AF756103

Key Challenges in Geography

EUROGEO Book Series

Series Editors

Kostis C. Koutsopoulos, European Association of Geographers, National Technical University of Athens, Pikermi, Greece
Rafael De Miguel González, University of Zaragoza & EUROGEO, Zaragoza, Spain
Daniela Schmeinck, Institut Didaktik des Sachunterrichts, University of Cologne, Köln, Nordrhein-Westfalen, Germany

This book series addresses relevant topics in the wide field of geography, which connects the physical, human and technological sciences to enhance teaching, research, and decision making. Geography provides answers to how aspects of these sciences are interconnected and are forming spatial patterns and processes that have impact on global, regional and local issues and thus affect present and future generations. Moreover, by dealing with places, people and cultures, Geography explores international issues ranging from physical, urban and rural environments and their evolution, to climate, pollution, development and political economy.

Key Challenges in Geography is an initiative of the European Association of Geographers (EUROGEO), an organization dealing with examining geographical issues from a European perspective, representing European Geographers working in different professional activities and at all levels of education. EUROGEO's goal and the core part of its statutory activities is to make European Geography a worldwide reference and standard. The book series serves as a platform for members of EUROGEO as well as affiliated National Geographical Associations in Europe, but is equally open to contributions from non-members.

The book series addresses topics of contemporary relevance in the wide field of geography. It has a global scope and includes contributions from a wide range of theoretical and applied geographical disciplines.

Key Challenges in Geography aims to:

- present collections of chapters on topics that reflect the significance of Geography as a discipline;
- provide disciplinary and interdisciplinary titles related to geographical, environmental, cultural, economic, political, urban and technological research with a European dimension, but not exclusive;
- deliver thought-provoking contributions related to cross-disciplinary approaches and interconnected works that explore the complex interactions among geography, technology, politics, environment and human conditions;
- publish volumes tackling urgent topics to geographers and policy makers alike;
- publish comprehensive monographs, edited volumes and textbooks refereed by European and worldwide experts specialized in the subjects and themes of the books;
- provide a forum for geographers worldwide to communicate on all aspects of research and applications of geography, with a European dimension, but not exclusive.

All books/chapters will undergo a blind review process with a minimum of two reviewers.

An author/editor questionnaire, instructions for authors and a book proposal form can be obtained by contacting the Publisher.

Gerry O'Reilly
Editor

Place Naming, Identities and Geography

Critical Perspectives in a Globalizing and Standardizing World

eur●
ge●

Springer

Editor
Gerry O'Reilly
School of Geography and History
Dublin City University
Dublin, Ireland

ISSN 2522-8420 ISSN 2522-8439 (electronic)
Key Challenges in Geography
ISBN 978-3-031-21509-4 ISBN 978-3-031-21510-0 (eBook)
https://doi.org/10.1007/978-3-031-21510-0

© The Editor(s) (if applicable) and The Author(s), under exclusive license to Springer Nature Switzerland AG 2023
This work is subject to copyright. All rights are solely and exclusively licensed by the Publisher, whether the whole or part of the material is concerned, specifically the rights of translation, reprinting, reuse of illustrations, recitation, broadcasting, reproduction on microfilms or in any other physical way, and transmission or information storage and retrieval, electronic adaptation, computer software, or by similar or dissimilar methodology now known or hereafter developed.
The use of general descriptive names, registered names, trademarks, service marks, etc. in this publication does not imply, even in the absence of a specific statement, that such names are exempt from the relevant protective laws and regulations and therefore free for general use.
The publisher, the authors, and the editors are safe to assume that the advice and information in this book are believed to be true and accurate at the date of publication. Neither the publisher nor the authors or the editors give a warranty, expressed or implied, with respect to the material contained herein or for any errors or omissions that may have been made. The publisher remains neutral with regard to jurisdictional claims in published maps and institutional affiliations.

This Springer imprint is published by the registered company Springer Nature Switzerland AG
The registered company address is: Gewerbestrasse 11, 6330 Cham, Switzerland

Place names are reflections of the varied attitudes, aspirations, and identities that have sought to appropriate, control, 'civilize' and 'make' places through time

(Alderman, 2008).

Conflictory nature of naming and renaming points to the centrality of place names as expressions of hegemony over space and ideologies

(Nash, 1999; Berg & Kearns, 2009).

'Naming places is a primary act of geographical appropriation, a demonstration of control over nature, the landscape and everything in it'

(Duffy, 2007, 63).

Street names 'are embedded into the structures of power and authority' and in street naming 'political regimes and elites utilize history to legitimate and consolidate their dominance and reinforce their authority'

(Azaryahu, 1997, 480).

Acknowledgments

This book is dedicated to those practitioners, researchers, staff, and organizations that do work concerning place names as without their efforts it is most likely that there would be greater voids in our understanding of the significance of toponymic studies that facilitate cognizance of their multi-layered heritages and histories, practical applications, and importance to supporting the UN Sustainable Development Goals and conflict resolution.

I would particularly like to thank all the authors listed below who have collaborated and contributed chapters to this Place Naming Project and book. Their diversity, originality, and scope have enhanced the work, counteracting any echo-chamber syndrome effects regarding perspectives, research approaches, and case studies.

Thanks to the academics who engaged in communications, abstract writing, and discussion but who could not contribute to finalized chapters for various personal, family, and pandemic-related reasons including Derek Alderman—Department of Geography, University of Tennessee, Apostolia (Lia) Galani—National and Kapodistrian University of Athens, Ali Aldosari—Saudi Arabia Geographical Society, and colleagues at the Russian Academy of Sciences—Institute of Geography and Institute of Linguistics.

During the pandemic lockdowns, virtual discussions and meetings were held, as well as international class linkups, especially for guest lectures and discussions for students concerning geography, identity, and place naming with Maoz Azaryahu, University of Haifa, and also Arnon Medzini, and Tal Yaar-Waisel—Oranim College of Education, Israel. A special word of appreciation to Sungjae Choo—Department of Geography, Kyung Hee University, South Korea and UNGEGN, for his collaboration in discussions and place naming conference sessions in Seoul (2019), Gangneung (2020), and Madrid (2021).

Colleagues in the UNGEGN (United Nations Group of Experts on Geographical Names) and ICA (International Cartographic Association) were generous with their time answering questions, and making introductions to people with different areas of expertise in place naming. These included Paulo Márcio Leal de Menezes—Geography Department, Federal University of Rio de Janeiro, and UNGEGN, who hosted the virtual Pan American International Symposium on

Toponymy, in November 2021, introducing me to some great researchers and their innovative Big Data management and GIS work in Latin American countries.

On my life path with its toponymic milestones, gratitude is due to my parents Kevin and Kitty who introduced me to the multi-layered place names and lore of the Boyne Valley in Ireland where I was brought up. There, are located magic places, like the megalithic sites at Newgrange and Brú na Bóinne, and the mythical Tara that does exist in the landscape as well as in imaginations, but also the hundreds of named townlands there, i.e. ancient Celtic territorial units still in use as the basis for geocoding for the postal, census, and electoral systems. The Valley is replete with place names etymologically of Celtic Gaelic, Irish-Latin, Viking, Anglo-Norman French, and English origins. Hence, the same place was known by several names, just like me—Gearóid Ó Raghallaigh while in class in primary school and on special national day occasions, but Gerard at home or Church events, and Gerry O'Reilly with friends and in everyday public places. So my interest in eventually unravelling this was fostered by my secondary school teacher, Gerard Rice—teaching history, geography, and local studies (or civics as he sometimes called it) including toponymy. Later at university, Professors Patrick Duffy, William J. Smith, and William Nolan always integrated place naming into their lectures, research, and conversations, as did Anne Buttimer while I was carrying out post-doctoral research on sustainability with her.

So decoding the place names of wherever I was living or working, and to see where I was, was instinctive as in Oued Zenati, Wilayat de Guelma in the Atlas Mountains, Constantine region of Algeria, or in coastal Annaba City ("Place of the Jujubes" in Arabic; Berber languages: Aânavaen) called Bône during the French colonial era (1830–1962), or Hippo Regius at the period(s) of the Roman Empire, and a central place in the story path of Saint Augustin, who wrote The City of God, that inspired utopian concepts of the Ideal City. Similarly, numerous colleagues, friends, and students nurtured my interest in toponymy while working in Tunisia, France, the USA, England, and elsewhere.

Encouragement and support in completing the book came from Michael Leuchner—Geography Department, Aachen University, former Publishing Editor at Springer. A special thanks to Doris Bleier—Editor, Earth Sciences, Geography, and Environment, Springer Publications, for her patience and encouragement, especially during lockdowns. The help of Sanjievkumar Mathiyazhagan, Springer Project Coordinator Books Production has been invaluable. Colleagues in EUROGEO (European Association of Geography) were most supportive, including Rafael de Miguel González, Karl Donert, Daniela Schmeinck, Maria Luisa de Lazaro Torres, and Kostas Koutsopoulos. And, of course, my family—Valérie, and daughters Aisling, Alannah, and son Shane. My wider family must be included here, as they kept me going virtually during lockdowns—my sisters Olive Doody and Catherine Farley, and especially Teresa Kelly (1944–2021), and lifelong friend Kevin O'Connor (1956–2020).

I would like to thank those people responsible for the various EU Erasmus programs that afforded me opportunities for staff mobility visits and projects that enabled me to travel and research, so putting the pieces of the place name narratives together. Thanks are also due to Dublin City University for conference and research travel funding.

Dublin, Ireland
June 2022

Gerry O'Reilly

Contents

Editor and Contributors

About the Editor

Gerry O'Reilly is an Associate Professor in Geography, and International Coordinator for the School of History and Geography, Dublin City University (DCU). His research and teaching interests are in geopolitics, human-made catastrophes, humanitarian action, cultural geography, places of memory, sustainable development, and education. He obtained his Ph.D. from Durham University UK, MA from the National University of Ireland (University College Cork), HDipEd and BA from Maynooth University. Post-doctoral research was undertaken in political geography and sustainable development at University College Dublin. Before joining DCU in 1997, Gerry held lectureship and research posts at UCD, Universities of Durham, Tunis, and Algeria-Annaba, and Visiting Professorship at the Ohio State University, Columbus. He was Erasmus Mundus Visiting Fellow at the Western Cape University (2009), Toronto York University (2008), and Columbia University NY (2007). He is the Vice President of EUROGEO (European Association of Geographers).

Contributors

Marianna Ács University of Pécs, Pécs, Hungary

Alpaslan Aliağaoğlu Balıkesir University, Balıkesir, Turkey

Yılmaz Arı Department of Geography, Bandırma Onyedi Eylül University, Bandırma, Balıkesir, Turkey

Boris Avdić University of Sarajevo, Sarajevo, Bosnia and Herzegovina

Titi Bachtiar Bandung Basin Research Group, Bandung, Indonesia

Mar Batlle Institut d'Estudis Catalans, Barcelona, Spain

Úna Bhreathnach Fiontar & Scoil na Gaeilge, DCU, Dublin, Ireland

Laura Cassi University of Florence, Florence, Italy

Jonathan Cherry School of History and Geography, DCU, Dublin, Ireland

Sungjae Choo Kyung Hee University, Seoul, South Korea

Manoel do Couto Fernandes Federal University of Rio de Janeiro (UFRJ) - Department of Geography–Cartography Laboratory (GeoCart), Rio de Janeiro, Brazil

Nusret Drešković University of Sarajevo, Sarajevo, Bosnia and Herzegovina

Ritienne Gauci University of Malta, Msida, Malta

Matjaž Geršič ZRC SAZU Anton Melik Geographical Institute, Ljubljana, Slovenia

Frédéric Giraut Geography Department, UNESCO Chair "Naming the World", University of Geneva, Geneva, Switzerland

Jussi S. Jauhiainen Department of Geography, University of Turku, Tartu, Estonia; Institute of Ecology and Earth Sciences, University of Tartu, Tartu, Estonia

Velid Jerlagić Sarajevo Historical Archives, Sarajevo, Bosnia and Herzegovina

Elša Turkušić Jurić Faculty of Architecture, University of Sarajevo, Sarajevo, Bosnia and Herzegovina

Robin Kearns School of Environment, The University of Auckland, Auckland, New Zealand

Helen Kerfoot Honorary Chair UNGEGN, Ottawa, Canada

Máté Kitanics University of Pécs, Pécs, Hungary

Marko Krevs University of Ljubljana, Ljubljana, Slovenia

Multamia R. M. T. Lauder Universitas Indonesia, Depok, Indonesia

Neil Lindsay School of Environment, The University of Auckland, Auckland, New Zealand

César López-Leiva Universidad Politécnica de Madrid, Madrid, Spain

Tendai Mangena University of the Free State, Bloemfontein, South Africa

Ruth McManus School of History and Geography, Dublin City University, Dublin, Ireland

Paulo Márcio Leal de Menezes Federal University of Rio de Janeiro (UFRJ) - Department of Geography–Cartography Laboratory (GeoCart), Rio de Janeiro, Brazil

Ranko Mirić University of Sarajevo, Sarajevo, Bosnia and Herzegovina

Brian Ó Raghallaigh Fiontar & Scoil na Gaeilge, DCU, Dublin, Ireland

Ferjan Ormeling University of Amsterdam, Amsterdam, The Netherlands

Gerry O'Reilly School of History and Geography, Dublin City University (DCU), Dublin, Ireland

Taavi Pae Department of Geography, University of Tartu, Tartu, Estonia

Norbert Pap University of Pécs, Pécs, Hungary

Jennifer Ballantine Perera Gibraltar Garrison Library, Gibraltar, Gibraltar; University of Gibraltar, Gibraltar, Gibraltar

Drago Perko ZRC SAZU Anton Melik Geographical Institute, Ljubljana, Slovenia

Péter Reményi University of Pécs, Pécs, Hungary

Ats Remmelg Department of Geography, University of Tartu, Tartu, Estonia

Ayar Rodríguez de Castro University of Alcalá, Alcalá de Henares, Spain

John A. Schembri University of Malta, Msida, Malta

Cece Sobarna Universitas Padjajaran, Bandung, Indonesia

Joan Tort-Donada Universitat de Barcelona, Barcelona, Spain

Charles Travis Department of History, University of Texas, Arlington, USA; Trinity Centre for Environmental Humanities, Trinity College Dublin, Dublin, Ireland

Abdullah Uğur Pamukkale University, Denizli, Turkey

Adriana Vescovo Former Responsabile of the Department of Geographical Information, Instituto Geográfico Nacional (Ign–Argentina), Ciudad de Buenos Aires, Argentina;
Corresponding Member (Argentina), Joint ICA/IGU Commission of Toponymy, Ciudad de Buenos Aires, Argentina;
Treasure and Coordinator of the Commission of Toponymy, Centro Argentino de Cartografía (CAC), Ciudad de Buenos Aires, Argentina

Tal Yaar-Waisel Oranim College of Education, Kiryat Tiv'on, Israel

Pier-Giorgio Zaccheddu Head of Executive Unit—International Affairs, Federal Agency for Cartography and Geodesy (BKG), Frankfurt am Main, Germany

Matija Zorn ZRC SAZU Anton Melik Geographical Institute, Ljubljana, Slovenia

Introduction: Approaching Place Naming Narratives

Gerry O'Reilly

Concerning place naming this introduction aims to brainstorm ideas, concepts, perceptions, attitudes, oral and written language and cultural approaches, research, tools used in toponymic studies, agencies, actors, approaches to standardization, range of geographical scales and phenomena, power-relations, and the (geo)politics of naming. Literature reviews are embedded in the respective chapters of the individual authors. The introductory discussion is followed by a research synopsis of the respective authors including interdisciplinary and multilingual perspectives from different parts of the world ranging from Brazil to southern Africa, New Zealand, South Korea, Israel, Turkey, Balkan and Central European countries, Italy, France, Ireland and so forth.

Toponymy has a long tradition in the sciences, humanities, and arts with a renewed critical interest in geography and associated disciplines in recent decades.[1] Place location, cartographical aspects, naming, etymology, geo-histories, and narratives so salient in past studies are now being enhanced from a range of perspectives including radical approaches, especially in a globalizing, standardizing world with Googlization and attempted normalization and homogenization of many place names, perceptions, and images worldwide including those for branding and marketing purposes. Globalization flows literally interconnects all named place, by geocoding using selected algorithms Googlization interrelates all these place names.

Nowadays we often 'google a place' to find out how to get there, or to find more information about a place. Whether to search for directions including the distance, means of transport, and time factors involved; to locate a store that sells a particular product; plan a trip; or vacation to some 'place.' At another level, we

[1] John Everett-Heath. 2017 The Concise Dictionary of World Place-Names. Oxford University Press.

G. O'Reilly (✉)
School of History and Geography, Dublin City University (DCU), Dublin 9, Ireland
e-mail: gerry.oreilly@dcu.ie

© The Author(s), under exclusive license to Springer Nature Switzerland AG 2023

G. O'Reilly (ed.), *Place Naming, Identities and Geography*,
Key Challenges in Geography, https://doi.org/10.1007/978-3-031-21510-0_1

have our own long experience i.e. personal archaeology with places, perceptions of them from earliest childhood onto adulthood—home, farm, street, village, town, family or friend's milieus, shops, school—as lived; or even heard about from family, community, school, religion, TV, movies and various media. Therefore, knowing your space, your place in it, how to navigate it, and even to reach places beyond the immediate physicality and perceptions are important. But going beyond one's own space, and coping with other places physically, psychologically, name-wise, language/linguistically is crucial in avoiding being topographically disoriented.

Place names can be based on a literalist approach, denotational or descriptive, as being derived from physical geography or the natural environment, i.e. hills, valley (e.g. Glen, from the Gaelic-Irish word Glenn meaning valley, Glendubh—literally Dark Valley), marshes (e.g. the city and county of Cork gets its name from the Irish for word marsh—*Corcaigh*), minerals (e.g. Argentina, literally *Argento* meaning silver as in Spanish and Italian from Latin), plants (phytonyms), birds, animals, colours (e.g. Mont Blanc, Blue Danube, Black Forest) and so forth. Place names can be tangible and/or measurable in some way including the cardinal directions N, E, W, and S (e.g. country names - North Sudan, East Timor, South Africa). Many subdivisions within categories exist. Toponymic awareness enhances geographical and spatial literacy. Due to increasing environmental consciousness, research on traditional place names referring to phenomena such as springs, lakes, flora, or fauna that no longer exist in the area has become a source for further investigation due to factors such as climate change.[2]

Place names derived from human geographical processes in specific contexts and histories, often pertain to socio-cultural constructivism, that over time may be less evident, more intangible, or even existential to a person in 'that place' now. Unquestionably, changing linguistics and languages, and cultural succession with new arrivals of settlers, all impact original names and renaming over time. Sometimes, the physical descriptor is linked with a cultural qualifier i.e. syncretic place names that have a mixture of both physical and cultural aspects e.g. Mont Sainte-Odile, a mountain in the Vosges, overlooking the Alsace plain. Also Skellig Michael, literally St. Michael's Islets, off the southwest coast of Ireland, was brought to world audiences in the Star Wars films—The Force Awakens and The Last Jedi. The Gulf of St. Lawrence (Golfe du Saint-Laurent), is the outlet of the North American Great Lakes via the St. Lawrence River into the Atlantic Ocean. Koenigsberg/Königsberg (King's Mountain) is the old Germanic name for the present-day Russian city of

---

[2] See: UNGEGN (UN Group of Experts on Geographical Names) https://unstats.un.org/unsd/ungegn/.

Review of Historical Geography and Toponomastics—RHGT. https://doaj.org/toc/2393-4255?source=%7B%22query%22%3A%7B%22bool%22%3A%7B%22must%22%3A%5B%7B%22terms%22%3A%7B%22index.issn.exact%22%3A%5B%221842-8479%22%2C%222393-4255%22%5D%7D%7D%5D%7D%7D%2C%22size%22%3A100%2C%22sort%22%3A%5B%7B%22created_date%22%3A%7B%22order%22%3A%22desc%22%7D%7D%5D%2C%22_source%22%3A%7B%7D%2C%22track_total_hits%22%3Atrue%7D.

Shaun Tyan Gin Lim, and Francesco Perono Cacciafoco. (2020). Plant and Place Names: A Case Study of Abui Toponymy. Review of Historical Geography and Toponomastics, 15, 29–30: 121–142.

Kaliningrad (Kalinin - Russian surname (of a Soviet politician), etymologically derived from the word Kalina meaning 'guelder rose' and grad—town, city or castle) on the Baltic.

Due to the immense number of languages, dialects, and permutations thereof that have existed throughout the world, alongside linguistic evolution, innovation, and diffusion, place names embrace multiple heritages and narratives, much appreciated by storytellers for 'explanation' and 'locations', especially in lore, which can be a rich data source. For instance, *Dindshenchas* meaning "lore of places" (the modern Gaelic-Irish word for topography), is a class of onomastic text in early Irish literature, recounting the origins of place names, traditions concerning events, and characters associated with places.

People like a good narrative, and they come in all forms with their toponyms. Obvious examples include Homer's Odyssey, the holy books that are replete with stories such as Noah's ark and where it landed, Moses leading the Jewish people out of Egypt to Israel, the story of Jesus Christ the Nazarene and his travels throughout Judea and Samaria, Jerusalem, and the place now called Yardenit in Hebrew where Jesus was baptized, where the Jordan River leaves the Sea of Galilee. All these places can be geocoded with precise latitude and longitude. However, reading the culture and political codes embedded in the names can be more challenging due to language(s), etymology, histories, and current power dynamics at all scales right down to street levels in Jerusalem, Nicosia, Diyarbakır or Derry/Londonderry and hence the need for critical perspectives.[3]

Locations in the Koran include Al-Aḥqāf (The Sandy Plains, or the Wind-curved Sand-hills), Iram dhāt al-ʿImād (Iram of the Pillars), Al-Madīnah (Yathrib), ʿArafāt' so important in the life path of the Prophet Mohamed, and the Hajj pilgrimage. And Al-Ḥijr (Hegra), Badr (the place of the first battle of the Muslims), Sabaʾ (Sheba), and of course Makkah (Mecca) Al-Balad al-Amīn (Arabic: ٱلْبَلَد ٱلأَمِيْن, the secure land) and so forth. Both Sunni Muslims and Shia Muslims agree on the three holiest sites in Islam being, respectively, the Masjid al-Haram (including the Kaaba), in Mecca; the Al-Masjid an-Nabawi, in Medina; and the Al Aqsa Mosque compound, in Jerusalem. Najaf in present-day Iraq is of particular importance to the Shia that is often the target of extremist groups, especially since 2016, aiming to deny and wipe out footprints of Shia tradition and culture. Similar narratives and associated place names exist in other religious systems.

Besides religious and/or other etymology of secular place names in India, toponymy there holds many remarkable narratives. For instance, the name Bombay derived from the Portuguese/Galician words "Bom Bahia," meaning 'Good Bay', was later anglicized to Bombay, by the British, as seen in Lendas da India (Legends of India, 1508). Since 1995 the official name is now Mumbai, etymologically meaning

[3] See: Reuben S. Rose-Redwood; Derek H. Alderman; Maoz Azaryahu. Geographies of Toponymic Inscription: New Directions in Critical Place-Name Studies. August 2010. Progress in Human Geography 34(4):453–470. Also note the extensive bibliographical references for the geographical naming work carried out by the respective authors.

'mother' in the Marathi language, which has an associated epic political tale.[4] Toponymy holds a vital place in the long and short cycles of history, with ever-increasing impacts due to cyber histories and the information revolution, juxtaposed with the individuals' experiences of place names.

Space is location, presence, and physical geography, while the place and its name give meaning, personality, and a connection to cultural and/or personal uniqueness, but also shared experiences and identities. It is the culturally ascribed meaning given to a space. Having a meaning may counteract anomie and its adverse implications. The cultural geographer Yi-Fu Tuan (1977; 2001) speaks of the dichotomy of space and place when he says what begins as undifferentiated space becomes place as we get to know it better and endow it with value. Nonetheless, he poses the question of whether endowment is the beginning of dissolution into a utility of representation and value with the act of naming.[5] He emphasizes space and the individual human body, followed by space and the group, and also mythical-conceptual spaces. His research strongly underlined the importance of the *habitus* for the individual, and group i.e. spatial habitus, exploring the socio-cultural, historical, and environmental factors of feeling at home. He argues that Place both carries a sense of location and a social position and speculates which one is the main meaning. In short, a place is whatever location makes social and geographical sense. This refers to the human ability to have a sense of place that is associated with a form of knowledge and memory of place, as much as symbolic projection. Hence, a place is not merely a location but has to be a location "in itself," a "small world." The place is linked to usage, and therefore there are types of places.

Tuan's humanistic approach was influenced by reactions to the quantitative revolution (the 1950s–1960s) in the social and political sciences that had marked a strong change in the approaches underlying geographical research, from an integrated holistic regional geography to a functionalist spatial science. Quantitative approaches led to a shift from descriptive ideographic geography including place and name narratives to an empirical law-making (nomothetic) science. Tuan and other geographers such as Anne Buttimer (1938–2017) promoted humanistic approaches that fed into the post-modernist revolution (the 1970s–1990s) i.e. a general distrust of specific types of reason or what was projected to be absolute logic—the truth. This impacted researchers such as Derek Gregory and others with a critical, sometimes radical sensitivity to the role of ideology in asserting and maintaining political and economic power and the multi-layered cultural geographies.[6]

Post-modernist philosophers like Jacques Derrida the Algerian-born French philosopher is best known for developing a form of semiotic analysis i.e. deconstruction, while Michel Foucault, French philosopher, historian of ideas, writer,

---

[4] See: Anu Kapur 2019. Mapping Place Names of India. Routledge India.

[5] Tuan, Y.-F. (1977, 2001). Space and place: the perspective of experience. University of Minnesota Press.

[6] Gregory, Derek. 1998 Power, Knowledge and Geography: An Introduction to Geographic Thought and Practice. Maiden, MA and Oxford, UK: Blackwell Publishing.

Gregory, Derek 2005 The Colonial Present. Derek Gregory. Maiden, MA and Oxford, UK: Blackwell Publishing.

and political activist, also engages with deconstructionist approaches illustrating the centrality of power relations within society and its social constructs, but likewise between groups and societies, and naturally the citizen within those spaces and places and consequently place names attributed.

In Geography this was manifested with a transition to critical geography and sometimes led to radical approaches to concepts of place and naming, with research being enhanced by interdisciplinary approaches, especially linguistics enriched by the work of salient intellectuals like Noam Chomsky—American, linguist, political philosopher, and activist. Research in critical geopolitics underscored that power (from and to) relationships exist at all spatial scales from local to state and trans-state scales. Critical approaches facilitated the deconstruction of received frameworks, methodologies of the 'normalized' ideologies, and power constructs like imperialism, and its systemic embeddedness in institutions and mindsets. Radical approaches came with an engagement to use knowledge to redress grievances, socio-political, racial, or gender spatial apartheids and particularly to create a more ethical and just world.[7]

From an etymological perspective, the word geopolitics refers to 'geo'—earth, territory, place, while politics designates power; thus the influence, impact, and control in place naming. Historical examples include the cadastre system introduced by the bureaucracy in the Roman Empire and its polity, for usage in enumeration, including for taxation. When Christianity became the official religion of the Empire (4th Century), the Church used the existing cadastre system as the basis for the creation of the parish unit, a subdivision of the diocesan system with the associated tithe taxes, especially in Europe between the 8th–12th Centuries, and readjusted many times after that. This system of parishes was expanded by the Church with its universalizing missionaries as witnessed in the Americas, stretching from Argentina to California for instance. In this general example, toponyms formed a vital part in constructing the 'authorities' geographical knowledge of their expanding world—all parish roads in Christendom lead back to Rome with flows of information, and wealth. Besides spiritual and theological aspects of disputes and wars witnessed during the Protesters' Reformation (16th–17th centuries), strong 'national' states were emerging around powerful royal nodes in their quest to build strong states that eventually created 'national churches' as in England and Sweden for instance. Significantly with the 'nationalization' of culture, the bible was translated from Latin into 'national' languages, with Latin being replaced in ceremonies and rituals, and impacting place names.

Such experiences of people in the nation-state building processes in parts of Europe with centralizing government activities in the UK, Germany, Sweden, France,

---

[7] Seminal publications in critical and radical geography include Antipode—Radical Journal of Geography; Hérodote—Review of Geopolitics and Geography, founded by Yves Lacoste, a French geographer born in Morocco.

See D. Fuller and R. Kitchin (2004) for an exploration of the role of the radical academic in society, and V. Lawson (2007) AAAG 97, 1 on geographies of care and responsibility.

Other influential critical and radical geographers in the Anglophone world include Simon Dalby, Gerard Toal (AKA Gearóid Ó Tuathail), Karen E. Till, Klaus Dodds, Gerry Kearns, Doreen Massey, David Newman, and John O'Loughlin to name but a few.

and Spain for example, often included internal colonialism as evidenced in the regions of the Spanish state, noticeably until after the death of General Franco in 1975. Contested place naming existed in Scotland, and particularly Ireland, especially in the nineteenth century with the English/British strategies to create the state of the United Kingdom of Great Britain and Ireland. Place naming and languages used have remained a contentious issue in Northern Ireland. Similarly, in the imperial project outside Europe, the ruling Powers left their foot and place name prints. Such as the toponymic prints of Cecil Rhodes, imperialist and supremacist lingered on in Southern Africa and Britain in various iterations; not being removed in a similar manner to (Adolf) Hitler and the NAZIs after 1945 with applied denazification programs. Countries once named after Rhodes and protectorates of the UK—Northern Rhodesia and Southern Rhodesia are now officially Zambia and Zimbabwe, while a small village still has the name Rhodes, located in the Drakensberg mountains Cape province, in the Republic of South Africa. Controversy remains regarding spaces, places, and monuments dedicated to Rhodes in public and university spaces in Cape Town. The Rhodes Memorial on Devil's Peak is deemed particularly offensive by many citizens, as is his statue at Oxford University 90 km northwest of London, largely due to the struggle of raising citizen consciousness by such bottom-up groups as BLM (Black Lives Matter).

In many West European countries, cadastration, geographical naming, and classifications of place names became very intensive, scientific, technology-driven, and highly state-controlled as of the Napoleonic era that built on the long 18th Century Age of Reason. Enlightenment intellectuals drew on classificatory concepts taken from the Encyclopaedist approaches, eventually feeding into the creation of toponymy gazetteers. In short, such work and scientific vision contributed to the creation of (toponymic) Wikipedia which was founded in 2001, and of course, the cadastral system is the basic building unit of GIS (Geographical Information System), with its associated algorithms, so central to Google Maps. Place names provide the most useful geographical reference system, especially in the virtual-sphere, where consistency and accuracy are essential in referring to a place in order to prevent confusion in everyday business and recreation. However, here is a kernel naming issue, are the 'insider' and 'outsider' oral and written place names in harmony, agreed or contested, and the issue of multiple names of course; and who controls the algorithm today.

Cognizance of the endonym is essential. It entails the common or native name for a geographical place or group of people. It is their self-designated name for themselves, their homeland, or their language. This contrasts with an exonym i.e. designating a common non-native name for a place, people, language, or dialect. Conflict can exist at any scale as in the territories between the core area of Catalonia and other regions of Spain, i.e. within states, but also between states regarding endonyms. Exonyms exist with multiple explanations, and not just for historical–geographical reasons such as cultural succession, and/or imperialism; difficulties when pronouncing 'foreign' words must also be taken into consideration. A classic example is provided by the country endonym Deutschland, and its exonym in neighbouring France with 'Allemagne' derived from Latin, but also Alemania in Spanish; Niemcy in Polish (for designating both Germany and people), etymologically meaning "mute, people who can't speak" language. Variations of this exist in Russian and other Slavic languages and of course, the exonym for Deutschland is Germany in English. Nonetheless, without going into linguistic theory, including orthoepy (the science of pronunciation), phonology, and phonetics, a 'foreigner' may be dismissive of other languages due to their own ethnocultural and language-centrism as witnessed during the colonial era, and more recently in some mass tourism areas and ex-pat (expatriate) 'colonies' as observed in Spain, Cyprus, and elsewhere.

Therefore, place naming and renaming must be read with a certain understanding of 'the spirit of the time' i.e. the Zeitgeist, defining the spirit or mood of a particular period of history as shown by the ideas and beliefs of the time. In hindsight, researchers can investigate toponymic heritage from past eras as discussed above, but as we emerged from the postmodernist epoch in the early twenty-first century, it is challenging to define which Zeitgeist we are living in research-wise, and the effects of this on contemporary place naming studies. Can we speak of Neomodernism or New Modernism that has legitimized an outlook of comprehensive individualism and relativism![8] Has there been a movement from post-modernism to post-humanism with trans-paradigms? Have we gone from invention to deconstruction and on to Reconstruction? And especially issues concerning reconstructing identity, values, and meaning—with some ethical urgency.

Above was mentioned people and students' personal archaeologies regarding place names which can whet their interest in toponymy. Sometimes there can be a fun side regarding toponymy. I find Hollywood a very good place for student geography field trips including exploring its toponymy, which goes back there to the sixth century in lore and recorded history. That is—Hollywood in County Wicklow, Ireland—50 km from Dublin. Sometimes students are disappointed by the rural one-street village encountered. Much further away is Baltimore (literally in Gaelic-Irish - Baile (Bally) Mór (Big), or Big Town in County Cork, also a good place for field trips, while the journey can include visiting the Barak Obama Plaza and Visitor Centre. This is a service area just off the M7 motorway on the borders of Counties Offaly and Tipperary, beside the village of Moneygall (from the Gaelic-Irish word 'Muine' meaning shrubbery), where the former US President's third great grandfather, Falmouth Kearney, lived before emigrating to the US in 1851, and that was visited by the Obamas in 2011 while reconnecting with some distant cousins in the area.

Despite the so-called normalization and homogenization of places, names, and the driving forces of neoliberal capitalism, we are often sharply reminded that there exist conflicting voices regarding toponymy, whether it be for a local development project like an area name with a disputed Postal Address concerning perception and property prices, or the naming of a housing estate, or a retail mall, or even a hospital in Dublin. A largely bottom-up grassroots reaction campaign in Ireland rejected the proposed name for a new hospital in 2017, rebuffing the government's proposed name

[8] Faye, J. (2012). Neomodernism—A New Approach to Humanistic Science? In: After Postmodernism. Palgrave Macmillan, London. https://doi.org/10.1057/9780230355484_8.

'Phoenix' to be replaced by Kathleen Lynn (i.e. Lynn Hospital) to commemorate a social activist, suffragette, medical doctor, republican, and armed battalion officer during the revolutionary years 1913–1923.

On another level, issues regarding place naming came sharply into focus in the USA and worldwide on 25 May 2020, occasioned by the killing of George Floyd, an African-American citizen by a white policeman in Minneapolis, when global audiences were reminded of the legacies and systemic wrongs of the colonial project. This occasioned protesters demanding the removal of historical place names and monuments associated with colonial supremacist projects. Unofficial and official naming or renaming occurred since then in many areas of the USA. For instance, on 5 June 2020, an area close to the White House in Washington DC was (officially) renamed BLM Plaza to the north of Lafayette Square, after the Department of Public Works had painted the words "Black Lives Matter" on 11 m of yellow capital letters, during the series of George Floyd marches taking place in the city.[9]

On another scale, we see the contested use of the country name "Macedonia" that was disputed between Greece and the Republic of Macedonia (FYRM – The Former Yugoslav Republic of Macedonia), now officially the Republic of North Macedonia since 2019. Between 1991 and 2019, the name was a source **of** causing instability in the volatile Western Balkans. Interestingly, news reports in 2022 stated that the Turkish government wants to change the name of country internationally, so that it be spelled and called by its true endogenous name "Türkiye" as opposed to Turkey—the Anglicized version of the name is associated with Western culture.

On another level, on 24 February 2022 global citizens witnessed the Russian invasion of Ukraine, also bringing to attention naming issues regarding Kyiv (Ukrainian) Versus Kiev (Russian), and Ukraine (Ukrainian) Versus The Ukraine—due to historical Russian legacies. Once brought to global consciousness, the preferred Ukrainian spelling and pronunciation were adapted by most international media, businesses, and institutions. Undoubtedly, toponymic narratives including those in transitions in eastern and southern Ukraine will be much researched in the coming decade.

Box 1: Kyiv, Ukraine

How to pronounce and spell Kyiv is not a matter of personal, but political choice. Some years ago most Western media outlets change their spelling thanks to the "KyivNotKiev" online campaign. Airports around the world did likewise to respect Ukraine's autonomous national identity decisions supported by the vast majority of the population. After independence in 1991, the Ukrainian nation dropped the "the" and just called the country "Ukraine" corresponding to their vernacular. Therefore, the Ukrainian constitution states definitively that the name of the country is Ukraine. Etymologically, the name

[9] Nubras Samayeen, Adrian Wong and Cameron McCarthy. Space to breathe: George Floyd, BLM plaza and the monumentalization of divided American Urban landscapes. Educational Philosophy and Theory. Published online: 23 Jul 2020. Taylor and Francis online.

Ukraine means "borderland" so using "the" before the country's name may leave room for ambiguity regarding Russia's perception of its hinterlands. Essentially, it can suggest a subordinate status to a neighbouring country.

Regarding hydronomy, some commentators in Pakistan have called for a renaming of the 'Indian' Ocean, but this has received little support from other coastal or riparian countries in the region. Sea naming matters are especially important to Indonesia, also known as the Indonesian Archipelago or Nusantara. The exact number of islands comprising Indonesia varies depending on definitions and sources. According to Law No 9/1996 on the Maritime Territory of Indonesia, 17,508 islands are listed within the territory of the Indonesian state. This includes the Natuna Islands, an archipelago of 272 small islands that lie in a part of the sea close to China. That whole area is generally known as the South China Sea. However, the Indonesian authorities have reiterated that the water area near their islands should now be called the North Natuna Sea. China rejects this. In another area, China started creating artificial islands in the South China Sea in 2014, building them on top of rocks or reefs which were close to the water's surface at high tide. Dredging ships were used to scoop up the seafloor to build up the islands on top of the rocks. These 'construction' activities have been disputed by neighbouring countries like the Philippines, as have their names based on the respective historical nomenclature. In 2022, it was reported that China has fully militarized at least three of these artificial islands.[10]

And what about the naming of dream paradisal islands offering sunshine and wealth. The World Islands project was launched in 2003, with plans for constructing 300 islands, i.e. an archipelago of small artificial islands constructed in the shape of a world map, located in the waters off the coast of Dubai, United Arab Emirates. This project consists of seven sets of islands, named after the continents of Europe, Africa, Asia, North America, South America, Antarctica, and Oceania. Each island is named for its representative country, landmark, or region e.g. France, California, Mount Everest, Australia, New Mexico, Upernavik, Buenos Aires, New York, Mexico, St. Petersburg, India, and so forth.[11]

Hopefully, the above brainstorming approach in the introduction will give rise to reflection and debate.

---

[10] Rebecca Strating, Joanne Wallis, Maritime sovereignty and territorialisation: Comparing the Pacific Islands and South China Sea, Marine Policy. Volume 141, 2022, https://doi.org/10.1016/j.marpol.2022.105110.

[11] Ahmad Jalinoosi and Saeedeh Moradifar. Analysing the Consequences of the UAE Creating Artificial Islands in The Persian Gulf. Journal of World Sociopolitical Studies, Vol. 1, No. 2, October 2017, pp. 281–313. Web Page: https://wsps.ut.ac.ir// Email: wsps@ut.ac.ir. eISSN: 2588-3127 Print ISSN:2588-3119. https://doi.org/10.22059/wsps.2017.242117.1030.

Methodology

In the context of the above strands of discussion, I floated the idea of a toponymic book project, encouraging researchers from varied academic backgrounds, languages, or cultures interested in place naming to send a proposal abstract, especially during the pandemic lockdown(s) period, in order to gauge what's currently being done regarding place name studies. Hence, this book presents research on geographical naming on land and sea from a range of standpoints on theory and concepts, case studies, and education. The choice of broad benchmarks is deliberate, as it allows for a cross-sectional approach and conclusions. The interdisciplinary research is enhanced with authors from various universities, and regional, national, and international toponymy-related institutions and organizations including the UNGEGN (UN Group of Experts on Geographical Names), IGU (International Geographical Union), ICA (International Cartographic Association), and so forth.

Essentially, for a toponymist, through well-established local principles and procedures developed in cooperation and consultation with the United Nations Group of Experts on Geographical Names (UNGEGN), applies the science of toponymy to establish officially recognized geographical names, but this entails cognizance of all existing names and any conflictual issues. A toponymist relies not only on maps and local histories, but interviews with local residents to determine names with established local usage. The exact application of a toponym, its specific language, pronunciation, and origins and meaning are all essential facts to be recorded during name surveys. The United Nations Group of Experts on Geographical Names (UNGEGN) is one of the nine expert groups of the United Nations Economic and Social Council (ECOSOC) and deals with national and international standardization of geographical names. Every five years they hold the UNGEGN conference, and also publish international guidelines.

Significantly 48 people answered the call for the 'Topo Book Project.' Some abstracts were deemed not suitable for various reasons. While over 30 proposals approached toponymy from multiple perspectives with rich insights. It must be noted that four excellent chapters could not be completed on time due to a range of personal, family, and pandemic-related issues. These include toponymy and legal perspectives from Japan; feminist and racial analyses from the USA; a four-person team with toponymic expertise in Russia; and a geographer in Saudi Arabia exploring toponymy and creation of dream niche tourism places in isolated coastal areas for the international jet set.

The book is broken down into four main sections:

Part I: Challenging conceptual and theoretical approaches to place naming

Part II: Approaches to implementing standardization of place names

Part III: Geo-histories, legacies, and toponymy transitions

Part IV: Toponymy: narratives, languages, culture, and education

Part V: The Relationship Between Geographical Naming and Cultural Politics

Part I: Challenging conceptual and theoretical approaches to place naming

In Part I the authors explore research trajectories in critical toponymy and standpoints that contribute to understanding different perceptions of places. For instance, numerous layers of memories in the name of the city of Stalingrad, which was renamed Volgograd in 1961. Due to globalization and toponymic standardization, in dynamic landscapes, the cartography-digitalization nexus is continuously evolving. Hence Google's power is increasing in place naming. For the benefit of standardization and 'geocode name' efficiencies, it is essential that there be cognizance of the geopolitics of emotions within places, and between them including endonyms and exogamy as seen in the Volgograd narratives, but also in the place of toponymy and hydronyms in Japan-Korea relations that take into account how place names are linked to perceptions, territorialization(s), historical legacies, and sovereignty claims.

Changing global contexts affecting geographical information must be borne in mind, including the management of toponymy as part of the Geospatial Big Data. This question is explored in the context of Argentina, including the problems encountered and actors involved. From a technical perspective, this poses the question of to what extent is the meaning of a toponym conditioned by space, and should the etymology of a place name be known in order to interpret its 'geographical meaning' correctly and represent it accurately. Concerning Spain, it is posited that specific problems in the mapping process related to the scale factor are encountered. Therefore, is it possible to speak of an ideal scale, in the cartographic representation of the place names? On what criteria can the distinction between "macrotoponym" and "microtoponym" be based, and what difficulties does microtoponymy pose in the mapping process. Diverse aspects and modalities of toponymic research related to cadastral toponymy are highly relevant as a source of research. The scale, and change of scale, are critical factors in cadastral toponymy, as illustrated in research presented. Greater amounts of toponymic research have been carried out on land than on sea for a variety of reasons, yet hydronomy is of major significance as reviewed in research regarding northeast Asia, and also on a different scale in the Land of Israel as with relevance to the overt and hidden meanings of sea names and water sources there. This begs the question of the names used, *de facto* as opposed to those employed *de jure*.

A core remit of the UN is to attenuate contentions and prevent violence and war between countries and within states in the interest of peace, as formulated in the Sustainable Development Goals. Manifold strategies are employed in this through various UN agencies and partners, including UNESCO (United Nations Educational, Scientific and Cultural Organization) and especially UNGEGN (United Nations Group of Experts on Geographical Names).

Perspectives, narratives, and analyses presented include:

Sungjae Choo: Critical toponymy perspectives have contributed to understanding different identities which are sometimes vigorously contested. Conflicts often arise when naming a place or geographical feature, reflecting diverse human perceptions.

The validity of critical toponymy in elucidating human perceptions of places draws on three elements to design an analytical framework: formation of memories; development of discourses; and paths of exercising politics and power. The Russian city of Volgograd was renamed from its former name of Stalingrad, but still experiences intense pressure to return to its former name. Here, different memories that are accumulated in a name are exposed and developed into discourses and supported by different groups of people. Sungjae Choo illustrates how to recognize the validity of critical toponymy perspectives in revealing human perception of places; understanding contrasting memories contained in the name of Stalingrad; addressing power groups in Russia that exhibit contrasting views on the renaming of Volgograd. He discusses the pluralism of memory and commemoration contained in a toponym and the accumulation of different senses of place.

Gerry O'Reilly: Surveys toponymic legacies through perspectives from Korea and Japan. The cartography-digitalization nexus is appraised in globalization processes, including Google Map's power in 'naming', alongside the geopolitics of emotions as regards spaces and hydronyms in Japan-Korea relations. This reflects perceptions, territorialization(s), historical legacies, and claims to sovereignty, alongside peaceful aspirations for the future. Google's rationale for selecting geographical names is reviewed as are possible alternatives from international institutions. Like analyses of historical maps, created in specific political time-slice contexts, many international associations were founded in the colonial era and (re)invented after the World Wars and Great Power reconfigurations, influencing international law. Inevitably, such holds embedded legacies with systemic perspectives. Complex post-colonial and post-WWII heritages continue to be shared by Japan and Korea. So disputed toponymies cannot be read in isolation from changing geopolitical relations between neighbouring states in Northeast Asia including China. Essentially, the cartography-digitalization nexus is reviewed in globalization processes alongside Google's potential power in place naming. The geopolitics of emotions vis-à-vis place names in Japan-Korea affairs are evaluated, in order to elucidate perceptions, and sovereignty claims in the post-Colonial context.

Frédéric Giraut: Social, political, cultural, and economic dimensions of toponymy are expressed in the act of naming, that is, in the contradictory production of toponyms. Place naming is thus an object of toponymy distinct from the place name itself. Its study from the perspective of motivations and representations has been developed in the social sciences; particularly in cultural and political geography. The study of the naming of places presupposes observation and documentation of process, and therefore focuses on neotoponymy. This is both the study of new place names and the corpus formed by them. In addition to collections of names whose meaning and origin can be worked on, it offers, by definition, information on its mode of production, actors, and possible controversies. Neotoponymy is the result of a process of substitution (renaming, plurinaming) or of nominations of emerging places. The contemporary history of France since the Revolution offers an abundance of neotoponymies from successive or simultaneous contexts that allow us to test a theoretical framework on place naming. Frédéric Giraut asks, what are the different toponymic techniques applied to local and provincial government territorial names; what political challenges do they address in different contexts; what are the successive toponymic experiments that France has investigated to name its administrative territories (departments, regions) and communes; how to define neotoponymy and how it can be applied.

Adriana Vescovo: Geographical names in Argentina in the present, and challenges are appraised by exploring spaces to discuss toponymy both as geospatial data and for its value as intangible cultural heritage. Its vitality grows, not only in the public and academic spheres, but also accompanying social, political, economic, and cultural processes around the world. In recent decades its visibility has increased, as well as the need for its local, national, and global standardization. At the same time, new problems and questions have arisen. The aim of this research is to reflect on this particular moment in the evolution of place names in Argentina from a geographical perspective including factors and characteristics in this territory; it's wide multicultural richness; the impact of emerging new technologies and paradigm shifts in geospatial information management; reactions and involvement of multiple stakeholders; its central role in SDIs and in the fulfillment of SDGs; experiences, projects, and steps taken on the road to standardization; and challenges posed by the current global environmental, technological, and cultural context.

Joan Tort-Donada and César López-Leiva: Toponymy, scale, change of scale—provide geographical and linguistic challenges in analyses. The aim is to correlate three concepts—toponymy, mapping, and scale—which have an affinity with each other, yet this is not readily resolved in GIS application, both theoretically and from a technical applied perspective. Firstly, basic concerns associated with the problem are presented: from the concept of the toponym/place name itself to the difficulties representing its spatial content, with an appraisal of questions as yet not fully resolved. This is followed by analyses of the spatial representation of specific toponyms and associated difficulties posed i.e. changing of scale. The point of reference is the system of toponymic representation currently used by the cadastral survey in Spain. The theoretical analysis, is juxtaposed with a case study concerning forests and vegetation in the Autonomous Community of La Rioja. Conclusions target how to coherently improve toponymic mapping.

Tal Yaar-Waisel: The Mystery of Hydronomy in the Land of Israel appraises how history, religion, and politics have been used interchangeably in selecting toponyms in the state of Israel from the beginning of the Zionist movement and more explicitly, after the establishment of the state of Israel. This research examines hydronomy and the direct and indirect place name meaning, paying special attention to the role of the education system in shaping the perception of the country's future citizens. In the Land of Israel, there is (i) the Gulf of Eilat, known globally as the Gulf of Aqaba; (ii) The Salt Sea, but known worldwide as The Dead Sea; and Israel has only one lake i.e. Lake Kinneret which for Arabic speakers is called Lake Tiberias, and the Sea of Galilee by Christians. This investigation into hydronomy postulates that in contrast to the early decades of the state of Israel, nowadays economic potential and political interests in naming are gaining a certain predominance.

Ferjan Ormeling, Helen Kerfoot, and Pier-Giorgio Zaccheddu: The United Nations Group of Experts on Geographical Names (UNGEGN), one of the nine permanent expert bodies of the Economic and Social Council of the UN, promotes the national standardization of geographical names, as an essential building block for international toponymic standardization and production of gazetteers, and creation of geographical names databases, all necessary for location-based services. This requires specialized training, unavailable in most countries, and that is why UNGEGN has developed training material, online modules, and contact courses in order to build on a national level the required capacities for geographical names collection and processing, creation of names databases, and their validating, and distribution of the standardized names.

Since 1982 UNGEGN has engaged in international training courses in toponymy, especially in developing countries. It focuses on the necessary training material to be made available, structure of courses, required expertise of lecturers, and teaching conditions. Included are criteria for selecting suitable fieldwork areas, and necessary hardware and software for student participants. The whole process from collecting the names in the field to their incorporation into databases and their publication on maps and in web applications is described by the lecturers/authors engaged in these courses.

Part II: Approaches to implementing standardization of place names

In Part II, the authors discuss standardizing geographical names for each country, nation, or language, and complications that need to be addressed. Familiarity with the most important and common terms associated with names and standardization is imperative as argued by Slovenian researchers emphasizing that cognizance of current levels of toponymic and hydronymic standardization are cardinal. Differences between endonyms and exonyms can exist from the perspective of an individual nation or ethnic group, hence standardization in bilingual and multilingual areas is particularly significant in Slovenia, Croatia, and Italy, which is legally regulated. These themes are further elucidated by research from the New Zealand Geographic Board presenting reasons why some place names have been changed there to reflect *Māori* rights in a post-colonial world. From a contrastive geographical zone, toponymic multiplicity in southern Africa demonstrates transitions in place naming. Assessing the toponymic palimpsest concept in Zimbabwe and the metaphor of the rainbow nation in the RSA is critically relevant to the (geo)politics of legacies in the creation and interpretation of multiple toponyms. In the colonial project, existing African human geographies, polities, boundaries, frontiers, and trading patterns were 'structured' within a European mental framework administrative map, reconfiguring, and (re)naming places. The inherited 'nation-state' building model today with its toponymy impacts international trade, tourism, and Googlization.

As elsewhere, this is witnessed in Spain where centrifugal/centripetal factors regarding toponymy are being played out. The Spanish state invests much in toponymic normalization but continues to face defies of standardization including mechanisms for the resolution of toponymic conflicts, heritages of historical–geographical aspects - legal, jurisdictional, administrative, and socio-economic. Official normalized toponymy databases can pose challenges as conflicts may emerge in adding new names or changing existing ones. The role played by the toponymic authorities, advisory bodies, and broader society in their resolution is vital. Spain being a multilingual country, with Spanish being the official language has regions with a local language that is also official in given zones e.g. Catalonia, Basque Country (Euskadi), and Galicia. Autonomous regions have local governments with jurisdiction over local toponymy. Linguistic policy as applied to toponymy in Spain is not uniform. Often the interest to standardize place names is essentially related to the promotion of a language itself, opening up larger geopolitics of polity and sovereignty as witnessed in Catalonia, Corsica and Northern Ireland.

Like Spain, Turkey has a rich heritage of toponymy including hodonyms. Historical and ethnocultural group successions have affected urban toponymy in Turkey as elsewhere. However, the most frequently used names in urban environments since the early years of the Republic created in 1923 stand out. The Ottoman heritage continues to affect urban toponymy. However, political parties make use of toponyms in 'owning' spaces that impact the image and electorates. Throughout the Balkans, Ottoman/Turkish culture left a toponymic heritage print juxtaposed with a predominantly Slavic culture, but also Germanic prints as witnessed elsewhere.

Drago Perko, Matjaž Geršič, and Matija Zorn: Standardization of geographical names on land and sea began in Slovenia after its independence in 1991. In 1995, it established the Commission for the Standardization of Geographical Names of the Government of the Republic of Slovenia, which, among other tasks, gradually began to standardize Slovenian geographical names based on databases of country names, settlement names, and Slovenian exonyms, all three of which are maintained and updated by the ZRC SAZU Anton Melik Geographical Institute, as well as a database of geographical names on national maps maintained by the Surveying and Mapping Authority of the Republic of Slovenia. It also prepared two gazetteers of standardized geographical names on national maps. These four databases and both gazetteers include over 220,000 Slovenian geographical names. Barely one percent of all geographical names in Slovenia are marine names, but they are already almost completely standardized.

Neil Lindsay and Robin Kearns: The New Zealand Geographic Board appraises the contested nature of toponymy in New Zealand and how processes of naming the landscape are inherently political as how, why, and what names are used, can create conflict between competing groups, beliefs or ideas. Here is analysed how toponymy is implemented and changed. Detail is presented of the work of the New Zealand Geographic Board Ngā Pou Taunaha o Aotearoa[12] and how decisions for

[12] 'Ngā Pou Taunaha o Aotearoa' metaphorically means 'The Memorial Markers of the Landscape'.

place names are made. It is postulated that naming practices in New Zealand have evolved in response to social and political shifts since the 1970s, particularly with changes in attitude towards restoring indigenous *Māori* rights and the importance of place names as 'taonga' (treasures), giving practical and tangible effect to the Treaty of Waitangi principles of participation, protection, and partnership.[13] Key issues surrounding the politics of toponymy in New Zealand, using case studies of recent name changes are detailed. The work of the Geographic Board reflects changing New Zealand society and culture.

Tendai Mangena: Theorizing multiple place names in southern Africa uses the metaphors of palimpsest and rainbow nation to demonstrate the unique features of toponymic multiplicity. Zimbabwe fits into the frame of the argument that many places in southern Africa are palimpsests, whose toponyms have been altered many times but still bear visible traces of earlier toponymic forms. South Africa evokes the metaphor of the rainbow nation and the ambivalent coexistence of colonial and indigenous place names, especially with its Apartheid heritages. Essentially, both metaphors are used to appraise the coexistence of toponyms that originated and are negotiated in different historical periods, and in the current situation. Traces of past toponyms bear competing histories and discourses.

Ayar Rodríguez de Castro: Conflicts and challenges in the standardization of geographical names examines toponymic prerogatives that are largely delegated to regional authorities in Spain. If a locality wants to change its name, it must request permission from the regional government. In order to have changes made official, local authorities must follow several steps and argue the need for the change, for which they can request a report from different consultative institutions on matters of toponymy such as the Royal Geographical Society of Spain (RGSS). The complexity of Spanish toponymy, including names that can be traced back to antiquity, or a multilingual scenario with a complicated regulatory framework, leads to frequent toponymic conflicts between various social groups and governments. Clarification of the protocols involved in standardization is enhanced by the RGSS as one of the toponymic advisory bodies.

Mar Batlle: Language policies in the field of toponymy in Spain are of much interest to the governance institutions and citizenry alike. Spain is a multilingual country in which Spanish is presently the official language of the entire state. However, there are regions (i.e. legal autonomous regions) with a local language that is also official in the given zone. Examples are Catalonia, Basque Country (Euskadi), and Galicia. Autonomous regions are administrative divisions, and have local governments that have jurisdiction over specific aspects, like the management of local place names. Therefore, linguistic policy as applied to toponymy in Spain is not uniform. In some bilingual regions, such as Catalonia, Galicia, and the Balearic Islands, toponyms are only set in the local language, in accordance with legislation and ordinary use. In other bilingual regions, toponyms can be set in one language (either Spanish or the local

[13] 1840 agreement between Māori and the British Crown, with articles and principles to preserve and respect Māori rights and culture.

one) or in both languages, such as in Basque Country or Valencia. In this scenario, involving several languages with different statuses and levels of bilingualism, with all the movement back and forth that they have had over the centuries, and with the subsequent diversity of cultural identities that form Spain, the interest to standardize place names is related to the promotion to the language itself.

Alpaslan Aliağaoğlu and Abdullah Uğur: In Turkey, urban toponymy has been studied by various sciences. The act of naming is the bearer of meanings. Urban toponymy addresses place names, including street, square, park, school, and bridge names, at all scales in urban spaces. Hence 25 published articles are examined by using content analysis. Research reveals an increase in the number of studies done on the origin of urban place names. These are especially related to the classification of street and neighbourhood names. The effect of Turkey's recent history of naming is elucidated by using 'republic' streets, referring to governance and those who founded the Turkish republic in 1922. Human characteristics of the geographical setting are important in the analysis. Due to the upsurge in interest in cultural geography, urban toponymy is appraised here with a critical approach. Local administrators with different political perspectives have tried to project their ideologies onto spaces by naming places. Also, urban names are used to create more prestigious urban spaces for economic gain.

Boris Avdić, Ranko Mirić, Nusret Drešković, and Marko Krevs: A major cultural crossroads in toponymy is provided in a case study of Bosnia and Herzegovina exploring a country that for centuries has been in an area where the influences of large cultural and civilizational spheres are intertwined. This type of impact has left a significant mark on local toponymy. As the Slavic population has played a dominant cultural role in this area, since the Great Migration Period (sixth–seventh centuries), it is not surprising that most of the local toponyms have Slavic origins. However, influences of other languages, such as Illyrian, Latin, Italian, Hungarian, German, and especially Turkish (in combination with Persian and Arabic), are also evident. The Ottoman Empire era—from the fifteenth to nineteenth centuries, provided a formative period for most of today's towns in Bosnia and Herzegovina. Sarajevo as the capital city is a well-known example. Confessional differences arose then among the domestic population that has persisted to this day. Implications of these variances are partly visible through toponymy. This became emphasized in some cases in contemporary contexts when toponyms gained their ethnopolitical and nationalistic dimensions, especially in the 1990s and attempts since then to build sustainable peace in the Balkan region.

Part III: Geo-histories, legacies, and toponymy transitions

In Part III, human geo-histories and territorialities are debated starting with the hypothetical question—can or should a toponym really be owned or deleted by state authorities; especially if it is centuries or even millennia old, and therefore, a heritage for all citizens. This is explored in toponymy in Anatolia holding multi-layered heritages. It is argued that the Ottomans did not really change many existing place names there until the final decades of the declining Empire. Major motivations for toponymic changes came in the era of the Turkish Republic attempting to create a greater centripetal national identity. Hence, researchers must be cognizant of the geographies, histories, chronology, and motivation for (re)naming. This is also evidenced in the toponymic geo-history of the Mohács plain, in Hungary, a multicultural Central European crossroads area for millennia experiencing population successions, but also state-driven ethnolinguistic homogenization policies that impacted toponymy. In the Mohács region, toponyms remain highly significant in the cultural landscape of central Europe, with narratives of religions, empires, Hungarian nationalism, world wars, Cold War antagonisms, and independence struggle to leave the Soviet Bloc in 1989. Mohács was embedded in a conflicted Cold War borderscape (1947–1991) and similarly impacted during the Yugoslav wars of the 1990s. Significantly, it is where the Hungarian authorities erected the 'anti-migration fence' in 2015 to prevent migrants and asylum seekers from entering Hungary and the EU.

Contemporary toponymic perspectives from Sarajevo in the new districts of New Sarajevo and New City give insights into the nature of ideological legacies, but more especially current socio-economic and global influences that affect toponymic meanings. Different social groups—ethnic, gender, political, and religious—are represented in the local toponymy, as illustrated by the driving factors for place naming since the 1990s. Historically such politico-religious ideological struggles for territorialization polity have left toponymic heritage not only in the Balkan countries but in other places including Malta, a crossroads of civilizations located in the Mediterranean Sea, with distinctive toponymic layers in its urbanscapes as also seen in the polyglot Maltese people, culture and language. While Gibraltar (BOT) shares a certain British colonial heritage with Malta, its toponymic history is unique regarding historical memory of territory and heritages. Due to the long linguistic, cultural, economic, and political linkages that Malta held with Sicily and Italy. Italian remained an official language of Malta until 1934 and is still one of the languages spoken by a majority of the population there. Malta was a British Crown Colony as of 1814 and declared independence in 1964 becoming a republic in 1974. Strong political voices demanding British withdrawal within Malta called for the islands to become part of the Italian state in the years leading up to its independence.

Such polyglot populations and toponymy are somewhat different in Brazil due to the characteristics of Portuguese colonial projects. This is seen by the toponymic study of the map of New Lusitania, a Portuguese cartographic product from the eighteenth century, that is enhanced by the techniques and skills used in comparing and contrasting different versions of such historical maps with the Toponymic Extraction Methodology and GIS applications. The present-day importance of the New Lusitania map is not only for toponymic research in Brazil, and Latin America but for world heritage. GIS remains crucial in furthering such research. Reflecting on pre and post-colonial toponymic heritages, in Brazil, Zimbabwe, South Africa and New Zealand and contemporary naming (geo)politics is enriching knowledge and skills.

Throughout different toponymic analyses, there is a strong strand in place naming linked to ideological territorialization and geopolitics, as with religion, monarchism, nationalism, republicanism, capitalism, and political-economic struggles of the twentieth century including totalitarian systems like fascism, Nazism, communism, and (economic) liberal democracy. From a critical perspective, place naming and the creation of large Soviet-style agricultural farms connect to each other in the narratives of Estonia before, during, and after the Soviet occupation era leaving toponymic legacies today is insightful.

Due to greater access to place naming material and archives from the twentieth century, it is easier to research names linked to totalitarianism such as Fascism, Nazism, and communism. In contrast, it is not always obvious the deeper normalized naming layers related to capitalism. Such naming phenomena are now part of the globalizing economic processes, enhanced by Hollywood images and place names.

Yılmaz Arı: Giving an identity to space through (re)naming, the practice of village renaming places in the period of the Republic of Turkey discusses the Ottoman Empire that lasted until 1923, including multi-ethnic groups that often lived in close proximity simultaneously or consecutively. The Ottomans frequently continued to use the toponyms that existed prior to their incorporation into the Empire. This situation began to change when the Union and Progress Party promoted renaming the villages in the 1910s. While some official decisions were taken in this regard, there was no significant development with this within the Empire that witnessed a period of unstable administration, chaos, and wars. The founders of the new Republic of Turkey (1923) attempted to build a nation-state and considered name changes as part of this effort. Several studies have been done on the Turkishization of toponymy, focusing on different aspects of renaming implemented by the republicans. The renaming efforts included the decisions in the 2010s, allowing the use of some previously banned names. Evaluation of the multiple dimensions of these initiatives is interesting regarding the motivations to change place names, criteria used, connection or continuity between new and previous names, and how ordinary people perceived name changes. Research is primarily based on the laws, regulations, guidelines, previous studies on renaming, and the author's first-hand observations.

Norbert Pap, Máté Kitanics, Marianna Ács, and Péter Reményi: The geo-history of the toponymy of Mohács plain in SW Hungary emphasizes that Mohács is geostrategic and of symbolic significance both during the Ottoman occupation, and 17th-century liberation wars. Decisive battles and military routes crossed the plain, transforming it into a memorial landscape. By the twentieth century, it became one of the most conflicted Cold War border-scapes, where the Hungarian 'Maginot-line' was constructed against the Yugoslav army in the late 1940-ies. The Homeland War of Croatia was fought just south of the plain in the 1990s and the anti-migration fence of Hungary was installed here in 2015. The region, which for centuries was controlled by the bishops of Pécs and Abbots of Szekszárd played an outstanding role in the Hungarian Reformation and is inhabited by Serbs, and Croats, Germans, and Gipsies alongside the majority Hungarians. The geographical names of the plain are well documented from the early Middle Ages, and reflect the different migration waves, military operations, and state policies. Mohács is an ideal research area which represents the different origins and bottom-up changes of Hungarian place names as well as top-down policies influencing them. This has to be read in the greater toponymic heritage narratives of Europe, without blurring the actual geographical scales of Mohács.

Elša Turkušić Jurić and Velid Jerlagić: Recreating the future is explored by architects, in the modern residential neighbourhood and existing toponyms in Sarajevo. Two municipalities in Sarajevo, Novo Sarajevo (New Sarajevo) and Novi Grad (New City), were built in the second half of the twentieth century, as a result of ambitious and numerous modern urban planning projects. Because pragmatic and rational processes treated the existing rural landscape as a tabula rasa, original toponyms were often formally retained. In the perceptions of present-day residents, the urban areas in these two municipalities are an integral embedded part of their identities and memories. The analysis starts from the supposition that, after several decades, these formerly new residential areas have given new meanings to pre-existing toponyms. It illustrates both the spatial and cultural characteristics of these meanings, and recent instances of their "re-decomposition", as a consequence of globalization. The research is based on comparison of six modern housing estates with analysis of the typological toponyms of Sarajevo's historic districts that gives greater insight into these urban and social transformations. Reflecting on this analysis in contrast to places outside the Balkan region offers interesting perspectives.

John Schembri and Ritienne Gauci: Street-naming in Malta provides a geo-cultural and political kaleidoscope, that can be seen as a reflection of colonial and post-colonial influences. Geographical scales here include the national (country), regional (towns and villages), and local (streets). A significant factor influencing street naming is the language used in identifying their location with a spray of vernacular Maltese (Semitic), Italian (Romance), and English. Although the Maltese Islands were occupied by a succession of rulers who were intent on exerting their influence in the central Mediterranean it was the Arabs (870–1090), Knights of St. John (1530–1798), and British (1800–1964) who left their major marks on the written and spoken word. The description and critique of the initiation, evolution, change, and purposes of street naming in Malta use three main sources. (i) Maps and sketches drawn by the official designate such as the Public Works Department (Malta) and the Ordnance Survey (UK); (ii) the minutes and proceedings of the street-naming committees in both Malta and Gozo; and, (iii) a ground-truth exercise at selected sites. A significant conclusion is that geopolitical influences exist at local, regional, and international levels, together with the transition from socio-religious to secular attitudes as reflected in street naming.

Paulo Márcio Leal de Menezes and Manoel do Couto Fernandes: Toponymic study of the map of New Lusitania, a Portuguese cartographic monument from the eighteenth century appraises a set of four versions of the map "Carta da Projecçaõ Geographica Espherica Orthogonal da Nova Luzitania or America Portugueza and Estado do Brazil." This provides a wealth of detail about Portuguese America, Spanish America, and other European domains that is impressive. For instance, toponymic content, presented in colonial and indigenous languages illustrates a hitherto unique set of the location of indigenous groups and that is considered one of the first ethnographic maps of the Americas or perhaps even worldwide. The research focus is on comparative studies of the toponymy used in each version, and the documents that allowed its creation. A database is being created for the identification of historical toponymy, and extraction of cultural elements, areas, and behaviour at spatial levels. Extraction of toponyms from the 1798 copy revealed the identification and classification of 4,750 toponyms, according to groups, linguistic and geographic motivations. GIS-based holistic research on toponymy and hydronymy including historical geography and cartography, linguistics, anthropology and heritages being carried out in Brazil and Argentina offers great insights for reflection, approaches and skills that are adaptable and applicable to other areas of the world. In the toponymic narratives of Latin America, we witness the major impact of the ideology of religion and particularly Christianity. In contrast, Communist ideology tried to eradicate religious ideologies and associated toponyms including odonyms at all geographical scales.

Taavi Pae, Ats Remmelg, and Jussi S. Jauhiainen: Research in the naming of collective farms in Soviet Estonia illustrates an innovative approach to appraising toponymy as place naming is part of national and local identity-making. Names of important territorial objects are often changed when state ideologies fundamentally change. Names are ideologized for the inhabitants' officially approved (re)education. However, some older names might remain in use despite being dubious after the changes in political power. Some citizens refuse to use new names or else create nicknames based on these. Analyses of farm names and their naming process in Estonia during the Soviet occupation era after World War II offers understandings into socio-political ideological constructs. The focus here is both on names and naming processes: the principles and practices of farm names, their connections to ideologies, and curious international, national, and local features, and contradictions to these.

Following World War II, Estonia became part of the Soviet Union. The ideological organization of the political economy included agriculture. Privately owned small farms were collectivized into large units that were important locations in the everyday lives of people in rural Estonia. Thousands of *kolkhozes* (cooperative collective farms) and *sovkhozes* (state-owned farms) needed Soviet names. Most names reflected directly the new political and economic ideology. These names were meant to honour major communist leaders, such as Lenin, Stalin, Kirov, Kalinin, and so forth; or the communist system with such farm names as *Uus tee* (New road), *Ühistöö* (Collective work). Nonetheless, names deriving from Estonian origin were also permitted if they outwardly fitted in with the new state ideology. However, the authorities overlooked some interpretations of national romantic names, such as those from the epic poem *Kalevipoeg*. In the early stages, politically oriented individuals involved in the farms independently suggested names. Later, an administrative regulation was introduced for approving names. Throughout the decades, the share of place-connected names increased and the number of names inspired by the cult referring to a person, Soviet patriotism and national (political) romanticism declined. Following the restoration of Estonia's independence in 1991, farmlands were privatized again, and large farm units were broken up. At the same time, almost all of the Soviet-era names vanished—but not entirely. Much place-naming research remains to be done at all geographical scales throughout the former Soviet and Socialist republics since 1991, including their ideological constructs. Besides old or former names being reintroduced such as Saint Petersburg for Leningrad, an analysis of new or innovative toponyms would be interesting especially, to gauge their ideological etymologies.

Part IV: Toponymy: narratives, languages, culture, and education

Part IV presents research regarding toponymy-related narratives, language, culture, and education. Concepts are elucidated including issues like identity, legislative framework, and multi-linguism. Analyses of colonial imprints on toponymy in Ireland are developed in two chapters; one from the standpoint of human geo-histories including the partition of the island in 1921, and respective place naming experiences there since then. Another chapter considers Brian Friel's play *Translations* written in 1980, while the play is set in 1833 rural Ulster when Ireland's toponymy was being (re)mapped, and 'standardized' by the British administration. Perspectives on the British Overseas Territory of Gibraltar (BOTG) (1704-present) share a certain English-British colonial heritage with Ireland and many places including Malta, a Crown colony from 1814 to 1964. Nonetheless, the respective toponymic narratives are quite distinct. In BOT Gibraltar, a dual place nomenclature developed, that of the top-down military and British administration, but also another in the areas where civilian migrants from Mediterranean countries created the proto-Ethno Gibraltarian population alongside street developments in their urban areas, quite removed psycholinguistically and cartographically from the British administration and population. Initially, many trading migrants came from the Genoa region in Italy.

Regarding Italy's place naming, the concept of the historical memory of a territory is debated, as is the definition of cultural heritage and expression of perception by generations throughout the centuries. Arguments are made why place names are one of the most important components of Italy's intangible cultural heritage, but in the contemporary world certification and standardization are all-important so as not to lose this. Challenges exist due to the complex links between place names, individual landscapes, and territorial categories. Strong examples are presented both nationally and regionally. This local–regional-national interplay and toponymic nexus emphasizes the need to preserve toponymic heritages. This is explored from a similar yet contrasting perspective from Indonesia. It is argued that toponymic studies require a multidisciplinary approach, that includes physical geomorphological narratives, while hydronyms are important cultural artefacts, highly relevant to education in disaster prevention, and mitigation as seen in the Bandung Basin region. Collective memories in the form of myths, legends, or oral traditions there help to preserve geographical names that local residents can learn from regarding contemporary natural emergencies.

By reflecting on the multifaceted perspectives on place naming presented, this provides a rich corpus of knowledge and skills from different areas of the world adding to original critical toponymic debates regarding heritage, culture, environment, geographical scales, and legislation about top-down authorities and institutions, alongside grassroot bottom-up toponymic realities. Such research supports the Sustainable Development Goals, including SDG 16 Peaceful, inclusive societies, and institutions, and especially SDG 4 Quality Education. The innovative and array of perspectives offered on place naming should feed into applied high-quality teaching and learning.

Jonathan Cherry, Brian Ó Raghallaigh, and Úna Bhreathnach: Reading Ireland's colonial and postcolonial toponymic landscapes posits that one of the most basic, yet powerful and symbolic acts of geographical appropriation is the naming of places. The research presents the diverse toponymy as an integral component of the cultural landscape capable of providing unique insights into social, political, and cultural attitudes and perceptions of those who have both named and renamed places through time. In illustrating the themes of the politics of naming, appropriation, conflict, and identity in both historical and contemporary urban and rural settings, a range of examples, from both the Republic of Ireland and Northern Ireland and a case study from Dublin have been deployed. An overview of the various bodies involved in managing and promoting Ireland's place names as well as the legislative framework within which it operates is also evaluated.

Charles Travis: In Translating Topographies: Brian Friel's approach to language, landscape, and toponymy in Ireland is considered. Friel's play *Translations* takes place in 1833 and is set in a "Hedge School" near the village of Baile Beag, in Donegal, in Ulster during the period Ireland was being mapped by the British Ordnance Survey. The tension between Gaelic *Dinnseanchas* (Lore of Place Names) perceptions of place and 'Anglo-Cartesian' appellations of space suffuse the drama. Yolland, the English soldier—the topographical engineer is befriended by Owen, the urbane son of Hugh the Hedge Schoolmaster as he seeks to translate Gaelic place names into English words and labels for the survey's new maps. Analyses of *Translations* references past debates between Friel's 'fictive character' and geographer John Andrew's 'historic' perspectives on the impact of the British Ordnance Survey not only in Ulster but on a wider linguistical landscape. It also contextualizes *Translations* within other literary geographies detailing global 19th Century industrialization processes in addition to the current post-Brexit milieu of Northern Ireland as a means to address the complexity of landscapes, identities, and senses of place at play in twenty-first century Ulster. Contrastive references to the diffusion of British toponymic standardization strategies in Ireland and BOT Gibraltar give insights into top-down and bottom-up processes in place naming.

Jennifer Ballantine Perera: The Overlaid Past: The Politics of Space and Memory in Gibraltar's 'Doubling' Street Naming Principle engages with the resilience of the use of traditional street names in Gibraltar together with the resistance they posed to the implementation of official street names. The permanence of traditional names raises questions regarding Edward Said's theory of imaginative geographies, which whilst active in Gibraltar, discursively and colonially in mapping out the territory, proved less effective when it came to the urban centre. In fact, this space had become imaginatively mapped by Gibraltarians through their local knowledge of these streets and through the use of a linguistically and culturally codified naming principle that Services personnel and English settlers found quite impenetrable. By way of unpacking the tensions generated through the (co)existence, two different sets of nomenclature invariably stem from a similar imaginative process. The research draws upon a number of street name lists produced in different historical eras. Some of these became necessary as navigational guides to assist non-locals in finding and therefore accessing streets in Gibraltar's city centre. They also draw attention to the limitations inherent in a colonially informed mapping when overlaid by the supposedly liminal yet powerful imaginative mapping held by Gibraltarians. This is a process that not only led to a doubling of names but also to the presence of two Gibraltars. The toponymic narrative of the BOTG is so different and yet so geographically close distance-wise to Italy helps to illustrate the importance of place name research from the grass-roots bottom-up in each specific milieu.

Laura Cassi: From historical to new place names, the case study of Italy explores toponymy, and the need to identify the specific setting so as to proceed to the toponymic organization, which is a basic component of the historical memory of a territory. Thus they are a valuable part of the cultural heritage and expression of perception by generations throughout the centuries of their environment. This is all the more important in today's world which is subject to enormous processes of certification and standardization. They deserve to be protected and foregrounded by way of careful data collection and specific inquiry into the complex links between place names and individual landscape and territorial categories—waterscapes, vegetation, shapes, and characteristics of the land, settlements, historical road systems, and human activities—as well as the complex relationship between the whole place name corpus of a territory and its landscape, and the birth of new place names, originating, for example, in the tourist industry. After the general geographical lines of development of place name research in Italy, an outline is presented of a number of examples both nationally and regionally relevant to the above-mentioned themes.

Multamia RMT Lauder, T. Bachtiar, and Cece Sobarna: Geographical names represent a memory of places as illustrated in the Bandung Basin, West Java, in Indonesia which is a vast country with 17 thousand islands covering three time zones and with the cultural diversity of 718 languages. Hydronymy and toponymy in the Bandung Basin offers a rich achieve of geological, physical geographical, anthropological, linguistic, and associated data that is useful today in human survival in the region. The Basin is home to the Sundanese ethnic group and is the base of the Ancient Bandung Lake which was formed due to the eruption of *Mount Sunda* 105,000 years ago. The Ancient Lake collapsed around 16,000 years ago, gradually receding, leaving behind a vast lake and wetlands. Archaeological findings around the lake suggest that the surrounding area once supported early human habitation with sustainable water and food. The Sundanese have a legend called Sangkuriang that mentions the existence of the lake, as well as the mythical origin of the lake's creation. This may suggest the collective memories of the lake's existence, transmitted by oral tradition through the ages. In this Basin, we found an abundant variety of generic terms for hydronyms. However, business and tourism interests appear to be less compliant with the UNGEGN resolution to use local names. Changing geographical names actually eliminates the accumulated knowledge capital about natural conditions including earthquake mitigation.

Ruth McManus: Exploring place names in teaching and learning provides a rich source for education at all levels including activities, and questions for reflection on multiple disciplinary scales. To achieve this, it is imperative that sources provide resources to encourage both self-directed readers and educators to reflect on the themes presented in this volume where a wide range of sources and methodologies is presented. Combined with this are both chapter-specific and broader questions, project exemplars, and suggestions for extending learning by embracing the concepts of place, naming, identity, education, heritage, and the SDGs.

Part V: The Relationship Between Geographical Naming and Cultural Politics

Part V reviews the standpoints and conclusions of forty-eight multidisciplinary researchers working on geographical naming in different parts of the world. They analyze from different perspectives the state of the art in place naming due to its multilevel significance in issues of identity, perceptions, culture, polity, sovereignty, geopolitics, and GIS database creation and management. This includes not only the technical geospatial perspectives alongside the imperative of name standardization in a globalizing world but also the dynamic aspects of intangible cultural heritages embedded in names and cultural politics. It is highly significant in supporting the objectives and targets of the UN SDGs. The researchers work in numerous universities, regional, national, and international geographical naming-related institutions, and organizations including the UNGEGN (UN Group of Experts on Geographical Names), IGU (International Geographical Union), and ICA (International Cartographic Association) among others.

Conclusion

In conclusion, by brainstorming the topic of place naming this produced a rich range of ideas, concepts, themes and methods, and interdisciplinary experts contributing to toponymic studies. This illustrates the tensions between citizens' everyday lived realities on the ground in situ with multi-layered human geohistories, alongside the current organic characteristics of a place and name(s). This is in contrast to the continuous efforts of governance power brokers, epitomized by the State with its political-culture agenda, interconnected with global governance institutions. Hence approaches, actions and structures facilitating the continuous peaceful resolution of these tensions is vital as reflected in the work of UNGEGN and supported by education and research. Based on this, the research methodology chosen with its broad benchmarks is deliberate, as it allows for a cross-sectional approach and conclusions. This standpoint is supported by the evidence presented by the respective authors with theory juxtaposed with case studies from a wide range of countries. Engaging with this, geographical education as discussed in the final part of the book draws together and underscores the multiple aspects of place naming, but also the emerging common themes that educators must engage with.

Challenging Conceptual and Theoretical Approaches to Place Naming

Assessing the Validity of Critical Toponymy Perspectives for Understanding Human Perception of Places: An Analytical Framework

Sungjae Choo

Abstract The research field of critical toponymy has been established as a platform to investigate political elements, power relations and economic, cultural, and linguistic motivations that are involved in adopting, using, and changing place names. Critical toponymy perspectives have contributed to understanding different identities which are sometimes vigorously contested. Conflicts can arise when naming a place or geographical feature, reflecting diverse human perceptions of places. This chapter revisits the validity of critical toponymy perspectives in elucidating human perceptions of places. It draws on three elements to design an analytic framework: formation of memories; development of discourses; and paths of exercising politics and power. It then examines the case of the Russian city of Volgograd, which was renamed from its former name of Stalingrad, but still experiences intense pressure of return to its former name. A discussion is presented regarding how different memories that are accumulated in a name are exposed and developed into discourses, and supported by which groups of people.

Keywords Critical toponymy · Place perception · Pluralism of memories · Volgograd · Stalingrad

Objectives

- To illustrate how the research field of critical toponymy has been established as a platform to investigate political elements, power relations and economic, cultural, and linguistic motivations that are involved in adopting, using, and changing place names.
- To explore how critical toponymy perspectives have contributed to understanding different identities which are sometimes vigorously contested.
- To show how conflicts can arise when naming a place or geographical feature, reflecting diverse human perceptions of places.

S. Choo (✉)
Kyung Hee University, Seoul, South Korea
e-mail: sjchoo@khu.ac.kr

© The Author(s), under exclusive license to Springer Nature Switzerland AG 2023
G. O'Reilly (ed.), *Place Naming, Identities and Geography*,
Key Challenges in Geography, https://doi.org/10.1007/978-3-031-21510-0_2

- To apply the three main elements of the design of an analytical framework: formation of memories; development of discourses; and paths of exercising politics and power.
- To examine the case of the Russian city of Volgograd, which was renamed from its former name of Stalingrad, but still experiences intense pressure to return to its former name.
- To discuss how different memories that are accumulated in a name are exposed and developed into discourses, and supported by various stakeholders.

Introduction

Naming a newly-built bridge connecting the land and an island on the southern coast of Korea in 2016 produced a major controversy. Two names were proposed: Paryeongdaegyo, named after Paryeongsan (*daegyo* and *san* are generic terms for bridge and mountain, respectively) by the municipality on the western side; and Jeokgeumdaegyo, named after Jeokgeumdo (*do* is a generic term for island) by the municipality on the eastern side.

Each side possessed its own perception of the feature, i.e., a bridge leading to a landmark mountain of the area versus a bridge connecting to an island located at the center of a picturesque landscape. As neither perception would be abandoned by its corresponding its advocates, the two proposals were forced to be put into a vote at the national names committee meeting. The committee sided with Paryeongdaegyo, and placed more importance on the mountain name. The municipality who lost the vote, however, refused to abandon its perception. Indeed, the name Jeokgeumdaegyo was later adopted for the bridge connecting the other side of the island to the next one.

Different perceptions commonly emerge, and not just in situations of new naming. It is often the case that certain names are used with more approval than others, even standardized ones, and express the affinities and memories of specific groups of people. In addition, for a single geographical feature, multiple groups can exist, e.g., nations, ethnic and language groups, and regional communities, who want to implement their chosen term which is replete with their specific perceptions, memories, and identities. Various dimensions of perception could also lead to disputes over a name of a location to be adopted for common use.

Revealing the power relationship and political elements involved in the establishment, use and change of place names, sometimes combined with economic motivations, constitutes a crucial trend in current toponymic research. The field of *critical toponymy* focuses on whom, with what motivation and place perception, and through what path and method, names are proposed and promoted.

The perspective of critical toponymy contributes to understanding how human perception of place is formed and developed around place names. This chapter establishes a theoretical framework for this purpose, mainly based on current trends of critical toponymic research. It then examines, with this framework, the case of naming a Russian city Volgograd, which was renamed from Stalingrad, but which is still under pressure to return to its previous name.

The Perspectives of Critical Toponymy

The emergence and development of critical toponymy as a research field is summarized well by a series of prominent research reviews (Berg and Vuolteenaho, eds 2009; Rose-Redwood et al. 2010; Rose-Redwood and Alderman 2011). Inspired by critical studies of space and place, or critical geography, a shift occurred from etymological and taxonomic foci to place-naming as a "political practice par excellence of power over space" (Pinchevski and Torgovnik 2002).

Selective genealogy, as termed by Rose-Redwood et al. (2010), has been deepened and diversified by numerous theoretical and empirical accomplishments. Choo (2019) identified three pertinent lines of inquiry: representation of ideologies by place-naming and the process of involving politics and the exercise of power; possession of power and its relationship with the commodification of place names; and issues of language use and place names as cultural heritage. It is worthwhile to discuss them in order to further elucidate the elements of perception of places.

Representation of Ideologies and Commemorative Naming

Place-naming accomplishes the symbolic function of connotation and the utilitarian function of denotation (Azaryahu 1996). As the connotation encompasses cultural values, social norms and political ideologies (Eco 1986), conflicts inevitably arise when different or contrasting elements are revealed in the process of representing ideologies, identities, and other symbolic meanings. The primary concern of critical toponymic research is the path through which place names, as strong semiotic texts rooted in larger meanings and discourses, are read and interpreted differently by each group (Rose-Redwood et al. 2010).

Regime transformations have been specifically identified as an important factor that influences naming or renaming places. As an efficient and familiar media to infuse new ideology, political and social systems, and practice of power, place names have served a potent tool for justifying and strengthening such changes (Rose-Redwood 2008a, b; Berg and Vuolteenaho, eds 2009; Azaryahu 2012). Name changes for these purposes were commonly observed in pre-modern and modern history, which experienced the turbulence of colonial and post-colonial, communist and post-communist, and dictatorial and post-dictatorial regimes (Palonen 2008; Nash 2009; Kim 2010; Lam 2016; Sartania et al. 2017). In a case of contemporary communist society, Kim (2014) shows how directly the names of streets and subway stations in Pyongyang, North Korea, support the socialist revolution and its corresponding ideological images.

Indeed, name changes that take place in accordance with the orientation of regimes are subject to diverse facets of development. Renaming may occur selectively in some symbolic places as "part of a larger project of nation-building" (Karimi 2016, 740), or could take the form of "toponymic cleansing" (Rose-Redwood et al. 2010, 460) in a radical context. It is evident that the retreat of former powers or changes in ruling ideologies require the restoration of original place names or the adoption of new names, which creates a controversy in place-naming practices. Two perspectives continually collide: one that regards the remaining past place names as irrelevant remnants to be eradicated; and another that views them as political or cultural heritage that is clearly worth preserving.

Among critical toponymy researchers, commemorative naming has attracted relatively greater attention. Commemorating a historical figure or an event is normally motivated by the desire to ensure identification with the object, to remember or represent the person associated with the place, or to honor the spirit or victims of the place-based event. Debatable issues arise, however, when a commemorative name is given or imposed by a power group without an appropriate procedure of consensus-building.

Commemorative naming appeals to political powers because it is able to "transform an official discourse of history into a shared cultural experience that is embedded into practices of everyday life" (Rose-Redwood et al. 2010, 459). This advantage exerts a strong influence when the historical symbolism represented by the person or the event, and/or the memory of his or her life or the development of the event, are embodied as a common experience of the name users. Instances have often appeared in history, in which a tragic accident occurred to a national leader, which then led to strong solidarity among the people. Commemorative naming could be adopted as an efficient tool to support this process. However, as shown in the case of naming after Israeli Prime Minister Yitzhak Rabin after he was assassinated, too much naming could diminish such influence (Azaryahu 2012).

It is also the case that commemorative naming sometimes diffuses to names of other generic types, and may even survive after the original name has disappeared, e.g., Avenue de Stalingrad in Paris, which still remains after Stalingrad was renamed to Volgograd. Consequently, the object of commemoration may not attract the desired attention. For example, when people call one of the main streets in Seoul *Chungmu-ro* in their everyday lives, they generally just take note of the location, and not the historical symbolism or accomplishments associated with the sixteenth century marine admiral Yi Sun-sin (Chungmu is his *ho*, a kind of pen name, and *ro* is a generic term for street). Rose-Redwood et al. (2010, 459) refers to this as when "history becomes geography." This street was given a Japanese name, *Honmachi*, in the colonial period, but renamed to Chungmu-ro after liberation, and developed into a center of the Korean film and publishing industries. When this narrative is added, the meaning of 'who' is further diluted. This situation could be regarded as "geography becomes storytelling" (Choo 2019, 454).

Place Names with Economic Value and Commodification

Critical toponymic perspectives have been extended to the topic of commodification of place names, the nature and formation of naming rights, and the logic of capital and power surrounding them. Research is actively underway to investigate the characteristic, in which a place name has a potential exchange value that exceeds its use value, and to explore its essence (Medway and Warnaby 2014; Light and Young 2015; Karimi 2016; Medway et al. 2018; Rose-Redwood et al. 2018).

Choo and Kim (2015, 434) introduces three aspects of how the economic value of place names as commodities has increased: a general trend of emphasizing economic value and pursuing commercialization; developing concepts of branding and image-making in place-naming as a means of value creation; and increased motivation to utilize place names in promoting regionalism.

A related issue on a macro-level is that urban naming rights programs have emerged as a neoliberal urban policy idea (Rose-Redwood et al. 2018) in response to increased inter-urban competition. Although a theoretical discussion on naming rights has yet to be elaborated, a few case studies present the process, results, and meanings of transacting naming rights: those of subway stations (Rose-Redwood et al. 2018) or sports stadiums (Medway et al. 2018) that accompany money flows; and those of streets (Karimi 2016) or general urban facilities (Light and Young 2015) that remain as symbolic transactions.

The transfer of naming rights has been shown to be a process in which a public territory transforms into a private space, which blurs the boundary between the public and the private (Rose-Redwood et al. 2010, 457; Karimi 2016, 741). In addition, the influence of the private sometimes extends to general economic and political realms. For instance, the Etihad Stadium, home of the Manchester City F.C., and named after its sponsor Etihad Airways by a naming right contract, has left the name in nearby streets and facilities, and thus influenced the entire urban landscape (Medway et al. 2018).

It should also be noted that transacting naming rights may not be welcomed by local residents as users, and may even experience strong resistance due to their cherished values. In the case of Winnipeg, Canada, new commercialized names for sports facilities were viewed by the local community as "cheapening the symbolic value of urban toponymic landscapes" by overshadowing "historical commemoration and honorific recognition of real service, sacrifice and achievement" (Rose-Redwood et al. 2018, 13). It is highly reasonable that local communities will have strong emotional ties with the location of their local teams and possess salient memories and experiences connected to their names. Consequently, renaming of such a place could produce a major shock by imposing a contextual mismatch (Medway et al. 2018).

Place Names as Cultural Heritage and Language Issues

Critical toponymic issues extend beyond political and economic, into cultural and linguistic, spheres. Initiated by the United Nations Group of Experts on Geographical Names (UNGEGN), two topics have been identified: the perspective of seeing geographical names ('place names' in this chapter) as cultural heritage; and the discussion of the nature and usage of exonyms. These two themes possess a commonality, in which they aim to achieve the basic value of respecting the cultural elements and practices that ethnic groups and language areas have maintained through place names. In direct contrast to such aim are the initiatives of power groups to control the names of minority groups and a series of economic logic that obscures the cultural realm (Choo 2019).

The effort to respect place names as cultural heritage often causes conflict by attempting to give names from political or economic motives. The power struggle that exists between the two sides is diametrically opposed: one side is concerned with place names adopted as a means of solidifying political authority and symbolic domination; and the other side is concerned with names that represent legends, myths, cultural history and identity, linguistic changes, and embodied lifestyles. On the one hand, a function exists that produces brand value through place names as a commodity. On the other hand, a function exists which creates landscapes with social inclusion that accommodate minorities, women, and the relatively weak.

UNGEGN's concern for place names as cultural heritage was initially motivated by the need to preserve minority or indigenous language place names. This focus of language use has great relevance in terms of investigations of place names from a critical perspective. As identities accumulated in place names should be conveyed in their source language, resistance occurs when the names are replaced with those in other languages. For instance, place names in the Hawaiian language represented the 'spiritual ecology' of the indigenous people, and thus it was very important for them to restore their names that were destroyed by invaders, in spite of all difficulties and controversies (Herman 2009). Specific language conventions and lexical relationships with their original pronunciations are also regarded as bases of perceiving places and their names (Vuolteenaho and Berg 2009; Kearns and Berg 2009).

Furthermore, the issue of language use is closely connected with practices of exonym use. An exonym, i.e., a name with a form that is different from the language of the area where the feature is located, expresses the history of invasions, exchanges, and/or cultural diffusion (Choo 2018a, 143). As a consequence, its use could sometimes produce problems between countries. Moreover, problems have occurred which have been followed by diplomatic requests: that of Cote d'Ivoire in 1986 to use this name in French, and not the English exonym Ivory Coast; and that regarding Mongolia and the South Korean government when establishing diplomatic relations in 1990 concerning using 몽골[mɔ:ŋ gɔl] rather than 몽고[mɔ:ŋ gɔ], in which the Korean exonym had a negative meaning of 'stupid and old-fashioned.' Indeed, UNGEGN recommends reduction of the use of exonyms that could give rise to international problems.

Which language is chosen constitutes a highly important political decision in multilingual areas or on occasions of language conversion. Indeed, it sometimes results from a power struggle and can be immersed in conflicts. The political motivation and elements of conflict development that intervene in this decision process are essential aspects for critical investigations of language use.

Understanding Place Perception from Critical Toponymy Perspectives

Place names are the product of human perceptions of places, features, and facilities. As no place name is assigned randomly, identification of what kind of human perception played a central role in the naming process is a crucial component of toponymic research. Critical perspectives are expected to have the capacity to investigate the process of competing or contrasting perceptions regarding a name in order to substantially enrich the discussion. This section uses three elements to design an analytic framework: formation of memories; development of discourses; and paths of exercising politics and power.

Memory

Place names contain layers of memories inherited from the past, and are sometimes differentiated by time period. Variegated memories sometimes appear through a name, resulting from different feelings and experiences related to the object of naming. Certain conflictual memories, and subsequently names, cannot be accepted by specific groups. As a consequence, opposing groups may propose a different name that is more acceptable to them. Critical toponymy has contributed to understanding how the competing or contrasting memories are evoked, and how they are developed with conflicting arguments of power groups.

First, it should be noted that differentiated memories can be formulated and inherited from the same historical event, and thus markedly dissimilar feelings can exist for the name associated with this event. Such occurrences as colonial rule, war, dictatorship, civil movements, etc., could be perceived differently, depending on where the individual was situated and how he or she was influenced by the turbulence of these historical flows. For instance, for those who experienced severe sufferings from dictatorial practices, commemorative naming after the dictator cannot be accepted. In contrast, for those who benefitted economically under an authoritarian regime, it may be favorably perceived.

Varied views on historical events are associated with positions taken concerning the nature of cultural heritage, particularly contrasting thoughts on painful histories or unacceptable social norms. In the case of place names in Korea adopted during the

Japanese colonial period, vigorous debates are ongoing. Some argue for renaming, as they felt that colonial traces should be eliminated; while others advocate for their maintenance, as they are cultural heritages that contain memories of users for about one hundred years (Choo 2018). Furthermore, certain names that possess implications of stereotyped gender roles have been renamed, but arguments exist to maintain those names, as they reflect the social reality of the time in which the name was assigned. Similar debates continue regarding seemingly derogatory names.

The issue of language use is also concerned with how names are perceived and become accumulators of memories. For instance, memories are expected to be enlarged in the name Cote d'Ivoire, not Ivory Coast, and in Aotearoa as 'land of the long white cloud' in Maori. Dual naming in a native language area constitutes a way to respect memories that have long accumulated in their native names. Renaming Mount McKinley as Mount Denali in 2015 after almost one hundred years of its standardization can be partly interpreted as a restoration to the historical container of memories of the natives, as 'the high one.'

Naming practices in recent years have aimed to increase the value of places or facilities to be named, and attempted to create new or updated memories associated with those names. Cases of renaming administrative units in Korea with this motivation have fostered debates: Hanbando-myeon, taken from a specific feature shaped as the Korean Peninsula (*Hanbando* in Korean, *myeon* is an administrative unit of the small locality); Kimsatgat-myeon, named after the wandering poet Kim Byeongyeon (satgat is his pen name). The renaming was driven by the desire to have their locality better be better known by tourists and buyers of their products. Opposing opinions were voiced that advocated for the cultural heritage values of those replaced names. What memories coincide with new names after 12 years of adoption have yet to be investigated.

Finally, memories accumulated in a place name through the attachment of individual feelings develop into collective values, and then proceed into the sphere of discourse. Individual emotions are solodified, reinforced, and sometimes influenced by outside social or political factors, and enter into the realm of conviction, belief, and social engagement. A series of discourses are addressed and formed with involvement of expert insights.

Formation of Discourses

Place names representing ideologies and commemorative names are inevitably followed by formal or informal discourses. One track of discourse is formulated around the function of boosting solidarity among community members or nationals. Examples of such discourses are: place names that reflect the communist ideology, e.g., Victory Street in Pyongyang; and a commemorate patriotic figure or event, e.g., Ho Chi Minh City, named after the Vietnamese communist leader.

More conspicuous discourses are those that focus on universal human values, such as peace, social justice, reconciliation, human rights, inclusiveness, resistance to human oppression, etc. These discourses firmly contend that unique place names containing users' identities are an important means of guaranteeing human rights, and thus achieving social justice. Accordingly, giving due consideration to minority language names and preserving them is highly recommended. Place names that recall inhuman events disparage the socially disadvantaged should be excluded. In cases of controversial names between regions or countries, a substantial path to bring peace or reconciliation is to seek naming practices that accommodate perceptions and identities of all users, and thus resolving conflicts (Choo 2018, 171).

Respecting the African American legacy in North America has occupied an important position in the advancement of place names discourse, as well as diverse empirical studies. For instance, one way is commemorative naming of African-American rights activists, such as Martin Luther King, Jr, (MLK) (Alderman 2000, 2008), and another is the renaming of streets, schools, and urban facilities that commemorate racists or slavery advocates (Smith 2018; Thornton 2019; Tucker 2019). MLK naming is understood as 'symbolic resistance' to the cultural appropriation and exclusion of African-Americans. However, it is still restrained by the interests of real estate stakeholders who may be highly concerned with the possible resultant decreasing market value of their assets (Alderman 2008). Anti-discourses can also emerge, e.g., "All Lives Matter" in response to the "Black Lives Matter" movement (Smith 2018, 114).

Discourses sometimes also evolve in a more precise way. The issue of social justice in place-naming is elaborated into the aspects of procedural and distributive justice (Rose-Redwood et al. 2010, 465). Respecting identities contained in each name, for example, preserving place names of minority ethnic or language groups, is regarded as realization of procedural justice; whereas, designating those names to proper populated locations or urban facilities in relevant contexts belongs to the sphere of distributive justice. One of the measures to achieve gender equality in place-naming is the support given to commemorative naming after women in Austria (Jordan 2016).

There are also discourses aimed to increase the brand value of place names and corresponding places. Participating in the globalization trend and upgrading the status of place names could constitute valid logics. As a pertinent case, regarding the restructuring of street names in Kabul, Afghanistan, such phrases as 'democracy,' 'human rights,' and 'freedom of expression' were introduced to show that their society was becoming democratic, progressive, and cosmopolitan (Karimi 2016). However, this kind of economic discourse is not substantially externally addressed, possibly due to the stronger orientation toward cultural and humanities aspects.

Finally, it is worth noting that discourses on the methodologies of respecting place names also exist. One of them are arguments surrounding the proposal to use multiple names for one feature or place. The proposal to use the name East Sea together with the Sea of Japan has been made with layers of discourse, including benefits, norms, examples, practical methods, etc. (Stoltman 2017; Choo 2018). Very exceptionally, temporal solutions can also be offered and accepted, e.g., the nine-day official use of the name Stalingrad for Volgograd in Russia.

Politics and Power Exercise for Discourses

Every toponymic discourse is argued for, and supported by, a certain group of people. Collective movements develop to advocate for their name based on their beliefs and interests, sometimes through physical demonstrations or reactions in social network services. When deciding on a standardized name of the bridge cited in the introductory part, members of the names committee received visits from municipality representatives and heard the shouting voices of local residents who were brought by chartered buses. These movements are more pronounced when multiple names are in close contention. Indeed, politics and power are often exercised in these discourses and disputes.

In most cases, multiple groups become visible, irrespective of the type of naming dispute: when multiple names are fiercely advocated for by different parties; when attempts are made to remove existing names, e.g., removing racists' or dictators' names; or when monopolizing names are opposed, e.g., Sobaeksan-myeon in Korea, which was taken from a nearby mountain, but blocked by a neighboring municipality that shared the feature. Arguments are made to support each of the competing names or to oppose specific names and prevent them from being accepted.

Then, what groups appear more conspicuously? Within a nation, several political or interest groups extend various and/or contrasting views on place-naming based on their particular beliefs and positions. Conservative parties are normally apt to adopt names that represent national developments, regarding either a person, an event or an ideology; whereas, progressive parties express strong desire to move away from names associated with encroaching on human rights. Furthermore, as the economic aspects of place names are emphasized, interest groups other than political ones emerge, which are often segmented by economic sectors, municipalities, or even groups of citizens. They are mostly motivated to increase, or at least maintain, the value of their places by including corresponding place names or excluding devaluing names.

Religious groups sometimes also become active in such debates. For example, proposals of the Bongeunsa Station and the Sudeoksa Interchange in Korea (*sa* is a generic term for Buddhist temple) faced strong oppositions from Christian groups. Some family groups in Korea have also attempted to inscribe their name, e.g., Songchon, meaning village of the Song family (*chon* is a generic term for village), and Yichon as village of the Yi family. Interestingly, the name Songchon was adopted 200 years after the Song family migrated to the area (Kwon 2004).

Naming disputes between countries are generally handled at the national level. Each government, normally dominated by diplomatic channels, is endowed with the mission to address its national position and interests regarding naming. Governments are supported and given consultations by expert and civil groups, as well as ordinary citizens. Although these support groups usually maintain the same position as their governments, their ways of working may sometimes differ, i.e., strongly, weakly, rigidly, or flexibly.

The media is also an important channel for exercising politics and power. Private media companies establish their own policies for place-naming, but are sometimes strongly influenced by their stakeholders, e.g., readers, listeners, advertisers, and/or governments. The media can also become an influential actor in national interests. Indeed, when its voice is strong enough to induce high expectations of the people, the government's position can also be affected. Consequently, it is anticipated that the increased influence of social media will attract more attention from critical toponymy researchers.

For third-party name users, controversial names could be objects of political or economic deals. It is commonly found that place names representing relatively stronger power occupy more accessible and exposed locations. For example, in street naming in Toronto, Canada, names originating from the First Nations were given to less accessible streets, while names that corresponded to the imperialistic legacy occupied the main streets of the city (Casagranda 2013). MLK street naming is also concentrated in urban undervalued areas (Alderman 2000). This is an issue of 'placing names' as compared to 'naming places' (Azaryahu 2012, 74).

As an important name-receiver, map-makers are sometimes positioned at the center of a 'cartographic deal.' They design maps and adopt names according to their policies, and considerable room may exist to apply them flexibly. For instance, multiple names could be chosen with an order of priority between them, e.g., using a bracket or a slash. In online maps, a decision should be made concerning in which steps of zoom-ins the second name will be shown. In addition, sometimes no name is inscribed, realizing 'cartographic silencing' (Smith 2018, 113). When two sets of naming disputes exist between two countries, a decision could be made to take sides one by one, thus realizing 'toponymic barter.'

The Case of Stalingrad Versus Volgograd

This section will apply the framework of understanding human perception of places drawn from critical toponymic perspectives to the case of Stalingrad versus Volgograd. For this purpose, reports, interviews, and contributions in the Russian media since 2002, along with extant literature, are examined and reviewed, starting from when the issue of renaming Volgograd to Stalingrad began to emerge. A preliminary summary is provided in Choo and Jin (2020).

Background of the Name Changes

The rapid transformation from despotism to a socialist state in Russia, achieved through revolution and civil war, motivated Bolsheviks to inject their revolutionary ideology into members of society. Various state reorganization works based on the Bolshevik ideology were implemented in a radical way, of which the ideological transformation of living space occupied an essential part. Renaming places was one of the effective ways that this goal was achieved, in addition to the construction of city monuments. Initiated by the renaming of Taldom (Тáлдом) in Moscow Oblast to Leninsk (Ленинск) in 1918, cities and towns which had been named after emperors, members of the noble class, and biblical or nationalist figures were given new names of revolutionaries, politicians, war heroes, or aspects of communist ideology (Sartania et al. 2017; Demyanov and Ryzhenko 2017; Choo and Jin 2020).

Tsaritsyn (Царицын) was one of the cities that experienced such renaming. Located downstream of the Volga and near the Don River, the city established itself as a transportation hub that connected the Caucasus, Central Russia, and the Caspian and the Black Sea (Fig. 1). It was through its connection to Stalin in the Russian Civil War (1918–22) that Tsaritsyn emerged in history. As the head of the Communist Party's military service, Stalin defended the city to prevent the White Army on the Black Sea coast and in the Caucasus and Central Russia from connecting with each other. He was considered to have established the foundation for the city to serve as a strategic hub by cultivating the military power of the Red Army through his effective management of food and materiel supplies (Institute of Military History 1984).

Tsaritsyn became a space that symbolized Stalin's military career and state management capabilities, and simultaneously, a place that was closely related to the myth of his personal heroism. It was for this reason that the proposal to rename

Fig. 1 Location of Volgograd

Tsaritsyn into Stalingrad (Сталинград), meaning 'Stalin's city,' was passed in 1925 by the regional Soviet committee. However, it was known that Stalin himself was against this proposal, and suggested that the city be named after Sergey Minin, a revolutionary (Komsomolskaya Pravda 2018).

The geopolitical importance of Stalingrad, which developed into a heavy industrial city with Stalin's special focus, was reaffirmed through World War II. The center of military production attracted worldwide attention by becoming the stage of the Battle of Stalingrad (1942.8–1943.2), which was considered the biggest battle of the war. This fierce battle continued for six months to secure a bridgehead connecting the Volga River and the Caucasus, as well as to occupy the rail transportation hub. After a close push-and-pull battle, the Soviet army finally defeated the German army. Hundreds of thousands of lives, however, were lost on both sides, and the name Stalingrad remained as a symbol of this indomitable struggle of the Soviet military. It can be reasonably asserted that the symbolism of being a city with the name of Stalin, the leader representing the Soviet Union, led to a greater fierceness in the Battle of Stalingrad to both protect it and take it over (Beevor 1998).

The hero city of Stalingrad experienced a major change with Stalin's death in 1953. In de-Stalinization, led by Nikita Khrushchev, the name of Stalingrad was a core aspect. The direction of condemning Stalin, which began with the 1961 Soviet Communist Party Convention, was developed by local authorities and institutional media in Stalingrad to form public opinion to change the city's name. It is crucial to note that opinion was solicited only from members of the industrial class, e.g., factory workers and managers, and those of agricultural labor, education and culture personnel, and even veterans who fought in the Battle of Stalingrad and participated in urban restoration afterwards, were ignored (Lipatov 2017). The name adopted in 1961 after a naming contest was Volgograd (Волгоград), representing "the city at the Volga River."

Despite the usual challenges involved in the name adoption process, the new name of Volgograd was instituted without much conflict. It was after the dissolution of the Soviet Union, however, that a serious change occurred. Specifically, political and interest groups in Russia began struggling for power and over ideology, and a strong demand for honoring Stalingrad's memory increased. This was attributed to the emergence of the desire to reevaluate the role of the leader Stalin and the importance of the Battle of Stalingrad in modern and contemporary Russian history. It also involved an existing trend, in which markedly diverse political and social groups in Russia had created a complex environment by presenting views based on their own interests (Kangaspuro and Lassila 2017).

The issue reappeared from 2002 to 2003, marking the 60th anniversary of the Battle of Stalingrad. Since the Communist Party members of the Volgograd City Council proposed the return to the name Stalingrad, the renaming issue expanded to include the greater aggregate Russian historical memory, and not just the Communist Party's political struggle. As previously mentioned, this was based on discourses of the historical significance of the Battle of Stalingrad as "the Great Patriotic War" and reevaluation of the historical figure of Stalin (Terekhova and Steglenko 2015; Glushchenko 2016).

Table 1 Days of the year when the name Stalingrad is officially used

Day	Celebration
February 2	The day of the defeat of the Nazi troops in the Battle of Stalingrad
February 23	The day of Guardians of the Country
May 8	The day that Stalingrad was awarded the title of hero
May 9	The victory anniversary (Victory Day)
June 22	The day of the outbreak of the Great Patriotic War (Day of Memory and Sorrow)
August 23	The day of the massive bombing of Stalingrad by Nazi aircraft (Day of Remembrance)
September 2	The day of the end of World War II
November 19	The day that the counterattack began near Stalingrad
December 9	The day of Russian heroes

It was 10 years later, January 2013, that another noticeable change took place. The Volgograd City Council made the decision that the name "Hero City Stalingrad" be officially used for six days to celebrate its patriotic or war-related significance, to which three more days were added in December of the same year (Table 1). Although this measure was a decision that considered both sides, neither was satisfied. This problem has since emerged on almost every anniversary of the Battle of Stalingrad.

Conflicting Memories

The name Stalingrad began as a commemoration of Stalin, who led the victory of the Russian Civil War and showed his great ability to manage the country through industrial development. This new name, which replaced the name of Tsaritsyn which was more than three hundred years old, expressed strong praise of the leader who spearheaded communist development.

The battle against Germany that occurred after 27 years of its use had a profound influence on its connotation. Initially, the name Stalingrad functioned as denoting the location of the critical spot of the war. However, as it became the largest battlefield in World War II, with more than two million casualties in six months, the place called Stalingrad was given a corresponding identity. The victory that followed fierce battles, named "the Battle of Stalingrad" during the war, played a crucial role in enriching the connotations of symbolism represented by victory, patriotism, and national pride in the name Stalingrad. For individuals who participated in the battle and for all of the groups that were involved, these connotations became power that supported them and a meaning that they felt should absolutely never be forgotten.

Memories of Stalingrad advocates are currently centered on this glorious victory in the war. For them, Stalingrad, and not Volgograd, "became a symbol of their military art and the highest courage, and known to the peoples of the world as a turning point in the fight against fascism" (comment by Patriarch Kirill, head of the World Russian People's Council, report by Forbes Russia, 2013. 2. 5). However, it might be difficult to contend that the symbolism implied by Stalin be excluded from the name of Stalingrad when the Battle of Stalingrad, and not Stalin himself, should be commemorated. The fact that a strong motivation and urgency existed that the place named after the leader could never be taken away in the course of the battle shows that Stalin occupied a great portion of the symbolism of the name Stalingrad. Indeed, it is highly likely that the victory of the Battle of Stalingrad was written and spoken about in close connection with Stalin's heroic actions. The resistance that arose at the time of changing to the name of Volgograd also reflected this sentiment.

This is where conflicting memories collide. The Battle of Stalingrad brought the glory of victory, but simultaneously instilled the trauma of the war that caused an immense number of casualties. The victorious leader Stalin evoked memories of victory, but the dictator Stalin prompted memories of horrific oppression and brutality. The following comment illustrates this emotion: "Imagine that you are from Volgograd and your relatives were repressed and died because of Stalin. How can these people live in this city that bears the name of the murderer of their relatives?" (Comment by Vladimir Vorsobin, journalist, talks in Radio Komsomolskaya Pravda, 2013. 2. 6).

These conflicting memories are understood as "commemorative pluralization" (Kangaspuro and Lassila 2017, 153) that emerged after the dissolution of the Soviet Union, with various narratives about the leader Stalin and the historic Battle of Stalingrad, differentiated by the activities of individuals and civic groups.

Discourses

Khrushchev'S first attempt in 1956 to remove the name Stalingrad failed due to resistance from groups who clearly remembered and honored the heroic accomplishments of Stalin. The situation in which the memory of the leader who had died just three years ago was respected by the people caused a passive attitude of Communist leaders who wanted to evade responsibility (Serebryakov 2019). This eventually caused Khrushchev to fail due to overwhelming opposition in votes at the Communist Party Convention. Indeed, military groups, which accounted for the majority of the voters, and veterans of the Battle of Stalingrad, could not be persuaded.

The success of the renaming in 1961 was the consequence of systematic discourse development, centered on the trauma of Stalin's regime. The anti-theses to Stalin narrated five years prior to 1961, such as "damage of the Marx-Leninist principles" and "suppression and execution of innocent people," changed into more traumatic ones, such as "a murderer of competent commanders of the Red Army" and "a criminal who oppressed and murdered many innocent people" (Gradoboev 1962). As Khrushchev expected, Stalin became a name that evoked the trauma of illegal massive purges and oppressive dictatorship for many. This trend naturally led to the rejection of Stalingrad.

The extension of the discourse of trauma also had a great influence on those who later insisted on returning the name Volgograd to Stalingrad. They want to focus on the memory of victory by emphasizing the Battle of Stalingrad, and exclude the trauma imposed by the dictator Stalin. They expanded the connotation of the name Stalingrad, claiming that it symbolizes the glory of the most brilliant victory in world war history, as well as in Russian history, and commemorates everyone who sacrificed himself or herself for the victory. In addition, they insist on restoring Stalingrad before generations change and memories of victory disappear (Ozerov 2018).

Therefore, the issue became a question of whether the person Stalin, as the origin of the city's name, could be separated from the place name Stalingrad. One side stresses that the name of Stalingrad has already become a proper noun that exists separately from the person Stalin. The opposite side, however, argues that the restoration of Stalingrad, which is inexorably associated with Stalin, will absolutely express his terrible crimes. The former position is sometimes supported by the contention that Stalin's achievements were enormous, and thus must be acknowledged. The opponents, on the other hand, vigorously argue that, irrespective of how much the victorious memory of the Battle of Stalingrad is emphasized, ultimately, the name Stalingrad intrinsically refers to Stalin, his nefarious deeds, and the dark past of the Soviet Union. For them, the return to Stalingrad is a profound insult to those who were killed and suffered during Stalin's tyranny and in the war.

The return to Stalingrad, along with the statues of war heroes, weapons, and war-related street names currently existing throughout the city, is believed to constitute a peak of war semiotics (Yanushkevich 2014) that create a patriotic feeling and atmosphere in everyday life. However, there is also a strong rejection of the history of war. Decorating the present with events that have already become history, and continuously evoking the memories of war as a means, are contrary to future-oriented development. For some Russians, "the city will turn into a monument to the war if Volgograd becomes Stalingrad" (comment by Mikhail Turetsky, conductor, report by Komsomolskaya Pravda, 2015. 3. 18).

Finally, it is worth noting that the Stalingrad discourse is expanding from the political to the economic and cultural realms. A trend exists that emphasizes the brand value of the name Stalingrad. Stalingrad can be viewed a brand that symbolizes war and victory, and can be an important factor in attracting capital and tourists. Supporters of this point focus on the benefits that can be obtained by the citizens: "I never doubted the need to return to the great name Stalingrad. Not for the sake of Stalin, but of the people of Stalingrad" (comment by Dmitri Rogozin, former Deputy Prime Minister, report by Komsomolskaya Pravda, 2015. 3. 18). Another view exists, as well, in which the benefits of using the brand will exceed the administrative costs of the renaming (Kangaspuro and Lasila 2017, 160). However, it is also possible that a continuing controversy without a resolution could cause greater interest in the city, instead of ending with a name change.

In terms of the cultural domain, a discussion about returning to the city's first name, Tsaritsyn, has attracted great attention. The name was mentioned as a candidate when the name of Stalingrad was removed in 1961, but was rejected because of its association with the old system that suppressed workers (Serebryakov 2019). However, Tsaritsyn is known to have received more than 10% support in a poll conducted in April 2019 (report by Komsomolskaya Pravda, 2019. 4. 12). This fact shows that, even after nearly one hundred years after its removal, the old name could be revived, along with the sense of place and the cultural heritage residing within the name.

Exercise of Politics and Power

The demand for renaming or restoration to Stalingrad, which has been underway over the past 20 years, has not converged in one direction, and is developing in a complex manner. This may be largely ascribed to the contemporary situation, in which each component of Russian society vigorously expresses its opinions based on its own understanding and interests. The different thoughts, interests, and political tendencies of varied groups and generations, as well as the active use of social network services that convey them, deepen this complexity.

Contrasting memories and commemorations, and the resultant development of discourse, appeared differently depending on the nature of the group. The restoration movement advocating returning to the name of Stalingrad was initiated by the Communist Party, which had little political influence in Russian politics after the 1993 constitutional crisis. They wanted to demonstrate their existence and lionize the Soviet Union to the public. Stalingrad was also supported by veterans and conservative citizens, many of whom were members of the war generation, who insisted on commemorating the Battle of Stalingrad and returning to the name that contained that sense of place.

The discourse against the name was raised by many groups in January 2013 after the Volgograd City Council decided to officially use the name of the Hero City of Stalingrad on war and military anniversaries. The first reaction was made by the Justice Russian Party, which aimed to achieve social democracy. Its chairman, Nikolai Levichev, stated at a press conference that Volgograd should not be renamed to the former name related to the tyrant, even for a moment, in ceremonies commemorating the victory of the Battle of Stalingrad (report by Ria Novosti, 2013. 1. 31).

The contrasting position was expressed by the Russian Orthodox Church. The church itself declared that it would not take an official position, but Patriarch Kirill, a bishop of the church, emphasized that Stalingrad should be viewed separately from Stalin (report by Forbes Russia, 2013. 2. 5). The head of the church's public relations department, Vsevolod Chaplin, added that the word Stalingrad is associated with the celebration of a well-known victory (report by NTV, 2013. 2. 6). This seems to reflect the pride of the Russian Orthodox Church that it contributed to the victory of the war by declaring itself to be a patriotic religion.

The vociferous arguments for returning to Stalingrad produced great repercussions. The opposition was led by those who valued human rights, based on the inseparability of Stalingrad from the person Stalin. The main actors were the Presidential

Council for Human Rights, civic groups on human rights, and the Yabloko Party, which advocated for market liberalism. For them, "the glorification of the tyrant has nothing to do with the memory of the Great Patriotic War and its victories" (comment by the Yabloko Party, report by NTV, 2013. 2. 6), and "the public celebration of Stalin will be perceived as a positive attitude of Russian people towards this person, which is impossible" (Arseny Roginsky, head of Memorial, a human rights society, report by Interfax, 2015. 2. 25).

Another important aspect to note is the position of Putin, Russia's highest leader. In 2002, he had opposed the restoration of Stalingrad on the grounds that it could raise the suspicion that he was attempting to return to Stalinism. However, in a conversation with veterans of the Normandy Landings in 2014, he suggested the possibility that local residents could change the city's name through the process of voting, based on the Russian Federal Constitution. This can be evaluated as that the Putin government remained apart from the issue, albeit supporting the principle of local autonomy, and refused to mediate the confrontation between political and social groups over the naming issue. The government's non-interference should be viewed as the Putin administration's calculated action to unite the entire people and strengthen nationalism by continuously evoking the issue of patriotism (Choo and Jin 2020).

No clear difference in opinion seems to exist by generation; in general, however, the progressive younger generation opposes commemorating the Battle of Stalingrad excessively, due to its nature of returning to the past. For the generation born after Volgograd was adopted, the sense of place contained in this name is more important. However, teenagers and those in their 20s who are familiar with using social network services express pros and cons about the renaming. This is interpreted as a growing interest in the conservative trend resulting from the entire environment of Russia, as well as a lack of sufficient awareness of the impact of historical choices (Terekhova and Steglenko 2015).

A Summary

Volgograd-Stalingrad naming debates began with conflicting opinions on whether commemorating Stalingrad could be viewed solely as the Battle of Stalingrad while the person of Stalin was excluded. Plural memories contained in one name were developed into tracks of discourse, based on the perception of the place as victory, sacrifice and patriotism on the one hand, and on the perception that rejected the hero Stalin's city on the other.

Interestingly, conflicting objects of memory and commemoration, and differentiated discourses that subsequently developed, appear differently depending on the nature of the power groups with distinct political tendencies. On one side are the Communist Party, veterans, conservative citizens mainly of the war generation, and the Russian Orthodox Church. On the other side are opposition parties and organizations aimed at protecting human rights. The Putin regime claims to respect the choices

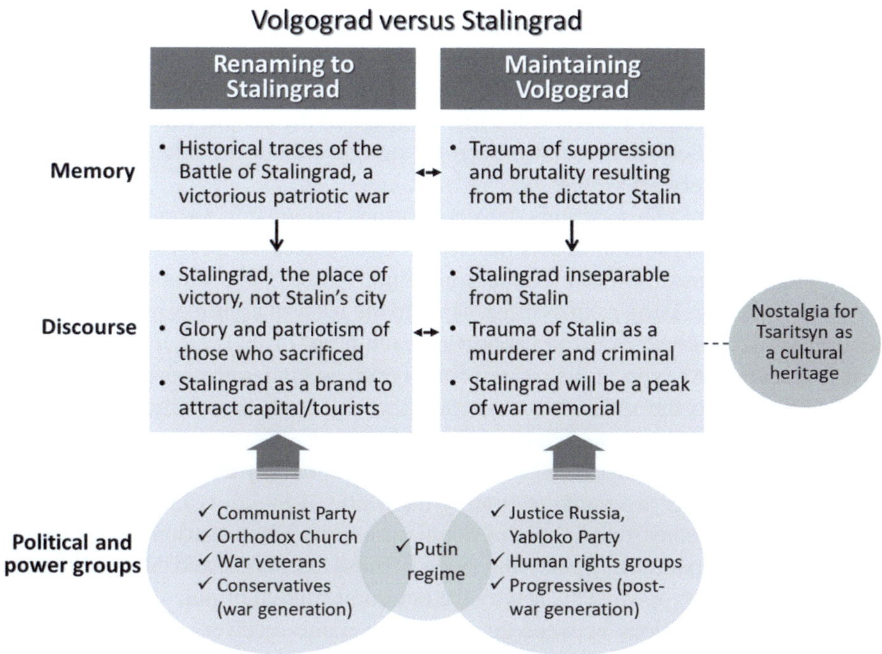

Fig. 2 Pluralism in the Volgograd-Stalingrad naming disputes. *Source* revised from Choo and Jin (2020)

of local residents, but implicitly tolerates the Stalingrad argument to strengthen its governance through national unity and the solidification of nationalism. Figure 2 shows this comprehensively.

Conclusion

The perspectives of critical toponymy have contributed to understanding human perception of places, features and urban facilities by examining the conflicting thoughts surrounding the naming, the extension of its arguments, and the path of exercising politics and power to carry them out. The motivation of the claim to use or not to use a name evolves from the realm of political ideology to economic, cultural, and linguistic values and heritage. Crucial to elucidating naming arguments is how layers of memories have been accumulated, how discourses centered on each memory have developed, and what political and power groups have advanced their views based on the respective discourses.

The movement of renaming Volgograd to Stalingrad provides an exemplar case for investigating how individuals or groups of a diversified society exhibit differentiated orientations of memory, and how contrasting discourses are formed and developed. Stalingrad advocates argue for the Battle of Stalingrad as the object of commemoration, centered on memories of victory, sacrifice, and patriotism. Those who oppose this emphasize the trauma of Stalin's dictatorship and oppression, and the much larger need to acknowledge this trauma and the people who suffered from it. Religious, political, and social groups are on one side or the other, according to their identity, beliefs, and interests.

The pluralism of memory and commemoration contained in a single place name, and the resultant accumulation of different senses of place, is expected to be a subject of continuous interest in future critical toponymic studies. The points of debate that appear in the progressing of this difference into the form of conflicts or disputes, and the process of various groups in supporting each point of their arguments, demand further research. As seen in the case of Stalingrad, an elaborate approach is needed to find answers to the question of whether the object of commemoration can be separated into more than one sense of place.

The debate about the name of Stalingrad is already deeply embedded in the political realm. Whenever an event celebrating the Battle of Stalingrad is held, the issue will reappear and constitute a topic of popular interest. This situation is expected to persist until the generations possessing either of those memories, and those influenced by them, almost disappear. This is why the temporal context of place perception and differentiated perceptions between generations as name-receivers are so important.

Questions

1. Why was the research field of critical toponymy established?
2. How have critical toponymy perspectives contributed to understanding different identities which are often contested?
3. How can conflicts arise when naming a place or geographical feature?
4. What are the three main elements of the research analytical framework used in this chapter?
5. Why and how was the Russian city of Volgograd renamed from its former name Stalingrad?
6. How have different memories accumulated in a geographical name such as Stalingrad been exposed and developed into discourses, supported by various stakeholders?

References

Alderman D (2000) A street fit for a king: naming places and commemoration in the American South. Prof Geogr 52:672–684

Alderman D (2008) Place, naming, and the interpretation of cultural landscapes. In: Graham B, Howard P (eds) The Ashgate research companion to heritage and identity. Ashgate, Aldershot

Azaryahu M (1996) The power of commemorative street names. Environ Plann D: Soc Space 14:311–330

Azaryahu M (2012) Rabin's road: The politics of toponymic commemoration of Yitzhak Rabin in Israel. Polit Geogr 31:73–82

Beevor A (1998) Stalingrad. Viking, New York
Berg LD, Vuolteenaho J (eds) (2009) Critical toponymies: the contested politics of place naming. Ashgate, Surrey
Casagranda M (2013) From empire avenue to Hiawatha road: (Post)colonial naming practices in the Toronto street index. In: Proceedings of the international conference on Onomastics "Name and Naming", vol 2, pp 291–302
Choo S (2019) Towards a multi-dimensional critical toponymic research. J Korean Geogr Soc 54(4):449–470
Choo S, Jin S (2020) Pluralism of memories and commemoration and the accumulation of placeness contained in place names: the case of Volgograd/Stalingrad. J Korean Geogr Soc 55(4):409–426
Choo S, Kim H (2015) The brand value of place names: topics in economic geography. J Econ Geogr Soc Korea 18(4):431–449
Choo S (2018a) People, places and place names. Hanul M Plus, Paju
Choo S (2018b) The essence and practicality of dual naming: considerations toward the East Sea/Sea of Japan dual naming. J Korean Cartogr Assoc 18(3):33–43
Demyanov KV, Ryzhenko GV (2017) Ideology, toponymy and politics of memory: On the massive renaming of cities in the USSR. Bull Omsk Univ. Ser: Hist Sci 4:153–160
Eco U (1986) Function and sign: semiotics of architecture. In: Gottdiener M, Lagopoulos A (eds) The city and the sign: an introduction to urban semiotics. Columbia University Press, New York
Glushchenko GY (2016) Identity of Volgograd: between the USSR and Orthodox church, culture, politics and understanding. In: Philosophical and political problems of identity: Russia and the modern world, pp 187–194
Gradoboev N (1962) Destalinization: exposures 22nd congress
Herman D (2009) The Aloha state: place names and the anti-conquest of Hawai'i. In: Berg LD, Vuolteenaho J (eds) Critical toponymies: the contested politics of place naming. Ashgate, Surrey
Institute of Military History (1984) Ministry of Defense USSR: Military Encyclopedic Dictionary
Jordan P (2016) Commemorative naming: Austrian experiences. In: Presentation at the Workshop on Implementing Resolutions: Experiences of Commemorative Naming, 29th Session of UNGEGN, 27 April
Kangaspuro M, Lassila J. From the trauma of Stalinism to the triumph of Stalingrad over Volgograd: the toponymic dispute. In: Fedor J et al (eds) War and memory in Russia, Ukraine and Belarus. Palgrave Macmillan, Cham
Karimi A (2016) Street fights: The commodification of place names in post-Taliban Kabul City. Ann Am Assoc Geogr 106(3):738–753
Kearns R, Berg L (2009) Proclaiming place: towards a geography of place name pronunciation. In: Berg LD, Vuolteenaho J (eds) Critical toponymies: the contested politics of place naming. Ashgate, Surrey
Kim K (2014) Changes of urban structure of Pyongyang represented in road names. J Cult Hist Geogr 26(3):34–55
Kim J (2010) A critical assessment on the colonial dual city of Gyeongseong (Keijo). J Seoul Stud 38:1–68
Komsomolskaya Pravda (2018) Dolmatov. Stalin: 1878–1953. Main documents
Kwon S (2004) The social construction of place-names as a landscape text. Geogr J Korea 38(2):167–181
Lam PTT (2016) (Re)naming Hanoi streets after historical people from colonialism to post-colonialism: creation and recalling collective memories. Onoma 51:25–44
Light D, Young C (2015) Toponymy as commodity: exploring the economic dimensions of urban place names. Int J Urban Reg Res 39(3):435–450
Lipatov AB (2017) Stalingrad-Volgograd: the history of the renaming of the hero city on November 10, 1961. News Volgograd State Pedag Univ 10(123):209–214
Medway D, Warnaby G (2014) What's in a name? Place branding and toponymic commodification. Environ Plan A 46:153–167

Medway D, Warnaby G, Gillooly L, Millington S (2018) Scalar tensions in urban toponymic inscription: The corporate (re)naming of football stadia. Urban Geogr 1–21
Nash C (2009) Irish place names: post-colonial locations. In: Berg LD, Vuolteenaho J (eds) Critical toponymies: the contested politics of place naming. Ashgate, Surrey
Ozerov V (2018) Why Volgograd needs to be renamed Stalingrad. Actual Comments 2. 1
Palonen E (2008) The city-text in post-communist Budapest: street names, memorials, and the politics of commemoration. GeoJournal 73(3):219–230
Pinchevski A, Torgovnik E (2002) Signifying passages: the signs of change in Israeli street names. Media Cult Soc 24(3):365–388
Rose-Redwood R, Alderman D (2011) Critical interventions in political toponymy. ACME: An Int EJ Crit Geogr 10(1):1–6
Rose-Redwood R, Alderman D, Azaryahu M (2010) Geographies of toponymic inscription: new directions in critical place-name studies. Prog Hum Geogr 34(4):453–470
Rose-Redwood R, Sotoudehnia M, Tretter E (2018) Turn your brand into a destination: toponymic commodification and the branding of place in Dubai and Winnipeg. Urban Geogr 1–24
Rose-Redwood R (2008a) Indexing the great ledger of the community: urban house numbering, city directories, and the production of spatial legibility. J Hist Geogr 34:286–310
Rose-Redwood R (2008b) From number to name: symbolic capital, places of memory, and the politics of street renaming in New York city. Soc Cult Geogr 9:431–452
Sartania D, Nikolaishvili D, Ujmajuridze A (2017) Soviet toponymy: the history and the present. Earth Sci 6(5–1):49–55
Serebryakov A (2019) "Many voted against": Stalingrad was renamed Volgograd 58 years ago—chronicle of events. V1.ru. 11. 10. https://v1.ru/text/culture/2019/11/10/66301870/
Smith B (2018) Engaging geography at every street corner: using place-names as critical heuristic in social studies. Soc Stud 109(2):112–124
Stoltman JP (2017) Geographical naming: Reflections on peace, cooperation and social justice. In: The society for east sea (ed) Achieving peace and justice through geographical naming: proceedings of the 23rd international seminar on sea names. Seojeon Tech, Seoul
Terekhova SA, Steglenko EV (2015) Volgograd or Stalingrad? Dispute about the names in the context of Russian politics and historical memory. Topical Issues Soc Sci: Sociol, Polit Sci, Phil, Hist 5–6(46):83–90
Thornton MP (2019) Geographical naming: the case of Hugh MacRae Park. In: Paper presented at the 10th international conference on geographic naming and geographical education. Northeast Asian History Foundation, Seoul, Korea
Tucker R (2019) Civil war: community reactions to the renaming of Robert E. Lee Elementary School in San Diego, California. In: Paper presented at the 10th international conference on geographic naming and geographical education. Northeast Asian History Foundation, Seoul, Korea
Vuolteenaho J, Berg LD (2009) Towards critical toponymies. In: Berg LD, Vuolteenaho J (eds) Critical toponymies: the contested politics of place naming. Ashgate, Surrey
Yanushkevich I (2014) Semiotics of social memory in urban space: the case of Volgograd(Stalingrad). Int J Cogn Res Sci, Eng Educ 2(1):43–50

Websites Visited for Interviews Cited in the Text

https://www.forbes.ru/news/233918-vsemirnyi-russkii-sobor-podderzhal-pereimenovanie-volgograda-v-stalingrad
https://www.kp.ru/daily/26028.4/2945777/
https://www.volgograd.kp.ru/daily/26355/3237712/
https://www.volgograd.kp.ru/daily/26965/4020726/
https://www.ntv.ru/novosti/457340/
https://www.interfax-russia.ru/south-and-north-caucasus/main/memorial-protiv-pereimenovaniya-volgograda-v-stalingrad-i-pamyatnika-stalinu-v-moskve

Legacies and Place Naming: Perspectives from Korea and Japan

Gerry O'Reilly

Abstract This chapter appraises the cartography-digitalization nexus in globalization processes, including Google's power in place naming, alongside the geopolitics of emotions regarding toponymy and hydronyms in Japan-Korea relations. This reflects perceptions, territorialization(s), historical legacies, and claims to sovereignty, alongside aspirations for the future. Google's rationale for selecting geographical names is reviewed as are alternatives from international institutions. Like analyses of historical maps, created in specific political time-slice contexts, many international associations were founded in the colonial era and (re)invented after the World Wars, presenting standpoints influencing international law. Inevitably, such holds embedded legacies and systemic perspectives. Complex post-colonial and post-WWII heritages continue to be shared by Japan and Korea; so disputed toponymies cannot be read in isolation from geopolitics and changing relations between neighbouring states in Northeast Asia including China.

Keywords Cartography · Digitalization · Place-naming · Territoriality · Japan · Korea · Colonialism · Legacies

Objectives

The objectives of this chapter are to:

1. Appraise the cartography-digitalization nexus in globalization processes and to assess Google's potential power in place naming.
2. Evaluate the geopolitics of emotions regarding place names in Japan-Korea relations, and elucidate perceptions, territorialization(s), historical legacies, and claims to sovereignty in the post-Colonial context.
3. Review processes whereby toponyms are selected or prioritized at home and in the international community, and evaluate the use of historical maps in disputed place-naming.

G. O'Reilly (✉)
Dublin City University, Dublin, Ireland
e-mail: gerry.oreilly@dcu.ie

© The Author(s), under exclusive license to Springer Nature Switzerland AG 2023
G. O'Reilly (ed.), *Place Naming, Identities and Geography*,
Key Challenges in Geography, https://doi.org/10.1007/978-3-031-21510-0_3

4. Appraise how post-colonial and post-WWII legacies continue to be shared by Japan and Korea regarding disputed toponymies.
5. Explore the role of toponymy regarding geopolitics and international relations in Northeast Asia.

Introduction

This chapter reflects on the changing world of cartography and digitalization and the subsequent implications of this alongside GIS and globalization processes. The power potential of the Google corporation in setting mapping agenda is explored, as with place names used on maps and the Japan Korea labeling dispute. This reflects perceptions, territorialization(s), emotions, and claims to sovereignty. Google's rationale for choosing geographical names is reviewed as are possible alternative policies from international institutions such as the UN, International Hydrological Organization (IHO), and international and national Geographical Associations, including Geographical Societies in Japan and Korea.

Regarding analyses of historical maps dating back centuries, created in specific cultural, and political-economy contexts, it should be noted that many international associations and authorities were founded during the colonial era, being reinvented after the world wars presenting 'new' perspectives and laws, but often with embedded legacies and systemic views.

Whatever the map image presented, there exists a realpolitik standpoint on the ground as becomes evident in the naming of the sea area between Korea and Japan, and also islands. Besides the physicality of these spaces, human psychological and cultural associations and consequent geopolitics of emotions exist as illustrated by the numbers of people visiting museums in Japan and Korea dedicated to place-naming and territorial claims. In this context, relationships between the former imperial power Japan having to adapt to the realpolitik post-WWII environment, post-colonial heritage, and reinventing itself in the twenty-first century have been challenging. The Republic of Korea has faced epic challenges in redefining itself nationally, regionally, and internationally since the mid-twentieth century facing the challenges of political stability, economic development, relations with North Korea, and of course heritages of colonization, annexation, and war impacting on collective landscapes, language, memories, and grievances. The geographical naming issues cannot be read in isolation from the geopolitics of emotions (Moïsi 2010).

Reflections on the Cartography-Digitalization Nexus

Essentially, a map depicts geography—the science that interconnects data and interprets the inter-relationships between physical environments and human phenomena. Embedded in this is human territoriality i.e. control and defense of territory—resources, people, culture, and identity. Much debate has surrounded 'territoriality' concepts since the nineteenth century; whether it is genetically innate, or socially constructed learned behaviour. Many researchers have argued that it is a combination of both. Kernel to territoriality is geopolitics (territory/earth and power), where geopolitical processes can exist from the smallest local scales to wider and global levels. However, in certain historical disciplinary contexts, the term geopolitics has sometimes been over-simplistically restricted to a synonym for statecraft or international relations.

Obvious expressions of human territoriality include the state itself with boundaries, defence structures, and flags delimiting a territory, with an organized political community under one government and recognized as such by a majority of other states. The modern 'state concept' is based on the Westphalian nation-state model acting as the basis for organization in the international community and upheld by the UN Charter. Culture, including place-naming and psycho-geography acts as the human-environmental interface reinforcing attachment to place and identity, especially in a globalizing world epitomized by the digisphere.

Technology has revolutionized cartography with major changes to map creation, diffusion, and usage open to people worldwide. Due to digitalization, map production is no longer the preserve of a limited number of individuals, geographers, cartographers, or civil servants, nationally or internationally. Any map is only as good as the data input i.e. available, selected for presentation. Hence the accelerated need to (re)appraise endogenous and exogenous geographical names as a basic element in map making and updating. This is nothing new in itself with a myriad of examples worldwide throughout history and especially in colonization-decolonization processes and conflictual cultural territorializations, but also less subtly in propaganda cartography. However, what is different now is the scale and compression of time–space geography in the cartographical process and map diffusion:

(i) Precision in mapping physical phenomena—due to digitization there is ever greater exactitude down to the smallest scales; onshore physical and human phenomena included e.g. offshore islets, outcrops, continental shelf features. This is enhanced by GIS map layering. Besides the changing physical geography of such being naturally subject to erosional and accretion processes, there's also the biota that can be monitored with updating in real-time. Therefore, geographical phenomena on the sea surface, in the water body, on the ocean floor, and under it can be mapped with ever-greater accuracy. This affords more economic opportunities for coastal states especially, in their EEZ (Exclusive Economic Zone) up to 200 nm as measured from the coastal baseline (UNCLOS 1982) but also has implications for the sea areas beyond the 200 nm i.e. common heritage of humanity.
(ii) Computerized speed in creating a map can now be globally diffused in the digisphere within seconds, and a picture with selected toponyms paints a thousand words (or worlds). However, with the proliferation of fake news, there are also fake maps that can lead to conflict.

In the colonial scramble of the imperial powers to carve up the world's territories and seas, and to avoid inter-empire armed conflict between themselves, many international organizations were developed from the 1880s on, promoting their concepts of international law; while other national and international associations were created such as the International Geographical Union and associated cartographical societies. Since then, many scientific data have been accumulated attempting standardization, depending on a range of historical factors, but one must remain cognizant of systemic attitudes, mind-sets, and portrayals of places and people, particularly the exogenous perceptions. On 25 May 2020, tragically occasioned by the killing of George Floyd, an African-American by a white policeman in Minneapolis, USA, the American and global public was reminded of the legacies and systemic wrongs of the colonial project, with street protesters demanding the removal of historical monuments and place names associated with colonial supremacist projects, arguing that they legitimated imperial and racist heritages being perpetuated in contemporary landscapes, place naming, and institutions in a normalizing manner, that is offensive and dangerous for many citizens. On 5 June 2020, an area close to the White House in Washington DC was (officially) renamed BLM Plaza to the north of Lafayette Square, after the Department of Public Works had painted the words "Black Lives Matter" in 11 m yellow capital letters, during the series of George Floyd protests taking place in the city.[1]

(iii) Intensification of pressures to standardize geographical names is often driven by global consumers and organizations, that may not be fully appraised of cultural geopolitical sensitivities regarding controversial nomenclature at local to regional and international levels, and the cultural importance of endogenous nomenclature. Standardization presents opportunities, along with challenges depending on mind-sets and willingness to collaborate but also how to deal with the burdens of the history, whether it be dual naming or the creation of a shared name symbolizing a new era in cooperation, or more so a shared realpolitik taking cognizance of changing geopolitical realities and necessities. Here it could be noted that in 2006, Korea 'informally' proposed the usage of the hydronym "Sea of Peace" or "Sea of Friendship" to Japan regarding their sea naming differences; this was rejected.[2]

On many geographical scales from the smallest to larger, sensitivities regarding naming 'places' exist as they are symbols of ownership, polity, and especially those concerning national (nation-state) pride; they are making a statement within and between countries. Noticeable examples include the Persian/Arabian Gulf naming disagreement between the Arab states and Iran; and the (Korean) East Sea Versus Sea of Japan; while Dokdo Versus Takeshima Island naming issues exist between Japan and Korea. Both are cognizant that the choice of toponyms and hydrosphere appearing on maps is of high significance. For the ROK (South Korea) and Japan,

[1] Nubras Samayeen, Adrian Wong and Cameron McCarthy. Space to breathe: George Floyd, BLM plaza and the monumentalization of divided American Urban landscapes. Educational Philosophy and Theory. Published online: 23 Jul 2020. Taylor and Francis online.

[2] Shiozaki: No need to change name of Sea of Japan. The Japan Times. 10 January 2007. https://www.japantimes.co.jp/news/2007/01/10/national/shiozaki-no-need-to-change-name-of-sea-of-japan/.

the names express ownership (polity and sovereignty), territoriality, and identity and consequently feed into subliminal geopolitics of emotions emanating from historical-cultural experiences of the respective populations and states. Names express 'interest' and balance of power in a particular space.

The North Korea (DPRK) stance favours the exclusive use of the hydronym "Korean East Sea" or "East Sea of Korea" according to sources that remain yet to be verified. By 2020, no known published maps on behalf of the North Korea (DPRK) asserting claims to such labelling have been located. Neither international academic scholarship nor organizations have been able to independently verify these 'naming' claims. In short, North Korea has not followed an aggressive policy on the hydronym issue, but still making it clear that it does not support the 'Sea of Japan' nomenclature. DPRK policy has rejected Japan's sovereignty claims over the Dokdo/Takeshima islands and expressed support for South Korean control of the 'Dokdo' islands. According to the CIA World Factbook (2020): North Korea supports South Korea in rejecting Japan's claim to the Liancourt Rocks (Dokdo - Tok-do/Take-shima).[3]

With this in mind alongside dilemmas for international regulation in the digisphere, some companies like Google LLC have articulated their position on mapping and toponymy, and their obligations to adapt to a changing world and markets (Google Public Policy Blog 2008). At a different level, those involved in Geography—mapping and education, have also to make informed decisions on materials that they use, create and present.

Googlization and Cartography

As with historical processes regarding economic territorialization practices embedded in globalization projects, people are familiar with such emblematic terms as Cocacolonization, Hollywoodization and Disneyfication that attempt to encapsulate and normalize 'reality' for global consumers; now Googlization must be added to this lexicon. Today, people don't just search for something online—they GOOGLE it.

Box 1 Google Statistics and Facts

IN 2022, GOOGLE IS THE THIRD MOST VALUABLE COMPANY IN THE WORLD. (1) Over 70,000 Google searches each second, 227 million an hour, 5.4 billion per day. (2) Some 90% of the global search engine market. (3) Has 4.3 billion users worldwide. Based on 4.72 billion internet users globally. (4) Earned $116.3 billion in advertising revenue in 2018; i.e. pre-Covid pandemic. (5) Mobile devices account for 63% of all Google searches. (6) Over 2.5 billion

[3] North Korea. CIA World Factbook. https://www.cia.gov/library/publications/the-world-factbook/geos/kn.html#Issues.

Android-powered devices worldwide. (7) GOOGLE MAPS—Five million live websites worldwide use Google Maps. (8) Over 154.4 million monthly users. (9) By 2018, 67% of all mapping app users relied on Google Maps. (10) Over 23 million downloads in 2020. (11) Google Map users contribute over 20 million pieces of information per day. (12) Has a 63.76% market share. (13) Over 624,000 US websites embed Google Maps. (14) Provides about 1 billion kilometers of alternative routes daily. (15) Uses 5 MB of data per hour. Source: Review 42. Deyan Georgiev. March 7, 2022 https://review42.com/resources/google-statistics-and-facts/.

Google is the world's predominant source of cartographic knowledge (Google 2011; The Economist 2014). Over a billion people use Google Maps every month and 5 million active apps and websites use Google Maps Platform core products every week. In 2022, there were over 154.4 million monthly users. (Ethan Russell, Google Product Director, Sep. 30, 2019). It could be argued that Google has taken over map production from sovereign states and is undermining their authority in mapping their territory. Google has much autonomy in making decisions with respect to marking borderlines and place name labelling. According to the Google Approach/Policies: to naming bodies of water (hydronym), place names, border locations, the content of placemarks generated by the Google Earth Community, and reasons for the blurred imagery that appears in a number of locations[4]:

- Like any cartographic publisher, Google policies have come under scrutiny, particularly when multiple countries disagree about place naming … with its broader political, historical, or cultural attributes e.g. the body of water between the Japanese archipelago and Korean peninsula i.e. The "Sea of Japan"/"East Sea".
- As the publishers of a geographic reference tool, Google should not choose sides in international geopolitical disputes and has chosen to implement a uniform policy of *Primary Local Usage*. Under which the English-language Google Earth client displays the primary, common, local name(s) given to a body of water by the sovereign states that border it. If all bordering countries agree on the name, then the common single name is displayed e.g. Caribbean Sea (English), Mar Caribe (Spanish). If different countries dispute the proper name for a water body, Google policy is to display both names, with each label placed closer to the country or countries that use it.

Hence, Google provides a clickable text box that gives more detailed explanatory text. For language clients other than English, Google displays only the preferred name in the relevant language e.g. the Japanese client of Google Earth shows: "Sea of Japan" in Japanese (日本海), while the Korean version shows "East Sea" in

[4] Google Public Policy Blog. Updates on technology policy issues. How Google determines the names for bodies of water in Google Earth. Tuesday, April 8, 2008. Posted by Andrew McLaughlin, Director of Global Public Policy. https://publicpolicy.googleblog.com/2008/04/how-google-determines-names-for-bodies.html.

Korean (동해). Google includes *both* labels in the click-box political annotation to facilitate "users in each language by presenting the name they expect to see, but without sidestepping the existence of a disputed alternative name. In that way, we (Google) provide more, rather than less, information while maintaining a good user interface…" according to Google.

Hence Google's stated policy is to display the "primary, common, local" names for a body of water with:

- "primary"—to include names of dominant use, rather than having to add every conceivable local "nickname" (?) or variation.
- "common"—to include names that are in widespread daily use, rather than giving immediate recognition to any arbitrary governmental re-naming.
- "local"—aim to reflect the primary common names used by countries that border the water body, as they are the countries recognized under international law as having a special sovereign stake in it.[5]

Concerning the disputed naming of the islands (Dokdo/Takeshima) the political geographer, Arnon Medzini (2017) analyses different 'naming' practices used by the digital companies and apps on the maps they produce regarding disputed areas as well as the respective relationships and reactions of the governments of Japan and Korea in his seminal article on *The role of geographical maps in territorial disputes between Japan and Korea*, in the European Journal of Geography (EJG) (Medzini 2017). By comparing and contrasting maps from Google Korea, Google Japan, Google International, and Bing Maps, he concludes that, unlike Google maps, Microsoft's search engine Bing Maps attempts to overcome the dispute and maintain neutrality by labeling the islands with all three names i.e. including the exogenous name Liancourt Rocks given by a French whaling ship crew in 1849 and often used by Europeans after that in their international maps.[6]

Alternative Policies for Naming Considered by Google and Others

Different guidelines for geographical nomenclature are considered by Google and others as with the use of names from: (1) authoritative international institutions such as UN agencies—e.g. UNGEGN (UN Group of Experts on Geographical Names) including the Cartographic Sect. (2) The International Hydrological Organization. (3) International and National Geographical Societies and Associations.

---

[5] Google Public Policy Blog. Updates on technology policy issues. How Google determines the names for bodies of water in Google Earth. Tuesday, April 8, 2008. Posted by Andrew McLaughlin, Director of Global Public Policy. https://publicpolicy.googleblog.com/2008/04/how-google-determines-names-for-bodies.html.

[6] Arnon Medzini (2017) The Role of Geographical Maps in Territorial Disputes Between Japan and Korea. European Journal of Geography Volume 8, Number 1:44–60, February 2017. Association of European Geographers.

Names from Authoritative International Institutions

Issues in deciding geographical toponyms and hydronyms may be approached by deferring to the determinations of an existing authoritative, multilateral or multistakeholder institution. However, Google argues that simply adopting the naming choices set by an authoritative international body runs the danger of losing independent judgment, arguing that:

- Publications and documents that use the UN Cartographic Section as the authoritative reference for naming bodies of water are considered. But Google states that the UN Cartographic Section's publications do not provide the level of coverage and detail that "(Google) hope to achieve for Google Earth."
- The UN as an institution does not take official positions on geographical names (competing claims of two or more member states), but instead the Cartographic Section issues guidance in the form of "informational practices for use in UN documents/publications." Essentially, the issue is bounced back on the competing claimants and their diplomatic skills to resolve the issue.
- The UN is viewed by some as a politicized organization, favouring the claims of some countries and regions over others according to Google.[7]

Remarks on the Google Narrative

Circumstances have forced Google's mappers to offer multiple interpretations of the Earth's geography and to adapt these to global geopolitical sensitivities. Google's policy attempts to avoid taking sides in disputes by labelling places using local names and displaying the maps in each country in accordance with its citizens' perspectives. Thus, in the dispute over the names of the seas and islands (East Sea/Sea of Japan; Dokdo/Takeshima), Google offers three different versions of the map, stating that each version is appropriate to the residents of the country that use the digital map. Yet the solution of providing several versions of the map is not acceptable to the countries claiming the territory. Japan and Korea object to Google's cartographic approach to place naming. Both fear that conceding on the naming of the hydrosphere would suggest compromise on the actual national and international perceptions of the '*de jure*' control of this area; while compromise on the naming of the island could undermine the '*de facto*' control of the ROK's sovereign claims. The '*de facto*' occupation, organization, and control of Dokdo (Takeshima) reside with Korea. To date, a relatively peaceful modus operand is in operation despite some diplomatic spats, but there is always the danger of escalation if exploited by populist or nationalist politicians.

---

[7] Google Public Policy Blog. Updates on technology policy issues. How Google determines the names for bodies of water in Google Earth. Tuesday, April 8, 2008. Posted by Andrew McLaughlin, Director of Global Public Policy. https://publicpolicy.googleblog.com/2008/04/how-google-determines-names-for-bodies.html.

However, those people involved in Geography—mapping and education, must make informed decisions concerning the material that they select, create and present which is not facilitated by naming disputes that may foster international ignorance feeding into partisan standpoints as digital maps have a greater impact on people's worldviews and opinions (UNESCO International Bureau of Education Achieves 2020). Digital maps have made people more aware of their geographical surroundings and so increased their regional geopolitical consciousness.

International Hydrological Organization (IHO)

The IHO aims to standardize nautical charts and documents, and was created in 1919–1921 initiated by the UK and France, during a period of major changes to geopolitical maps of the world. Japan joined the IHO in 1919 and the ROK in 1957, with the ROK commencing participation in meetings in 1962. North Korea (DPRK) became an IHO member in 1987. Hence it could be argued that the Japanese viewpoint and labelling of places became embedded and systemic in IHO material and maps at an early date while Korea was under Japanese control. In the historical context, the IHO became associated with the League of Nations (1920–1946) and its successor the UN in 1945. In 2022, the IHO had 97 member states out of the 193 UN member states. IHO membership includes national hydrographic offices, but with a representation of less than 50% of all countries. The IHO's naming work is now focused essentially on: (a) undersea features, and (b) setting boundaries and limits of oceans and seas. Resolution of current geopolitical disputes is not part of its remit.

International and National Geographical Societies and Associations

According to Google, the company considered adopting the naming conventions of one or more widely-respected national-level geographic organizations e.g. US National Geographic Society (founded 1888) and the UK Royal Geographical Society (founded 1830). But, "They … occasionally reach differing conclusions on names and naming conventions, and it would be difficult to set a neutral, objective rule for deciding which organization to follow." Also they "exist only in a handful of large, rich economies, and many believe they do not represent the views and values of other parts of the world."[8] Regarding professional Geographic associations, it is significant that the most reputed of them, the International Geographical Union

[8] Google Public Policy Blog. Updates on technology policy issues. How Google determines the names for bodies of water in Google Earth. Tuesday, April 8, 2008. Posted by Andrew McLaughlin, Director of Global Public Policy. https://publicpolicy.googleblog.com/2008/04/how-google-determines-names-for-bodies.html.

(IGU) was founded in 1922 while the global geopolitical map was being re-drawn by the WWI victors.

Japan and Korea—Geographical Societies

The Association of Japanese Geographers was founded in 1925, during the era of extreme nationalism and industrialization that followed Japan's victories in WWI, which was feeding into the imperial project, with the Association adapting to the respective regimes in control there before and after 1946. In 1998, the Japanese Geographical Union was founded, which is an umbrella organization for various Geo-type associations in Japan. Following WWII, geographical studies and curricula in Japan were greatly influenced by American trends, emphasizing physical geography and political economy. Japan is an active member of the IGU (International Geographical Union).

In South Korea (ROK), the Chosun Geographical Society was founded in September 1945 and renamed the Korean Geographical Society (KGS) in 1949 (Yu, Woo-ik and Son Ill 2000).[9] In post-WWII ROK, there was an extensive influence of the traditions of American Geography on South Korean education—curricula, and research, especially from the 1970s emphasizing physical and economic geography (Park Sam Ock 2004). Regarding North Korea (DPRK), after 1946, Chinese and Soviet influences impacted Geography research and curricula taking a different turn; Geography as an individual discipline was not supported by communist regimes with the precedent having been set in the USSR. The different subfields of Geography were blended into other research, studies, and institutions including the military rather than University Geography departments as such. The ROC is an active member of the International Geographical Union, since 1960, while the DPRK is not affiliated with the IGU (Lee and Butt 2014).[10]

The International Geographical Union (IGU) was formally established in Brussels in 1922, but the history of international meetings of geographers is longer dating back to 1871 in Belgium, preceding the infamous Berlin Congress (1878) organized by the imperial powers. Regarding the Korean Geographical Society and IGU, a book on the history of geography in Korea was published in both Korean and English

---

[9] Korean Geography and Geographers. The Organizing Committee of the 29th International Geographical Congress. 2000. Yu, Woo-ik (Seoul National University) Son, Ill (Gyeongsang National University) Korean Geography and Geographers (Editors). Hanul ACADEMY Publishing Co. Seoul, Korea 2000.

[10] Sam Ock Park. The Influence of American geography on Korean geography. GeoJournal Vol. 59, No. 1, Some International Reflections on American Geography on the Occasion of the Centenary of the Association of American Geographers (2004), pp. 69–72 (4 pages). Springer. https://www.jstor.org/stable/41147795.

Jongwon Lee and Graham Butt (2014) The reform of national geography standards in South Korea—trends, challenges and responses, International Research in Geographical and Environmental Education, 23:1, 13–24, https://doi.org/10.1080/10382046.2013.858404. https://www.tandfonline.com/doi/abs/10.1080/10382046.2013.858404.

in commemoration of the 29th IGU Congress held in Seoul in 2000. In 2015, in celebration of the 70th anniversary of the foundation of the KGS, a book in Korean was published summarizing the history of KGS.[11] Further research regarding the history of geography associations, societies, and school curricula in Korea would enhance insights into geographical issues including toponyms and hydronyms.

During the colonial era, evidently, the discipline of Geography as such was constructed by the imperial powers for their own geostrategic geopolitical agenda. As in the UK, France, Belgium, and the Netherlands for instance, there is a long tradition in Geography—education, research, and curricula especially dating from the latter half of the nineteenth century. Such geographical awareness and how it is framed includes the use of text, images, and maps as in official documents, manuals, and curricula, constructing the psychogeography, mental maps, self-image, and perceptions of the 'home country looking at the 'others.' There is the worldview at home, with embedded geographies of 'the others'—boundaries, sovereignty, lands, people, toponyms, and cultures. In the post-colonial era, attempts were made by newly independent countries to 'decolonize' Geography in their nation-state building projects that have had varying degrees of success, while in former colonial countries the process may be said to be ongoing, in their archaeology of knowledge, and systemic institutional and cultural legacies. It could be argued that despite the work done by UNESCO and its Bureau of Education, imperial pasts continue to cast shadows over many former colonies (UNESCO International Yearbook of Education 1948).[12]

Maps and Realpolitik

Donghae: East Sea Versus Sea of Japan

Extensive research on historical maps has been carried out by Korean and Japanese scholars, on the naming of the 'Sea' and changing meaning of "Donghae" (East Sea) over time. East Sea has been referenced for centuries, the Sea to the East of China and Korea, corresponding to Sea No. 52 as used in the third edition of the IHO Limits of the Seas and Oceans (1953) indicating the sea to the east of the Eurasian continent. The hydronym "Donghae" has a long history in cartography in Asia, and Korea promotes that international maps that label the sea area as such between Korea and Japan, rejecting the standpoint that Japan promotes in naming it as the "Sea of

[11] Email communication from: Prof. Chul Sue Hwang, Department of Geography, Kyung Hee University, Vice President of the Korean Geographical Society. http://www.kgeography.or.kr.

[12] See: UNESCO International Bureau of Education Achieves. https://unesdoc.unesco.org/search/N-EXPLORE-54271774-c980-4fca-b432-1e2d4381a0ef.

UNESCO. International Yearbook of Education. V, 10, 1948 https://unesdoc.unesco.org/ark:/48223/pf0000132903.

Japan" only. Research on historical maps of the region includes a range of sources.[13] In brief, different names were used for the "East Sea" before the nineteenth century; hydronyms referencing Korea were most frequent in the eighteenth century. Naming processes were influenced by European 'discoverers, explorers, merchants' while in 1849 the name Liancourt Rocks started appearing on maps, named by French whalers.

The number of maps that chose the name "Sea of Japan" for 'the Sea' between Korea and Japan grew from the early nineteenth century. This was reinforced during the European and Japanese colonial era from the late nineteenth century on and the founding of international institutions and associations such as the IHO. Many Korean observers emphasize the shift from using the name Sea of Korea and East Sea to the "Sea of Japan" as late as possible, whereas some Japanese commentators try to set it as early as possible (Tanabe et al. 2010). Japan highlights that this change of name falls into the time of the Japanese "period of national isolation" (1639–1835), prior to the colonial era. Korea emphasizes the contact between Japan and European imperial countries as impacting the names on maps.

Usage of the name 'Sea of Japan' is most noticeable on maps created by Europeans in the first half of the nineteenth century, reflecting their perceptions of the area. The move to using 'Sea of Japan', on maps created in Japan becomes most obvious in the second half of the nineteenth century, reflecting Japanese horizons 'adapting' the European names. The name is used on maps created in Korea from approximately 1910 on with the Japanese annexation of Korea. With opposing viewpoints, the Sea was to the West of Japan, but also an inland sea within the Japanese regional worldview. Whatever the standpoints, Japan adopted the name Sea of Japan attempting to normalize it regionally and internationally.[14]

Many people in Korea associate the name Sea of Japan with Japanese occupation (1910–1945) and feel uncomfortable with the sole use of the name "Sea of Japan". According to UN Res. III/20 "Names of Features beyond a Single Sovereignty," published at the third UNSCGN, 1977, recommends: "countries sharing a given geographical feature under different names should endeavour, ... to reach agreement on fixing a single name for the feature concerned. It further recommends that when countries sharing a given geographical feature do not succeed in agreeing on a common name, it should be a general rule of international cartography that the name used by each of the countries concerned will be accepted. A policy of accepting only one or some of such names while excluding the rest would be inconsistent in principle as well as inexpedient in practice. Japan denies that both resolutions to which the ROK refers can be applied to the Sea between Korea and Japan, stressing that the sea areas between them do not fall under the sovereignty of some states, but

[13] See: Historical map data from the Dokdo Museum, Seoul and the National Museum of Territory and Sovereignty, Tokyo. Also a wide range of historical maps, material and analyses, including references is available in Wikipedia, Sea of Japan naming dispute. https://en.wikipedia.org/wiki/Sea_of_Japan_naming_dispute.

[14] Tanabe, Hiroshi / Yaji, Masatki / Takizawa, Yumiko/Watanabe, Kohei (2010) Origin and Function of Geographic Names. Study on the geographical name "Japan Sea". Teikyō University research group on geographical names. Tōkyō: Teikyō University.

exist in the zone of the high seas. In 2006, the ROK president informally proposed to Japan's PM that an agreed compromise on the hydronym would be to rename it with a neutral name such as the "Sea of Peace" or "Sea of Friendship". This was rejected by the Japanese authorities.[15]

Islands: Dokdo Versus Takeshima

Dokdo/Takeshima consists of two main islands with approximately 30 smaller rocks; the total surface area of the islets is 0.187554 sq. km., the highest elevation being 168.5 m. A South Korean coastguard detachment has been stationed there since 1954 and it's reported that the last survivor of a small fishing community also lives there. The geographical distances: Dokdo (Korean name), Takeshima (Japanese name), and Liancourt (European name dating from the nineteenth century)—lie at 116 nm (215 km) from mainland Korea and Korea's Ulleung-do Island at 47 nm (87 km); while it is 135 nm (250 km) from Japan proper. Dokdo/Takeshima is 85 nm (158 km) from Oki island (Japan) and 313 nm (580 km) from Liancourt to Japan (NEAHF 2020).[16] The waters surrounding the islands are rich in fish, and it is believed that gas deposits exist offshore. To date, verifiable data regarding the quantity and quality of such has not been provided.

Official websites of the Ministries of Japan and Korea Foreign Affairs claim sovereignty over the islands, stressing historical narratives and interpretations of the post-WWII San Francisco Peace Treaty (1951), aimed at re-established peaceful relations between Japan and the Allied Powers on behalf of the UN to officially end hostilities and seek redress for actions up to and including World War II. This impacted the countries that Japan had colonized, annexed or invaded. In the turmoil caused by Japan's policies in Asia and its fallout after 1945, alongside the evolving Super Power Cold War and Korean War (1950–1953), the isolated small islets of Dokdo (Takeshima) did not really figure on the international agenda (Achieves/Web MOFA ROC; MOFA Japan).[17] As early as 1954 Japan proposed that the issue be brought to the International Court of Justice (ICJ), but this has been rejected by

[15] Shiozaki: No need to change name of Sea of Japan. The Japan Times. 10 January 2007. https://www.japantimes.co.jp/news/2007/01/10/national/shiozaki-no-need-to-change-name-of-sea-of-japan/.

[16] A Visual Study of Dokdo and Ancient Territorial Perceptions. https://www.dokdo-takeshima.com/a-visual-study-of-dokdo.html.

NEAHF (Northeast Asian History Foundation). Dokdo in the East Sea. http://contents.nahf.or.kr/english/item/level.do?levelId=eddok_003e_0010_0030.

[17] See: Ministry of Foreign Affairs. Republic of Korea. Dokdo: Documents and Official Position - http://search.mofa.go.kr/search/search_en.do.

Ministery of Foreign Affaires of Japan (MOFA). An Outline of the Japanese Position on Sovereignty over Takeshima and the Illegal Occupation by the Republic of Korea. https://www.mofa.go.jp/region/asia-paci/takeshima/position.html.

MOFA: The Issue of Name 'Sea of Japan' https://www.mofa.go.jp/policy/maritime/japan/index.html.

Korea, underlining that such could lend some sort of credence to Japan's claims, which Korea completely rejects (Van Dyke 2003, 2007).[18]

The Geopolitics of Emotions

Museums: Territorial Narratives, Projections, and Mind-Maps

The Dokdo Museum (opened on August 8, 1997) in Seoul, South Korea, has the stated aim to collect and research Dokdo and the East Sea; exhibiting and managing these materials. Korea aims to reinforce its claims on what it considers its sovereign territory. There is also a Dokdo Museum on Ulleungdo Island, 127 km east of the Korean peninsula.[19]

In Japan, the National Museum of Territory and Sovereignty is located in the Toranomon Mitsui Building, Tokyo. It covers Takeshima, Senkaku Islands, and the Northern Territories including the Kuriles, essentially entailing territorial disputes between Japan and its neighbours—Russia, the ROK, North Korea, PR China, and Taiwan. It was opened on January 25, 2018. Its aim is to: raise national and international awareness regarding these sovereignty issues. The museum was relocated to the Mitsui Building, Tokyo in January 2020.[20]

The new exhibition area is seven times larger than before, and exhibits on the Northern Territory were increased significantly, as well as space for Takeshima (Dokdo) and Senkaku Islands as grounds for Japan's territorial claims—historically and internationally. There is a strong rebuttal of the sovereignty claims of the ROK, PRC, and RC (Taiwan), but it is stated that: 'they (Japan) are in a position to solve problems through dialogue.'

Clashing standpoints projected by these state-sponsored museums give the government positions on respective sovereignty claims, and illustrate how state-construction nationalism attempts to connect with the respective national toponym narratives and collective subliminal experiences of the former 'colonizer' and the 'colonized'. Both Korea and Japan underline that their differences on sovereignty issues must be settled through diplomacy. Nonetheless, it would be reasonable to say that an impasse has been reached, that this could be exploited by populist politicians,

[18] Van Dyke J.M. 2003 The Republic of Korea's Maritime Boundaries. The International Journal of Marine and Coastal Law 18(4):509–540.

Van Dyke J.M. 2007 Legal Issues Related to Sovereignty over Dokdo and Its Maritime Boundary. Ocean Development and International Law 38(1):157–224.

[19] See: Dokdo Museum, Seoul. http://www.dokdomuseumseoul.com/en/.

See: Dokdo Museum, Ulleungdo. http://www.dokdomuseum.go.kr/eng/page.htm?mnu_uid=751&.

[20] Japan: National Museum of Territory and Sovereignty https://www.cas.go.jp/jp/ryodo_eg/tenjikan/outline.html.

somewhat reminiscent of the 'invented tradition' as of 2005 in Japan, of Takeshima Day, to celebrate, or to commemorate, which remains open to interpretation.

Takeshima Day

On March 16, 2005, in Matsue, Japan, the Shimane Prefectural Assembly members voted a bill designating February 22 as 'Takeshima Day'. The so-called ceremonies have continued annually since then, leading to street skirmishes and official protests from the ROK.[21]

Historically, 2005 references the centenary of the 1905 Japan war with Russia and victories at Port Arthur and declaring Korea a Japanese Protectorate as a prelude to its annexation as a Colony (1910) and imperial expansion onto the Asian mainland, while occupying militarily Dokdo in the process in 1905.[22] For the Korean authorities and citizens, this is perceived as a provocation inflaming emotions and grievances.

Korean Grievances

Japan's imperial project impacted Korea long before the annexation in 1910. Imperial cultural destruction policies targeted the Korean language, as well as Korean toponyms, culture, and architecture, for instance, the enforced prayer for Koreans at Shinto shrines.[23] Other examples are well illustrated by author and journalist Min Jin Lee, in her Pachinko (2017) novel, a saga about four generations of a poor Korean family, spanning the twentieth century up to the 1980s. It depicts the narratives of the

---

[21] Happy Takeshima Day…?" Shimane Declares Takeshima Day as February 22nd – Japan Adds Insult to Injury. https://www.dokdo-takeshima.com/happy-takeshima-day.html#:~:text=These%20rocks%20located%20between%20Japan,22%20as%20'Takeshima%20Day'.

[22] Shimane Prefecture / Shimane Prefecture Board of Education the Shimane Citizen's Conference Supporting the Return of Takeshima and the Northern Territories. Takeshima: Japanese Territory. https://www.pref.shimane.lg.jp/admin/pref/takeshima/web-takeshima/takeshima06/pamphlet/index.data/takeshima-leaflet-eigo.pdf?site=sp.

Bukh, Alexander 2015 Shimane Prefecture, Tokyo and the territorial dispute over Dokdo/Takeshima: regional and national identities in Japan. The Pacific Review. Volume 28, 2015 - Issue 1: Identity Politics in Japan's International Relations.

[23] Taehoon Kim. The Place of "Religion" in Colonial Korea around 1910: The Imperial History of "Religion". Journal of Korean Religions. Vol. 2, No. 2, Korean Religions in Inter-Cultural Contexts (October 2011), pp. 25–46 (22 pages). Published By: Institute for the Study of Religion, Sogang University, Korea. https://www.jstor.org/stable/23943381.

Suga Kōji and. A Concept of "Overseas Shinto Shrines": A Pantheistic Attempt by Ogasawara Shōzō and Its Limitations. Japanese Journal of Religious Studies. Vol. 37, No. 1, Religion and the Japanese Empire (2010), pp. 47–74 (28 pages). Published By: Nanzan University. https://www.jstor.org/stable/27822899.

characters' experiences of colonialism at home, and then abroad in Japan as immigrants. Sharp insights are given into relationships and places, as well as the enforced usage of names, instead of their own personal Korean ones for officialdom.

Part of the WWII memory legacy for Koreans and others as in the Philippines is the so-called 'Comfort women' i.e. sex slavery whereby Koreans were forced to 'work' in military brothels. Survivors and descendants continue to protest. In 2007, the US House of Representatives and EU Parliament called on Japan to make a genuine apology and reparations. In 2015, Japan agreed to some compensation for the surviving women, but the dispute is not really fully resolved.[24]

Grievances include that in Japan, government-approved textbook revisionism (1982) was attempted regarding the imperial era that was condemned by neighbouring countries and the international community. While in 2006, Japanese textbooks stated that: the 'Liancourt Rocks' is Japanese territory. Controversial visits have continued by some Japanese politicians including PMs to the Yasukuni Shinto shrine and graveyard, with its war cemetery. The controversy began especially in 1978 when 14 Japanese civilian and military leaders convicted by a post-WWII international tribunal and known as "Class A War Criminals" were enshrined there. Nonetheless, Yasukuni remains sacred to Japanese nationalists; but is contested by other countries especially China, South and North Korea, and Taiwan. Some Japanese citizens reject the use of this place for ceremonies and commemorations arguing that the temple glorifies Japan's former militarism and colonialism.[25]

Korea like many former colonies demands the return of stolen artefacts. Many remain in the Tokyo National Museum and private collections in Japan. According to the ROC: 75,311 cultural artefacts were taken from Korea. Japan has 34,369,

---

[24] Sonya Kuki. The Burden of History: The Issue of "Comfort Women" and What Japan Must do to Move Forward. Journal of International Affairs. Vol. 67, No. 1, The Gender Issue: Beyond Exclusion (FALL/WINTER 2013), pp. 245–256 (12 pages). Published by: Journal of International Affairs Editorial Board. https://www.jstor.org/stable/24461685?seq=1.

Lee, C., and Crowe, J. (2015). The Deafening Silence of the Korean "Comfort Women": A Response Based on Lyotard and Irigaray. Asian Journal of Law and Society, 2(2), 339–356. https://doi.org/10.1017/als.2015.9. https://www.cambridge.org/core/journals/asian-journal-of-law-and-society/article/deafening-silence-of-the-korean-comfort-women-a-response-based-on-lyotard-and-irigaray/495BD182C34D74E086B8C673A9BC393C.

[25] Ryu, Yongwook. "The Yasukuni Controversy: Divergent Perspectives from the Japanese Political Elite." *Asian Survey*, vol. 47, no. 5, 2007, pp. 705–726. *JSTOR*, www.jstor.org/stable/10.1525/as.2007.47.5.705. Accessed 22 Oct. 2020.

Aljazeera. Yasukuni: caught in controversy as Japan struggles with history: As Yasukuni marks 150 years since it was founded, the decision to enshrine war criminals casts a shadow over the shrine. By Kelly Olsen. 18 Oct 2019. https://www.aljazeera.com/economy/2019/10/18/yasukuni-caught-in-controversy-as-japan-struggles-with-history.

BBC News. Yasukuni Shrine: Japan's ex-PM Abe visits controversial memorial. https://www.bbc.com/news/world-asia-54216632.

BBC News. Yasukuni Shrine: Chief priest to quit after criticising Japan's emperor. 11 October 2018. https://www.bbc.com/news/world-asia-45821700.

Ryu, Yongwook. "The Yasukuni Controversy: Divergent Perspectives from the Japanese Political Elite." *Asian Survey*, vol. 47, no. 5, 2007, pp. 705–726. *JSTOR*, www.jstor.org/stable/10.1525/as.2007.47.5.705. Accessed 22 Oct. 2020.

the USA has 17,803, and France has several hundred, seized by France (1866) but loaned back to Korea in 2010. In 2010, Japan's PM expressed: "deep remorse" and arranged an initial plan to return the Royal Protocols of the Joseon Dynasty and 1,200 other books, which was carried out in 2011. Korea has pursued active policies in the recuperation of its cultural heritage.[26] Due to the nature and trauma of war and colonial projects, grievances can remain strong over generations in former colonies, and resurface especially when it is perceived that the former colonizing power is attempting to lay claim to its culture and any part of its sovereign territory.

In 2022 both Japan and China commemorated the 50th anniversary of their normalization of relations after a difficult history. This included a series of wars (1880–1945), with Japan seizing Taiwan, Manchuria, and most of coastal China. While Japan was defeated in 1945, tensions remained due to the Korean and Cold Wars. Grievances endured regarding Japan's war crimes in China, such as the Nanjing Massacres (1937–1938). Japan, Asian states, USA and Russia continue to monitor Chinese naval activities in the region, deemed as expansionist by many observers.

Conclusions

Digital cartography is embedded in globalization processes along with Googlization and the power of Google in attributing 'names' to geographical places such as those claimed by Japan and Korea reflecting disputed perceptions, territorialization(s), and sovereignty. Regarding analyses of historical maps, created in specific contexts, many international associations were founded in the colonial era, reinvented after the World Wars presenting worldviews, and 'updated' international law, with legacies and systemic perspectives. Intricate post-colonial legacies continue to impact Korea and Japan, so disputed toponymy and hydronomy cannot be read in isolation.

Both parties to the 'naming' disputes present a variety of maps, some hundreds of years old, to support the justice of their claims, despite their understanding that these maps in essence have no legal value today. Both parties understand that modern perceptions of sovereignty and borders cannot be based upon maps from former eras. International and arbitration courts are aware that modern maps are not always objective in presenting reality because cartography has the ability to represent conveyed reality. Courts may disregard maps when discussing countries' territorial claims. Of

[26] History Today. Do historical objects belong in their country of origin? Four historians consider one of the most contentious questions facing the West's museums and galleries. Volume 69, Issue 3, March 2019. https://www.historytoday.com/archive/head-head/do-historical-objects-belong-their-country-origin.

Atkins, E. Taylor (2010). *Primitive Selves: Koreana in the Japanese Colonial Gaze, 1910–1945'*. Berkeley: University of California Press.

Geoffrey R. Scott. 2008 Spoliation, Cultural Property, and Japan. https://www.law.upenn.edu/journals/jil/articles/volume29/issue4/Scott29U.Pa.J.Int'lL.803(2008).pdf.

course in the background looms the 1982 UN Convention on the Law of the Sea—Exclusive Economic Zone (EEZ) possibilities and associated sovereignty issues (UNCLOS 1982).[27]

Finding an agreed name for the Sea between Korea and Japan is more than a simple question of name changing. Further engagement is needed to 'deconstruct' systemic reactions, mindsets, and attitudes giving credence to post-colonial and post-WWII apologies and building sustainable futures. Korean efforts to erase the legacy of the colonial past include: redress the negative effects of Japanese imperialism and its embedded 'sense of entitlement'; questioning systemic negative legacies in Japan, and inherited international approach to Korean grievances. Significantly, the ROK—has actively promoted its stance since the 1970-80s in international fora; with growth in self-confidence shedding the 'colonial complex' often evident in former colonies. The emotional weight that the ROK attributes to naming and territorial issues attests to the strength of historical collective memory. Significantly both North and South Korea agree on this.

Japan, like other former nationalist colonial powers, is having to (re)assess and (re)adjust to the loss of empire, post-WWII US occupation, re-configuration of dynamic relationships with neighbouring states; and political-economic experiences within the US sphere of influence, as well as contentions with the PRC and its geo-strategy regarding maritime territorial zones in North-East Asia. Japan's geopolitical issues with neighbouring states include: (i) agreement on maritime boundary delimitation; and (ii) the PRC's active geo-strategy to establish 'sovereign' dominance in Asia's seas and oceans. Japan is attempting to reinvent itself, like other former imperial powers such as the UK and France, but also powerful countries which were not victors in WWII like Germany. If Japan agrees to the 'naming concessions' with Korea, this could encourage a domino effect with other neighbouring states impacting on claims and the consequences thereof. Nevertheless, appealing to extreme nationalism may feed into old imperial reflexes from a bygone era, as witnessed with the Russian invasion of Ukraine in 2022, undermining sustainable futures. Creative place name solutions have to be worked on by both Japan and Korea, including the possibility of a neutral name for the dispute regarding Sea hydronomy.

Questions

1. Define what the cartography-digitalization nexus means, and Google's power in place naming?
2. Assess the geopolitics of emotions regarding toponymy and hydronyms in Japan-Korea relations, taking into account how place names can be linked to perceptions, territorialization(s), historical legacies, and claims to sovereignty.

[27] NRB (National Bureau of Asian Research). The Law of the Sea and South Korea: The Challenges of Maritime Boundary Delimitation in the Yellow Sea by Seokwoo Lee and Clive Schofield. April 23, 2020. https://www.nbr.org/publication/the-law-of-the-sea-and-south-korea-the-challenges-of-maritime-boundary-delimitation-in-the-yellow-sea/.

UN Law of the Sea. https://www.un.org/Depts/los/index.htm.

3. What are the processes whereby toponyms are selected or prioritized at home and in the international community, and evaluate the use of historical maps in this?
4. Can it be argued that post-colonial and WWII legacies, like disputed toponymies, continue to impact international and geopolitical relations in Northeast Asia countries such as Japan - Korea?

References

CIA World Factbook. North Korea. https://www.cia.gov/library/publications/the-world-factbook/geos/kn.html#Issues

Google Public Policy Blog (2008) Updates on technology policy issues. How Google determines the names for bodies of water in Google Earth. Tuesday, April 8. Posted by Andrew McLaughlin, Director of Global Public Policy. https://publicpolicy.googleblog.com/2008/04/how-google-determines-names-for-bodies.html

Lee J, Butt G (2014) The reform of national geography standards in South Korea—trends, challenges and responses. Int Res Geogr Environ Educ 23:1, 13–24. https://doi.org/10.1080/10382046.2013.858404. https://www.tandfonline.com/doi/abs/10.1080/10382046.2013.858404

Medzini A (2017) The role of geographical maps in territorial disputes between Japan And Korea. Eur J Geogr 8(1):44–60, February. Association of European Geographers

Moïsi D (2010) The geopolitics of emotion: how cultures of fear, humiliation, and hope are reshaping the world. Anchor Books, NY

MOFA ROC - Ministry of Foreign Affairs. Republic of Korea. Dokdo: Documents and Official Position. http://search.mofa.go.kr/search/search_en.do

MOFA Japan - Ministry of Foreign Affairs of Japan (MOFA). An Outline of the Japanese Position on Sovereignty over Takeshima and the Illegal Occupation by the Republic of Korea. https://www.mofa.go.jp/region/asia-paci/takeshima/position.html

MOFA: The Issue of Name 'Sea of Japan' https://www.mofa.go.jp/policy/maritime/japan/index.html

NEAHF (Northeast Asian History Foundation) (2020) Dokdo in the East Sea. http://contents.nahf.or.kr/english/item/level.do?levelId=eddok_003e_0010_0030

Park SO (2004) The influence of American geography on Korean geography. GeoJournal 59(1). Some International Reflections on American Geography on the Occasion of the Centenary of the Association of American Geographers, pp 69–72 (4 pages). Springer. https://www.jstor.org/stable/41147795

Tanabe H, Masatki Y, Yumiko T, Watanabe K (2010) Origin and function of geographic names. Study on the geographical name "Japan Sea". Teikyō University research group on geographical names. Teikyō University, Tōkyō

UNESCO (1948) International Yearbook of Education. V, 10. https://unesdoc.unesco.org/ark:/48223/pf0000132903

UNESCO International Bureau of Education Achieves. https://unesdoc.unesco.org/search/N-EXPLORE-54271774-c980-4fca-b432-1e2d4381a0ef

Van Dyke JM (2003) The Republic of Korea's maritime boundaries. Int J Marine Coastal Law 18(4):509–540

Van Dyke JM (2007) Legal issues related to sovereignty over Dokdo and its maritime boundary. Ocean Dev Int Law 38(1):157–224

Yu W-I, Son, Ill. Korean (2000) Geography and geographers. The organizing committee of the 29th international geographical congress. Yu, Woo-ik (Seoul National University) Son, Ill (Gyeongsang National University) Korean Geography and Geographers (Editors). Hanul Academy Publishing Co. Seoul, Korea

Place Naming and Neotoponymy: French Experiences Through the Lens of a Theoretical Framework

Frédéric Giraut

Abstract The social, political, cultural, and economic dimensions of toponymy are expressed in the act of naming, that is, in the contradictory production of toponyms. The place naming is thus an object of toponymy distinct from the place name itself. Its study from the point of view of motivations and representations has been developed in the social sciences, and in particular in cultural and political geography. The study of the naming of places presupposes the observation and documentation of the process, and therefore focuses on neotoponymy. Such neotoponymy is both the study of new place names and the corpus formed by them. In addition to collections of names whose meaning and origin can be worked on, it offers, by definition, information on its mode of production, its actors, and possible controversies or debates related to it. Neotoponymy is the result of a process of substitution or addition (renaming, plurinaming) or of nominations of emerging places. Depending on the context, the type of places, and the motivations linked to it, issues and techniques will be different. The contemporary history of France since the Revolution offers an abundance of neotoponymies from successive or simultaneous contexts that will allow us to test a theoretical framework on place naming.

Keywords Place naming · Neotoponymy · Political toponymy · Critical toponymy · Theory · French place names · Territoriality

Objectives

- To have an overview of the French experience of naming political territories (administrative units and local governments units).
- Compare the experiences and initiatives of political place naming in different historical contexts.
- Use the notion of neotoponymy in its different aspects.

F. Giraut (✉)
Geography Department, UNESCO Chair "Naming the World", University of Geneva, Geneva, Switzerland
e-mail: frederic.giraut@unige.ch

© The Author(s), under exclusive license to Springer Nature Switzerland AG 2023

G. O'Reilly (ed.), *Place Naming, Identities and Geography*,
Key Challenges in Geography, https://doi.org/10.1007/978-3-031-21510-0_4

Introduction

This chapter deals with neotoponymy, i.e. toponymy resulting from the voluntary naming of new geographical objects or the renaming of old ones. The French experience over more than two centuries provides material to illustrate the issue and also to consider its logics since experience over the long term has a heuristic value.

We will focus in particular on the names of the units of territorial organization: both public administration and local government. They have been crossed during the period by several major changes linked first to the establishment of a new revolutionary order, then during the XXth and XXIst centuries, by the new regionalism and its municipal declensions.

A toponymic production is an object in itself (Amilhat Szary 2008; Berg and Vuolteenaho 2009; Boyer and Cardy 2011; Giraut and Houssay-Holzschuch 2008a, b; Guillorel 2008; Pourtier 1983; Rose-Redwood et al. 2010). It is the result of a complex interplay of actors and takes place in specific contexts which determine the issues, objectives, and techniques available (Giraut and Houssay-Holzschuch 2016; Puzey and Kostansky 2016). The result of this production is the corpus of new toponyms that can be called neotoponymy (Giraut and Houssay-Holzschuch 2022). The field of study of this neotoponymic production can be described as Place naming studies. It is at the heart of critical toponymy, which focuses on the stakes, the logic, and the effects of place naming and renaming. It includes the contemporary processes of commodifying naming rights and place branding (Light and Young 2015: Medway and Warnaby 2014; Rose-Redwood et al. 2022).

We have recently developed a framework dedicated to issues and rationales in place naming (Giraut and Houssay-Holzschuch 2016). We will first outline it. The second part will then be devoted to the French revolutionary experience of setting up departments and communes in place of the units of the *Ancien régime*: provinces and parishes, with a system of names that broke partially with the previous toponymy. The third part will be devoted to the longer-term experience of adapting the territorial system to a new regionalism, to decentralization, and to metropolisation. The introduction of new toponymic layers is then marked by local geopolitics and territorial marketing. We will thus be able to qualify the successive logics that can intervene in the setting up of the territorial neotoponymy.

Neotoponymy: A Framework

This chapter is based on a theoretical article published in the journal Geopolitics called "Place Naming as *Dispositif*: Toward a Theoretical Framework". Together. with Myriam Houssay-Holzschuch, we suggest that renaming practices (neotoponymy), as varied as they are throughout the world, present some commonalities. These practices and processes can be broken into three elements: contexts, actors, and technologies.

These elements are constitutive of a theoretical framework (Fig. 1) inspired by the Foucauldian notion of *dispositif* as a complex and historically contextual arrangement of norms, knowledges and technologies of power. All elements of contextualization not only have the same potential for generalization, but both the geopolitical context and the objectives of renaming point to more general trends that can help theorize the renaming process, as actors are more context specific.

We argue that four types of political contexts are especially conducive to neotoponymic production: those of **Revolution** (radical change of political regime), **Conquest** (colonial and imperial), **Emergence** (urbanisation and territorial restructuring) and **Commodification** (competition between places in territorial marketing and branding). In turn, renaming practices aim at four main objectives turned into toponymic power technologies: **Cleansing**, **Restoring**, **Founding**, and **Promoting/Branding**.

While the link between a type of context and the objectives of renaming is not straightforward, less alone deterministic, preferential combinations or nexuses can be identified. We will now illustrate this framework with the French territorial restructuring experience.

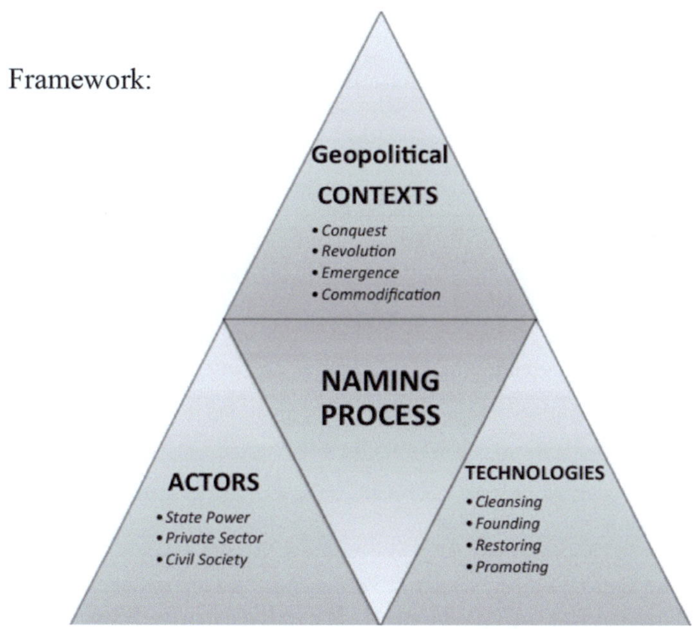

Fig. 1 Framework showing the main components (*dispositif*) of Place naming processes namely contexts and technologies. *From* Giraut and Houssay-Holzschuch (2016), p. 8

Casestudy Box

Cleansing, Restoring, Founding, and Promoting/Branding

- **Cleansing** aims at discarding the toponymic imprint inherited from a culture, a language, and/or a previous political order. Cleansing an existing toponymic landscape can be done by erasing place names, renaming, or translating them, and removing them from all official language and legal documents.
- **Founding** inscribes cultural and political references in the toponymy in order to create, legitimize, and, ultimately, sustain a new political and cultural order at the local, or the national, level. Naming places after founding fathers, ideological values, or founding events shapes the toponymic landscape and helps "legitimate existing power structures by linking the regime's view of itself, its past, and the world, with the seemingly mundane settings everyday life."
- **Restoring** strives to reinstitute ancient, or dominated, memories and cultures by deploying previous toponyms from such a culture in order to atone for (newly considered) historical injustices, or to legitimize territorial claims.
- **Promoting** is the way to brand a place, a development, a resort, a territory, or a city through its name (or nickname), which is used as valuable and marketable symbolic capital. It is an attempt to attract investors and consumers, and to be well placed in international, or national, rankings in the context of places' fierce competition between places for economic and political gain. At its most extreme, promoting technologies include speculating on a place name's economic value. For instance, private companies create a new toponym to market a leisure destination, or acquire an existing toponym by renting it temporarily, or by purchase. Whatever the case, promoting considers a place name as both symbolic and economic capital.

From Giraut and Houssay-Holzschuch (2016), p. 9

The Neotoponymy of the French Revolutionary New Territorial Order

As early as 1789, the revolutionary project was to be embodied in a radical territorial reform on two levels. On the one hand, the Constituent Assembly decided to generalize the status of municipality to more than 40,000 localities, which became communes, and on the other hand to create 83 *départements*. This replaced the heterogeneous system of *provinces* and local entities: villages, towns and parishes with various types of franchises.

The Communes: Permanence and Toponymic Selection

If the municipal system is revolutionary from a political point of view, it takes over the jurisdiction of the basic localities, i.e. the villages and towns, which were instituted both as parishes and as tax-collecting localities (Nordman and Ozouf-Marignier 1989). Also, the major institutional change that endowed these entities with municipal assemblies did not lead to any change in the jurisdictions (still not precisely established until the generalization of the cadaster under the Napoleonic Empire) or in the names, apart from a few rare cases. Thus, when the communes were created, the heterogeneous toponymic layer of locality names was preserved. However, the revolutionary craze was to result for a few years, from *An II* (1793) to the fall of the Napoleonic Empire (1814), in radical name changes concerning about 10% of communes (Maréchaux 2016). These changes aimed to erase some of the mainly religious and secondarily aristocratic or monarchical traces. Thus hagiography (names after Saints) is particularly targeted and many qualifiers of Saint disappear temporarily (*Ste-Croix-à-Lauze* becomes *La Lauze*; *St-Geniez-de-Dromon* becomes *Dromon*); sometimes to promote an entirely new name, thus *Saint-Ouen* is renamed *Bains-sur-Seine* and *Saint-Etienne* becomes *Ville-d'Armes*. References to religious places in the toponymy of localities (church, chapel, basilica, monastery…) are also deleted (*Les-Églises-d'Argenteuil* becomes *Argenteuil*). Aristocratic and feudal references are also removed. Thus *Aisey-le-Duc* became *Aisey-sur-Seine*, *Pagny-le-Chateau* was renamed *Pagny-Egalité*. Sometimes the change affects borderline cases of homonymy, e.g. *Grenoble* is renamed *Grelibre* when its ending has nothing to do with the word *noble*. It is interesting to note that of the nearly 10% of renamed communes (about 3000) that had their names changed, about 90% regained their names from 1814 onwards, but several hundred retained their revolutionary names. As Xavier Maréchaux (2016) shows, this contributed to the partial de-Christianization of French toponymy according to a national political geography that largely overlapped with that of the radical secular and republican strongholds.

If we want to interpret this process of limited implementation of a communal neotoponymy in terms of the proposed framework (Table 1), we should first emphasize the revolutionary context of the French **Revolution**. Its actors are the public authorities that emerged from this revolution and certainly committed segments of civil society that carried the revolutionary process locally. As for the toponymic technologies, it is mainly a question of **Cleansing** *Ancien Régime* references with the promotion of revolutionary references and values. Conversely, the (partial) end of this toponymic parenthesis in an equally revolutionary context (**Revolution** in the sense of a radical change in the political order) at the end of the Empire involved the technology of **Restoring**.

Table 1 Contexts and technologies involved in French municipal neotoponymies

Neotoponymies	Contexts	Technologies
Revolutionary municipalities	Revolution	Cleansing/Founding
Post-imperial municipalities	(counter-)Revolution	Restoring

Source Giraut (2002)

The Naming of Revolutionary **Départements***: Erasure and Neutralization*

The creation of the *départements*—a new institution in a completely new territorial framework—is exactly contemporaneous with the creation of the institution of municipalized communes on a village basis. As early as 1789, the French Revolution, through its *Comité de Constitution*, set out to eliminate the complex map of the *provinces*. Territorially and culturally, they embodied the history and power of the *Ancien Régime*, which was at once royal, religious and aristocratic, and even bourgeois through certain franchises and assemblies. These provinces are made up of a number of area based institutions with different jurisdictions: *Généralités* (fiscal function), *Gouvernments* (military function), *Parlements* and *Baillages* (judicial function), ecclesiastical *Provinces* and *Diocèses* (religious function). They had to give way to a new and unique institution, which was both a framework for exercising the authority of the new state and a set of new assemblies for expressing regional wills. The idea was to free oneself completely from the old order and to produce a new system, so the temptation of pure abstraction was exercised. The initial projects are thus strictly speaking utopian.

A first map is proposed made up of 81 squares, themselves subdivided into 9 entities. A standard and replicable module was therefore envisaged, based on logical nesting and a dimension that ensured average internal accessibility in line with the travel capabilities in one day (Meuriot 1917; Ozouf-Marignier 1988). But no centrality is predefined and the function of the chief town is potentially rotating. The highly critical reception on the ground given to these utopian proposals is territorial in nature, highlighting functionalities and identities with their own history and spatial framework. The critique calls for the recognition of historical social entities, natural entities (hydrographic, topographic) and their boundaries linked to natural discontinuities.

The final draft of the new *Comité de Division* retained the standard approximate module (83 *départements* instead of the 81 proposed) but adapted the definition of the contours to the inherited entities, natural discontinuities and the jurisdiction of certain cities. However, the system of naming which intervenes in fine (February 26, 1790) will be without compromise contrary to the cartography.

While all the work of the committee had been carried out on entities bearing the names of potential chief towns or parts of dismantled provinces, the final names refuse all references to cities and provinces,[1] retaining only natural references, mainly

[1] Garat Aîné, D., P. Samary, J. X. Bureaux de Pusy, L. m. d. Foucault de Lardimalie, G. J.-B. Target, J.-B. d. b. d. Pinteville de Cernon, J. A. E. Fos de Laborde, H.-G. R. c. d. Mirabeau and J. S. Maury

hydrographic: 59 *départements* out of 83 are thus named after watercourses. The name of a river alone in 23 *départements* and in combination with another one in 11 cases. It may also be the subject of a demultiplication with precision of the part of which it refers (*haute, basse, inférieure*…) in 6 cases. Thus the referent *Seine* is found in three initial *département* names, once in combination with *Marne*, once with *Oise* and once specified as *Seine-inférieure*. It will soon (1795) be found alone in place of the referent *Paris*. *Paris* which was until then the exception by referring to a city, the capital city of the Kingdom. Secondly, the topography and orography are mobilized through references to mountain ranges alone (Jura; Vosges; Ardennes) or multiplied and specified when the main ranges are involved (*Hautes* and *Basses -Alpes*; *Hautes*, *Basses* and *Orientales Pyrénées*). Finally, certain geographical specificities (river mouths, estuary, strait, finistere/lands' end, island, source, volcano, geological formation, etc.) are promoted as names of *départements* in about fifteen situations, often in coastal situations, less marked by the presence of major rivers.

Apart from a few departmental creations named according to the same logic (two in 1794, one in 1808), a few ephemeral changes will occur, as for the communes, from the An II of the new Republican Calendar (1793). The royalist *Vendée* and the protesting *Gironde* momentarily lost their names, which had become synonymous with regional opposition to the Jacobin revolutionary doctrine. They then took the names of *Bec-d'Ambes* and *Vengé* (an allusion to the republican revenge following the treason) respectively. This was only for two years.

If we disregard the imperial *départements* in the first decade of the nineteenth century, then the colonial departments, particularly in Algeria, the main change in the rationale of naming the *départements* will come with the annexation of new territories in the context of the *Second Empire*. In 1860, Savoie was integrated into France, and this territory retained the reference to the historical entity of *Savoie* by giving the two names of *Savoie* for the *département* formed in the south around *Chambéry*, and *Haute-Savoie* in the north around Annecy with the French *Genevois*. At the same time, the entry of the County of *Nice* into the national territory resulted in the creation of a *département* with a classic name: les *Alpes-Maritimes*.

In the second half of the twentieth century, there were several new names. On the one hand, some *départements* were partially renamed, and on the other, the new *départements* resulting from the reconfiguration of the *Paris* region had to be given new names. The renominations, which were staggered from 1941 to 1990, therefore concerned a few cases, six in all. The aim was to rid them of qualifiers that were considered negative and stigmatizing: *Basse* and *Inférieure*, and to replace them with the qualifiers of *Maritime* or *Atlantique* for river names (*Charente, Loire, Seine*) or mountain names (*Pyrénées*). However, two *départements* reverted to names evoking either a former *province* (*Alpes-de-Haute-Provence* replaced *Basses-Alpes* in 1970) or a regional language, in this case the Breton language, for *Côtes-d'Armor* in place of *Côtes-du-Nord* in 1990 (Le Bart and Procureur 2011).

(1880). Discussion sur la dénomination des départements, lors de la séance du 26 février 1790. Archives Parlementaires de la Révolution Française: 711–711. https://www.persee.fr/doc/arcpa_0000-0000_1880_num_11_1_5850_t1_0711_0000_16.

The territorial restructuring of the Paris region in the 1960s introduced hitherto exceptional referents into the new *département* names and broke with the logic of toponymic neutrality of the preferential use of hydronymy and oronymy. The re-creation of a *département* of *Paris*, this time cantered on the city alone, in place of the *département* of the *Seine*, definitively establishes the possibility of an exception for the capital city. Until then, only *Paris* (in a short-lived manner) and the *Territoire-de-Belfort* (created between the two World Wars after the return of Alsace to France) constituted a reference to a capital city or chief town. A principle which derogates from that of the original nominations. This practice is found again in one of the six new *départements* created under this reform on the former departments of the *Seine* and *Seine-et-Oise*. This is the *département* of *Seine-Saint-Denis*, which has a name that combines the name of a river and an historical royal city, and which is also hagiographic (the name of a saint) and therefore doubly taboo in relation to the original revolutionary naming principles. The other names are either the classic use of a river name alone (*Essonne*), or a river name associated with a landscape and territorial evocation (*Val-de-Marne*; *Val-d'Oise*; *Hauts-de-Seine*). Finally, the name *Yvelines* refers to a forest name which is also that of a local *pays* of the *Ancien Régime*.

We can see (Table 2) that the initial revolutionary context generated a neotoponymy motivated by **Cleansing** (of *Ancien Régime* references), **Neutralising** (of cities) and the **Ordering** of modules on an egalitarian basis of alphabetical order. The subsequent conquest of new territories anchored them in the departmental system (**Founding**) while opening the possibility of **Restoring** the provincial reference. In the twentieth century, the changes in names considered as derogatory must be interpreted as taking place in a context of **Commodification** of territories. Questions of image have an economic value and not only an identity value (Grégory 2011). Promotional technologies may therefore involve the restoration of *Ancien Régime* references or regional languages (**Promoting** and **Restoring**). Finally, the creation of new *départements* to accompany metropolisation takes place in a dual context of the **Emergence** of new functional entities and the relative marketing of territories (**Commodification**). This leads to a neotoponymy that borrows from the technologies of **Promoting** and **Founding** a new metropolitan device.

Table 2 Contexts and technologies involved in French *départements* neotoponymies

Neotoponymies	Contexts	Technologies
Revolutionnary *départements*	Revolution/Emergence	Cleansing/Neutralising/Ordering
Annexed *départements*	Conquest	Founding/Restoring
Renamed *départements*	Commodification	Promoting/(Restoring)
Restructured *départements*	Emergence/Commodification	Founding/Promoting/(Ordering)

Source Giraut (2002)

The Neotoponymy of the Contemporary French Territorial Restructuring: Administrative Regions and New Communes

The second part of the twentieth century and the beginning of the twenty-first will see the emergence in France of two new types of political and administrative units: the administrative regions and the jurisdictions of inter-municipal cooperation and mergers of communes. (Giraut 2002; Offner 2006; Vanier 2008).

The Naming of French Administrative Regions: The Return of the Repressed

In 2016, brand new French regions were given names. These new names complete the process of simplifying the map from 22 to 13 regions. The toponymic equation seemed difficult, but clearly identified: (a) to ensure internal cohesion, i.e. to overcome internal geopolitical tensions, and (b) to ensure a regime of visibility at the international level by valorizing identifiable regional resources (Giraut 2016). Does the new regional naming meet this double challenge?

The map of the 13 regions called for new names for at least the 7 new merged regions. Two chose to remain as the sum of their former merged regions: *Auvergne-Rhône-Alpes* and *Bourgogne-Franche-Comté*, while 5 have adopted a new name, possibly supplemented by the names of the former regions. We thus have the regions: *Hauts-de-France*; *Grand-Est* (with *Alsace-Champagne-Ardennes-Lorraine* as a secondary name); *Nouvelle Aquitaine* (with *Aquitaine-Limousin-Poitou–Charentes* as a secondary name), *Occitanie* (with *Pyrénées-Méditerranée* as a secondary name) and *Normandie.*

As far as the toponymic positioning of these new entities is concerned, they reveal the aspirations of the regional government who steered the operation and of the citizens who took part in the various types of prior consultation. These were organized in a participatory democracy perspective that is now systematic for questions of neotoponymy. On this occasion, the limits of the exercise became apparent, with consultations intended mainly to bring out or validate proposals that were filtered and completed by ad hoc committees under the control of the regional presidencies. Thus, the name *Grand-Est*, which was ultimately retained, was introduced by the regional government into the proposals submitted to the vote, even though it did not appear on the list drawn up by the committee of experts and citizens.

In order to appreciate this neotoponymy, we must first go back to the previous map drawn up at the end of the 1950s, that of the so called *regions de programme* which were first endowed with economic and social councils and then, during the decentralisation process in the 1980s, became full-fledged territorial authorities, with elections, budget and taxes (Masson 1983). The introduction of this regional level and the technocratically produced map constituted a break, including a toponymic one, with the exclusive territorial system of the *départements.* The latter, as we have seen, had been substituted for the tangle of different jurisdictions of the *Ancien Régime.* Their names did not recall any of them. In this sense they were neutral because they related to water or orography or geographical position, but never to history nor regional language.

The names of the 21 and then 22 regions were based on a number of references, in particular those of the *Ancien Régime*, alongside references to the major rivers and mountain ranges (*Rhône; Loire; Alpes; Pyrénées*) and to the position in the national ensemble (*Nord*; *Centre*; *Midi*). It is well known that the notion of *province* in the *Ancien Régime* is rather overused because it corresponds to multiple and changing institutions and maps (*Généralités*, *Gouvernements*, *Diocèses*, *Parlements …*) and that only the jurisdictions of the *États* and *Parlements* were intended to embody territories truly endowed with autonomy and recognized particularities, yet these so called *Pays d'État* existed essentially on the periphery of the Kingdom (*Bretagne*, *Languedoc*, *Provence*, *Bourgogne*, *Franche Comté*, *Dauphiné*) and not in the vast Parisian and Ligerian areas. In any case, the map of the 1950s invoked a number of *provinces*, preferring instead to the map of the Governments which subdivided the whole of the Kingdom. In a number of cases it made the contemporary administrative region a subdivision of the reference entity (the two Normandy regions), or on the contrary an entity encompassing more than the reference Province (*Provence-Alpes-Côte-d'Azur*; *Languedoc-Roussillon*; *Poitou–Charentes*).

While the new map and names recreate an integral *Normandie*, they do away with *Alsace*, *Lorraine*, *Bourgogne*, *Limousin*, *Poitou*, *Languedoc*, *Champagne* and *Picardie*, and do not satisfy aspirations for an integral *Bretagne* despite the name. Among the toponymic inventions, some allow for reference to historical and cultural spaces while adapting to the new geography. Thus the *région Centre* did not change its perimeter as a result of this reform, but its name. This was done by adding a reference to the *Val-de-Loire*, a real touristic brand that makes the name less banal.

The Occitan reference is in fact a hold-up (Aigouy-Campoy et al. 2019). Indeed, *Languedoc* referred to a well circumscribed former *Pays d'Etat* centered on *Montpellier*, whereas the Occitan belonging, or the Occitan speaking area, could also be claimed by *Aquitaine*, the former *Limousin*, *Provence*, and part of *Auvergne-Rhône-Alpes*. But this successful hold-up by the new region resulting from the merger of *Midi-Pyrénées* and *Languedoc-Roussillon* is paid for on another scale. It is *Roussillon*, in other words French Catalonia, that is paying the price with its toponymic erasing. This is a rule of political geography: integration on one scale generates losses of recognition and peripheralization on another. In the end, it is mainly a question of territorial marketing, despite the pretense of participatory management of the naming process. What do the new names say from this point of view?

Apart from the Occitan power grab and the Norman evidence, only *Nouvelle-Aquitaine* does not sacrifice the historical reference for internal cohesion. But this is done with the use of the pirouette that is the addition of the qualifier *nouvelle*. Taken independently, it is a positive keyword that would naturally carry the new entity towards the future and progress, but in toponymy this construction is rather reminiscent of colonial names that transferred the names of European regions to the other side of the world. There is therefore a great risk that *Nouvelle-Aquitaine* will be seen internationally as a region distinct from *Aquitaine*.

This is a problem that is even more acute with the two new regions of *Hauts-de-France* and *Grand-Est*. In both cases, the reference point is strictly national. The challenge of creating large regions was explicitly to position them in international

Table 3 Contexts and technologies involved in French *régions* neotoponymies

Neotoponymies	Contexts	Technologies
First *régions*	Emergence	Founding/Restoring
Restructured *régions*	Commodification	Promoting/Participating

Source Giraut (2002)

competition. In the case of *Hauts-de-France*, it is not so much the direct reference to *France* that poses a problem, but the absurd height claimed by transposing the position 'at the top' on the map of France. It ignores the qualifier *septentrional* (North) and the pure convention of orienting the North *en haut* in cartography! As a result, the whole is situated in an exclusive national framework, its name already has no meaning in itself, but no meaning at all if one leaves the image of the oriented hexagon (shape of the French state). The *Grand-Est* is a case of the same confusion. The reference to *grandeur* allows one to try to mobilize in territorial competition on the basis of supposed power, but this is done at the expense of heritage resources.

To summarize and interpret the two phases of contemporary French regional naming (Table 3), it can be said that they were first carried out in the context of the **Emergence** of a new functional scale, and then in the context of a competition between territories seeking to enhance their symbolic and economic resources (**Commodification**). From the point of view of the toponymic technologies deployed, the foundation of the new entities involved recourse to references to the restored *provinces* of the *Ancien Régime* (**Restoring**). In a second phase, the new names of the merged regions were essentially a matter of **Promoting** and branding. This was achieved at the cost of erasing certain historical references, but also by spectacularly promoting the reference to *Occitanie*.

The Naming of the New Communes: From Localities to Territories

Finally, let us return to the narrow municipal grid inherited from the Revolution. This was initially made up of more than 40,000 entities. In the nineteenth and twentieth centuries, it underwent a number of mergers, which brought it down to about 36,000 entities. These mergers, often linked to urbanization (Edelblutte 2000), mostly taking the form of absorption by the central city, which gave its name to the new entity.

Also during this long period name changes occurred (Guerrin 1998). They were relatively few: 1,863 from 1943 to 2006 (Delattre 2007). They mainly allowed the names of communes to be specified in their location with the mentions *sur* (on - most of the time a river) or *en* (in - most of the time an old rural *pays* name[2]) or *les* (*pays* or town name). This is to avoid confusion with other French communes of the same name. From the end of the nineteenth century onwards, these changes were also motivated by economic and notoriety reasons. Thus, many localities tried and often succeeded in distinguishing themselves by adopting a name complement in *les-bains* or *sur-mer* or *...-plage*.

[2] Regarding the historical and geographical aspects of these French rural so called *pays* see Gallois (1908) and Chamboredon (1988).

The recent movement to create the *nouvelles communes* through mergers is well underway since the 2010 law. At the end of 2019, more than 800 entities were recorded, grouping together more than 2,500 former communes for nearly 2.5 million inhabitants (Bideau 2019). The choice of name, often presented as a secondary detail, turns out to be an issue in itself that focuses the debate during the project phase, just like the fiscal effects of the merger. The choice of name also often appears to be an opportunity to enhance some specific resources. The debates on the choice of the toponym, often relayed by the regional press, can therefore reveal many burning questions of *glocal* geopolitics: positioning of the constituent localities in terms of centre/periphery relations, and positioning of the new entity in a regional and global landscape.

This new process of merging French communes must be placed in the broader international dynamics of territorial restructuring related to metropolization and critical local government size (Brenner 2004; Barlow and Wastl-Walter 2004; Bulkeley et al. 2016; Cardoso and Meijers 2017). It follows, in France, the systematic creation of a generalized double intercommunality layers: project intercommunality, and service management intercommunality which is also a project for metropolises and urban communities (Booth 2009; Estèbe 2008; Gerbaux 1999; Guichard 1976). This creation of intercommunal entities has already generated an intensive production of new names, less supervised by the state, which have been the subject of analyses (Bailly 2008; Brunet 2021; de La Soudière 2004; Delfosse 1997; Giraut and Lajarge 1996; Landel and Senil 2008; Quillot 2018), as well as those of the French cantons following their redrawing (Brunet 2015). The first generations of inter-municipal entities were rather endowed with multi-barreled names or with the names of former rural *pays*, allowing the component communes not to be symbolically dissolved in the reference to the central town or city alone. Subsequently, the issue of economic promotion, and in particular tourism, through a name that is gratifying or evokes a regional resource as landmark, became the main focus. This is part of a territorial branding and marketing approach (Alaux et al. 2015) which is also connected to metropolitan new regionalism at international level (Guo and Zhang 2019; Lu et al. 2018).

In particular, the new movement of mergers of French communes is part of the fusions that have taken place in Europe since the middle of the twentieth century, mainly in countries where, like France, the municipal grid was based on that of the parishes. Thus, Switzerland, in a very different institutional context, has been experiencing a significant movement of mergers of communes for the last thirty years (Destrem 2016; Raffestin 2000). Elsewhere, a movement of creation of communes by progression of the so-called incorporated domain to municipalization, or by process of decentralization, has produced a very large number of new communes in the last decades in very different contexts (Antheaume and Giraut 2005; Baldersheim and Rose 2010; Meligrana 2004). These processes of massive creation of new communes have been the subject of analyses, including the toponymic component (Adam 2008; Boujrouf and Hassani 2008; Destrem 2016; Hoffman 2000; Ji 2018; Lehr and McGregor 2016; Lima 2008; Masuda and Bookman 2018; Soguel and Silberstein 2015; Verdeil 2011; Zimmerbauer and Paasi 2013). A certain number of neotoponymical stakes and challenges run through the great diversity of situations (Giraut 2005; Giraut and Houssay Holzschuch 2008a). The main one is the question of the always complicated toponymical relationship with the city-centre, which is both a pole of centrality and an unavoidable magnet for any territorial reform, and a dominating and predatory chief town that threatens the identity of the peripheries with its hegemony. There is also a tendency for all local authorities, and in particular those administering small or medium-sized towns and cities, to capture by naming the resources, including heritage resources, of a hinterland for which the city would constitute a gateway (Giraut 2005). This is expressed in the neotoponymy resulting from territorial recompositions (Giraut and Houssay-Holzschuch 2008a, b).

In the case of France and the creation of *nouvelles communes*, which are not immune to these trends, it is necessary to take into account national specificities as well as previous experiences of naming and renaming concerning communes and their groupings. The issue of naming these entities has not escaped the attention of the central government (*Commission Nationale de Toponymie* and *Direction Générale des Collectivités locales*). The latter issued a circular on 18 April 2017,[3] addressed to the *préfets* responsible for validating the names and to the municipal councils responsible for drawing up the project including the name proposal. This circular recalls the rules that should govern their choice and determination. In addition to a reminder about spelling (hyphens, capital letters, accentuation), which was abused in many of the names of the first generation of *nouvelles communes*, the circular and a more recent guideline advise[4] the use of the initial names of the merged communes or (where appropriate) 'an old name with which the commune can identify'. Furthermore, it recalls that the *Conseil d'État* does not accept 'names based on considerations of simple tourist or economic publicity'.

The objectives of the project leaders of new communes may be significantly different and include questions of plural internal identities and external identification, coupled with the valuation of resources, particularly touristic ones. Finally, the process of choosing a name appears to be a means of involving people around the project. Thus, in 2017, in the public presentation of the (aborted) merger project between *Doussard* and *Faverge-Seythenex* in *Haute-Savoie*, we learn that the name can and must be "Support for a common territorial project, a vector of image for the inhabitants and the outside world, evocative of the territory and its assets. (…) The question of the name constitutes a challenge which must at the same time seek to respect each individual and each inhabitant in his or her history and identity, but also enhance the value of this exceptional territory in the tourist sense (one of the only French communes which will benefit from a beach and a ski resort as well as a certain number of high-quality outdoor activities or sites) or even try to identify the new commune geographically, quite simply (…) The search for the name of the new

---

[3] Direction Générale des Collectivités Locales, 2017, «Fixation du nom d'une commune nouvelle», Online: http://cnig.gouv.fr/wp-content/uploads/2017/04/Indications_nom-commune-nouvelle_18042017.pdf.

[4] Commission Nationale de Toponymie 2021, *Décider du nom d'un lieu. Guide pratique à l'usage des élus*, Délégation Générale à la Langue Française et aux Langues de France.

commune can be a formidable vector of solidarity and joint construction between all the economic, associative and political players and with the inhabitants. In order to build, through the choice of a name, the beginning of a common identity, the elected representatives have decided to engage in a very broad consultation process with the inhabitants and local players, representatives of associations, entrepreneurs, shopkeepers, partners, etc."

Initial observations on the new names issued from *nouvelles communes* show that they often mention a reference to a former rural pays (territory or area), a landscape feature or a province of the *Ancien Régime*. The prepositions *en*, *au* and *de*, but also *les*, *lès* and *sur*, are over-represented among the names of *nouvelles communes* already in existence. The use of *en* followed by the name of a rural *pays* or *province* thus rises from less than 2% among the names of merged communes to almost 7% among the names of new communes. And this more than doubles (to nearly 15%) if we add the *de* and *au* followed by a rural *pays*. This is in line with the observations already made on the numerous requests (not always validated) to modify the names of communes over the last few decades, with a clear tendency to add hydronyms and especially regionyms (Guerrin 1998), among other examples: *Châlons-en-Champagne* and *Saint-Dié-des-Vosges*.

We also note the frequent assimilation or association of *nouvelles communes* with a *val* or *vallée*, which avoids referring to one of the constituent localities and refers to a topographical and possibly historical and economic territory. Thus 85 *vaux*, *vallées* or *vallons* are included in the names of the 810 *nouvelles communes* created between 2015 and 2019, i.e. almost 11%, whereas only 10 occurrences existed among the names of the 2525 initial communes (less than 0.5%), five of which were already the result of previous mergers. Thus, the names chosen in the end may be fairly neutral compromise names that are the result of neutralizing initial proposals that enhance the value of a particular constituent locality (the city centre in particular) and/or of highlighting a generic resource, for example *Entre-Vignes* between *Montpellier* and *Nîmes* or *Valforêt* in rural *Côte-d'Or*. There is therefore a tension around the reference to the city-centre, between internal fear and external need. There is also a tendency towards toponymic neutralization, excluding the names of localities in favour of generic references to the landscape, the topography and/or the local and regional environment in the form of references to *pays* and *provinces* (typically: *Val-en-* …; *Monts-du-* …; *Terres-de-* …).

The very few changes in the recent names of *nouvelles communes* that have already taken place could indicate a certain return to the reference to the central town after a period of toponymic neutralization. In fact, in 2017, the *nouvelle commune* formed by the communes of *Plémet* and *La Ferrière*, freshly named *Les Moulins*, abandoned this generic referent to take on the name of the largest former commune: *Plémet*. At the end of 2021, *Capavenir-Vosges*, with the marketing name of an ex-commune of 3 communes, became *Thaon-les-Vosges* with the name of its ex-centre commune; and *Le Hom*, from 5 communes, became *Thury-Harcourt-le-Hom* by adding the name of the local hydrographic figure common to that of its ex-centre commune.

At the same time, there is a culture issue with the fear of losing heritage and local identity to the imposition of generic or off-ground or regional references for

Table 4 Contexts and technologies involved in French municipal cooperation and mergers neotoponymie

Neotoponymies	Contexts	Technologies
Urban municipal merger (XIX-XX)	Emergence (agglomeration)	Absorptioning
Municipal cooperation areas 1st generation	Emergence	Neutralising
Municipal cooperation areas 2d generation	Emergence/Commodification	Neutralising/Promoting
Communes nouvelles	Emergence/Commodification	Neutralising/Promoting/Participating

Source Giraut (2022)

territorial marketing purposes. More specifically, concerns about the conservation or loss of onomastic heritage are expressed. This is true from three points of view:

- linguistic with, in particular, the risk of erasing Breton for French in the new names (Bourges 2017; Larvor 2017). However, this has not been proven for the moment (Bideau and Giraut 2022). There is even a case of neo-Breton in Ille-et-Vilaine: *Maen Roch*;
- religious: hagiographic toponymy with the *Saint.es* would disappear; indeed, from the beginning of 2015 to the end of 2019, it was present in 18.5% of the names of merged communes to 4.3% of the names of new communes, a trend that increases even more if we take into account the disappearance of references to churches, crosses, chapels, abbeys, etc.;
- historical: disappearance of references to localized events or to specific historical statuses of localities or to historical personalities….

These fears and denunciations are mainly expressed in the regional press. They will have to be verified precisely, but it can be assumed that while some generic names are appearing, a certain erasure of local linguistic and historical specificities is at work. This is true of the new names, since those of the former communes may remain and be displayed in the signage if they become *communes déléguées*. More generally and more profoundly, there would be a shift from communal references linked to localities (ex-parishes) to words linked to landscape and territorial environment.

In terms of interpreting the reasons for the neotoponymy linked to the *nouvelles communes* (Table 4), it can be said that the new names linked to inter-communal cooperation and the merging of communes take place in a context of the **Emergence** of new functional entities linked in particular to urbanization. Since the end of the nineteenth century, the symbolic valorization of resources for notoriety and economic development has also been an element of context that is similar to toponymic **Commodification**. The combined and successive technologies may have involved the neutralization (of the domination of the chief towns) (**Neutralizing**) and the promotion of the image and value of the new entities (**Promoting**).

Conclusion

With the three French territorial restructurings of the last centuries, it appears that radical political changes and territorial conquests are particularly favourable contexts for the eradication of old territories (the *provinces* of the *Ancien Régime*), the production of new ones (the *départements*) and the change of status and partial renaming of inherited entities (the parishes and cities). It also appears that the redistribution of the population as a result of urbanization has led to the emergence of new territorialities and related denominations either through the constitution of new types of entities (*régions*, inter-municipal cooperation areas) or through mergers (new large *régions*, *nouvelles communes*). Finally, a context of marketing and competition between territories is likely to generate renaming. The toponymic goals and technologies that govern the choice of new names are diverse and can be combined. Thus, *cleansing* consists of not including in new names toponymic references to former entities belonging to a bygone political order or a denied tradition (in the rarer contexts of our case of conquest). *Neutralising* will generate names which do not recall those of former entities or which do not give primacy to a central place in a territorial entity. Its implementation favours generic names referring to the environment and landscape or possibly to a specific regional environment or heritage. They erase references to localities to favour references to local territories marked by landscape or heritage sites or areas. The promotion will consist in the valuation of symbolic or material resources. These can range from the notoriety of a site, a town or an environmental element to the evocation of the greatness of the entity or its capacity to provide access to a whole region.

Finally, it appears that contemporary naming or renaming practices are increasingly using the technology of public participation in the choice of name. This can range from elementary and superficial modalities such as a vague online consultation based on a pre-selection, to the constitution of real workshops to elaborate an initial list which is then submitted to a wide consultation and debate. The neotoponymy linked to territorial restructuring is therefore marked by a succession of toponymic technologies that are recombined according to the context and the issues at stake.

Questions

- What are the different toponymic techniques applied to local and provincial government territories names?
- What political challenges do they address in different contexts?
- What are the successive place-naming experiments that France has experimented with to name its administrative territories (departments, regions) and its communes?

References

Adam F (2008) L'autorité et l'autre. Parcours toponymiques et méandres linguistiques au Québec. L'Espace politique (2):31–39
Aigouy-Campoy P, Marchand P, Ratinaud P (2019) Resémantisations du toponyme «Occitanie» dans la presse quotidienne régionale. Pôle Sud 50(1):67–88
Alaux C, Serval S, Zeller C (2015) Le marketing territorial des Petits et Moyens Territoires: identité, image et relations. Gestion et management public 4/2(4):61–78
Amilhat Szary A-L (2008) Des territoires sans nom peuvent-ils être sans qualité ? Réflexions sur les modifications de la carte administrative chilienne. L'Espace Politique (2)
Antheaume B, Giraut F (eds) (2005) Le territoire est mort, Vive les territoires ! IRD Editions, Paris
Bailly G (2008) Nommer les espaces de coopération intercommunale. L'espace Politique 2:79–96
Baldersheim H, Rose L (eds) (2010) Territorial choice: the politics of boundaries and borders. Springer
Berg LD, Vuolteenaho J (eds) (2009) Critical toponymies: the contested politics of place naming. Aldershot Ashgate
Barlow M, Wastl-Walter D (ed) (2004) New challenges in local and regional administration. Ashgate
Bideau G (2019) Les communes nouvelles françaises (2010–2019): une réforme territoriale silencieuse. Annales De Géographie 728(4):57–85
Bideau G, Giraut F (2022) (forthcoming) Faire territoire: logiques de la dénomination des communes nouvelles françaises. l'Espace Politique
Booth P (2009) Planning and the culture of governance: local institutions and reform in France. Eur Plan Stud 17(5):677–695
Boujrouf S, Hassani EM (2008) Toponymie et recomposition territoriale au Maroc: Figures, sens et logiques. L'espace Politique 2:40–52
Bourges A-Y (2017) Les «communes nouvelles» et la néo-toponymie en Bretagne: le cas du Coglais. Blog: Variétés historiques. http://varietes-historiques.blogspot.ch/2017/01/les-communes-nouvelles-et-la-neo.html
Boyer H, Cardy H (2011) Localiser, identifier, valoriser. Mots, Les langages du politique (97):5–13
Brenner N (2004) New state spaces: urban governance and the rescaling of statehood. Oxford University Press, Oxford
Brunet R (2015) Nouveaux cantons en France: jeux de noms. L'espace Géographique 44(1):73–80
Brunet R (2021) Nouveaux territoires, nouveaux noms de la France. Hermann, Paris
Bulkeley H, Luque-Ayala A, McFarlane C, MacLeod G (2016) Enhancing urban autonomy: towards a new political project for cities. Urban Stud 55(4):702–719
Cardoso RV, Meijers EJ (2017) The metropolitan name game: the pathways to place naming shaping metropolitan regions. Environ Plan A 49(3):703–721
Chamboredon J-C (1988) Carte, désignations territoriales, sens commun géographique: Les "Noms de pays" selon Lucien Gallois. Études Rurales 109:5–54
de La Soudière M (2004) Lieux dits: nommer, dé-nommer, re-nommer. Ethnol Fr 34(1):67–77
Delattre E (2007) Le changement de nom des communes françaises aspects économiques, marketing et stratégiques. Revue D'économie Régionale & Urbaine 2:269–291
Delfosse C (1997) Noms de pays et produits du terroir: enjeux des dénominations géographiques. L'espace Géographique 26(3):222–230
Destrem L (2016) Un état des lieux des noms de communes fusionnées en Suisse. Blog Neotoponymie. https://neotopo.hypotheses.org/473
Edelblutte S (2000) Réflexion sur les modifications du maillage communal français depuis 1790. Revue Géographique de l'est 40(4):21p
Estèbe P (2008) Gouverner la ville mobile: intercommunalité et démocratie locale. France, Presses universitaires de France, Paris
Gallois L (1908) Régions naturelles et noms de pays. Armand Colin, Paris
Gerbaux F (ed) (1999) Utopie pour le territoire: cohérence ou complexité ? La Tour d'Aigues, Ed. de l'Aube

Giraut F (2002) L'"exception territoriale française" sous influence européenne? L'information Géographique 66(2):133–161
Giraut F (2005) Fabriquer des territoires. Utopies, modèles et projets, Habilitation à diriger des recherches. Université de Paris I
Giraut F (2016) Nouveaux noms de régions françaises et marketing territorial, ou quand le grand, le haut et le nouveau ratent ce que l'Occitan réussit. Neotoponymie/Neotoponymy, En ligne. https://neotopo.hypotheses.org/625
Giraut F (2020) Pour une théorie de la nomination des communes nouvelles françaises: des localités aux territoires. Néotoponymie/Neotoponymy, En ligne. https://neotopo.hypotheses.org/2526
Giraut F, Houssay-Holzschuch M (2008a) Au nom des territoires! Enjeux géographiques de la toponymie. L'espace Géographique 37(2):97–105
Giraut F, Houssay-Holzschuch M (2008b) Néotoponymie: formes et enjeux de la dénomination des territoires émergents. L'espace Politique 2:5–12
Giraut F, Houssay-Holzschuch M (2016) Place naming as Dispositif: toward a theoretical framework. Geopolitics 21(1):1–21
Giraut F, Houssay-Holzschuch M (eds) (2022) The politics of place naming. Wiley, Hoboken
Giraut F, Lajarge R (1996) "Les "pays-tests", quelles configurations pour quels projets?" Montagnes Méditerranéennes: Le pays: échelle d'avenir pour le développement territorial?:19–27
Grégory M-A (2011) Changer de nom pour changer d'image. Le cas des modifications de dénomination de département. Mots (97):15–29
Guerrin C (1998) «Étude socio-toponymique des changements intervenus dans la nomenclature officielle des communes françaises entre 1943 et 1993». Actes des colloques de la Société française d'onomastique 8(1):133–140
Guichard O (1976) Vivre ensemble: rapport de la Commission de développement des responsabilités locales. Paris, France, la Documentation française
Guillorel H (ed) (2008) Toponymie et politique: les marqueurs linguistiques du territoire. Bruxelles, Bruylant
Guo J, Zhang Z (2019) Does renaming promote economic development? New evidence from a city-renaming reform experiment in China. China Econ Rev 57:101344
Hoffman KE (2000) Administering identities: state decentralisation and local identification in Morocco. J North Afr Stud 5(3):85–100
Ji X (2018) City-renaming and its effects in China. GeoJournal 83(2):381–397
Landel P-A, Senil N (2008) Les nouveaux territoires et leurs noms entre projet et compétitivité: le cas des "pôles d'excellence rurale". L'Espace politique (2):66–78
Larvor R (2017) Toponymie bretonne. «Un trésor en danger...». Le Télégramme. http://www.letelegramme.fr/bretagne/toponymie-bretonne-un-tresor-en-danger-12-09-2017-11659329.php
Lehr JC, McGregor B (2016) The politics of toponymy: naming settlements, municipalities and school districts in Canada's Prairie provinces. Prairie Perspect: Geogr Essays 18:78–84
Le Bart C, Procureur T (2011) Quand les Côtes du Nord sont devenues les Côtes d'Armor. Le département entre identité et attractivité. Mots. Les langages du politique (97):31–44
Light D, Young C (2015) Toponymy as commodity: exploring the economic dimensions of urban place names. Int J Urban Reg Res 39(3):435–450
Lima S (2008) L'émergence d'une toponymie plurielle au Mali. L'espace Politique 2:13–30
Lu S, Wu Y, Xie X (2018) Benefit of historical reputation: evidence from the city renaming reforms in China. China Econ Q 17(3):1055–1078
Maréchaux X (2016) République et «laïcité»: le nom révolutionnaire des communes de France sous le Consulat et l'Empire (et au-delà), essai de recensement et d'interprétation. Napoleonica. la Revue 25(1):94–122
Masson J-L (1983) Provinces, Départements, Régions: l'évolution des circonscriptions administratives de la fin de l'Ancien Régime à nos jours. Doctorate Thesis in Law, Paris II University
Medway D, Warnaby G (2014) What's in a name? Place branding and toponymic commodification. Environ Plann A: Econ Space 46(1):153–167

Meligrana J (ed) (2004) Redrawing local government boundaries. An international study of politics, procedures, and decisions. Vancouver, University of British Columbia Press
Meuriot P (1917) Pourquoi et comment furent dénommées nos circonscriptions départementales? Auguste Picard, Paris
Nordman D, Ozouf-Marignier M-V (1989) Atlas de la Révolution française: Le territoire. Éd. de l'École des Hautes Études en Sciences Sociales, Paris
Offner J-M (2006) Les territoires de l'action publique locale. Fausses pertinences et jeux d'écarts. Revue Française de Science Politique 56(1):27–47
Ozouf-Marignier M-V (1988) La formation des départements: la représentation du territoire français à la fin du 18e siècle. Paris, France, Éd. de l'École des hautes études en sciences sociales
Pourtier R (1983) Nommer l'espace. Émergence de l'État territorial en Afrique noire. L'espace Géographique 12(4):293–304
Puzey G, Kostanski L (eds) (2016) Names and naming: people, places, perceptions and power. Bristol, Multilingual Matters
Quillot E (2018) Néologie toponymique des nouveaux groupements intercommunaux: l'innovation langagière est-elle au rendez-vous?, 4ème Congrès international de néologie des langues romanes 4–6 juil. 2018 Lyon (France) Lyon
Raffestin C (2000) Les projets de fusion territoriale: entre politique et économie. Bulletin de la Société Neuchâteloise de Géographie 44:9–18
Rose-Redwood R, Alderman D, Azaryahu M (2010) Geographies of toponymic inscription: new directions in critical place-name studies. Progr Human Geogr 34:453–470
Rose-Redwood R, Vuolteenaho J, Young C, Light D (eds) (2022) Naming rights, place branding, and the cultural landscape of neoliberal urbanism. Routledge, London
Soguel N, Silberstein J (2015) Welfare loss with municipal amalgamations and the willingness-to-pay for the municipality name. Local Gov Stud 41(6):977–996
Vanier M (2008) Le pouvoir des territoires: essai sur l'interterritorialité. Paris, Anthropos
Verdeil É (2011) Délimiter les banlieues de Beyrouth. Des noms et des périmètres. Des banlieues à la ville, espaces et acteurs de la négociation urbaine, Presses de l'IFPO
Zimmerbauer K, Paasi A (2013) When old and new regionalism collide: deinstitutionalization of regions and resistance identity in municipality amalgamations. J Rural Stud 30:31–40

Geographical Names in Argentina: Present and Challenges

Adriana Vescovo

Abstract Spaces to discuss toponymy both as geospatial data and for its value as intangible cultural heritage, are expanding. Its vitality grows, not only in the public and academic sphere, but also accompanying social, political, economic, and cultural processes around the world. In recent decades its visibility has increased, as well as the need for its local, national, and global standardisation. At the same time, new problems and questions have arisen. The aim of this chapter is to reflect on this particular moment in the evolution of place names in Argentina from a geographical perspective including: factors and characteristics in this territory; its wide multi-cultural richness; the impact of emerging new technologies and paradigm shifts in geospatial information management; reactions and involvement of multiple stake-holders; its central role in SDIs and in the fulfillment of SDGs; experiences, projects and steps taken on the road to standardisation; and challenges posed by the current global environmental, technological, and cultural context.

Keywords Argentine toponymy · Place names standardisation · SDI · SDG · Decision-makers · Intangible cultural heritage

Objectives

- By the end of this chapter, readers should be able to recognise the impact that technological revolution and new cultural paradigms have had in the global management of geospatial information.

A. Vescovo (✉)
Former Responsabile of the Department of Geographical Information, Instituto Geográfico Nacional (IGN- Argentina), Avenida Cabildo 381, (C1426AAD), Ciudad de Buenos Aires, Argentina
e-mail: adriana.vescovo2012@gmail.com

Corresponding Member (Argentina), Joint ICA/IGU Commission of Toponymy, Ciudad de Buenos Aires, Argentina

Treasure and Coordinator of the Commission of Toponymy, Centro Argentino de Cartografía (CAC), Avenida Cabildo 381, (C1426AAD) Ciudad de Buenos Aires, Argentina

© The Author(s), under exclusive license to Springer Nature Switzerland AG 2023
G. O'Reilly (ed.), *Place Naming, Identities and Geography*,
Key Challenges in Geography, https://doi.org/10.1007/978-3-031-21510-0_5

- By the end of this chapter, readers should be able to recognise the social actors involved in the management of geographical names as part of geospatial Big Data, and actions that have arisen in Argentina as a response to this new scenario.
- By the end of this chapter, readers should be able to recognise the main problems and challenges that management of geographical names poses in Argentina, considering its own characteristics and those of the current global technological, cultural, and environmental context.
- By the end of this chapter, readers should be able to recognise the importance of geographical names in SDIs and in compliance with the SDGs -globally and in Argentina- and the urgent need for their standardisation.
- By the end of this chapter, readers should be able to apply the concepts developed in the analysis of cases offered as examples.

Introduction

This chapter aims to offer a summary of the current situation of the management of geographical names in Argentina, its progress, its problems, and its future challenges.

In this sense, its purpose is not to develop a complete and deep analysis of Argentine toponymy, but to characterize the current scenario of its treatment according to its own territorial, demographic, political organization, and idiosyncratic characteristics, and as a response to the present global technological, cultural, and environmental context.

In this regard, geographical names will be considered from their dual approach: as geospatial data and as intangible cultural heritage.

Before closing, a tour of some Argentine toponyms will be offered, as selected examples of the application of the themes developed.

The permanent intention of the chapter is to promote a more conscious participation in the management of geographical names by all social actors, as part of a broader global commitment.

In no case will it attempt to be an exhaustive description and it will only cite background or examples that may be considered important, interesting, or closer to the understanding of non-local readers.

Global Context of the Geospatial Information

As part of geospatial information and as signifiers of identity, geographical names were affected by the profound technological changes and by cultural paradigms that transformed the world and geographical science from the mid-twentieth century to the present.[1]

[1] If you are interested in this theme, please see Suggested reading.

Technological Context, New Paradigms, and Trends

To synthesise this context, a summary of its main characteristics is included in Table 1.

Geographical Names in the Current Global Scenario

Toponyms identify most of geographical features. In the current technological context, they are a basic and fundamental element of geospatial information and accompany almost all geographic features throughout their management cycle: from their capture to their processing and dissemination.[2]

From this place, they were affected by the digital and technological revolution already described and are currently immersed in the global scenario in which all geospatial information is developed, sharing the same challenges as part of the geographic big data and artificial intelligence approaches.[3]

Need to standardised geospatial information to serve its digital interoperability under the collaborative model and Open Data policy has reinforced the need for the application of principles and guidelines that regulate the standardisation of geographical names at the national and international level. This circumstance has renewed the role and work that the United Nations Group of Experts on Geographic Names (UNGEGN) has been developing since 1967.[4]

Planetary awareness and actions aimed at achieving the goals of the SDGs[5] have focused on the need for managing geospatial information at a global level. In this sense, the creation of UN-GGIM, the recognition of geographical names as the Global Fundamental Geospatial Data Theme in the SDI[6] and the drafting of the Integrated Geospatial Information Framework (IGIF)[7] are clear responses to these requirements posed by the twenty-first century.

In this regard, the UN Resolution VIII/6 of the 8th UN Conference on the Standardization of Geographical Names in 2002, had recommended that *"standardized geographical names data should be considered in the establishment of national and*

---

[2] If you are interested in this theme, please, see Suggested reading.

[3] About applications of AI in geospatial information, please read about ARIES for SEEA, on https://seea.un.org/content/aries-for-seea.

[4] To access a comprehensive store of information of geographical names standardisation please, visit the UNGEGN website: https://unstats.un.org/unsd/ungegn/.

[5] For complete information about SDG please visit: https://sdgs.un.org/2030agenda.

[6] UN-GGIM includes geographical names as one of the 14 basic and fundamental geospatial information layers. And it recognizes its link with all the SDGs. https://undesa.maps.arcgis.com/apps/Cascade/index.html?appid=4741ad51ff7a463d833d18cbcec29fff.

[7] The Integrated Geospatial Information Framework (IGIF) "provides a basis and guide for developing, integrating, strengthening, and maximizing geospatial information management and related resources in all countries. It will assist countries in bridging the geospatial digital divide, secure socio-economic prosperity, and to leave no one behind." Read more in: https://ggim.un.org/IGIF.

Table 1 Global context of geospatial information

Technological context The irruption of the digital age meant one of the most revolutionary events that occurred in geographic science. It fueled technological innovation, improved accuracy, and exponentially expanded the volume, speed, and formats in which geospatial information is captured, managed, processed, and disseminated	Its main characteristics are: • Digital and global management of information • Exponential development of technologies, platforms, and devices • Extensive incorporation of private capital into ground and space technology • New products, services, and applications • Incorporation of artificial intelligence (IA)[a] • Global and permanent access to information on the planetary system • Change of paradigm in the official cartographic production • Development and dissemination of GPS use	GEOSPATIAL BIG DATA Need for norms and standards of quality and interoperability
Cultural context Geographic science and geospatial information management were also affected by the new paradigms that emerged in the cultural changes of the twenty-first century	Its main characteristics are: • Georeferencing of everyday life: everything is geolocated • Empowerment of the user in the private and public spheres • Emergence and development of the collaborative model • Participation of all social actors in the construction of knowledge and information • Information as a citizen's right • Emergence and dissemination of the OPEN DATA policy • Affirmation of awareness of planetary challenges • Emergence of global and integrated responses (SDG-Agenda 2030) • Recognition of the strategic, social, and economic value of geospatial information • Emergence of social movements on gender and anti-discriminatory issues, recovery and revaluation of native and local cultures, and historical review of colonization and decolonization processes	SDIs AND INTEGRATED GEOSPATIAL INFORMATION Need to have accessible, up-to-date, and reliable georeferenced geospatial information as an essential ingredient of public, regional and global policies

Developed by the Author

[a]For a definition of Artificial Intelligence (AI), please read: https://ec.europa.eu/futurium/en/system/files/ged/ai_hleg_definition_of_ai_18_december_1.pdf

regional spatial data infrastructures and included in their design, development and implementation" (United Nations 2002, 37–38).

More recently, in its Strategic Plan 2021–2029, "*UNGEGN recognizes the UN Sustainable Development Goals (SDGs) are the blueprint to achieve a better and more sustainable future for all. They address global challenges, including those related to poverty, inequality, climate, environmental degradation, prosperity, and peace and justice*" and adds that "*monitoring of SDGs should be adequately supported by thematic analyses based on a set of geospatial data, of which geographical names are a core element*". Finally, the document discussed and approved during the last Virtual Session held in April 2021 closes by saying that the "*use of nationally standardized geographical names applies to all seventeen SDGs.*" (UNGEGN 2021, 13).

Likewise, geographical names have renewed their vitality as part of the intangible cultural heritage. This growth of interest in toponyms as a reflection of identity is not only reserved for the academic field. A greater participation of the community in the choice of new names is observed, as well as an increase in the recovery of names in indigenous languages, an increase in discussions regarding the use of geographical names linked to historical revisions in colonized territories, as well as other related to gender issues. Use of geographical names in the official and business sphere has also expanded, as country and city brands, in tourism, and in certifications of origin in local food production.

In short, geographical names have notably renewed their vitality within the framework of the new trends that characterise the current global scenario.

Geographical Names in the Argentine Republic

Like what happens in other geographical spaces, Argentine toponymy reflects an intimate link with its landscape considered—in the broadest sense of the term—as "*an area, as perceived by people, whose character is the result of the action and interaction of natural and/or human factors;*" as defined by the Council of Europe Landscape Convention (Council of Europe ETS 176, 2000, 2).

According to this definition, the scenario of the management of geographical names in Argentina will be analysed, considering the responses of the different actors involved, achievements, challenges, and a selection of some cases, as an application of the concepts developed.

Argentine Local Context: Territorial Characteristics and Spatial Occupation Processes

Naming places establishes a link between man and his geographical space. In this sense, with the aim of making this relationship clearer and giving a framework to the current situation of the local management of the geographical names, a very synthetic characterisation of the territory, the processes of its occupation and their political organization are included in Table 2.

These varied physical characteristics are associated with multiple natural hazards such as earthquakes, volcanism, fires, and floods. Climate change affects the Argentine territory in different aspects, areas, and severity.

According to the last population census 2010[8] 2.4% of the population recognized themselves as descending from indigenous or native peoples and 0.4% as Afro-descendants. 97.2% comes from the primitive Creoles of the times of the Conquest, their miscegenation, and the following migratory waves. Italians and Spanish were the majority, but diversity includes a long list of communities from all continents.[9]

Management of Geographical Names in Argentina: Response of the Actors Involved

In summary, the current panorama of geographical names management in Argentina is the response to the exceptional changes in global technological and cultural paradigms already described, applied to its own territorial characteristics, its demographic history, its form of government and its idiosyncrasies.

In this new scenario, toponyms ceased to be a static element of the map on paper, to become digital information with multiple applications in everyday life, and a basic and fundamental datum of geographic Big Data, and of the new geospatial engineering in all its formats and supports.

Emerging from multiple sources and disseminated at great speed, geographical names have gained everyone's attention.

In the official sector, daily work experience pointed to the urgency of having standardised geographical names to respond to the need for accurate geospatial information, updated and in real time.

Likewise, within the current context, the geospatial information management has gone far beyond the scope of official cartographic production, reaching the most varied social actors.

---

[8] INDEC. National Population and Housing Census 2010. Statistics. Population Groups. https://www.indec.gob.ar/indec/web/Nivel3-Tema-2-21.

[9] This vast genealogy of migratory origin was so significant that it is usually synthesised with the widespread phrase: "Argentines descend from ships". For more details about demographic composition, distribution, stories, etc. please, visit: https://culturaargentina.org.ar/colectividades/index.html.

Table 2 Argentine local context

 *Territorial characteristics*[a] The situation, extension and multiplicity of physical factors define the diversity of its natural environments	• BICONTINENTAL SITUATION A portion at the southern tip of the American continent and another on the Antarctic[f]continent, linked by an arc of oceanic islands • EMERGED SURFACE[b]: 3.761.274 km^2 South American: 2.791.810 km^2 Antarctica: 965.597 km^2 Oceanic islands: 3.867 km^2 7th place in the world order and 4th in America • LENGTH OF COASTS[c]: 15.960 km Atlantic coastline: 4.725 km Antarctica coastline and islands coastline: 11.235 km • SUBMERGED SURFACE[d]: 6.581.000 km^2 Between the baselines and the outer limit of the continental shelf (175% of the emerged territory) • NORTH–SOUTH EXTENSION[e]: 3.600 km In the South American area. Represents a difference in latitude of 30° • GREAT DIVERSITY OF TYPES, FORMS AND HEIGHTS OF RELIEF Different origins and geomorphological processes • MULTIPLICITY OF CLIMATES: WARM–COLD/HUMID-ARID Originated by the great extension in latitude, atmospheric circulation, height, and layout of the relief • MORE THAN 100 BASINS AND SURFACE WATER REGIONS Rivers, wetlands, salt flats, lakes, and lagoons of very diverse flow, extension, shape, and depth • AROUND 17.000 BODIES OF ICE According to the first National Inventory of Glaciers[g]

(continued)

Table 2 (continued)

Territory occupation processes[h] The great cultural diversity is the main characteristic of its social composition	• The human occupation of the current Argentine territory dates back 10.000 years • Currently there are 39 original peoples[i] distributed in 1.770 indigenous communities • The most numerous indigenous groups are the Mapuches (especially in Patagonia), the Guarani (in the NE), the Wichis and the Qom (in the N), and the Quechuas (in the centre and NW) • 16th century: European settlers and explores arrive, mostly Spanish, as part of the process of conquest and formation of the Viceroyalty of the Río de la Plata (1776–1814) • Expeditions from England, France and the Netherlands also toured the Patagonian coasts and islands • From the beginning of the conquest, several religious orders settled in the current Argentine territory and established links with the indigenous population (Jesuits, Franciscans, Salesians, Anglicans, etc.) • The slave trade introduced the black population • 19th century: Europeans of various origins (French, English, Irish) will participate in the struggles for independence and will precede future migratory currents • On July 9, 1853, Argentina declared independence from Spain and adopted a representative, republican and federal system of government • The incorporation to the armies, the diseases, the definition of the borders and the political organization of the territory generated the displacement, the decrease, and the extinction of great part of the black and native population • Towards the end of the 19th century and the beginning of the 20th, an intense immigration policy promoted the largest wave of immigration[j] • The two great world wars generated new waves of migration • More recently, the population from neighbouring and Asian countries is added

(continued)

Table 2 (continued)

	• The immigrants were distributed throughout the territory • The current distribution of the Argentine population is very irregular: a third of the total population is settled in the Metropolitan Area of Buenos Aires (AMBA) with a maximum density of more than 14.000 inhabitants per km^2. Almost 60% of the population is concentrated in the mid-latitude fringe, with large gaps in mountainous and arid areas • From a political point of view, Argentina has a federal government system: it includes 23 provinces and a federal district

Developed by the author from the cited sources

[a]Official map. Please, visit: http://www.plataformaargentina.gov.ar/en/mapaPlataforma_i

[b]*Source* IGN. Límites, Superficies y Puntos Extremos | Instituto Geográfico Nacional (ign.gob.ar)

[c]*Source* IGN. Límites, Superficies y Puntos Extremos | Instituto Geográfico Nacional (ign.gob.ar)

[d]For more information, maps, and official documents, please visit: http://www.plataformaargentina.gov.ar/en/%28copla_i%29

[e]*Source* IGN: Límites, Superficies y Puntos Extremos | Instituto Geográfico Nacional (ign.gob.ar)

[f]Since 1904, Argentina has been constantly and uninterruptedly present in Antarctica, which constitutes the longest continuous presence in such continent. Argentina is one of the twelve original signatories of the Antarctic Treaty. For more information about Argentina in Antarctica, please visit: https://www.cancilleria.gob.ar/en/foreign-policy/antarctica

[g]https://www.argentina.gob.ar/ambiente/agua/glaciares/inventario-nacional

[h]If you are interested in this issue, please see Suggested Readings

[i]For the Map of the Distribution of the indigenous people, please see: https://www.argentina.gob.ar/sites/default/files/2020/11/mapapueblos_112021.pdf

[j]If you are interested in this issue, please visit: https://www.argentina.gob.ar/interior/migraciones/museo-de-la-inmigracion

In fact, the new trends mobilized the relationship between geographical names and the community, as an active manager of its cultural heritage.

In this sense, the management of geographical names at local level will be analysed in relation to the impact of current paradigms and the response of the most active actors involved.

Management of Geographical Names in the Official Sphere

New trends and the need of access to interoperable geospatial information have generated a remarkable impulse to the standardisation of geographic names in recent years. Before beginning the analysis of the current situation, it is important to mention some significant antecedents.

Background: Pioneering Projects in the Geographical Names Management

Until recent decades, collection projects and efforts at national standardisation of geographical names were erratic and not always complete, both in terms of their continuity over time, as well as the coverage of the territory or the approach of the entire toponymic cycle. The available technology at that time did not allow great precision either.

However, some works and projects meant valuable contributions for the future path of the management of geographical names in Argentina.

In this sense, the following initiatives were selected and include in Table 3 as pioneering precedents for introducing methodologies, criteria and principles that were innovative at the time of their elaboration; for the depth of the research; for demonstrating a broad and long-term vision; or for having the commitment of numerous and representative social actors in pursuit of a national toponymic policy.

Geographic Names in the Current Local Geospatial Scenario

Transcendental changes in technology and cultural paradigms revolutionised the field of geospatial information in Argentina. At official level, this scenario promoted tasks aimed at adapting geographic information to these new trends. Consequently, the standardisation work of all stages of the cartographic production process was required, as well as the creation of Institutional Geographic Databases (BDGI), official SDIs and numerous platforms for the exchange of information, products, and services in the IGN[10] and in the respective official agencies responsible for the different areas of competence.

---

[10] The National Geographic Institute (IGN) of the Argentine Republic is the official mapping agency whose mission is to understand in the official representation of the national territory and the development of regulatory frameworks for its realization by obtaining precise geographic information,

Table 3 Pioneer projects in the management of geographical names

Project	Values and characteristics
1935–1942. Gazzetteer of National Governorates [a]	• First Official Compilation Project of Argentine Toponymy • Objective: collection, dissemination, and preservation of thousands of toponyms of the country and their meaning. Aimed at understanding toponyms and consolidating the sense of belonging • Execution: local work of teachers and evaluation of linguistic experts • Publication: covered about half of the existing political jurisdictions at the time of publication
1942 Meaning of the nomenclature of railway stations in the Argentine Republic[b]	• Timeliness and quality of information • Quantity and criteria of the selected sources • Application of pioneering principles and methodologies in the standardisation of geographical names at the time of its execution: – cultural importance of toponyms – encouragement to use names of local flora and fauna – need to set rules for imposition and change of names – involvement of the local community – creation of a Station Nomenclature Commission The railway infrastructure is one of the few toponymic topics studied methodically at national level. Current interest in civil associations of former workers and fans of this transport
1970. Toponymy of the Argentine Antarctic Sector[c]	• Dictionary of toponyms corresponding to the Antarctic area and the Argentine sub-Antarctic islands (South Georgia and South Sandwich) • Development: more than ten years of in-depth research on the origin of the geographical names of this continental portion • Sources analysed: more than 350 publications, national and polar institute archives, logbooks, and officers' reports from Argentine and foreign ships that sailed the area between 1675 and 1956 • Toponymic information: specific name; description of the accident; minute coordinates; naval cartographic source; name evolution; corresponding versions in other languages; origin and reason for the name • Importance: work of international value due to the special historical and current characteristics of the continent • Toponymic value: as a strategic datum: in logistics, salvage, transport, and tourism activities, and for its • Scientific value in all branches of research, with special interest in atmospheric and climate change studies, geodetic, geological, biodiversity, oceanographic, etc. And as cultural heritage: the origin of each of its toponyms highlights the link between man and his environment, his history, and his identity

(continued)

Table 3 (continued)

Project	Values and characteristics
1982 Toponymy of the Argentine Republic[d]	• First Project of the IGM inspired by the spirit of the first United Nations International Conferences on the Standardisation of Geographical Names • Project with a clear vision of national toponymic policy • Execution: institutional team and more than 200 experts from different disciplines belonging to official organizations at all levels, national and private universities of all political jurisdictions, National Academies, and scientific institutions dedicated to related topics • Tasks developed: research, classification, organization, registration, and processing of the data received • Toponymic information: location, coordinates, official cartographic sources, origin, evolution, meaning and historical and linguistic background • Processing Methodologies: application of early advances in digital technology • Objectives: creation of a digital Data Bank for the preparation of toponymic lists and searches; publication of a volume dedicated to each political jurisdiction • Implicit recognition of toponymic jurisdiction of provincial jurisdictions; methodologies and guidelines
2011. Malvinas Islands Gazetteer. Phase I[e]	• Objective: to generate a single official Toponymic Database of the Malvinas Islands • Methodology: reconcile the information generated by the two national bodies with competence in the elaboration of the cartography of the archipelago • Execution: Working Group made up of representatives from the SHN and the IGN • Result: Gazetteer with 687 geographical names and four fields: Specific Name, Generic Name, Geographical Coordinates to the minute and Description of its normalised location in order of accident hierarchy • Final version approved by the Ministry of Foreign Affairs, International Trade and Worship in 2010, and available on official sites[f] • Dissemination: as of 2011, the Nomenclátor Islas Malvinas Fase I was disseminated at International and National Conferences and Congresses and received comments from UNGEGN authorities and other stakeholders • Updating: Project foresees the elaboration of future updating PHASES that allow improving the precision and quantity of toponyms based on the use of the best available technologies and other sources of consultation on a larger scale

Developed by the author from the cited sources

[a]In English: Gazetteer of National Governorates. Please, see Consejo Nacional de Educación (1941, 1942)

[b]In English: Meaning of the Nomenclature of Railway Stations in the Republic of Argentina. Please, see Udaondo (1942)

[c]In English: Toponymy of the Argentine Antarctic Sector Please, see Armada Argentina (1970)

[d]In English: Toponymy of the Argentine Republic. Please, see Ejército Argentino. Instituto Geográfico Militar (1982)

[e]Project developed within the framework of the agreement between the governments of Argentina and the United Kingdom of Great Britain and Northern Ireland. For the document, please read: https://www.cancilleria.gob.ar/userfiles/ut/1999_-_declaracion_conjunta.pdf. For more details and documents of the question of the Malvinas Islands please read: https://www.cancilleria.gob.ar/en/foreign-policy/question-malvinas-islands

[f]https://www.ign.gob.ar/NuestrasActividades/Geografia/Nomenclador/Malvinas

- Geographical Names in the Processes of Standardisation

According to the protagonists themselves, this new paradigm meant *a new interpretation of the Institute's missions* that to its *role as the governing body in the field of geodesy and cartography…* must add *…the growing demand of geospatial information by all citizens and, by State agencies dedicated to planning public policies.* In the case of the IGN, the BDGI *emerged from the agreement of all the sectors involved in production and attending to the multiple demands of the organization…* and *…the decision was made to adopt and adapt widely accepted standards at the international level…according to the normative antecedents of the IGN itself and of the other cartographic agencies of the region, to establish a better integration at the regional level* (Castellaro y Cimbaro 2016, 49).

Geographic names are included in this new production model:

Management Plan: As part of its institutional change,[11] IGN prepared the Management Plan 2015–2020.[12] Place names were implicitly included in all geospatial information management and were explicitly mentioned among its Specific Objectives and Goals.

Feature Catalogue: As a result of numerous exchanges between multiple producers and users of official geospatial information, the IGN defined its Feature Dictionary and Catalogue. The latest version[13] includes 10 Classes, 47 Subclasses and 227 Features. Geographic names are attributes of almost all these Features.

Standardisation of Geographical Names: As part of this task of normalisation the production processes of official cartography, the IGN developed the following projects and official documents aimed at the specific management of geographical names:

a. Manual for National Standardisation of Geographical Names.[14] Its objective is to serve as a Guide for the implementation of a Program for the national standardisation of geographical names in the Argentine Republic. Supported within the framework of the UNGEGN principles, it aims to reach place names throughout their management cycle: from their capture to their dissemination; and in its two aspects: as geospatial data and as intangible cultural heritage.

---

timely and concise, essential for the comprehensive development of the country: https://www.ign.gob.ar/AreaInstitucional/Mision.

[11] In 2009, by Decree No. 554/09, the Geographical Institute changed its designation from Military to National and became part of the Ministry of Defence's scientific and technological development policy as a decentralized agency. Based on this change, the IGN began a new stage focused on modernizing its production process, strengthening its federal presence, and updating its technical and operational capabilities.

[12] New Plan is in the process of being elaborated. Complete document in: https://static.ign.gob.ar/plan_de_gestion/Plan_de_Gestion_del_IGN_2015-2020.pdf.

[13] To Access to the document Catalogue of Geographic Features version 2.0, please read: https://www.ign.gob.ar/NuestrasActividades/InformacionGeospatial/catalogo-de-objects-geographic.

[14] IGN, Manual for the National Standardisation of Geographical Names. Version 0.1 Argentine Republic 2021.

b. Principles and Guidelines for the Definition of Geographical Names of the Argentine Republic.[15] Prepared within the framework of a national programme for the standardisation of geographical names, in accordance with the above-mentioned Manual. Its objective is to serve as a guide for definition of the written form of the toponyms of the Argentine territory, considering its own characteristics and problems. It is intended especially for those who oversee standardisation tasks. It also aims to guide institutions and individuals in the proposal or change of toponyms.
c. Standardisation and Validation of Geographical Names of the Bicontinental Map of the Argentine Republic Scale 1: 5.000.000.[16] The objective of this Project was to standardize and validate all the geographical names included in the Bicontinental Map at a scale of 1: 5.000.000, to update its 2021 edition.

   It included: analysis of 2.395 Geographical Names. They were ordered according to the Classes and Subclasses of the Feature Catalogue. The document includes Classification, Methodology and Reference Sources. This task will be taken as the basis for the future Concise National Gazetteer of Geographical Names of the Argentine Republic.

- Geographical Names in Integrated Geospatial Information

According to this new paradigm, geographical names are currently integrated into multiple digital platforms. They accompany the geographical features that are ordered and disseminated in geoportals, databases, products and services provided by the agencies responsible for cartography and official geospatial information (IGN and SHN)[17]; as well as by numerous public bodies responsible for thematic information at all levels of government: national, provincial and municipal.

Through the digital format, geographical names have increased accessibility, volume, and speed of exchange. Consequently, visibility and application.

In this sense, the Human Settlements Base of the Argentine Republic (BAHRA, for its acronym in Spanish) can be cited.[18] It is the first official and standardised database on this subject, aimed at contributing to the knowledge and integration of geospatial and statistical information.

It is an important initiative and a case to be analysed. In this sense, it shows the possibilities and, also the problems and challenges those new technologies and paradigms open.

---

[15] ["Principles and Guidelines for the Definition Of Geographical Names of the Argentine Republic"] Version 1.0 Argentine Republic 2021. Published in: https://unstats.un.org/unsd/ungegn/nna/.

[16] It included toponyms corresponding to the American continental, insular, oceanic, and Antarctic portions of Argentina; definition of the form of writing of the GN of the rest of the Antarctic space; and of all those belonging to neighboring countries.

[17] According to Law 19.922 "Hydrographic Law", Art. 12. (The Naval Hydrography Service will oversee studying the toponymy in the areas of its jurisdiction to establish the names to be adopted and the modifications that were appropriate: http://www.hidro.gov.ar/Institucional/Normativa.asp).

[18] It is the result of an interministerial agreement celebrated in 2011. It was launched in May 2014 and currently has 14.673 records, including all 13 Antarctic bases.

In this regard, BAHRA shows joint work, and the social and economic value of geographical names as basic and fundamental data to be integrated into multiple themes and projects. In another sense, it raises the need to have standardised toponymic bases; to define toponymic competencies; and to apply homogeneous criteria for the definition of data quality.[19]

Geographic Names in SDIs

The Spatial Data Infrastructure of the Argentine Republic (IDERA, for its acronym in Spanish) *is a geospatial information community that aims to provide the publication of data, products and services, in an efficient and timely manner as a fundamental contribution to the democratisation of access to information produced by the State and various actors, and support in decision-making in different activities of public, private, academic, non-governmental and civil society spheres.* (Infraestructura de Datos Espaciales de la República Argentina [IDERA], s.f.)

Based on initiatives that have arisen since 1999, IDERA was officially born in 2007. Today it has almost 200 members from the national, provincial, municipal, academic and NGO levels.

As already mentioned, geographic names identify most geographic features. In this regard, IDERA included geographic names as Basic and Fundamental Data along with nine other layers of information.

The Basic and Fundamental Information Layers of IDERA were selected after the consensus analysis between producers and users of Geographic information. In this sense, geographical names meet the following conditions:

a. they are among the most relevant geographical features for the development of the country
b. coincide with the topics of greater consumption by the institutions
c. provide the basis on which users supplement specific thematic information
d. they are essential for carrying out projects that involve geographic information generated by institutions of each specific sector
e. they are produced by different organizations or institutions involved in decision-making aimed at promoting sustainable development
f. they allow the spatial location of data, the consolidation of reference base maps, and the best analysis of statistical data (IDERA 2016).

In summary, as a Basic and Fundamental Data of an SDI, geographical names must provide a common, homogeneous, reliable, and accessible base on which the users-producers of thematic geospatial information can integrate and share their specific data, in a way that allows them to focus their efforts, energy, time and human and economic capital to achieve their own competency issues.

---

[19] Argentina participated in the Project "Diagnosis of the current situation of methodologies and procedures used for GI Quality Assessment in the PAIGH Member States". For more information, please see Suggested Reading.

In this sense, IDERA should serve as a guarantee of reliability through accessing standardised geographical names.

Geographic Names and New Cultural Trends

Especially in the last decades, geographical names have been involved in the different cultural trends that affect geospatial information. In this sense, a very active participation by the most diverse social actors is verified, affecting the toponymy of the entire territory. Here are just few examples:

Geographic Names and Gender Perspective

- Odonyms: In 1995, by citizen proposal to the Permanent Commission of Urban Nomenclature[20] the City of Buenos Aires baptised streets of Puerto Madero with the names of notable women from Argentine history.[21] This trend is completed with other initiatives, such as the "Women who open the way" Program presented in 2021 that seeks to incorporate the gender perspective into urban nomenclature and make visible female personalities from history. More recently, the City Government, with the participation of social organizations and the support of UN Women, called on residents to choose women's names for the new underground stations.[22]
- Antarctica: The SHN approved in 2018 the incorporation of four new place names corresponding to the first women scientists who participated in an Antarctic campaign between 1958 and 1959. Their surnames name two inlets and two bays on the west and east coasts of the peninsula Antarctica and were presented to SCAR. An article in the UNGEGN Bulletin gives details of this decision.[23]

Geographic Names as a Brand

Use of geographical names in Trademarks, Geographical Indication and Designation of Origin registrations continues to expand in Argentina: Country Brand; City Brand and use of place names in the registers of touristic areas, wines, cheeses, meats, and fruit and vegetable production, among others.

---

[20] http://www2.cedom.gob.ar/es/legislacion/normas/leyes/ley83.html.

[21] https://turismo.buenosaires.gob.ar/es/article/las-mujeres-de-puerto-madero.

[22] https://www.buenosaires.gob.ar/laciudad/noticias/vecinas-Y-vecinos-eligen-nombres-de-mujeres-para-las-nuevas-estaciones-de-la-linea.

[23] Known as "the four of Melchior", his works meant notable contributions to the knowledge of the biology of the continent. The details of this tribute, history, photographs, and designation were published on official sites and in UNGEGN Bulletin. https://unstats.un.org/unsd/ungegn/pubs/Bulletin/UNGEGN_bulletin_no.56_rev.pdf.

Presence and Revaluation of Native Languages

Every geographical name expresses the empirical image that people and communities have of the space they inhabit. In this sense, this first link was established through the place names with which the first occupants of the current Argentine territory identified the elements of their environment.

Native names survive widely in almost the entire country, especially rooted in the areas where these ethnic groups still have a presence and a living language. Essentially, three American languages meet these conditions: Guaraní, Quechua and Mapuche.[24] As Berta Elena Vidal de Battini analysed: *beneath that apparent simplicity, we must consider an intricate knot of differentiated languages and dialects that have disappeared or live in reduced human groups … and are in the process of disappearing …* (Vidal de Battini 1960, 330).

Native toponymy is present in all geographical features and in all their categories, including many of the names of the first and second order political divisions.

Because they are languages without writing,[25] indigenous names were commonly deformed in their transcription into Spanish. This complicates their collection and standardisation.

In another sense, indigenous toponyms can provide important indicators of the landscape, its resources and its evolution, and serve as a tool for land management, as expanded upon in Table 4.

In the same sense, it is important to highlight the value of the documents prepared by the religious orders that maintained long contact with these communities. Beyond the conflictive aspects of these links, the compilation work in bilingual dictionaries meant a great contribution to the preservation of geographical names as intangible cultural heritage. In the case of the Jesuits, the toponymy was preserved in extensive cartographic collections of appreciated international value, such as the Atlases of the Blaeu family.

In this aspect, must be added the life and testimonies of missionaries of very different nationalities: the work of Thomas Bridges,[26] Alberto de Agostini[27] and

---

[24] Please, see Table 4.

[25] Using of ISO 639 language-code is important to identify toponymys´ languages in Gazetteers. Internationally recognized codes for the representation of more than 500 languages or language families. For ISO language-codes, please, see: https://www.iso.org/iso-639-language-codes.html.

[26] Thomas Bridges (1842–1898) wrote the first Yamana-English dictionary. He shared his childhood with Yamana children in the Malvinas Islands, where he learned their language. He was an Anglican pastor and founded the first rural establishment in Tierra del Fuego, which he named Estancia Harberton in honour of his wife's birthplace in Devon.

[27] Geographer and Salesian missionary, he explored, documented, and photographed the south of the continent. To expand this information, we suggest reading the tribute to Father Alberto de Agostini (1883–1960) with interesting current comments, in: https://www.researchgate.net/publication/282251794_EDITORIAL_Un_homenaje_al_Padre_Alberto_Maria_De_Agostini_1883-1960.

Table 4 Descriptive aspects of indigenous toponyms

Indigenous toponymy very clearly summarises the main functions of geographical names: identification, location, description, and appropriation of the place. In this sense, native names offer a great wealth of meanings. When transparent, they can provide excellent descriptions of local flora and fauna: sizes, brightness, colors, sounds, tastes, locations, and many other descriptive characteristics of the geographic features they designate. Beyond its richness and linguistic beauty, this descriptive character transforms the indigenous toponyms into indicators of the evolution of the landscape, of environmental changes and of the occupation of the territory. They can collaborate in the management of natural disasters; or provide indications of potential water, mining, or energy resources. Guaraní toponymy dominates the Northeast and East of the territory, applied especially to watercourses and terrain features. Their names describe sizes and aspects, such as: *Iguazú* (big water); *Iberá* (bright water), *Paraná Guazú* (big river); o *Paraná Miní* (small river). Quechua toponymy is numerous in the Northwest, West, and center of the country, coinciding with the occupation and influence of the Inca empire. They are recognisable by suffixes such as *huasi* (house): as in *Carahuasi*, *Pumahuasi, Incahuasi, Condorhuasi* and *Intihuasi; gasta*(town): as in *Nonogasta* and *Cochagasta*; or *yoc* (place of): such as *Vicuñayoc, Vizcachayoc* and *Yaretayoc.* Mapuche toponymy is characteristic of the center and South of the territory, except in Tierra del Fuego. It is very numerous, coinciding with the presence of communities of this origin. Very recognisable in prefixes and suffixes associated with types of geographic features and their characteristics of size, color, number, shape, and applications. Examples are: *Pehuen* (pine) as in *Pehuenche* (people of the pines); *Cari* (green) as in *Cari Lauquen* (green lagoon); *Buta or Futa* (large), as in *Futalaufquen* (large lake); *Pichi* (small); Choique (ñandú, rhea); Mahuida (hill, mountain): as in Pichi Choique Mahuida (hill of the little ñandú). The suffix Co (water) is also very common. In the indigenous toponymy of Tierra del Fuego, the ending derived from *waia* (in yamana, bay) is common, as in *Ushuaia* (deep bay); and *Lapataia* (wood bay or wooded bay).

Developed by the author from the cited sources

Martin Gusinde[28] represent an exceptional contribution to the appreciation of the indigenous cultures and to the geographical names' standardisation.

To toponyms that resulted from the presence of native peoples, must be added those that are currently included as part of the trend towards the recovery of their languages and cultures. Here are three examples:

- Application in the Nomination of New Features: *Amutui Quimei* was the name chosen to baptise the reservoir that was formed as result of the construction (1971–1976) of a hydroelectric plant in Los Alerces National Park (Chubut), in Andean Patagonia. As consequence of the works, a 9.250-hectare water mirror was formed, which determined the disappearance of the associated natural landscape. A series of four natural lakes, their connecting rivers, and their successive waterfalls, as well as the surrounding forest were left under water. *Amutui Quimei* means "lost beauty" in the Mapuche language.
- Legal Revaluation of Original Names: In 1984, the province of Córdoba legally imposed [29] the use of the native names of the main rivers of its territory. The law requires its use in posters, cartography, public documents, and its dissemination in education. These names -of origin associated with Quechua- should be included in parentheses after the toponyms in the Spanish language, which name -from the first to the fifth- the main rivers of the province, according to the order in which they were discovered by the Spanish colonizers.
- Alternative Official Language: In 2004, the province of Corrientes legally established [30] Guarani as an alternative official language, as well as its teaching, preservation, and dissemination; and topographic signage in both languages.

Geographic Names and Participatory Model

Access to technology is enhanced with the participatory model. In this sense, in recent decades, citizen participation in the survey, construction and dissemination of geospatial information and in the nomination of places has increased. In Argentina, participatory model's examples are:

- Geolibres[31]: It is a civil association that emerged in 2019. Its main objective is to promote collaborative production and free access to open license geospatial information.
- Citizen participation: The very positive response of citizens to government calls for the nomination of new features in both urban and natural areas is

[28] Austrian priest. He dedicated many years of his life to documenting his experiences and the culture of the last Yámanas in an extensive body of work. To expand this information, we suggest reading the anthropologist Anne Chapman's comments on the work of Martín Gusinde (1886–1969) and its editions, in: https://www.persee.fr/doc/jsa_0037-9174_1984_num_70_1_2246_t1_0199_0000_1.

[29] SAIJ Id: LPO0007105.

[30] SAIJ Id: LPW0005598.

[31] For more details of this Project please, visit: http://geolibres.org.ar/.

Table 5 New names for new places

A recent well-publicized case corresponds to the contest launched in 2021 by the local municipal government to choose the name of three new islands that emerged in the Paraná River delta, due to the acceleration of the natural sedimentation process. Due to its extension, the SHN demanded the nomination of the islands for their inclusion in the official nautical cartography. 1900 students from 87 courses participated in the contest.

Developed by the author from the cited source
For more information about the experience please read: https://www.sanisidro.gob.ar/novedades/alumnos-de-san-isidro-bautizaron-las-nuevas-islas-del-delta-y-fueron-conocerlas.

observed. Local communities, schools and associations actively participate in baptising unnamed geographic features or new ones. The most prominent inclusion of these issues in the media is also notable, as the case descripted in Table 5.

Likewise, the initiatives of proposals and citizen participation in conflicts arising from the imposition of place names not consulted are maintained. All the cases mentioned confirm the close relationship and strong identification between the local community and its geographical names.

Geographic Names and Planetary Awareness

The importance of geographic names as basic and fundamental data of SDIs and their recognition by the main international organizations in the management of geospatial information and geographic names -UN-GGIM and UNGEGN- have been previously mentioned. In the case of Argentina, the reasons for a similar rating by IDERA have been analysed.

In this sense, the presence of geographical names as an information layer integrated into all natural resource inventory projects, as well as in the monitoring of the SDGs, is very clear. Among the most recent, we can mention:

- National Inventory of Glaciers (2010)[32]: Of great importance as freshwater reserves, aesthetic, heritage, tourist, and scientific value associated with monitoring climate change. As a first result, a total of 16.968 ice bodies emerged. Most of them, without nominating.

[32] For more details and information about the law, the inventory, and the Atlas, please visit: https://www.argentina.gob.ar/ambiente/agua/glaciares.

- Wetlands Inventory[33]: The Argentine´s territory characteristic already mentioned (see Table 2) determine the presence of a great number and diversity of wetlands. Important for its surface and multiple ecosystem services; or for its local value in the development of arid zones. The specific names of these water bodies also are a georeferenced data from the inventory, included as an attribute. Its elaboration is a task of enormous difficulty, especially in the definition of its limits and sources of access to information. Both aspects are linked to the extension of the feature and the scope of its geographical names. The methodological document does not include a special section for the treatment of names.[34]
- The Committee for the Sustainable Development of the Mountainous Regions of the Argentine Republic [35] is an area for policy articulation, which brings together public sector institutions linked to the issue of mountains from multiple aspects: environmental, economic, and social. The management of these projects requires the unambiguous identification of the associated mountainous features with information coming from very diverse organisms. In this sense, the standardisation of its geographical names emerges as an urgent need to allow the interoperability and management of geospatial information.

Glaciers, wetlands, and mountains are just examples of the multiple features that must be inventoried and integrated for the management of public policies for territorial and environmental development. Its effective management requires unique and standardised geographical names. Also, to be integrated in the management of regional and global projects.

- Geographical Names in the SDGS in Argentina[36]

Geographic names are closely associated with compliance with the SDGs. In this sense, UNGEGN stated geographical names data *"has to be integrated into national, regional and international spatial data infrastructures (SDIs). SDIs shall constitute the centrepiece and main geospatial data and services' platform to be used for monitoring the SDG indicator framework under the geospatial lens as well"* (8th *UN Conference on the Standardization of Geographical Names, 2002, 37–38).*

Due to their broad focus, the SDGs include the management of environmental, economic, and social information. Multiple layers of geospatial information are integrated in the monitoring of the goals associated with climate change (SDG 13); to

---

[33] For more details about this theme, please visit: https://www.argentina.gob.ar/ambiente/agua/humedales/inventarionacional.

[34] For more details about this issue, please read: https://www.argentina.gob.ar/ambiente/agua/humedales/inventarionacional/marco-conceptual-metodologico.

[35] For more details about this project, please visit: https://www.argentina.gob.ar/ambiente/ordenamiento-territorial/comite-montana.

[36] For more information about Argentine environmental policy and management of SDG in Argentina, please visit: https://www.cancilleria.gob.ar/en/foreign-policy/environment and https://www.argentina.gob.ar/politicassociales/ods/institucional.

the management of water resources (SDG 6), ecosystems (SDG 15), or human settlements (SDG 11). In this sense, in Argentina, its compliance is associated with the multiple layers of thematic information of the IDERA and its corresponding toponyms. In all cases, geographical names identify each of these features. And they require their standardisation for more efficient management.

Associated Actions

Other actions also support the management of geographical names.

International Participation: UNGEGN and Others

Especially in the last decade, Argentina has regularly and actively participated in international efforts aimed at the standardisation of geographical names.

Parallel to the actions carried out at national level, Argentina has participated in sessions, meetings, symposiums, and training webinars organized by UNGEGN, UN-GGIM, and other organizations of the United Nations system and cooperation institutions on toponymic issues. Such as the PAIGH, ICA, IGU and ICOS.

In this sense, it has submitted its national reports,[37] submitted the list and recorded register of Argentine geographical names to the UNGEGN website, contributed to the UNGEGN Bulletin[38] and to the comments to the UNGEGN Strategic Plan 2021–2029, and exchanged communications and consultations with its experts and Working Groups.

Awareness, Dissemination, and Training of Toponymic Topics

As a National Authority on geographical names, the IGN disseminates the work of the UNGEGN on its official site[39] and its documents of interest, such as the updated version of the official names of countries[40], the synthesis of their participation in the International Sessions and Conferences, presentations, etc.

It also carries out institutional training actions to reinforce the importance of the standardisation of geographical names in all areas of production. Develops institutional exchanges and presentations at international and national conferences related to geospatial information. Likewise, it disseminates toponymic themes

[37] For the complete report, please read: https://unstats.un.org/unsd/ungegn/sessions/2nd_session_2021/documents/2021_70-Report_Argentina_S.pdf.

[38] "Towards the standardisation of geographical names in Argentina", in: https://unstats.un.org/unsd/ungegn/pubs/Bulletin/UNGEGN_bulletin_no.57_rev.pdf.

[39] Please, read: https://www.ign.gob.ar/AreaInstitucional/UNGEGN.

[40] Please, read: https://www.ign.gob.ar/NuestrasActividades/Geografia/Nomenclador/Paises.

and their normalisation through its Institutional Magazine aimed at the public, scientific-technical, academic, educational sphere, and the community in general.[41]

It is not easy to record all the activities that take place in the country linked to toponymic themes. However, it is worth noting the dissemination actions developed by some Universities, and the Centro Argentino de Cartografía (CAC, for its acronym in Spanish).[42] In this sense, the CAC has included the subject of geographical names and the need for their standardisation in its annual meetings and International Congresses, as well as in other scientific activities of national and international scope. Likewise, it promotes the inclusion of geographical names in the curricular designs of all careers related to geosciences.

Problems and Challenges of Toponymic Management in Argentina

The problems of managing geographical names in Argentina stem from two groups of factors:

(a) Those derived from the own territorial, environmental and linguistic characteristics; and of the local idiosyncrasy itself and its consequences in government management.
(b) Those derived from the technological revolution and new trends on the global stage.

According to the experience gathered, the main problems and challenges are summarised in Table 6.

Conclusions

As a component of geospatial information and as intangible cultural heritage, geographical names were affected by the revolutionary technological and cultural paradigm shifts that transformed the world and geographical science from the mid-twentieth century to the present.

Technology, Open Data policy and the participatory model have generated a profound synergy resulting in global access and production of geospatial information by all social actors.

This Geospatial Big Data offers enormous possibilities of application for decision making in the private, governmental, and academic sphere, and society in general.

In the current context of global challenges: social, economic, and environmental, geospatial information has revitalised its permanent strategic value.

---

[41] Please, visit: https://www.ign.gob.ar/AreaServicios/Publicaciones/RevistaOjoCondor.

[42] Scientific-technical association, member of the ICA. https://centroargentinodecartografia.org/.

Table 6 Problems in the management of geographical names in Argentina

Factor	Problem
Large bicontinental extension, including its territorial, insular, and oceanic portions	Difficulty in the direct collection of geographical names and their permanent updating
Large population gaps	Numerous unnamed geographic features and difficulty in naming them
Extensive presence of toponyms in languages without writing, their dialects and their transcriptions into Spanish produced at different historical moments, without the application of homogeneous linguistic criteria	Multiple versions and complex standardisation task
Easy accessibility of digital devices; speed of information exchange; multiple user-producers; multiplicity of platforms, products and services; widespread use of global platforms; inclusion of geospatial information from collaborative platforms	Multiple official and unofficial versions online; difficulty in identifying and reliable sources; lack of application of quality criteria and use of standardised geographical names
Lack of awareness about the benefits of standardisation of geographical names	Lack of incentives to develop standardisation programs and actions
Lack of knowledge about the concepts of standardisation of geographical names; and toponymic principles and guidelines	Difficulties and absence of its application in multiple public and private spheres and in their respective integrated geospatial information projects
Lack of standardisation of geographical names	Unreliable quality of geospatial products and services; losses of capital resources, time, and human lives
Lack of awareness and knowledge about standardisation concepts by non-specialized users	Lack of interest in the use of official sources and standardised geographical names
Lack of further development of the SDIs culture and its metadata	Lack of use of standardised geographic names in integrated geospatial information projects in both the public and private spheres
Lack of knowledge about artificial intelligence tools applied to geospatial information and their own risks	Lack of quality control of geospatial information from this source
Ignorance and dispersion of toponymic works carried out in the country	Documents and unused efforts in the construction of the toponymic heritage and in the efficient management of geographical names
Federal system and characteristics of the local idiosyncrasy	Overlapping competences and complex bureaucratic networks

Developed by the Author

In this scenario, geographical names have been recognised as basic and fundamental data of SDIs, and as a core element in the fulfilment of the SDGs and the integrated management of geospatial information.

As intangible cultural heritage, geographical names have achieved remarkable visibility as part of new cultural trends: revalorisation of indigenous cultures, gender perspective, historical revisions, and an active citizen participation.

However, along with this considerable growth in vitality and visibility, geographical names face the same challenges of geospatial information in the era of Big Data: to offer reliability of information in a scenario of online, permanent, and fast exchange of data from millions of user-producers.

Standardisation of geographical names is a clear response to this need.

In this sense, Argentina is immersed in the same global scenario, and all its social actors have responded to these new goals.

In addition to these global challenges, there are those arising from its own characteristics: a large territorial extension, very irregular distribution of the population, the presence of geographical names in unwritten languages, its federal form of government and the overlapping of competences, the need to manage standardised geographical names and to increase the culture of SDIs as a response to the management of reliable and integrated geospatial information.

In this regard, it has been one of the essential objectives of this chapter to stimulate joint reflection on the strategic importance of geographical names, the need for their standardisation and to achieve a greater awareness of these challenges in all actors involved in the use and production of geospatial information and in its treatment as intangible cultural heritage, as active participants in its management.

Argentine Toponymy: Selected Cases

To close this chapter, a selection of geographical names is included for the application of the topics developed.

1. Argentina

Argentina derives from the Latin "argentum" silver, by association with this metal, which was supposed to be abundant in these regions of America. Two images linked to this geographical name have been included in Fig. 1. Figure 1a shows the title page of the epic poem where the name Argentina was first used (Lisboa, 1602) to designate approximately the current Argentine territory. Figure 1b shows the image of the logo of Argentina as a Country Brand, officially established in 2021.

2. Indigenous Toponymy

2.1 Humahuaca

Humahuaca is the name of the creek, in the NW of Argentina, declared a World Heritage Site by UNESCO. Occupied for more than 10.000 years, it was a strategic

Fig. 1 Argentina

axis of population, trade, and production. Dominated since the fifteenth century by the Inca empire and then by the Spanish conquest. This geographical name derives from Omaguacas, one of their native peoples at the arrival of the Spaniards. Toponyms of the area reflect the evolution of human occupation over millennia. In Fig. 2, three images illustrate this toponym: In Fig. 2a, the famous geological formation with more than 30 shades of colours. In Fig. 2b view of the pre-Hispanic archaeological site; and in Fig. 2c, a sixteenth century church (Photos: Martín García Moritán).

Fig. 2 Humahuaca

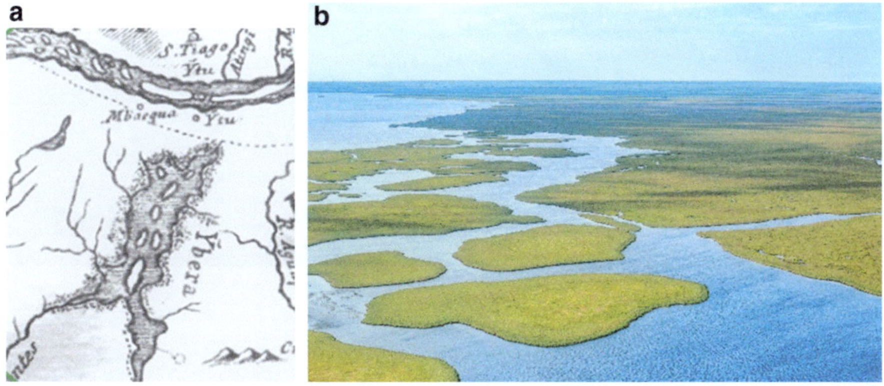

Fig. 3 Iberá

2.2 Iberá

Toponym that currently designates one of the most important tropical wetlands in the world in terms of extension and biological richness. It was registered in the Jesuit cartography since the seventeenth century with different indigenous and Spanish names. Finally, Yberá, "bright water" in the Guarani language, prevailed. Officially *Iberá*. Figure 3 shows two images linked to this place name: in Fig. 3a, the toponym Yberá in the Jesuit cartography (1732). In Fig. 3b Aerial view of the Esteros (Photo: Matías Rebak/Fundación Rewilding Argentina).

2.3 Ushuaia

This toponym in the Yamana language designates a deep inlet on the coast of the Canal Beagle. Derived from Ushu: at the bottom, and Waia: entrance, inlet. According to Thomas Bridges "inland port to the west". Today it is also the name of other nearby natural and artificial features: the bay, the harbour, a peninsula, and a rural settlement; one of the administrative divisions of the Province of Tierra del Fuego, and of the capital city of the province, the southernmost city in the world, at 54º48′30″S/68º19′25″W. In Fig. 4, two images associated with this place-name: in Fig. 4a, view of the bay, harbour, and city of Ushuaia. In Fig. 4b, image of the toponym, in one of the coasts of the bay (Photos: Marisol Vereda).

3. European Toponymy Over Time

3.1 Isla de los Estados

Isla de los Estados is located to the East of Tierra del Fuego. Its current name derives from *Staten Landt* (Land of the States) as it was baptised by the Dutch expedition of Le Maire and Schouten in 1616, in homage to the States of the Netherlands. In search of a third alternative pathway on the spice route, the expedition named several accidents on the Patagonian coast. The original name of the island was *Chuanisin*, in Yamana language: region of abundance. Figure 5 includes two maps associated

Fig. 4 Ushuaia

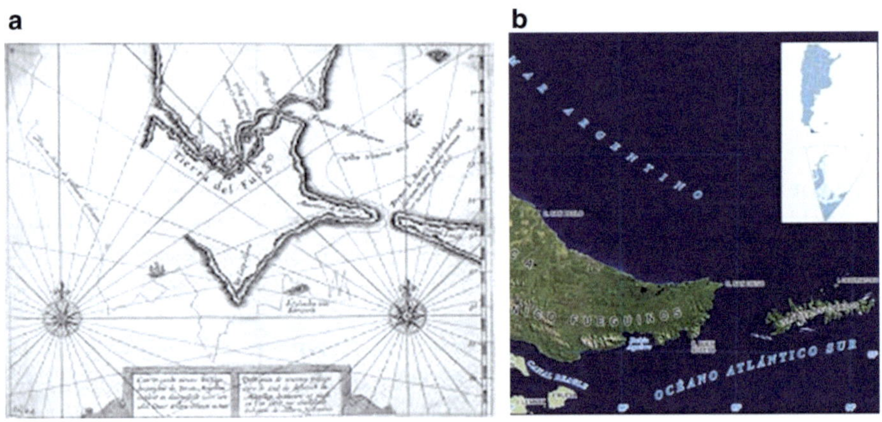

Fig. 5 Isla de los Estados

with this place name. In Fig. 5a, a map that shows the route of Willem Schouten and the original toponymy (Willem Janszoon Blaeu, 1618). In Fig. 5b, a satellite chart (detail of Argentine cartography Source: IGN).

3.2 Islas Malvinas

The toponym *Malvinas* derives from Malouines that refers to the port of Saint-Malo, in French Brittany, which in the 18th century established a colony of fishermen in the South Atlantic archipelago. Officially, *Iles Malouines* was the place name that Louis-Antoine de Bougainville assigned to the islands in the document in which he described them before France recognised Spain's rights over the islands in 1767. They were mentioned in historical cartography with different names as Islas de San Antón or Sebaldinas among others. Figure 6 shows two images associated with this place name: In Fig. 6a, detail of the Map of Iles Malouines in the Atlas of Martin de Moussy (1873). In Fig. 6b, current view of Saint-Malo, Bretagne, France.

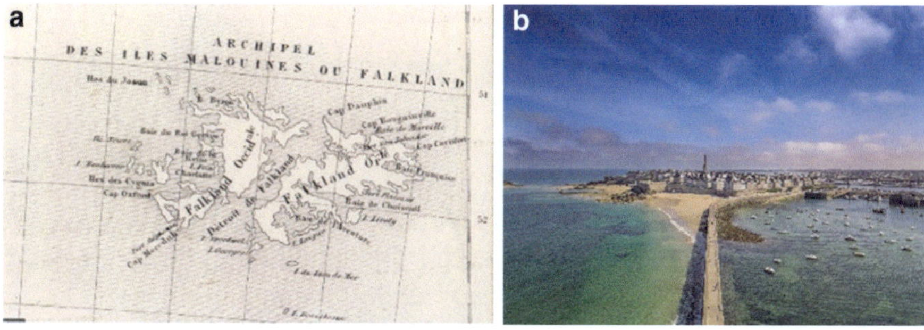

Fig. 6 Islas Malvinas

3.3 Welsh Toponymy

Welsh colonisation arrived in Argentine Patagonia in 1865. It was established first on the East coast and then in the Andean region. Culture and ties to the land of origin are very much alive in both Argentina and Great Britain. Welsh toponymy is observed in the two colonisation areas: such as *Puerto Madryn, Gaiman, Trelew*, and *Dolavon*, in the East, and *Trevelin*, in the West. Figure 7 includes three images associated with Welsh place names in Argentine Patagonia: Fig. 7a shows the Official Signposting of Madryn, important port and touristic centre (Photo: Martín García Moritán). Figure 7b shows Welsh symbology on the coat of Trevelin (Image: Municipalidad de Trevelin), and, in Fig 7c, image of double road nomenclature with the name that recalls María Humphreys, the first Welsh baby born in Chubut (Photo: unknown author).

3.4 Nueva Lubecka

Settlement founded by the German immigrant Juan Plate in 1895 in the West of the Province of Chubut. He arrived in Argentina in 1880 as part of the great migratory wave. He developed the cattle activity in Patagonia and was one of the first importers and traders of wire fences in the country. The rural settlement was baptised by Plate as Estancia Nueva Lubecka in honor of Lübeck, the hometown of his wife. Currently, *Nueva Lubecka* is a place on the strategic Route 40. Figure 8 includes two images associated with this toponym. Figure 8a is an historical image of the settlement

Fig. 7 Welsh toponymy

Fig. 8 Nueva Lubecka

(in Hardt 1972), and Fig. 8b shows the main building of the farm (Photo: Alberto Azparren, 2013).

3.5 Monte Fitz Roy o Chaltén

Toponym of the pike located in the Andean Mountain range. Baptised in 1877 as *Fitz Roy* with the name of this English sailor in homage to his explorations of areas and rivers in southern Patagonia. Then, the original name *Chaltén* was added, which in the Tehuelche language means smoking mountain due to the cloud formations that regularly surround its top. In Fig. 9, four images associated with this mountain. In Fig. 9a, the easily recognisable profile of this mountain, considered a summit of extreme difficulty by climbers from all over the world (Photo: Carlos Nelson). The intense link between the mountain and community explains the inclusion of its image in the coats of the province (Fig. 9b), of the new neighboring town of *El Chaltén* (Fig. 9c), and of the Los Glaciares National Park, where it is located (Fig. 9d).

Fig. 9 Monte Fitz Roy o Chaltén

Fig. 10 Lago Roca o Acigami

4. New Trends and Last Inmigrations Waves in Argentine Toponymy

4.1 Lago Roca o Acigami

Acigami is the new name of the *Lago Roca*, as part of the process of recuperation of the original geographical names in indigenous languages in Tierra del Fuego. Both names are official. Figure 10 shows the register of this trend. In Fig. 10a, signposting in the National Park. In Fig. 10b, interesting detail with the meaning of the name: long bag/basket in Yamana language (Photos: Marisol Vereda).

4.2 Women in the New Odonyms in the City of Buenos Aires

Puente de la Mujer (Woman Bridge) is the name of the bridge of the Spanish architect Santiago Calatrava that joins the banks of one of the five dams on the Río de la Plata, in Puerto Madero, the newest neighborhood in Buenos Aires. In 1995, by citizen proposal, its streets were baptised with the names of notable women in Argentine history. Figure 11 shows two images linked to the place names of this part of the city. In Fig. 11a, a view of this bridge (image from the Official Website of the Government of City of Buenos Aires), and in Fig. 11b, signposting in a cross of streets (Photo: Adriana Vescovo).

Fig. 11 Women in the new odonyms in the City of Buenos Aires

Fig. 12 Paseo Beijing and streets of Chinese Quarter

4.3 Paseo Beijing and Streets of Chinese Quarter

Chinese odonyms and double nomenclature in the train station and streets sign in the "Chinese Quarter", in the NE of the City of Buenos Aires, as it is showed in three images in the Fig. 12. It shows the impact in the toponymy of the latest immigration from Southeast Asia in Argentina, which included the Japanese, Chinese, Korean and Vietnamese communities, among others. Chinese immigration is one of the most recent and numerous, along with others from neighboring countries (Figs. 12a, b, and c) (Photos: Adriana Vescovo).

Questions

1. What do you consider to be the main problem of the current management of geographical names as part of the geospatial Big Data in Argentina? Justify this answer based on the analysis of the new global context affect geospatial information.
2. Cite the actions developed in Argentina to solve this problem and its responsible actors.
3. What are the challenges of Argentina in the management of geographical names considering its territorial extension and the toponymy of the original peoples?
4. Justify the importance of standardised geographical names in SDIs and their link to compliance with the SDGs

Acknowledgements To the authorities, colleagues, and staff of the IGN; to authorities and members of the CAC; special thanks to Helen Kerfoot and Peter Jordan; to colleagues, students and friends from Universities and Institutions in Argentina and abroad. To Eva Szabo for the suggested source on the European communities in Argentina. To Horacio Speri, Pablo Losada, Erwan Nérambourg (from the Ville de Saint -Malo), Roberto Liebenthal and Monica Bader (from Forschungs- und Dokumentationszentrum der Deutschsprachigen Immigration nach Argentinien (DIHA)-Universidad Nacional de San Martín (UNSAM) who helped me to find photographs to illustrate some of the cases, and to Ernesto Friederichs (Juan Plate's great-grandson) for his availability, important information and sources. To Martín García Moritán, Carlos Nelson, Marisol Vereda, and Matías Rebak/Fundación Rewilding Argentina who generously offered their photographs. To all those known and anonymous people who inspired, supported and maintain with passion and interest the link with the geographical names in different parts of Argentina and around the world.

References

Armada Argentina (1970) Toponimia del Sector Antártico Argentino. Servicio de Hidrografía Naval. [Toponymy of the Argentine Antarctic Sector] Publicación H-914. Texto elaborado por el Capitán de Corbeta (RS) Enrique Jorge Pierrou. Buenos Aires

Castellaro H, Cimbaro S (2016) Cambio de paradigma en la producción del IGN, [Paradigm shift in IGN production]. El Ojo del Cóndor 7:49–51. https://www.ign.gob.ar/descargas/elojodelcondor/Ojo_del_Condor_07.pdf

Consejo Nacional de Educación (1941) Diccionario Geográfico de las Gobernaciones Nacionales. [Gazetteer of National Governorates] 1er Tomo. Misiones, Chaco, Formosa, Los Andes. Buenos Aires

Consejo Nacional de Educación (1942) Diccionario Geográfico de las Gobernaciones Nacionales [Gazetteer of National Governorates] 2° Tomo. La Pampa, Río Negro, Neuquén, Chubut, Santa Cruz y Tierra del Fuego. [Gazetteer of National Governorates] Buenos Aires

Council of Europe, ETS 176. 20.X.2000. Council of Europe Landscape Convention, Chapter I—General provisions Article 1 – Definitions (2) Florence. https://www.coe.int/en/web/landscape

Ejército Argentino. Instituto Geográfico Militar (1982) Toponimia de la República Argentina. (Toponymy of the Argentine Republic) Volumen I. Territorio Nacional de la Tierra del Fuego, Antártida e Islas del Atlántico Sur. Primera Parte Tierra del Fuego. Buenos Aires

Hardt J (1972) Juan Plate. UN PIONERO PATAGÓNICO (Los alemanes en el Sur) 1859–1938. [Juan Plate. A PATAGONIAN PIONEER (The Germans in the South) 1859–1938] Universidad Nacional del Comahue

IDERA. Infraestructura de Datos Espaciales de la República Argentina. ¿Qué es IDERA? [What is IDERA?] https://www.idera.gob.ar

IDERA (2016) Descripción de Datos Básicos y Fundamentales. Documento Técnico Versión 2.0. https://www.idera.gob.ar/images/stories/downloads/catalogo/DescripcinDByF_V2.0_IDERA.pdf

Udaondo E (1942) Significado de la nomenclatura de las estaciones ferroviarias de la República Argentina [Meaning of the nomenclature of railway stations in the Argentine Republic] Buenos Aires

UNGEGN (2021) Strategic Plan and Programme of Work 2021–2029. Alignment of Strategies to the UN Sustainable Development Goals. https://unstats.un.org/unsd/ungegn/documents/UNGEGN_Strategic_Plan_2021_V4.pdf

United Nations (2002) Eight United Nations Conference on the Standardization of Geographical Names Berlin, 27 August-5 September 2002. E/CONF.94/3.ISBN 92-1-100915-4. https://unstats.un.org/unsd/geoinfo/ungegn/docs/8th-uncsgn-docs/8th_UNCSGN-E.pdf

Vidal de Battini B (1960) Toponimia, en La Argentina. Suma de Geografía. Tomo VIII. Capítulo III, 273–458. Buenos Aires

Suggested Reading

Buzai G (1999) Geografía Glob@L. El paradigma geotecnológico y el espacio interdisciplinario en la interpretación del mundo del siglo XXI. [Glob@LGeography. The geotechnological paradigm and interdisciplinary space in the interpretation of the world of the 21st century] Lugar Editorial

Buzai G (2011) La Geotecnología: ¿Nuevo paradigma de la geografía o paradigma geográfico de la ciencia? [Geotechnology: New paradigm of geography or geographical paradigm of science?] Revista Catalana de Geografía IV XVI, 42. http://www.rcg.cat/articles.php?id=187

González D et al (2010) Línea de tiempo en la ocupación del espacio argentino. [Timeline in the occupation of Argentinean space] Coordinadores: María Cecilia Velázquez y Edmundo Szterenlicht. Consejo Federal de Inversiones. ISBN 978-987-510-085-5

Manduca F, Mansilla R (2018) Proyecto "Diagnóstico sobre la situación actual de las metodologías y procedimientos empleados para la Evaluación de calidad de la IG en los Estados Miembros del IPGH" Participación of Argentina. [Project "Diagnosis of the current situation of methodologies and procedures used for GI Quality Assessment in the PAIGH Member States". Participation of Argentina] El Ojo del Cóndor 9:48–50. ISSN 1853 9505. https://www.ign.gob.ar/AreaServicios/Publicaciones/RevistaOjoCondor

Martínez C (2017) Revelaciones de un manuscrito francés sobre las islas Malvinas: Louis-Antoine de Bougainville en los contornos de la América hispana (1767). Revelations from a French manuscript on the Malvinas Islands: Louis-Antoine de Bougainville in the contours of Hispanic America (1767) Corpus [En línea] 7(2). http://journals.openedition.org/corpusarchivos/1928. https://doi.org/10.4000/corpusarchivos.1928

Murray C et al (2008) Tras la estela del Hoorn. Arqueología de un naufragio holandés en la Patagonia. [In the wake of the Hoorn. Archaeology of a Dutch shipwreck in Patagonia] Coordinador Ricardo Bastida. 1ª. Edición, Buenos Aires. ISBN 978-987-9132-19-7 CDD 930 1021 804

Ruiz I Almar E (2010) Consideraciones acerca de la explosión geográfica: geografía colaborativa e información geográfica voluntaria acreditada [Considerations about the geographic explosion: collaborative geography and accredited voluntary geographic information]. GeoFocus-(Artículos) 10:280–298. ISSN: 1578-5157. http://www.geofocus.org/index.php/geofocus/article/view/201

United Nations Education, Scientific and Cultural Organization (UNESCO). Intangible Cultural Heritage. https://ich.unesco.org/en/

Vescovo A (2019) Geographical names in the era of big data. A necessary updating in the curriculum design of the geosciences. Micro, Macro & Mezzo Geo Information (Issues) 13:16–28. ISSN: 1857-9000 (printed version) EISSN: 1857-9019 (electronic version) UDC: 528:004. https://mmm-gi.geo-see.org/journal/volume-13/

Zaccheddu P (2018) Do geographical names data play a specific role for the SDGs under a geospatial lens? UNGEGN Information Bulletin No. 54, May 2018. SPECIAL FEATURE. Geographical names supporting sustainable development. https://unstats.un.org/unsd/ungegn/pubs/Bulletin/UNGEGN_bulletin_54_finalver.pdf

Toponymy, Scale and the Change of Scale. A Geographical and Linguistic Challenge

Joan Tort-Donada and César López-Leiva

Abstract We propose a reflection that seeks to correlate three concepts—toponymy, mapping and scale—which have an enormous affinity with each other (yet one that is not readily resolved), from both a theoretical point of view and from a technical or applied perspective. In the first part, we address some of the basic concerns associated with the problem as presented: from the concept of the toponym/place name itself, to the difficulties encountered in representing its spatial content, and a discussion of questions as yet unresolved. In the second part we examine the problem of the spatial representation of the toponym and of the specific difficulties posed, in this regard, by a change of scale. To do so, we take as our point of reference the system of toponymic representation currently used by the cadastral survey in Spain, and which we analyse, by means of different examples concerning place names related to forests and vegetation in the Autonomous Community of La Rioja. The conclusions we draw enable us to summarise some of the key issues raised by the debate and to discuss how they might be addressed more coherently—especially, in terms of their toponymic mapping.

Keywords Toponym · Spatial meaning · Scale · Mapping · Cadastral survey · Phytotoponymy · La Rioja (Spain)

Objectives

- Clarify the fundamental ideas about the concept of toponym and about its theoretical dimension (as a geographical proper name, endowed with a meaning related to space), on the one hand, and its practical dimension (with regard to its application in cartography and in public signage), on the other hand.

J. Tort-Donada (✉)
Universitat de Barcelona, Barcelona, Spain
e-mail: jtort@ub.edu

C. López-Leiva
Universidad Politécnica de Madrid, Madrid, Spain
e-mail: cesar.lopez@upm.es

© The Author(s), under exclusive license to Springer Nature Switzerland AG 2023

G. O'Reilly (ed.), *Place Naming, Identities and Geography*,
Key Challenges in Geography, https://doi.org/10.1007/978-3-031-21510-0_6

- Recognize the problem of the cartographic treatment of toponymy, and especially the issues that affect the scale and the change of scale.
- Capturing the diversity of interactions and correspondences that occur between 'macrotoponymy' and 'microtoponymy', taking into account that these are the two great levels or scales into which toponymy is conventionally divided.
- Consider, from a practical point of view, the problems of cartographic representation of the spatial meaning (or semantic content) of toponyms, taking as an example the phytotoponyms of the Spanish region of La Rioja as they are represented in the cartography of the cadastre.

Introduction

The chapter addresses an issue of relevance in toponymy, namely the concept of *scale*, and examines the key problems that derive from it both from the point of view of mapping and from the more specific perspective of the *change of scale* (which, in addition to the technical difficulties it poses, is of great importance in all matters concerning the determination of the spatial scope, or "meaning",[1] of the toponym).

In line with this approach, the text is organized in two main sections. The first takes an eminently theoretical stance and seeks to outline a general framework for the study, placing particular emphasis on the four main fields in which we detect a specific need for scientific clarification of the underlying question: that is, the concept of the toponym itself and its delimitation within the broader group of proper names; the "meaning" of the toponym in strictly spatial terms (in other words, the problem of the *spatiality* of the toponym); the particular relationship established between the toponym and scale when it comes to mapping, and, finally, and closely related to this last point, we specifically address the difficulties (both conceptual and technical) that derive from the application of a change of scale—essential in cartography as well as in the more general field of the geographical sciences.

The second section of the text takes a more applied approach and reports the findings of a specific case study. It constitutes a broad-based analysis of the use of scale in the maps of the official cadastre in Spain, taking as its reference the territory of the Autonomous Community of La Rioja and its phytotoponyms, that is, place names that allude to plants and vegetation. Here, by focusing on a specific case we are able to limit to some extent the complexity inherent in the representation of space and of the toponyms on the cadastral maps and, consequently, we are able to identify the problems that are of most interest to our discussion here. Moreover, this study of cadastral mapping provides an opportunity to examine one of the main problems that has always presented itself when considering place names from a geographical perspective, that is, the "connection" between the name and the geographical space

---

[1] We put the word *meaning* between inverted commas when we apply it to a place name, given the reservations generated by this concept linguistically speaking when applied to something other than common nouns.

that the name identifies. The recent digitization of cadastral toponymy, and its inclusion in large databases, highlights the great potential for future development offered by the study and management of toponyms from this new perspective in almost any part of the world.

Finally, in the conclusion, we are able to summarise some of the key issues raised in the article and to underline, from our particular perspective, the extent to which they represent both a geographic and a linguistic challenge.

Toponymy, Mapping and Scale: A Complex Relationship

The Toponym, a Concept of Imprecise Delimitation

From the general perspective of knowledge, it is apparent that the concept of the *toponym* (and of its synonyms, *place name* and *geographical name*) is somewhat nebulous; that is, it lacks precision and is not easily defined. Indeed, to date, linguists have been unable to agree on a meaning for the term or on its exact semantic scope. Non-specialist dictionaries usually define the toponym quite simply as: "a proper name of place" (RAE 1984: 1320; IEC 1995: 1780). As is evident, this is an especially broad definition that Moreu-Rey seeks to limit to some degree, yet without deviating from the essential idea that it seeks to convey, as: "A proper name that serves to distinguish a precise, unique place in a given context" (Moreu-Rey 1995: 45). What is significant about this definition, in our opinion, is that it puts the emphasis on a place name's spatial quality: the toponym's "object of reference", that is, the *place*, individualized (or distinguished, or singularized) by means of its name. By so doing, space is conceived as a prior and necessary condition for the existence of a toponym. And, ultimately, the pre-eminence afforded space allows us to speak of the *geographical* dimension of toponymy. Also the most recent perspectives of linguistics affect the need to approach the place name as a "form" of representation of space (Reszegi 2020).

We consider Moreu-Rey's proposal interesting insofar as it offers an initial, conceptual delimitation and, at the same time, it draws our attention to the very broad scope that the notion of the toponym can have when considered in extenso. Indeed, we might even conclude that the concept hints at a universality of meanings because, in fact, it has the potential to embody the "geographical environment" understood in its globality. But problems arise when researchers, keen to get started on their analyses, are faced with the inescapable need to limit their field of study. It is at this point that we believe it is worth coming back to the thinking of Moreu-Rey, who notes that in the set of place names of any specific geographical area we will find two main groups of toponyms:

(a) place names whose meaning is clear and understandable—and which usually constitute the largest group, and

(b) place names that apparently have no meaning "because they do not correspond to any word or expression in the language spoken within the geographical and linguistic framework studied" (Moreu Rey 1982: 13). Moreu-Rey adds that "these toponyms, in reality, used to have a meaning in languages long disappeared or in concepts long abandoned. They are known as 'fossils' because they are crystallized or petrified ancient common names, some of which have been preserved for millennia" (Moreu Rey 1982: 13).

In practice, one thing is clear: for those interested in the study of toponymy from the point of view of philology—and, more strictly, of etymology—the names of greatest interest lie in this second group. These are the names that pose the great problems as regards their evolution and/or transformation over time, and those that, ultimately, give rise to the formulation of hypotheses that specialists seek to confirm or refute. This does not necessarily mean that the names in the first group are devoid of interest, but it does mean that they are interesting in a very different way—a way that has little to do with etymological research in its strictest sense which tends to be based in the study of documented historical forms.

However, the names in the first group—those of "clear, understandable meaning", which we can refer to as being "transparent"—are, in a certain way, those that most interest researchers seeking to undertake studies from the perspective afforded by geography: that is, when their meaning refers to the appearance and the constitutive characteristics of a certain place on earth, however small or large it might be. Geography, insofar as it is a body of knowledge that possesses a very important descriptive component, considers these "transparent" names an essential bedrock of the discipline, insofar as they provide the scholar with a highly valuable descriptive picture of the territory to which they are applied. Yet, this does not mean that the geographer has no interest in the "unintelligible" place names—rather their interest is conditioned by the fact that, first, these names constitute a fairly small subset within the overall set of place names; and, second, because, as a general rule, the geographer does not have the training to engage in the field of etymological debate, and he or she can only fully exploit the potential of these names when their etymology has been well established and is not prone to conflicting interpretations.

The Geographical Perspective: The Spatiality *of the Toponym*

In this section, our aim is to focus in more detail on the question of the spatial dimension, or *spatiality*, of the toponym, an aspect we consider fundamental when undertaking an onomastic study from the specific perspective of geography. In this regard, it appears to us essential, from the very outset, to bear in mind that toponymy, as a multidisciplinary field of knowledge, must always be considered in the context of its two fundamental coordinates: namely, space and time (Tort-Donada 2010). Indeed, these two coordinates allow us to speak, according to Henri Dorion, of a "toponymic function", when the main vector of reference is space, and of a "toponymic memory", when the main vector is time:

> Let us first recall that toponymy, in common with many of the human sciences, is inscribed within the dual dimensions of space (the toponymic function) and time (the toponymic memory). Toponymy has, therefore, an essential relation with geography (place names being its basic vocabulary) and with history (to the extent that the names are witnesses over time to a given man-place relationship) (Dorion 1984: 103).

In this context, the need is apparent to defend the idea, as geographers (that is, as experts in the consideration of the spatial dimension of terrestrial phenomena), of the importance of the role corresponding to the geographical perspective in relation to toponymy. We refer, in the general framework of the interrelationships established between "territory" and "language", to the qualitative contribution that can be made from an in-depth appraisal of the "spatial content" of any given toponym (Tort-Donada 2020). This potential contribution—which we believe could help clarify many of the problems posed by the etymological study of toponyms—remains today largely unknown to many of the experts who work in the field of toponymy from non-spatial perspectives (for example, etymologists and experts in diachronic linguistics). It is our belief—contrary to the opinion (frequently held among linguists) that it is not possible to speak of "semantic content" in relation to place names—that, from a geographical point of view, any toponym alive, or in current use, has semantic content; that is, a meaning that corresponds with the concretion in space that the name in question has at this particular moment in time. "Meaning" and "spatial dimension" (or spatial or geographical transcendence) are, from this point of view, equivalent expressions (Tort 1999, 2001). Studying the spatial dimension of a place name means, from this perspective, analysing the individuality of this space, its "specific differentiation" from the rest of the space. The sole prerequisite for the analysis of this "individualized space" is the clear specification of the temporal coordinate: that is, an explicit identification of the point in time to which the reference is made. For the etymologist, this moment is a remote point in time (and difficult to determine): the moment in which the word was originated. For the geographer, in contrast, the moment of reference is very much "now", the present moment: that is, the present reality constitutes the starting point of any situation that we propose addressing as a geographical problem.

In line with this reasoning, we can conclude that, for the etymologist and for the geographer, space has a different temporal dimension. The former is interested in space as situated in a more distant moment of time: it is the space that will allow him or her to discover the reason that originated the word, its genesis. The latter, in contrast, is interested in space as a current, discernible, perceptible reality. The two 'spaces' may or may not coincide: it is more than normal that, over time, the significant entity, the "individualized space" to which any toponym alludes, changes. But, even though they might not coincide, there is no doubt that both spaces are of interest to the aforementioned scholars—albeit, that is, from the different perspectives that we have just described.

We opt, at this juncture, to illustrate our argument with a consideration of a question raised which, although it originates from a linguist, is of particular value inasmuch as it finds support in geography. We refer to the specific defence of space as a vector made by the Catalan etymologist Joan Coromines; a scholar that we do not hesitate to label as "transversal" (Tort-Donada 2021) due to his high competence in linguistics but also in geography. This author, when proposing a detailed methodology for the etymological study of toponymy (Coromines 1972, I: 130–156), calls on the scholar to place a particular emphasis on the geographical data and on the observation of the reality associated with the toponym. Moreover, at one point, and in another of his methodological studies (Coromines 1965), he offers an explicit defence of his methodological prudence, providing by way of example (from Catalan toponymy) the Latin root, SPATHA (meaning "abyss, precipice", and which gave rise to the vernacular form *espadat* [sword], today fallen into oblivion), which, as he explains, gave rise to numerous toponyms in the Catalan geography that maintain, in their origin, the aforementioned idea of an "abyss":

> Often the appearance of the place will guide us towards its meaning. It can be no coincidence that all the places I mention below are located near a great precipice or cliff: l'*Espà* [and the author refers, with great precision, to the location of each of the toponyms], Sant Salvador de les *Espases*, the Torre de l'*Espada*, Castells*espasa*, *Espades*, the Barranc de l'*Espada*, Coma-l'*espada*, the Serra d'*Espadà* (…). It is, therefore, a prudent conclusion to believe that the Latin SPATHA, of which in Catalan we know the variants *espasa*, *espaa* and *espada*, had the same meaning as its derivative *espadat*[2] (Coromines 1965: 18).

On the Particular Interaction Between Toponyms and Cartography: The Scale Factor

When it comes to delimiting the concept of the toponym in a more rigorous fashion, we have already mentioned the difficulties involved in defining the fragment of space to which the toponym alludes with any degree of precision. It is here that the concept of *scale* is especially useful as applied to the field of toponymy. Scale is a technical concept, usually defined as the ratio between a distance measured on a map and the corresponding distance on the ground (Monkhouse 1978: 172), but which also concerns, in a highly specific way, what is usually an inseparable ingredient of any cartographic representation: its toponymy.

In this regard, an important consideration needs to be stressed from the outset: if, on the one hand, the scale is a critical factor when conceiving the goodness of fit of a given toponym to a specific cartographic representation, it can, on the other hand, also be a factor of great utility and interest when we speak of the toponym itself—in other words, when it is approached as a concept that necessarily alludes to a specific space (or "semantic field" in a geographical sense), and which, therefore, is capable of being understood as well, like any toponym that appears on a map at different scales. In fact, we would maintain that it is possible to speak of an *ideal scale* for any toponym, that is, of a "proportion" (or a "relationship of proportionality") between

[2] Italics is ours.

the name and the concrete space alluded to, outside of which the toponym ceases to make sense because its meaning is distorted in terms of space.

If the ultimate basis for the concept of an *ideal scale* lies in an individualized consideration of the toponym (which implies, as a theoretical principle, that "each toponym has a specific scale"), from a practical point of view—specifically, in relation to the mapping process—it is usually accepted, as a general rule, that there is a relationship of correspondence between the specific scale used on a map and the type of toponym appropriate to that scale. For the territory of Catalonia, and as far as its official cartography is concerned, the general scheme of correspondence between scales and types of toponyms is as follows (in summary form, based on Moll and Tort 1985):

- Scales smaller than 1:50,000 (that is, 1:100,000, 1:250,000, 1:500,000, etc.): When referring to large territorial areas, these scales are, in general, only suitable for higher-ranking toponyms, that is, for what is known as *macrotoponymy.*
- Medium scales (from 1:20,000 to 1:50,000): The level of detail allows such maps (above all, those with a scale closest to 1:20,000), to widely incorporate (albeit subject to a process of selection, depending on the scale) the minor toponymy, or *microtoponymy.*
- Detailed scales (from 1:5,000 to 1:20,000): Placing ourselves in this range of scales (especially around 1:10,000 and higher), there is an almost exhaustive inclusion of the microtoponymy on the maps (the only exceptions being in particularly dense urban areas).
- The larger scales, above 1:5,000, allow, a priori, a complete, unrestricted incorporation of the entire toponymy of a given territory (be it urban or rural). Their drawbacks lie in the difficulty a map at these scales has for providing a global overall view of the territory represented.

It should be stressed that the goodness of fit of the scale of the toponyms to the map is always a highly complex task. Typically, this complexity derives from the intangibility of the toponym itself or, as discussed in the section above, it may be a consequence of the difficulties encountered when seeking to delimit, as accurately as possible, the "spatial dimension" of the toponym. Yet, it should be recognised that this complexity is, as a rule, inherent in any operation of *generalization* conducted in the mapping process (and that it affects the toponymy just as much as it affects the other elements involved in this process). In this regard, we should not lose sight of the fact that the purpose of any generalization "is to reduce what is in reality a very complex visual-intellectual process to manageable proportions", and that, ultimately, "generalizations [in the cartographic process] are always necessary" (Robinson et al. 1987: 126).

The Change of Scale in Toponymy: Fundamental Questions

To complete our analysis, we seek to present, within the framework of the fit or correspondence between "toponyms" and "scale" described up to this point, some of the problems that affect toponyms when their mapping is subject to a change of scale. Such a change can, in fact, give rise to marked distortions in everything that impacts the meaning and *raison d'être* of the toponym in relation to the map: whether the change is from the small to large scale, or vice versa. We seek to illustrate this by means of two types of example: the first concerns the large toponymy, or the *macrotoponymy* (or, specifically, a subtype of these toponyms, which we refer to as 'topocultism') and the second concerns the very broad field of small toponymy, that is, the *microtoponymy*.

Among the macrotoponymy, there exists a type of toponym, of scholarly or erudite creation, known as topocultism (Tort 1999: 63). This type of toponym only achieves its full spatial meaning at a certain scale, one that is generally small in relation to its natural scope of application.[3] Examples of topocultisms in Catalan geography include, for example, *Depressió Central* (Catalan Central Depression), *Serralada Prelitoral* (Catalan Pre-Coastal Range) and *Prepirineus* (Pre-Pyrenees). In all three cases, we are dealing with toponyms that are widely recognized, despite their specificity, and which are very broad in their geographical scope (if we take the whole of Catalan territory as our point of reference) and, to a certain extent, well delimited. Although we might imagine that in their origin, these names were proposed from highly sectoral and specific perspectives, and within a strictly erudite register, they have achieved widespread acceptance. These names, however, only make sense within what is their "ideal scale"; that is, the scale at which it is possible to infer their specific spatial scope in relation to the whole of Catalonia. Outside of this, it is impossible for these names to fulfil the function of identification inherent to toponyms (and proper names in general). In fact, these place names can be considered "invented" toponyms: in other words, they arose out of the need to name large territorial areas that were originally nameless and, moreover, they cease to have any meaning when considered from a more detailed perspective or scale or from the angle of popular perception. Consequently, we can conclude that while such names are toponyms in the *spatial* sense of the concept, they are not, however, in relation to what would be their *roots* (or "feelings of identification") among the population directly concerned with these place names.

The problem outlined is also present, and with a similar degree of complexity, at the inverse scale—that is, in relation to what we might refer to as the toponymy of detail, or *microtoponymy*, a type of name that acquires its full meaning, unlike topocultisms, at scales of detail (that is, at the large scale) in relation, once again, to its natural scope of application.[4] Microtoponyms normally comprise, in relation to

[3] If we take Catalonia as our territory of reference, then it would tendentially be between the scales of 1:800,000 and 1:1,500,000.

[4] In the case of microtoponyms, the scales that are considered best suited to their mapping usually range between 1:5,000 and 1:25,000.

a given human community (grouped in a population nucleus or in a municipality), the universe of names used, on a daily basis, by the individuals of that community, be it to identify a *specific* geographical element (for example an individual house or building, a spring, a mountain peak, a monument, a bridge, etc.), be it *linear* (such as a road, a street, a highway, a river or river course in general, etc.) or *extensive* (for example a field of crops, a mountain, a plain or a territorial demarcation). A problem that often occurs in relation to toponyms of this type is that of the *concurrence* of different names for the same geographical concept. This typically occurs in relation to linear toponyms, such as the names of river courses of a lower rank[5] or the names of secondary roads and tracks. But it also occurs in relation to extensive toponyms: above all, those that are applied to places or demarcations that have a large surface area and which are often used by a range of different owners or users (which, once again, can result in the concurrence of different names for the same geographic element or concept).

In the cases described, if a change of scale is proposed, the inconveniences derived from abandoning the "ideal scale" once again appear: the concurrence of names for a linear geographical element can make it impossible to choose a single toponym when a geographic reality is represented in a simplified fashion; or, if there is a concurrence of names for broad, extensive toponyms, it might be equally impossible to determine which name should prevail if the change in scale involves a radical simplification of the reality represented.

This problem of the change of scale is both a geographical and linguistic challenge, which, given the complex nature of the factors that originate it, is not easily resolved. However, the example that we present below by way of a case study (see, section "A Case Study: The Mapping of Toponyms Alluding to Plants and Vegetation in the Cadastral Cartography of La Rioja") offers us the possibility of addressing the problem from a different angle—that is, we focus our attention on a type of cartography and a type of toponymy (associated with that cartography) that has its own specific parameters for addressing the problems of spatial delimitation associated with the meaning of toponyms.

[5] An interesting case in point in Catalan microtoponymy is that of the hydronym *la Riera*, in the municipality of Reus (Tarragona). It is a ravine which, over its ten kilometres of length, changes name nine times (Amigó 1989: 66–67).

A Case Study: The Mapping of Toponyms Alluding to Plants and Vegetation in the Cadastral Cartography of La Rioja

Preliminary Notes: The Scale Factor in Toponymy Related to Vegetation

In the geo-botanical interpretation of toponyms related to plant communities, the first question that needs to be addressed is what we should understand by the concept of a 'plant community'. According to Gleason (1917), a plant community is essentially the sum of its constituent parts: the species distribute themselves over the territory as a continuum, and their distribution responds *individualistically* to the variation in environmental factors—factors that vary continuously in both space and time. The mixture of species in an enclave can be attributed to the ecological conditions and their possibilities of chorological concurrence; consequently, plant communities are, fundamentally, the fruit of chance. And this variation in environmental factors impacts on both the composition of these communities and the abundance or scarcity of the species of which they are composed. Similarly, communities are not, as a rule, stable. They are only ever temporary.

In stark contrast, Clements (1916, 1928) considered plant communities to be clearly recognizable and definable entities (Alcaraz 2013) and, therefore, capable of being named. Likewise, he claimed, they tend to repeat themselves with great regularity over a given territory. According to this conception, the species that make up a community constitute an organic system characterized by mutual interaction between all of its elements.

The cognitive intuition of the inhabitants of a territory tends to resolve this duality, since in practice a plant community is recognized as a unique element of the landscape. And, based on this recognition, it can be named without further theoretical debate; for example: in Spanish, *Hayedo* is a collective of 'haya' (beech, *Fagus sylvatica*) or *Robledal* meaning an oak forest (some species of *Quercus*). A plant community is a functional set, insofar as it serves to locate, demarcate and identify a place. Indeed, the geo-botanical toponymy (or phytotoponymy) provides a set of data in which it is possible to detect the regularities and singularities presented by the plant landscape of a territory. And, therefore, in practice, it facilitates a typification of the vegetation cover that extends over that territory.

However, the precise location and delimitation of the area that corresponds to the named element has not been the object of detailed reflection in much of the traditional toponymic literature.

It is apparent that, in cartography, especially in the case of maps drawn at more general scales (above all, scales greater than 1:10,000), higher referential precision is needed (Vázquez Hoehne and Rodríguez de Castro 2012). This means that the area corresponding to a toponym is usually subject to the interpretation of the reader (and, therefore, to that particular reader's map-reading skills) and so the limits of the place to which the place name corresponds are uncertain and ambiguous. Yet, the explicit delimitation of an area is not only an imprecise activity because of problems associated with its spatial representation (albeit that, today, these are readily overcome thanks to developments in geographic information systems), but also because it is inherent to the nature of the named geographic element. Indeed, the delimitation of a geographic element can, in practice, also be diffuse, vague and variable over time.

In the case of toponyms related to vegetation types, the scale factor brings with it a notion of spatial variation or change, known technically as *gradient*. The problem lies in the fact that the precise delimitation of plant communities is, in most cases, complex (except in cases related to sudden changes in environmental factors or the soil structure). The limits of a place identified with a particular toponym are also usually diffuse and vague, with introgressions of one type of vegetation cover in another together with transition or ecotonic bands (Van der Maarel 1990; Ruiz de la Torre 1990). For this reason, the scale ultimately determines that the delimitation of a place or territorial entity is imprecise. Drawing, in part, on the work of Vázquez Hoehne and Rodríguez de Castro (2012), we can say that the place or territorial area has a *core* or *area of certainty*, that can be delimited, and an associated *marginal* or *diffuse area*. Naturally, both are subject to change over time, which means in works of toponymy at the *meso* and *micro* scales it is possible to draw conclusions of a chorological (i.e. determine areas of distribution of plant groups) or dynamic type (i.e. the substitution of some plant communities in the core of certainty by others, in a diachronic process). However, the imposition of a name on the core of certainty is conditioned by factors related to historical perception and even to psycholinguistics—factors, nevertheless, unrelated to a purely geo-botanical interpretation, beyond that provided by the intuition of the inhabitants of the rural forest environment. Hence, the scale is connected with the most typical *perspective* of observation of a plant community—so, an area or piece of land that people can access freely is not the same as an area that has to be observed from a certain distance because of difficulties of access.

A Way of Representing a Toponym Based on the "Territorial Area" of Its Meaning

A generally accepted principle is that the progressive improvement of spatial information should be oriented to the precise determination of what is known as the *toponymic area*: that is, the area of territory ascribable to a toponym—in other words, its "spatial meaning". There are still many questions to be addressed in this regard, but, for the time being, it is apparent that the information provided by the Cadastre in Spain has improved greatly in recent decades and, that this improvement has, by and large, impacted the spatial dimension of toponyms too.

In the systematic collection of toponymy for the creation of a *corpus* of present-day geographical names and its subsequent transfer to a database, digitized cadastral information can be useful. This alpha-numerically referenced information is originally represented in *shape* format (the latter can be downloaded and processed with software created by the Cadastre office) and the resulting information gathered by the corresponding provincial land register service can be combined with the vector information (*shape*). This operation allows us to obtain a corresponding table of attributes, containing a field with the name of the place associated with each cadastral parcel.

This collection of cadastral toponymy needs to be followed up with a process of data purification, consisting of the elimination of duplicates and the correction of entries (a task that is currently highly necessary because of the high proportion of transcription errors presented by toponyms from digital cadastral information). Data correction should involve crosschecking with data from other sources (for example, the transcripts of topographic maps at the scales of 1:50,000 and 1:25,000 to confirm the correct version of names that appear doubtful). Exceptionally, a certain number of incomplete or incorrectly recorded toponyms that appear in the digital cadastre can be corrected using thematic maps. This occurs when the appellative value of a name to be verified is associated with an area in which there is express evidence of the presence of the type of plant cover on which the phytotoponym is based.

With this information, it is possible to graphically represent the surface area to which, according to the Cadastre, each toponym is assigned.

Toponyms that refer to verifiable characteristics of the territory, such as those of a geo-botanical type, have the advantage of facilitating the delimitation sought. However, the lack of definition, discussed above, of the limits between plant communities must still be taken into consideration. This vagueness, which usually results in different degrees of composition and structure of the communities, frequently prevents the clear tracing of dividing lines that separate *areas* assigned to different vegetation units.

Some Technical Characteristics of Cadastral Toponymy and a Recent Innovation: The Creation of Spatially Referenced Toponymic Databases

Often, the cadastral toponymy turns out to be the most abundant and diverse. These place names originate from several different data banks:

(a) Historical cadastre: This is especially valuable for its planimetric precision and the rigour of the toponymic collections and transcriptions, the result of meticulous field work. The historical cadastre is made up of:

  - Land maps or *catastrones*. In the territory of La Rioja, for example, there are some 2,700 land maps at a scale of 1:20,000 (approximately one for each cadastral polygon), produced between 1928 and 1969. In the area of La Rioja where agriculture is the predominant land use (that is, the northern part of the region), some 43,000 toponyms have been identified and digitized based on these land maps. They are currently deposited in the Provincial Historical Archive of La Rioja.

- The 1957 land parcel topographic cadastre and the planimetric evaluation records on photographs of the terrain. The land parcel boundaries of the rustic property cadastre were drawn on aerial photograms. The "cadastre files" were, from that date on, considered to be the ideal basis for all consultations. These files contain the reference and, sometimes, the description of the elements used for the delimitation of the toponym, for each municipality and each cadastral polygon (with its corresponding number). And, of even greater relevance from that moment on, in terms of toponymic studies: they were based on the so-called "parajes", rustic estates or places, i.e. spatial units of reference for each cadastral polygon, identified by means of their corresponding toponym.

(b) Current cadastre. The computerization of cadastral data which began in the 1990s meant a loss of precision in the recording of place names. The lack of rigour employed in this process led to defective transcriptions and the introduction of changes in spelling, aggravated by the absence, at that time, of clear criteria for the standardization of spellings. At times, the problem lay in the excessive length of the names in the databases. Subsequently, all these issues have gradually, and in part, been corrected.

There are two types of toponymic data in digital cadastral mapping:

Those referred to as *labels* and which provide an approximate location of the place name using the coordinates of the centroid point of this label.

Those referred to as *cadastral areas* (land parcels and polygons), associated with vector layers. When undertaking spatial analyses and when crosschecking with other areal-type thematic information (such as vegetation mapping), they are an especially useful tool, insofar as toponyms always refer to surface entities of a certain area.

The location (or georeferencing) of toponyms in GIS and their mapping can be carried out in detail in the polygon maps corresponding to the cadastral sites. In general, these sites can be considered as having a precise location and surface area, with the exception of certain theoretical considerations regarding the arbitrariness of their delimitations.

The mapping material that can be downloaded from the Spanish Cadastre server consists of two parts. On the one hand, graphic information in *shape* format of the cadastral polygons with the delimitation of land parcels and sub-parcels. On the other, subsequent alpha-numerically referenced information (each identified by means of a common field, the cadastral reference). Among the data of this alpha-numeric information, we find a field that holds the name of the principal site assigned to each parcel. This field is the one that provides us with the toponym with which the rural estate was officially registered. One sole name corresponds to each cadastral parcel.

All neighbouring cadastral parcels with a common toponym can therefore be integrated. Thus, the polygons on the map correspond to differentiated areas of the contiguous parcels and are identified with the same name (regardless of whether the polygons were originally subdivided into different parcels). Cadastral parcels may also be disjointed cadastral parcels, separated from each other but lying close together, but which belong to the same toponymic area.

Thus, at the end of the process, we obtain a map of polygons representing the parcels of the rustic property cadastre that have been assigned a place name which refers to one category of meaning.

Auxiliary Procedures in Cadastral Mapping for the Analysis of Phytotoponymy: Some Examples

The applications of toponymic mapping are highly diverse. Having a surface area reference—the "toponymic area"—for each geographical name means toponyms can be crosschecked with other thematic maps.

This type of toponymic mapping is particularly useful for studying the evolution of a territory's vegetation cover over time, with the toponyms being used as indicators of the presence or absence of the plant communities or botanical species to which the names allude. As such, it is a question of identifying actual correspondences between the toponym and the presence of present-day species or units of vegetation.

The scales of the different maps however need to be compatible to ensure the coherence of such analyses.

Chorological maps or maps of the geographical distribution of vegetation are used in conjunction with the toponymic mapping for comparative purposes and in this way the presence or absence of the plant species/cover alluded to by the toponym at each location can be analysed (López Leiva and Cuevas Moreno 2013; López Leiva 2016, 2018). If a geographical name, in conjunction with its corresponding toponymic area, mentions a plant element currently absent from the area of the polygon it represents, it is considered an "external toponym".

When working with cartography whose characteristics are similar to those of the Spanish cadastral maps, the procedure for detecting external toponyms with a botanical reference involves selecting polygons that do not intersect with any subplot related with that meaning and which appear on the chorological or plant distribution map.

To clarify, "non-external" means that all or part of a polygon overlaps with the area of distribution of a plant community or a subplot in which that community is the dominant or main botanical species. Yet, "non-external" might also imply that this polygon only lies "close" to an area in which there is a high probability of the presence of this plant community: it may, for example, partially overlap with areas in which the botanical species alluded to is only *accessory*, with a less abundant or even token presence. This is the case of a plant that is a simple companion of another dominant species; or, when the allusion is to a plant community, the case of a small inclusion that cannot be represented at that particular map scale. Once again this highlights the fact that the scale of representation of a map is fundamental for the

Table 1 Typology of "external" and "non-external" toponymic polygons according to the cadastral parcel, and the main problems detected

External	Problems
All or one of the parcels is external a. Pluri-parcels b. Mono-parcels	A fragment of one pluri-parcel plot might indicate that the toponym is external when most of it is not
Non-external	
All the parcels are non-external (in contact or minimum degree of overlap) a. Pluri-parcels b. Mono-parcels	A minimum degree of overlap might suggest that a toponym is non-external when most of it is in fact "close" lying

adequate use of toponymic mapping. It should be noted, here, that in the case of the Spanish Cadastre the most usual scale of reference employed is 1:25,000.

The different possible outcomes described above are summarised in Table 1.

A "non-external" toponym can occur in one of the following two situations: (a) When it is divided in several spatially independent fragments and all these fragments lie in the area of distribution or one or more fragments overlap with a part, however small, of an area outside that distribution; (b) when the toponym is formed by a single polygon, the polygon must overlap with a part, however small, of the area of distribution. A toponym is considered as being "external" when its area of representation is not in contact with the area of distribution, or when its surface area is divided into several fragments and one of these fragments is external (Fig. 1).

The area occupied by each polygon or "locatable" (i.e. spatially georeferenceable) toponymic record can be matched with information identifying its plant cover, as can their *buffer* area, which, to some extent, can be considered to correct any inaccuracies in the delimitation of the place and to broaden the definition of the closest lying area of distribution of the species or group to which the toponym alludes.

Thus, whatever the specific characteristics of the vegetation mapping being used, by adopting the concept of the "external toponym" we dispose of highly relevant qualitative information: that is, a set of locations identified by a geographical name that alludes to a botanical species or a plant population that no longer exists, or whose presence is reduced to a mere vestige or which is unlikely. Such an analysis can be fundamental, or at least of some interest, for the reconstruction of the landscape, the choice of species for landscape restoration, the management and planning of the territory and for an accurate estimate of the intensity and duration of human action in the past.

For a study of this nature, it is essential that the geobotanical information be as accurate as possible. Detailed scale maps exist in which information is provided (typically in associated data tables) not only about the most abundant plant elements (i.e. the dominant or main species) but also about the subordinate elements (i.e. isolated examples, small stands, patches or tesserae of dimensions barely representable at the particular map scale).

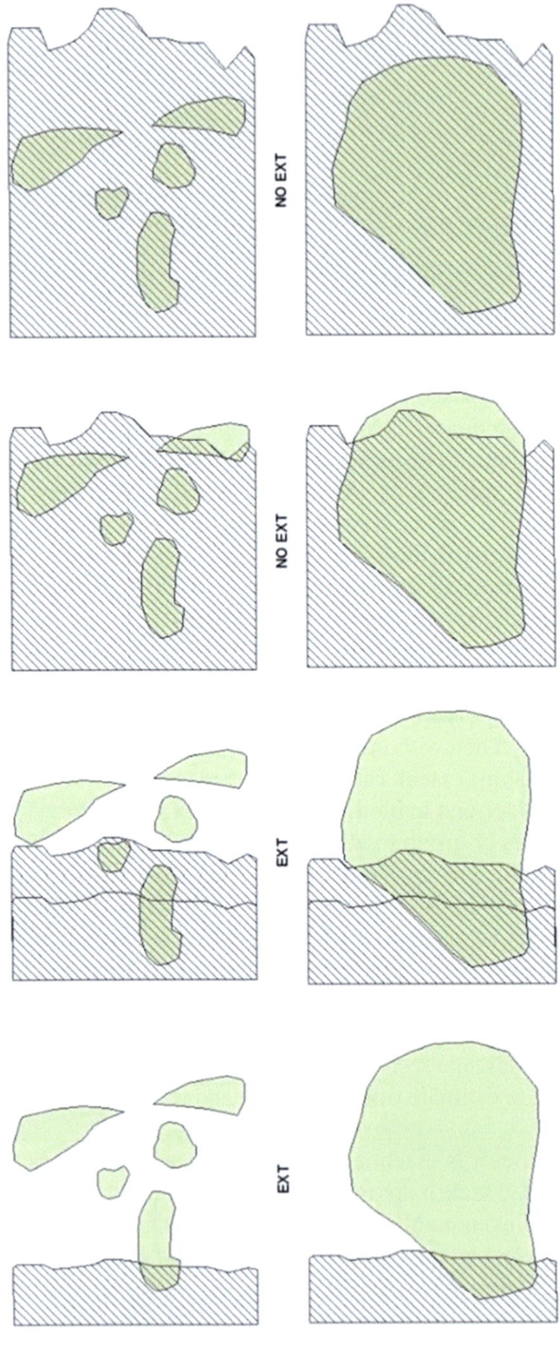

Fig. 1 Possible scenarios involving the detection of "external" toponyms. *Note* The toponymic mapping is shown in green and the hatched lines show the area of distribution of the plant species or communities. *Source* Own elaboration

At this juncture, we need to consider the parameters that need to be taken into account when comparing the toponymic areas that are represented and the areas of distribution of the plant species.

One such parameter is the distance (in units of length) to either the dominant or co-dominant vegetation, on the one hand, or to the non-main or accessory vegetation, on the other. In itself, however, distance serves merely as an indicator, since a species or community's physiographic characteristics (altitude, slope, and exposure) and opportunities for dispersal have a greater influence on their actual distribution. Obtaining physiographic information for all the toponymic areas related to a semantic group facilitates its subsequent analysis. Similarly, it also allows the statistical analysis of the most usual ranges—in relation, that is, to the location's altitude, slope, and exposure—for a set of toponyms, provided these place names are related to a specific species or plant group.

The calculation of distances can, however, be very useful when *assigning species to generic toponyms*, that is, those, with an imprecise meaning, alluding to different species related by their taxonomy or physiognomy. In this way, the most likely taxon can be assigned, taking into account the current degree of chorological proximity of the species whose vernacular names coincide.

Another auxiliary procedure relies on quantification and the location of external toponyms, that is, of place names whose presence provides confirmation of the disappearance of previously existing locations of a species. Figure 2 presents an example of the location of external toponyms in La Rioja (Spain), using geographical names from the Cadastre whose meaning is unequivocal and the current area of distribution of the reference species. Here, we use the vegetation maps available in Spain's *Mapa Forestal*, available at scales of 1:200,000 and 1:50,000. A specific example of one of these polygons is shown on Fig. 3.

Conclusions

The basic issue addressed throughout this article, above and beyond that is the theoretical considerations raised in the first section and the discussion developed in the second centred around a specific case study, is very simple. Essentially, our aim has been to understand the *raison d'être* of the toponym from the perspective of its spatial dimension, and to consider, on this basis, all the questions posed by its cartographic representation—questions that range from the inherent difficulties involved in its definition and its intangibility as a linguistic phenomenon, to its fit in the technically complex system of representation of the earth's surface that is cartography—and, in particular, the problems related to the scale factor and its derivations, especially, those related to a change of scale: an operation categorically linked to the process of producing a map but with a transcendence that is projected over all its elements but, in particular, its toponymy.

In this context, the consideration of the methods and techniques of representation developed by cadastral cartography, which, by definition, places a particular emphasis

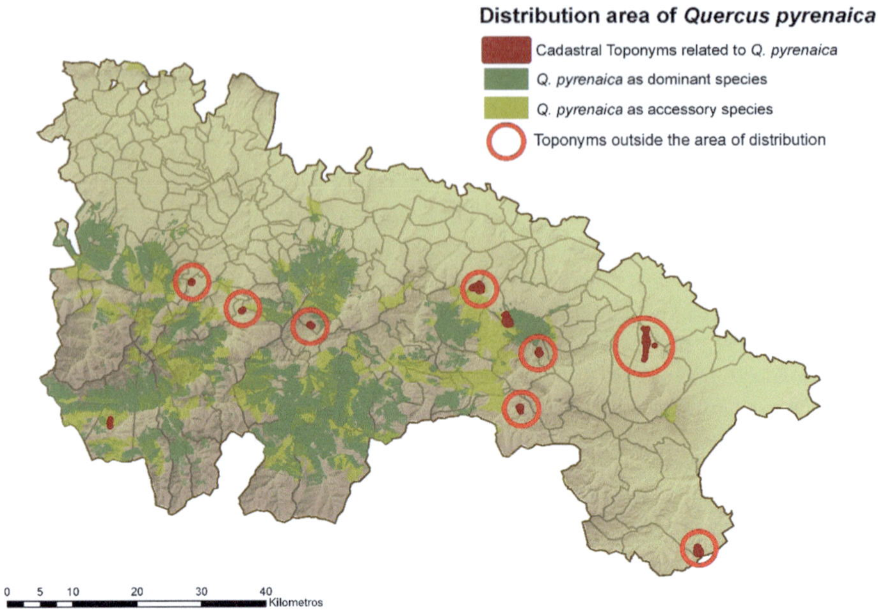

Distribution of cadastral toponyms

Unequivocal	Flora Vascular de La Rioja
Uncertain	Main
	Accessory
	Toponyms outside the area of distribution

Fig. 2 Cadastral toponyms related to Quercus pyrenaica outside the distribution area of this species. Prepared by the authors from information provided by the Spanish Cadastre and the Forest Map of Spain at scales 1:200,000 [MFE200 CC-BY miteco.gob.es] and 1:50,000 [MFE50 CC-BY] miteco.gob.es]; regional and municipal boundaries from centrodedescargas.cnig.es CC-BY. The Spanish Cadastre provides a license in a governamental legal settlement [Res. 23/03/2011, Dirección General del Catastro, http://www.catastro.minhap.gob.es/documentos/res_230311.pdf]

on the value of *precision*, has furnished us with a practical point of reference when it comes to appraising the strengths and weaknesses of mapping of this kind. This is true, above all, when we are faced with complex questions such as the definition of the actual space of the parcels of land, the links between these parcels and certain toponyms and the problems posed by the passage of time and the transformation of the natural environment; especially, when seeking to establish clear correlations between the names and the specific places to which these names are applied. The case study presented, despite its intrinsic complexity, allows us to verify the great technical advances that have taken place in this field in recent decades (particularly, in all aspects concerning the systematization and the quantitative and qualitative treatment of the registered data).

Fig. 3 Example of graphic output of an external toponym of the Spanish Cadastre in La Rioja, related to the species Quercus pyrenaica (a submediterranean oak). One of its possible collective place names is "Marojal" from the name 'marojo' or 'roble marojo'. The polygon shows its location together with an interpretable image of the vegetation cover in this and the surrounding area. *Source* Own elaboration based on polygons from the Spanish Cadastre over ortophotography [PNOA ign.es CC-BY]. The Spanish Cadastre provides a license in a governamental legal settlement [Res. 23/03/2011, Dirección General del Catastro, http://www.catastro.minhap.gob.es/documentos/res_230311.pdf]

A little over seventy years ago, Bertrand Russell raised the question as to whether the method of referring to elements of reality by means of proper names—and therefore, by means of toponyms—might one day be replaced by a much more precise and efficient system of co-ordinates. And he concluded that while such a system could perhaps reduce their number, it could never eliminate our need for them: "Without proper names we can express the whole of theoretical physics, but no part of history or geography" (Russell 1948: 90). Despite their intangibility, and their uncertain concretion, toponyms today remain essential elements for the construction of territorial identity.

Questions

1. To what extent is the meaning of a place name conditioned by space? Should the etymology of a place name be known to interpret its 'geographical meaning' correctly and represent it accurately?

2. Which specific problems in the mapping process does the scale factor raise with place names? Can we speak of an "ideal scale", in the cartographic representation of the place names?
3. On what criteria can the distinction between "macrotoponym" and "microtoponym" be based? What difficulties does microtoponymy pose in the mapping process, and how can they be resolved?
4. Why does toponymy related to plant communities exemplify some of the main problems raised by mapping toponyms? To what extent may phytotoponyms, and particularly microphytotoponyms, be used as indicators of biogeographical traits and landscape dynamics?
5. In what aspects and in what modalities of toponymic research can cadastral toponymy be relevant as a source of study?
6. To which extent the scale, and the change of scale, are critical factors in cadastral toponymy?

Acknowledgements Sincere thanks are due to Iain Kenneth Robinson for his linguistic assistance in the text.

References

Alcaraz F (2013) Fundamentos de la clasificación de la vegetación. Curso de Geobotánica. Universidad de Murcia, Murcia

Amigó R (1989) Sobre inventaris de noms de lloc. Introducció metodològica. Centre de Lectura, Reus

Clements FE (1916) Plant succession. An analysis of the development of vegetation. Carnegie Institute, Washington

Clements FE (1928) Plant succession and indicators. H.W. Wilson, New York

Coromines J (1965) Introducció a l'estudi de la toponímia catalana. In: Estudis de toponímia catalana, vol I. Editorial Barcino, Barcelona, pp 7–30

Coromines J (1972) Tópica hespérica. Estudio sobre los antiguos dialectos, el substrato y la toponimia romances, vol 2. Gredos, Madrid

Dorion H (1984) Les relations entre la toponymie et les autres sciences sociales. In: 450 ans de noms de lieux français en Amerique du Nord. Les Publicacions du Québec, Québec, pp 103–108

Gleason HA (1917) The structure and development of the plant association. Bull Torrey Bot Club 43:463–481

Institut d'Estudis Catalans (IEC) (1995) Diccionari de la llengua catalana. Enciclopèdia Catalana, Barcelona, p 1780

López Leiva C, Cuevas Moreno J (2013) Valor de la toponimia riojana como indicador de la historia de los aprovechamientos forestales y usos del suelo. Cuadernos de la Sociedad Española de Ciencias Forestales, p 38

López Leiva C (2016) Onomástica, ecología y territorio. La toponimia de La Rioja como indicador biogeográfico y de la dinámica del paisaje forestal. Tesis doctoral, ETSI Montes, Forestal y del Medio Natural. Universidad Politécnica de Madrid, Madrid

López Leiva C (2018) Fitotopónimos de La Rioja (España): indicadores de distribución y evolución de las cubiertas vegetales. Onomàstica. Anu Soc d'Onomàstica 4:63–122

Moll J, Tort J (1985) Toponímia i cartografia. Assaig de sistematització. Institut Cartogràfic de Catalunya, Barcelona
Monkhouse FJ (1978) Diccionario de términos geográficos. Oikos-Tau, Vilassar de Mar
Moreu-Rey E (1982) Els nostres noms de lloc. Moll, Palma de Mallorca
Moreu-Rey E (1995) Tipologia toponímica. In: Rosselló Verger VM, Casanova E (eds) Materials de toponímia-I. Generalitat Valenciana-Universitat de València, València, pp 45–52
Real Academia Española (RAE) (1984) Diccionario de Lengua Española. Espasa Calpe, Madrid, p 1320
Reszegi K (2020) Toponyms and spatial representations. Onomastica LXIV:23–39
Robinson AH, Sale RD, Morrison J, Muehrcke PC (1987) Elementos de cartografía. Omega, Barcelona
Ruiz de la Torre J (1990) Mapa Forestal de España a escala 1:200.000. Memoria General. ICONA-Ministerio de Agricultura, Pesca y Alimentación, Madrid
Russell B (1948) Human knowledge. Its scope and limits. Allen and Unwin, London
Tort J (1999) Algunes reflexions sobre el concepte de topònim. Butll Inter Soc d'Onomàstica 77:59–72
Tort J (2001) Els noms de lloc i la geografia. La toponímia com a eina per a la conceptualització de l'espai. Butll Inter Soc d'Onomàstica 83:86–98
Tort-Donada J (2010) Some reflections on the relation between toponymy and geography. Onoma 45:253–276
Tort-Donada J (2020) On the connection between the physical environment and the toponym. Geography's contribution to clarifying the problem. In: Balode L, Zschieschang C (eds) Onomastic investigations II. Latvian Language Institute, Riga, pp 286–299
Tort-Donada J (2021) Names and naming in the Iberian Peninsula. Joan Coromines' intercultural approach in onomastics. In: Felecan O, Bugheşiu (eds) Names and naming. Intercultural aspects. Palgrave Macmillan, Chan, pp 295–314
Van der Maarel E (1990) Ecotones and ecoclines are different. J Veg Sci 1:135–138
Vázquez Hoehne A, Rodríguez de Castro A (2012) Metodología para la investigación de las áreas de referencia de los topónimos en cartografía. In: Actas del X Congreso Topcart. Congreso Iberoamericano de Geomática y CC de la Tierra, Madrid

The Mystery of Hydronomy in the Land of Israel

Tal Yaar-Waisel

Abstract History, religion and politics have been used interchangeably in the selection of place names in the State of Israel from the beginning of the Zionist movement and even more explicitly, after the establishment of the state of Israel; this trend continues today. Similar factors are involved in naming water reservoirs. This chapter examines the history of sea names and their direct and indirect meaning, paying special attention to the role of the education system in shaping the perception of the country's future citizens. In the southern part of the Land of Israel there is the Gulf of Eilat, known globally as the Gulf of Aqaba. Also, there is The Salt Sea which is the lowest place below sea level on earth and is known worldwide as *The Dead Sea*. Israel has only one lake, known as lake *Kinneret*, while for Arabs it is Lake *Tiberias* and the Sea of Galilee by Christians. This chapter illustrates that in contrast to the early decades of the state of Israel, nowadays economic potential and political interests in naming are gaining precedence.

Keywords Gulf of Aqaba/Eilat · Dead Sea · Sea of Galilee · Google maps · Geography education

Objectives

To introduce the reader to the toponymy and hydronomy in the Land of Israel from historical, political, geopolitical and economic perspectives.

To understand the complexity of the names of places in Israel due to the uniqueness of Israel in the Middle East, and the rich history of the region religiously and politically.

To recognise the ambiguous situation of local geopolitics.

T. Yaar-Waisel (✉)
Oranim College of Education, Kiryat Tiv'on, Israel
e-mail: Yaar.tal@gmail.com

© The Author(s), under exclusive license to Springer Nature Switzerland AG 2023

G. O'Reilly (ed.), *Place Naming, Identities and Geography*,
Key Challenges in Geography, https://doi.org/10.1007/978-3-031-21510-0_7

Introduction: 'Whose Gulf is It?'

It was only when I started my geography studies at the university that I first heard about the existence of the Gulf of Aqaba. Of course, I knew well the Gulf of Eilat, the coral reefs and the beautiful mountains that rise above the bay. I knew the city of Aqaba that can easily be seen from Eilat, including the Jordanian king's palace and a huge waving flag, but the gulf was the Gulf of Eilat, without any connection to the city of Aqaba. How can this be? Is Israel so disconnected from the world around it? Google also knows the Gulf of Eilat—but only if one speaks Hebrew.

Naming of Places and Geographical Elements in Israel

Naming of places and geographical elements in Israel is a complicated geo-political issue, as are economic, touristic, or commercial concerns in Israel. A lot has been written on this subject either from an objective-scientific point of view, or from a national—Zionist perspective (Kadmon 2000; Azaryahu and Golan 2001; Medzini 2012).

Zionist nation-building aimed at restoring the Hebrew toponymy of the land. The creation of the national Hebrew map of Israel was designated to assert the Jewish identity of the State of Israel in terms of a conflation of cultural and territorial aspects of Jewish sovereignty (Azaryahu and Golan 2001). History, religion, and politics have been used interchangeably in the selection of names in the State of Israel from the beginning of the Zionist movement (Carta 1953–1985), and more explicitly, after the establishment of the State of Israel in 1948 (Kadmon 2000), this trend continues today.

Place names—foundation upon religious narratives

The history of names goes far back to biblical times and even earlier. Names of places that were preserved unchanged, were re-raised after the resurrection of the Jewish nation in the lands of Israel. Thus, the city of Jericho goes back more than 11,000 (!) years, as do *Beit-Lechem* or *Hebron*. All three are now Palestinian cities in the West Bank, and have also Israeli, Hebrew names. Many biblical names have religious significance for Judaism, Christianity and Islam due to their shared geographies in the development of monotheistic narratives over thousands of years. Sometimes, one single place can be powerfully meaningful for all three religious traditions; then, the name is, of course, of special significance as it is with Jerusalem—Yerushalayim (Hebrew: יְרוּשָׁלַיִם) and Al-Quds (Arabic: اَلْقُدْس).

The name Temple Mount demonstrates the above significance: as *Har ha-Moria* (Moriah Mountain), Temple Mount, contains the rock from which the world was allegedly created, where the sacrifice of Isaac is assumed to have taken place, and thus is the holiest place for both Jews and Christians (see Fig. 1). Yet, the same place is also named *El Karam el Sharif* (the Holy Place), or *El Aksa* (the Edge) from where

Fig. 1 Ghiberti, 1401–1402, Sacrifice of Isaac (Florence Baptistery)

Muhammad ascended to heaven. It may be one the smallest of places, yet with such a deep meaning, attracting geopolitical controversy (O'Reilly 2019).

Another salient example is Nazareth, the place where the Annunciation was given to Mary, mother of Jesus, and the beginning of Christianity. In Hebrew, *Nazareth* means Christian.

Place names—foreign rule

The group refers to naming after local names, in use for 2000 years of foreign rule, from the Romans to the British; a salient example is Caesarea, built in the first century BC and named after Julius Caesar. The name remains the same even today. Two thousand years later, the two King George Boulevards in Tel-Aviv and in Jerusalem were named after the British King—George V, in whose reign the Balfour Declaration was made in 1917 regarding 'a national home for the Jews' in Palestine. Significantly, Balfour, is the name of street in Jerusalem where officially the Israeli prime minister resides.

Place names—Arab civilization

The Arab/Palestinian inhabitants in the Land of Israel are the third and very significant source of naming. For instance, *Ramla* was founded by Arabs in the eighth century. The name of the city originates in Arabic, which translates as "the city of sand," because it was built on dunes, while *Hadera* follows the Arabic word for "the green" (*El Hader*).

There are names that combine biblical history with a very similar Arabic name. This makes the (geo)politics of official place naming even more complicated, when the question arises as to which historical or ethnic tradition the place/name can be ascribed or concerns territorial and identity-related emotions linked to politics and sovereignty in the region over the past three millennia, at least.

Shefar'am (Hebrew), or, in Arabic: *Shafa'amar* sound quite similar but have a different history: while the first is originally Hebrew and dates back to the 200 AD, the Arabic name origin dates from the time of Othman, son of Daher al-Omar, who ruled in the second half of the eighteenth century. Both narratives claim the place is originally theirs, although it has been inhabited by Arabs for hundreds of years.

Finally, there are modern Zionist-Hebrew names that are sometimes rooted in old history or cultural bases; the most famous one is Tel Aviv, adopted from Theodor Herzl's book *Altneuland*[1] in the Hebrew translation.

Who is Responsible for Naming in Israel?

The Government Naming Committee is a public scientific group composed of scientists from various fields,[2] natural scientists, geographical historians, archaeologists, scholars of the bible, scholars of Hebrew and Arabic, representatives of government ministries and other ministries: the Prime Minister's Office, and the Ministries of: Interior, Transport and Road Safety, Construction and Housing (The Israel Mapping Center), the Jewish National Fund and Nature and Parks Authority. The committee has three subcommittees: Geographical, Historical, and Settlement Names.

The committee's sources of authority and functions are:

1. The committee operates by virtue of Government Resolution No. 258 of March 8, 1951, which documents the appointment of the Government Naming Committee. This decision stipulates that the committee will operate in the Prime Minister's Office and that its decisions will be binding on state institutions. This decision therefore united the committee for determining geographical names in the Negev, along with the committee for determining names of localities of the Jewish National Fund.
2. The committee is the only authorized body for deciding upon names for various localities and sites, including intersections, interchanges, tourist sites, nature and landscape, industrial and employment sites, etc. in the State of Israel. It was emphasized that "it is inconceivable that every single person and even a community in a certain place be allowed to name on its own."[3]

---

[1] Theodor Herzl, Altneuland, 1902. On the book cover is written: "Wenn ihr wollt, ist es kein Märchen", If you want, it's not a fairy tale, became one of the cornerstones of Zionism.

[2] https://www.gov.il/he/departments/units/names_committee.

[3] There: https://www.gov.il/he/departments/Units/names_committee.

3. The ultimate goal of the committee is to decide upon Hebrew names, on Israeli maps, as far as possible.
4. The committee's decisions are made in its plenum and are published officially.

Having explored the rich historical and contemporary palimpsest of toponymy in the Land of Israel, and the official aspects of place naming in the State, the focus turns to hydronomy. Here is explored the complexity of naming a world-renowned water body—the Dead Sea/Salt Sea; an area of major international significance—the Sea of Galilee/*Kinneret*/Lake *Tiberias;* and the geostrategic Gulf of Eilat/Aqaba offering access to the waters of the Indian Ocean and world seaways.

The Gulf of Eilat/Aqaba—Geography and History

The Gulf of Aqaba, which is known by Israelis as the Gulf of Eilat, is located at the northern tip of the Red Sea, east of the Sinai Peninsula and west of the Arabian Peninsula. Its coastline is divided among four riparian countries: Egypt, Israel, Jordan, and Saudi Arabia. The gulf measures 24 km (15 mi) at its widest point and stretches some 160 km (99 mi) north of the Straits of Tiran (*Wikipedia*, Gulf of Aqaba, Brawer 2014). Geologically, the Gulf forms the southern end of the Dead Sea Transform, which is the reason for the Gulf's depth.

At its very tip, in the north, there are two cities: Aqaba in the north-east (Jordan), and Eilat in north-west (Israel)[4] (see Fig. 2).

This location has a long history as a transfer point between Africa and Asia; ancient Egyptians, Romans and Ottomans used this route, Goes back 4000 years BC. Known as the Nabataean way at the point where Aqaba was established (second century BC), and *Eilot* is where King Solomon built his navy and copper mines (Tenth century BC). Aqaba is famous for being the location of the last battle of the Ottoman Empire against the British, in 1917.

In 1948, with the establishment of the State of Israel, the small place named *Um-Rashrash* was conquered in one of the first military operations of the Israel Defense Forces. The city of Eilat was established in 1949. Both Aqaba and Eilat are port cities and touristic destinations (see Fig. 3). Hence Eilat is part of the southern Negev desert, at the southern end of the Arabah, adjacent to the Egyptian resort city of Taba to the south, the Jordanian port city of Aqaba to the east, and within sight of Haql, Saudi Arabia, across the Gulf to the southeast.

The gulf is well known as one of the most beautiful coral reservations and famous for its rich underwater life.

---

[4] A tour in Eilat
https://www.youtube.com/watch?v=NrCNWFLLRdY
A tour in Aqaba
https://www.youtube.com/watch?v=RgsNq58sCKs.

Fig. 2 At the northern end of the Gulf are located the cities of Eilat (foreground) and Aqaba (background). The photo was taken from the Eilat mountains and looks to the east, towards Edom—Trans-Jordan mountains

Fig. 3 The border area between Israel and Jordan as seen from the Gulf, looking to the north

Managing the Gulf

Although the Gulf is not a big water body, four different states have rule it. In practice, the Kingdom of Jordan and the State of Israel are actually in charge of the management.

The maritime border crosses the Gulf at its median line, from the point where the *Arava Wadi* enters the Gulf to the very south; Egypt lies to the west and Saudi-Arabia to the east. The southern part of the Gulf is wider, and thus allowing non-territorial waters for access to the high seas and the associated international legal regime (See UNCLOS 1982).

It is interesting that even in its official documents, the State of Israel uses the name Gulf of Aqaba, whereas Gulf of Eilat is not really in use; as seen in Fig. 4, the word Aqaba is used in the very early publications of the State of Israel.

The name, Gulf of Aqaba appears once again in the Peace Treaty with Egypt in 1979, as Egyptian officers did not accept the Eilat Gulf term, and former Prime Minister Menachem Begin agreed to use the term Aqaba (Serbro 2021). Begin well understood the historic opportunity and greatness of the event and did not want to jeopardize the historic peace agreement over such a naming issue.

Israel and Jordan signed a Peace Treaty, which has a Marine Appendix, in 1994.

A state must follow several aspects in managing the sea: International law (UNCLOS 1982), the Charter of the Seas, principles, international maritime courts and more. Since 1994, with the signing of the Maritime Annex to the 1994 Peace Treaty with Jordan, shared management based on mutual trust has been practiced, and indeed, cooperation in the Gulf is very good (Shahar 2021). Beside everyday

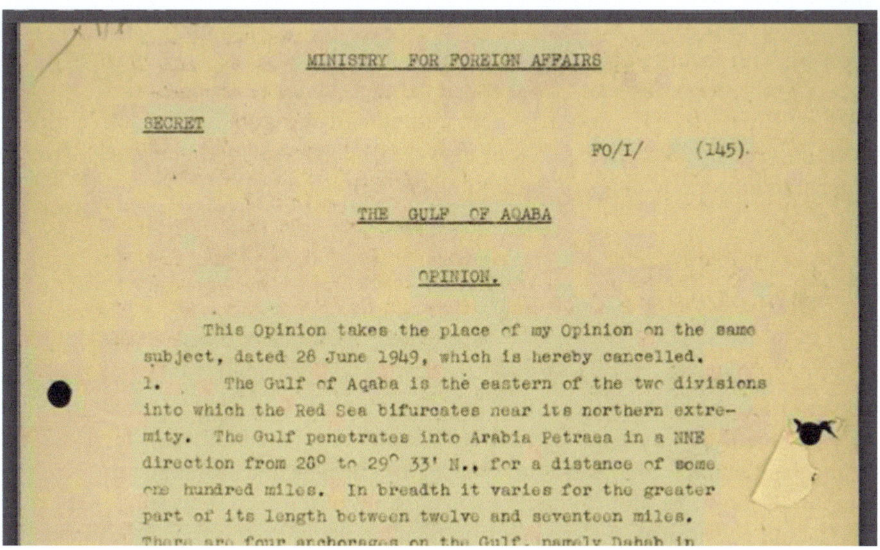

MINISTRY FOR FOREIGN AFFAIRS

SECRET

FO/I/ (145).

THE GULF OF AQABA

OPINION.

This Opinion takes the place of my Opinion on the same subject, dated 28 June 1949, which is hereby cancelled.

1. The Gulf of Aqaba is the eastern of the two divisions into which the Red Sea bifurcates near its northern extremity. The Gulf penetrates into Arabia Petraea in a NNE direction from 28° to 29° 33' N., for a distance of some one hundred miles. In breadth it varies for the greater part of its length between twelve and seventeen miles.

Fig. 4 Document of the Ministry of Foreign Affairs, 1951 contact with Jordanians, there is constant security cooperation in the borders' triangle, together with Egyptians.

Most ongoing connections and cooperation are between military forces from both sides of the international border. For the military, it is easier to communicate because they hold similar attitudes and duties regarding the work to be done. Most unexpected or planned events are managed by the two navies (Benjy 2021).

There are several bodies in the Red Sea that have taken security goals upon themselves, to allow safe sailing and crossing. There are places where the UN or NATO have taken part in the process. Ad-hoc teams are dealing with challenges, like of pirates trying to operate in the area. It is common, that a central international management operates when there is a specific mission, as happens if a fishing boat is swept away by sea currents across the border (Shahar 2021).

When there is "diplomatic sensitivity" involved, the Israeli forces prepare for it in advance. For cross border delegations there is a need to be prepared, they are the ones to determine the terms that be used in the specific negotiation. Since some of the maps in which the name Aqaba Gulf appears are British, there is no doubt about the term being in use. Besides, everyday connection is rather simple, and has dozens of instances when cooperation is needed. In practice, it is the commanders of the two naval forces, the military who conduct cooperation. Various cooperation exercises, for example readiness for an ecological hazard, are part of the Gulf management routine (Benjy 2021). Marine communications are conducted worldwide as usual, on the international marine communications channel.

The Israeli Ministry of Environmental Protection is also cooperating with the Jordanians, and the Egyptian consul, based in Eilat is aware of this. The main issue discussed in the environmental context is docking in the bay. Ships carrying oil demand great care; there is awareness of this concern on both sides of the Gulf. The introduction of an oil tanker into the port of Eilat, a sensitive issue today, is following with security care; the Navy is in charge of the process. Only when a ship has received a permit to enter the port is it handled by the port authorities.

When Bresler (2021) asked Israelis if they call the gulf "Eilat Gulf" in their meetings with the Jordanians, he was answered: "This is a good question … we do not want to hurt the feelings of the Jordanians" (Bresler 2021).

History of Naming the Eilat Gulf

It is difficult to know who was the first to call the site Eilat Gulf. In the past, Hebrew speakers used the biblical term *Yam-Suff*, meaning the Red Sea, to which the Gulf is geographically connected. There is no written evidence of a formal decision, just to the establishment of the city of Eilat in 1952; then, the Geographical Naming Committee chose the biblical name Eilat. It is not unfounded to assume that David Ben-Gurion was involved in the decision to name the gulf after the city and thereby do geographical justice to its being a gulf, and not part of the Red Sea, just as the

Gulf of Suez, the "twin brother" from the west, got its own name. Ben Gurion uses the term "Eilat Gulf" already of 1949 at the government meetings.[5]

The formal list that included the name "Eilat Gulf" is the work of the Geographical Names Committee appointed by the Prime Minister Ben Gurion already in July 1949 to "determine Hebrew names for all the mountains, valleys, springs, valleys in the Negev from Eilat Bay to the Ein Gedi-Gaza line." It was first published in May 1950, and Hebrew names added to the official 1:250,000 map (GovNamesCommittee 29 May 1950).

International agreements use the "Common name" (a term in international law) and its local pronunciation. In Israel's it is the British common name (Gulf of Aqaba) (Biger 2021). Gulf of Aqaba is the name that appeared on British Mandate maps. The name Gulf of Eilat is new; the connection between the name of the sea and Eilat is neither biblical nor has historical roots (Biger 2021).

In the peace agreement with Egypt (1979) Prime Minister Begin agreed to register Gulf of Aqaba, regarding freedom of navigation in the gulf (Serbro 2021). In the peace agreement with Jordan (1994), Amnon Rubinstein, the head of the Israeli peace delegation, following Prime Minister Rabin's directive, agreed to use the term Gulf of Aqaba because it had already been adopted in the agreement with Egypt. Despite this, Dr. Serbro had made three copies of indexes of maps. Gulf of Aqaba was written in the Jordanian copy, Gulf of Eilat in the Israeli one, and both—in the American. Three different copies were signed: In the maritime border agreement with the Jordanians, the term Aqaba Gulf was used again. This time the Jordanians insisted, and it was not allowed to list the Gulf of Eilat in the border documentation either. But in the sea map of *Rosh Hamifratz* (the gulf head) both terms were used (Serbro 2021).

Naming of the Gulf in Maps

Place naming and cartography

As is well known, many people regard maps as sacrosanct products of scientifically accurate work (Medzini 2012) that assumes cartography, based upon satellite photos, mathematical and engineering calculations—an exact science. The public has faith in maps and considers them reliable representations of reality. As political and social scientists know that belief is not true: maps are subjective and interpretive. They may present a pre-ordered reality, which allows their use as an extremely meaningful political tool. Maps express national and local pride and help create national narratives. Cartographers cannot avoid introducing their own subjective point of view, cultural background, or target audience into their maps. Many maps attempt to conceal the existence of the 'other side.' In effect, they represent ideological landscape more than they do the actual one (Medzini 2012). People trust maps and see them as true

[5] Documentation of government Meeting, State Archives.

representations of reality because cartography has an aura of scientific accuracy, stemming from the notion that map and territory are identical. In fact, map drawing is interpretation rather than an exact copying of reality to paper (Medzini 2012). Presence and absence of places on a map are of political significance.

Maps are products of their creators' interests, and they reflect "hidden assumptions of mapmaking" (Wood 2010). This is the case all over the world, especially when it comes to the Land of Israel; in this case the use of maps is a means of establishing the presence of the Zionist state and determining its borders (Leuenberger and Schnell 2010). Ability to use maps to represent an elected reality that goes beyond the physical one turns maps into an extremely significant political tool (Collins-Kreiner et al. 2006). An examination of Israeli and Palestinian maps shows how each side tells its narrative of the conflict from its point of view, while disregarding that of the other side (Medzini 2012).

Transition from paper and atlas maps to digital maps has expanded and deepened the cultural-sociological significance of maps (Meishar-Tal 2014); cartographers are not traditional cartographers employed by governments, and it is not just countries that produce maps. Thus, map production is motivated not by nationality but rather by economics.

Map naming in Google Earth reflects broader political, historical, or cultural disputes (McLaughlin 2008). McLaughlin claims that Google should not choose sides in international geopolitical disputes. For this reason, they "have chosen to implement a uniform policy of Primary Local Usage." Google determined that "if different countries dispute the proper name for a body of water, [Google's] policy is to display both names, with each label placed closer to the country or countries that use it" (McLaughlin 2008). This may be true for Google Earth, but it is not always reflected in Google Maps, at least not in this case. For language clients other than English, Google "displays only the preferred name in the relevant language." "When [their] policy says that [they] display the 'primary, common, local' names for a body of water; each of those three adjectives has an important and distinct meaning. By saying 'primary,' [they] aim to include names of dominant use, rather than having to add every conceivable local nickname or variation" (McLaughlin 2008). Many negative posts against Google, mostly Iranian, question this description. It seems that "this explanation reflects a poorly thought-out and ill-conceived excuse after-the-fact to sweep the issue under the (Persian) rug and to provide Google executives with a face-saving PR talking-point" (Google 26 April 2008). Google uses publications and documents of the United Nations Cartographic Section as authoritative references for naming bodies of water. One source for Google Maps' names is the International Hydrographic Organization (IHO) (Google 26 April 2008, Boorstin 2009).

Similarly, for traditional maps, which have political significance, Google Maps have political expression. This is reflected in many places in the world in general, and in seas and water bodies in particular. This is the case, for instance, in the controversy over the name Sea of Japan, called either the Sea of Korea or the East Sea. Similarly, the Persian Gulf—the Arabian Gulf, received much resonance on websites and incited responses from writers (Taylor 2013; Claughlin 2008, and others).

Israelis and Palestinians use Google Earth to express their spatial narratives and their interests in the conflict. Thus, Google Earth is used as an arena for political activism and as a bridge between the parties. Findings also indicate that Google is becoming a third party in the conflict by using its ability to determine location names and boundaries, and by taking the role of a censor of information published in their domain (Meishar-Tal 2014).

Researchers Rodrigo Ochigame (MIT) and Katherine Ye (Carnegie Mellon University) explain[6] that "Different results from region to region show that the very idea of neutral search engines is a myth. Any attempt to quantify the relevance of results through code entails moral and political priorities." They contend this is the "divide and rule" method of Google searches (Leshem 2021, p. 1972). *Algorithms of Oppression: How Search Engines Reinforce Racism* is the title of a book published in 2018 by Safiya Umoja Noble, a senior researcher at UCL, who sought to prove that Google's search engine offers results that increase social bias (Leshem 2021).

If one searches for Eilat Gulf from Google in North America or in Europe, the result is Gulf of Aqaba; Gulf of Eilat cannot be found on Google. It happens that when looking for Aqaba Gulf, the city of Eilat does not appear at all.

It becomes even more complicated, if one uses Google Earth in Israel; it offers three different options: The English version indicates only Aqaba Gulf, the Hebrew one suggests only Eilat Gulf, even if we search for Aqaba Gulf, and at last, if you do your search in Arabic, there is no Eilat. By the way, the city of Aqaba does not exist even in the Arabic version; this might be just a mistake.

"Trusting Google is understandable. Google tries to create the impression of a benevolent company. Search is not just a technical matter, but a political one … with financial interests" (Ye and Ochigame 2021). Google Maps show different maps depending on the location of the viewer (Ye and Ochigame 2021).

The Dead Sea

The Dead Sea is well known around the world as the lowest place on earth, more than 436 m below sea level. It is a salty inland water body, approximately 80 km long and 12 km wide, at maximum. It is part of the border between the States of Israel and Jordan. The border crosses it right in the middle (see Fig. 5).

No one in Israel calls it the Dead Sea, even though for many years it was common to think that nothing is alive in this sea, due to the high percentage of minerals and salts, which do not allow any flora or fauna to survive. Today, a few species are known to live in the shallow parts of the sea despite the high concentration of salts. In Hebrew, the sea is called *Yam Hamelach* (The Salt Sea) but in the online translation, even if you ask for a translation of Salt Sea, the result is Dead Sea. The Dead Sea has different Arabic names.

---

[6] Ye C, Ochigame R (2021) Search Atlas—visualizing divergent search results across geopolitical borders. https://dl.acm.org/doi/pdf/10.1145/3461778.3462032.

Fig. 5 The Dead Sea

Is it a sea? It is not connected to the world-wide ocean system, it is not a part of seas around the world, its salt percentage is much higher and finally, it is not absolutely "dead".

In various languages, the Dead Sea is referred to by different names, all reflective of its characteristics. The earliest known name is recorded in the Hebrew Bible, where it is referred to as The Salt Sea, due to its significantly high salt content. In the later Roman era, salt was a very highly valued commodity—Roman soldiers were paid in salt rather than money! This is also the source of the Latin word *salary*, which comes from salt.[7] Early translations into foreign languages continued to use the original name Salt Sea; by the Roman era, visitors to Judea began calling it The Dead Sea, as they noticed that the water is devoid of all life forms, whether plants or animals. There are quite a few references to the Dead Sea in the Bible, and it has long been associated with mysticism, wonderment, and religious significance (Dead Sea 2021-(www.deadsea.com)).

The Dead Sea belongs to two countries—Israel and Jordan; the border line crosses the sea in the middle.

The Dead Sea, like the Gulf of Eilat/Aqaba, is located on the Israeli Jordanian border. In both cases, the maritime border poses common challenges for the two countries at the civilian and military levels. For instance, two "Dead Sea factories" in Jordan and in Israel use the same minerals of the sea. The Dead Sea has argent ecological issues and heavy economic interests. Cooperation is mainly not published.

Jordanians called this sea "Dead Sea" (*البحر الميت, albahr almayit*), that's how it is been presented in Google Maps.

[7] https://www.deadsea.com/articles-tips/interesting-facts/why-is-the-dead-sea-called-the-dead-sea/#name.

Fig. 6 Sea of Galilee

The Sea of Galilee

Yam Kineret (Kineret Sea) or *Ha-Kineret* (The Kineret) as known in Hebrew, is the Sea of Galilee for English speakers, and called in Arabic the Sea of Tiberias. It is the one and only lake with fresh drinking water in the land of Israel, located in the northeast (see Fig. 6). Its maximum length is 21 km, and its maximum width is only 13 km[8] (during drought years it becomes even smaller). As it is too small to be called a sea, and its water is not salty as in seas, the Sea of Galilee is not really a sea in any way. It is a small lake (less than 170 km^2). However, throughout history it was called a sea, probably because it is the only sweet water lake in the whole area.

People when asked what the meaning of *Kineret* is, are likely to answer that *Kineret* in Hebrew comes from the word *Kinor* (a violin), due to its shape. In fact, the name appears several times in the bible, but even earlier it was mentioned in Egyptian sources that go back 15,000 years, before the violin in its known shape was invented; thus, this is only a late interpretation.

The Sea of Galilee has a wide meaning in Christianity: Various miracles are reputed to have happened to Jesus and his believers in the Sea of Galilee: "Shortly before dawn Jesus went out to them, walking on the lake. When the disciples saw him walking on the lake, they were terrified. 'It's a ghost,' they said, and cried out in fear" (for the full episode, see Matthew 14:22–33. Compare John 6:16–21). In the Gospels, the Sea of Galilee is often mentioned by several names: Galilee Sea, Ginosar Sea and Tiberias Sea (Barkai and Schiller 1999).

The city of Tiberias is located on the western shore of the lake, was founded by the Romans, and has significance for Christians, Muslims and Jews. The city walls were built by the Arab ruler Dahar al-Omar during the eighteenth century, and his rule at that time made Arabs call the lake after the city. Since it is an ancient Roman

[8] http://kineret.org.il.

name (after emperor Tiberius) that had been in use for thousands of years, before Hebrew gave the lake its name, why is it not in use by the Christian world?

In the 1994 Peace Agreement between Israel and Jordan there is a water chapter and a special annex regarding water affairs. The sources of water mentioned in both relate to the Jordan River, the Yarmouk River and the underground water in the *Arava Valley*. There is no mention of the Sea of Galilee, and probably not by chance (Serbro 2021).

In the southern Jordan River and the Yarmouk River there is basic cooperation between the parties because the international border passes through them. There is no sharing in the Sea of Galilee and "there was no desire to create such a link." Therefore, it was not mentioned. Although there is a reference to the transfer of additional water to Jordan, the name of the Sea of Galilee does not appear as a source of water (Serbro 2021).

Until 1967, the Israeli border line with Syria was near the east bank of the Sea of Galilee. Since then, Syria is only about 12 km away, but it has nothing to do with the water agreements signed between Israel and Jordan.

The Education System's Role in Using Geographical Names

What is the role of the education system in strengthening national feelings? Education is a widespread means for shaping the consciousness of the young generation, the future citizens of the country. The geographical map of the State of Israel must hang in every classroom, by law. Students are familiar with the common geographical terms from their early childhood; this map tells the Jewish narrative of the land; it appears in the names of landscape units, cities, and seas. Arab schools have the same map, although it is translated to Arabic. Sometimes the atlases used in Arab schools were bought from foreign Arab publications, which are significantly cheaper than the atlases of Israeli publications; thus, terms appearing in the atlas are not Israeli.

It may not be surprising that maps and geography textbooks represent the national narrative; yet here the case is different, since the national narrative is not always in use by the State, but, nonetheless, it is strongly preserved in the educational system. It is also surprising that in the translation to Arabic not all the terms have been translated consistently; Gulf of Eilat is written in its Hebrew term while the Sea of Tiberias and the Dead Sea are presented in their common Arabic form. Why do these differences occur? As explained above, the maps are being used to confirm Israel's reassurance of its sovereignty in these two places, but the attitude to the third one is not the same.

The use of different terms for the same sites has been well researched and published (Yaar-Waisel 2021). When there are conflicting political approaches, teachers use different terms at the same time; for example, the terms Judea and Samaria and 'the occupied territories' are used interchangeably. Confusion and use of different terms were found to characterize geography education regarding the Land of Israel's map also in Israeli-Arabic schools, that use Sea of Tiberias and the *Kineret* Lake, at same time.

Discussion and Summary

Names of places in Israel are complex, loaded with historical, religious, national, and political interpretations. Names of water bodies in Israel do not make things simpler. On the contrary, every water body has an ancient historical source and a not-so-simple contemporary reality, with different interpretations. It can be said, with a bit of humor that The Gulf of Eilat is not Eilat's gulf, the Dead Sea is not dead, and the Sea of Galilee is not a sea.

The reasons for the complexity of names are varied: The revival of Hebrew, Zionist settling in the Land of Israel and the desire to preserve the history of the Jewish people in its territory and cultural landscape are the primary factors in the preservation of ancient Hebrew names. This is why the name Eilat was given to the Gulf, although the name was not accepted anywhere else in the world. The same is true for the naming of the northern lake *Kineret*, even though the Sea of Galilee is common throughout the world. Lastly, a mixture of the Israeli-Arabic every-day language is reflected in geography teaching, too.

Despite implied rigidity, Israel has often been found to be flexible, patient with its neighbors and refraining from hurting their feelings, because of a desire for good neighborliness and at the same time fear for the fragility of peace agreements. Even when signing peace agreements, and in day-to-day conduct, both in the military and civilian spheres, Israelis refrain from calling the gulf Eilat Gulf, to avoid issues with the neighbors and use the term that is common outside of Israel.

Military responsibility for managing the sea may seem unique, since Jordan and Israel have been in a peace relationship for many years. Nonetheless, there are different seas around the world managed by security units like coast guards, in charge of preventing maritime terrorism, stopping immigration, blocking smuggling and so on. It happens in the Mediterranean, the Aegean and elsewhere.

Use of a name is influenced by geopolitics and, conversely, affects it vice-versa. Cooperation in the Red Sea would not have been possible without open or covert recognition of the importance and status of Aqaba as the important port city in the Gulf. The volume of economic activity is significant on a day-to-day basis, whether it is political strategic thinking or understanding of the people on the ground. This every-day conduct is manifested first of all in economic activities, in each of the seas we dealt with. Economic interests affect Israeli publications to encourage tourism in the Sea of Galilee: Pilgrimage has a large share in tourism in the area and is also the one that leaves more money. Israeli use of the Christian name Sea of Galilee is a clear outcome.

Today's technology makes it possible for any person, or any interested party to produce and design maps and to distribute them easily across the globe via the Internet (Sungjae 2021) In this reality, it is very easy for any organization to use maps to shape political attitudes (Medzini 2012). Atlases have always been shaped by the assumptions and interests of their makers. Today's digital maps are no different (Ye and Ochigame 2021).

Google as one of the world's leading online web mapping services, is also trying to walk the impossible tight rope between political bickering of countries (Dempsey 2012). Concentration of power in technological infrastructures has become a matter of public concern. Such infrastructures, including search engines, seem to play a key role in the spread of false information and hate speech (Ye and Ochigame 2021). Ye and Ochigame are developing an atlas for web search, which will show results as they are obtained for the same search in different places around the world. At the end of the process, the atlas will probably enable to identify that an existing Eilat Gulf does not exist, except in the Hebrew language.

It was in 2010, after the widely published controversy following the naming of the Persian Gulf/Arab Gulf by Google, that Robert Boorstin, the director of Google's public policy team states, "We work to provide as much discoverable information as possible so that users can make their own judgments about geopolitical disputes" (Google Public Policy World 2008). Yet, despite this approach, Google has been targeted by various nations over the designation of disputed geographic areas (Dempsey 2012). The use of Google names results from the number of users and the number of estimated 'shouters'. In the Israeli case, there are few users in relation to the number of Arab users and the number of protesters is therefore lower (Biger 2021). It can be concluded that the naming of Eilat Gulf is just one more maritime dispute and in a small scale and low flames, mainly because Israel is not interested in causing problems for Google.

The UN Geographical Names Committee is the one that determines, but its decisions are significant only for UN member states. The decisions of this committee do not bind individuals who suggest a name, prepare a map or an atlas and they can choose any term they want (Biger 2021).

The educational message derived from this study is not unequivocal. There is no consistency in naming in textbooks and in maps translated into Arabic. The determination of the education system to preserve Hebrew terms is in conflict with the tendency of the Foreign Ministry and military personnel that incline to display more flexibility, so as not to hurt the neighbors' feelings. The Foreign Ministry, which has a broad, global outlook, is not delighted with changing names to Hebrew terms (Biger 2021).

Perhaps the controversy is not important at all. It seems to Israel that the continuation of cooperation with Jordan and Egypt is more significant, and therefore Israel is less concerned about the name of the Gulf. Historically, until 1949 this Gulf was not "Israeli", although today Israeli military superiority in it is evident to all. Yet, from an economic-maritime perspective Israel has much to learn from its neighbour Jordan; "recognition" or "honour" of joint ownership of the Gulf is not important enough to harm the delicate and fragile situation in this area.

This study indicates that economic and political interests are stronger these days than the meaning of words. In contrast to the first decades after the establishment of the State of Israel, the name of a geographical place is less meaningful, while the economic potential matters more, as it is much more important to ensure stability.

So, whose gulf is it? History, economic activity today, the large population of the city of Aqaba and above all, geopolitical conceptions of non-governmental (especially Google) and governmental organizations today, are what determine that the Gulf of Eilat is not really the gulf of Eilat.

Questions

- What do the names of the seas in the Land of Israel mean?
- What are the overt and hidden meanings of sea names and water sources in Israel?
- How are religions, politics and nationalism involved in naming places in Israel?
- What are the names used, de facto as opposed to de jure, in Israel and why?

References

Azaryahu M, Golan A (2001) (Re)naming the landscape: the formation of the Hebrew map of Israel 1949–1960. J Hist Geogr 27(2):178–195

Barkai G, Schiller E (eds) (1999) The Sea of Galilee and its surroundings in Christian Sources. Ariel - Encyclopedia for Israel, Jerusalem

Benjy R (2021) Lieutenant Colonel, Israeli Navy (retired). An interview, 2 July 2021

Biger G (2021) Professor (Emeritus) of Historical Geography. An interview, 15 July 2021

Boorstin B (2009) Director, Google's Public Policy Team, when sources disagree: borders and place names in Google Earth and Maps. Google Public Policy Team Blog, 4 Dec 2009. https://publicpolicy.googleblog.com/2009/12/when-sources-disagree-borders-and-place.html

Brawer M (ed) (2014) University Atlas, Yavne

Bresler G (2021) Former Environmental Protection Officer. An interview, 3 July 2021

Claughlin A (2008) How Google determines the names for bodies of water in Google Earth. https://publicpolicy.googleblog.com/2008/04/how-google-determines-names-for-bodies.html

Collins-Kreiner N, Mansfeld Y, Kliot N (2006) The reflection of geographical conflict in mapping: the case of Israel's borders. Middle East Stud 42(3):381–408

Dead Sea (2021) A tourism site. https://www.deadsea.com/articles-tips/interesting-facts/why-is-the-dead-sea-called-the-dead-sea/#name

Dempsey C (2012) The politics of Google's mapping. GIS Lounge, 18 May 2012

Fine T, Segev M, Lavie R (2007) Israel: man, and space. Tel Aviv: Centre of Technological Education, Tel Aviv (Arabic version)

GovNamesCommittee (29 May 1950) *Rashimon*—list of geographical names in the southern issue of the map of Israel (Gulf of Eilat to the Gaza line—Ein Gedi), scale 1: 250,000, ISA-PMO-GovNamesCommittee-000dgna

Kadmon N (2000) Toponymy: the lore, laws and language of geographical names. Vantage Press

Leshem E (2021) Why does the same search word get different results in Google around the world? Haaretz, 12 July 2021 (Hebrew)

Leuenberger C, Schnell I (2010) The politics of maps: constructing national territories in Israel. Soc Stud Sci 40(6):803–842

McLaughlin A (2008) How Google determines the names for bodies of water in Google Earth. Google Public Policy Blog. https://publicpolicy.googleblog.com/2008/04/how-google-determines-names-for-bodies.html

Medzini A (2012) The War of the maps: the political use of maps and atlases to shape, national consciousness—Israel versus the Palestinian Authority. Eur J Geogr (EJG) 3(1):23–40

Meishar-Tal H (2014) Google Earth as a reflection of cultural changes and socio-spatial processes in the digital age. Int J Inf Syst Soc Change (IJISSC), IGI Glob 5(3):1–11

O'Reilly G (2019) Aligning geopolitics, humanitarian action and geography in times of conflict, vol 2. Springer International Publishing, Switzerland, pp 9–27
Rose-Redwood R, Alderman D, Azaryahu M (2009) Geographies of toponymic inscription: new directions in critical place-name studies. https://doi.org/10.1177/0309132509351042
Serbro H (2021) Head of the Joint Committee of Experts on the Israel-Jordan Border, Director General of the Israel Mapping Center (2003–2012). An interview, 13 June 2021
Shahar Z (2021) Colonel (retired), former Red Sea Arena Commander. An interview, 12 July 2021
Sungjae C (2021) The name East Sea in the digital era. In: Geographical naming in the digital era II proceedings of the 27th international seminar on sea names, Gangneung, Korea, 14–16 Oct 2021
Taylor A (2013) The 11 most controversial places on Google maps. 16 May 2013. https://www.businessinsider.com/most-controversial-places-on-google-maps-2013-5
Wikipedia (2021) Gulf of Aqaba. https://en.wikipedia.org/wiki/Gulf_of_Aqaba
Wood D (2010) Rethinking the power of maps. The Guildford Press, New York
Yaar-Waisel T (2021) Teaching geographical boundaries: from challenge to model—the case of the Israeli education system. Bord Glob Rev 2(2):52–60
Ye C, Ochigame R (2021) Search Atlas—visualizing divergent search results across geopolitical borders. https://dl.acm.org/doi/pdf/10.1145/3461778.3462032

List of Interviews

Benjy R (2021) Lieutenant Colonel (retired) Israeli Navy, 2 July 2021
Biger G (2021) Professor (Emeritus) of Historical Geography, 15 July 2021
Bresler G (2021) Former Environmental Protection Officer, 3 July 2021
Shahar Z (2021) Colonel (retired), former Red Sea Arena Commander, 12 July 2021
Serbro H (2021) Head of the Joint Committee of Experts on the Israel-Jordan Border, Director General of the Israel Mapping Center (2003–2012), 13 June 2021

United Nations Capacity Building in Toponymy

Ferjan Ormeling, Helen Kerfoot, and Pier-Giorgio Zaccheddu

Abstract The United Nations Group of Experts on Geographical Names (UNGEGN), one of the nine permanent expert bodies of the Economic and Social Council of the UN, promotes the national standardization of geographical names, as an essential building block for international toponymic standardization and the production of gazetteers, the creation of geographical names databases, all necessary for location-based services. This requires rather specialized training, unavailable in most countries, and that is why UNGEGN has developed training material, on-line courses and contact courses in order to build on a national level the required capacities for geographical names collection and processing, the creation of names databases and their validating, and the distribution of the standardized names. Since 1982 UNGEGN has regularly engaged in international training courses in toponymy in developing countries. This chapter focuses on the necessary training material to be made available, the structure of these courses, the required expertise of the lecturers and the teaching conditions, the criteria for selecting suitable fieldwork areas, and the necessary hardware and software for student participants. The whole process from collecting the names in the field to their incorporation into databases and their publication on maps and in web applications will be described by the lecturers who have recently been engaged in these courses.

Keywords Geographical names standardization · Toponymic teaching material · UNGEGN · Spatial Data Infrastructures · SDI · INSPIRE

F. Ormeling (✉)
University of Amsterdam, Amsterdam, the Netherlands
e-mail: f.j.ormeling@uva.nl

H. Kerfoot
Honorary Chair UNGEGN, Ottawa, Canada

P.-G. Zaccheddu
Head of Executive Unit—International Affairs, Federal Agency for Cartography and Geodesy (BKG), Frankfurt am Main, Germany
e-mail: pier-giorgio.zaccheddu@bkg.bund.de

© The Author(s), under exclusive license to Springer Nature Switzerland AG 2023
G. O'Reilly (ed.), *Place Naming, Identities and Geography*,
Key Challenges in Geography, https://doi.org/10.1007/978-3-031-21510-0_8

Objectives

- To show how the educational infrastructure of UNGEGN developed and evolved over time.
- To display what UNGEGN considers the necessary content of its teaching programmes, and what expertise is to be transferred.
- By the end of this chapter readers should be able to infer how the UNGEGN Working Group on Training Courses in Toponymy operated until now (2020) and what students can expect when enrolling in one of its programmes.

United Nations Capacity Building in Toponymy

Objectives

This chapter will show how the educational infrastructure of UNGEGN developed and evolved over time, what is considered the necessary content of its teaching programmes, and what expertise is to be transferred. By the end of this chapter readers should be able to infer how the UNGEGN Working Group on Training Courses in Toponymy operated until now (2020) and what students can expect when enrolling in one of its programmes.

UNGEGN Set Up Training Programmes

Starting from the conviction that geographical names play an essential role in spatial information transfer and navigation, the need for standardization of the spelling of toponyms is obvious. It is also obvious that the task of standardizing a nation's toponyms falls to some national institution. But where should this national institution get the expertise to do this? The task to collect toponyms in most countries falls to the national mapping agency, which usually collects all the other spatial data. But in the curriculum for those who collect and process spatial data, toponymy is seldom to be found, as it falls outside the scope of the technical expertise transferred.

With a few exceptions, it is only in large countries that dedicated training facilities, such as the former WTUSM (now part of Wuhan University in China), ENSG in Paris, or the Moscow-based MIIGAiK, could emerge, and also could provide specialized training in toponymy. For other countries, toponymy very much was a subject either neglected or taught in in-house training, because, first, it is a very specialized subject, and secondly, each country would only need new toponymists once in a decade, and that is no basis for setting up a specialized training programme.

The lack of training programmes brought with it a lack of teaching material. Apart from the general introduction to the subject of toponymy by the Australian geographer Aurousseau (1957), which resulted from his wartime experience in map production, there were no manuals, if we disregard the in-house lecture notes produced by training institutions like ENSG. Aurousseau's introduction was only updated and extended in

2001 by Kadmon (2001). So in the absence of training facilities and training material, UNGEGN organized them itself.

How Did UNGEGN Training Evolve Until 2020?

Contact Courses

When UNGEGN became operational the need for capacity building was immediately felt and at the Second United Nations Conference on the Standardization of Geographical Names (UNCSGN) in 1972 the resolution II/18, *Technical training of personnel*, was adopted (see Ormeling 2017). It called for pilot training courses in toponymy, and at the 5th UNGEGN session in 1973, a draft programme for such a course was presented. A working group was set up to realize such a course, and its convenor was able to engage the support of the government of the Netherlands, which proved to be willing to pay half the costs, provided the other half would be paid by another party, such as the UN. A try-out for the course was held in Suriname that year, including fieldwork in the upstream Tapanahony area, chosen in order to test operational procedures in the field collection of names. As the UN proved unable to finance the lecturers and their travel costs the working group was back to square one.

The working group was reinstated for the next 1977–1982 term, and changed its approach: instead of a course in the Netherlands, it decided either to develop a programme of travelling experts that would go on a lecturing trip through countries eager for support, or to hold courses outside Europe and North America, on a regional basis, in the centre of a group of countries in need of toponymic training. The working group opted for the latter model, and was able to find a country willing to host a course, providing the meeting facilities and accommodation, while the lecturers covered their own travel costs.

The country that was able to break the financial and organizational deadlock after 15 years was Indonesia—its national mapping organization Bakosurtanal consented to organize a course for the UNGEGN Asia South-East and Pacific South-West Division in June 1982, at no costs to the UN, as the seven foreign lecturers were paid for by their own departments. This was the Cisarua Pilot Course in Toponymy (see Box 1), held in 1982 just before the Fourth UNCSGN. In its resolution IV/6, the Conference recommended holding similar courses in other UNGEGN divisions, asking the UN to contribute financially, in order to get more potential participants to these courses. The UN reacted by budgeting $25 000 for a second course, preferably a course held in the Arabic countries.

Box 1: Programme of the 1982 Cisarua pilot course in toponymy, following the guidelines agreed to during UNGEGN's 6th session

June 7: Opening,
Toponymic terminology
Functions of geographical names
Global distribution of languages and scripts
UN activities on the standardization of geographical names
June 8: General problems of national names authorities
Writing systems
Conversion on non-written languages
June 9: Geographical name problems in Indonesia
Exonyms
Glossaries and national gazetteers
June 10: Technical visit to Hydrographic office and the Topographic office at Jakarta and to Bakosurtanal in Cibinong
June 11: General and specific problems encountered in field collection of names
Field collection of names in Indonesia
The role of office procedures in standardization of geographical names
June 12: Standardization of geographical names in Australia
Automatic processing of geographical names
June 14–16: Practice: field collection of geographical names
June 17: Discussion: field collection of geographical names. Diplomas; Closure

In 1985 a second chance presented itself: it was Morocco that consented to organize a course for the Arabic Division in Rabat, December 1985, at the premises of the Département de la Conservation Foncière et des Travaux Topographiques, with the help of the renowned Moroccan toponymic expert Dr Abdelhadi Tazi. The programme was comparable to that of Cisarua, albeit without fieldwork.

In 1986 the new convenor of the working group was made aware of a parallel network of toponymy courses, given in Spanish, organized by the Pan-American Institute of Geography and History, so the next year he participated in the PAIGH Toponymy course held in Panama, and at the Fifth UNCSGN was instrumental in having resolution V/10, recommending collaboration between PAIGH and UNGEGN, accepted. He was also able to organize a French-language international toponymy training course in Québec, Canada, the next year. This course differed from the two previous ones by the inclusion of topics such as language contact, relationships between toponymy and history, name changes and the creation of new names in 'empty' spaces. Generally, the programme was more culturally oriented, as opposed to the more practically-oriented Spanish-language PAIGH courses and the more theoretical and technically oriented English-language UNGEGN courses. But in all three course types, the inventory of geographical names in the field and

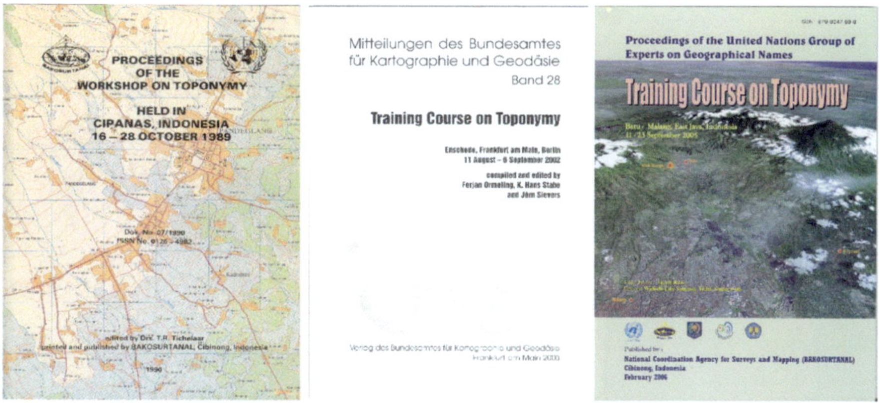

Fig. 1 Examples of proceedings of UNGEGN toponymy training courses in Indonesia (1989, 2005) and Germany (2002). *Source* Ferjan Ormeling

their processing in the office formed the central focus. For many UNGEGN courses proceedings have been published (see Fig. 1).

Apart from these toponymy courses for people from developing countries, the UNGEGN Working Group also was engaged with promoting university courses in toponymy. Resolution IV/5 already had called for university courses in toponymy, and universities in Vienna, Zürich, Dresden and ITC-Enschede had complied and provided such courses. After resolution V/30 calling for more university courses was accepted in 1987, more universities started creating such programmes, for example Université Laval in Canada, and universities in Jerusalem and Nanjing. In the discussions at the 14th UNGEGN session in 1989, there was a difference in opinion between those advocating a more practical, technical approach and those who wanted to include more cultural aspects.

At the 15th UNGEGN session a new convenor was chosen, and Canada reported on the kit it had developed for training courses. It consisted of fieldwork manuals and rules and regulations used in Canada and has been a most welcome addition handed out at these courses. In 1992, just before the Sixth Conference, UNGEGN and local institutions organized a toponymy course for Southern Africa in Pretoria, the first one in Africa south of the Sahara. This was more scientifically oriented than the previous ones, with emphasis on etymology, semantics and research. At the 1992 Sixth Conference, again a resolution was accepted (VI/13) that called for financial assistance from the UN for toponymy courses. It had the desired effect as since then the working group has been able to offer a regular programme of toponymy courses all over the world (see Fig. 2), supported locally by **universities**, like in Pretoria, Enschede, Bathurst, Timişoara, or by **national mapping organizations**, such as in Indonesia, Germany, Burkina Faso, Brazil, Tunisia, Madagascar, Austria, Mozambique and the Philippines, or by **Ministries**, like in Sudan, or by **research institutions** like in Kenya. The working group has held these courses in 12 of the 24 divisions of UNGEGN, and has had a fair coverage of the globe, with the exception

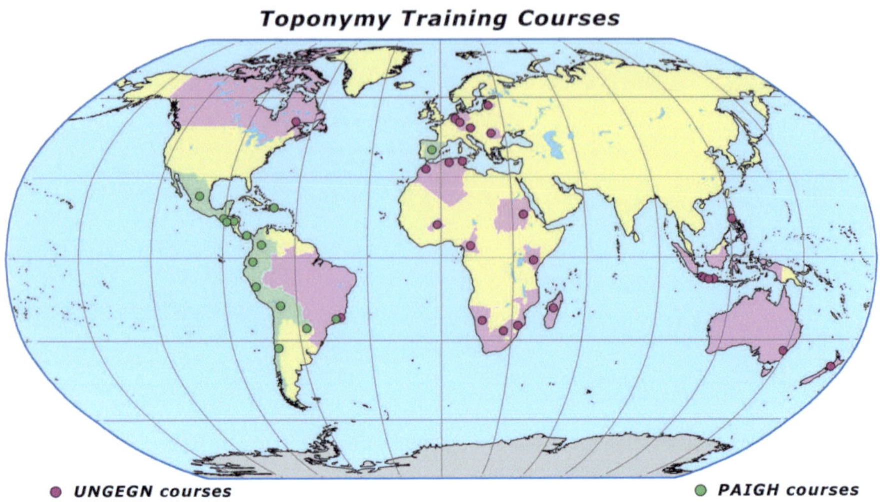

Fig. 2 Locations of UNGEGN and PAIGH international training courses on toponymy, as of 2022 (https://unstats.un.org/unsd/geoinfo/UNGEGN/docs/Training/Manila/day%202/01_KERFOOT_Philippines_UNGEGN_HK%20final.pdf)

of Central America and Andean South America (an area served by PAIGH) and continental Asia. The toponymy course organized in Vienna, however, especially targeted Central Asian countries.

Web Course

As early as 1998, in resolution VII/9, the Working Group was urged to consider the development of an internet-based training course in toponymy, and this was reiterated in resolution X/11 (2012) *Web-based course in toponymy*. The reason for developing this web course was the reality that only a limited number of people could be reached through the contact toponymy courses; moreover, job switching by employees of names bureaus to other departments meant a further reduction in the usefulness of these courses. That is why some recent in-person courses were directed at training the trainer. Developing a web course could also play a role in preparing course participants for in-person courses, so that they all would have a common starting level. So, from 2002 onwards, a web course was being developed, first with the support of Eötvös Lorand University in Budapest, and consecutively with support of Utrecht University and of the International Cartographic Association. As always, the problem has been procuring the funds for its production, and here the United Nations Statistics Division has supported the working group again . So, since 2012 the course has been

Fig. 3 Toponymy training manual and other resources provided by UNGEGN (available at https://unstats.un.org/unsd/ungegn/pubs)

up and running (https://unstats.un.org/unsd/geoinfo/UNGEGN/docs/_data_ICAcourses/index.html). It needs regular maintenance, though, and funds are needed for that as well.

Manuals

The 2002 conference resolution VIII/15 *Support for training and publications*, called for the production and publication of a manual on Romanization systems (see UNGEGN 2007) and of a basic manual on geographical names standardization (UNGEGN 2006). These were realized prior to the Ninth Conference in 2007 and form a perfect combination with the web course. As a matter of fact, the web course frequently refers to these manuals, as well as to the *Glossary of Terms for the Standardization of Geographical Names* (UNGEGN 2002, 2007).

To provide further reading for those engaged in the standardization of geographical names, since the Tenth Conference a so-called advanced toponymy manual has been produced by an international team of experts. It is called the *Toponymy Training Manual* (Ormeling et al. 2017, see Fig. 3), and can be downloaded from the UNGEGN website as an e-book, the chapters of which can be downloaded individually as well. The format has the advantage of being easily updated.

With What Aspects of Names Collecting and Processing Must Toponymy Staff Be Familiar?

Fieldwork Preparation

Fieldwork is a labour-intensive and expensive aspect of names standardization. In principle each area is only visited once for names collecting by topographers or

Fig. 4 Village elders in Mozambique, 2006 (left) and in Madagascar, 2013 (right). *Photos* Ferjan Ormeling

toponymists, so during this one time visit all pending names issues should be solved, and all spelling uncertainties should be eliminated. So, through study of all sources relating to the fieldwork area differences in spelling should be ascertained, and names claims should be studied. Fieldwork forms an essential part of toponymy courses, as it confronts the participants directly with the inhabitants who use the collected names on a daily basis. They will find that different subgroups of the local population (based on gender, age, social status) might use different name forms for the same object, that not all the features to be named on the map bear local names, and that many local names exist for feature categories that will remain unnamed on the map.

The selection of the fieldwork training area aims at finding places with sufficient population density to find local informants and with sufficient variation in hydrography and orography. It would be a bonus to find an area where discrepancies exist between local and official names, so that the students can identify pending naming issues.

The visit to the fieldwork area should be planned in advance, by identifying and contacting persons knowledgeable about the area's history, and people in public administration, who deal with the area on a daily basis (see Fig. 4). Prior warning of the arrival of a names collecting party intent on recording the people's cultural heritage, might also induce locals to react in a more friendly way and keep their dogs at bay.

Fieldwork

For the actual fieldwork one needs to have available maps and aerial photographs of the fieldwork area, a GPS for assessing coordinates of topographic objects, a mobile device for recording pronunciation by the local population, and notebooks. A toponymy course could also include some guidelines for interview techniques. Course participants should never ask leading questions, such as 'what is the name of that mountain' but rather refer to the various elevations around the persons interviewed and leave it to them whether to assign the status of hills or mountains or peaks, etc. Timing the interviews is also an important aspect. It would be essential to reach those most likely to be knowledgeable about names of surrounding features, so harvest times or religious ceremonies should be avoided for the interviews.

The toponymist must learn how to indicate the names for features on the map, so that cartographers will later link the proper names to the named features. For the final map, a predefined density of named features will be sought; in most cases the small scale of the final map would prevent all collected names being inserted.

Frequently, the interviews will have to take place with the help of interpreters speaking the local language or dialect. Toponymists must also assess whether the opinions regarding names are carried by more than one person: preferably a group of local persons would be interviewed simultaneously so that any differences in opinion would be solved locally at the time.

Fieldwork in areas that have been surveyed already in the past, also has the function to check whether the names on the map are still valid, so older editions of topomaps should be brought along.

The result of the fieldwork would be lists of names with attribute data (location, type of feature, orthographic variants, language, meaning, etc.) or preformatted files in which, for every toponym, attribute data have to be filled in. Audio files with the pronunciation of the toponyms by locals can be added. These lists or files will be turned into name records in the next phase, office treatment.

Office Treatment

As part of the process of collecting, processing, storing and disseminating geographical names data, "office treatment" is an essential step in establishing accurate and unambiguous toponymic records.

Emphasis is put on care and consistency to establish data that will become authoritative and ready for further use. At this time students are also introduced to terminology relevant to standardization of spelling and application of the toponyms to the landscape.

Results from fieldwork may be integrated with other records, such as cartographic materials, archival records, existing office information and administrative region data (see Fig. 5). Depending on the country and the language issues involved, specialists in writing and pronunciation of particular languages and dialects may need to be consulted to interpret, clarify or correct information during the time records are being reviewed for permanent inclusion.

Before individual records are reviewed for accuracy, clarity and completeness, discussion takes place with students about the purpose of the database and the fields of information that should be included. For example, at the very least a record would include: the toponym, the feature type, the coordinates of at least one point, and the status of the record. This will involve a determination of feature types (such

Fig. 5 Course participants (Vienna, 2006) preparing records. *Photo* Helen Kerfoot

as mountain, river, swamp, town) suitable for a particular region, and whether they would be better labelled more generally as highlands, flowing water, vegetation, populated place, etc. The assignment of coordinates will also be questioned, leading to discussion as to where the point will be selected, and the possibility of multiple coordinates and determination of the extent of a named feature. Status, in its simplest form, could be "approved", "not approved" and "formerly approved" (historic), and will link to the concept of authorizing the records and including a field for approval date. Students usually suggest a number of other fields they think to be suitable (for example, other names for the same feature—allonyms, administrative region, language, meaning and origin of the name). In addition, inclusion of photos and pronunciation (spoken or IPA) are often requested.

In reviewing the individual records that students have collected and possibly elaborated in the office, the main aim is to make them consistent and complete for presentation to a names board for a decision on their suitability for approval. For example, the geographical name should be written appropriately in lower-case letters, with capital (upper-case) letters, hyphens and diacritics as in use, or as required by language rules. Alternative spellings of the toponym, or other names (in the same or different languages) supplied or found on maps should be added to the record (in an Excel file or in the existing database). Each place or feature should be illustrated appropriately on an existing official topographic map. In some cases, the names will be newly recorded, in other cases they may be recommendations for a name change to match current local usage. (In addition, names on a map that are no longer found to be in use may be listed for the names board to consider rescinding them.)

Current trends to crowd-sourcing data (volunteer geographical information) may provide advantages in gathering toponyms, but review of the information and its reliability will increase the workload of the toponymists in the office.

Processing in the office, preferably by an established Secretariat, includes preparation of documents for the meetings of a names board, but also involves steps after the meetings. This may include preparing documents and summary lists for the signature of a Minister of a government department, listing the board decisions for inclusion in a gazette service, or preparing notifications for public circulation. The work of a Secretariat and the advantages of having one are presented and discussed with students.

It is key that office treatment progresses in a timely way, taking into account the need for accuracy in geo-referencing and paying careful attention to the cultural value of names. Good records are beneficial for decision-making and for the use of future generations.

Authorizing Geographical Names

To authorize standardized spelling and application of a toponym, a decision-making process is necessary, with robust principles and policies to follow. In many countries, the national mapping authority fills this role, but usually only focuses on areas where maps are being updated. At the same time, military mapping or hydrographic charting may be making independent name decisions. To have authoritative nationally-accepted names available for government and public use, UNGEGN recommends that some form of national names authority be established. There is no universal standard for this, as its form will vary depending, for example, on a country's size, population distribution, languages, administrative units and political arrangements. Basically, however, we review a centralized approach with a names board, supported by a secretariat, and possibly with various sub-committees and advisors. Examples include Estonia, New Zealand, and Hungary. We also discuss an alternative of a decentralized approach, where various regional committees have the decision-making responsibilities and feed their data up to a national board. Australia, Canada and Malaysia illustrate this type of process (see Fig. 6).

During the courses, time is spent discussing possible steps in establishing a board, the need for the board (or "committee", "commission" or "council") to have a legal framework, how broad is the mandate, and the possibility of having a names law (as, for example, in Norway). The importance of a Secretariat to a board is emphasized and its functions discussed. Various details about a board and its responsibilities are looked at more closely, such as how many members are required and what sort of backgrounds they should have, the terms for the chair, the frequency of meetings, volunteer versus paid participants, types of advisors that could be needed, and so on.

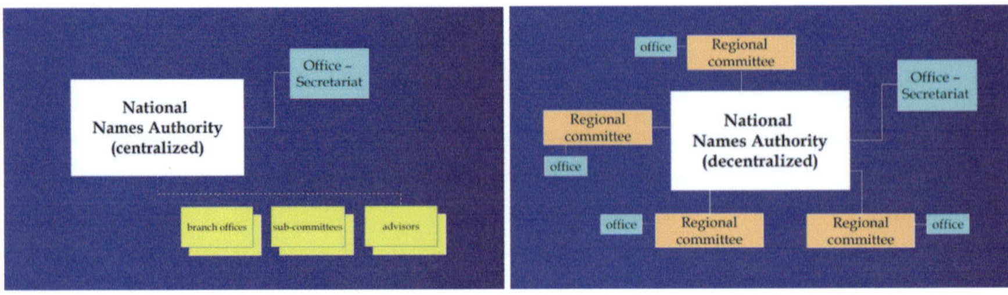

Fig. 6 Structures of geographical names boards: centralized (left) and decentralized (right) (https://unstats.un.org/unsd/geoinfo/UNGEGN/docs/Training/Manila/day%201/04_KERFOOT_Philippines_National%20agencies_final.pdf)

An important step in establishing a names authority is the development of policies and procedures to be followed. This will include the possibility of authorizing names from different language groups. Will more than one name be recognized for a single entity? Will they have equal status, or will there be different occasions to use particular language names? The concept of local usage of toponyms (rather than government-imposed names) is important to address, as are questions of pejorative or other types of unacceptable names.

In some courses, a mock board meeting has been organized. From the fieldwork exercises, teams have selected a few names they collected and have prepared and presented them to a board for possible authorization. Questions from board members often underline the need for names records and their submission to be clear and complete!

Other questions arise regarding boards. For instance, how much evidence should there be for approval of a toponym? When is a name change justified? Our board exists but it is not functioning, how can we fix this? Of course, answers depend on the situation and it becomes clear that there is no one solution that can be applied universally. What is very important is an appreciation that a board should be providing unbiased judgement and the decisions that it makes reflect its integrity and effectiveness.

Today's issues of public resistance to names considered to be colonial in origin, or those not reflecting local usage or respecting minority languages make the need for a board more significant, particularly to take the public pressures off the Government ministers.

In this section, or elsewhere in the schedule, the challenges of setting up a national geographical names programme are considered, which leads to presentation and discussion of the benefits of nationally standardized toponyms and examples of problems that have occurred when standardized names are not available in emergency situations.

With What Aspects of Databases, Gazetteers and Name Servers Must Toponymy Staff Be Familiar?

Theoretical and Practical Aspects of Database Training

After the geographical name information has been gathered in the field and authorized in the office, this information has to be stored and prepared for dissemination. Several methods exist for storing the data, from the paper cards in the past to the advanced digital techniques today using databases.

This section of the training was structured into a theoretical "teaching" part and into a "workshop (hands-on-experience)" part, where the students

- learned the basics about the layout, structure and characteristics of a geographical names database;
- inserted the data from the field collection into a (simple) database;
- performed queries in the database;
- created a gazetteer from the database; and, if applicable,
- created a map with the geographical names in a Geographical Information System (GIS).

Generally, before the design of a geographical names database is started, the following questions have to be answered:

- What is the purpose of your database?
- What tables do you need in the database?
- What (attribute) fields do you need in the tables?
- What are the fields in the tables with unique values?
- What are the relationships between tables?

After the purpose of the geographical names database has been determined, the structure of the database becomes crucial. This means that the database schema or model, the tables and fields have to be selected and their relationships to be set.

After the creation of the database, students need to evaluate and assess the structure and content of it and—if needed—refine the design, add data and create other database tables and fields.

The "workshop (hands-on-experience)" was dependent on the technical IT infrastructure available at the training locations. The exercises with the database application software aimed at demonstrating how the database could be used and how the database content could be imported into other systems, like GIS. Furthermore, it was demonstrated how to display names from the database by using Google Maps/Earth or other applications. In a nutshell, with the Google Maps application fieldwork results can be quickly and easily displayed without the need to use any database software or GIS tool. Alternatives might be Bing Maps or Yahoo Maps. Most often 'OpenStreetMap (OSM)' is used to publish a map which shows data collected by others in a geographical context.

Aspects of Designing a Geographical Names Database

Today, geographical names can be gathered into sub-national, national, regional or global databases to make them accessible to the widest possible audience, often through spatial data infrastructures (SDIs). A geographical names database should serve many purposes, particularly if linked to other spatial data, as a geographical name is very often the first entry point to search for geospatially referenced information, for example, pertinent to natural disasters, and is of key importance for socially relevant issues such as stable living conditions, climate change and land use.

In addressing geographical data needs to analyze those relevant issues, our focus has been put on the geographical names data itself: What attributes should be included? How will data from different sources be combined? What data quality (e.g. accuracy, consistency, completeness) will be guaranteed? How will the data be kept up-to-date? How will the data be accessed?

With technology development, local authorities have started to create their own geographical names databases to meet/fulfil their specific local needs. In the courses the students reflect on the issues for designing a national geographical names database, which will be accessible for international purposes, too.

When databases were first created, it was invaluable for the production of printed 'gazetteers' and for map production, streamlining the creation of both products. Today, geospatial databases serve a broad set of needs and the national data is included in many applications (e.g. GPS and smart phone applications). Key to the transformation from one comprehensive attribute-based file to a geospatially-based, relational system of files—thus using a database system—was the inclusion of unique feature/object identifiers (UID). This means that in the database system one 'named place'/geographical object may be associated to one or more geographical names (endonyms) in different languages. An example from Germany is the named place of the 'City of Bautzen' which is associated with the geographical names 'Bautzen' (in the German language) and 'Budyšin' (in the Sorbian language). Both names have equal official status and are, therefore, treated equally in the German geographical names database. The UIDs unambiguously combine the information.

Concerning the content of the database, the following object/feature types might be included: settlements, transport, vegetation, water bodies, relief areas, etc. This is related to the decision about the respective attributes (see Fig. 7) associated with the objects/features. Examples are: administrative division (statistical key number), stream system (hydrological area code number), population size, status of the geographical name, language of the geographical name, height (height in metres above sea level), etc.

Within UNGEGN's *Technical reference manual for the standardization of geographical names* (UNGEGN 2007) a data transfer standard and format is mentioned, providing tables of Roman characters used in different languages, and a 'toponymic data exchange standard'. Although the manual was published in 2007, the toponymic data exchange standard still provides a "core content" as it has been used and adapted to country-specific discussions and requirements within several

Fig. 7 UNGEGN Data model content (status: 2016)

Field Name	Data Type
RecordID	Index
UID	Number (long integer)
Name	Text (50 char.)
Latitude	Number (double)
Longitude	Number (double)
FeatureCode	Text (50 char.)
AdminUnit	Text (50 char.)
Language	Text (50 char.)
Description	Text (255 char.)
VariantName	Text (50 char.)
MapSheet	Number (long integer)
Source	Text (255 char.)
Status	Text (50 char.)
Pronunciation	OLE-Object
Location information	OLE-Object

toponymic training courses from 2007 until today. The country-specific training material is available through the UNGEGN website: https://unstats.un.org/unsd/ungegn/working_groups/wg6.cshtml.

Usually, a national feature classification is available and can be used for describing and classifying the objects/features in the database. A national feature classification recognizing the situation in a particular country and with feature code names in one (all) official language(s) as well as translations into other language(s) (e.g. English) should be a requirement in order to use the database content for international purposes.

In 2010, the European Commission's implementing rule for interoperability of spatial data sets and services of the European geospatial data infrastructure was published (INSPIRE 2010). The INSPIRE data specification on geographical names is a technical guidance document associated with the implementing rule that was prepared following the participative principle of a consensus-building process, also involving UNGEGN experts (INSPIRE 2014). In INSPIRE the concept that the same named place can be referred to by several names was reflected. Figure 8 shows part of the INSPIRE geographical names application schema.

The INSPIRE specification of geographical names provides technical guidance and can serve as a general data model reflecting today's complexity and the multi-purpose-requirement of databases. As the INSPIRE specification is not restricted to a European approach it can be used for modelling geographical names with a sub-national, national, regional or global perspective.

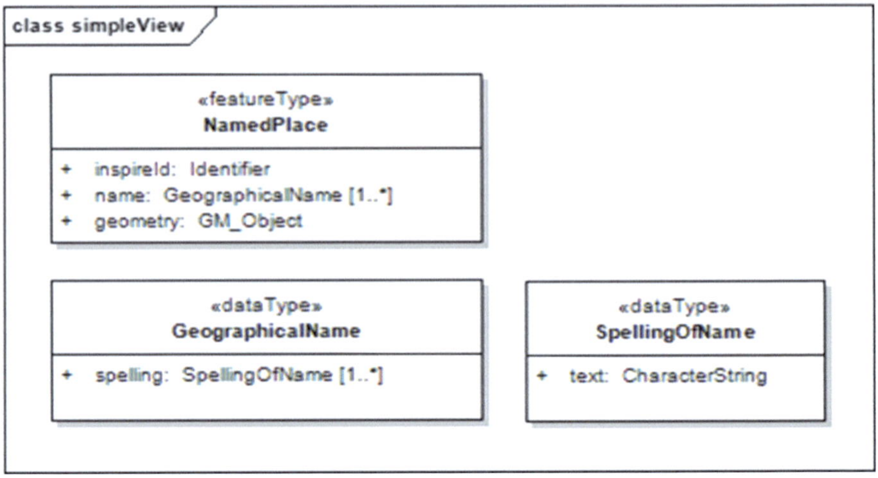

Fig. 8 Core of the INSPIRE geographical names application schema (status 2009)

A more comprehensive content on the issues related to the design of databases can be obtained from UNGEGN's *Toponymy Training Manual* (2017). The manual was designed to bring together a collection of articles on a variety of toponymic/geographical names' issues—amongst others, advanced technical ones on databases and their management. Therefore, this reference also elucidates the following two sections on database management and publication.

Requirements for Efficient Database Management and Updating

Geographical names database management should allow flexibility. Methods of geographical names data management must keep pace with technology developments. Computers, the internet, and geographical information systems have all changed the way data is managed.

The database schema should explain whether the geographical names will be used as an attribute of a spatial object/feature ("geospatial-based"), or whether the geographical names entries are the main database entries ("attribute-based"). In the geospatial community the first option is favoured, whereas the geographical names and language community very often model the databases while focusing on the names entries.

Using a database management system to store and manage data has become state-of-the-art and comes with advantages. One of the biggest advantages of using a database management system is that it lets end users and application programmers access and use the same data while managing data integrity. The latter is the maintenance and the assurance of the accuracy and consistency of data over its entire life-cycle. Data is better protected and maintained when it can be shared using a database management system instead of creating new iterations of the same data stored in new files for every new application. The database management system provides a central store of data that can be accessed by multiple users in a controlled manner.

Within any database design and management, it should be considered as well that geographical names data is used at all levels of resolution. The spatial resolution of a geographical names data set is typically described by the scale of the map from where, or for which, it has been captured.

If geographical names data from multiple sources are to be included in the database, some issues may be encountered and have to be documented: standardization requirements, handling of features/objects that span jurisdictional boundaries, quality of the data and frequency of data updates. This information can be stored as metadata information which describes the content of the database or as direct attribute to the geographical name entity in the database.

From a spatial data infrastructure's (SDI) point of view, the connection of geographical names with a geographical feature/object is only one of many attributes to the entity 'name'. The same geographical feature/object may be described by many different names in different languages, each one of which may again be pronounced, transcribed, transliterated or otherwise rendered graphically in different ways, which may or may not be considered 'official' (see example 'Bautzen' in section "Aspects of Designing a Geographical Names Database").

No matter whether local, provincial, sub-national or national, all geographical names databases should be designed and maintained with the purpose of publishing their content and linking to other spatial data within a national SDI. Therefore, it is crucial to make the databases compatible with the requirements and international standards for data integration in SDIs. A wide range of different applications (e.g. geoportals) can be built up using the SDI and by that providing discovery functionality for spatial data sets and services.

Publication of Derived Geographical Names Products, Servers and Applications from the Database

Today, a patchwork of heterogeneous international, regional, national and sub-national geographical names related products, web services and applications exist and geographical names are easily findable and accessible for national and international use, and the ways in which geographical names can be made accessible continually evolves.

The potential for standardising geographical names utilizing the internet was first recognized in resolution VII/9 of the Seventh Conference on the Standardization of

Geographical Names in 1998, along with a recommendation that data on geographical names' websites be provided free of charge in the interests of international standardization. That positive trend continues and is also fully compliant with, and recognizes, resolution VIII/6 of the Eighth Conference in 2002, which recommends integrating geographical names better into national, regional and international spatial data infrastructures (SDIs). Those recommendations have been widely adopted and the internet has been imperative to the dissemination of geographical names data and information for some time (Birtles 2021).

One way of publishing geographical names data from the database is through a 'gazetteer'. A gazetteer can be defined as a list, a report or a repository of location information that is used to search for specific locations. The term 'gazetteer' in an SDI context is considered as "any geospatial dataset which contains 'spatial identifiers'". These can be geographical names, postal codes or other indexes for indirect spatial referencing. This technical SDI view on 'gazetteers' is different from the much simpler view of the United Nations Group of Experts on Geographical Names (UNGEGN) on 'gazetteers': "List of toponyms arranged in alphabetic or sequential order, with an indication of their location and preferably including variant names, type of (topographic) feature and other defining or descriptive information." A special Internet-/web-service, addressing a specific profile or use case for publishing geographical names data is the so-called 'gazetteer service'.

Web services are the vehicle for accessing geographical names database content through an application, for integrating it in a Geographical Information System (GIS) or a map production software or just using them as essential parts of SDIs. A geoportal usually reveals the contents of the SDI, it is the central point of access to the data and services of the SDI. Users can search within the central search engine, which contains sets of metadata on geospatial data sets and services from across all levels of public administration in a country. The example of the German national geographical names database is shown in Fig. 9. A gazetteer service usually is a core search component of a geoportal.

Where the WWW is a source of human readable information, the Semantic Web conceptualizes a web of machine-actionable data. As part of the broader vision of the Semantic Web, the so-called 'Linked Data' method makes standardized, tagged and machine-actionable data from heterogeneous sources available online for linking. For data publishers, it aims to efficiently optimize the capacity for interoperability and correct interpretation of published data. For data consumers it aims to optimize the efficient and correct re-use of data. In this way, it facilitates inter-organizational data sharing, linking and the development of context-driven applications that make use of data from various sources (Birtles 2021). Considering Linked Data approaches in relation to geographical names data management, this new data integration method can make geographical names database content like 'gazetteers' and related information more interoperable and re-usable in the WWW.

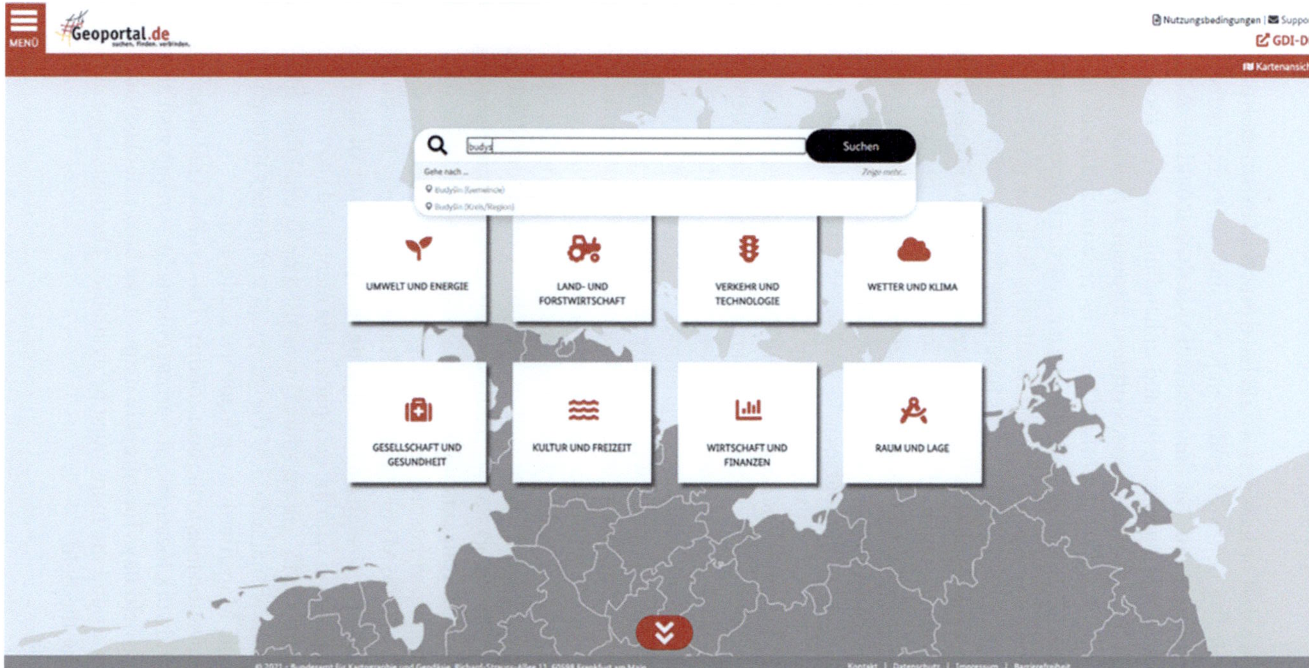

Fig. 9 The German national geographical names database (GN-DE) published as a web service as part of the national spatial data infrastructure (GDI-DE) and visualized through the Geoportal application (www.geoportal.de) (status 2022)

Qualifications for Lecturers, Participants, and Teaching Conditions

Preferably, teaching staff or lecturers would have practical experience themselves in collecting names during fieldwork and/or in processing names in a names bureau or secretariat for a names board. On the technical side they should be able to instruct course participants in reference systems and in handling receivers for Global Positioning Services (GPS). For the courses offered through UNGEGN, teaching staff should be familiar with the work of this Group, its recommendations and available materials. Local teaching support is very important, particularly to provide linguistic experience with the local language(s) and familiarity with the physical and cultural landscape of the area. Course participants have come from a variety of backgrounds and with differing levels of skill in the language of the course. Primarily the courses are aimed at individuals involved (or to be involved) in aspects of toponymy in mapping or with a national names authority. Also those with local government, in hydrography, in database management or taking university geography courses have participated in a number of the courses offered. For the course participants, computer literacy is required, as well as expertise in map reading.

Regarding the teaching conditions, UNGEGN courses have usually been held in air-conditioned class rooms within a university or government department. For at least part of the course computer access was required, with PCs that had been uploaded beforehand with database programmes or software packages. As for the hardware no specific requirements were set beforehand, because the software requirements provided the requirements for the hardware, too. Concerning the software for the exercises, 'Microsoft Windows' (proprietary) was used as the operating system and 'Microsoft Access' (proprietary) and 'PostgreSQL' (open source, with 'PostGIS' as spatial database extender) was used as the object-relational database system. The data used for the "workshop" (hands-on-experience) comprised any records from field collection or from existing national geospatial/topographic data in digital format from the countries hosting the training, if available, or open data available from OpenStreetMap (OSM) or the Shuttle Radar Topography Mission (SRTM3). For the creation, editing, visualizing, analyzing and publishing of the geographical names data in combination with the geospatial data mentioned above, the Geographic Information System 'QGIS' (open source) was used or, if available, ESRI ArcGIS (proprietary). For the visualization of the geographical names through the web the applications Google Maps/Earth were made available, too.

The local organization falls to the host country. This includes pre-determining suitably accessible fieldwork locations, pre-arrangements for a group visit and provision of transport on the field day. Topographic maps of the area and classroom space for group discussions around the maps must be available. The hosts have also provided GPS equipment for the field.

What Must Toponymy Lecturers Know About the Functioning of UNGEGN?

UNGEGN Structure and Resources

The Economic and Social Council (ECOSOC) of the United Nations passed resolution 715 A (XXVII) in 1959 establishing a group of experts to provide advice on the handling of geographical names in UN maps and documents. The first small meeting was held in 1960 and has led to the 11 conferences (UN Conferences on the Standardization of Geographical Names) held every five years from 1967 to 2017. Two sessions of UNGEGN were convened between conferences and one with every conference, giving 30 to 2017. At that time the conferences and UNGEGN were discontinued and a "new" UNGEGN established, with sessions held for a week every two years.

For the second session of the new UNGEGN held in 2021, a strategic plan and programme of work had been compiled and was approved at the session. The five strategies presented are: Technical expertise; Relationships, Links and connections; Effective work programmes; Culture, heritage and language recognition; and Promotion and capacity building. This document is available for download from the UNGEGN website and will be the backbone of UNGEGN's work through until 2029. (https://unstats.un.org/unsd/ungegn/sessions/2nd_session_2021/documents/UNGEGN_Strategic_Plan_final_5May2021.pdf).

During its 60 years, UNGEGN has followed up on conference resolutions and functioned between conferences through linguistic or geographical divisions (for example, Portuguese-speaking Division; Africa East Division) to address shared issues at a more grass roots level. UN Member States can join one or more divisions to assist with their work; currently there are 24 such divisions. Also, UNGEGN has established working groups to address particular geographical names topics, such as romanization systems, terminology, database management, cultural heritage, etc. When their assignment is complete, a working group is disbanded; at present there are 10 working groups. Additionally, a special Toponymic Task Team for Africa was established in 2004, and a coordinated project to produce national toponymic guidelines for the benefit of international cartographers was started in 1979. Details of all these aspects of UNGEGN are presented to students and their relevance explained. The structure of UNGEGN is presented in Fig. 10.

In the past, UNGEGN published its work through reports of conferences and UNGEGN sessions as well as through the UN publication *World Cartography*. Since 2001, UNGEGN has prepared two publicity brochures (translated by some countries into their own languages), manuals (on national standardization, technical and romanization issues, and toponymic training), a terminology glossary, and a media kit.

In 2002 the UNGEGN website was created, so today all publications and conference/session documents are online for free public access. A World Geographical Names Database was created in 2004 to respond to public enquiries and to

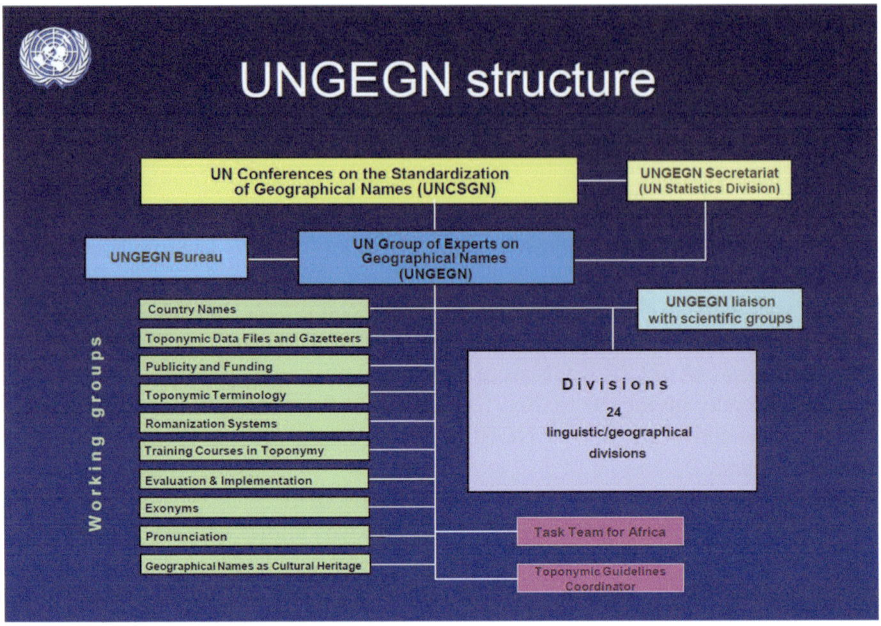

Fig. 10 Diagram of structure of UNGEGN until 2017, after which the UNCSGN element ceased to exist (https://unstats.un.org/unsd/geoinfo/UNGEGN/docs/Training/Manila/day%202/01_KERFOOT_Philippines_UNGEGN_HK%20final.pdf)

encourage the use of nationally standardized toponyms. Online access is available to this multilingual, multiscriptual data set of names of countries, capitals and major cities/towns. The website has a complete set of the *UNGEGN Information Bulletin* issues since 1988 and links to national standardization materials of individual countries. Although access to the website is in English, some documentation is available in other languages. For example, conference resolutions are posted as compendiums and in searchable database format in several UN and other languages.

Upcoming events, tweets on toponymy, and since the Covid-19 pandemic curtailed in-person meetings, resource materials from webinars have been uploaded to the UNGEGN website.

During the toponymy courses, students are presented with various aspects of the UNGEGN website and are encouraged to explore it for themselves. *UNGEGN website:* https://unstats.un.org/unsd/ungegn/.

How Can UNGEGN Help?

Many individuals involved with geographical names administration and standardization in their country work by themselves or as part of a small group. Consequently, a wider view of the challenges they face on geographical names issues is not always available, and networking with offices in similar fields in other countries is minimal. UNGEGN provides an opportunity to expand one's horizons, to link up with others, to find examples of good practices in standardization topics and to discuss challenges. Although in geographical names standardization there is no one right way of treating names, often difficulties faced by one country have already been encountered by another, and helpful ideas can be exchanged.

The UNGEGN website, as described above, provides various tools to assist in standardization and all files included can be downloaded free of charge. The reports and most of the documents presented at former UN Conferences (UNCSGN) and UNGEGN sessions are available, as are the manuals, publicity pamphlets and media kit. Ideas to promote programmes, to respond to management about benefits of toponymic standardization, to answer questions from the media, and much background documentation are all there. Links provide access to geographical names authorities, publications and geographical names databases in many countries (see Fig. 11).

The *UNGEGN Information Bulletin* has been produced twice a year since noted in resolution V/24 at the Fifth Conference in 1987. Recently there have been thematic issues (for example, "Legislation on Geographical Names" in Vol. 55, December 2018), and for any country the opportunity exists to showcase their activities through contributions, while at the same time learning from the work of others.

The World Geographical Names Database (currently under reconstruction in 2021) provides the opportunity for countries to contribute the standardized forms of their major cities/towns, in their official languages and in their original scripts together with Romanized forms. In addition, pronunciation sound files can be provided for upload. This database allows for the promotion to the public of nationally standardized names.

Fig. 11 Home page of the UNGEGN website, https://unstats.un.org/unsd/ungegn/ (as of 2021)

Training courses have in the past been offered through UNGEGN Divisions in various parts of the world. Countries wishing to host such training can apply to the UNGEGN Secretariat for consideration. Also available for training are the online course, the *Toponymy Training Manual*, materials from past courses, and offerings of recent webinar sessions.

UNGEGN has a variety of working groups that address thematic issues concerning geographical names. Participation in working groups is open to interested individuals and benefits can be gained by joining relevant groups. For example, the working groups addressing Geographical Names Data Management and Geographical Names as Cultural Heritage provide the environment for exchange and learning from leaders in these fields. The 2021–2029 Strategic Plan and Programme of Work shows specific projects that fall under the auspices of the various working groups.

The other sub-families of UNGEGN are the divisions. (Figure 12 shows the initiation of the Portuguese-speaking Division in 2007.) A country, through a division based on geography or language, has the opportunity to communicate with countries having similar standardization challenges. Networking can be at a less formal level than at the UNGEGN sessions.

Upcoming events are shown on the UNGEGN website, and when an UNGEGN session is scheduled every two years, all the details needed by participants from Member States are provided ahead of time.

Fig. 12 Portuguese-speaking Division start-up in 2007 at the UNCSGN, where capacity building was also discussed. *Photo* Helen Kerfoot

Outlook

The Covid-19 pandemic in 2020/21 forced us to actively seek other means of communication that could at least temporarily replace contact training courses. So during 2021, the new convenor of the UNGEGN Working Group on Training Courses in Toponymy (elected in 2017) organized a webinar on the significance and benefits of geographical names standardization, particularly for Caribbean countries (https://unstats.un.org/unsd/ungegn/working_groups/webinars/docs/geographical-names-standardization-supporting-national-development.pdf). The webinar was popular and attended by students from around the world. Similar webinars are planned on a regional or local basis.

During the Second UNGEGN Session in 2021 the Working Group made the proposal to produce a pilot streamed toponymy course, consisting of four basic lectures on toponymy that can be accessed at will from all over the world, via the Working Group's website (https://unstats.un.org/unsd/ungegn/working_groups/wg6.cshtml). The idea is that this pilot streamed toponymy course will be presented to the Third UNGEGN session in 2023 and, if deemed useful by the Session experts, a full course will hopefully be funded by the UN Statistics Division.

Questions

- What factors could influence the membership and structure of a national geographical names board?
- What are some principles that you would consider useful for a geographical names board?
- What is the purpose of your database?
- What tables do you need in the database?
- What (attribute) fields are needed in these tables?
- Which of the fields in the tables have unique values?
- What are the relationships between the tables?
- On accessing the UNGEGN website, what publications can be found by the general public to learn about toponymic standardization?

References

Aurousseau M (1957) The rendering of geographical names. Hutchinson, London

Birtles S (2021) Linked Data considerations for geographical names standardisation. Report submitted by Australia to the second UNGEGN session. https://unstats.un.org/unsd/ungegn/sessions/2nd_session_2021/documents/2021_6-Linked_Data_E.pdf

INSPIRE (2010) Implementing Directive 2007/2/EC of the European Parliament and of the Council as regards interoperability of spatial data sets and services (COMMISSION REGULATION (EU) - No 1089/2010 of 23 November 2010). http://eur-lex.europa.eu/LexUriServ/LexUriServ.do?uri=OJ:L:2010:323:0011:0102:EN:PDF

INSPIRE Thematic Working Group Geographical Names (2014) Data specification for geographical names. European Commission Joint Research Centre. https://inspire.ec.europa.eu/Themes/132/2892
Kadmon N (2001) Toponymy: the lore, laws and language of geographical names. Vantage Press, New York
Ormeling F (2017) 50 years trying for training courses in toponymy. Presentation during 11th UNCSGN/30th session UNGEGN. United Nations, New York. https://unstats.un.org/unsd/geoinfo/UNGEGN/docs/11th-uncsgn-docs/history_training50years.pdf
Ormeling FJ, Kerfoot H, Zaccheddu P-G (eds) (2017) Toponymy training manual. United Nations, New York. https://unstats.un.org/unsd/ungegn/pubs/documents/Training%20Manual.pdf
UNGEGN (2006) Manual for the national standardization of geographical names. United Nations, New York. https://unstats.un.org/unsd/ungegn/pubs/
UNGEGN (2007) Technical reference manual for the standardization of geographical names. United Nations, New York. http://unstats.un.org/unsd/geoinfo/UNGEGN/publications.html
UNGEGN (2002, 2007) Glossary of terms for the standardization of geographical names. United Nations, New York. https://unstats.un.org/unsd/ungegn/pubs/

Approaches to Implementing Standardization of Place Names

Standardization of Geographical Names on Land and Sea in Slovenia

Drago Perko, Matjaž Geršič, and Matija Zorn

Abstract Slovenia began to regulate geographical names after its independence in 1991. In 1995, it established the Commission for the Standardization of Geographical Names of the Government of the Republic of Slovenia, which, among other tasks, gradually began to standardize Slovenian geographical names based on databases of country names, settlement names, and Slovenian exonyms, all three of which are maintained and updated by the ZRC SAZU Anton Melik Geographical Institute, as well as a database of geographical names on national maps maintained by the Surveying and Mapping Authority of the Republic of Slovenia. It also prepared two gazetteers of standardized geographical names on national maps at scales of 1:1,000,000 and 1:250,000 issued by the Surveying and Mapping Authority of the Republic of Slovenia. As part of these four databases and both gazetteers, which include more than 220,000 Slovenian geographical names, the commission has standardized more than 10,000 geographical names as of the beginning of 2022. Barely one percent of all geographical names in Slovenia are marine names, but they are already almost completely standardized.

Keywords Geographical name · Toponymy · Hydronomy · Endonym · Exonym · Standardization · Slovenian

Objectives

- After reading the chapter, readers should understand the importance of standardizing geographical names for each nation or language, how standardization takes place, who participates in it, and what problems need to be addressed.

D. Perko · M. Geršič · M. Zorn (✉)
ZRC SAZU Anton Melik Geographical Institute, Ljubljana, Slovenia
e-mail: matija.zorn@zrc-sazu.si

D. Perko
e-mail: drago.perko@zrc-sazu.si

M. Geršič
e-mail: matjaz.gersic@zrc-sazu.si

© The Author(s), under exclusive license to Springer Nature Switzerland AG 2023

G. O'Reilly (ed.), *Place Naming, Identities and Geography*,
Key Challenges in Geography, https://doi.org/10.1007/978-3-031-21510-0_9

- Readers should be familiar with the most important and most common terms associated with geographical names and their standardization.
- Readers should understand the current level of standardization of geographical names in Slovenia, which started being standardized only in 1995, with special emphasis on marine names.
- Readers should understand the basic differences between endonyms and exonyms from the perspective of an individual nation.
- Readers should be aware that standardization in bilingual or multilingual areas is especially important, and they should understand how bilingual geographical names are regulated in Slovenia.

Introduction

There are at least a few hundred thousand geographical names in Slovenia (Perko and Geršič 2021). Like elsewhere across the globe, these names are of exceptional importance at all levels of communication because Slovenians encounter them daily on identification documents, postage stamps, road signs, maps, the internet, and television, as well as in newspapers and other media. Lack of uniformity in using geographical names impedes communication, which is why it is imperative to make their usage uniform. This process is known as standardization of geographical names.

Standardization must take into account that geographical names are a vital part of the world's cultural heritage as well as that of each nation or language. They imbue spaces with meaning, develop spatial identities, and are the fundamental building blocks of any civilization or culture (Perko et al. 2017). As a young country, Slovenia also consolidates its identity through geographical names because foreigners often first encounter Slovenia, Slovenians, and the Slovenian language through the geographical names they see on maps or hear in the media.

A geographical name refers to a strictly defined geographical feature that it unambiguously defines and individualizes (Furlan et al. 2000). It takes shape in a specific linguistic area at a specific time (Šivic-Dular 1988). In areas of contact between various cultures, geographical names can overlap, and thus the same geographical features may have different names in different cultures and languages. This also applies to Slovenia, which, albeit small, is a geographically and culturally very diverse country (Perko et al. 2020). On the one hand, the Italian and Hungarian ethnic minorities living in Slovenia have their own geographical names and, on the other, Slovenians living in the border regions of Slovenia's neighboring countries have given Slovenian names to the geographical features there.

Geographical names can also serve as a window into the past (Perko et al. 2017). Slovenian geographical names reflect the influence of various peoples and nations that, during different periods, lived in the Alps, the Mediterranean, the Balkans, and the Pannonian Basin or at their meeting point at the heart of Europe in what is now Slovenia.

Types of Geographical Names

In terms of standardization, it is sometimes important to know the type of geographical name because individual languages may apply different normative rules to them, and individual countries may apply different legal frameworks.

For example, in Slovenian different capitalization rules apply to the names of settlements compared to other types of geographical names. In simplified terms, all the words in the names of settlements (except for prepositions and conjunctions) are capitalized (with a few exceptions), whereas in all the other types of geographical names only the first word is capitalized, unless the other words are themselves proper nouns. For example, in the name Kranjska Gora (literally 'Carniolan mountain'), referring to a settlement with a well-known Slovenian ski resort, both words are capitalized but, if there were also a mountain with the same name, the second word would not be capitalized (i.e., the correct Slovenian spelling would be Kranjska gora).

Slovenian legislation also distinguishes settlements from other geographical features, prohibiting two settlements in the country from having the same name (this applies to all new or altered names of settlements from 2008 onward), whereas it allows any two other types of geographical features, such as rivers, to have the same name.

Geographical names or toponyms can be divided in a number of ways, but especially in terms of the following (Kladnik et al. 2020):

- The location of a geographical feature (a distinction is made between terrestrial names or geonyms, which designate geographical features on Earth, and extraterrestrial names or cosmonyms, which designate all other features in outer space);
- The spatial scope of a geographical feature (a distinction is made between macrotoponyms and microtoponyms);
- The settlement of a geographical feature (a distinction is made between anoikonyms, which refer to unsettled geographical features, and oikonyms, which refer to settled geographical features and are divided into astionyms for the names of cities or towns and geographical features in towns, and comonyms for the names of villages and geographical features in the countryside);
- The native character of names (a distinction is made between endonyms, which designate geographical features in one of the languages spoken in the territory of that feature, and exonyms, which designate the same geographical feature in a language not spoken in the territory of that feature if it differs from the endonym for that feature);
- The type of a geographical feature (a distinction is made between various -onyms, such as hydronyms for the names of waters, or choronyms for the names of spatial units).

With regard to the location, type, and settlement of a geographical feature, at the highest levels toponyms can be hierarchically categorized as follows (Gundacker

2014; Backus Borshi 2015; Urazmetova and Shamsutdinova 2017; Bijak 2019; Perko 2022):

- I cosmonyms (extraterrestrial names);
- II geonyms (terrestrial names);
- IIA anoikonyms (names of unsettled geographical features);
- IIA1 oronyms (names of landforms);
- IIA2 hydronyms (names of waters);
- IIA3 choronyms (names of regions);
- IIB oikonyms (names of settled geographical features);
- IIB1 astionyms (names of cities or towns and geographical features in them); and
- IIB2 comonyms (names of villages and geographical features in them).

Most of these basic types of names also have several subtypes. Thus hydronyms, for example, are at least further divided into okeanonyms for the names of oceans, pelagonyms for the names of seas, limnonyms for the names of lakes, potamonyms for the names of rivers, rheithronyms for the names of streams or creeks, and helonyms for the names of wetlands (Kladnik et al. 2020).

Standardization of Geographical Names Introduction

Standardization denotes a planned establishment of harmonized rules to regulate a specific area in the best possible way. It is carried out by a standardization authority composed of experts, users, and decision-makers in this area. In turn, the standardization of a geographical name is a process that uses selected criteria to define a uniform and binding written form of the name of a geographical feature (Dobrovoljc 2013) or a procedure that establishes and defines uniform written forms of the names of geographical features based on selected criteria.

As a rule, the standardization authority selects only one name for every geographical feature, defines the linguistically and geographically correct spelling of that name, and, if needed, also its spelling and usage in foreign languages (Kladnik and Perko 2013). It is only a standardized geographical name, with all its components and their fixed order defined, that makes it possible to unambiguously identify the geographical feature it refers to because it is the only one, without doublets, and hence invariant (Furlan et al. 2000).

Slovenia began standardizing geographical names after its independence in 1991. Because UN resolutions encourage countries to establish national geographical names authorities, on September 14th, 1995 the Slovenian government established a national standardization authority as its permanent working body and named it the Slovenian Government Commission for the Standardization of Geographical Names. It is composed of geographers, linguists, cartographers, and representatives of relevant ministries (Kladnik et al. 2020). The commission operates at several levels:

- The **global level**, where it represents Slovenia's interests in the UN names authorities;
- The **regional level**, where it takes an active part in the East Central and South-East Europe Division as one of the twenty-three regional linguistic-geographical divisions of the United Nations Group of Experts on Geographical Names (UNGEGN), and collaborates with similar commissions in other countries where members of the Slovenian minority live; and
- The **national level**, where, in accordance with the law, it approves changes to geographical names, standardizes endonyms in Slovenia and Slovenian exonyms outside Slovenia, and issues standardization documents (Kladnik et al. 2020).

Linguists, geographers, and decision-makers at various levels play a key role in standardization because geographical names may generally be problematic in three ways—that is, in terms of:

- **Language**, especially when names deviate from the normative rules;
- **Geography**, especially when names deviate from geographical reality; and
- **Legislation**, especially when names deviate from legal acts (Perko and Geršič 2021).

Also problematic is the use of the same name for different geographical features. Because of the unmanageable number of names used across the globe, it is practically impossible for every geographical feature to have its own individual or unique name in every possible language. What can at least be achieved through the standardization of geographical names is that the same names are not used for two or even more different geographical features within the same country or type of geographical names. For example, with regard to the names of settlements, this means that two settlements within the same country cannot have the same name, but a settlement can have the same name as a different type of geographical feature, such as a river or a hill (Perko and Geršič 2021); a good example in Slovenia is *Vipava*, which refers to both a town and a river. However, the problem in Slovenia is that, for example, there are several settlements called *Črni Vrh* and several elevations called *Črni vrh* (literally, 'black peak'), and as many as eight villages named *Pristava*.

In Slovenian, the problem of identical geographical names, especially the names of settlements, is most often solved by adding a left or right modifier. For example, in the name of the settlement *Črni Vrh v Tuhinju* (literally, 'black peak in the Tuhinj Valley') the adjective *Črni* is a left modifier of the noun *Vrh* and the word combination *v Tuhinju* a right modifier of the name *Črni Vrh*.

The commission seeks to fully adhere to the UN resolutions on geographical names, the Slovenian normative guide, Slovenian toponymic guidelines, and Slovenian legislation.

The most important legal instrument governing geographical names in Slovenia is the Act Regulating the Determination of Territories and the Naming and Marking of Settlements, Streets, and Buildings (ZDOIONUS) adopted on February 29th, 2008, which, among other things, defines the procedure for naming and renaming settlements, and specifies the government authority that approves the naming or renaming—that is, the Slovenian Government Commission for the Standardization of Geographical Names. The act does not affect previously established names of settlements, even though they may be problematic in several ways, but it does ensure that all new names and re-namings conform to this (Perko 2022).

The act's most important requirements are as follows:

- Every settlement must have a name;
- A settlement's name can refer to a geographical feature or to its history and cultural heritage;
- The new name of a settlement must differ from the names of other settlements in Slovenia;
- The name of a settlement or its renaming is specified by the municipality through a special ordinance;
- The name of a settlement must be in Slovenian and, in official bilingual areas, also in Italian or Hungarian, which must be approved by the minority representatives on the municipal council;
- The written form of the name of a settlement must agree with the opinion by the government working body; and
- Geographical names must be protected.

The toponymic guidelines were published by the Slovenian Surveying and Mapping Authority in 1995 in two volumes: *Toponimska navodila za Slovenijo* in Slovenian (Radovan 1995) and *Toponymic Guidelines for Slovenia* in English (Radovan and Majdič 1995). They are composed of ten sections. The first sections present the population of Slovenia, the alphabet, dialects, and normative rules for writing geographical names. They also contain some of the main features of spelling Italian and Hungarian geographical names. This is followed by sections dealing with names and standardization authorities, toponymic sources, and descriptive labels and abbreviations on maps. They also include a dictionary of common nouns and a section presenting the administrative division of Slovenia into municipalities. Considering that the guidelines were prepared twenty-five years ago, they are outdated in many respects and should therefore be updated.

Standardization of Geographical Names on National Maps

Several UN resolutions encourage national standardization commissions to compile registers of geographical names and publish lists of geographical names (i.e., gazetteers) in various languages as part of their standardization procedures (Kladnik et al. 2020).

To date, the Slovenian Commission has participated in the compilation of four registers (the register of country names in 1994, the register of settlement names in 1995, the register of geographical names in 1997, and the register of Slovenian exonyms in 2010) and two gazetteers (a gazetteer for a 1:1,000,000 map in 2001 and a gazetteer for a 1:250,000 map in 2007).

The commission first tackled country names, which it has already standardized twice to date. It is also gradually standardizing Slovenian exonyms, the names of settlements, and geographical names in the Register of Geographical Names, which is the largest collection of geographical names in Slovenia. It has fully standardized the geographical names in Slovenia on the 1:1,000,000 and 1:250,000 national general maps (Perko 2022).

Gazetteer for a 1:1,000,000 Map

In 2001, the commission compiled a gazetteer for the 1:1,000,000 *General Map of Slovenia* and standardized all the geographical names on it. That same year, the map and gazetteer were published in a bilingual English–Slovenian publication issued by the Surveying and Mapping Authority of the Republic of Slovenia as part of the *United Nations Series of National Gazetteers.*

The introductory part of the publication describes the characteristics of Slovenian regions and the Slovenian language, with the most common generic expressions used in Slovenian geographical names presented in a table in English, German, French, and Spanish. There is a two-page map (Fig. 1) with standardized names in the middle of the book, and the second half of the publication features a gazetteer of all names with the type of geographical feature that they designate, their position on the map grid, and their latitude and longitude (Perko 2001). The latest edition of the map was published in 2017 (Fig. 1).

There are a total of 843 names on the map, of which 464 or 55.0% are in Slovenia. Based on the type of geographical names in Slovenia, 8.6% are oronyms, 11.6% hydronyms, 7.1% choronyms, and 8.4% astionyms—and a full 64.2% are comonyms (Kladnik et al. 2020).

Bilingual variants are provided for seventy-eight Slovenian endonyms (separated from the endonyms in the majority language with a slash) in cross-border Slovenian ethnic territory (e.g., *Trieste/Trst* for the port in northeastern Italy) and for twenty-two Slovenian exonyms outside Slovenian ethnic territory, with the Slovenian name provided in parentheses (e.g., *Rijeka (Reka)* for the port in Croatia). There are twelve bilingual names written in the two bilingual areas within Slovenia: four are Slovenian–Hungarian (e.g., *Lendava/Lendva*) and eight are Slovenian–Italian (e.g., *Piran/Pirano*).

On the 1:1,000,000 map, the commission has thus standardized 452 Slovenian geographical names, four Slovenian–Hungarian names, and eight Slovenian–Italian names within Slovenia plus twenty-two Slovenian exonyms outside Slovenia, adding up to a total of 486 Slovenian names standardized.

The standardization authorities in Austria, Hungary, Croatia, and Italy are responsible for standardizing the seventy-eight Slovenian endonyms outside Slovenia.

Fig. 1 Detail from the 1:1,000,000 General Map of Slovenia, which contains 464 standardized geographical names in Slovenia and twenty-two standardized Slovenian exonyms outside Slovenia. *Source* © Surveying and Mapping Authority of the Republic of Slovenia

Gazetteer for a 1:250,000 Map

In 2007, the commission also produced a gazetteer for the 1:250,000 *General Map of Slovenia* and standardized the geographical names on it. A year later, the names were published in the Slovenian–English *Gazetteer of the National General Map of the Republic of Slovenia at the Scale 1:250,000* issued by the Surveying and Mapping Authority of the Republic of Slovenia.

The publication has a map on the front (Fig. 2) and a list of names on the back with the type of geographical features they designate, their grid locations on the map, and their easting and northing UTM coordinates in meters (Furlan et al. 2008). The latest edition of the map was published in 2019 (Fig. 2).

There are a total of 8,203 names on the map, of which 4,273 or 52.1% are in Slovenia. According to the type of geographical names in Slovenia, 8.1% are oronyms, 6.5% hydronyms, 1.0% choronyms, 0.9% astionyms, and 83.5% comonyms (Kladnik et al. 2020).

Bilingual variants are provided for 613 Slovenian endonyms (separated by a slash) in cross-border Slovenian ethnic territory (e.g., *Klagenfurt/Celovec* for the capital of Austrian Carinthia) and twenty-six Slovenian exonyms (provided in parentheses) outside Slovenian ethnic territory (e.g., *Udine (Videm)* for the largest city in the Italian region of Friuli). There are fifty-two bilingual names written in both bilingual areas within Slovenia: thirty-two are Slovenian–Hungarian (e.g., *Dobrovnik/Dobronak*) and twenty are Slovenian–Italian (e.g., *Izola/Isola*).

On the 1:250,000 map, the commission thus standardized 4,221 Slovenian geographical names, thirty-two Slovenian–Hungarian names, and twenty Slovenian–Italian names within Slovenia plus twenty-six Slovenian exonyms outside Slovenia, adding up to a total of 4,299 Slovenian names standardized.

The standardization authorities in Slovenia's neighboring countries are responsible for the standardization of the 613 Slovenian endonyms outside Slovenia.

Register of Geographical Names

The Register of Geographical Names (*Register zemljepisnih imen*, REZI) was established in 1997 by the Surveying and Mapping Authority of the Republic of Slovenia, which updates it regularly. The register contains geographical names from national maps at four scales: 1:5,000 (or 1:10,000 for sparsely populated areas), 1:25,000, 1:250,000, and 1:1,000,000. For each geographical name, data are provided on the type of name, its position on the map, and its link to other national spatial data registers. The register distinguishes thirty-seven types of geographical names: nine types of oronyms (e.g., mountains or caves), ten types of hydronyms (e.g., rivers or waterfalls), three types of choronyms (e.g., countries or provinces), four types astionyms (e.g., cities or streets), and eleven types of comonyms (e.g., villages or farms).

Fig. 2 Detail from the 1:250,000 General Map of Slovenia, which contains 4,273 standardized geographical names in Slovenia. *Source* © Surveying and Mapping Authority of the Republic of Slovenia

Among these, eleven are micronyms (e.g., churches or solitary rock formations) and twenty-six are macronyms (e.g., mountain ranges or railway lines).

That same year, the commission also produced theoretical guidelines for standardizing geographical names in the register, which it later published in both an abridged form (Furlan et al. 2000) and a full edition (Furlan et al. 2001). By the end of 2021, the register included 216,638 entries. According to the type of geographical names in Slovenia, 7.4% are oronyms, 8.1% hydronyms, 0.9% choronyms, 1.1% astionyms, and 82.5% comonyms (Kladnik et al. 2020).

The commission has already standardized all the geographical names at the 1:1,000,000 and 1:250,000 scales because these names are the same as in both printed gazetteers. The only difference is that the basic unit of the gazetteers is the geographical name, and the basic unit of the register is the form (entry) of the geographical name provided on the map. Because a name can be written multiple times, bilingually, or in several variants, the number of entries in the register is greater than the number of names in the gazetteers (Table 1).

The number of names in the gazetteer for the 1:1,000,000 map is 843, and the number of entries in the register for the same scale is 1,043, or 25.1% more. For the 1:250,000 scale, the number of names in the gazetteer is 8,203, and the number of entries in the register is 8,811, which is 7.4% more.

Geographical names at the 1:25,000 scale (68,467 entries) have only been toponomastically verified, which means they are only at the initial stage of standardization, and not even this has been done for the names at the 1:5,000 and 1:10,000 scales (130,317 entries). A detailed toponomastic examination should be performed in cooperation with local residents and include field and desk collection, recording, processing, research on the origin, spelling, and pronunciation of dialect and standard Slovenian variants, and a basic spelling conformity check.

Of all the names in the register, 9,854, or 4.5%, have already been standardized and 31.6% have been toponomastically reviewed (Kladnik et al. 2020).

Marine Geographical Names

Marine names have been neglected in Slovenia so far, and this contribution breaks new ground in this respect.

There are several reasons for this neglect. Slovenia's coast is only 47 km long. Its border is 1,370 km in length, of which its maritime border accounts for only 48 km or 3.5% (29 km with Italy and 19 km with Croatia). Slovenia's territory on land covers 20,271 km^2 and its territorial waters in the Gulf of Trieste cover 231 km^2 or nearly two-fifths of the entire gulf, which is nonetheless barely 1.1% of the country's total land and sea area (Perko et al. 2020).

A similar disproportion can be observed between the number of Slovenian geographical names for terrestrial features and marine features. The Register of Geographical Names includes 202 marine entries, which corresponds to only 0.9% of all entries. In terms of country, 151 are in Slovenia, thirty-one in Italy, and twenty

Table 1 Number and share of geographical names in the gazetteers and the number and share of entries in the Register of Geographical Names by map scale and type of geographical name in and outside Slovenia

Type	Gazetteer				Register									
	1:1,000,000		1:250,000		1:1,000,000		1:250,000		1:25,000		1:5,000		Total	
	n	%	*n*	%	*n*	%	*n*	%	*n*	%	*n*	%	*n*	%
Oronyms	58	6.9	559	6.8	122	11.7	655	7.4	15,314	22.4	16,095	11.6	32,186	14.8
Hydronyms	91	10.8	612	7.5	132	12.7	717	8.1	7,703	11.3	13,233	9.6	21,785	10.1
Choronyms	55	6.5	64	0.8	26	2.5	76	0.9	10,228	14.9	54,843	39.7	65,173	30.1
Astionyms	85	10.1	86	1.0	97	9.3	97	1.1	823	1.2	976	0.7	1,993	0.9
Comonyms	554	65.7	6,882	83.9	666	63.8	7,266	82.5	34,399	50.2	53,170	38.4	95,501	44.1
Total	843	100.0	8,203	100.0	1,043	100.0	8,811	100.0	68,467	100.0	138,317	100.0	216,638	100.0

in Croatia. In terms of language, 103 are Slovenian, eighty Italian, and nineteen Croatian. In terms of map scale, six are found at a scale of 1:1,000,000, nineteen at a scale of 1:250,000, eighty-nine at a scale of 1:25,000, and eighty-eight at a scale of 1:5,000.

Only three types of marine names are found in the register: the name of one sea (twelve entries), the names of parts of that sea (mostly various types of gulfs; 149 entries), and the names of salt pans (forty-one entries).

Within Slovenia, as already mentioned, there are 151 entries referring to marine names:

- In terms of language, ninety are in Slovenian and sixty-one in Italian (because a native Italian minority lives on the coast alongside Slovenians);
- In terms of map scale, four can be found at a scale of 1:1,000,000, six at a scale of 1:250,000 scale, fifty-three at a scale of 1:25,000, and eighty-eight at a scale of 1:5,000.

Again, only three types of names are found: the name of one sea (three entries), the names of parts of that sea (108 entries), and the names of salt pans (forty entries).

Fig. 3 The town of Piran with the Bay of Piran (photo by Andrej Šafarič, https://www.shutterstock.com)

Fig. 4 The town of Koper with the Bay of Koper (photo by Cortyn, https://www.shutterstock.com)

The 1:1,000,000 Scale

The Register of Geographical Names contains six marine names from the 1:1,000,000 map, which accounts for only 0.6% of all entries. Of these, four are in Slovenia and two in Italy, three are written in Slovenian and three in Italian, and two are the name of a sea and four are names of parts of that sea.

Within Slovenia, there are four entries referring to marine names. Of these, three are written in Slovenian and one in Italian, one corresponds to the name of a sea and three to the names of parts of that sea. These four entries in fact refer to three geographical features; specifically:

- The Slovenian entry *Jadransko morje* refers to the Adriatic Sea, named after the town of Adria north of the Po Delta in Italy;
- The Slovenian entry *Tržaški zaliv* refers to the Gulf of Trieste, named after the largest Adriatic port;
- The Slovenian entry *Koprski zaliv* and the Italian entry *Vallone di Capodistria* refer to the Bay of Koper, named after the largest port in Slovenia (Fig. 4).

The 1:250,000 Scale

The Register of Geographical Names contains nineteen marine names from the 1:250,000 map, or only 0.2% of all entries recorded at this scale. The number of entries is three times the number of entries at the 1:1,000,000 scale, but their share is only a third of what is found at that scale. Of these nineteen entries, six are found in Slovenia, ten in Italy, and three in Croatia. In terms of language, eight are provided in Slovenian, eight in Italian, and three in Croatian. Three correspond to the name of a sea and sixteen to the names of parts of that sea.

Within Slovenia, six marine names are recorded: four in Slovenian and two in Italian, with one corresponding to the name of a sea and five to the names of parts of that sea. These six entries in fact refer to four geographical features; specifically:

- Four entries (Sln. *Jadransko morje*, *Tržaški zaliv*, and *Koprski zaliv*, and Ital. *Vallone di Capodistria*; Fig. 4) refer to the same three geographical features as at the 1:1,000,0000 scale above;
- Two entries (Sln. *Piranski zaliv* and Ital. *Golfo di Pirano*) refer to the Bay of Piran, named after the picturesque little Mediterranean coastal town in Slovenia (Fig. 3).

The 1:25,000 Scale

The Register of Geographical Names contains eighty-nine marine names from the 1:25,000 map, corresponding to 0.1% of all entries recorded at this scale. The number of entries is five times the number of those at the 1:250,000 scale, but their share is only half of what is found at that scale. Of these eighty-nine entries, fifty-three are in Slovenia, nineteen in Italy, and seventeen in Croatia. Forty are written in Slovenian, thirty-three in Italian, and sixteen in Croatian. Seven correspond to the name of a sea, seventy-one to names of parts of that sea, and eleven to names of salt pans.

Fifty-three marine names are recorded within Slovenia. Of these, thirty-one are written in Slovenian and twenty-two in Italian. One refers to a sea, forty-two to parts of that sea, and ten to salt pans.

Again, the same names appear as at the 1:1,000,000 and 1:250,000 scales. There are forty-four bilingual (Slovenian–Italian) entries referring to twenty geographical features (e.g., *Sečoveljske soline/Saline di Sicciole* 'Sečovlje Salt Pans'), with two pairs of bilingual names referring to the same features appearing twice (i.e., *Piranski zaliv/Golfo di Pirano* and *Zaliv Valdoltra/Baia di Valdoltra* 'Bay of Valdoltra').

There are eight entries in Slovenian only (e.g., *Žumrove kotanje* 'Žumer's Hollows', referring to an underwater karst feature with springs west of Izola discovered by the geography teacher Jože Žumer), but one name (i.e., *Tržaški zaliv*) appears twice and so actually only seven different names are recorded. Accordingly, the fifty-three names in Slovenia recorded at the 1:25,000 scale actually refer to twenty-seven different marine geographical features.

The 1:5,000 Scale

The Register of Geographical Names contains only eighty-eight marine names from the 1:5,000 map, which is negligible compared to the total number of entries recorded at this scale. All the names are found within Slovenia. Fifty-two are written in Slovenian and thirty-six in Italian. Fifty-eight refer to parts of the sea and thirty to salt pans.

All the names from the 1:1,000,000 and 1:250,000 maps also appear here, except for *Jadransko morje*. Compared to the entries from the 1:25,000 map, only seven are new and four are missing, but there are more repetitions of the same name.

There are seventy bilingual (i.e., Slovenian–Italian) entries referring to nineteen different geographical features (e.g., *Portoroški zaliv/Golfo di Portorose* 'Bay of Portorož'); among these, one pair of names (for the Bay of Koper) appears five times, one pair appears four times, two pairs appear three times, and five pairs appear twice.

Eighteen entries are provided in Slovenian only (e.g., *Stare soline* 'Old Salt Pans'), but one (*Tržaški zaliv*) appears four times, and so in fact there are only fifteen different names. Accordingly, the eighty-eight names in Slovenia recorded at the 1:5,000 scale actually refer to thirty-four different marine geographical features.

The Surveying and Mapping Authority of the Republic of Slovenia no longer updates the printed versions of the 1:25,000 maps, on which only the names of settlements are provided in their bilingual form (Fig. 5), nor the printed editions of the 1:5,000 maps, on which practically no bilingual forms are provided (Fig. 6). Instead, it covers and updates the geographical names at these two scales within the Register of Geographical Names. It published the last printed series of the 1:25,000 maps for all of Slovenia in 1995.

Standardization of Country Names

The oldest of the gazetteers mentioned is the register of country names, which was created in 1994 as a basis for the first standardization of Slovenian names of countries and has been regularly updated ever since.

As toponyms, country names can be categorized as:

- Geonyms based on the location of the geographical feature they refer to;
- Macrotoponyms based on the spatial scope of the geographical feature they refer to;
- Oikonyms based on the settlement of the geographical feature they refer to;
- Choronyms based on the type of the geographical feature they refer to; and
- Mostly exonyms based on the name's origin.

The names of countries are the best-known and most frequently used geographical names at the global level. The Slovenian normative guide requires them to be Slovenianized, and so nearly all Slovenian names of countries are Slovenian exonyms with

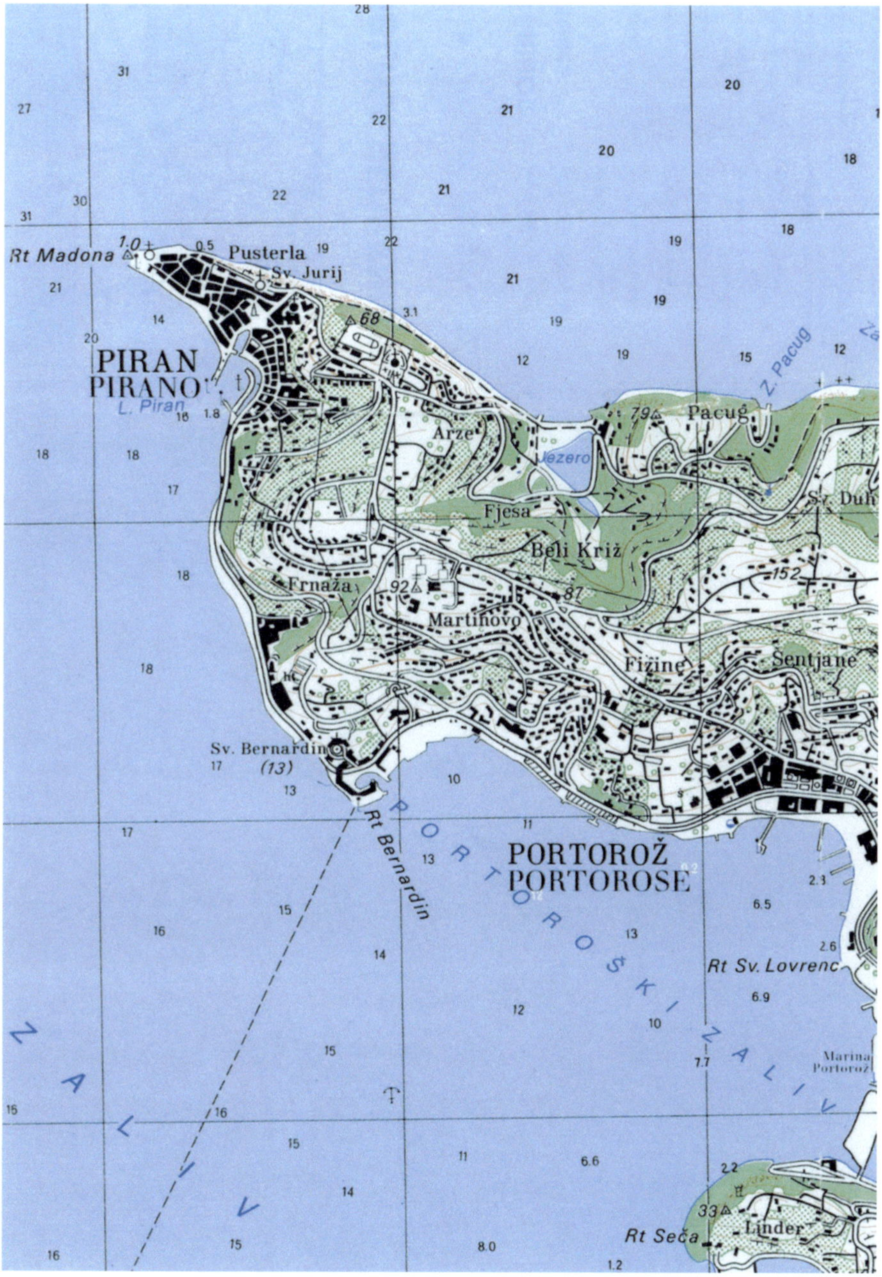

Fig. 5 In the Register of Geographical Names, most names at the 1:25,000 scale are written bilingually, whereas on the printed 1:25,000 *National Topographic Map of Slovenia* bilingual forms are only provided for the names of settlements. *Source* © Surveying and Mapping Authority of the Republic of Slovenia

Fig. 6 In the Register of Geographical Names, nearly all names at the 1:5,000 scale are written bilingually, in contrast to the printed 1:5,000 *National Topographic Map of Slovenia*, which has been replaced by more modern cartographic representations. *Source* © Surveying and Mapping Authority of the Republic of Slovenia

various degrees of Slovenianization (Kladnik and Perko 2013, 2015a, b). This is the only group of Slovenian geographical names that has been fully standardized.

The commission first standardized Slovenian country names in 1996 and published them in the Slovenian standard SIST ISO 3166:1996, which distinguishes between three versions of country names: the Slovenian short name, the official Slovenian short name, and the official Slovenian full (or long) name (Perko 1996a, b; Kladnik et al. 2020).

Every country suggests or changes its own short and formal names in its official language(s), alongside English and French, and reports them to the UN. If all other UN members agree with its suggestions, the UN approves them and the International Standardization Organization includes them in its ISO 3166 standard, thereby validating them internationally. Short names of countries, which may differ from their official short names, are formed within an individual language and apply at the national level. The Slovenianization of official short and full names of countries is thus based on the names accepted by the UN (e.g., *Ruska federacija* 'Russian Federation', and *Demokratična ljudska republika Koreja* 'Democratic People's Republic of Korea'), and the Slovenianization of short names of countries follows the Slovenian language tradition (e.g., *Rusija* 'Russia' or *Severna Koreja* 'North Korea'). With most countries and languages, the short name of a country is the same as its official short name.

In defining all three versions of Slovenian country names, the commission combined the normative rules into a single rule, according to which the names of countries are translated into Slovenian, whereby only those components of country names that are considered less well-known geographical or personal names are left in their original form and possessive forms are used for adjectives based on personal names (Perko 1996b). The commission fully Slovenianized the 185 names of countries at that time, except for the names of two countries that contained less well-known geographical names (i.e., *Gvineja Bissau* and *Sierra Leone*) and the names of eight countries that contained the names of persons, including saints (e.g., *Saint Kitts in Nevis* 'Saint Kitts and Nevis').

Due to many political changes after 1996, in 2004 the commission set up the Sub-Commission for Country Names composed of geographers and linguists, which again systematically reviewed all the names and introduced a uniform use of the common noun *otoki* 'islands' instead of *otočje* at the end of multiword names and a more consistent Slovenianization of the names of countries named after saints and noble or royal families. In 2007, the commission unanimously standardized this coordinated proposal. In this way, the Slovenian names of countries were standardized for the second time.

Compared to the names standardized in 1996, the commission really only introduced two new changes: for Congo and Papua New Guinea. To distinguish between the two countries with the same name *Kongo* 'Congo', the commission decided to use the short forms *Zahodni Kongo* 'West Congo' and *Vzhodni Kongo* 'East Congo' based on the model of *Južna Koreja* 'South Korea' and *Severna Koreja* 'North Korea', and the two countries' location in relation to one another. For the name *Papua Nova*

Gvineja 'Papua New Guinea', whose syntax is problematic in Slovenian, the commission standardized the form *Papuanska Nova Gvineja* 'Papuanian New Guinea' (just like *Ekvatorialna Gvineja* 'Equatorial Guinea'), which follows the French official name *la Papouasie-Nouvelle-Guinée* (Kladnik and Perko 2007, 2015b).

Since 2007, the commission has regularly been standardizing the changes to country names after their publication in the UN bulletin. The last change to the Slovenian standard was made in 2020, when Swaziland (Sln. *Svazi*) changed its name to Eswatini (Sln. *Esvatini*).

On the current list of 198 standardized country names, twenty-nine names in Slovenian are equivalent to the endonyms and the remaining names are Slovenian exonyms with various degrees of Slovenianization, ranging from minor changes, such as for Poland (*Polska* in Polish and *Poljska* in Slovenian), to full translations, such as for Ivory Coast (*Côte d'Ivoire* in French and *Slonokoščena obala* in Slovenian). There is only one true Slovenian name (i.e., not a translation): *Nemčija* 'Germany' (*Deutschland* in German).

By the end of 2021, the register of country names included 250 units (198 countries and fifty-two dependent territories with a high degree of independence) containing numerous data, including all versions of names in Slovenian, English, French, and the original language. Theoretically, the register encompasses 750 standardized geographical names, but because the short name is the same as the official short name for 230 units (e.g., *Ruska federacija*) and seventy-six units do not include the official full name at all (e.g., *Irska* 'Ireland'), the actual number of different standardized names is 426. Three units (i.e., *Dominikanska republika* 'Dominican Republic', *Srednjeafriška republika* 'Central African Republic', and *Združeni arabski emirati* 'United Arab Emirates') even have all three forms the same.

Standardization of Settlement Names

The register of settlement names, which was created only a year after the register of country names, was compiled in 1995 as a basis for the standardization of Slovenian geographical names or, in this case, the Slovenian and bilingual names of settlements.

As toponyms, the names of settlements can be categorized as:

- Geonyms based on the location of the geographical feature they refer to;
- Macrotoponyms based on the spatial scope of the geographical feature they refer to;
- Oikonyms based on the settlement of the geographical feature they refer to;
- Astionyms based on the type of the geographical feature they refer to; and
- Endonyms based on the name's origin.

The names of settlements in Slovenia are mostly problematic in linguistic terms because many are written in their dialect form, as well as from a legal point of view because, due to their repetition, they are not unambiguously defined.

At the initial stage of standardizing settlement names, the commission defined the following three rules:

- Linguistically or geographically problematic names of settlements need not be changed because they have already become established;
- Legally problematic names of settlements need not be changed in full, but only by adding a left or right modifier to the existing name;
- When a settlement is renamed or newly named, the linguistic, geographical, and legal aspects must strictly be taken into account.

Based on these rules, in 1995 the commission reviewed the names of 5,972 settlements known at that time (as of January 1st, 2022, there were 6,035). Of these, 1,216 were problematic and 4,756 were unproblematic (Perko and Orožen Adamič 1995). The commission produced a gazetteer of all settlements (Gabrovec and Perko 1997) and prepared a renaming proposal for the problematic names (Gabrovec and Perko 1996).

The names of settlements were largely problematic for three reasons:

- Abbreviations in the name; e.g., *Gradišče (K. o. Št. Lovrenc)*;
- The word *del* 'part' in the name, which occurred due to the division of a settlement located in two neighboring municipalities into two settlements; e.g., *Čeče (del)*; and, first and foremost,
- The existence of several identical names referring to different settlements (e.g., the name *Pristava* appeared eight times).

The renaming proposals were based on the following rules:

- No settlement in Slovenia can have the same name as another settlement;
- If there are several settlements with the same name, only the largest settlement retains the simple name and a left or right modifier is added to the names of smaller settlements;
- No abbreviations may be used in names;
- A hyphen with spaces before and after is the only punctuation mark that may be used in a name.

With bilingual endonyms and exonyms, the following are also used:

- A slash without spaces, which replaces the conjunction *ali* 'or' in the official bilingual names of settlements; e.g., *Piran/Pirano* instead of "*Piran* ali *Pirano*" and *Hodoš/Hodós* instead of "*Hodoš* ali *Hodós*" (the draft version of the new normative guide also envisages a slash with spaces before and after, such as in *Spodnje Škofije/Valmarin* or *Dolga vas/Hosszúfalu*);
- Parentheses for endonyms and exonyms referring to the same geographical feature; e.g., *Wien (Dunaj)* 'Vienna' for the German endonym and the Slovenian exonym, or *Ljubljana (Laibach)* for the Slovenian endonym and German exonym (Perko et al. 2017).

The commission thus added a right modifier to the identical names of different settlements, except for the 437 names of the largest settlements sharing the name.

The renaming proposals for the most frequently repeated name, *Pristava*, included: *Pristava nad Borovnico* (i.e., 'above Borovnica'), *Pristava pri Ljutomeru* ('near Ljutomer'), *Pristava pri Novi Gorici* ('near Nova Gorica'), *Pristava pri Podgradu* ('near Podgrad'), *Pristava pri Štjaku* ('near Štjak'), *Pristava pri Vojniku* ('near Vojnik'), and *Pristava v Halozah* ('in Haloze'). The first six modifiers are the names of nearby settlements and the last one is the name of a region. The settlement of Pristava with the largest population retained its name unmodified.

All the names of settlements that contained an abbreviation were written out in full or otherwise modified; e.g., *Sv. Ema* was changed to *Sveta Ema*, *Št. Jurij* to *Šentjurij*, and *Razbore (K. o. Ježni vrh)* to *Razbore nad Kostrevnico*. For settlements that included the word *del* 'part' in their name, the commission suggested that every part of the divided settlements obtain the status of an independent settlement and hence a new name. For example, based on the 1996 proposals, the settlement *Zaplana (del)* in the Municipality of Logatec was to be renamed *Log pri Logatcu* because the hamlet of Log is the central part of Zaplana in that municipality, and *Zaplana (del)* in the Municipality of Vrhnika was to be changed to *Zaplana* because there is no other settlement with the same name in Slovenia and therefore a left or right modifier is not required (Gabrovec and Perko 1996).

New names were proposed for a total of 779 settlements, or 13.0% of all settlements at that time: 698 for settlements with the same names, thirty-five for settlements with an abbreviation in their names, and forty-six for settlements with the word *del* in their names.

The commission sent the renaming proposal (Gabrovec and Perko 1996) to the municipalities, which are responsible for naming settlements in Slovenia. However, only a few of the proposed new names have been implemented to date (Perko and Geršič 2021).

By the end of 2021, the register of settlement names comprised 6,035 names, of which the commission has standardized 3,605, which corresponds to 59.7% of all settlement names in Slovenia. The register is regularly updated following the division or merger of existing settlements and the establishment of new ones.

Standardization of Exonyms

Slovenian exonyms are Slovenian geographical names outside Slovenian ethnic territory that differ from endonyms. In the narrow sense, exonyms include only Slovenian geographical names that differ completely from the original endonyms (e.g., *Nemčija* for *Deutschland* 'Germany') and, in the broader sense, they also include Slovenianized geographical names (e.g., *Pariz* for *Paris*) and translated geographical names (e.g., *Rumena reka* for *Huang He* 'Yellow River').

As toponyms, exonyms can be categorized as:

- Both geonyms and cosmonyms based on the location of the geographical feature they refer to;

- Macrotoponyms and only rarely microtoponyms based on the spatial scope of the geographical feature they refer to;
- Oikonyms and anoikonyms based on the settlement of the geographical feature they refer to;
- The opposites of endonyms based on the name's origin; and
- Various types are based on the type of geographical feature they refer to.

The UN opposes the excessive use of exonyms and prioritizes endonyms. However, well-established exonyms are part of cultural heritage and should therefore be suitably protected (Kladnik et al. 2020).

To facilitate the standardization of exonyms, a gazetteer of Slovenian exonyms was compiled, which roughly reached its present size in 2010. It includes more than five thousand of the most frequently used exonyms that were collected from more than fifty thousand various forms of these types of geographical names appearing in the fourteen Slovenian world atlases (Kladnik et al. 2013), including *Atlant*, the first Slovenian world atlas, which was published in individual sheets between 1869 and 1877 (Urbanc et al. 2006), several important encyclopedias, and *Slovenski pravopis* 2001 (2001 Slovenian Normative Guide). For every exonym in the gazetteer, data are provided in thirty-five categories (the nominative, genitive, and adjectival forms, the original endonym, etymology, type, and so on). Important among these is the "recommended use" category, which distinguishes between the necessary, recommended, less recommended, not recommended or unnecessary, and inappropriate use of the exonym, which is crucial in deciding whether to standardize a specific exonym or not (Kladnik and Geršič 2014).

The analysis of the gazetteer shows that, as expected, the number of Slovenian exonyms across the globe decreases with distance from Slovenia. The only exception is North America, which is connected with Slovenian emigration to this continent over the past centuries and with its prominent global role over the past century. In the open-access Slovenian digital corpus Gigafida, which contains 1.2 billion words from texts published between 1990 in 2011, the most frequent Slovenian exonyms in North America refer to the names of countries and their administrative divisions. Exonyms that occur over five hundred times include *Misisipi* 'Mississippi' (510 times), *Aljaska* 'Alaska' (572), *Kolorado* 'Colorado' (726), *Teksas* 'Texas' (1,597), *Havaji* 'Hawaiian Islands' (1,819), *Kolumbija* 'Colombia' (1,899), *Kalifornija* 'California' (2,136), *Kanada* 'Canada' (9,174), and *Združene države Amerike* and *ZDA* 'United States' (over 200,000 times). In terms of frequency, these are followed by *Dolina smrti* 'Death Valley' (133), *Veliki kanjon* 'Grand Canyon' (121), *Niagarski slapovi* 'Niagara Falls' (109), *Skalno gorovje* 'Rocky Mountains' (96), and *Aleuti* 'Aleutian Islands' (57) (Perko and Kladnik 2017).

The average density of Slovenian exonyms per million km^2 is 477 in Slovenia's four neighboring countries (966 in Austria, 435 in Italy, 301 in Croatia, and 279 in Hungary), 103 in Europe, eighteen in Asia, fourteen in Africa, eight in North America, six in Australia, five in South America, and four in Central America (Perko and Kladnik 2019).

By the end of 2021, the gazetteer comprised 5,044 units. The use of 544 of these was defined as necessary, and the use of a further 2,154 is recommended. To date, the commission has standardized 219 Slovenian exonyms (197 country names and twenty-two other types of geographical names), which corresponds to 4.3% of all the exonyms in the gazetteer and 40.3% of the exonyms whose use was defined as necessary.

Standardization of Bilingual Geographical Names

Minority languages or bilingualism, which includes the bilingual names of settlements, often constitute a complex political issue that reflects the diplomatic relations between individual countries. The importance of the use of bilingual names was also recognized by UNGEGN, which adopted Resolution II/36: Problems of Minority Languages at its second conference, held in London in 1972. It was agreed that, where possible and in consultation with native speakers of the minority language, the countries in question should adopt a common orthography for all geographical names in the minority language, use that orthography for standardization, and publish the standardized names on their official maps and in national gazetteers (Raper 1996).

Because of historical developments, especially the border changes after the First and the Second World Wars, there are two officially recognized ethnic minorities in what is now Slovenia: the Hungarian minority in the northeast and the Italian minority in the southwest. In the bilingual area, Italians account for around 5% of the population. Most live in the coastal towns of Koper (Ital. *Capodistria*), Izola (Ital. *Isola*), and Piran (Ital. *Pirano*). Hungarians mostly live in the countryside, with Lendava (Hung. *Lendva*) as the local center (Geršič et al. 2017).

The earliest regulations on writing geographical names in bilingual areas in Slovenia can be traced back to the days of the former Yugoslavia, when the Socialist Republic of Slovenia was allowed to independently create and adapt its legislative framework, although this had to comply with the Yugoslav constitution (Geršič et al. 2017). At the end of April 1980, the Decree on Writing Geographical Names in Ethnically Mixed Areas of the Socialist Republic of Slovenian Plans and Maps was adopted. Article 3 of the decree provided that in ethnically mixed areas geographical names on plans and maps had to be written bilingually (i.e., in both languages), with the Slovenian name preceding the Italian (Figs. 7 and 9) or Hungarian (Figs. 8 and 10) name. Article 4 required that in ethnically mixed areas the names of settlements, hamlets, streets, regions, waters, peaks, mountain ranges, and microtoponyms had to be written in both languages. The typography and size of letters had to be the same in both languages (Geršič et al. 2017).

Today both minorities are, first and foremost, legally protected by the Slovenian constitution, which mentions both ethnic communities in Article 5 and Italian and Hungarian as the official languages in areas where the two communities live in Article 11. Article 64 of the constitution grants both communities special rights,

Fig. 7 Bilingual Slovenian–Italian sign in a linguistically mixed area near the Italian border in southwest Slovenia (photo by Matija Zorn, GIAM ZRC SAZU Archive)

Fig. 8 Bilingual Slovenian–Hungarian sign in a linguistically mixed area near the Hungarian border in northeast Slovenia (photo by Tatjana Kikec, GIAM ZRC SAZU Archive)

Fig. 9 Slovenian–Italian names of settlements in a bilingual area along the Italian border in southwest Slovenia

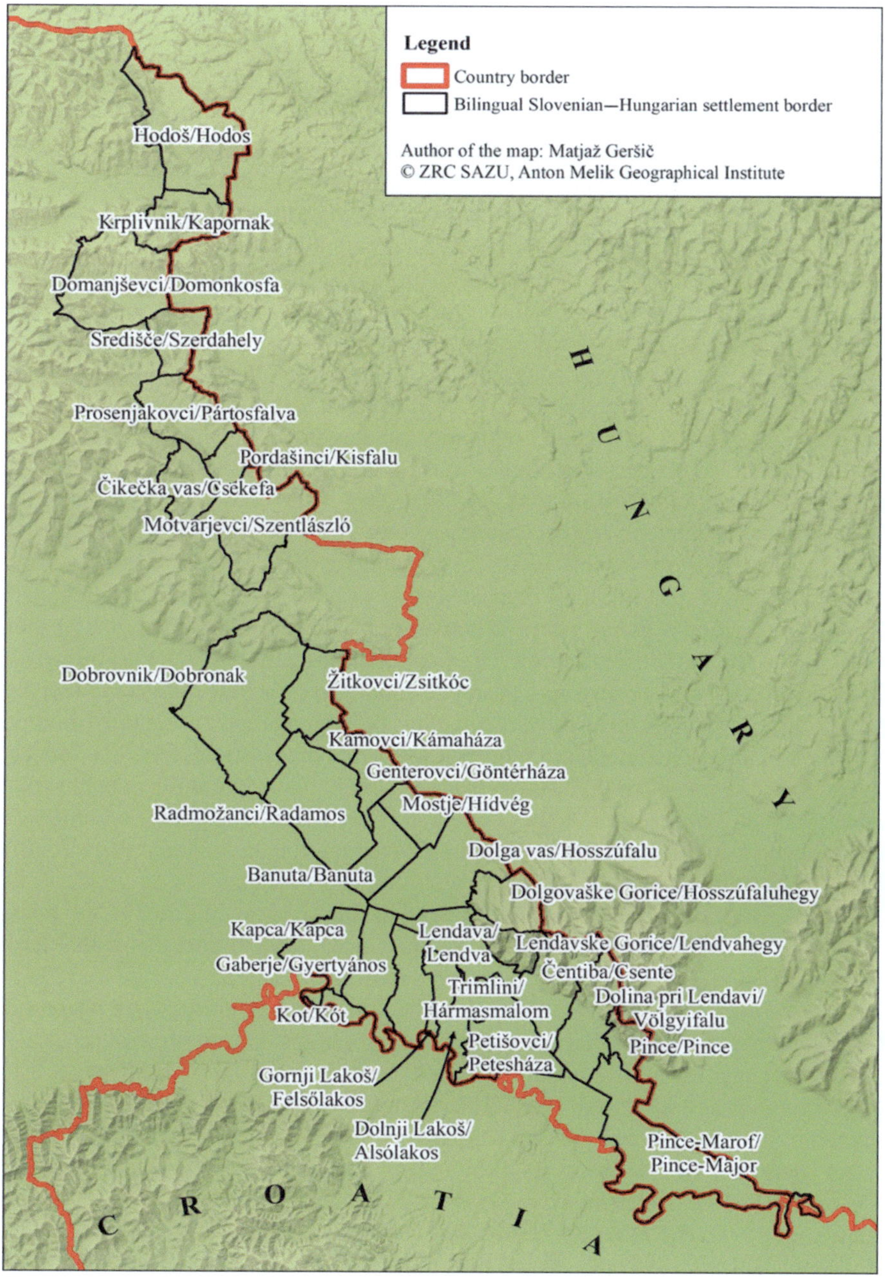

Fig. 10 Slovenian–Hungarian names of settlements in a bilingual area along the Hungarian border in northeast Slovenia

but the manner of exercising them is defined by individual laws or acts. The Self-Governing Ethnic Communities Act, which entered into force in 1994, does not stipulate any special protection regarding the use of language. The written form of geographical names in bilingual areas is defined by two legal documents: The Decree on the Transcription of Geographical Names on National Maps in Ethnically Mixed Areas in the Republic of Slovenia, adopted in 2014, and the Act Regulating the Determination of Territories and the Naming and Marking of Settlements, Streets, and Buildings, adopted in 2008. Article 9, Paragraph 3 of this act, which defines the naming of settlements and documenting their names, stipulates that "in the territories of municipalities in which Italian or Hungarian are used as official languages alongside Slovenian, the names of settlements are provided in Slovenian and Italian or Hungarian." As already evident from its title, the aforementioned decree refers to the written form of geographical names on maps. It also mentions the Register of Geographical Names, in which geographical names from bilingual areas are to be written in both languages. This applies to the names of municipalities, settlements, and streets, and to other geographical names only if they are standardized by the responsible Slovenian government working body.

Following the decree from 2014, the representatives of both ethnic minorities spent much more time preparing their respective lists of names than we had anticipated. Cooperation with the Italian ethnic community has been incomparably better than that with the Hungarian community. Throughout the preparation of the list of names, the Italian community worked with the representatives of the Surveying and Mapping Authority of the Republic of Slovenia, whereas the Hungarian community did not, even though it received much more input material for preparation of the list. Indeed, the commission compiled the list together with the Italian community, which meant any errors could be resolved as they appeared (Geršič et al. 2020). The commission standardized a coordinated list of bilingual Slovenian–Italian names at the 1:25,000 scale in 2020 (Table 2).

No cooperation took place with the Hungarian community while the list was being prepared. The list eventually submitted is made up of several documents that have not been harmonized one with another, and therefore it is full of mistakes and issues that remain unclear. Moreover, a number of non-uniform solutions were proposed. A detailed review showed that various types of problems arose with more than half the Hungarian geographical names proposed (Geršič et al. 2020).

The commission still has to standardize the names in the Register of Geographical Names at the 1:5,000 scale, which is anticipated in the coming decades.

An interesting case of observing bilingualism with regard to a road tunnel recently occurred in the ethnically mixed area with Slovenians and Italians in the southwest of the country (Geršič et al. 2017). In 2015, the newly built Markovec Tunnel (under Markovec Hill) was opened. At the beginning, the sign in front of it was only in Slovenian (Markovec; Fig. 11), whereas all of the other road signs in this area are bilingual. This issue was even discussed by the European Commission, which agreed that the sign in front of the tunnel must include both the Slovenian and Italian names. After this and a concurring judgement by the Slovenian Government Commission

Table 2 List of Slovenian–Italian bilingual names in the Register of Geographical Names at a scale of 1:25,000, which the commission standardized in 2020 in cooperation with the Italian minority

Slovenian name	Italian name
Arce	Arze
Bared	Baredo
Belveder	Villaggio turistico Belvedere
Bolnišnica Valdoltra	Ospedale di Valdoltra
Cerkev Marijinega vnebovzetja	Chiesa della Beata Vergine Assunta
Dolina svetega Jerneja	San Bortolo
Fontanige	Fontanigge
Gravisijev grad	Castello Gravisi
Jagodje	Saleto
Jurijev hrib	Monte San Giorgio
Kanal Giassi	Canal Giassi
Kanal Kurto	Canal Curto
Kanal Piketo	Canal Picchetto
Kanal svetega Jerneja	Canal San Bortolo
Kažadjevolo	Casa del Diavolo
Kažanova	Casa Nova
Kortina	Cortina o Valle Piccola
Križišče Sečovlje	Incrocio Sicciole
Loncan	Lonzano
Marcane	Marzanedo
Mednarodni mejni prehod	Valico internazionale
Mejni prehod Čampore	Valico internazionale di Chiampore
Mejni prehod Kaštelir	Valico internazionale di Santa Barbara
Molin	Mulino
Pomorski mednarodni mejni prehod Izola	Valico marittimo internazionale di Isola
Pomorski mednarodni mejni prehod Koper	Valico marittimo internazionale di Capodistria
Pomorski mednarodni mejni prehod Piran	Valico marittimo internazionale di Pirano
Pri baronici	Casale della Baronessa
Pri Totu	Cortivo Totto
Rt svetega Lovrenca	Punta San Lorenzo
Segadisi	Segadissi
Slami	Slami

(continued)

Table 2 (continued)

Slovenian name	Italian name
Sodniki	Sonnichi
Sveta nedelja	Santa Domenica
Tri kaže	Tre case
Valeta	Valetta
Zaliv Svetega duha	Golfo di San Spirito

Fig. 11 The sign in front of the Markovec Tunnel was initially only in Slovenian (left). Later it was replaced by a bilingual Slovenian–Italian sign (right) because the tunnel is in a bilingual area (photo by Tjaša Škamperle, GIAM ZRC SAZU Archive)

for the Standardization of Geographical Names, the Italian name Monte San Marco was added (Fig. 11).

Slovenian Geographical Names as Exonyms

The most frequent Slovenian geographical names in other languages were collected based on two sources: Geonames and Wikipedia. Despite many errors, both offer a sufficiently large range of Slovenian geographical names in other languages to reveal specific principles of their use. The following groups of geographical names are presented in addition to the name of the country: the names of settlements, rivers and lakes, hills and mountains, (current) regions, and historical regions. Geographical names were analyzed based on the degree of exonymization—that is, the ratio between the number of exonyms of a selected geographical name and the maximum number of language variants of all the selected names in the group (Kladnik et al. 2020).

Ljubljana (Fig. 12) stands out in terms of exonymization of **settlement names**. This is not surprising because Ljubljana is Slovenia's capital and in many respects the most important Slovenian city. It is followed by Maribor (the second-largest Slovenian city), Ptuj (the eighth-largest city), Nova Gorica (twelfth), and Novo Mesto (seventh). It is already clear with the third name that the degree of exonymization and size are not closely connected. Reasons for this can be ascribed to various factors. It is certainly not insignificant that, at the time when exonyms were created more intensely, the ranking of settlements by population was different than today; in addition, some settlements also had a different function.

The names of the twenty-nine largest Slovenian settlements appear in thirty-seven languages. The majority of exonyms are in German. This is hardly surprising because the German linguistic environment was one that coexisted in and dominated today's Slovenian environment the longest, and German exonyms in particular are a reminder of historically strong Germanic influence. It is also important that the official language of Slovenia's northern neighbor Austria is German. German is followed by Latin. Latin exonyms in Slovenian territory are also not surprising. Many

Fig. 12 Slovenia's capital, Ljubljana, has the most exonyms among all Slovenian endonyms; for example, German Laibach, Italian Lubiana, Czech Lublaň, Polish Lublana, and Spanish and Portuguese Liubliana (photo by Matic Štojs, https://www.shutterstock.com)

settlements go back to Roman antiquity, which is also reflected in their names. In addition, from an ecclesiastical administrative point of view, Slovenia was under the jurisdiction of Rome, where Latin was the language of the Church, including liturgy, and of science and art. There are also a considerable number of Italian exonyms. This is not surprising, considering that this is the official language of Slovenia's western neighbor and that the western part of today's Slovenia was part of Italy during the interwar period and its southwestern part belonged to the Republic of Venice for centuries. Somewhat more surprising is the small number of exonyms in the languages of Slovenia's other two neighbors, Croatia and Hungary (Kladnik et al. 2020).

With regard to the **names of rivers**, the highest degree of exonymization among major Slovenian rivers is achieved by the Sava, Slovenia's longest river, followed by the Soča and Drava rivers. The Soča is not among the six longest Slovenian rivers, but it is important in terms of tourism and hydroelectric power (Komac and Zorn 2021), and it is even better known for being the Austrian–Italian front line (the Soča/Isonzo Front; Kumer et al. 2020) in the First World War. The Drava is Slovenia's second-largest river, important especially because of its hydroelectric power plants (Zorn 2018). All the rivers mentioned above flow through different countries. Rivers with basins only in Slovenia have quite a low degree of exonymization. The degree of exonymization of the names of rivers also shows that it does not depend on the river length and that it is influenced by several factors (Kladnik et al. 2020).

With the **names of lakes**, Lake Bled on the eastern edge of the Julian Alps stand out in terms of the degree of exonymization. Bled is one of the most important Slovenian tourist destinations, and in the past it was a fashionable tourism resort that was frequented by many wealthy foreign visitors. A somewhat lower degree is typical of Lake Bohinj in the Julian Alps and an even lower one for Lake Cerknica in the classical karst area of southwest Slovenia. The low degree of exonymization of Lake Cerknica is a little surprising considering that it was described as an example of an intermittent lake as early as the second half of the seventeenth century (Fig. 13; Valvasor (1689) 2009–2013). Judging from the examples presented, it seems that tourism plays a greater role than research, which is understandable considering the difference between the number of tourists versus researchers. Most exonyms for hydronyms are in English, and German stands out among the languages of Slovenia's neighboring countries (Kladnik et al. 2020).

In terms of degree of exonymization, Slovenia's highest peak, Mount Triglav (2,864 m; Fig. 14)—also considered a German or Italian peak in the past—stands out among the **names of mountains** (Mikša and Zorn 2018). This is followed by Mount Krn (2,244 m), which is not among the highest Slovenian mountains, but it played an important role on the Austrian–Italian front line during the First World War and it belonged to Italy during the interwar period. Elevation is thus not a key exonymization factor. Italian predominates among the exonyms, which is hardly surprising considering that western Slovenia, which includes the majority of the highest mountains, belonged to Italy during the interwar period (Zorn and Mikša 2018; Kladnik et al. 2020), when Italy's Italianization efforts also included actively changing geographical names.

Fig. 13 Map of intermittent Lake Cerknica from the second half of the seventeenth century, with German geographical names (Valvasor 1689; *Source* GIAM ZRC SAZU Archive)

Fig. 14 The name of Mount Triglav, Slovenia's highest peak, is the most common exonym among the names of Slovenian mountains. It was first written in Slovenian as Terglou in the middle of the eighteenth century (photo by Miha Pavšek, GIAM ZRC SAZU Archive)

Slovenia is a very diverse country in terms of landscapes (Ciglič and Perko 2013; Perko et al. 2020). Landscape diversity also results in a number of **choronyms, or names of regions**. English names for forty-eight Slovenian regions are presented in Table 3 and Fig. 15.

Among choronyms, the Julian Alps (Sln. *Julijske Alpe*) have the highest degree of exonymization. This is the highest and largest mountain range in Slovenia, extending across the border to Italy. The reason for the high degree of exonymization is definitely their exceptional natural diversity and appeal for both winter and summer tourism. A relatively high degree of exonymization (0.38) is also typical of two mountain ranges, the Karawanks (Sln. *Karavanke*) and the Kamnik–Savinja Alps (Sln. *Kamniško-Savinjske Alpe*). Second place on this scale goes to the Gulf of Trieste (Sln. *Tržaški zaliv*). It is part of the larger Gulf of Venice (Sln. *Beneški zaliv*) and its territory is shared by Italy, Slovenia, and Croatia. The ports of Trieste and Koper are located in its hinterland, providing access to the open sea to many countries in central and eastern Europe. The gulf's tourism role is also important. A high degree of exonymization is also typical of the Slovenian choronym *Kras*, which refers to the Karst Plateau rising steeply from the Gulf of Trieste (Kladnik et al. 2020). The Slovenian name *Kras* is especially important in terms of describing the type of landscape that was scientifically studied for the first time in this very region (Ferk and Zorn 2015). This scholarly importance is clearly reflected in the degree of the name's exonymization (Kladnik et al. 2020).

When comparing the number of exonyms for these choronyms by language, most can be found in English. This is followed by German, which is hardly surprising given the historical circumstances. The same applies to Italian (Kladnik et al. 2020).

The **names** of Slovenian **historical regions** or lands are among (the most) frequent Slovenian geographical names used in other languages (Fig. 16). The highest degree of exonymization is typical of Istria (Sln. *Istra*). Most of the former land of Istria now lies in Croatia. Croatian Istria is an important European tourism region, which might be one of the reasons for the high degree of exonymization of its name. However, the historical aspect seems more important because this was already an important region in antiquity, and later on it became a strategically important border area between the Republic of Venice and the Habsburg Monarchy. Carniola (Sln. *Kranjska*) dominates among the other names, which is not surprising considering that it was the central part of Slovenian ethnic territory. When comparing the number of exonyms by language, English predominates, followed by Italian and Latin (Kladnik et al. 2020).

Exonyms are also used for the **name of the country**, *Slovenija*. Sixty-four exonym versions were identified for it (Table 4 and Fig. 17). The name written as Slovenia is the most common, followed by *Eslovenia*, *Slowenien*, and *Sloveniya*. The frequency of individual exonym versions reflects the colonial past of third-world countries, and thus English, German, and Spanish name equivalents predominate (Kladnik et al. 2020).

Taking into account all the geographical names discussed, the majority of exonyms referring to Slovenian territory are in German, followed by English and Italian. The reason for the great number of German exonyms in Slovenian territory lies in history: Slovenian ethnic territory, part of which now belongs to the Republic of Slovenia, was

Table 3 English and Slovenian names of macroregions and (mezzo)regions (for location see Fig. 15; Perko and Ciglič 2020)

English names	Slovenian names
Macroregion	***Makroregija***
Region	*Regija*
Alps	***Alpe***
Western Karawanks	*Zahodne Karavanke*
Eastern Karawanks	*Vzhodne Karavanke*
Kamnik–Savinja Alps	*Kamniško-Savinjske Alpe*
Julian Alps	*Julijske Alpe*
Cerkno, Škofja Loka, Polhov Gradec, and Rovte Hills	*Cerkljansko, Škofjeloško, Polhograjsko in Rovtarsko hribovje*
Sava Hills	*Posavsko hribovje*
Velenje and Konjice Hills	*Velenjsko in Konjiško hribovje*
Pohorje, Strojna and Kozjak	*Pohorje, Strojna in Kozjak*
Ložnica and Hudinja Hills	*Ložniško in Hudinjsko gričevje*
Sava Plain	*Savska ravan*
Savinja Plain	*Savinjska ravan*
Pannonian Basin	***Panonska kotlina***
Goričko	*Goričko*
Lendava Hills	*Lendavske gorice*
Slovenian Hills	*Slovenske gorice*
Dravinja Hills	*Dravinjske gorice*
Haloze	*Haloze*
Mount Boč and Macelj	*Boč in Macelj*
Voglajna and Upper Sotla Hills	*Voglajnsko in Zgornjesotelsko gričevje*
Central Sotla Hills	*Srednjesotelsko gričevje*
Krško, Senovo, and Bizeljsko Hills	*Krško, Senovsko in Bizeljsko gričevje*
Mura Plain	*Murska ravan*
Drava Plain	*Dravska ravan*
Krka Plain	*Krška ravan*
Dinaric Alps	***Dinarsko gorovje***
Kambreško and Banjšice Plateaus	*Kambreško in Banjšice*
Trnovo Forest Plateau, Mount Nanos, and Hrušica Plateau	*Trnovski gozd, Nanos in Hrušica*
Idrija Hills	*Idrijsko hribovje*

(continued)

Table 3 (continued)

English names	Slovenian names
Javornik Hills and Snežnik Plateau	*Javorniki in Snežnik*
Pivka Lowland and Mount Vremščica	*Pivško podolje z Vremščico*
Inner Carniola Lowland	*Notranjsko podolje*
Krim Hills and Menišija Plateau	*Krimsko hribovje in Menišija*
Bloke Plateau	*Bloke*
Big Mountain, Mount Stojna, and Mount Gotenica	Velika gora, Stojna in Goteniška gora
Ribnica–Kočevje Lowland	*Ribniško-Kočevsko podolje*
Little Mountain, Kočevje Rog Plateau, and Mount Poljane	*Mala gora, Kočevski rog in Poljanska gora*
Velike Lašče Region	*Velikolaščanska pokrajina*
Ljubljana Marsh	*Ljubljansko barje*
Novo Mesto Region	*Novomeška pokrajina*
Lower Carniola Lowland	*Dolenjsko podolje*
Radulja Hills	*Raduljsko hribovje*
Dry Carniola and Dobrepolje	*Suha krajina z Dobrepoljem*
White Carniola	*Bela krajina*
Gorjanci Hills	*Gorjanci*
Mediterranean	***Sredozemlje***
Gorica Hills	*Goriška brda*
Vipava Valley	*Vipavska brda*
Karst Plateau	*Kras*
Brkini Hills and Reka Valley	*Brkini in dolina Reke*
Podgorje Karst Plateau, Čičarija Plateau, and Podgrad Lowland	*Podgorski kras, Čičarija in Podgrajsko podolje*
Koper Hills	*Koprska brda*

under strong German influence in the past (it was part of the Habsburg Monarchy for centuries). English, which follows German, is in second place primarily because of the use of the definite article in the English versions of choronyms and hydronyms—which, however, is not the case with settlement names. The high position of Italian as the official and majority language of neighboring Italy on the scale is not surprising. It can primarily be ascribed to strong Italianization carried out in western Slovenia during the interwar period, when this part of the country was part of Italy. In addition, part of Slovenian ethnic territory belonged to the Republic of Venice for several centuries (Kladnik et al. 2020).

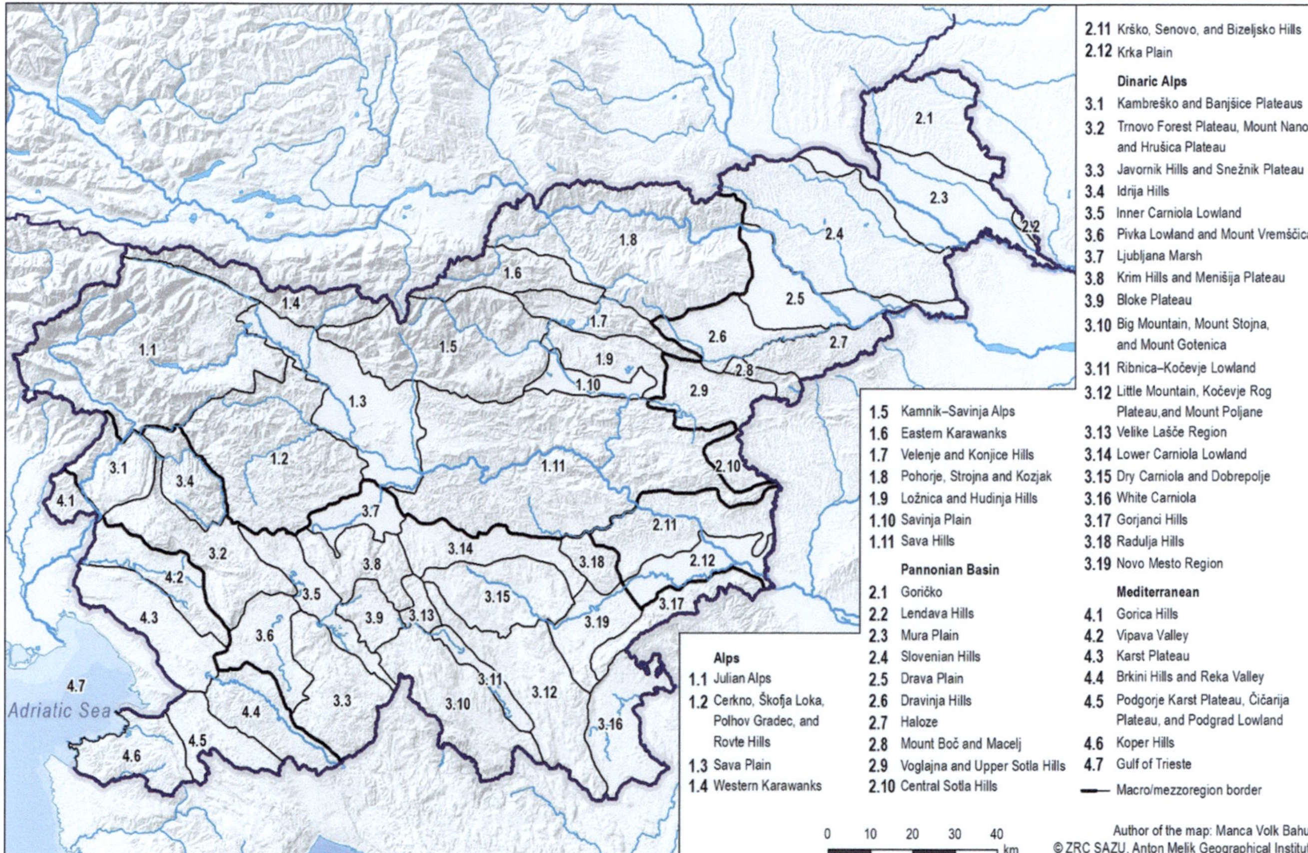

Fig. 15 English names for Slovenia's regions (Perko and Ciglič 2020; *Source* GIAM ZRC SAZU Archive)

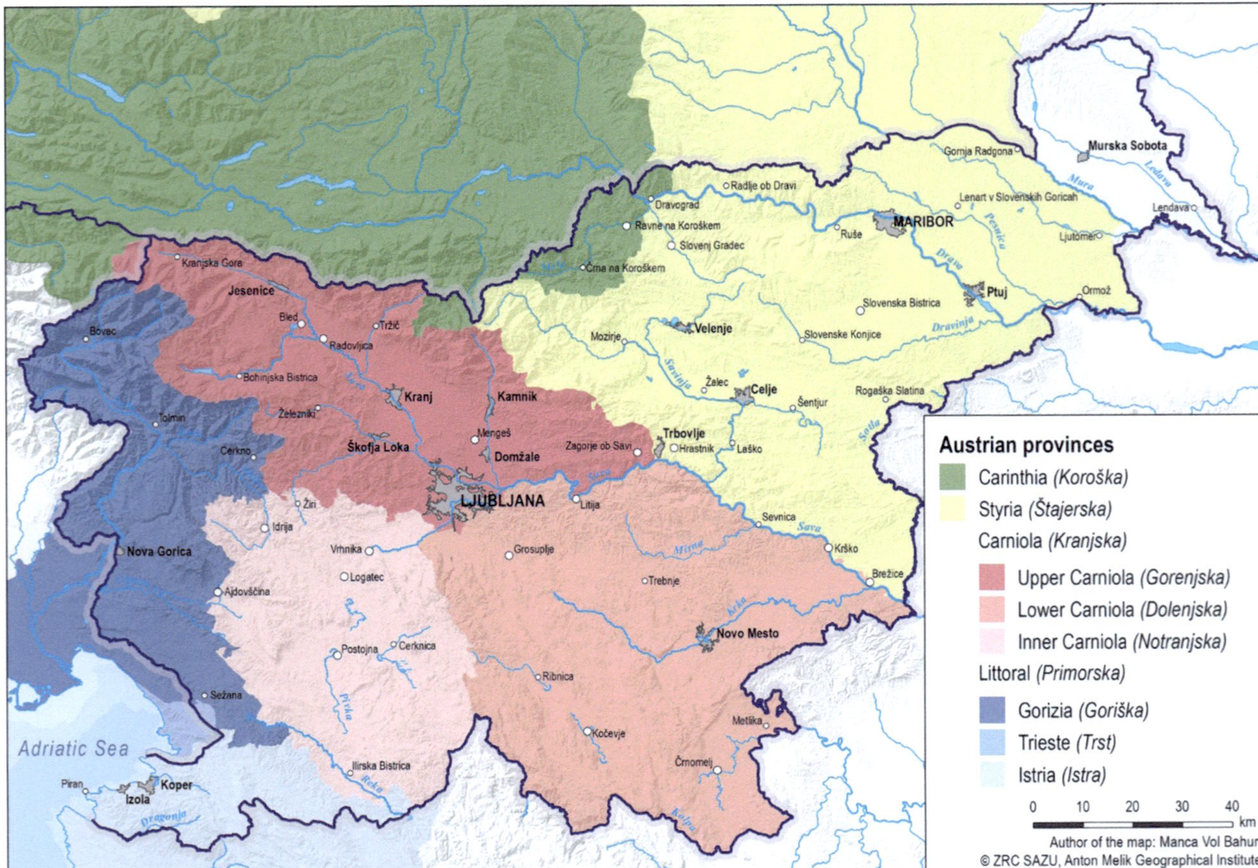

Fig. 16 A map of the Austrian lands showing the administrative division of what is now Slovenia in 1914, before the First World War. The legend contains English names with the original Slovenian forms in parentheses (Perko and Ciglič 2020; *Source* GIAM ZRC SAZU Archive)

Conclusion

In accordance with the UN resolutions on geographical names and Slovenian legislation on geographical names and their linguistic, geographical, and legal aspects, as of January 1st, 2022 the Slovenian Government Commission for the Standardization of Geographical Names has standardized:

- A total of 464, or 100%, of the 464 names in Slovenia and twenty-two exonyms outside Slovenia in the gazetteer of the 1:1,000,000 map of Slovenia;
- A total of 4,273, or 100%, of the 4,273 names in Slovenia and twenty-six exonyms outside Slovenia in the gazetteer of the 1:250,000 map of Slovenia;
- A total of 9,854, or 4.5%, of the 216,638 entries referring to names in Slovenia in the Register of Geographical Names;
- A total of 426, or 100%, of the 426 names in the Register of Country Names;

Table 4 Slovenia's name in various languages

Language	Slovenia's name
Scottish Gaelic	An t-Slòbhain
Irish	an tSlóivéin
Iloko	Eslobenia
Tagalog	Eslobenya
Aragonese	Eslovenia
Asturian	Eslovenia
Spanish	Eslovenia
Basque	Eslovenia
Galician	Eslovenia
Quechua	Eslovenia
Galician	Eslovenia
Portuguese	Eslovénia
Tetum	Eslovénia
Catalan	Eslovènia
Occitan	Eslovènia
Portuguese	Eslovênia
Waray	Eslovenya
Wolof	Esloweeni
Wolof	Esloweni
Maori	Horowinia
Somali	Islofeeniya
Somali	Isloveeniya
Zulu	i-Slovenia
Sardinian	Islovénia
Quechua	Isluwiniya
Quechua	Isluwinya
Maltese	is-Slovenja
Malagasy	Silaovenia
Yoruba	Silofania
Lingala	Siloveni
Luba-Katanga	Siloveni
Tongan	Silōvenia

(continued)

Table 4 (continued)

Rundi	Siloveniya
Ganda	Sirovenya
Albanian	Slloveni
Albanian	Sllovenia
Navajo	Słobíín Bikéyah
Bicolano	Slobinya
Welsh	Slofenia
Yoruba	Slofénìà
Gaelic	Sloibhìnia
Estonian	Sloveenia
Võro	Sloveeniä
Western Frisian	Sloveenje
Latgalian	Sloveneja
Walloon	Sloveneye
Bambara	Sloveni
Cornish	Sloveni
English	Slovenia
Breton	Slovenia
English	Slovenia
Finnish	Slovenia
Faroese	Slovenia
Interlingua	Slovenia
Indonesian	Slovenia
Igbo	Slovenia
Ido	Slovenia
Italian	Slovenia
Kikuyu	Slovenia
Kalaallisut	Slovenia
Latin	Slovenia
Malagasy	Slovenia
Malay	Slovenia
Norwegian Bokmål	Slovenia
North Ndebele	Slovenia
Norwegian Nynorsk	Slovenia
Norwegian	Slovenia
Pampanga	Slovenia
Piemontese	Slovenia
Romansh	Slovenia
Romanian	Slovenia
Northern Sami	Slovenia
Shona	Slovenia
Swahili	Slovenia
Javanese	Slovénia
Icelandic	Slóvenía
Ewe	Slovenia nutome
Friulian	Slovenie
Western Frisian	Slovenië
Limburgish	Slovenië
Dutch	Slovenië
French	Slovénie
Narom	Slovénie
Arpitan	Slovènie
Danish	Slovenien
Swedish	Slovenien

(continued)

Table 4 (continued)

Dutch	Sloveniën
Veps	Slovenii
Bosnian	Slovenija
Serbo-Croatian	Slovenija
Croatian	Slovenija
Lithuanian	Slovėnija
Latvian	Slovēnija
Esperanto	Slovenio
Azerbaijani	Sloveniya
Uzbek	Sloveniya
Volapük	Sloveniyän
Esperanto	Slovenujo
Kurdish	Slovenya
Turkish	Slovenya
Samogitian	Slovienėjė
Akan	Slovinia
Czech	Slovinsko
Slovak	Slovinsko
Polish	Słowenia
Afrikaans	Slowenië
Bavarian	Slowenien
German	Slowenien
Luxembourgish	Slowenien
Low German	Slowenien
Fulah	Slowenii
Turkmen	Sloweniýa
Uighur	Slowéniye
Kashubian	Sloweńskô
Pitcairn-Norfolk	Slowenya
Upper Sorbian	Słowjenska
Lower Sorbian	Słowjeńska
Silesian	Słowyńijo
Lombard	Sluenia
Corsican	Sluvenia
Sango	Solovenïi
Hawaiian	Solowenia
Hausa	Sulobeniya
Hakka	Sû-lo̍k-vì-nì-â
Fulah	Suloweniya
Hungarian	Szlovénia
Nauru	Tsirobeniya
Vietnamese	Xlô-ven-ni-a
Vietnamese	Xlô-ven-ni-a
Manx	Yn Clovean
Greek	Σλοβενία
Belarusian	Славенія
Chechen	Словени
Mongolian	Словени
Ossetic	Словени
Macedonian	Словенија
Serbian	Словенија
Bulgarian	Словения
Kazakh	Словения
Kyrgyz	Словения

(continued)

Table 4 (continued)

Russian	Словения
Tajik	Словения
Tatar	Словения
Ukrainian	Словенія
Armenian	Սլովենիա
Georgian	სლოვენია
Japanese	スロベニア
Yiddish	סלאוועניע
Hebrew	סלובניה
Persian	اسلوونی
Kashmiri	سَلووینیا
Persian	سلوانیا
Pashto	سلوانیا
Arabic	سلوفينيا
Urdu	سلووینیا
Uyghur	سلوۋېنىيە
Sindhi	سلوينيا
Kurdish	سلۆڤێنیا
Uyghur	سلوۋېنىيە
Aramaic	ܣܠܘܒܢܝܐ
Nepali	स्लोभेनिया
Hindi	स्लोवेनिया
Marathi	स्लोव्हेनिया
Assamese	শ্লোভেনিয়া
Bangla	স্লোভানিয়া
Punjabi	ਸਲੋਵੇਨੀਆ
Gujarati	સ્લોવેનિયા
Odia	ସ୍ଲୋଭେନିଆ
Tamil	ஸ்லோவேனியா
Telugu	స్లోవేనియా
Kannada	ಸ್ಲೋವೇನಿಯಾ
Malayalam	സ്ലോവേനിയ
Sinhala	ස්ලෝවේනියාව
Thai	ประเทศสโลวีเนีย
Thai	สโลวีเนีย
Burmese	ဆလိုဗေးနီးယား
Tibetan	ས་ལོ་ཝེ་ནི་ཡ།
Dzongkha	སུ་ལོ་བི་ནི་ཡ
Lao	ສະໂລເວເນຍ
Khmer	ស្លូវេនី
Amharic	ስሎቬኒያ
Tigrinya	ስሎቬኒያ
Korean	슬로베니아
Chinese	斯洛文尼亚
zh-hant	斯洛維尼亞

Fig. 17 Selected names for Slovenia across the globe

- A total of 3,605, or 59.7%, of the 6,035 names in the Register of Settlement Names; and
- A total of 219, or 4.3%, of the 5,044 names in the Register of Slovenian Exonyms.

Among the marine geographical names, the commission has standardized:

- Three, or 100%, of the three marine names in Slovenia in the gazetteer of the 1:1,000,000 map of Slovenia;
- Three, or 100%, of the three marine names in Slovenia in the gazetteer of the 1:250,000 map of Slovenia;
- Three, or 4.5%, of the sixty-seven names (202 entries) in the Register of Geographical Names.

Nearly all the registers and gazetteers are freely available on the official website of the Slovenian Government Commission for the Standardization of Geographical Names (www.gov.si/teme/standardizacija-zemljepisnih-imen).

Currently, the most important source of standardized Slovenian endonyms is the gazetteer of the 1:250,000 map of Slovenia, which includes all the standardized names from the register of settlement names, the gazetteer of the 1:1,000,000 map of Slovenia, and the Register of Geographical Names at scales of 1:1,000,000 and 1:250,000.

Hence, since its establishment in 1995, the commission has standardized over ten thousand Slovenian geographical names, or four hundred a year. It has standardized just under 5% of the approximately two hundred thousand Slovenian endonyms and just under 5% of the approximately five thousand Slovenian exonyms. This means it still needs to standardize the majority of Slovenian geographical names, which will take a lot of time to complete, especially because it will have to once again review the names it has already standardized due to envisaged changes to the Slovenian normative rules for writing geographical names.

Questions

1. Why is the standardization of geographical names important for everyone?
2. Who is responsible for standardizing geographical names at the global and national levels?
3. Which experts should participate in standardizing geographical names?
4. Which punctuation mark is recommended in bilingual endonyms to separate names in two languages?
5. Why does Slovenia have so few marine names?

References

Backus Borshi O (2015) Formal expression of definiteness in Albanian n: a description based on comparison with English. PhD thesis, Charles University, Prague

Bijak U (2019) The development of Polish toponomastic terminology. Ann Univ Mariae Curie-Skłodowska, Sect FF: Philol 37(1):91–103. https://doi.org/10.17951/ff.2019.37.1.91-103

Ciglič R, Perko D (2013) Europe's landscape hotspots. Acta Geogr Slov 53(1):117–139. https://doi.org/10.3986/AGS53106

Dobrovoljc H (2013) Predgovor (Preface). In: Kladnik D, Perko D (eds) Slovenska imena držav. Geografija Slovenije 25. Založba ZRC, Ljubljana, pp 8–9. https://doi.org/10.3986/9789612546052

Ferk M, Zorn M (2015) Kras - večpomenski izraz z mednarodno veljavo (Kras - a polysemous term with international recognition). Kronika 63(3):367–378

Furlan M, Gložančev A, Kladnik D, Perko D, Šivic-Dular A (2008) Imenik zemljepisnih imen Državne pregledne karte Republike Slovenije v merilu 1 : 250.000 = Gazetteer of the National general map of the Republic of Slovenia at the scale 1:250,000. In: Državna pregledna karta Republike Slovenije 1 : 250.000: standardizirana slovenska zemljepisna imena = National general map of the Republic of Slovenia 1:250.000: standardized Slovene geographical names. Geodetska uprava Republike Slovenije, Ljubljana

Furlan M, Gložančev A, Šivic-Dular A (2000) Pravopisna ustreznost zapisa lastnoimenskega gradiva v registru zemljepisnih imen in registru prostorskih enot (Orthographically correct representation of proper names in the Register of Geographical Names and Register of Spatial Units). Geogr Vestn 72(1):73–86

Furlan M, Gložančev A, Šivic-Dular A (2001) Pravopisno ustrezen zapis zemljepisnih in stvarnih lastnih imen v Registru zemljepisnih imen in Registru prostorskih enot (Orthographically correct representation of proper names in the code catalogues Register of Geographical Names and Register of Spatial Units). Geodetska uprava Republike Slovenije, Ljubljana

Gabrovec M, Perko D (1996) Seznam predlogov novih uradnih imen za sporna uradna imena naselij v Sloveniji (List of proposals for new official names for disputed official names of settlements in Slovenia). Unpublished project report. ZRC SAZU Geografski inštitut Antona Melika, Ljubljana

Gabrovec M, Perko D (1997) Imenik uradnih imen naselij v Sloveniji (Gazetteer of official names of settlements in Slovenia). Unpublished project report. ZRC SAZU Geografski inštitut Antona Melika, Ljubljana

Geršič M, Kladnik D, Brnot M (2020) Standardization of geographical names in officially recognized bilingual areas of Slovenia. Onomàstica 6:145–167

Geršič M, Kladnik D, Repolusk P (2017) Geographical names in the languages of official minorities in Slovenia. Onomàstica 3:105–126

Gundacker R (2014) The significance of foreign toponyms and ethnonyms in old kingdom. In: Höflmayer F (ed) The late third millennium in the Ancient Near East: chronology, C14, and climate change. Oriental institute seminars 11. The University of Chicago, Chicago, pp 333–426

Kladnik D, Ciglič R, Hrvatin M, Perko D, Repolusk P, Volk M (2013) Slovenski eksonimi (Slovenian exonyms). Geografija Slovenije 24. Založba ZRC, Ljubljana. https://doi.org/10.3986/9789612546021

Kladnik D, Geršič M (2014) A gazetteer of Slovenian exonyms. In: Jordan P, Woodman P (eds) The quest for definitions: proceedings of the 14th UNGEGN Working Group on Exonyms Meeting. Verlag Dr. Kovač, Hamburg, pp 223–247

Kladnik D, Geršič M, Perko D (2020) Slovenian geographical names. Acta Geogr Slov 61(3):7–190. https://doi.org/10.3986/AGS.9394

Kladnik D, Perko D (2007) Problematična imena držav v slovenskem jeziku (Problematic country names in Slovene language). Geogr Vestn 79(2):79–95

Kladnik D, Perko D (2013) Slovenska imena držav (Slovenian country names). Geografija Slovenije 25. Založba ZRC, Ljubljana. https://doi.org/10.3986/9789612546052

Kladnik D, Perko D (2015a) Družbena občutljivost standardizacije imen držav na primeru Južne Afrike in Moldavije (Reception of the standardized names of countries – the delicate case of South Africa and Moldova). In: Dobrovoljc H, Lengar Verovnik T (eds) Pravopisna razpotja: razprave o pravopisnih vprašanjih. Založba ZRC, Ljubljana, pp 251–259. https://doi.org/10.3986/9789610504429_20
Kladnik D, Perko D (2015b) Problematika poimenovanja držav in odvisnih ozemelj (Names of countries and dependent territorial units). In: Dobrovoljc H, Lengar Verovnik T (eds) Pravopisna razpotja: razprave o pravopisnih vprašanjih. Založba ZRC, Ljubljana, pp 235–249. https://doi.org/10.3986/9789610504429_19
Komac B, Zorn M (2021) Stoletje (ne)participativnosti: gradnja hidroelektrarn v Posočju (A century of (non)participation: construction of hydroelectric power plants in the Soča Valley). In: Nared J, Polajnar Horvat K, Razpotnik Visković N (eds) Regionalni razvoj včeraj, danes, jutri. Regionalni razvoj 8. Založba ZRC, Ljubljana, pp 163–188
Kumer P, Zorn M, Žorž G, Gašperič P (2020) From the Great War to interwar fortifications: changing narratives attached to the military landscape in Western Slovenia. In: O'Reilly G (ed) Places of memory and legacies in an age of insecurities and globalization. Springer, Cham, pp 247–263. https://doi.org/10.1007/978-3-030-60982-5_11
Mikša P, Zorn M (2018) The, "battle" for the mountains: Germans versus Slovenes in the South-Eastern Alps in the late 19th and early 20th century. Adv Environ Res 65:199–227
Perko D (1996a) Sporna imena držav v slovenskem jeziku (The problemacy in translating foreign countries' names into Slovene). Geogr Obz 43(3):20–27
Perko D (1996b) Standardizirana imena držav v slovenskem jeziku (Standardized names of countries in Slovenian language). Geogr Obz 43(4):18–26
Perko D (2001) Zgoščeni imenik zemljepisnih imen Slovenije = Concise gazetteer of Slovenia. Geodetska uprava Republike Slovenije, Ljubljana
Perko D, Ciglič R (2020) Slovenia's regions. In: Perko D, Ciglič R, Zorn M (eds) The geography of Slovenia: small but diverse. Springer, Cham, pp 227–243. https://doi.org/10.1007/978-3-030-14066-3_15
Perko D, Ciglič R, Zorn M (2020) Slovenia: a European landscape hotspot. In: Perko D, Ciglič R, Zorn M (eds) The geography of Slovenia: small but diverse. Springer, Cham, pp 1–20. https://doi.org/10.1007/978-3-030-14066-3_1
Perko D, Geršič M (2021) Sporna imena naselij v Sloveniji (Disputed names of settlements in Slovenia). Georitem 32. Založba ZRC, Ljubljana. https://doi.org/10.3986/9789610505365
Perko D, Jordan P, Komac B (2017) Exonyms and other geographical names. Acta Geogr Slov 57(1):99–107. https://doi.org/10.3986/AGS.4891
Perko D, Kladnik D (2017) Slovenian exonyms in North America. Acta Geogr Slov 57(1):119–139. https://doi.org/10.3986/AGS.4777
Perko D, Kladnik D (2019) Slovenski eksonimi v sosedstvu Slovenije (Slovenian exonyms in Slovenia's neighborhood). Geogr Vestn 91(2):139–160. https://doi.org/10.3986/GV91208
Perko D, Orožen Adamič M (1995) Uradna imena naselij v Sloveniji (Official names of settlements in Slovenia). Unpublished project report. ZRC SAZU Geografski inštitut Antona Melika, Ljubljana
Perko D (2022) Standardizirana zemljepisna imena v slovenskem jeziku (Standardized geographical names in the Slovenian language). In: Lengar Verovnik T, Vranjek Ošlak U (eds) Pravopis na zrnu graha. Založba ZRC, Ljubljana, pp 181–197
Radovan D (1995) Toponimska navodila za Slovenijo (Toponymic guidelines for Slovenia). Geodetska uprava Republike Slovenije, Ljubljana
Radovan D, Majdič V (1995) Toponymic guidelines for Slovenia. Surveying and Mapping Authority of the Republic of Slovenia, Ljubljana
Raper EP (1996) United Nations documents on geographical names. Names Research Institute, Pretoria

Šivic-Dular A (1988) K normiranju slovenskih zemljepisnih imen (Towards the standardization of Slovene geographical names). In: Pogorelec B (ed) XXIV. seminar slovenskega jezika, literature in kulture. Filozofska fakulteta, Ljubljana, pp 55–66

Urazmetova AV, Shamsutdinova JH (2017) Principles of place names classifications. XLinguae 10(4):26–33. https://doi.org/10.18355/XL.2017.10.04.03

Urbanc M, Fridl J, Kladnik D, Perko D (2006) Atlant and Slovene national consciousness in the second half of the 19th century. Acta Geogr Slov 46(2):251–283. https://doi.org/10.3986/AGS 46204

Valvasor JV (2009–2013) Čast in slava vojvodine Kranjske (Die Ehre dess Hertzogthums Crain). Zavod Dežela Kranjska, Ljubljana

Zorn M (2018) The economic role of the Drava River in Slovenia: from navigation to hydropower. Podravina 17(33):32–62

Zorn M, Mikša P (2018) The Rapallo Border between Italy and Yugoslavia after the First World War. Hist Alpes 23:165–181

Suggested Reading

Kladnik D, Geršič M, Perko D (2020) Slovenian geographical names. Acta Geogr Slov 61(3):7–190. https://doi.org/10.3986/AGS.9394

Perko D, Jordan P, Komac B (2017) Exonyms and other geographical names. Acta Geogr Slov 57(1):99–107. https://doi.org/10.3986/AGS.4891

Perko D, Geršič M, Zorn M (2022) Endonyms and exomyns. In: Richardson D (ed) The international encyclopedia of geography: people, the earth, environment, and technology. Wiley, Chichester. https://doi.org/10.1002/9781118786352.wbieg2094

The *New Zealand Geographic Board* and the Contested Nature of Place Names in New Zealand

Neil Lindsay and Robin Kearns

Abstract Processes of naming the landscape are inherently *political* as how, why, and what names are used can create conflict between competing groups, beliefs or ideas. This chapter examines place names in New Zealand and how they are implemented and changed. In doing so, we detail the work of the New Zealand Geographic Board Ngā Pou Taunaha o Aotearoa and how decisions for place names are made. We assert that naming practices in New Zealand have evolved in response to social and political shifts since the 1970s particularly with changes in attitude towards restoring indigenous Māori rights and the importance of place names as 'taonga' (treasures), giving practical and tangible effect to the Treaty of Waitangi principles of participation, protection and partnership. We detail the key issues surrounding the politics of place names in New Zealand, using case studies of recent place name changes. We also examine how the New Zealand Geographic Board reflects broader changes to New Zealand society.

Keywords New Zealand · Place names · Toponyms · Critical toponymies · Naming politics · Indigenous naming

Objectives

By the end of this chapter, readers should be able to recognise:

1. The role of the New Zealand Geographic Board and how place names are implemented, removed or changed.
2. The reasons why some place names in New Zealand have been changed in recent decades.

N. Lindsay · R. Kearns (✉)
School of Environment, The University of Auckland, Auckland, New Zealand
e-mail: r.kearns@auckland.ac.nz

N. Lindsay
e-mail: neil.lindsay@auckland.ac.nz

© The Author(s), under exclusive license to Springer Nature Switzerland AG 2023

G. O'Reilly (ed.), *Place Naming, Identities and Geography*,
Key Challenges in Geography, https://doi.org/10.1007/978-3-031-21510-0_10

3. How naming practices in New Zealand have evolved in response to shifts in cultural politics since the mid twentieth century.

Introduction

The process of naming any town, beach, mountain, or feature on the ocean floor involves a deeply entrenched set of ideas around the identity of a place and how its identity is represented. Toponyms (place names) are intrinsically linked with our attachment to particular places, our sense of self, and our identity within a larger collective.[1,2] Often, these names have symbolic meaning that represents our past, present, or future. When two groups have different names for a place it can create conflict over what a particular geographic feature represents. In these instances, a choice to use a particular name over another is usually a *political* process, in which numerous pieces of evidence—history, culture, linguistics or emotional connection—are used to assess which name should be given. In many countries a singular governmental organisation, group or figure is called on to assign place names and mediate these perspectives. In New Zealand the *New Zealand Geographic Board* fulfils this role, making place names official to enable "clear and unambiguous communication and connection between people and place" (New Zealand Geographic Board 2021, p. 2). But how are these decisions made, and how does the Board manage competing views?

The following chapter examines place names in New Zealand[3] and how they are implemented and/or changed. To do so, we introduce the processes behind how the New Zealand Government makes place naming decisions, referring to the work of the New Zealand Geographic Board and how it balances competing perspectives for naming places. We begin by briefly outlining the historical context behind place names in New Zealand, particularly drawing on the differences between indigenous Māori place names and introduced European place names. We then touch on recent naming issues in New Zealand, detailing how broader cultural expectations around Māori rights and calls for de-colonisation have come to alter how place naming is practised and governed by the New Zealand state. In describing these issues, we identify how the New Zealand Geographic Board plays a key—if understated—role in New Zealand's cultural challenges.

[1] 'Ngā Pou Taunaha o Aotearoa' metaphorically means 'The Memorial Markers of the Landscape'.

[2] 1840 agreement between Māori and the British Crown, with articles and principles to preserve and respect Māori rights and culture.

[3] We use the English name *New Zealand* rather than the Māori name *Aotearoa* for clarity. Nevertheless, the Māori name is regularly used both informally and formally to describe the name of the country. This is both used as a standalone name and combined as *Aotearoa New Zealand.* Recent debates, as examined later in this chapter, have called for the country to be renamed to use either the Māori or dual name.

Context

The first settlers migrated from eastern Polynesia to what is now New Zealand sometime between 800 and 1300 AD. These indigenous people, later collectively referred to as *Māori*, developed a strong connection to the natural landscape (Murton 2012). In Māori culture, people are seen as being deeply connected to the land and have ancestral relationships to it, therefore developing a strong place identity. The original naming of New Zealand places therefore "played an important role in the [Māori] way of seeing the world" (Berg and Kearns 1996, p. 107), and recorded past events, ancestors, or whoever controlled a particular territory. In effect, place names symbolically represented (and still reflect) a strong relationship with land, place and *te ao Māori* (the Māori world view).

While Māori were seen as a singular group upon the arrival of European settlers, it is important to note that Māori are not a homogenous group and arrived in New Zealand at different times and places, developing different dialects and customs. Prior to colonisation, Māori never referred to themselves as a singular group, but rather individual *iwi* (tribes), *hapū* (sub-tribes), and *whānau* (family). Accordingly, each group had different names for places, reflecting their cultural differences. The South Island, for example, is officially and alternatively named *Te Waipounamu* ('The Waters of Greenstone') in te reo Māori (the Māori language), but several iwi prefer to use the name *Te Waka-a-Māui* ('The canoe of Māui') or *Te Waka-a-Aoraki* ('The canoe of Aoraki'). As New Zealand was not a collective cultural entity until European colonisation, no Māori name for New Zealand existed until European arrival. The name *Aotearoa* ('land of the long white cloud') was originally used in reference to the North Island but has since, for many, come to refer to the entire group of islands and as a commonly and popularly used term for New Zealand, even among non-Māori.

After Māori, the islands were next explored by Dutch seafarer Abel Tasman in the seventeenth century, who named the country *Staten Landt*, believing it was connected to another island near Cape Horn in South America. Discovering that the lands were not connected, Dutch cartographers later renamed the islands *Nieuw Zeeland*, after the Dutch province of Zeeland. This name was then anglicised to *New Zealand*. It became commonly used by Europeans following James Cook's circumnavigation in 1769–1770 and the corresponding publication of his journals. Cook named much of what he found on his voyages, naming his first site of landfall *Poverty Bay* because "it afforded us no thing we wanted" (Mackay 1949). Many names originally recorded by Cook remain in widespread and official use today.

The British Crown formally declared sovereignty over New Zealand in 1840 with the signing of the Treaty of Waitangi. The Treaty (*Te Tiriti* in te reo Māori) is a culturally important document that plays an important role in New Zealand political debates today. The Treaty promised Māori control of land, full citizenship and protection, while Māori gave the British Crown *kāwanatanga* ('right of governance') (Stokes 1992). Mistranslations between the English and Māori versions of the Treaty, as well as a general failure to uphold the Treaty by the British Crown, resulted in conflict, land confiscation, and a loss of authority for Māori. In contemporary times, the New Zealand Government aims to uphold the principles of Te Tiriti in all decision-making practices, although the extent to which this is successful is regularly challenged. Issues of whether the principles of Te Tiriti are upheld are central to cultural politics within New Zealand and are implicated in contemporary discussions around place naming.

With European settlers came a 'naming as claiming'—whereby new place names were imposed on the landscape. While the majority of these names were of British origin, earlier application of French (*d'Urville Island*), Dutch (*Cape Maria van Diemen*), and Austrian (*Franz Josef Glacier)* names reflect the origins of other explorers. During this colonial period, many Māori names were replaced by those chosen by European settlers, in part due to not having a writing system until the early nineteenth century. There was also a desire to make New Zealand more culturally aligned to Britain. Many European names were applied in the nineteenth century by surveyors working for colonising associations or provincial governments (McKinnon 2009). These names often followed a regional pattern, with much of Otago and Southland, for example, being given Scottish place names (e.g., *Dunedin*) in light of the region's primarily Scottish settlers. While many Māori names survived, many were also bastardised by misspellings, several of which survive to this day. Many instances of renaming in recent years can be attributed to instances of misspelled names of Māori origin. Popular etymologies of European place names include military heroes (*Wellington),* royalty (*Queenstown*), politicians (*Gisborne*), ships (*Coromandel*), and sponsors/benefactors (*Ashburton).*

As English became the main language of New Zealand from the mid-nineteenth century, te reo Māori became increasingly confined to Māori communities. Te reo was suppressed in schools and wider society to ensure Māori assimilated into the now European-dominated society, and Māori were punished for speaking the language up until the 1970–1980s. Many elderly Māori still recall these punishments today. Despite this, most Māori spoke te reo as their first language up until World War II, after which there was a period of mass urban migration. As many inner-city schools did not teach te reo, fewer than 20% of Māori were considered native speakers by 1980. Urbanised Māori became largely alienated from their language and culture, much of which is passed on orally from generation to generation. Importantly, many Māori became disconnected from their iwi and ancestral land, and so lost their sense of belonging and identity. Many Māori have since made efforts to reconnect with their ancestral *marae* (meeting house) and *whenua* (land) to reaffirm their identity.

During the second half of the twentieth century, Māori activists increasingly began to assert their linguistic and political rights under Te Tiriti, in what was labelled the Māori renaissance (Moon 2009). Following this work, an increasingly popular commitment was made to strengthen *Māoritanga* (Māori culture), from which major language recovery programmes were established. Increasingly, Māori words were heard on radio and TV, with announcers commonly beginning their shows by saying "Kia ora" (an informal greeting), although not without pushback. Efforts to secure the survival of the language gained momentum in 1985. At this time, the Waitangi Tribunal (an ongoing commission of inquiry investigating claims brought by Māori relating to actions which breach the Treaty of Waitangi) agreed that the government was obliged to protect the language under the Treaty. Consequently, Māori was made an official language of New Zealand under the Māori Language Act 1987. Throughout this period, significant political shifts were made towards recognising Māori culture and worldviews as an integral part of the New Zealand identity.

Through this renaissance, the ongoing importance of place names to Te Ao Māori became increasingly realised. To Māori, place names are *taonga* (cultural treasures) that tell stories of important events, identify prominent landmarks, and lie at the heart of Māori identity. In many places, both Māori and English names are given equal status and separated by a forward slash (*Aoraki/Mount Cook).* While the practice of giving dual names began in the 1920s, dual names became more common following the 'Māori renaissance'. By convention Māori names are generally put first.

As part of this political shift, the New Zealand Government provided an opportunity for Māori to seek redress for breaches of Te Tiriti made by the Crown through the Treaty of Waitangi settlements process (Wheen and Hayward 2012). Māori became able to negotiate historical claims to land, resources, and taonga and be financially and culturally compensated by the New Zealand Government. Since the first claim settlement in 1992, many settlements have included the renaming of places as a form of cultural redress. The earliest and most prominent of these changes was the renaming of New Zealand's highest mountain Mount Cook to Aoraki/Mount Cook.

Case Study: Mount Taranaki

Mount Taranaki (Fig. 1) is a prominent example of how multiple naming and renaming has occurred in the New Zealand context. According to Taranaki-born historian Danny Keehan, the mountain was originally called *Pukehaupapa* (Ice Hill), or *Pukeonaki*, by the original occupants of the region (Coster 2016). The primary *tupuna* (ancestor) during this period was called Rua Taranaki, who arranged for his son to be the first person to climb the mountain and who subsequently named it *Taranaki* to affirm his authority over the area.

During James Cook's exploration of the region in 1770, he named the mountain *Egmont* after John Percival the 2nd Earl of Egmont, who had supported the expedition to New Zealand but had never set foot in the country. Mount Egmont became the commonly used name by the European majority in New Zealand and appeared on maps also giving the same name to the encircling national park. Use of the Mount Egmont name was, and still is, considered culturally insensitive to local Māori (Ngarewa 2022).

Following years of debate and controversy, the New Zealand Geographic Board decided to formally change the name of Mount Egmont to Mount Taranaki in 1983. Following pushback, the Board altered this decision in 1985 to Mount Taranaki (Egmont), with the English name as secondary for mapping purposes. In 1986, the Minister of Lands officially modified this decision to make the English name equal to the Māori name, deciding on alternative

Fig. 1 Mount Taranaki from Pouakai Tarns

names—*Mount Taranaki* or *Mount Egmont*. This decision came under considerable criticism from nearby residents, the majority of whom supported using *Mount Egmont*. However, local Māori supported the decision. The Minister's decision, obviously in response to objecting submissions to the inclusion of the Māori name, highlights the politics surrounding place naming and how competing ideas of place identity must be weighed up by naming authorities.

Since this decision, Mount Taranaki has become the more popularly used name. As part of a Treaty settlement with Ngā Iwi o Taranaki, the mountain will be renamed to just *Taranaki Maunga*, a process expected to be completed within the next few years.

Recent Trends of Place Naming in Aotearoa New Zealand

Since the Māori renaissance and subsequent revival of te reo Māori, indigenous names for places are being increasingly used, reflecting an openness to recognising the impacts of colonisation and the Māori view that place names are taonga (treasures). In many instances, including by local and central government and media outlets, Māori place names are increasingly being used in preference over non-Māori names. Tāmaki Makaurau (Auckland), Te Whanganui-a-Tara (Wellington), Ōtautahi (Christchurch) and Otepoti (Dunedin) are, for example, increasingly used on promotional websites, in news broadcasts and academic texts. Despite this ongoing use, non- Māori place names are still far more commonly used by the general public. A distinction between official and common naming practices prevails, however. The foregoing names of the nation's major cities are not official nor do they necessarily reflect the traditional names applied by local iwi to what are now large urban areas.

Attempts to change non-Māori place names often attract stern criticism and objection, particularly among local residents who identify specifically with commonly used names and oppose attempts to reinstate original Māori (but often underused) place names. As Berg and Kearns (1996) note, these views reflect *identity politics* which involve often barely conscious constructions of race, culture, and the nation. In their example of naming places in the coastal part of the Otago region, Berg and Kearns (1996) argue that objections to renaming popularly used (but offensive to Māori) names such as Murdering Beach (referring to an incident where whalers from the Tasmanian Brig *Sophia* were killed by Māori after an arguably provoked altercation) act to *legitimise* the dominant role of a Pākehā (non-Māori, of European descent) power politics in New Zealand. Local residents commonly claim that Māori names have been reinstated by top-down rather than bottom-up, local-led decision making. This, of course, neglects the role of local iwi who are largely supportive of reinstating names but may be comparatively low in number or excluded from local discussions.

One prominent and recurring debate is around the name of the country itself. In 2021, Te Pāti Māori (the Māori Party) launched a petition to restore all Māori place names, including renaming New Zealand as Aotearoa (Whyte 2021). Polls indicate this to be an unpopular move with several factors complicating the debate (NZ Herald 2021). Firstly is discussion of whether the country should be completely renamed as Aotearoa or have the alternative names, either Aotearoa or New Zealand, an idea polls suggest is more popular. Secondly, the historical name of Aotearoa is itself disputed, as pre-European Māori had no name for the New Zealand archipelago as a whole. Accompanied by a widespread belief that such a large name change would have to occur through public referendum and be finally decided by Parliament, the Aotearoa debate showcases the complexities in naming politics and how such decisions have to account for multiple sets of often conflicting, linguistic, historical and cultural evidence. As of 2021, no referendum has been called and public support for the idea is limited, although Prime Minister Jacinda Arden has stated that she believes the use of Aotearoa as an interchangeable name is a "positive thing" (One News 2020).

In recent years, a small number of settlements have had their original Māori name restored (e.g. Benneydale to Manaiti/Benneydale). To rename a New Zealand city, town, village, or suburb, the New Zealand Geographic Board Ngā Pou Taunaha o Aotearoa (Board) requires proposers to have consulted with local government. Even if support is gained, attempts to restore original Māori place names often see considerable pushback from locals fearful of losing their place-identity. Resistance is evident even when only a small change is made, such as the addition of a macron (as evidenced by angry reactions to changing Otumoetai to Otūmoetai). To supporters of these name changes, correcting or renaming is an important aspect of decolonisation that involves recognising the influence of colonisation and contributes to the positive normalisation of te reo Māori. Many supporters decry the continued use of non-Māori names as a continuation of the colonial state—as Māori academic Pare Keiha states, 'we continue to delight in historical cleansing by continuing to deny Māori [their] own history in [their] own land… it says much about a country still wedded to colonisation, still wedded to its own racism' (Thornber 2020). The strong feelings and mobilisation that is triggered in support of, or in reaction to, a proposed name change highlights the complex linguistic politics surround place renaming.

Not all efforts to change the names of places invoke bicultural politics and the restoration of original Māori names to the landscape. By way of example, Blenheim is a town of 28,000 at the north of the South Island or Te Waipounamu and is the service centre for the grape-growing Marlborough region, famous for its sauvignon blanc wines. In 2016, a brief media campaign led by winemakers sought to change Blenheim's name to 'Marlborough City'. This idea was short-lived and never reached the stage of a formal proposal. However, it does indicate the potential links between naming and branding, in this case an attempt to better highlight the international reputation of the Marlborough region's wines (Kearns and Lewis 2019). The defeated proposal speaks to at least aspirational efforts to link place names with products and, in overlooking the original Māori name, threatened to over-write the established settlement history. Ironically, Blenheim and Marlborough are both colonial names that displaced a long-established Māori name, Te Waiharakeke. Kearns and Lewis (2019) note that this attempt at 'toponymic commodification' can be read as a neoliberalised place-making project in the sense that it was being presumptuously advanced by particular for-profit interests.

Case Study: Tāmaki Makaurau Auckland

The Auckland isthmus was first settled by Māori in approximately 1350 AD for its strategic location, fertile soils and harbour access. Numerous iwi settled in the area, calling it *Tāmaki* (of disputed meaning, possibly referring to battle; heavily populated; or successful attack by Maki, a local chief) or *Tāmaki Makarau* (meaning *Tāmaki* desired by many).

Following the signing of the Treaty of Waitangi in 1840, the Governor of New Zealand, William Hobson, chose the Waitematā Harbour as the location of New Zealand's new capital, constructing a new settlement and naming it *Auckland* after patron George Eden, the 1st Earl of Auckland. Due to colonial settlement, the European population of Auckland began to rapidly outnumber Māori and so Auckland became the dominant name of the new city and surrounding area. The name *Tāmaki* later became applied to a small suburb in the city's east. The city experienced significant Māori in-migration in the second half of the twentieth century, and the current Auckland population is 13% Māori with 24% of all Māori in New Zealand living there (Auckland Council n.d.). The wider demographic context is that almost 40% of the city's population was born overseas, with Auckland being listed as having the fourth largest overseas-born population of any major city in the world (Pio 2016).

In recent years, ongoing shifts to recognise indigenous Māori rights have led to the growing use of *Tāmaki Makarau* instead of Auckland, particularly among governmental institutions and media. Both local and central government often use the names interchangeably rather than use one or the other, reflecting the changing recognition of indigenous place names. Neither name is official.

The Role and Scope of the New Zealand Geographic Board Ngā Pou Taunaha o Aotearoa

The New Zealand Geographic Board is New Zealand's national place naming authority, responsible for official place names in New Zealand, its offshore islands, its continental shelf, and the Ross Dependency of Antarctica. Its role is set out in the New Zealand Geographic Board (Ngā Pou Taunaha o Aotearoa) Act 2008, and includes

- deciding on proposals for new, altered or discontinued place names, including for Crown protected areas, Antarctic, and undersea features
- approving and adopting existing recorded place names as official
- meeting Tiriti o Waitangi commitments by providing advice on place names in the cultural redress part of settlements, and engaging with post-settlement governance entities about processing other Māori place names
- validating Antarctic and Crown protected area names
- maintaining a publicly available, accessible, searchable and re-usable New Zealand Gazetteer of place names.

The Board has ten members. The Minister for Land Information appoints eight of them—four are each nominated by Te Rūnanga o Ngāi Tahu,[4] the New Zealand Geographical Society, the Federated Mountain Clubs of New Zealand and Local Government New Zealand; two are recommended by the Minister for Māori Development; and two others are the Minister for Land Information's choice. Two members are ex-officio from Toitū Te Whenua Land Information New Zealand (LINZ), one of whom is the Surveyor-General who acts as Chairperson, and the other being the LINZ official with primary responsibility for setting hydrographic information standards for New Zealand. The Board's work is supported by dedicated administrative and advisory staff within LINZ.

The legislation that underpins the Board's activity includes a requirement to use official place names in all official documents, including on road signs, maps and charts, and in scientific publications and information for tourists. This is a large and complex task, and the Board is involved with education through outreach, engaging

---

[4] A Māori iwi that holds mana whenua over much of the South Island or Te Waipounamu.

with media, responding to complaints, responding to incorrect use of place names, and checking for compliance in agency's databases and publications.

Establishing place names in New Zealand occurs through a formal consultation process, outlined in Fig. 2.

A place name can also be made official without public consultation through a **fast-track process** if there is no other recorded name for the feature and if public objection is considered unlikely. A recorded name must be shown in at least two authoritative publications. Of the over 50,000 place names recorded in the *New Zealand Gazetteer* (New Zealand's record of all place names), only 39% of these are official, hence the need for a fast-track process.

The Board is statutorily required to encourage the use of original Māori place names where possible, with special importance given to their correct spelling and use of macrons. Despite the replacement or correction of many Māori place names during colonisation, often these original names endure through oral traditions and recall important ancestors and historical events. To the Board, officially restoring or correcting them is culturally significant as it publicly recognises the long-held association and connection to a specific area of tangata whenua[5] Māori.

Case Study: Maxwell to Pākaraka

Maxwell is a small town near Whanganui in the North Island (Te Ika-a-Māui). To Māori, the area was known as *Pākaraka*, meaning 'the pā (Māori village/defensive settlement) where the karaka trees grew'. Maxwell was first settled by Europeans in the mid-1800s and named Maxwelltown until 1927, after Sergeant George Maxwell, one of the founding members of the Kai-iwi Yeomanry Calvary Volunteers a settler militia, who died in battle nearby.

The volunteers were involved in an encounter called the 'Handley's Woolshed' incident in November 1868, during which the militia came across a group of Māori boys aged 10–12, hunting geese and pigs. The volunteers charged and according to historical accounts a number of boys were killed. Therefore, to Māori this name is considered highly offensive.

In 2019 local Māori approached the Whanganui District Council (WDC) to discuss a name change. The WDC supported the application following a period of consultation. While submissions mostly supported the proposal, submissions from a letter-drop in Maxwell were mostly against it. In 2021 a proposal was made to the Board, who decided to accept the name change based on: (1) its function to encourage use of Māori names, (2) the offensive nature of the name to local Māori, and (3) the support of WDC. The Board notified the proposal for three months so that the public could make submissions supporting or objecting to the name change, before making a final decision. In December 2021 the Minister for Land Information confirmed the change from Maxwell to Pākaraka.

While the Board has jurisdiction over the naming of entire cities, towns, localities and suburbs, the names of roads and streets remain the domain of local authorities.

[5] People of the place.

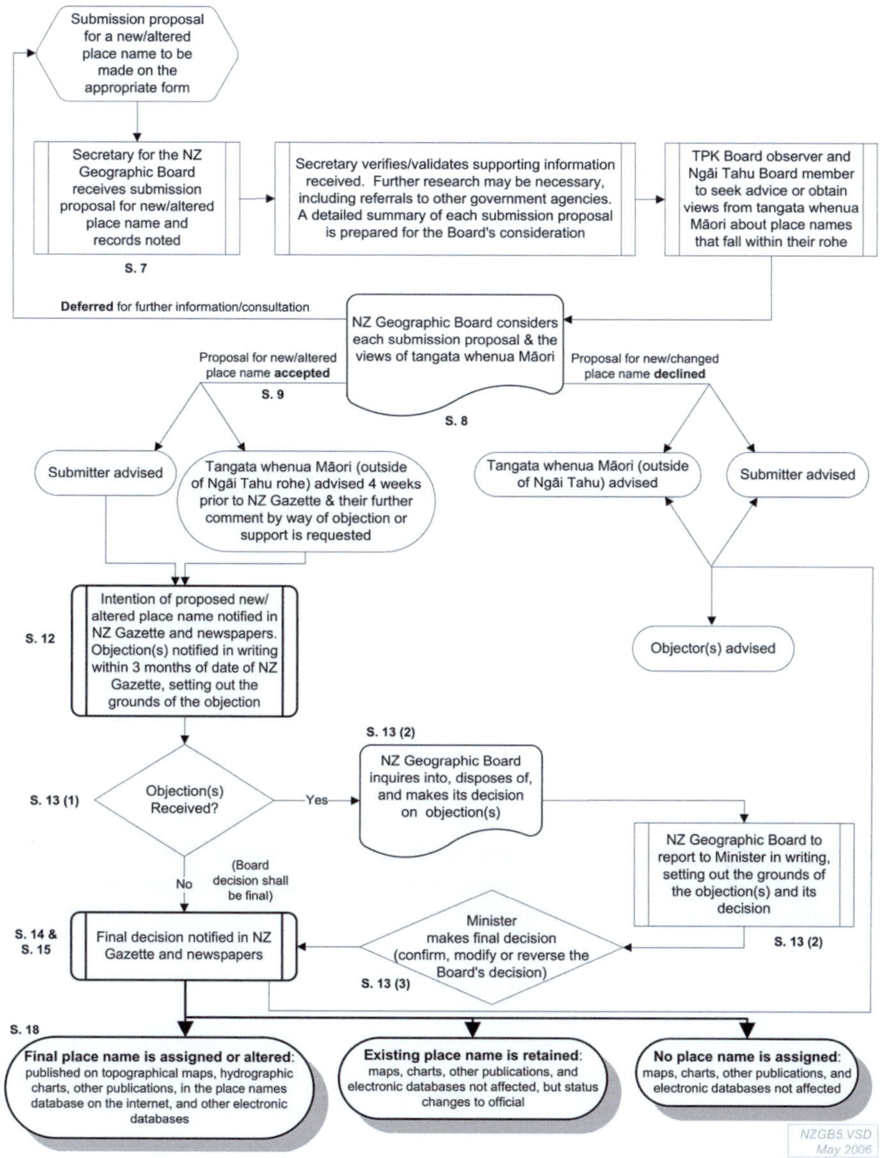

Fig. 2 New Zealand Geographic Board Ngā Pou Taunaha o Aotearoa process for New Zealand place names. From *New Zealand Geographic Board Ngā Pou Taunaha o Aotearoa process for New Zealand place names*, by New Zealand Geographic Board, 2006, Land Information New Zealand (https://www.linz.govt.nz/system/files_force/media/pages-attachments/nz-names-process-200811.pdf). Reproduced with permission

Therefore, there is an understandable variety in their appropriateness in light of good naming practice. At the level of suburbs, a recent example is a precinct of the capital city of Wellington. The area was officially named Paekākā in 2021. Wellington City Council proposed the name after mana whenua (Māori with authority over an area) Taranaki Whānui gifted it to them. Paekākā was considered to be an appropriate name for recognising Māori heritage in the area as it had been the traditional Māori name for nearby gardens that no longer exist. The name refers to the perch or realm of the native kākā, a large forest parrot which has, through predator eradication, been restored in the area. Paekākā is now the name of an area that spans multiple suburbs but does not replace the names of these suburbs, localities, or any local reserves.

However, place names are not only written. Although the legislative mandate of the Board focuses on inscribed toponomy (names that are recorded on road signs and maps, and in policy documents) te reo Māori is historically an oral form of communication. A recent concern—throughout public agencies, educational institutions, broadcast media as well as the Board itself—has therefore been the oral communication of place names, particularly their pronunciation. As Kearns and Berg (2002) argue, identity is at least in part narrated through place-referenced linguistic tactics such as pronunciation. From this perspective the efforts, or lack thereof, to emulate authentic pronunciation of a place name can speak to one's degree of solidarity with a land's first settlers. In the case of Aotearoa New Zealand, as Yoon's (1980) early analysis of Māori place name distributions highlighted, larger settlements are less likely to have retained a Māori name. Yet those names that have endured have often been mispronounced to the point of being almost unrecognisable compared to their original oral form. By way of example, the second author's childhood was spent in the nation's northernmost city, Whangārei. In his early days there it was referred to as 'Wong-gar-ray'. In Māori, however, the 'Wh' is a soft 'f' sound, and the macron indicates a long 'a' sound, so the name is correctly pronounced 'Fang-aa-ray'. This has only recently become a more common usage, aided by broadcast media and a generational change with more te reo learners. However, one of the challenges in promoting authentic proclaiming of place names is that while naming in print can be mandated by law, pronunciation cannot. Only education can be used as tool to achieve proper use and compliance.

Due to its large exclusive economic zone (EEZ), New Zealand is also responsible for naming undersea features and bodies of water (hydronomy) and is responsible for naming features and places in the Ross Dependency—New Zealand's part of Antarctica. Given that the Board has an official strategy to employ Māori names where possible, a challenge is in places like Antarctica where Māori did not travel to or explore. Unlike processes of renaming in which alternative names are often clear, many new features and places are named each year, raising additional challenges for what names to use. To Dodds and Yusoff (2005) who examined Antarctic features named by New Zealand, the use of Māori names in Antarctica often represents a 'fraught assimilation' and could be seen as an appropriation of a suppressed language for further colonial use. Moreover, New Zealand's Antarctic station is called *Scott Base*—after Robert Scott, a British citizen who has little to do with New Zealand history (Lindsay and Yoon 2021). Applying a new name to a feature represents a complex challenge, as the Board has to negotiate the possibility of using Māori names (without appropriation), historical figures (while using New Zealanders where possible) and maintaining relations with other countries involved in Antarctic science, protection and administration.

Part of the Board's role is accepting proposals from people asking for names to be changed. In areas with particularly strong colonial histories or a large population, reactions can be contentious. In recent years, proposals have been increasingly made by local Māori, who have required place name changes as part of a Treaty of Waitangi settlement redress or have sought to change place names considered offensive. In Berg and Kearns's (1996) work on New Zealand place names, they detail the renaming of *Murdering Beach* to *Whareakeake*, which was heavily opposed by residents. In many cases offensive names have become culturally embedded and are cemented in place-based identity, making name changes fraught. A number of high-profile cases involving the Board's decisions over the last few decades have been contentious but, in most cases objections diminish after a few years of the new name being in use.

Case Study: W(h)anganui

Whanganui, a city of approximately 50,000 in the North Island or Te Ika-a-Māui, was first named by Europeans as *Petre* after Lord Petre, an officer of the New Zealand Company, but was officially changed to Wanganui in 1854. This arose from a language issue—the term *Whanga Nui* refers to 'the long wait' but the linguistic challenges of pronouncing the 'wh' sound in the local dialect resulted in the name being mispronounced.

In 1991, the Board changed the name of the major river that flows through the city to Whanganui River after calls from local Māori, despite the city remaining spelled Wanganui. In 2006, a non-binding referendum was held, with 82% of the city voting to retain the h-less name based on a 55.4% turnout. In 2009, local Māori applied to the Board to change the name of the city to Whanganui. Public submissions were roughly equal, with a slim majority of submissions wanting to keep the status quo. The local mayor at the time, Michael Laws, was strongly against the change. Another Council referendum was held in May 2009, with 77% of residents voting to retain the name Wanganui based on a 61% turnout, reflecting the controversy.

With the decision being political in nature, the Board, as a requirement of the statutory process, handed over the decision to the Minister for Land Information in September 2009 but recommended the name be changed having considered the public consultation results. In December 2009, the Government decided that while either spelling was acceptable, government agencies would use the correct spelling Whanganui (Fig. 3). In 2015 and 2019, the Board further changed the names of the district and region to Whanganui District and Manawatū-Whanganui Region.

Fig. 3 Overwritten Whanganui road sign (changed from Wanganui). From *Whanganui road sign* [Photograph] by C. Stephens, n.d. Reproduced with permission

A number of other issues add to the broader context of the politics of place naming in New Zealand. As alluded to earlier, the news media play an influential role in naming disputes. Over recent years, New Zealand's major TV networks have made a newfound commitment to employing te reo Māori and Māori place names. During New Zealand's COVID-19 announcements, for example, Maori words like 'mahi' (work) and 'motu' (island or country) were regularly used. Place names in both English and Māori are often used interchangeably. In many cases this has seen a prominent uptake among many businesses, although limited use by the general public.

Issues of decolonisation and indigenous rights have also become increasingly important to many New Zealanders, particularly in the wake of the Black Lives Matter movement. There have been subsequent moves to remove offensive icons, including statues and place names. Accompanied by an increasing revival of te reo Māori, the Board is responding to this context through outreach into Māori communities, investigating ways to make the stories behind names available to the public, and collaborating with other governmental strategies of Māori language revitalisation.

Case Study: Spelling and Pronunciation

Much of the work of the Board involves correcting misspelled place names to their original Māori spelling. Upon European arrival in the nineteenth century, many Māori place names were recorded by European explorers upon meeting Māori. As te reo Māori was not a written language until the introduction of the Roman alphabet in the nineteenth century, these names were often misspelled. While place names like Waiotahe (corrected from *Waiotohi*) and Remutaka (corrected from *Rimutaka*) have been restored, many prominent misspellings (like the Lower Hutt suburb of Petone, originally *Pito-one)* are still recorded.

While many Māori place names are in widespread use, pronunciation of these names is often incorrect, generally stemming from the different use of vowel sounds and the use of macrons in te reo. While governmental figures and news presenters have made efforts to pronounce Māori place names correctly over recent years, numerous towns and cities with Māori names are mispronounced by the majority of New Zealanders—Taupō and Tauranga being typical examples. Such misspellings and mispronunciations are considered offensive to Māori and represent the continuation of colonial power over Māori worldviews.

Conclusions

This chapter has examined the political, historical and cultural context of place names in Aotearoa New Zealand and how renamings are implemented and changed. To do this, after cursorily reviewing the nation's history of colonisation, we examined the processes of the Board as New Zealand's place naming authority including how decisions are made over naming. Naming practices in New Zealand are a result of changes in attitudes towards the importance of te reo Māori (the Māori language) and of reaffirming original Māori naming practices. Drawing on selected case studies, the chapter outlined how and why places are named in New Zealand and the reasons behind place names being changed in recent decades. The chapter detailed the importance of understanding the cultural politics of place naming disputes, particularly through knowing the nation's settlement history and colonisation, how it has impacted Māori culture, and how place names act as a prominent symbol of Māori sovereignty and governorship. In part, this knowing reflects commitment towards decolonisation, although the extent to which this has occurred continues to be discussed.

What is important is that the official naming authority is not separate from the nation's cultural politics. Naming authorities both influence, and are influenced by, cultural debates surrounding the role of naming, who gets to name, and why places are named in such ways. In New Zealand the Board has an important role to play in shaping the nation's cultural politics by reaffirming the importance of Māori names, but also represents the public's views by way of the decisions that it makes.

Ongoing recognition of indigenous Māori rights, culture, and sovereignty within both governmental settings and popular culture prompts reflection over whether

New Zealand will ever be renamed Aotearoa. In a 2021 context the suggestion is currently unlikely considering its relative unpopularity in the eyes of the general public. However, what can be seen is that original Māori names, correct pronunciation, and te reo Māori have become increasingly important over recent decades reflecting a change in how the general public relates to cultural politics and the identity of the country. Any change to the country's name is likely to be a result of a public referendum and would require significant buy-in from the general public as well as an Act of Parliament. This is not to say that political considerations and toponymic governance do not have a say in the matter. Government naming conventions (both informally and formally) drive how people see and engage with place. As government (and media) increasingly use original Māori place names, so too will the general public. The Board has an important part to play in these discussions, despite its role being specifically excluded from engaging with the name of the country. By continuing to reassert Māori names, perhaps flow-on effects will prompt reflection over what the country's name is and will be.

Questions and Activities Regarding This Chapter

1. How does the New Zealand Geographic Board make a place name official?
2. In what ways is a political shift towards reflecting Māori rights changing how place name decisions are made?
3. Reflect upon your own country's practices of using indigenous names. What is your country's approach to using indigenous names, and how has this changed over time?

References

Auckland Council (n.d.) Māori in Tāmaki Makaurau. Auckland Council. https://www.aucklandcouncil.govt.nz/plans-projects-policies-reports-bylaws/our-plans-strategies/auckland-plan/about-the-auckland-plan/Pages/maori-tamaki-makaurau.aspx. Accessed 17 Jan 2022

Coster D (2016) History reveals Taranaki's peak was once known by a different name. Stuff. https://www.stuff.co.nz/taranaki-daily-news/news/78445421/history-reveals-taranakis-peak-was-once-known-by-a-different-name

Dodds KJ, Yusoff K (2005) Settlement and unsettlement in Aotearoa/New Zealand and Antarctica. Polar Rec 41(2):141–155

Kearns R, Lewis N (2019) City renaming as brand promotion: exploring neoliberal projects and community resistance in New Zealand. Urban Geogr 40(6):870–887

Lindsay N, Yoon HK (2021) Toponyms on the ice: the symbolic and iconographical role of Antarctic research base names. Polar Rec 57

Mackay JA (1949) Unhappy naming of Poverty Bay. In: Historic Poverty Bay and the East Coast, N.I., N.Z, pp 37–45. http://nzetc.victoria.ac.nz/tm/scholarly/tei-MacHist.html

McKinnon M (2009) Place names—naming the country and the main islands. Te Ara: The Encyclopedia of New Zealand. https://teara.govt.nz/en/place-names/page-1
Moon P (2009) A chequered renaissance: the evolution of Maori society, 1984–2004. Te Kaharoa 2(1)
Murton B (2012) Being in the place world: toward a Māori "geographical self." J Cult Geogr 29(1):87–104
Ngarewa A (2022) The naming of Taranaki Mounga. The Spinoff. https://thespinoff.co.nz/atea/17-01-2022/the-naming-of-taranaki-mounga
NZ Herald (2021) Aotearoa-New Zealand name change debate: new poll shows while majority in favour of status quo there is support for change. New Zealand Herald. https://www.nzherald.co.nz/nz/aotearoa-new-zealand-name-change-debate-new-poll-shows-while-majority-in-favour-of-status-quo-there-is-support-for-change/UUFJQB3XRUVC7H3SLPVAVXQ36E/
New Zealand Geographic Board (2021) Annual report of the New Zealand Geographic Board 2020/21. https://www.linz.govt.nz/about-us/publications/new-zealand-geographic-board-maps-and-publications/annual-report-new-zealand-geographic-board
One News (2020) Jacinda Ardern says changing NZ's name to Aotearoa not on her agenda, but wants people to use it. One News. https://www.1news.co.nz/2020/09/14/jacinda-ardern-says-changing-nzs-name-to-aotearoa-not-on-her-agenda-but-wants-people-to-use-it/
Pio E (2016) Discovering diversity. Employment Today, pp 26–28
Stokes E (1992) The treaty of Waitangi and the Waitangi tribunal: Maori claims in New Zealand. Appl Geogr 12(2):176–191
Thornber L (2020) Is it time to restore all of New Zealand's Māori place names?. Stuff. https://www.stuff.co.nz/pou-tiaki/te-reo-maori/300107523/is-it-time-to-restore-all-of-new-zealands-mori-place-names
Wheen NR, Hayward J (eds) (2012) Treaty of Waitangi settlements. Bridget Williams Books
Whyte A (2021) Over 10,000 sign Te Pati Māori's place name petition. One News. https://www.1news.co.nz/2021/09/14/over-10000-sign-te-pati-maoris-place-name-petition/
Yoon HK (1980) An analysis of place names of cultural features in New Zealand. NZ Geogr 36(1):30–34

Suggested Reading

Berg LD, Kearns RA (1996) Naming as norming: 'race', gender, and the identity politics of naming places in Aotearoa/New Zealand. Environ Plan D: Soc Space 14(1):99–122
Dodds KJ, Yusoff K (2005) Settlement and unsettlement in Aotearoa/New Zealand and Antarctica. Polar Rec 41(2):141–155
Kearns RA, Berg LD (2002) Proclaiming place: towards a geography of place name pronunciation. Soc Cult Geogr 3(3):283–302
Stokes E (1994) Geographic naming in New Zealand: the role of the New Zealand geographic board. NZ Geogr 50(2):51–54

Theorising Multiple Place Names in Southern Africa

Tendai Mangena

Abstract This chapter discusses multiple place names that are prevalent in southern Africa using specific examples from Zimbabwe and South Africa. It makes use of the metaphors of the palimpsest and the rainbow nation to demonstrate the unique features of place name multiplicity in the two selected countries of southern Africa. It specifically shows the ways in which particular examples of place name multiplicity in the case of Zimbabwe fit into the frame of the argument that many places in southern Africa are palimpsests, whose toponyms have been altered many times. Typical of palimpsests, which are written over several times, in most instances, such places still bear visible traces of 'earlier' toponymic forms. In the case of South Africa, the chapter evokes the metaphor of the rainbow nation to speak about and think through the ambivalent coexistence of colonial/apartheid and indigenous places names. The two metaphors are used to explain the coexistence of names of a place that originated and are negotiated in different historical periods. The chapter is particularly interested in what it means for places to be referred to through different names; how traces of past names bear competing histories and discourses which undergird (dis)continuities that play out through the prism of toponyms.

Keywords Zimbabwe · South Africa · Palimpsest metaphor · Rainbow nation metaphor · Multiple toponyms

Objectives

- By the end of this chapter readers should be able to give examples of place name multiplicity from Zimbabwe to demonstrate the point that place names in southern Africa are palimpsests whose toponyms have been altered many times.
- From this chapter, readers will learn about the significance of the metaphor of the 'rainbow nation' to an understanding of the co-existence of colonial/apartheid and indigenous names in post-apartheid South Africa.

T. Mangena (✉)
University of the Free State, Bloemfontein, South Africa
e-mail: tendaimangena@gmail.com

© The Author(s), under exclusive license to Springer Nature Switzerland AG 2023

G. O'Reilly (ed.), *Place Naming, Identities and Geography*,
Key Challenges in Geography, https://doi.org/10.1007/978-3-031-21510-0_11

- Readers will learn about what it means for places to be referred to through different names and how traces of past names bear competing histories and discourses in postcolonial countries.

Introduction

In this chapter, I discuss multiple place names that are prevalent in southern Africa using specific examples from Zimbabwe and South Africa (see Fig. 1, map of southern Africa). The two countries are chosen over others because over and above sharing a common history, borders and languages as well, toponomastic research is particularly more vibrant in both countries than in any other part of southern Africa (see Koopman 2012; Neethling 2016; Nyambi et al. 2016; Mangena 2020; Mamvura 2020). Specifically, renaming of places that takes or does not take place in political transitions has been an interesting subject of toponomastic research in these two countries (Klerk 2002; Guyot and Seethal 2007; Swart 2008; Ndletyana 2012; Duminy 2014; Nyambi et al. 2016; Mangena 2020; Mamvura 2020). This makes Zimbabwe and South Africa interesting sites in which to think about and through place name multiplicity typical in southern Africa. The specific examples of place name multiplicity from Zimbabwe are carefully selected to fit into the frame of the argument

Fig. 1 Map of southern Africa

that many places in southern Africa are palimpsests, whose toponyms have been altered many times. Typical of palimpsests, which are written over several times, in most instances, such places still bear visible traces of 'earlier' toponymic forms. This element of traces of the past call to mind Derrida's (1994, p. 5) specters and traces of, especially the imperial past, '*the* non-objects, *the* non-present present'. In the case of South Africa, I argue that the metaphor of the rainbow nation is equally relevant as an explanation of the coexistence of colonial and or apartheid and indigenous place names. I use both metaphors, of the palimpsest and rainbow nation, to explain the coexistence of names of a place that originated in different historical periods. From this, I argue that both metaphors offer a theoretical framework that accounts for the multiplicity of place names in this region of the continent. I am particularly interested in what it means for places to be referred to through different names; how traces of past names bear competing histories and discourses which undergird dis/continuities that play out through the prisms of toponyms.

The chapter consists of four sections. In the first section I discuss the concept of place name multiplicity. In the second section, I call attention to the metaphors of the palimpsest and rainbow nation and their usefulness to an understanding place name multiplicity in Zimbabwe and South Africa respectively. In the third section I employ specific examples from Zimbabwe to think through the various toponymic layers that are created in and through political transitions. In the last section, I use examples from South Africa to speak to and about the contested concept of the rainbow nation vis-a-vis the common coexistence of apartheid\colonial and vernacular/indigenous toponyms.

Place Name Multiplicity

Frederick Giraut's article 'Plural toponyms: When place names coexist' (2020), is useful and productive in the formulation of what constitutes toponym multiplicity. From the title of his article, Giraut conceptualizes place name multiplicity as when place names coexist. Multiplicity of toponyms, in other words, concerns the use of more than one name to refer to a place or simply more than one label for a single place. From this, one could be tempted to argue that what is generally at stake in multiple place names is the question of the endonym versus the exonym, the official versus the unofficial, the vernacular versus the formal. I give examples of place name multiplicity from my specific case study to illustrate the definition. From Zimbabwe I refer to the many names of what is popularly known as Victoria Falls, a famous waterfall which is one of the seven wonders of the world. From South Africa I highlight the example of Maloti\uKhahlamba versus Drakensburg place names.

As I show later in this chapter, the interpretation of place name multiplicity should naturally go beyond mentioning that names belong to the category of endonym or exonym to include a reflection on the different or conflicting registers in the different place names that are engendered by (post)colonial politico-cultural legacies. As it can be seen in the example of the Victoria Falls. The 'falls' were 'renamed' Victoria Falls after the British Queen Victoria by David Livingstone. I say they were 'renamed' and not 'named' after the British Queen Victoria to make the point that the place had a name already used by the indigenous people in and around the place. It is important to note that the falls are located at the border between Zimbabwe and Zambia, but the town of Victoria Falls itself is located on the Zimbabwean side. So, because of the 'falls' location, different linguistic groups from both countries relate to them using different names. According to Hang'ombe et al. (2019) the Lozi/Kololo referred to the Falls as Mosi-Oa-Tunya (the smoke that thunders); the Leya named it Syungu Namutitima (the mist that thunders); the Nambya called it Chinotimba (the thundering place), and the Ndebele called it aManzi aThunqayo (steaming or boiling water). Significantly, Mosi-Oa-Tunya is the most well-known indigenous name for Victoria Falls. So, Victoria Falls is a typical colonial place name, which however persists in the present. Reclaiming Mosi-Oa-Tunya in post-independence Zimbabwe as per calls made in 2013 by politicians would have represented 'a cultural restoration of the falls' original identity' (Nyambi and Mangena 2016, p. 5). The important point is, in as much as the world relates and may continue to relate to the falls as Victoria Falls, the native Lozi have always called the place using the language they know since the name Victoria Falls alienates them from the falls. In a way therefore, the many labels, specifically Mosi-Oa-Tunya and Victoria Falls seem to coexist at an official level (see Fig. 2: Victoria Falls and Mosi-Oa-Tunya). The argument that we make elsewhere is 'if the essence of tourism is encountering new places, then visiting Vitoria Falls is fundamentally a less enriching tour than a visit of Mosi-Oa-Tunya' (Nyambi and Mangena 2016, p. 5).

The same logic, to a certain extent applies to the South African example of the Drakensburg Mountain referred to in vernacular languages as Maloti\uKhahlamba (see Figs. 3 and 4). Drakensberg is an Afrikaans name meaning "Dragon's Mountain", the Zulu refer to the place as uKhahlamba ("barrier of spears") while in Sesotho the place is Maloti ("mountains"). But the notable difference in this case is in the existence, and popular usage of what I would call 'competing vernaculars', where over and above the 'apartheid' toponym Drakensburg, we have two vernacular labels, that coexist in identifying a similar place.

As many examples from Zimbabwe and South Africa show, places are palimpsests whose toponyms have been altered many times during political transitions and the long history of migration, with some old toponyms 'resurfacing'. In the context of South Africa, the concept of a rainbow nation has enabled an acceptable co-existence of colonial/apartheid toponyms in post-apartheid democracy. In the next section, I discuss the relevance of the palimpsest and rainbow nation metaphors to a conversation on and about place name multiplicity in Zimbabwe and South Africa respectively.

Fig. 2 Toponyms Mosi-Oa-Tunya and Victoria Falls coexisting

Fig. 3 The toponyms Maloti and Drakensburg coexisting

Fig. 4 The toponyms uKhahlamba and Drakensburg coexisting

Metaphors of Palimpsest and Rainbow Nation in Relation to Multiple Place Names

According to Dillon (2005, p. 244),

> palimpsests originated … in some parts of Europe, whereby writing material was recycled as a direct result of changing historical and cultural factors which rendered some texts obsolete either because the language in which they were written could no longer be read, or because their content was no longer valued. Palimpsests were created by a process of layering whereby the existing text was erased, using various chemical methods, and the new text was written over the old one.

Dillon explains further that most or basic definitions of 'palimpsest' do not explain that:

> the first writing on the writing material seemed to have been eradicated after treatment, it was often imperfectly erased. … and its ghostly trace then reappeared in the following centuries as the iron in the remaining ink reacted with the oxygen in the air, producing a reddish-brown oxide (2005, p. 244).

In this sense then, in palimpsests, any layer didn't fully erase its predecessors, so it was possible to always recognize the previous layers of the text written earlier (Mitin 2010). Characteristic or defining features of palimpsests: Of something reused or altered but still bearing visible traces of its earlier form; of layering and subsequent textual reappearance are important to my discussion. In the words of Mitin, 'these specific features have made a palimpsest an important metaphor used in social sciences and the humanities to stress multiplicity of a text or phenomenon, to witness its layering and to single out some—by chance partly hidden—layers of reality' (Mitin 2018, p. 2). The palimpsest then 'provides the possibility for erasure and overwriting and the co-existence of several different scripts, implying not just different historical eras, but several historical and contemporary actors as well' (Schein 1997, p. 662). Schein here evokes 'the very multiplicity of human interpretations and representations of a place'.

Using the metaphor of the palimpsest as conceptualized in this sense, in this chapter, I use examples from Zimbabwe to argue that the majority of places in southern Africa have been named many times, with toponyms erased to make way for new ones especially in and through political transitions. However, the erased toponyms often reappear in other forms; in other words, the erasures are 'imperfect', since they show ghostly traces of earlier toponyms. I find links between these characteristic features of the palimpsest and Derrida's (1994) concept on hauntology, which is an ontology based on haunting specters. The ghostly traces have largely to do with contestations in place naming and renaming. This supports the critical toponym view that toponyms, especially commemorative ones, are more than referential, that they are political entities as well.

South Africa's post-apartheid national identity has been conceived as that of a 'rainbow nation' (Baines 1998). Archbishop Desmond Tutu is believed to have coined the phrase 'rainbow nation' as a reinforcement of the vision of national building (Baines 1998). This 'rainbow nation ideology' was later espoused by Nelson Mandela, the first president of the 'new' South Africa to capture the African National Congress's non-racialism political approach (Myambo 2010). In the words of Habib, 'the notion of the rainbow nation projects an image of different racial groups coming together and living in harmony' (1997, p. 16), where the many colors of the rainbow are interpreted as symbols of South Africa's cultural diversity. In that sense 'the adoption of a radical multiculturalism … allowed for a reconciliation of opposed forces that were tearing the nation apart' (Myambo 2010, p. 95). The rainbow nation is, in this chapter found to be a productive metaphor that one can use to understand some of the meanings engendered in the co-existence of colonial/apartheid and indigenous/vernacular toponyms especially in the majority of the cities in South Africa.

Situations where places have been (re)named repeatedly abound in Zimbabwe and South Africa, as is typical in other parts of southern Africa. The 'toponymic palimpsests and rainbows', as I would want to call them, raise questions about 'competing toponymic corpuses' (Giraut 2020) created out of temporal and political changes. The existing palpable toponymic layers are then read, interpreted and related to differently by different groups of people. That is the typical environment that engenders the use, development or persistence of multiple names for a single place in most parts in southern Africa.

Politics of Legacies in Zimbabwean Toponyms

Using examples from Zimbabwe, I focus on the diverse toponymic layers created in and through political transitions, to speak about the relevance of what I call politics of legacies to the creation and interpretation of multiple toponyms.

I begin this section by invoking the many names for the International Airport in Harare, Zimbabwe's capital city. The airport's current name is *Robert Gabriel Mugabe International Airport*. Prior, it was *Harare International Airport*, the name which replaced the colonial, in this case, original name *Salisbury Airport*. Because of these changes, the airport in Harare has been described as having worn 'many hats'.

The transition from *Harare International Airport* to *Robert Mugabe International Airport* has not been smooth. In some quarters, the airport remains *Harare International Airport*. For example, an attempt to book a flight to or from *Robert Gabriel Mugabe International Airport* shows that it does not exist. So, the name *Harare Airport* persists on IATA even though the change of the name of the airport in Harare was effected in 2017. There have also been attempts to erase Robert Gabriel Mugabe International Airport signage in the transitional context of the coup in November 2017 (see Fig. 5) and after, when calls were made to rename the airport to Mnangagwa Airport. Both attempts at erasing RG Mugabe signage are some of the cultural markers of the transition from Robert Mugabe's regime to the current dispensation. What motivated the attempts at such an erasure are issues to do with resentment of Mugabe's perceived unwarranted fame (Mangena 2020).

This story about the airport in Harare, especially the image of 'wearing many hats', reveals that it has been susceptible to replacement within the context of regime changes (Azaryahu 2018). Its susceptibility to change, its openness "to a layering of different historical narratives" (Rose-Redwood et al. 2018, p. 8) transforms into a palimpsest. In particular, the airport in Harare could be taken to be a "palimpsest which has continuously been written and re-written by multiple 'authors' as well as reinterpreted by different 'readers'" (Rose-Redwood et al. 2018, p. 8). The example of Harare airport also shows that layers of place names normally take identifiable forms in postcolonial countries; colonial official layer, postcolonial official layer, and then at both levels, an addition of unofficial layers whose main function is to disrupt especially the commemorative official texts (see Giraut 2020; Mitin 2018).

Renaming of places does not mean that old names vanish. It is possible that at one point, *Salisbury Airport, Harare International Airport* and or *Robert Mugabe International Airport* coexist(ed). This persistence of erased toponyms has been vividly imagined in Zimbabwean literature. By way of example, Petina Gappah's character in one of the short stories in *An Elegy for Easterly* says the following:

> The government renamed the places that the whites had renamed, so Umvukwes became Mvurwi again, Selukwe became Shurugwi, and Marandellas became Marondera. Queque became Kwekwe. The changes did not affect people like my grandmother for whom independence was a reality that did not alter the memories. She continued to speak of Fort Victoria and not Masvingo, of visiting us in Salisbury when she meant Harare, and of my aunt who was married in Gwelo instead of Gweru. (2009, p. 156).

Fig. 5 Erasure of R. G. Mugabe from R. G. Mugabe international airport direction sign

Gappah's character grapples with a situation where many years after the end of colonialism, some people, especially the older generation, still called places using colonial toponyms. This persistence of colonial toponyms, which takes the form of what Stoler (2008) names 'colonial debris', connects with the concept of places as palimpsests, with older toponyms resurfacing or persisting in the present. This then means that old and new toponyms are able to coexist and often compete for existence. The coexistence is another way in which a place is known by more than one names.

A related example in Zimbabwe is that of a district which is known as Matobo. Similar to Harare Airport, Matobo has been renamed several times. Matobo's original name is Mwalindidzimu (which means sacred place of ancestors) given by the Kalanga people who lived there before Mzilikazi invaded the space in the early nineteenth century (Nyambi and Mangena 2016; Mangena 2020). Mzilikazi renamed the place Matobo. Matobo was later appropriated to Matopos by Cecil John Rhodes, the chief architect of colonialism in southern Africa. Upon the attainment of independence, the name Matobo was reinstituted. It's possible that different social groups relate to the place using either of the labels Mwalindidzimu, Matobo or Matopos since complete erasure in palimpsests is not always guaranteed. Specific to the change from Matobo to Matopos, one could argue that since Rhodes' bones remain in his grave in the Matobo Hills, there are chances that the place will continue to be associated with Rhodes, and Matopos will possibly linger on alongside that memory.

There are other examples of place name multiplicity from Zimbabwe, that involve the use of alternative or unofficial toponyms as protest. An apt example is that of the Tonga people in some parts of Matabeleland who use place names different from official ones as documented in Chabata et al.'s article 'Mangling of place names and identity corruption amongst the Tonga of Zimbabwe' (2017). The Tonga people largely use place names which are different from the official ones because most of the place names in their areas are not indigenous, rather they are corruptions of Shona and Ndebele, which are dominant languages in Zimbabwe (Chabata et al. 2017, p. 109). Instead of using the corrupted toponyms, which are official, the Tonga, in order to preserve their language (and culture,) and resist domination, continue to relate to such places using indigenous Tonga names which however do not appear on maps.

Examples: Unofficial/Official

Kamatibi \Kamativi.

Silupane\Lupane.
Siyakobo\Siyakobvu.
Kasambabezi\Zambezi.

These examples of multiplicity or possibly duality of place names in a predominantly Tonga area reveal the complexity of officialization and standardization of toponyms in a postcolonial country like Zimbabwe. The standardization fixed the forms in which the names were to be written on the maps. This happened in the colonial period at a time when there was no working or standard orthography in the Tonga language (Chabata et al. 2017). So, the already developed Shona, Ndebele and English orthographies corrupted otherwise Tonga names during this process. Struggles for ethnic recognition, in this case, the continued use of 'unofficial' labels by the Tonga undermines the linguistic domination of Ndebele and Shona.

There are other forms that fit in to such toponymic protests as in the case of the Tonga that also gave rise to a multiplicity of toponymy in colonial Zimbabwe that are worth mentioning. The British who occupied southern Africa, named and renamed places in order to tame the new space they occupied using their language. The renaming included new English and also transphonologised indigenous toponyms. Transphonologised toponyms included examples of *Umtali* for *Mutare, Gwelo* for *Gweru* etc. Suffice to say that all this aided the creation of a corpus of unofficial toponyms that were used alongside the official ones. However, what fits into toponymic protest in this case, is the manner in which, without outright rejection of English toponyms, the indigenous people subtly undermined them through 'Africanization' and or mispronunciation. The mispronunciation of, for instance, '*Top Area*' into '*Topola*', '*Damn Good Mine*' into '*Demgudu*', '*Glengarry*' into '*Ngerengere*', for instance, goes beyond being a display of lack of knowledge of the English language by indigenous people to become a clever way of refusing to use English otherwise foreign toponyms imposed on indigenous places (see Mangena & Ndlovu 2013). This form of linguistic innovation allowed marginalized people, in this example, the colonized blacks in Zimbabwe, to relate to places and their names in ways that were comfortable to them.

The last two examples that I want to draw in Zimbabwe speak to the politics of commemorative toponyms in post-independence. The first example relates to commemoration of the sitting president as provided in the country's by laws on naming of places. The second example allows a reflection on the conflicting ways of relating to space through use of different toponyms for a single place by the ruling party and opposition. The transition from Robert Mugabe to the current regime has seen a continuation of a damnation of colonialism through toponym changes. One of the major name changes in the new post-Mugabe commemorative space involves an attempt at immortalizing Emmerson Dambudzo Mnangagwa in the same way that Mugabe was immortalized during his reign. In every city in Zimbabwe there is a road or street that has been named after the current president (see Mamvura 2020). However, in certain instances, especially in Bulawayo, the second largest city in Zimbabwe, mainly occupied by the Ndebele ethnic group, the imposition of Mnangagwa's name on the city's landscape provoked backlashes from residents of that city, as could be seen in the following narrative reported in a newspaper:

> Some of us have bad memories of ED Mnangagwa. Having a busy road named Mnangagwa is obviously traumatic. … We have our own heroes like Lookout Masuku, Dr Dumiso Dabengwa, Mqondisi Moyo and others, (Sithole 2019, Zimbabwe Situation, November 24, 2019).

Emmerson Dambudzo Mnangagwa, just like Mugabe, is at the center of the traumas associated with the memory of *Gukurahundi* in Matabeleland and the Midlands where an excess of 20 000 predominantly Ndebele citizens were killed by state agents in the early years of Zimbabwe's independence. This opposition to commemoration of Mnangagwa in Bulawayo, arguably, may trigger parallel toponyms to the official yet unpopular and contested new place names. The parallelism may include a retention of the old toponym, the deliberate non-use the "official" name or appropriation of the meanings of the new place name. It is highly probable that when some Bulawayo residents see Mnangagwa signage on their roads, that would not make them think of a him as a hero, as suggested in the official script, but rather a perpetrator of violence who traumatized their people. This is one way in which place names, especially commemorative ones, gain multiple meanings.

There are other examples of place name multiplicity that fall outside the frame of palimpsest in Zimbabwe that I invoke here to argue that there are other ways of interpreting this phenomenon of use of more than one name for a place. This interesting example of how people may choose to use different names to refer to a single place in the Zimbabwean context is that of an open space opposite the Rainbow Towers Hotel (incidentally, people still call the hotel by its old name Sheraton Hotel) in Harare. ZANU-PF calls the open space *Robert Mugabe Square*, while the opposition Movement for Democratic Change refers to it as *Freedom Square*. None of the two names are considered legitimate by the City Council of Harare (NewsDay 16 December 2017). The City Council of Harare just dismisses the two names as unofficial without stating the official label. Maybe to them, it is one of those places that do not need to be named. This example shows how it is possible to think about toponymic layering not only as result of successive regimes (not only as relating to the past), but as triggered by different conflicting political ideologies between the ruling and 'opposition' political parties in the present.

The many examples of the toponymic layers in Zimbabwe evoked in this section of the chapter bear testimony to the role played by political legacies in engendering place name multiplicity. In cases of erasures, these are not always complete. Thus, the old toponyms persist and often compete with the new ones. The incomplete erasures of removed toponyms, otherwise bear testimony to the fact that place names are not easy to erase. With commemorative toponyms, memorialized persons' political legacies influence how the general public relates to such places of memory. Place name multiplicity therefore bring to the fore issues to do with contestations of memory.

Colonial/Apartheid and Indigenous Toponyms Coexisting in Post-apartheid South Africa

One could say that there are many toponymic layers on most places in South Africa. These include the Khoi San, the Bantu, Dutch, Portuguese, British and Afrikaans toponymic 'layers'. The Bantu layer has sublayers of toponyms from different ethnic groups (Xhosa, Ndebele, Tswana, Venda, Sotho, Zulu etc.). This analysis simplifies an otherwise complex political history that resulted in the creation of an equally complex linguistic landscape in South Africa. But simplifying the linguistic landscape is here a strategy meant to highlight the various toponymic layers that have the potential to enable a transformation of most places in South Africa into palimpsests that are or have been written and rewritten many times by different authors (Azaryahu 2018). The existence of the different toponymic layers noted here suggests that just like in the case of Zimbabwe, it is possible to interpret place name multiplicity in South Africa using the metaphor of the palimpsest. The many names of Durban are an apt example. The following excerpt from Koopman (2012, pp. 135–136) demonstrates how the place called Durban has been named and renamed several times:

> The earliest known name for the bay on which Durban is situated is the 1450 name given by Portuguese seafarers—Rio de Natal—and names for the Bay of Natal in French, Dutch, English, Italian and Portuguese are found on many early maps predating the first settlement by white explorers, traders and missionaries. Survivors from early shipwrecks left short-lived names in Dutch, English and other languages for the bay and other topographical features, but it was only from 1820 onwards when Europeans started to settle that more permanent names were given. The early name Port Natal, which referred to the bay as well as the young growing township, was officially replaced by D'Urban after the name of the then Governor of the Cape, and within a few years this had become Durban, still today the official name of the city, and the name found in gazetteers and atlases world-wide. The Zulu name eThekwini was also early recognized, and many other Zulu names have been used for Durban or significant parts of Durban over the years. Some of these, like kwaMalinde, isiBubulungu, iFenya and eGagasini …. Official names like Port Natal and Durban, used on document and title deeds, have always lived side by side with unofficial names like Durbs and eMdubane.

While in Zimbabwe, indigenous toponyms which were bastardized during colonialism were 'corrected' at post-independence, (at least for all cities and towns), the situation is quite different in South Africa. Many places were renamed following the end of apartheid. However, scholars have noted that, 'careful not to undermine his reconciliatory project, Mandela was wary of erasing place names that honoured apartheid figures and history, especially replacing them with names of individuals associated with the liberation struggle' (Ndletyana 2012, p. 92). I argue in this chapter that this 'carefulness' in renaming of places has resulted in many, if not the majority of cities in South Africa at the present having two officially coexisting toponyms, one from the apartheid/colonial era, the other indigenous and African. These have been made to coexist in a clever way. A distinction has been made between the city and its metropolitan (see examples in Table 1). So, most cities maintain the apartheid/colonial place name while the metropolitan names are a formalization of the indigenous African place name, which, in the majority of cases, is the original name to the place in which the city is located. This scenario, where indigenous and colonial/apartheid toponyms are made to co-exist officially in South Africa, can be understood within the broader metaphor of the rainbow nation used to describe the country's post-1994 national identity. In this sense, not replacing the apartheid or colonial toponyms was regarded as a promotion of social cohesion.

In post-apartheid South Africa, Durban city remains Durban, while the Municipality or Metro is now officially known as eThekwini (see Koopman 2012). Suffice to note that what is popularly known as Durban has always been referred to as eThekwini by the Zulu people (Guyot and Seethal 2007). Also Pretoria, the city, retains the apartheid name in post-apartheid South Africa, but the new name for Pretoria Municipality or Metro is Tshwane (see Raper 2008; Neethling 2016). 'Pretoria was named in honour of the Voortrekker leader Andries Wilhelmus Jacobus Pretorius' (Raper 2008, p. 221). The origins and meanings of Tshwane are contested (Raper 2008). However, the most accepted meaning is that it is a Tswana name for Apies River near where the city was established (Raper 2008). Pretoria and Tshwane, Durban and eThekwini, as in the rest of the examples, are usually used interchangeably. Interchanging the labels Durban and eThekwini, Pretoria and Tshwane, for instance, means that the respective place names coexist. Durban and Pretoria are remnants of apartheid/colonialism, while eThekwini and Tshwane are examples of decolonizing toponyms. The formalization of eThekwini and Tshwane, for instance, can

Table 1 Examples of colonial/apartheid city and indigenous metro toponyms

City name	Municipality\Metropolitan name
Durban	eThekwini
Pretoria	Tshwane
Louis Tritchard	Makhado
Pietersburg	Polokwane
Port Elizabeth	Gqeberha
Grahamstown	Makhanda

be framed or understood within the decolonial turn that has gripped especially the South African Universities. This decolonial turn is traceable to the Rhodes-Must-Fall campaign which was started in 2015 by a majority of black students at the University of Capetown who were infuriated by the slow progress at the university (Nyambi and Mangena 2016, p. 6). The Rhodes Must Fall was.

> intended to address *the* unequal vision of the world as it manifests itself within universities … mobilising for direct action against the reality of institutional racism at the University of Cape Town. The chief focus of this movement is to create avenues for REAL transformation that students and staff alike have been calling for (Chaudhuri 2016, no page).

In this case 'Rhodes is more than a name and a historical person. He is the symbol of colonial debris stifling transformation at the university'. However, in the naming of cities and metros in the manner shown above, the South African government is enabling a persistence of 'Rhodes' (used broadly here to capture the injustices of apartheid and colonialism).

The coexistence of typically colonial\apartheid and indigenous toponyms is tension-filled, and 'reflects a desperate attempt to integrate two worlds that hitherto conflicted on the fundamental question of black people's humanity. The names indicate a transition and the ambivalence caused in part by different sets of history' (Nyambi and Mangena 2016, p. 5). The question that one may pose is, is there real transformation in allowing a co-existence of apartheid/colonial and indigenous and original toponyms? In the South African context putting side by side colonial/apartheid and indigenous toponyms in this manner represents an attempt 'to create a sense of reconciliation among the people who identify with the names and the symbolic past they bring to bear in the forging of the present' (Nyambi and Mangena 2016, p. 6). In that sense, the colonial/apartheid toponyms are kept not as a 'preservation of reminders of former history' (Swart 2008, p. 112) but rather as a gesture of reconciliation. That is actually the essence of the metaphor of the rainbow nation, which is a theory of multiculturalism intended to bring people together (Myambo 2010). Nevertheless, the maintenance of colonial/apartheid toponyms in South Africa is impartial. The examples of the city-metropolitan toponyms are just a small portion, otherwise in other contexts places were renamed, in what Adebanwi (2018, p. 233) has called an attempt at 'metaphorically changing the colour of the street signs into black-only' that however goes against toponymic multiculturalism. This ambivalence also applies to the effectiveness of the 'rainbow nation' vision. The coexistence of colonial/apartheid and indigenous toponyms gives a semblance of 'equality', equal distribution in the cultural landscape, but so much else remains unequal in post-apartheid South Africa, what Musila (2014) terms the 'cracks' of the rainbow nation. For instance, according to Habib, 'the racial character of the ownership structure of the South African economy *continues* to be a stark reminder of apartheid and its inequities' (Habib 1997, p. 19). Myambo also asks

> Was the multiculturalism advocated by Archbishop Desmond Tutu to end the endemic epistemological and literal racial violence of the apartheid system ever anything other than a nation conceived as ultimate abstraction? Yet, coming from a historical context in which most of the country was/is literally owned by whites, is a feeling of metaphorical ownership

of the abstract Rainbow Nation enough for the disenfranchised masses crammed together in overcrowded townships and unsanitary "squatter camps"? (Myambo 2010, p. 95)

The coexistence of colonial/apartheid and indigenous names in South Africa takes more than one forms. First, some colonial/apartheid toponyms persist even if some of them were removed and replaced by indigenous ones at the end of apartheid in 1994. The second form, which has been the major concern of this section of the chapter, relates to what I would call an officialization of the persistence colonial/apartheid city toponyms by the post1994 African National Congress government. This officialization was facilitated within the framework of the rainbow nation through which colonial/apartheid city names were allowed to sit alongside indigenous toponyms. It has been my argument that because of such a policy that is hugely contested, some places are eventually known by more than one names.

Conclusion

In this chapter, I discussed the multiplicity of place names in southern Africa using specific examples from Zimbabwe and South Africa. I used the frame of the palimpsest to think through toponymy and political legacies in Zimbabwe and the rainbow nation metaphor to interpret the co-existence of apartheid\colonial' and indigenous place names in South Africa's cities.

As shown through some examples from Zimbabwe, the colonial toponyms are not the only ones that are problematic and have been contested, we also have to think about the hierarchies between and among indigenous languages and the attendant domination of hegemonic ethnic groups in that respect. This has been shown in this chapter through the example of the Tonga, who subversively use unofficial but otherwise indigenous names for places in their area because the official ones are distorted versions of the original.

I have also noted that there are relations between various layers of one and the same place. The different layers speak to the long history of migration and colonization which resulted in the creation of multiple meanings of toponyms and corresponding multiple memories. In particular, the coexistence of apartheid\colonial names and the original\vernacular names in post-apartheid South Africa represents a problematic coexistence of two distinct worlds that conflict\ed on the fundamental question of black people's humanity (Nyambi and Mangena 2016, p. 2). The toponyms which are allowed to coexist can be interpreted as symbolic of social cohesion but as I have argued the persisting apartheid\colonial place names symbolize colonialism\apartheid's enduring impact and visible continuities.

Questions

1. What does place name multiplicity concern?
2. Why are Zimbabwe and South Africa interesting sites of theorising place name multiplicity in southern Africa?
3. How useful is the concept of the palimpsest to a discussion of place name multiplicity in Zimbabwe?
4. In what ways is the metaphor of the rainbow nation productive in the accounting of the multiplicity of place names in South Africa?
5. Use at least two examples from either Zimbabwe or South Africa to demonstrate the relevance of the politics of legacies to the creation and interpretation of multiple toponyms.

References

Adebanwi W (2018) Coloring "rainbow" streets in South Africa: the struggle for toponymic multiracialism in urban post-apartheid South Africa. In: Rose-Redwood R, Alderman D, Azaryahu M (eds) The political life of urban streetscapes: naming, politics and place, 2018. Routledge, London and New York, pp 218–239

Azaryahu M (2018) Revisiting East Berlin and Haifa: a comparative perspective on renaming the past. In: Rose-Redwood R, Alderman D, Azaryahu M (eds) The political life of urban streetscapes: naming, politics and place, 2018. Routledge, London and New York, pp 56–73

Baines G (1998) The rainbow nation? Identity and nation building in post-apartheid South Africa. Mots pluriels, pp 1–10.

Chabata E, Mumpande I, Mashiri P (2017) Mangling of place names and identity corruption amongst the Tonga of Zimbabwe. In: Mashiri P, Chabata E, Mamvura Z (eds) Zimbabwe naming practices and patterns: a sociolinguistic approach, 2017. University of Zimbabwe Press, Harare, pp 109–127

Chaudhuri A (2016) The real meaning of Rhodes Must Fall. The Guardian, 2016, March 16

Derrida J (1994) Spectres of Marx: the state of the debt, the work of mourning and the new international. Translated by Peggy Kamuf, with an introduction by Bernd Magnus and Stephen Cullenberg. Routledge, London and New York

Dillon S (2005) Reinscribing De Quincey's palimpsest: the significance of the palimpsest in contemporary literary and cultural studies. Textual Pract 19(3):243–263

Duminy J (2014) Street renaming, symbolic capital, and resistance in Durban, South Africa. Environ Plan, Soc Space 32(2):310–328. https://doi.org/10.1068/d2112

Gappah P (2009) An elegy for easterly. Faber and Faber, London

Guyot S, Seethal C (2007) Identity of place, places of identities: change of place names in post-apartheid South Africa. S Afr Geogr J 89(1):55–63

Habib A (1997) South Africa-the rainbow nation and prospects for consolidating democracy. Afr J Polit Sci/Rev Afr Sci Polit 15–37.

Hang'ombe K, Chabata E, Mamvura Z (2019) Syungu namutitima or victoria falls? Contest for place and place naming. Nomina Afr: J Afr Onomast 33(1), 19–31

Klerk VD (2002) Changing names in the "new" South Africa: a diachronic survey. Names 50(3):201–221

Koopman A (2012) The post-colonial identity of Durban. Oslo Stud Lang 4(2):133–159

Light D, Young C (2014) Habit, memory, and the persistence of socialist-era street names in post-socialist Bucharest, Romania. Ann Assoc Am Geogr 104(3):668–685. https://doi.org/10.1080/00045608.2014.892377

Giraut F (2020) Plural toponyms: when place names coexist. Echo Géo 53:1–8. https://doi.org/10.4000/echogeo.20760
Mamvura Z (2020) Kutonga Kwaro Gamba': politics and the renaming of defence cantonments in the 'Second Republic' in Zimbabwe. Geo J 1–5
Mangena T, Ndlovu S (2013) Selected transphonologised Zimbabwean toponyms. In: Felecan O (ed) Name and naming: onomastics in contemporary public space. Mega, Baia Mare, pp 347–354
Mangena T (2020) Rhodes, Mugabe and the politics of commemorative toponyms in Zimbabwe. Geopolitics 25(4):1015–1036
Mitin I (2010) 'Palimpsest'. In: Warf B (ed) SAGE encyclopedia of geography. SAGE: Thousand Oaks
Mitin I (2018) Constructing urban cultural landscapes & Living in the Palimpsests: a Case of Moscow City (Russia) Distant Residential Areas. Belgeo Revue Belge De Géographie 4:1–15. https://doi.org/10.4000/belgeo.28126
Musila G (2014) Laughing at the rainbow's cracks: blackness, whiteness and the ambivalences of South African stand-up comedy. In: Obadare E, Willems W (eds) Civic agency in Africa: arts of resistance in the 21st century, 2014. James Currey, Woodbridge, pp 147–166
Myambo MT (2010) The limits of rainbow nation multiculturalism in the new South Africa: Spatial configuration in Zakes Mda's Ways of Dying and Jonathan Morgan's Finding Mr. Madini. Res Afr Lit 41(2):93–120
Ndletyana M (2012) Changing place names in post-apartheid South Africa: accounting for the unevenness. Soc Dyn 38(1):87–103
Neethling B (2016) Street names: a changing urban landscape. In Hough C (ed) The Oxford handbook of names and naming. Oxford University Press, Oxford, pp 144–160
Nyambi O, Mangena T (2016) Introduction: the way we name now. In: Nyambi O, Mangena T, Pfukwa C (eds) The postcolonial condition of names and naming practices in Southern Africa, 2016. Cambridge Scholars Publishing, Newcastle upon Tyne, pp 1–18
Raper PE (2008) Tshwane, a San name for Pretoria, South Africa. Names 56(4):221–230
Rose-Redwood R, Alderman D, Azaryahu M (2018) The urban landscape as political cosmos. In: Rose-Redwood R, Alderman D, Azaryahu M (eds) The political life of urban streetscapes: naming, politics and place. Routledge, London and New York, pp 1–24
Schein RH (1997) The place of landscape: a conceptual framework for interpreting an American scene. Ann Assoc Am Geogr 87(4):660–680. https://doi.org/10.1111/1467-8306.00072
Sithole P (2019) Byo residents bitter over selective renaming of roads. Zimbabwe Situation, 2019 November 24
Stoler AL (2008) Imperial debris: reflections on ruins and ruination. Cult Anthropol 23(2):191–219
Swart M (2008) Name changes as symbolic reparation after transition: the examples of Germany and South Africa. German Law J 9(2):105–121

Suggested Reading

Beyer H, Brenzinger M, du Plessis T (eds) (2017) Critical toponym: place names in political, historical and commercial spaces. In: Proceedings of the 4th international symposium of place names Windhoek, Namibia, 18–20 September 2017. Sun Media, Bloemfontein
Bigon L (ed) (2016) Place names in Africa: colonial urban legacies, entangled histories. Springer, New York City
Nyambi O, Mangena T, Pfukwa C (eds) (2016) The postcolonial condition of names and naming practices in Southern Africa. Cambridge Scholars Publishing, Newcastle upon Tyne

Conflicts and Challenges in the Standardisation of Geographical Names in Spain

Ayar Rodríguez de Castro

Abstract In Spain, toponymic prerogatives are largely delegated to regional authorities. If a locality wants to change its name, it must request permission from the regional government. In order to have changes made official, local authorities must follow several steps and adequately argue the need for the change, for which they can request a report from different consultative institutions in matters of toponymy such as the Royal Geographical Society of Spain (RGSS). The complexity of Spanish toponymy, including place names that can be traced back to antiquity, or a multilingual scenario and a complicated regulatory framework, leads to frequent toponymic conflicts between different social groups and governments. This chapter addresses the protocols involved in the standardisation and rubber-stamping of geographical names in Spain, the factors that may affect them, and the work carried out by the RGSS as one of the toponymic advisory bodies of reference. A number of examples are presented to illustrate the complexity of the Spanish toponymic system and strategies adopted to mitigate conflict.

Keywords Spanish toponymy · Standardisation · Official toponymy · Toponymic conflicts · Conflict mitigation

Objectives

- This chapter emphasises the importance of official and normalised toponymy databases, and will present the different challenges involved in their production, using Spain as a case study.
- By the end of the chapter, readers should be able to recognise the conflicts that may emerge in the process of making a new place name, or changing an existing one, and the role played by the toponymic authorities, advisory bodies and broader society in their resolution.

A. Rodríguez de Castro (✉)
University of Alcalá, Alcalá de Henares, Spain
e-mail: ayar.rodriguez@uah.es

© The Author(s), under exclusive license to Springer Nature Switzerland AG 2023
G. O'Reilly (ed.), *Place Naming, Identities and Geography*,
Key Challenges in Geography, https://doi.org/10.1007/978-3-031-21510-0_12

- Readers will learn to recognise the work undertaken by advisory bodies, understand their role, and identify the strategies put in place to solve toponymic conflicts.

The Importance of Toponymic Normalisation

Having normalised in as far as possible, official toponymic databases and geographical gazetteers are essential for all countries and regions for many reasons, including the following:

(1) official documents and maps must have a source that identifies different places without ambiguity (Fons 2015);
(2) toponyms, as geographical markers, play a crucial role in emergency management and in the transmission and dissemination of geographical information (Grace 2021);
(3) toponyms, as the proper noun of places, are part of the immaterial cultural heritage of a people and of the territory in which they are framed and controlled repositories are necessary for their preservation (IPGH 2004);
(4) official toponymic registers are a key instrument for the management of toponymic conflicts (Arroyo Ilera 2018);

Toponyms, as the emblem and display case of the place which they designate, or to which they belong (Riesco 2010; Tort-Donada 2006) can be the source of controversy and strife, so it is important to have protocols in place to administer and register them. Geographical names can cause a wide array of conflicts, the resolution of which involves comprehensive research and reliable and up-to-date toponymic sources.

Toponymic databases, however, are not always available. They are difficult and laborious to make, and pose significant administrative challenges, as they require a complex institutional structure to elaborate, normalise, endorse and publish them. Sometimes, simply, they are not high on the list of priorities. In any case, elaborating a geographic gazetteer should be in the to-do list of all countries and regions.

In Spain there is a substantial number of national and regional geographic gazetteers and the necessary administrative and legal framework is in place for their ongoing development and updating. As a result, toponymic conflicts are identified and managed with relative ease. Not all toponyms, however, are the same, and sometimes the normalisation and rubber-stamping of certain toponyms requires a finer, more careful approach, to prevent social, political, economic or administrative conflict.

This study analyses the conflicts that can emerge around the management of official toponymy and the strategies that can be deployed to resolve them, in Spain. This work focuses especially on the work undertaken by the RGSS, one of the main scientific and advisory bodies of reference for the Spanish toponymic authorities.

Toponymic Normalisation in Spain

In Spain, toponymic prerogatives are widely distributed across a constellation of toponymic authorities. This has posed considerable challenges to the elaboration of a common toponymic database. Despite this, the intense efforts begun in the 1990s have led to the compilation and publication, in 2016, of a *Nomenclátor Geográfico Nacional* (National Geographic Gazetteer of Spain) (Vázquez Hoehne et al. 2011) that lists and geo-references the standardised (but not necessarily official) denominations that must be used in the country's official cartography. To date, this geographic gazetteer includes a *Nomenclátor Geográfico Básico de España* (Basic Geographic Gazetteer of Spain, published by the Cartography Central Registry of the National Geographical Institute, which records the country's toponymy at a 1:25,000 scale and is matched with other official databases) and the regional geographic gazetteers.

The Spanish government holds toponymic prerogatives over a significant proportion of the country's place names, but most of these legal powers are devolved to the governments of the seventeen regions (*Autonomous Regions*) and two autonomous cities into which the country is divided.[1] Only in exceptional circumstances do institutions that depend on the central government such as the National Geographical Institute or Naval Hydrographic Institute, use their superior prerogatives to impose specific toponyms, for instance, in border denominations between regions, coastal names and nautical toponymy.

This wide array of toponymic authorities (Table 1) has not made the compilation, normalisation and standardisation of place names, undertaken by the National Geographical Institute for the Basic Geographic Gazetteer of Spain, an easy task, owing not only to the complex structure of prerogatives but also to the associated conflicts between different institutions.

Owing to the complex administrative and legal toponymic structure in place, standardising and making official a place name in Spain is a complicated process. When a social group or an institution applies for the recognition or change of place names, each case must be examined individually to determine not only the suitability of the application but also the institution to which it must be submitted.

The following diagram (Fig. 1) illustrates the process that a local council must follow to officially change the locality's denomination. This process cannot be directly extrapolated to other geographic entities, for which the protocol is less standardised. In any case, in most instances the first step is to submit the application to the local administration (council or equivalent local ruling body), which will redirect the application to the appropriate toponymic authority to decide, with the support of reports issued by one or more toponymic advisory bodies. Should the suggested place name lead to toponymic conflicts because of two or more toponymic authorities with prerogative over the name having different criteria, the final decision falls to the National Geographical Institute.

---

[1] Not all of them exercise these prerogatives fully, and few have official geographic gazetteers, although most have a legal framework in place to manage their toponymic responsibilities.

Table 1 Main bodies with toponymic prerogatives in Spain

Toponymic authority	Toponymic prerogatives
Comunidades autónomas (Autonomous Regions)	The seventeen regions and two autonomous cities exercise their toponymic prerogatives to different extents. Each of them have one or more toponymic bodies. Most establish a significant proportion of official toponyms
Instituto Geográfico Nacional (National Geographical Institute)	Official denomination of geodesic vertices and certain toponyms in the borders between regions, as well as the toponymy of regions that do not exercise their toponymic prerogatives
Registro de Entidades Locales (Register of Local Entities)	Official denomination of municipalities and other populated areas
Instituto Hidrográfico de la Marina (Naval Hidrographic Institute)	Coastal and nautical toponyms
Ministerio de Defensa (Ministry of Defence)	Military air bases
Ministerio de Agricultura, Alimentación y Medio Ambiente (Ministry of Agriculture, Food and Environment)	National parks
AENA	National airports and heliports
Ministerio de Fomento	Roads that are totally or partially managed by the central government
Puertos del Estado	National harbour facilities

This table is not exhaustive, and only presents major bodies with toponymic prerogatives in Spain, and those which have had a greater role in the elaboration of the Basic Geographic Gazetteer of Spain. Author's own, based on Basic Geographic Gazetteer of Spain (https://www.ign.es/web/ign/portal/rcc-nomenclator-nacional)

Challenges and Problems of Toponymic Standardisation

Establishing new place names or changing existing ones is a common practice, owing to the dynamic nature of toponymy and its sensitivity to current social needs and wants. Toponyms are, first and foremost, geographical markers that identify places and locate them in maps, but they are also proper nouns, which imbues them with an emotional signification that empowers them as identity markers, territorial landmarks and political and commercial labels, and as such they can attract the sympathy or enmity of users.

In this regard, the creation or modification of a toponym ideally requires full consensus among political institutions, scientific bodies and civil society. Should this not apply, toponymic changes can lead to a wide variety of conflicts, depending on the level of rejection that the new denomination inspires in one or more of the relevant stakeholders. Table 2 summarises the main conflicts that these changes can trigger, based on the 'change name' processes examined by the RGSS.

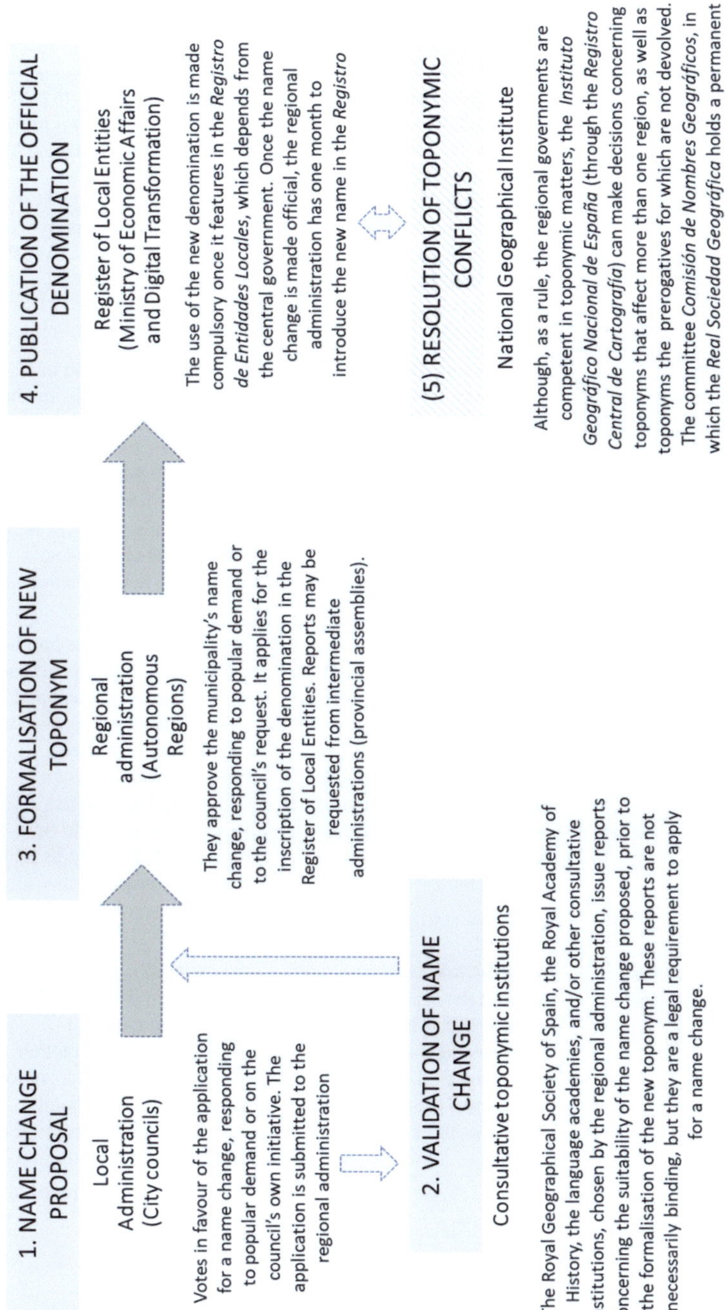

Fig. 1 Stages and agencies and authorities involved in name change proposals set forth by a Spanish municipality/locality. Author's own

Table 2 Main types of conflict caused by the creation/modification of place names

Main types of conflict detected in the creation/modification of toponyms	Description	Paradigmatic example
C1. Lack of agreement between two or more ruling bodies	Two or more toponymic authorities with prerogatives over the total or partial denomination of a given geographical entity disagree on the official or primary denomination. This problem often emerges around toponyms situated on administrative borders	A topographic feature shared (mountain range, bay, gulf…) by two administrative jurisdictions. Each administration espouses the use of their own toponym as a way to lay a claim over the feature
C2. Lack of clarity or discrepancy over the competent authority concerning the management of a given toponym	Two or more toponymic authorities, situated in the same or different administrative levels, claim the authority to establish official and/or primary denominations. This problem is often associated with historical territorial disputes and multilingual areas	The local population of a given territory vindicate the use of a toponym (for instance, of the locality in which they live) against the opinion of a higher administrative body, which rejects it or supports an alternative denomination
C3. Disagreement over the status of a toponym	Two or more toponymic authorities that hold total or partial prerogatives over a given toponym disagree about the official/unofficial, normalised/not-normalised status that said toponym and its variants must hold in the appropriate geographic gazetteers	No agreement exists around the status of a restricted-use endonym. Some users and/or institutions consider that said toponym can be used officially, like the main endonym, while others believe that it should be regarded differently, as a toponymic variant, a non-recommended place name, etc
C4. Conflicts around the consideration of exonyms	A toponymic authority, with or without the support of the civil society which it represents, proposes the use of an exonym that is rejected by other toponymic authorities and/or the scientific community	An exonym proposed to refer to a foreign country or city is rejected by the local population and/or another toponymic authority, and/or the scientific community, which regard the proposal as offensive or inappropriate

(continued)

Table 2 (continued)

Main types of conflict detected in the creation/modification of toponyms	Description	Paradigmatic example
C5. The toponym suggested/changed by a competent authority is rejected by the local population	A toponym suggested by the competent toponymic authority/ies is rejected by the inhabitants/users of the geographical entity so denominated	The creation or modification of toponyms to, for instance, pay homage to a historical character or politician, is not supported by the local population
C6. The toponym proposed/changed by the competent authority causes the enmity of a third party	A toponym proposed by the competent authorities to designate a given geographical entity, with or without the support of the local population, is rejected by a third party, which perceives the toponym as an offence	The change of a local toponym is perceived as an insult by members of a different community which are not, in principle, users of said toponym. For instance, reference to a given historical character can be perceived as an attempt to lay claim over his or her legacy, leading to the rejection of another people, on which the influence of said character is also significant
C7. The proposed toponym is not scientifically acceptable	A toponym proposed by one or more toponymic authorities and/or civil society is not considered appropriate on scientific grounds for whatever reason (e.g. semantic errors, linguistic imprecisions, etc.), leading to its rejection by some of the institutions involved in the process of toponymic formalisation	The scientific community believes that the toponym proposed by a toponymic authority and/or civil society is not scientifically acceptable, for instance, the appropriation of a toponym which is regarded as shared heritage or toponyms that lack sufficient scientific basis

Author's own after toponymic files in the RGSS

When one of these scenarios ensues, it is crucial for a mediating institution to act as arbiter. In Spain, the competent authority is ordinarily the National Geographical Institute, which has the support of the Comisión Especializada de Nombres Geográficos de España (Spanish National Commission on Geographical Names), an advisory collegiate body attached to the ruling Permanent Committee of the Geographic High Council (Consejo Superior Geográfico), which is the institution responsible for solving toponymic conflicts. This committee is constituted by members of national and regional toponymic bodies, representatives of scientific institutions and language academies (Spain has six official languages in addition to several dialects, whose toponymic role is officially recognised), and other institutions related to toponymy, such as the RGSS. In addition to participating in the committee, the RGSS monitors, and aims to prevent, the emergence of toponymic conflicts in its role as one of the advisory bodies of reference in the field of toponymy.

Advisory toponymic bodies play a central role in preventing conflict, by assessing the pertinence of toponymic changes, presenting proposals to resolve existing conflicts, and evaluating the risks of potential additional conflicts. The RGSS has been playing this role for over a century and has been involved in enormously relevant decisions for the Spanish toponymic corpus. For instance, in 1916 this institution fostered an initiative put forth by the Marquis of Foronda to revise the use of the same toponyms by different localities suggesting the use of toponymic surnames (Arroyo Ilera 2017) in order to, among other things, facilitate the task of the postal service.[2] In 1986, the Real Decreto (Royal Decree) 1690/1986 legally endorsed the Society's role as main advisory body for name changes, and since that date it has supported other toponymic authorities, local bodies and private groups in this capacity, generally by issuing reports about the relevance of specific name changes. The society manages these processes through a specific Comisión de Nombres Geográficos y Cambios de Denominación (Geographical Names and Toponymic Changes Commission), which also handles enquiries about geographical names and disseminates their scientific work through the Comisión Especializada de Nombres Geográficos.

The following section explains in detail, and provides illustrative examples of, the criteria used by the RGSS to resolve toponymic conflicts and the strategies it uses to tackle specific issues.

The Resolution of Toponymic Conflicts from a Geographical Perspective

Solving toponymic conflicts involves examining every proposed place name from a geographical perspective in full detail, that is, by taking into consideration all factors that can affect the projection of place denominations in their given geographical framework. This chapter goes over the geographical factors that need to be taken into account to evaluate toponymic changes, and provides examples to illustrate both the Society's methodology to identify and assess these conflicts and the strategies used to diagnose and resolve them.

Table 3 presents the factors that affect the resolution of toponymic conflicts from a geographical perspective. These factors need to be addressed to assess the relevance of a new toponym or the change of an existing one, as well as the possible cultural, social, economic and political conflicts resulting from it, and their resolution. These factors can have more or less impact on the diagnosis, and even contradict one another, so some of them can be used for, and some against the new denomination. Since every toponym is unique, they have all to be analysed individually and in detail,

---

[2] The Society revised a total of 570 toponyms (Arroyo Ilera 2017). This initiative had the full support of the National Postal Service and the toponymic authorities, although these not always took into account the interests and wishes of the inhabitants of the localities affected by the changes.

turning the resolution of toponymic conflicts into a challenging task in which every angle must be examined.

The following sections will present examples (3rd column) of the types summarised in the table. These are all recent cases in which local councils and toponymic authorities requested reports about a toponymic change from the RGSS. Details about the specific arguments used to assess each of these proposed changes and in the Society's rulings will be provided.

Table 3 Factors that affect the diagnosis of toponymic conflicts (i.e. they need addressing in order to assess any toponymic denomination)

Factors that affect the diagnosis of a toponymic conflict	Description and aspects to be examined by the toponymic advisory body	Example (RGSS case studies)
F1. Semantic and linguistic considerations of the toponym	The etymological suitability of the toponym must be analysed in relation to the designated geographical entity, ensuring that the toponym's history and meaning are correct and free from social and political implications that can lead to its rejection	Ría de Ribadeo, Higuera de Albalat, Navalmoral de la Sierra
F2. Coherence of the toponym with the geographical context in which it is used	The toponym must match its spatial, cultural and linguistic context. The risks include introducing an out-of-context toponym, in a language different from that of the territory within which it is framed, or making reference to cultural elements that are foreign to the local population	Montoro de Mezquita, Miera/La Cárcoba, Ría de Ribadeo, Higuera de Albalat, Navalmoral de la Sierra, Ancares
F3. Historical/geographical aspects	Toponyms may channel historical, political and geographical claims. The rigorous historical examination of a toponym and the sources used to justify its creation or recovery are decisive aspects in the assessment of said toponym's pertinence	Montoro de Mezquita, Ría de Ribadeo, Higuera de Albalat, Miera/La Cárcoba, Ancares

(continued)

Table 3 (continued)

Factors that affect the diagnosis of a toponymic conflict	Description and aspects to be examined by the toponymic advisory body	Example (RGSS case studies)
F4. Legal, jurisdictional and administrative considerations	Administrative considerations can play a significant role in the assessment of the pertinence and legitimacy of a toponym. When it comes to solving an existing conflict, the institution that holds the toponymic prerogative must be clearly established, and the status of the toponym (main toponym, alternative toponym, variant, non-recommended form, etc.) neatly defined. Similarly, the possibility of other geographical entities having the same toponym must be investigated	Ancares, Miera/La Cárcoba, Ría de Ribadeo
F5. Socio-economic considerations	Changes in place names can have direct or indirect social, identity and economic consequences, for instance, by incorporating the geographical entity designated into a territorial label (e.g. linking a place to a protected designation of origin or a touristic trademark etc.), or by flagging the political stance of a given social group (e.g. toponyms that pay homage to political figures). The analysis of these factors is essential to assess the motivations behind a change in place names	Ría de Ribadeo, Navalmoral de la Sierra, Montoro de Mezquita, Miera/La Cárcoba, Ancares
F6. Social and political conflicts	The creation and change of a place name can be rejected by some stakeholders, and this plays a major role in the assessment of the pertinence of a toponym. Although names are identity features, they can operate as conciliatory mechanisms, rather than the object of social and political dispute. As a result, it is desirable to find toponyms that meet as wide a consensus as possible	Ancares, Ría de Ribadeo, Miera/La Cárcoba

(continued)

Table 3 (continued)

Factors that affect the diagnosis of a toponymic conflict	Description and aspects to be examined by the toponymic advisory body	Example (RGSS case studies)
F7. Specific considerations of each individual toponym	Toponyms are idiosyncratic, and for this reason they must always be analysed individually, so that the specific circumstances are fully taken into account: e.g. the popularity of an unofficial toponym and generational differences in the use of toponyms, among many others	Miera/La Cárcoba, Higuera de Albalat, Ancares

Author's own after toponymic files in the RGSS

Semantic and Linguistic Considerations

Sometimes, toponyms are 'transparent' labels that describe a place (Jordan 2009). This occurs when the generic term, the specific term, or both carry a semantic or etymological signification that corresponds to the reality of the geographical subject or its past. If a toponym is regarded as inadequate because it conveys a wrong idea of reality or a conceptual error, it can cause its rejection or lead its users to errors concerning the scope and nature of the geographical entity it designates. In order to avoid this, it is crucial to examine the etymological origin and semantic implications of all proposed toponyms.

In two recent instances (2007 and 2021), the RSG was asked to analyze the toponym "Ría de Ribadeo" (*Ribadeo sea inlet*), which refers to a sea inlet situated at the border between the Spanish regions of Galicia and Asturias. In Galicia, the sea inlet is referred to by this name because of its close connection with the homonymous Galician municipality of Ribadeo. For years, the regional government of Asturias has tried for the sea inlet to be designated officially as "Ría del Eo" (*Eo sea inlet*), in reference to the Eo River, whose estuary converges with the sea inlet, supporting this petition with historical, cultural and economic arguments. From a geographical perspective, the sea inlet is the result of an elevation of the sea vis-à-vis the land following an instance of 'eustatic sea level change' (Comisión de nombres geográficos y cambios de denominación de la Real Sociedad Geográfica 2021a) and its origins are, in consequence, independent from the hydrographical network. It is thus understood that no sea inlet can be ascribed to a river flowing into it. Therefore, from a semantic perspective, it is geographically wrong to assign the sea inlet to the river; in fact, it is often the case that more than one river flows into the same sea inlet. In virtually every case, sea inlets are named after an important locality on its shoreline ("Ría de Navia" and not "Ría –of the river– Navia", "Ría de Arousa" and not "Ría –of the river– Tambre", etc.). Accordingly, "Ría del Eo" is not a geographically and semantically an appropriate toponym, unlike "Ría de Ribadeo". Based on this

Table 4 Case study: "Ría de Ribadeo"/"Ría del Eo"

<table>
<tr><td>Toponym:
"Ría de Ribadeo" (Ribadeo sea inlet)/"Ría del Eo" (Eo sea inlet)</td><td>Aspects to be considered: F1, F2, F3, F4, F5 and F6
Possible conflicts derived from an incorrect evaluation or unsatisfactory resolution of the issue: C1, C3, C4, C6 and C7</td></tr>
<tr><td rowspan="2">Location: estuary at the border between the regions of Galicia and Asturias

Author's own after Instituto Geográfico Nacional de España maps</td><td>Conclusion: the toponym "Ría del Eo" is semantically inappropriate</td></tr>
<tr><td>RSG's ruling: the RGSS proposes "Ría de Ribadeo" as only toponym to refer to the sea inlet, in the belief that "Ría del Eo" is inadequate because a sea inlet is independent from the rivers that flow into it. However, owing to the complexity of the toponymic conflict, fostering the recognition of "Ría del Eo" as a restricted-use variant or seeking other denominations to satisfy the Asturian population are not ruled out</td></tr>
</table>

Author's own. Every aspect taken into consideration by the RGSS is presented in each case study. These factors match those presented in Table 3, and the associated conflicts those presented in Table 2. More detail on these case studies can be found in the toponymic reports published by the *Boletín de la Real Sociedad Geográfica* (Comisión de nombres geográficos y cambios de denominación de la Real Sociedad Geográfica 2017, 2019, 2021a, b, c, d)

and other considerations, especially the potential jurisdictional and political conflict derived from the change name (see Table 3), the RGSS ruled against the toponymic suitability of "Ría del Eo" (Table 4).[3]

Coherence of the Toponym with the Geographical Context in Which It is Used

Toponyms are not isolated elements, and sometimes a toponym may be rejected, by the local population or the scientific community, for lacking coherence with the local toponymic network, for being foreign to the local territorial context, or for being at odds with nearby toponyms. The suitability of toponyms, therefore, must be evaluated with reference to the contexts in which they are placed.

In 2019, the RGSS was asked to issue a report concerning the change of denomination of "Montoro de Mezquita" (*Montoro of Mosque*), a hamlet in the province of Teruel, one of the most sparsely populated regions in Spain. The applying institution argued that the decades-old denomination was at odds with the cultural and

[3] This toponymic issue is at the centre of a heated social and political conflict between the regions of Asturias and Galicia. Asturias argues that "Ría del Eo" is a widely used toponym in Asturias, and for this reason the RGSS does not rule out alternative strategies/denominations that meet the wishes of the Asturians who live in the vicinity of the sea inlet.

Table 5 Case study "Montoro de Mezquita"/"Montoro de Maestrazgo"

Toponym: "Montoro de Mezquita" (*Montoro of Mosque*)/"Montoro de Maestrazgo" (Montoro of Maestrazgo)	*Aspects to be considered*: F2, F3, F5 *Possible conflicts derived from an incorrect evaluation or unsatisfactory resolution of the issue*: C5, C7
Localización: locality situated in the province of Teruel, in the southeast of the region of Aragón  Author's own after Instituto Geográfico Nacional de España maps	*Conclusion*: the origin of the extant toponym is uncertain, and is probably the result of an error. The toponym suggested is coherent with the locality's geographical and historical context *RSG's ruling*: the change of toponym is regarded as pertinent and generates no conflict, so its implementation is recommended

geographical context of the hamlet, since there are not Andalusian remains in the vicinity to justify the reference to a 'mosque'. It was suggested that this reference is the result of a semantic error which, although by now part of the collective imaginary, is geographically incorrect. Indeed, the reference "de Mezquita" (of Mosque) seems to spring from Pascual Madoz's *Diccionario geográfico-estadístico-histórico de España y sus posesiones de Ultramar* of 1848 (Madoz 1848), a recurrently used source for assigning place names in Spain in the early twentieth century. In this case, the *Diccionario* mentions the passing of a "Río Mezquita" near the hamlet, no further evidence for which exists.

The applying institution suggested renaming the locality as "Montoro de Maestrazgo", a reference to the district in which it is located, the Maestrazgo, a leading touristic area in the region, and an economic powerhouse for the province more broadly. Moreover, Montoro de Mezquita is a paradigmatic locality for the district's landscape. In this case the RGSS, after establishing that the etymological origin of "Montoro de Mezquita" could not be confirmed, decided to recommend the change of toponym for "Montoro de Maestrazgo" (Table 5), in the belief that the suggested denomination, as well as respecting the wishes of the local population, corrects the contextual inaccuracy of the previous toponym.

Historical/Geographical Aspects

Many new toponyms, and especially changes in place names, follow from historical claims and the interest of some social groups to recover old denominations with heritage value. Historical arguments must be examined with caution, and the sources used to be carefully studied and chosen, because any mistake can easily lead to toponymic conflict.

In October 2020, the municipality of "Higuera", in the region of Extremadura, decided to change its name to "Higuera de Albalat" (*fig of Albalat*), requesting a report from the RGSS. The toponymic surname "de Albalat" has its origin in the "Campana de Albalat", a group of localities formed by Higuera and another two municipalities situated in the former territory of the Andalusi city of Medina Albalat, which dates from the early tenth century. The fortress of Medina/Makhada Albalat was built during the reign of Abderraman III, and abandoned by the Muslims for good in the twelfth century; it is in a ruinous state and will likely disappear altogether soon (Hispania 2014). It is believed that the etymological origin of the denomination "Albalat", of obvious Arabic origin, is "paved road", perhaps in relation to the construction technique often found in Roman roads (Ríos Camacho 2010). Since Makhada means "pass" or "ford", the name of the Andalusi city refers to its location in the Roman road from *Toletum* to *Augusta Emerita*, at the point in which the road crosses the Tagus River. This is distant several kilometres from the three modern localities, perhaps an excessive distance to believe that this is the etymological origin of the denomination.

The Arabic toponym is amply documented in the town's historical records. Although cities with "Albalat" toponyms are common in the eastern coast of Spain (Albalat dels Sollers, Albalat de la Ribera, Albalat dels Taronchers, etc.), and although the etymological origin of these toponyms match that of Albalat in Cáceres—la) بلاط Jl-balāṭ), meaning slab, "paved gound" or "road" (Generalitat Valenciana 2009)—toponymic references that include the term "Albalat" are rare in the western regions. According to Ranz and López's (Ranz and López 2015) analysis of a toponym found in the central region of the country, "Albalate de Zorita", the Arabic term *Albalat-e* could be a corruption of the Celtic term *alp-alba*, meaning "elevation" or "slope" (Celdrán 2002). From this perspective, the toponymic surname "Albalat" would be etymologically incorrect for "Higuera de Albalat" if the aim is to recall the Andalusi legacy; the etymological origin of the term is unclear, and the toponym could lead to long-term archaeological error. However, the presence of the "Campana de Albalat" and the fortress of Medina Albalat, and, especially, the fact that the toponym is not only abundantly documented but also deeply ingrained in the collective imaginary, led the RGSS to support the change name (Table 6), also arguing that this would give the valuable archaeological remains additional visibility, despite the uncertain etymology of the toponym. The RGSS's proceedings and bulletin published the whole argument, to ensure that the etymological problem was sufficiently documented (Comisión de nombres geográficos y cambios de denominación de la Real Sociedad Geográfica 2021d).

Table 6 Case study "Higuera de Albalat"

Topónimo: "Higuera de Albalat" (*fig of Albalat*)	*Aspects to be considered*: F1, F2, F3, F7 *Possible conflicts derived from an incorrect evaluation or unsatisfactory resolution of the issue*: C5, C6, C7
Location: municipality in the northeast of the region of Extremadura  Author's own after Instituto Geográfico Nacional de España maps	*Conclusion*: it is unclear that the toponym suggested is etymologically correct, but it is suitable from the historical perspective, and can contribute to give visibility to the municipality's archaeological remains *RSG's ruling*: the change of place name was recommended on historical and social grounds, although further etymological investigations were considered desirable

Legal, Jurisdictional and Administrative Considerations

The legal and administrative mesh surrounding toponyms can be a decisive factor in the assessment of name changes. A wrong interpretation of the legal framework can result in a new denomination leading to conflict. Similarly, the jurisdictional pertinence of name changes must also be evaluated, and the suitability of the suggested name for its geographical scale and administrative framework assessed.

In May 2018, the RGSS was asked to issue a report about the change of denomination of "La Cárcoba", the capital of the municipality of "Miera" (region of Cantabria), for "Miera", that is, making the capital's and the municipality's toponyms coincide. As the petition argued, the toponymic change requested aimed to officialise the denomination "Miera" for the capital of the municipality, which, according to the application, was the popular and historical use of the term since its foundation, but this was opposed by part of the local population, which argued that there is no certainty with regard to the existence of a historical capital besides the modern neighbourhood. The proposal suggested for a group of neighbourhoods known as "los barrios altos" ('upper neighbourhoods') to be grouped under the toponym "Miera" and for this new entity to replace "La Cárcoba" as capital. "La Cárcoba" would be one of the locality's neighbourhoods, as it was further argued that a neighbourhood cannot hold administrative capital status.

This issue, therefore, raised not only a toponymic problem, but also an administrative one, namely the creation of an artificial population nucleus, brought about by bringing together a series of neighbourhoods that had never been united in this way. Although it was possible to attest that both the capital and the municipality had historically been referred to as "Miera", the name proposed involved the creation of an administrative unit with no historical precedents. The neighbourhood of "La Cárcoba" has held capital status in the municipality for over a hundred years, and this

Table 7 Case study "Miera/La Cárcoba"

Toponym: "Miera"/ "La Cárcoba"	*Aspects to be considered*: F2, F3, F4, F5, F6, F7
	Possible conflicts derived from an incorrect evaluation or unsatisfactory resolution of the issue: C2, C3, C4, C5, C7
Location: municipality in the interior of the region of Cantabria  Author's own after Instituto Geográfico Nacional de España maps	*Conclusion*: although the recovery of the historical denomination "Miera" is considered pertinent, the creation of an artificial locality with said name to replace the current capital cannot be justified
	RSG's ruling: it is recommended to postpone the name change until consensus exists concerning the population nucleus that is to hold capital status

Author's own after (Comisión de nombres geográficos 2019)

cannot be challenged on administrative grounds, as the Ley Orgánica de Régimen Municipal de Cantabria (Municipal Regime in Cantabria Act) establishes that population nuclei of every kind may hold municipal capital status (Ley Orgánica 8/1981). Since "La Cárcoba" is defined as a "neighbourhood", that is, "a group of houses or hamlet that depends on another locality" (Real Academia Española de la Lengua 2014) which, therefore, can legitimately act as the head of the municipality, being, to boot, a consolidated name and entity whose suppression for the creation of new ones is unadvisable, especially since no local consensus exists. For these reasons, the RGSS ruled that, although officialising "Miera" as the name of the municipal capital makes sense in terms of historical and current use of the name, attaching it to a geographical entity that does not yet administratively exist requires a wide consensus and in-depth scientific scrutiny. As such, the RGSS recommended against the name change until the administrative boundaries of the suggested capital were defined and the change was supported by a significant majority of the local population (Table 7).

Socio-Economic Considerations

Toponyms can be used as commercial trademarks, showcasing the place which they designate (Light and Young 2015). In principle, the creation or modification of a toponym for profit is only problematic insofar as it is contested by the local population or is in any way offensive. Yet, it must also be taken into account that geographical names are items of immaterial cultural heritage (UNGEGN 2006). Therefore, any

Table 8 Case study "Navalmoral de la Sierra"

Topónimo: "Navalmoral de la Sierra" (Navalmoral 'of the hills')	*Aspects to be considered*: F1, F2, F5
	Possible conflicts derived from an incorrect evaluation or unsatisfactory resolution of the issue: C6, C7
Location: municipality in the southeast of the region of Castilla y León  Author's own after Instituto Geográfico Nacional de España maps	*Conclusion*: no objection is found to the name place change
	RSG's ruling: the proposed change is considered legitimate, as the inclusion of the surname "de la Sierra" is geographically and semantically coherent, and seems to generate no conflict or enmity

Author's own after Comisión de nombres geográficos y cambios de denominación de la Real Sociedad Geográfica (2017)

initiative that undermines the heritage and identity value of a place name can lead to important toponymic conflict.[4]

It is seldom that the RGSS is asked to review this sort of case, but some of the toponymic issues it examines have important economic implications. In 2017, the city council of Navalmoral de la Sierra, in the region of Castilla y León, requested a report concerning the replacement of the locality's name, "Navalmoral", by the current "Navalmoral de la Sierra". It was attested that, since records exist (1842), the surname "de la Sierra" ('of the hills') has been incorporated to the place name of 41 localities (Ministerio de Asuntos Económicos y Transformación Digital 2008). Although in most instances the aim behind this change was to distinguish these localities from others with the same name, the toponymic embellishment of the place name to make reference to a "sierra", as a touristic resource, or to a protected designation of origin may have also contributed to the decision.

Had the addition of the toponymic surname resulted in commercial or social conflict, the case would have been examined in more detail. However, no such conflict ensured, and the RGSS ruled that no issues resulted from the toponym "Navalmoral de la Sierra" being made official (Table 8). Moreover, the new toponym allowed the locality to make a more explicit reference to its geographical location, increasing its touristic appeal.

[4] One notorious instance is "Wiesbauerspitze", when the meat company "Wiesbauer" persuaded a group of Austrian city councils to name a local mountain, the Mullvitzkogel, after the company's trademark, a publicity stunt that generated great controversy in these towns and elsewhere (Stani-Fertl 2014).

Social and Political Conflicts

As put forth by the school of critical toponymy (Berg and Voulteenaho 2009), place names can carry political and identity meanings which, occasionally, can lead to a wide variety of conflicts. Owing to the role of place names as identity markers, toponyms can be seen to reflect the place to which they are attached or be used to lay claim to it, pitching different social groups, and even scientific authorities, against one another. It is, for this reason, crucial to always assess the degree of consensus met by new toponyms or by the change of existing ones.

In January 2021, the mayor of Candín, in Castilla y León, asked the RGSS for a report concerning the change of the municipality's name to "Ancares", as requested by the residents. The municipality is situated in the district of "Comarca de Ancares"; this district borders with the Galician district of "Os Ancares", which shares the Castilian district's denomination and has strong historical ties with it. In addition, other toponyms in both regions use the term "Ancares", such as "Sierra de Ancares" ('mountains of Ancares') and "Río Ancares" ('Ancares river'). Several city councils in Galicia actively opposed Candín's proposal, arguing, among other things, that the proposed toponym "Ancares" implied that the town was the region's main locality, and that this was problematic on several counts.

The city council of Candín argued in turn that the locality used to be called "Valle de Ancares" ('valley of Ancares'), which means that this was not, strictly speaking, a toponym in dispute, but the use of a similar toponym in different areas. However, this argument was not accepted owing to the fact that even the "Sierra de Ancares" itself was situated across the border between both regions, and other evidence for a shared etymological network exists. In addition, owing to the area's growth as a touristic destination in recent years, the change of toponym could have significant economic consequences.

The proposal met the widespread rejection of toponymic authorities, other institutions and individuals (Lopez 2021), which was a strong argument against the place name change, at least before a serious debate was held by all the toponymic authorities involved. It is worth recalling that the UN's Group of Experts on Geographical Names (UNGEGN 2006) recommend that all toponymic institutions involved act in agreement, that "It is desirable for operating guidelines to be conservative in regard to changing official names" and that "unnecessary changing of names" is to be avoided.

In this way, the RGSS decided to rule against the change of place name, despite which, because the historical arguments for the new toponym were well grounded (the pre-existence of the toponym "Valle de Ancares" to refer to the municipality), it was also suggested that an alternative such as "Candín de Ancares" ('Candín of Ancares') or "Candín en Ancares" ('Candín in Ancares') could be used to meet the wishes of the local population (Table 9).

Table 9 Case study “Ancares”

Toponym: “Candín”/“Ancares”	*Aspects to be considered*: F2, F3, F4, F5, F6, F7
	Possible conflicts derived from an incorrect evaluation or unsatisfactory resolution of the issue: C2, C4, C6, C7
Location: municipality in the northwester of Castilla y León, in the border with Galicia  Author’s own after Instituto Geográfico Nacional de España maps	*Conclusion*: the proposed denomination was rejected owing to the opposition of other localities whose territorial mark could be undermined and to the fact that it used a toponym shared by different regions, among other reasons
	RSG’s ruling: an alternative denomination was suggested to avoid social and political conflict, because the toponym originally put forth lacked sufficient geographical and historical grounds (the municipality was never called “Ancares”, but “Valle de Ancares”, which restricts the toponym to a very specific area in the hill-range and district of “Ancares”)

Specific Considerations of Each Individual Toponym

Analysing the pertinence of toponyms involves evaluating specifically toponymic parameters, such as the intensity of use (it is not the same to change the name of a small rural village and that of a densely populated urban settlement), the use made of the name in the media and the collective imaginary, strictly local idiomatic and dialectal factors, which can affect the spelling of the toponym, and so on. Although the projection of these factors is more limited than that of the issues examined above and they rarely lead to conflict, they can play an essential role in the resolution of said conflict, owing to the fact that most toponymic conflicts are local disputes.

This can be illustrated with some of our previous case studies. In the example posed by “Higuera de Albalat”, for instance, we find that the current toponym has long been used indiscriminately in the media, websites and other unofficial sources to refer to the municipality of “Higuera”. The widespread social use of the toponym was traced back to the mid-nineteenth century. In addition, no evidence was found that this use ever triggered conflicts of any sort, and this became a decisive argument in support of the name change. With “Candín”, on the other hand, it was argued that having the name officially changed to “Ancares” would encourage new applications for toponymic changes in the area, destabilising its toponymic network and undermining the heritage value of the centuries-old denominations of many of the localities in the vicinity. Owing to toponymy’s potential as touristic showcase, in recent years “Ancares” has become a very valuable trademark. As such, it was likely that many other municipalities in the area would try to have their names changed accordingly, or even to dispute the use of the denomination “Ancares” on geographical and historical grounds. In fact, the municipality of Peranzanes, near Candín, floated the possibility of changing its name with the support of historical arguments soon after the city council of Candín put forth their own proposal (Varela 2021).

Conclusion

The management of toponymic conflict involves the detailed and individualised analysis of place names. For this reason, it is essential to have scientific bodies in place to advise the toponymic authorities about the risks that toponymic changes entail. These bodies must be as independent as possible and be capable of providing a comprehensive perspective on all angles of real or potential conflict.

In Spain, the existence of a consultative body (Comisión Especializada de Nombres Geográficos) and scientific institutions able to advise the authorities about the pertinence of toponymic changes greatly contributes to prevent most toponymic conflicts and to resolve them swiftly when they appear. Among these, the RGSS has been playing an essential role for over a century, as the toponymic institution of reference, not only helping to solve existing conflicts but also designing strategies to better address them in the future.

Questions

1. What aspects must be taken into account for the evaluation of a toponymic change or the creation of a new toponym?
2. What institutions must participate in the processes of toponymic standardisation and formalisation?
3. What kind of toponymic conflict can ensue when two or more toponymic and/or scientific and/or civil society disagree about the pertinence of a denomination?
4. How should toponymic conflicts be resolved?

References

Arroyo Ilera F (2017) Reforma de la nomenclatura municipal de 1916: proceso, criterios y consecuencias. Boletín de la Real Sociedad Geográfica 2017, CLII

Arroyo Ilera F (2018) La toponimia como patrimonio cultural inmaterial. Boletín de la real sociedad geográfica 2018, CLIII,pp 33–60

Berg LD, Vuolteenaho J (2009) Critical toponymies: the contested politics of place naming 2009. Ashgate Publishing, Ltd.

Celdrán P (2002) Diccionario de topónimos españoles y sus gentilicios, 2002. Ed. Espasa, Madrid

Comisión de nombres geográficos y cambios de denominación de la Real Sociedad Geográfica. Expediente de toponimia del cambio de denominación de Navalmoral por Navalmoral de la Sierra, 2017 (pending publication)

Comisión de nombres geográficos y cambios de denominación de la Real Sociedad Geográfica. Informe sobre la propuesta de cambio de denominación de la capital del municipio de Miera.

Boletín de la Real Sociedad Geográfica 2019, CLIV: 263–274. https://www.boletinrsg.com/index.php/boletinrsg/article/view/86/95
Comisión de nombres geográficos y cambios de denominación de la Real Sociedad Geográfica. Expediente de toponimia del caso Ría de Ribadeo-Ría del Eo, 2021a (pending publication)
Comisión de nombres geográficos y cambios de denominación de la Real Sociedad Geográfica. Expediente de toponimia sobre el cambio de denominación del municipio de Candín por Ancares, 2021b (pending publication)
Comisión de nombres geográficos y cambios de denominación de la Real Sociedad Geográfica. Informe sobre el cambio de denominación del municipio de Montoro por Montoro de Mezquita en la reforma de la nomenclatura municipal de 1916. Boletín de la Real Sociedad Geográfica 2021c, CLVI: 263–268. https://www.boletinrsg.com/index.php/boletinrsg/article/view/124
Comisión de nombres geográficos y cambios de denominación de la Real Sociedad Geográfica. Informe sobre el expediente del cambio de denominación del municipio y capitalidad de Higuera por Higuera de Albalat. Boletín de la Real Sociedad Geográfica, 2021d, CLVI: 263–268. https://www.boletinrsg.com/index.php/boletinrsg/article/view/124
Fons ME (2015) La gestió i el control dels noms geogràfics de les Illes Balears: necessitat de creació d'un nomenclàtor georeferenciat. Onomàstica i identitat: un repte universal i multiescalar. II Congrés de la Societat d'Onomàstica, XXVII Jornada d'Antroponímia i Toponímia de la UIB. Universitat de les Illes Balears
Generalitat Valenciana (2009) Gran Enciclopedia Temática de la Comunidad Valenciana, 2009. Ed. Prensa Valenciana
Grace R (2021) Toponym usage in social media in emergencies. Int J Disaster Risk Reduct 52:101923
Hispania N (2014) Fortaleza de Medina-Albalat. Lista Roja del Patrimonio Español 2014. https://listarojapatrimonio.org/ficha/fortaleza-de-medina-al-balat/
IPGH (Instituto Panamericano de Geografía e Historia-Pan American Institute of Geography and History) (2004) Report of the United States of America Delegation. XXXVII Meeting of the Directing Council of the Pan American Institute of Geography and History
Jordan P (2009) Some considerations on the function of place names on maps. In: Proceedings of the 24th international cartographic conference, 2009. ICC, Santiago, Chile, pp 15–21
Ley Orgánica 8/1981, de 30 de diciembre, de Estatuto de Autonomía para Cantabria. https://www.boe.es/diario_boe/txt.php?id=BOE-A-1982-635
Light D, Young C (2015) Toponymy as commodity: exploring the economic dimensions of urban place names. Int J Urban Reg Res 39(3):435–450
Lopez MF (2021) Los alcaldes gallegos recelan de Candín: «Ancares es comarca». Diario de León 6–1–2021. https://www.diariodeleon.es/articulo/bierzo/alcaldes-gallegos-recelan-candin-ancares-es-comarca/202101050232132075570.html
Madoz P (1848) Diccionario geográfico-estadístico-histórico de España y sus posesiones de ultramar, 1848, 4. Est. tip. de Madoz y Sagasti
Ministerio de Asuntos Económicos y Transformación Digital (2008) Variaciones de los municipios de España desde. Registro de Entidades Locales. https://www.hacienda.gob.es/Documentacion/Publico/SGT/CATALOGO_SEFP/100_Variaciones-INTERNET.pdf
Ranz JA, López JR (2015) La toponimia de Guadalajara: estado de la cuestión. Rev Folk 401:4–16
Real Academia Española de la Lengua. Diccionario de la lengua española 2014
Riesco P (2010) Nombres en el paisaje: la toponimia, fuente de conocimiento y aprecio del territorio. Cuad Geográficos 46:7–34
Ríos Camacho XC (2010) Mozarabismo en la Gallaecia Altomedieval. Estudios generales y análisis desde sus fuentes documentales monásticas (siglos VIII, IX, X y XI)
Tort-Donada J (2006) Els noms de lloc i el territori: La toponímia des de la geografia. Toponímia, paisatge i cultura: els noms de lloc des de la lingüística, la geografia i la història. Associació d'Història Rural de les Comarques Gironines p 4
UNGEGN (United Nations Group of Experts on Geographic Names) (2006) Manual for the national standardization of geographical names 2006, 88. United Nations Publications

Varela S (2021) Peranzanes, en Ancares, también se quiere cambiar el nombre: Valle de Fornela. La Voz de Galicia 26–3–2021. https://www.lavozdegalicia.es/noticia/lugo/2021/03/26/peranzanes-ancares-quiere-cambiar-nombre-real-valle-fornela/00031616754230823382103.htm
Vazquez Hoehne A, Rodríguez de Castro A, Luján Díaz A, Montilla Lillo M, Castaño Suárez A (2011) Propuesta metodológica para la elaboración del Nomenclátor Geográfico Básico de España a partir de la autocorrección de la Base Cartográfica Numérica con la información de las comunidades autónomas, 2011. Biblioteca Técnica de Política Lingüista

Suggested Reading

Stani-Fertl R (2014) Motifs for renaming geographical features. In: Proceedings of the XXIV ICOS 2014

Language Policies in the Field of Toponymy: Perspectives on Spain

Mar Batlle

Abstract Spain is a multilingual country in which Spanish is presently the official language of the entire state. However, there are regions (called autonomous regions) with a local language that is also official in the given zone. Some examples are Catalonia, Basque Country (Euskadi) and Galicia. Autonomous regions are administrative divisions, and they have local governments. These governments have jurisdiction over specific aspects, like the management of the local place names. Therefore, linguistic policy as applied to toponymy in Spain is not uniform. In some bilingual regions, such as Catalonia, Galicia and the Balearic Islands, place names are only set in the local language, in accordance with legislation and ordinary use. In other bilingual regions, toponyms can be set in one language (either Spanish or the local one) or in both languages, such as in Basque Country or Valencia. In this scenario, involving several languages with different statuses and levels of bilingualism, with all the movement back and forth that they have had over the centuries, and with the subsequent diversity of cultural identities that form Spain, the interest to standardize place names is related to the promotion to the language itself.

Keywords Spain · Geographical naming · Standardizing · Bilingualism · Regional languages

Objectives

- By the end of the chapter, readers will be able to get an understanding of the linguistic policy applied to the guidelines for place names in Spain, taking into account the diversity of languages spoken there.
- Readers will get insights into how the various actors and participants are involved in the place naming process.
- Readers will be exposed to the basic resources available on the internet regarding the gazetteers for place names in Spain.

M. Batlle (✉)
Institut d'Estudis Catalans, Barcelona, Spain
e-mail: mbatlle@iec.cat

© The Author(s), under exclusive license to Springer Nature Switzerland AG 2023
G. O'Reilly (ed.), *Place Naming, Identities and Geography*,
Key Challenges in Geography, https://doi.org/10.1007/978-3-031-21510-0_13

Introduction

Spain is a country with considerable linguistic diversity in which there are different situations and levels of bilingualism. Spanish or Castilian is the official language throughout the state. Some regions have another local language that is also official in that region. Thus in Galicia, Galician is spoken; Basque, in the Basque Country (Euskadi) and Navarra; and in Catalonia, the Balearic Islands and Valencia, Catalan is spoken, called Valencian in Valencia. In Catalonia, in addition, to Aranese, a variety of Gascon is spoken in the Val d'Aran and is also official. In other regions, the local language does not have official status, such as Bable in Asturias and northern Leon, and Aragonese in Aragon (and Catalan in the area called La Franja, next to Catalonia), although there is legislation regarding the use, and academies that regulate the corresponding language. As can be seen in the following two maps, sometimes the extension of the local language (Fig. 1) affects more than one administrative division (autonomous region) (Fig. 2).

The distribution of local languages presents different nuances depending on the area. In Catalonia, the Balearic Islands and Galicia the local language is spoken along with Spanish throughout the region. However, in the Basque Country, Navarra, Valencia and Aragon the corresponding local language and Spanish are distributed by linguistically differentiated areas (Figs. 3, 4 and 6). In the first case, Catalonia, the Balearic Islands and Galicia, there is language contact and bilingualism throughout

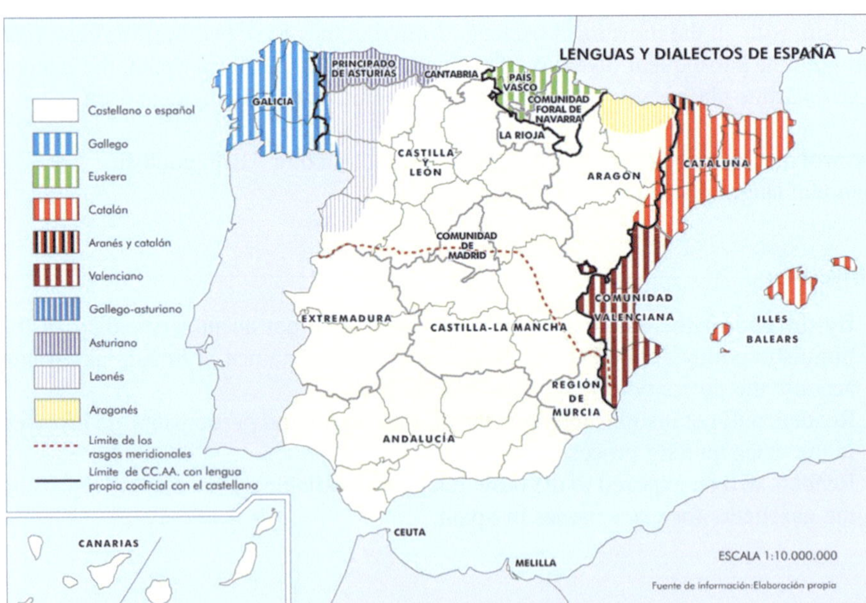

Fig. 1 Linguistic map of Spain (*Source Toponimia. Normas para el MTN25*, p. 19. © Dirección General del Instituto Geográfico Nacional; © Centro Nacional de Información Geográfica)

Fig. 2 Map of the administrative divisions of Spain. (*Source* https://www.ign.es/espmap/mapas_desadm_bach/pdf/DesAdm_Mapa_01_texto.pdf) © Dirección General del Instituto Geográfico Nacional; © Centro Nacional de Información Geográfica

the region; and in the second, only in the area where the local language is also spoken, which may have a greater or lesser extent of use.

In bilingual regions, toponymic standardization has always been considered a priority issue, and in monolingual regions it has hardly aroused interest. For this reason, bilingual areas, in general, have advanced more in standardization than monolingual ones, with the exception of Andalusia, which in recent years has made a notable effort to promote the standardization of its toponymy. Clearly, this situation is the result of the very unequal language policy that exists throughout the Spanish state.

It must be taken into account that the initiative for the toponymic standardization process depends on different entities depending on the area. Thus, in the case of the Basque Country, the law has recently been modified, and the city council is in charge of linguistic regulation. On the contrary, in the case of Catalonia, as a large part of the power of toponymy belongs to the autonomous government (the Generalitat de Catalunya), it is this body who decides. Hence, the *Nomenclàtor oficial de toponímia major de Catalunya* (Official Gazetteer of Major Toponymy of Catalonia) was launched in 1998. It is the same for Galicia.[1] For this reason, in the regions where the local government has more power and dedicates adequate funding, and so there has been greater standardized of the toponymy. It is also important that

[1] See: https://toponimia.xunta.gal/es/toponimia/normativa.

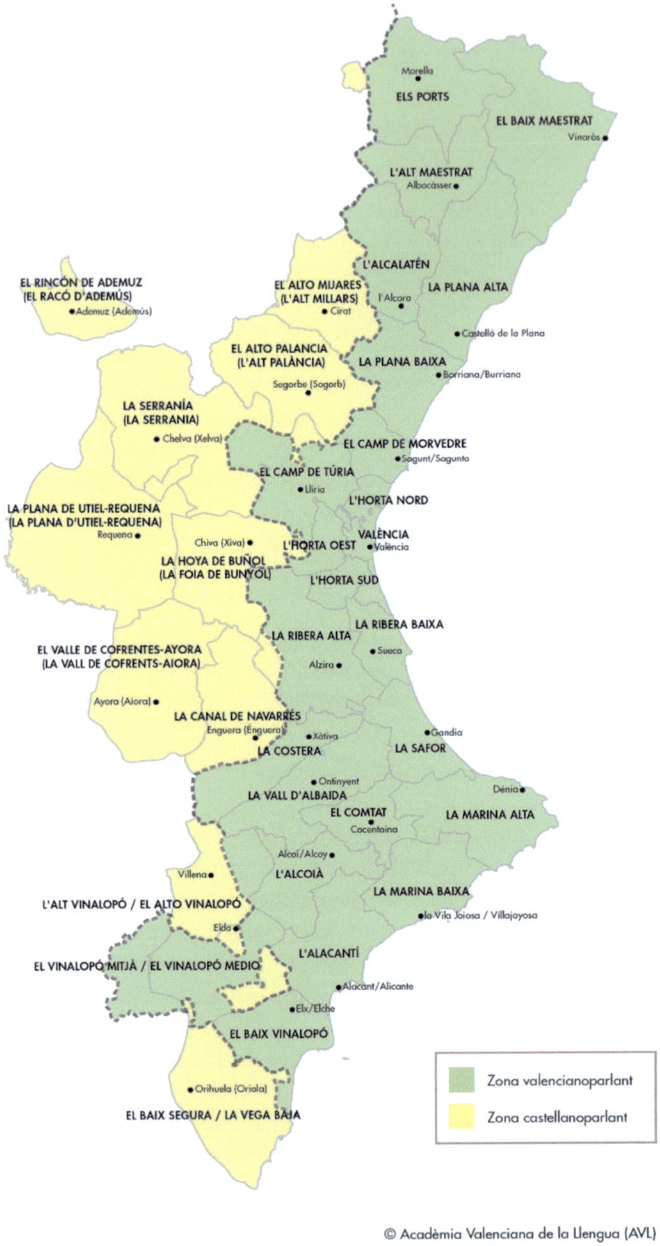

Fig. 3 Current geographical distribution of the Spanish (yellow) and Catalan (green) in Valencia. (©Academia Valenciana de la Llengua (AVL))

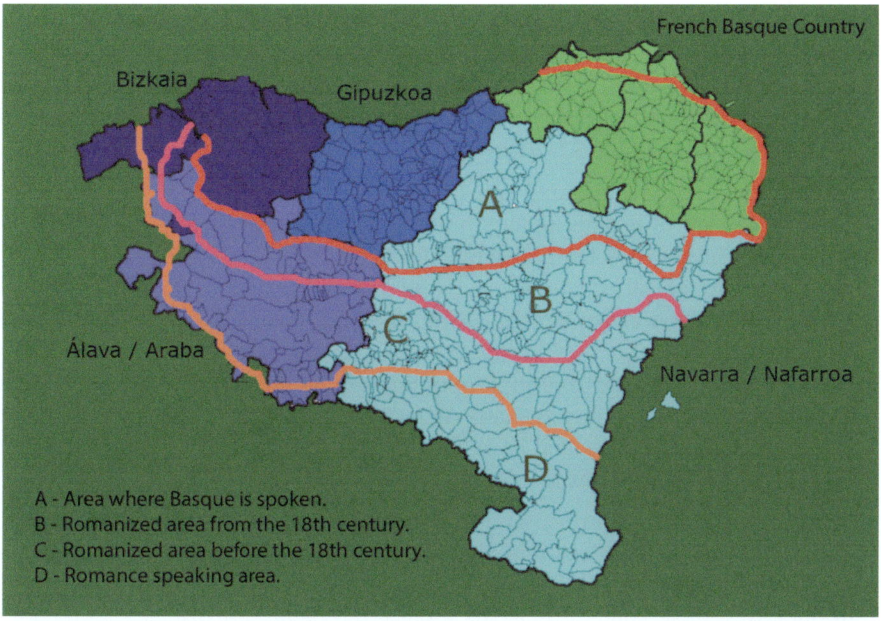

Fig. 4 Map showing the progressive decline of the Basque language. (© Mikel Gorrotxategi)

the government promotes not only the linguistic regulation of toponymy, but likewise its use in all spheres, as the government of Catalonia did, or that of Andalusia is doing. Otherwise, there is the situation like in Galicia, where despite having standardized names, their public use progresses slowly.

Regarding the toponymy of the monolingual regions using Spanish, Madrid and Andalusia stand out. In general, it is poorly standardized and works at a lower rate than in bilingual areas. For example, we have *Córdoba* in Andalusia, but *Cordovilla* or *Cordobilla*, probably its diminutive, is the name of other towns such as *Cordobilla* (Navarra), *Cordobilla de Lácara* (Badajoz), *Cordovilla* (Salamanca) or *Cordovilla la Real* (Palencia).

In contrast, the toponymy of the bilingual regions can be bilingual or monolingual, depending on the regulation that each community has selected to do. The official toponymy of Galicia, Catalonia and the Balearic Islands is, according to the legislation, only in the local language, not in Spanish. That is, in Galician in Galicia, and in Catalan in the Balearic Islands and in Catalonia, except for the part of Val d'Aran which is in Aranese. Although there are no official bilingual names in these areas, the names of prominent cities have names in Spanish, such as *Lleida* and *Gerona* (*Lleida* and *Girona*) in Catalonia; *La Coruña* and *Orense* (*A Coruña* and *Ourense*) in Galicia; or *Ibiza* (*Eivissa*) in the Balearic Islands. In the Basque Country, Navarra and Valencia, the official toponymy can be in Spanish, in the local language (Basque or Catalan depending on the region), or it can have a bilingual name in the two languages of the area. In many cases, even if the official name is in only one language, there is also the equivalent name in the other language. For example, *Ademuz* (Valencia) is the official name in Spanish, but there also exists *Ademús* in Catalan.[2]

Regarding the Basque Country and Navarre, the Basque language has declined over the centuries and its area of usage can be divided into three zones according to this regression (Nieto 2013: 82–84):

(a) The current Basque-speaking area (A zone on the map, Fig. 4). This area naturally has plenty of Basque toponyms. However, there are also Spanish toponyms, some of which are anciently documented.
(b) The area was recently Castilianized (B zone on the map, Fig. 4) but still possesses many Basque toponyms in good shape, although Basque speakers are not very common there.
(c) The zone was Castilianized in ancient times (C zone on the map, Fig. 4) such that the people living there usually do not speak Basque. In this area, there are some Basque toponyms that have been greatly transformed and there are a lot of toponyms set in Spanish.

The D zone on the map (Fig. 4) was always Romance language-speaking area. As can be seen on the map, the regression of the Basque language goes from south to north and throughout all the regions.

In big cities, there sometimes exists a preferred name; for instance, in the city named *Donostia* in Basque and *San Sebastian* in Spanish, it is quite frequent for Spanish-speaking people to use the Basque name *Donostia* instead of the Spanish name *San Sebastian*. Therefore, although the name of the municipality is officially bilingual, the local name *Donostia* is widely used, like in the name of the major hospital of the city *Hospital Universitario Donostia*. Even the adjective in Spanish is *Donostiarra*. Another salient example is found in the name *Vitoria-Gasteiz*, which is officially bilingual but the Spanish name *Vitoria* is used more than the Basque name *Gasteiz*. It is interesting to note that the official name is set as a compound name. Therefore, there is a clear intention to preserve the Basque name in formal scenarios regardless of the language used. According to Nieto (2013) in zones B and C there is a tendency to recover Basque names that have long disappeared.

In the case of autonomous communities with non-official local languages, the same situation occurs when the language is official. Therefore, the toponymy is bilingual or monolingual. For example, in Aragon there are names in Spanish and Catalan (*Benabarre/Benavarri*); or in Spanish and Aragonese (*Alcalá del Obispo/Alcalá d'o Bispe*). In Asturias, the standardization of the traditional names of Asturian toponymy is being carried out, which in many cases only consists of eliminating the Castilian spelling of the name and adapting the name to the current orthographic regulations (for example, *La Arena* is written *L'Arena*). Cases of double denomination, such as *Oviedo/Uviéu* and *Gijón/Xixón* are very few compared to other areas, but they are increasing, especially in main cities.

It should be noted that in areas with languages in contact, situations are generated that do not appear as frequently in monolingual areas, such as the duplication of

[2] See: https://www.avl.gva.es/web/publicacions/noms-dels-municipis-valencians.

names for the same entity. In these cases, if the toponym is within the autonomous community, this regulates the setting of the name. For example, in Valencia the bilingual name occurs in some hydronyms, such as *Riu Palància/Río Palancia* (Segorbe), which originates in a Spanish-speaking area (*Río Palancia*) and ends in the Catalan-speaking part (*Riu Palància*); or the *Riu Magre/Río Magro* (Macastre). The case of *Riu Millars/Río Mijares* is similar, with the difference that it starts in Aragon, enters Valencia through the Spanish-speaking part, and ends with the Catalan-speaking part.

If two or more communities share the geographical entity, it is the National Geographic Institute that sets the name in the official cartography. At the state level, therefore, the orography and hydrography of a supra-autonomous nature appear in a bilingual way in the official cartography. For example, in the *National Topographic Map 1*: 25,000 (MTN25) the Spanish name is used in its journey through the monolingual Castilian community and in the bilingual name in the community with its own co-official language, introducing the Castilian name first. Thus, we find *Río Ebro* in Aragon and *Río Ebro/L'Ebre* in Catalonia (Fig. 5), or *Río Júcar* in Castilla-La Mancha and *Río Júcar/Ríu Xúquer* in Valencia (*Normas Toponimia*, p. 26).

With the linguistic diversity described, the different nuances presented by the linguistic distribution in the communities, and the different degrees of vitality of the local languages, it is understood that in the Spanish environment it is not easy to regulate toponymy in a harmonized and coherent way. Especially if you only have in mind a system designed for a monolingual country, which has happened frequently. In addition, it must be taken into account that the jurisdiction of the toponymy is divided between different entities depending on the type of toponym. This is how the National Geographic Institute and other state bodies (Spanish Airports and Air Navigation, National Parks, State Ports, etc.), the autonomous communities and local administrations intervene. This diversity of actors implies, sometimes, the application of different criteria.[3]

The State Commission and the Regional Commissions

Since 2007, there is the Comisión Especializada en Nombres Geográficos (CENG, Specialized Commission for Geographical Names), integrated into the Consejo Superior Geográfico (Higher Geographic Council), which is the governing body of the National Cartographic System, and depends on the Ministry of Development. As you can see on their website,[4] «its mission is to promote the standardization of geographical names in Spain, in coordination with the competent bodies of the State and the Autonomous Communities, and to promote their knowledge, normalized use and value as cultural heritage by the different Public Administrations, private entities

[3] See: https://www.mitma.gob.es/organos-colegiados/consejo-superior-geografico/csg/comisiones/normativa.

[4] See: https://www.mitma.gob.es/organos-colegiados/consejo-superior-geografico/csg/comisiones/comisionespecializada-de-nombres-geograficos.

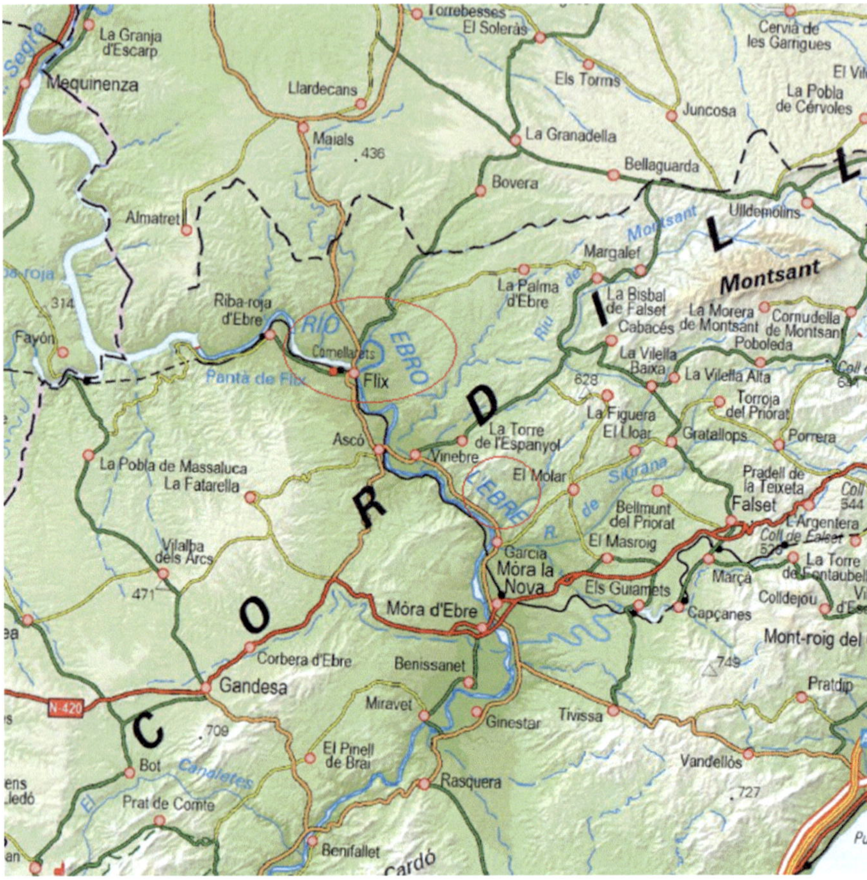

Fig. 5 National Topographic Map (MTN25) showing Catalan and Spanish names for the river Ebro (*Source* http://centrodedescargas.cnig.es/CentroDescargas/catalogo.do?Serie=MAMOV; © Dirección General del Instituto Geográfico Nacional; © Centro Nacional de Información Geográfica)

and the rest of society, having to channel their work into annual action proposals that will be submitted to the aforementioned Permanent Commission». Currently, the main activity carried out is the coordination for the elaboration of the *Nomenclátor Geográfico Nacional* (*National Geographic Gazetteer*).

The CENG is made up, among others, by representatives of the entities that participate in the standardization of toponymy at the regional level, who share their cartographic databases with the National Geographic Institute in order to facilitate state toponymic harmonization. Autonomous governments usually have specialized toponymic commissions to carry out better management, promote place names standardization and disseminate official names for public use. Salient examples of this exist in Galicia, Asturias, Aragon, Andalusia and Extremadura.

In Galicia, it is worth mentioning the enormous work of collecting and standardizing place names coordinated by the Comisión de Toponimia,[5] of the Consellería de Presidencia e Administración Pública de la Xunta, which has also made a toponymy search engine available to the public on the internet.[6] The commission was created in 1983 thanks to the Language Normalization Law. It is made up of representatives of different departments of the Xunta de Galicia, the local government, (Presidency, Language Policy, Local Administration, Urban Planning, Statistics and Regional Development); representatives of various scientific institutions; and experts in Galician place names and Galician language.

In Asturias, in 2004 was created the Xunta Asesora de Toponimia del Principáu d'Asturies (Toponymy Advisory Board of the Principality of Asturias), which promotes the standardization of place names of Asturias. It is made up of representatives of the local government (General Directorate competent in cultural matters; General Directorate for Territorial Planning); of the Academia de la Llingua Asturiana (Academy of the Asturian Language); the Real Institutu d'Estudios Asturianos (Royal Institute of Asturian Studies); experts in toponymy; and the University of Oviedo. It now has a digital place names search engine.[7]

In Aragon, there is the Comisión Asesora de Toponimia de Aragón (Aragon Toponymy Advisory Commission) since 2016 to coordinate the work of the various entities that participate in the management and regulation of toponymy. Representatives of different departments of the Aragonese government are part of this commission; of the Geographic Institute of Aragon; representatives of the University of Zaragoza, the Instituto de Estudios Turolenses, the Instituto de Estudios Altoaragoneses (Huesca), the Institución Fernando el Católico (Zaragoza) and experts in Aragonese place names. For some time now, *Nomenclátor geográfico de Aragón* has been carried out, which is updated annually and which already has more than 800,000 revised place names. Aragon has three different languages: Spanish, Aragonese and Catalan. The following map shows the Aragonese speaking area in green and the Catalan speaking area in orange (Fig. 6). However, in many areas near the linguistic border with Spanish, only the oldest speakers use Aragonese and this language is in decline.

In Extremadura, the entity that coordinates the management of toponymy is the Consejo de Información Cartográfica y Territorial de Extremadura (Extremadura Cartographic and Territorial Information Council),[8] which depends on the regional government. It also has the Comisión Temática de Toponimia de Extremadura (Extremadura Toponymy Thematic Commission), made up of personnel from different administrations in order to facilitate the exchange of information of interest in relation to the study and knowledge of place names. Among the various works they

[5] See: https://toponimia.xunta.gal/es/toponimia/comision.

[6] See: https://toponimia.xunta.gal/gl/visor?termino=.

[7] See: https://politicallinguistica.asturias.es/toponimia-buscador.

[8] See: http://sitex.gobex.es/SITEX/inicio.

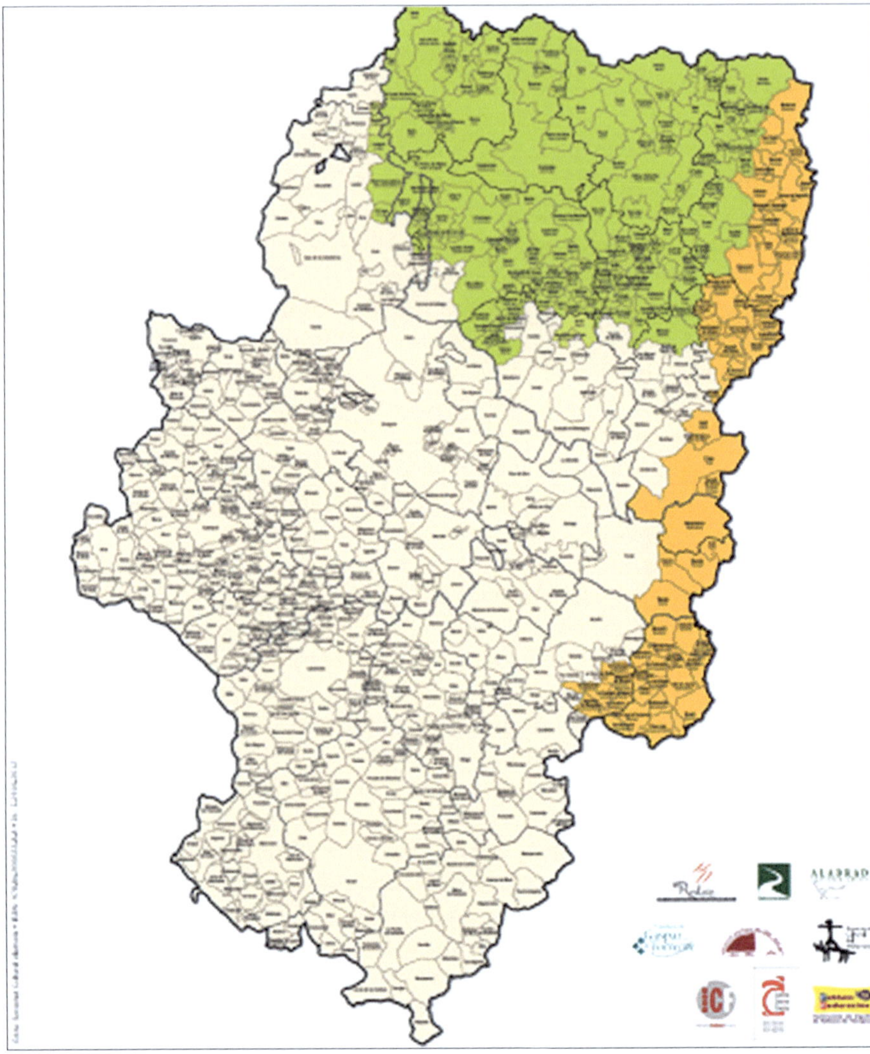

Fig. 6 Linguistic map of Aragon. The area where Aragonese is spoken appears in green. The area where Catalan is spoken appears in orange (*Source* https://www.roldedeestudiosaragoneses.org/mapa-linguistico-aragon-segun-anteproyecto-ley-lenguas-2001; © Rolde de Estudios Aragoneses)

carry out, the updating, revision and modification of place names and place names collection stand out. Digital cartography is available online.[9]

Unlike the previous cases, in Andalusia there is no commission. The different entities that have authority over toponymy coordinate with the Instituto de Estadística y Cartografía de Andalucía (Andalusian Institute of Statistics and Cartography),

[9] See: http://www.ideex.es/Geoportal/.

which is legally in charge of inventorying, normalizing and disseminating Andalusian toponymy. It depends on the Andalusian government, the Junta de Andalucía. It has public digital cartography,[10] such as the *Nomenclátor Geográfico de Andalucía*, which allows you to search for place names.

Basque Country (Euskadi) and Navarra

In the Basque Country, there is no type of commission that coordinates the management of toponymy. According to local legislation, toponymy is the responsibility of the municipalities, except if the toponym is shared by two municipalities or with other communities. By law, the advisory institution on place names is the Euskaltzaindia (Royal Academy of the Basque Language). As town halls do not usually have staff to manage toponymy, the mapping shows the toponymy proposed by the Eusko Jaurlaritza (the Basque government), except in cases where the town hall has made the corresponding regulation, which by law implies that the Department of Linguistic Policy of the government and the Euskaltzaindia have also intervened in the standardization process. The autonomous government has digital cartography that can be consulted.[11]

In Navarra there is also no toponymy commission, and this is managed in a similar way to the one explained above. As in the Basque Country, the Euskaltzaindia is the advisory body. They have an official, searchable digital cartography,[12] although incomplete, since the names of the villages or the municipal toponymy do not appear. The town councils request the regulation of urban toponymy from the language academy, the Euskarabidea (Instituto Navarro del Euskera), and the latter sends it to the Euskaltzaindia for review. Doubts are resolved by talking to the municipalities directly. Once reviewed, it is returned to the Euskarabidea for forwarding to the corresponding town hall. Recently, work is also being done to regulate the names of towns.

The Euskaltzaindia has a place name locator open to the public that collects place names from the two Spanish communities and the French Basque Country, in the department of the Atlantic Pyrenees.[13]

---

[10] See: https://www.juntadeandalucia.es/institutodeestadisticaycartografia/.

[11] See: https://www.geo.euskadi.eus/s69-15375/es/.

[12] See: https://administracionelectronica.navarra.es/toponimia/?lang=es.

[13] See: https://www.euskaltzaindia.eus/index.php?option=com_ecoeoda&task=lekuaPortada&Itemid=794&lang=es.

Catalonia, the Balearic Islands, and Valencia

Toponymy in Catalonia is exclusively in one language, Catalan, and it also has abundant legislation on which rests the policy of linguistic normalization promoted by the Generalitat de Catalunya, the local government. Very succinctly, it can be said that there are three important points in the legislation. The first, is the royal decree promulgated in 1976 that recognizes the Institut d'Estudis Catalans (Catalan Studies Institute, IEC) as an academic, scientific, and cultural institution. The second is the Catalan Language Policy Law approved in 1991, which recognizes that the IEC is the institution responsible for regulating the Catalan language. And the third, and most important for toponymy, the Law of Language Policy of Catalonia approved in 1998 which indicates that place names in Catalonia are only in Catalan (or in Aranese in Val d'Aran), according to the orthographic regulations of IEC.

Also that same year, 1998, the Comissió del Nomenclàtor (Nomenclator Commission) was set up, which initially had a temporary nature and its sole mission was to draw up the official *Nomenclàtor oficial de toponímia major de Catalunya.* Later, in 2001, in view of the usefulness of having a commission where the management of toponymy was centralized, it was established on a permanent basis and was renamed the Comissió de Toponímia de Catalunya (Toponymy Commission of Catalonia).[14] The commission is made up of all the institutions that in one way or another participate in the standardization and officialization of Catalan place names, or have some power over it, and they are:

- The Institut Cartogràfic i Geològic de Catalunya (Geographical and Geological Institute of Catalonia), which provides its database and is in charge of official cartography.
- The Institut d'Estudis Catalans, which by law is responsible for proposing the spelling of place names in accordance with current spelling regulations.
- The Conselh Generau d'Aran (General Council of Aran), in charge of fixing the spelling of the place names in the Aranese language located in Val d'Aran.
- Various departments of the government of Catalonia, since it has power over a large part of Catalan place names.
- Two associations of municipalities, representing the municipalities of Catalonia, which have authority over their name and that of their towns.
- The Consorci per a la Normalització Lingüística (Consortium for Language Standardization), which promotes the use of Catalan, included standardized place names.

The main functions of the commission can be summarized as: coordinating the entities involved in the standardization and officialization of place names; propose names and spellings that have to be approved by the municipalities or the government of Catalonia; disseminate the official Catalan place names; and promote the importance of Catalan toponymy as part of the cultural heritage. The most important

[14] See: https://llengua.gencat.cat/ca/direccio_general_politica_linguistica/02_organismes_vinculats/comissions-i-consells/comissio-de-toponimia/.

project carried out by the commission is the *Nomenclàtor oficial de toponímia major de Catalunya*, which includes some 53,000 standardized place names at a scale of 1: 50,000. Besides, it has also promoted other projects. In 2006 it published *Criteris per a la toponímia municipal* (*Criteria for the municipal toponymy*) in order to respond to the demand from the councils for advice on the standardization of toponymy at the municipal level.

In addition, the commission follows the recommendation of GENUNG to disseminate the importance of the cultural heritage of toponymy (Resolution VIII/9 Geographical Names as cultural heritage of 2002). In this sense, between 2005 and 2009, the commission organized a traveling exhibition entitled *Mots amb arrels: els noms de lloc ens parlen* (*Rooted words: place names speak to us*) to disseminate the heritage value of toponymy and the importance of its preservation. Currently the exhibition material is available on the internet.[15]

The most recent project of the commission is the *Nomenclàtor mundial* (*Gazetteer of the world*), a corpus of standardized exonyms that fixes the spelling of foreign toponymy in accordance with the recommendations of GENUNG (Resolution II/28 Lists of exonyms of 1972, among others). In this sense, it promotes the use of well-known names. These names may have been adapted over time (*Milà* for *Milano*) or used in the original language (*Oxford*). Guidelines are also given on some aspects that are related, such as the use of the article in lower case (if the place name is adapted to Catalan, as in *l'Índia* or *l'Iraq*) or initial capital (if the name is not adapted, as in *El Salvador* or *Le Mans*). In some cases, some flexibility is allowed depending on the context. Thus, you can choose between *Tromsø* or *Tromso* (Norway), depending on who the text is addressed to, and between *Regensburg* or *Regensburg* (Germany), depending on whether it is a text referring to the present time or a text in the field of history. For the place names of languages in non-Latin alphabet, the most suitable transcription for public use is indicated.

As for the Balearic Islands, on April 15, 2011, the Comissió de Toponímia de les Illes Balears (Balearic Islands Toponymy Commission) was created "as a body to promote the collection and study of the Balearic Islands toponymy and their labelling, and, in general, their georeferenced representation, setting the corresponding criteria and preparing the georeferenced gazetteers it deems necessary." It is attached to the Department of Education and Culture of the Government of the Balearic Islands and its functions are carried out "without prejudice to the functions that the University of the Balearic Islands has attributed in terms of linguistic advice, in accordance with Article 35 of the Statute of Autonomy of the Balearic Islands.[16] And it is made up of representatives of the various entities that intervene in the management of place names, such as various departments of the Government of the Balearic Islands (Language Policy, cartography, Balearic Institute of Statistics); the University of the Balearic Islands, the Federation of Entities Locals of the Balearic Islands; the Technical Commission for linguistic advice; the UIB Linguistic Service; the Geographic

---

[15] See: http://www.gencat.cat/toponimia/motsarrels.htm.

[16] See: https://notib.recerca.iec.cat/la-comissio-de-toponimia-de-les-illes-balears-2011/.

Information Systems Laboratory of the UIB; and the Territorial Information Service of the Balearic Islands (SITIBSA).

The project *Nomenclàtor Toponímic de les Illes Balears* (NOTIB) has been carried out by the University of the Balearic Islands and assumed by the Institut d'Estudis Catalans. It consists of the collection and presentation of the list of the current non-urban place names of the Balearic Islands. Currently they are working on the Nomenclàtor Geogràfic de les Illes Balears (Geographical Gazetteer of the Balearic Islands), already georeferenced, on the initiative of the Government of the Balearic Islands, through SITIBSA, which is the public company that deals with it.

Regarding to Valencia, the Acadèmia Valenciana de la Llengua (AVL) is the institution that has the competence to establish the linguistically correct forms of Valencian place names, in accordance with regional legislation, especially article 41 of the Statute of Autonomy of the Valencian Community (2006) and Article 7b of Law 7/1998, of September 16, on the creation of the Acadèmia Valenciana de la Llengua (Valencian Academy of the Language). Within this body, the Onomastics Section is in charge of the study, inventory, and standardization of geographical names. The excellent work carried out by this working group has been reflected in the *Corpus toponímic valencià* (50,000 standardized place names). Furthermore, the work carried out by the Institut Cartogràfic Valencià (Valencian Cartographic Institute) in the construction of the Valencian Spatial Data Infrastructure, together with the standardizing activity of the AVL, has resulted in the generation of the *Nomenclàtor Toponímic Valencià* (Valencian Toponymic Gazetteer), which contains 123,000 standardized and georeferenced place names. These place names are also official as of the promulgation of Law 2/2020, of December 2, of the Generalitat Valenciana, about the Geographic Information and the Institut Cartogràfic Valencià.

Unlike what happens in Catalonia and the Balearic Islands, in Valencia, there is no commission to coordinate the management and regulation of toponymy. However, there are two entities that coordinate the realization of the *Nomenclàtor Toponímic Valencià*, which contains the official toponymy of the Valencian community. They are the Acadèmia Valenciana de la Llengua and the Institut Cartogràfic Valencià. The first does an excellent job of normalizing the toponymy, already reflected in the *Corpus toponímic valencià*.The second elaborates the corresponding cartography. The Acadèmia Valenciana de la Llengua has also elaborated the *Criteris per a la fixació de la toponimia valenciana* to help to standardize the toponymy.

Linguistic Aspects

One Language and More Than One Academy

The various statutes of autonomy establish that in those communities that have a local language, the autonomous government has the power to regulate that language. This usually transfers the power to the corresponding language academies. For example, in Asturias, the Academia de la Llíngua Asturiana; in Galicia, the Royal Galician Academy; in Catalonia, the Institut d'Estudis Catalans; in the Basque Country and Navarra, the Euskaltzaindia; in Aragon, the Academia Aragonesa d'a Luenga, which is made up of two institutes: Instituto de l'Aragonés and Institut Aragonès del Català.

In the case of Catalan, even in the case of the same language, it presents certain phonetic, lexical and morphosyntactic variation in each of the areas where it is spoken (Catalonia, the Balearic Islands, Valencia, and La Franja). The entity that regulates the language is the Institut d'Estudis Catalans (IEC), which has been an academy of the language since its foundation (1907). However, since 1998 there is also the Acadèmia de la Llengua Valenciana, with authority, according to the Statute of Autonomy of the Valencian Community (2006) over the variety of Catalan spoken there. In the case of the Balearic Islands, is the University of the Balearic Islands that deals with the language, always in coordination with the Institut d'Estudis Catalans. It is worth noting that, despite the fact that the creation of the Acadèmia Valenciana de la Llengua caused some discomfort in the Institut d'Estudis Catalans, it has served to promote the use of the local language in the territory and has done an exceptional job in standardizing Valencian place names, that has been done without departing from the criteria recommended by the Onomastic Office of IEC. There is also the Institut Aragonès del Català (Aragonese Institute of Catalan), recently created, with the aim of promoting the use of Catalan and the standardization of place names in La Franja.

The fact that the standardization of toponymy in Catalan is carried out by different institutions depending on the administrative division, has certain implications. On the one hand, in the way of organizing the process, since each autonomous government is organized in the way that is most convenient for it and promotes standardization with more or less interest. Catalonia and Valencia have advanced much more than the Balearic Islands. On the other, in the standardization criteria applied, this may differ on some points. However, in recent years the coordination between the three entities has improved considerably to the benefit of the linguistic unit of regulated place names and the language itself.

Castilianization and Standardization

The standardization of toponymy in bilingual areas always presents the same problems regardless of the area, irrespective of those that each area presents in particular. The two most frequent are the presence of names altered by the influence of Spanish, and the need to adapt the spelling of names to the current spelling in each language.

The process of Castilianization started with the Nueva Planta decrees signed between 1707 and 1716, by which the laws and institutions of the Crown of Aragon were abolished. At that time, the Crown of Aragon was integrated by the Kingdom of Aragon, the Kingdom of Valencia, the Kingdom of Mallorca, and the Principality of Catalonia. Spanish became the only language of government, displacing other languages like Catalan or Aragonese, the regular ones until that moment. At the same time, the toponymy will appear little by little Castilianized (Barella 1996). The map of the Principality of Catalonia drawn by Josep Aparici (1654–1731) in 1720,[17] maintains the names of populated places in Catalan, except the best known in Spanish, such as *Lérida* for *Lleida*, or *Ebro* for *Ebre*. Some generic terms are also in Spanish, like *golfo* instead of *golf*.

The process of adapting to Spanish the toponymy present in the cartography will only be interrupted during the short time that Catalonia was part of the Napoleonic Empire (1812–1814). During this period, Catalan became the official language in Catalonia again, with the intention of gaining the sympathy of the population, heavily punished by the war it had lost against France. It is noteworthy that the maps produced at this time benefited from advanced French cartography, very different from the little evolved cartography of Madrid, characterized by the carelessness of toponymy and the little consideration for local languages (Barella 1996). It will not be until the establishment of the Mancomunitat de Catalunya (Commonwealth of Catalonia), 1914–1925, that appears the possibility of preparing the cartography of Catalonia in Catalan, but it was not meant to last. After the Spanish Civil war (1936–1939), Francoism will prohibit any use of a language other than Spanish. It will not be until the end of the Franco regime (1975) that the conditions begin to be met to use Catalan again in all areas (Tort Donada 2019).

The Catalan names that had been graphically Castilianized were corrected some time ago thanks to the work promoted by the Government of Catalonia. It has been a long and expensive process, but it has been worth it. Unfortunately, this process collides with the trend of customizing digital cartography on platforms such as Google Maps, where names are displayed adapted to the language of the browser. The Spanish-speaking user likes to see *Londres* and not *London*, or *Padua* and not Padova. And from this, Google deduces that they will also like to see a whole series of Castilianized names that were used in the Franco era, such as *Villanueva y Geltrú* instead of *Vilanova i la Geltrú*. In these cases, a balance must be sought between the adaptations that have a well-established use, such as *Gerona* and *Lérida* in Spanish, and the translations into Spanish that have no real use, which was very fashionable in Franco's time.

There are only a few cases where the local government does not have the power to change the name. Mainly, the name of the towns. In these cases, if the corresponding city council does not want to modify the name, it is not possible to rectify it. They are marked with an asterisk in the *Nomenclàtor oficial de toponímia major de Catalunya*. For example, *El Fresno*, *l'Hidalgo*, Los *Rosales*, *Buenos Aires*. However, it must be taken into account that, due to so many years of forced Castilinization of Catalonia, the presence of Spanish is reflected in microtoponymy. Thus, we essentially find two types of place names that contain names of Castilian origin (Batlle 2014):

a. Place names formed by an anthroponym in Spanish, since for many years it was mandatory to use only Spanish praenomens i.e. the names of people in the toponym. For example, *mas d'en Francisco* (*Francesc* in Catalan), *maset de*

[17] See: https://cartotecadigital.icgc.cat/digital/collection/catalunya/id/1775.

Santiago, *torre del Marcelino* (*Marcel·lí* in Catalan), *mas de Don Pedro* (*Pere* in Catalan).

b. Place names that include common names in Spanish. Most correspond to the nicknames of the owner. For example, *mas de Mancebo*, *mas del Sargento*, *cal Manco*, *torre del Señorito*, *mas del Ligero*, *molí del Blanquillo*.

It should be noted that the pronunciation of all these names corresponds to the phonetics of Catalan, not Spanish since they have been adapted according to the actual use of the speakers. In some cases, it is necessary to modify the spelling of the word of Castilian origin to reflect this phonetic adaptation. For example, *mas del Metxon* (in Spanish *Mechón*) or *cal Mutxatxa* (in Spanish *Muchacha*), *torre del Ranxet* (in Spanish *Ranchito*), *font del Sanxo* (in Spanish *Sancho*).

Furthermore, as the western area of Catalonia is close to the linguistic domain of Spanish, some Spanish generic names appear applied to the toponymy. For example, it is common to use the Spanish word *venta* instead of *hostal* (*Venta del Pinar*, *Venta del Ros*), the diminutive *casilla* (*Casilla del Canal*), with the typical Spanish suffix *-illa*, or the word *desaigüe* (*lo Desaigüe de les Olles*), which is an adaptation of the Castilian *desagüe* 'drain'. In these cases, the names are standardized by capitalizing the Castilian origin term to fully incorporate it into the proper noun and separate it from the common Catalan lexicon.

Regarding the adaptation of the archaic spelling to the current orthographic regulations, the vast majority of place names have been corrected thanks to the work of the Institut d'Estudis Catalans and the decisions of the Generalitat de Catalunya. One of the tasks of the Comissió de Toponímia de Catalunya is the social awareness of the patrimonial value of autochthonous toponymy and the promotion of the use of standardized toponymy in all areas. Only a few names of towns keep the traditional archaic spelling. For example, *el Figaró* (correct spelling, *el Figueró*), *Lladó* (correct spelling, *Lledó*). In the Val d'Aran area, it also occurs in the case of *Baqueira* (correct spelling, *Vaqueira*). The city council does not want to correct the spelling to keep the same name as the well-known Baqueira-Beret ski resort. There is no doubt about the relationship between the name of the place and the trademark nor of the interest of protecting any possible affectation of the economy of the company (and therefore of the municipality). Fortunately, other municipalities in the same situation have adopted the name to the current spelling regulations, such as *Arànser*, although the ski resort continues to use the archaic name *Aransa* as a trademark.

It should be noted that little by little, the work of social awareness continues to give results and some municipalities and towns are correcting the spelling, such as that of *Roda de Berà* in 2012, which was previously written *Roda de Barà*.

Changing the name (or the spelling) of a town carries a very clear social implication, and many times the fact of changing the name or not divides the opinion of the affected community. All these processes are usually reflected in the local press (Fig. 7) and in the graffiti on the road signs (Figs. 8 and 10). Sometimes the city council organizes a popular vote to decide whether to change the name. This was the case in the town of *el Figueró*, which in 2004 held a popular consultation to decide whether the graphic name of the name (*el Figaró*) should be adapted to the current

EL 9 NOU | POLÍTICA I SOCIETAT Temes de l'any | 19

Figaró: ancorats en l'error

Els veïns decideixen en referèndum mantenir el nom del poble mal escrit

Figaró-Montmany

Ferran Polo

Els veïns de Figaró van passar per les urnes el 17 d'octubre. Aquell dia, però, no van votar per escollir l'alcalde, el president de la Generalitat o el del govern espanyol. Van votar en una consulta popular per dir si volien que el nom del municipi seguís igual i, per tant, mal escrit –*Figaró-Montmany*– o bé s'adaptés a la forma lingüísticament correcta que proposava l'Institut d'Estudis Catalans: *el Figueró i Montmany*. Els de Figaró es van decantar per la continuïtat i per deixar les coses tal com estaven. 212 persones van mostrar la seva fidelitat a la lletra a i al guionet i 155 es van mostrar partidàries de la correcció del nom, que no es va materialitzar. Per tant, després del 17 d'octubre Figaró-Montmany segueix dient-se *Figaró-Montmany*.

La consulta popular va tenir una doble lectura: des de fora es va criticar que es permetés als veïns decidir sobre una qüestió científica: "Aquest no era un tema perquè el decidís la gent del poble", afirmava el director de l'Oficina d'Onomàstica de l'Institut d'Estudis Catalans, Josep Moran, que qualificava el resultat de "deplorable". "Trobo absurd que una qüestió ortogràfica s'hagi de decidir en un referèndum. Els únics que queden malament són els veïns de Figaró", deia Jordi Castellanos, professor de filologia catalana de la Universitat Autònoma de Barcelona. La directora del Centre per a la Normalització Lingüística del Vallès Oriental, Anna Maria Jansana, apuntava que "és important tenir l'Institut d'Estudis Catalans com a referència en aquests temes".

El cartell de l'entrada del poble encara conserva l'article –que es va descartar– i la forma amb *ga*

RAMON FERRANDIS

El recompte de vots va atreure una quarantena de persones a la sala cultural de l'antiga escola

JORDI SALAS

Des del poble, en canvi, es considerava que la correcció ortogràfica del nom no era només una decisió lingüística. "Els qui coneixem la realitat social de Figaró sabem que no es tracta només d'un tema lingüístic. També es tracta d'un tema social i identitari", considerava el regidor de Participació Ciutadana, Marc Parés. L'alcaldessa, Gemma Sànchez, explicava que un canvi de nom fet per decret "hauria generat molta més divisió al poble". Un 41% dels veïns amb dret de vot –375 de 905– van votar. La participació també va ser considerada positiva pel govern local. La campanya informativa que alguns veïns van

Des de fora del poble, es va criticar que els veïns decidissin sobre un tema científic

fer abans de la consulta també van servir perquè "ara tothom reconeix que la versió correcta és *el Figueró*", apuntava l'historiador Jaume Oliver.

Amb tot, pot ser que la decisió a les urnes no hagi tancat definitívament el debat sobre la correcció lingüística d'aquest topònim: la Generalitat es planteja, durant aquesta legislatura, fer un canvi del reglament de demarcació territorial i població dels ens locals que li permetria actuar d'ofici i adaptar a la normativa la forma *Figaró-Montmany*.

La consulta ciutadana sobre el nom del poble s'emmarca en el procés d'introducció de la democràcia participativa al poble: a partir del 2006 els veïns decidiran a què es destina part del pressupost municipal. Aquest mateix any també han pogut decidir quins usos es donen al Casino municipal després de la reforma que està a punt de començar.

Fig. 7 Local press commenting on the vote on the change of spelling of the name of the municipality. (© El 9 Nou)

orthographic regulations (*el Figueró*) or not. The majority of people decided to keep the archaic spelling of the name. In the following figure you can see the front page of the local press with the following headline: *Figaró, anchored in error*.

On other occasions, the name change not only affects a specific town, but also affects the other towns that make up the municipality, which have to agree. One example of the most difficult to solve was the one that took place in 2006 in the municipality of Avinyonet del Penedès. This municipality is made up of four towns: les Cabòries, les Gunyoles, l'Arboçar and Cantallops. The town named *les Cabòries* wanted to change its name because, due to popular etymology, *les Cabòries* acquired a very well-known meaning ('concerns'). However, as it belonged to a municipality

Fig. 8 Image showing a traffic signal with the Castilian name erased (*Source* https://www.rtve.es/noticias/20201007/castello-somos-nosotros-dos-municipios-luchan-toponimo/2043740.shtml) (Información de rtve.es © Corporación Radio Televisión Española)

made up of several towns, it was not easy to find a name that everyone agreed with. In the end, the proposal of the Comissió de Toponímia was accepted, and *les Cabòries* was changed to *Avinyó Nou* ('New Avinyó'). This name is related to the name of the municipality since Avinyó Nou is the most important town of the municipality. *Avinyonet* means 'Little Avinyó' and the diminutive was used to differentiate it from another town that is also called *Avinyó* in the Bages region.

The standardization of place names in bilingual areas can also present region-specific problems. This is the case of the Basque Country (Nieto 2013 and 2014). The fundamental criterion is respect for the rules of the creation language of the place name, something that is not always as clear as it might seem at first. The place names of Basque origin will be written in Basque and those of Latin origin in Romance. However, in areas where Basque is no longer the usual language, speakers have ended up modifying the name. For example, the Basque name *Muniziaga* (Galdames) has become popularly *Murcíelaga* ('feminine bat') by popular etymology; or *Larreineta* (Trapagaran) 'mineral fields' which has become *La Reineta* with a more Romanesque than Basque appearance and more sense (*reineta* is a type of apple). In other cases, the place name has been translated, such as *el Molino Viejo* 'the old mill' (Heredia), which in the old documentation was known with the Basque name *Errotazarra* 'the old mill'. The great migratory phenomena of the late nineteenth century also favored place name substitution. For example, *Atxúriaga* (Galdames, Bizkaia) became *La Aceña.*

Another interesting area is Extremadura (García 2022), where three other linguistic modalities coexist with Spanish, but in a state of diglossia so advanced that all of them are in serious danger of extinction. Pure speakers are scarce and of middle or advanced age —except for the case of the *fala*—, and generational transmission continues to occur generally in few municipalities. These minority languages are Rayan Portuguese (*português raiano*), the *fala* of Valle de Jálama (*fala do Val de Xálima*) and the language of Extremadura (*estremeñu*). As in the previous case, speakers have Castilianized or reinterpreted the name. For example, *Centieiras* is Castilianized in *Centeneras* and later, in an attempt to restore the original name, the hybrid name *Centeneiras* was generated; and similar in the case of *O Colmeal*, which was adapted as *El Comenar* and re-adapted as *O Colmenal*.

Finally, monolingual areas are not exempt from disturbances, although for different reasons. In the case of Andalusia, one of the most frequent problems is that pronunciation has influenced the spelling. In cases where the pronunciation responds to a general phenomenon of the language, usually the written name is amended. For example, it is written *Fuente el Collado* (and not *Fuente el Collao*) or *Fuente del Dornajo* (and not *Fuente del Dornao*) (Garrido et al. 2021: 114).

The Importance of City Councils

When talking about place name normalization, one cannot fail to mention the importance of municipalities in this process. In some areas, they carry more weight than in others in relation to microtoponymy. In addition, according to state legislation, they have power over the name of towns and urban toponymy.

In the case of Catalonia, when a municipality wants to change its name, or the name of a town, or simply wants to modify the spelling of the name, the plenary session of the City Council approves the new name and, once this is done, requests the change to the Generalitat de Catalunya. The latter requests a report from the Comissió de Toponímia and the latter entrusts it to the Institut d'Estudis Catalans, which prepares the report stating whether it approves or rejects the name. This report is advisory in nature and the Generalitat decides whether to allow the requested name change. Due to the social implication that this type of change entails, the Generalitat is usually sensitive to the reasons why a certain town wants to change its name.

The Institut d'Estudis Catalans usually accepts the majority of requested changes. Many requests are made in order to adapt the name of the municipality to the new reality after a segregation or aggregation of the towns that make up the municipality. Thus, for example, *Boadella d'Empordà* was renamed *Boadella i les Escaules* so that the name of the municipality would include the name of the town added to the municipality, or *Passanant* which was renamed *Passanant i Belltall*. Likewise, after the segregation of *Vila-seca*, the municipality was no longer *Vila-seca i Salou* and both towns were constituted as independent municipalities with the names of *Vila-seca* one and *Salou* the other.

On other occasions, the town wants to recover the traditional name of the municipality. For example, the municipality of Cabrera d'Igualada decided to regain the name of *Cabrera d'Anoia*, and Fogars de Tordera decided to regain the name *Fogars de la Selva*. Recently (October 2021), the Sant Carles de la Ràpita city council organized a consultation to recover the name used until the eighteenth century, *La Ràpita*, and the result of which did not reach the minimum of social acceptance to change the name. However, the name was changed. At the end of 2021, the Cabacés city council launched an informative campaign[18] to recover the normative spelling of the name, which is *Cabassers*, which was the one used before the Spanish Civil War and the one accepted by the Institut d'Estudis Catalans.

However, on other occasions the proposals for change are not well seen by the Institut d'Estudis Catalans. For example, when the change means losing a traditional name to put a commercial one (such as going from *Mas Boada* to *Tourist Club*), or when the change could lead to confusion between two different towns. For example, in Catalonia there are two very close towns that have the same name and are only differentiated by the determinative, *Vilassar de Mar* and *Vilassar de Dalt*. The town of Vilassar de Dalt wanted to be called only *Vilassar*, a fact that would have generated confusion regarding the referent. In these two and similar cases, the Institut d'Estudis Catalans prepares a report for the toponymy commission advising against the change.

In other areas, such as Aragon, the weight of the municipalities is even greater, since the legislation is much vaguer as regards the protection of bilingual toponymy. For this reason, the recent Commission Assessora de Toponímia of the Government of Aragon has tried to promote the officialization of the bilingual names of towns. In the Matarranya and Baix Cinca area, where Spanish coexists with the Catalan in the area, the town councils generally assume bilingual names and respect the Catalan names of minor toponymy; but in other areas they reject the names in the local language. For example, the Valderrobres city council rejects the Catalan name, *Vall-de-roures*, even though it has been very well documented since ancient times. In the area where Spanish and Aragonese coexist, minor place names in Aragonese are usually accepted without too much problem, and with more reluctance, the names of towns. In Aragon and in other areas, the power for changes rests with the town councils. It is very important to make them aware of the patrimonial value of the place names, since without their consent the Commission's proposals are not applied.

On some occasions, politics ends up having a prominent importance in the naming of towns. This is the case of some municipalities in Valencia, where the choice of the name in Spanish or Catalan is usually linked to the party that governs the town hall. Without a doubt, the most representative case is that of the Valencian municipality that since 2020 is named *Castelló*. The name of this municipality has been accepted by the Acadèmia Valenciana de la Llengua to avoid that every time there is a change of political party in the town hall, the name is changed, which to date went from *Villanueva de Castellón* (in Spanish) to *Castelló de la Ribera* (in Catalan) and vice versa. All this changes are usually reflected in the graffiti on traffic signs (Fig. 8).

Another worse case because it breaks the law, is the change that in 2012 the city council of Maó (Menorca) approved when the Popular Party governed. Infringing the Balearic language policy law that indicates that toponyms have only one name in Catalan, they adopted the Castilianized name *Mahón* as the official name of the

[18] See https://www.cabassers.org/.

Fig. 9 Signage with the two names of the city. (©Diari MENORCA)

city along with the Catalan name *Maó* (Figs. 9 and 10). In February 2021, with the town hall in the hands of the Socialist Party again, the legitimate name, *Maó*, was restored.

Conclusions

In accordance with current state legislation, the management of the co-official languages and a good part of the toponymy falls to the autonomous government. Hence, in the whole of the Spanish state, the linguistic policy applied to toponymy (and to the protection and promotion of co-official languages) is not one, but many and very unequal. Bilingual communities are those that, within the linguistic policy of preservation and promotion of the co-official language, have worked more on the standardization of toponymy, such as Catalonia, Valencia, the Balearic Islands, the Basque Country, Navarra and Galicia. However, in monolingual areas, such as Madrid or Extremadura, little work has been done on toponymy, since it is considered that "it is already fine" or non-problematic. In this case, the exception is Andalusia, where the differences between oral and written names have forced an effort to standardize the toponymy in order to avoid confusion. In addition, there are regional governments that dedicate more budget to the study of toponymy than others do, and

Fig. 10 Road signage with the Spanish name *Mahon* covered in paint. (©Diari MENORCA)

this is reflected in the advance of standardization. It is still worth mentioning that only in the areas in which the autonomous governments, in addition to promoting standardization, disseminate and boost standardized names, do they end up fulfilling the objective of their own legislation on language policy applied to toponymy.

In many areas, city councils have an important role in the management and standardization of toponymy, since they are the only ones that can promote the standardization of names and promote their public and social use. In these cases, it is important that the local government, in coordination with the corresponding language academy, provide the means so that they can carry out the task. In any case, it is necessary to promote awareness at all levels of the intangible heritage that the toponymy represents, whatever the language. Hence, the importance of collecting it, preserving it, and using it publicly standardized.

Questions

1. What institution standardizes toponymy in Spain?
2. In what language are the Spanish toponyms?
3. What are the two main problems when it comes to standardizing toponymy in Catalonia?
4. Can you find the influence of Castilian in Catalan toponymy?
5. What role do the municipalities play in the standardization of toponymy in Spain?

Acknowledgements I am grateful for the information provided by the experts Angélica Castaño (Instituto Geográfico Nacional, Madrid), Vicente Feijoo (Real Academia Galega, Galicia), Jairo Javier García Sánchez (Universidad de Alcalá, Extremadura), Maria Teresa Garrido Borrego (Instituto de Estadística y Cartografía de Andalucía, Andalusia), Javier Giralt (Universidad de Zaragoaza, Aragon), Mikel Gorrotxategi (Euskaltzaindia, Basque Country), Maite Mollà (Academia Valenciana de la Llengua, Valencia) and Joan Tort (Universitat de Barcelona).

References

Barella i Miró A (1996) La corrupció de la toponímia catalana en els mapes de finals del segle XVIII. Treballs de la Societat Catalana de Geografia pp 15–35. https://raco.cat/index.php/TreballsSCGeografia/article/view/157063. [Consulta: 22–02–2022]

Batlle M del Mar (2014) Políticas lingüísticas en el campo de la toponimia: una mirada general a España y Catalunya. In: Finco F, Iannaccaro G (eds) (2014) Nomi, luoghi, identità toponomastica e politiche linguistiche. Udine, Società Filologica Friulana, pp 37–49

Corpus toponímic valencià (2009) València: Acadèmia Valenciana de la Llengua. https://www.avl.gva.es/documents/31987/99450/CTV/c2a1b117-2df6-4c78-a736-0cf84b8150b1

Criteris per a la fixació de la toponímia valenciana. València, Acadèmia Valenciana de la Llengua, 2015. https://www.avl.gva.es/documents/31987/58581/MANUALS_03.pdf

Criteris per a la toponímia d'àmbit municipal. Barcelona, Generalitat de Catalunya, 2006. https://llengua.gencat.cat/ca/direccio_general_politica_linguistica/02_organismes_vinculats/comissions-i-consells/comissio-de-toponimia/criteris/

García Sánchez JJ (2022) El patrimonio toponímico de la raya entre España y Portugal. Estud Romànics 44:141–171

Garrido M, Torrecillas C, Benabad I, Cardenas L, Nicolás C (2021) Normalización de datos geoespaciales de fuentes y manantiales inventariados en el *Nomenclátor Geográfico de Andalucía* (España). Rev Cartográfica 103: 99–121. https://doi.org/10.35424/rcarto.i103.969

Gorrotxategi Nieto M (2013) Normativización, oficialización y normalización de la toponimia en Euskal Herria. In: Peral MDG (coord.) Lengua, espacio y sociedad. Investigaciones sobre normalización toponímica en España. De Gruyter, Berlín/Boston, pp 81–98

Gorrotxategi Nieto M (2014) La normalización de la toponimia del País Vasco, estado actual y problemas (políticos, sociales, lingüísticos…). In: Finco F, Iannaccaro G (eds) (2014) Nomi, luoghi, identità toponomastica e politiche linguistiche. Udine, Società Filologica Friulana, pp 199–210

Mapa Topográfico Nacional 1:25.000 (MTN25). Consulta: 2–11–2021. http://centrodedescargas.cnig.es/CentroDescargas/catalogo.do?Serie=MAMOV

Nomenclátor Geográfico de Andalucía. https://www.ideandalucia.es/portal/nomenclator

Nomenclátor geográfico de Aragón. https://idearagon.aragon.es/toponimia/cata.html

Nomenclàtor oficial de toponímia major de Catalunya. Barcelona, Generalitat de Catalunya (2009). https://www.icgc.cat/Descarregues/Llocs/Nomenclator-oficial-de-toponimia-major-de-Catalunya

Nomenclàtor Toponímic de les Illes Balears (NOTIB). https://notib.recerca.iec.cat/

Nomenclàtor Toponímic Valencià. https://icv.gva.es/va/nomenclator-toponimico-valencia

Pedregal Montes MA (2014) El proceso de oficialización toponímica en el Principado de Asturias. Dentro de Tort i Donada, Joan; Montserrat Montagut i Montagut (ed) El noms en la vida quotidiana. Actes del XXIV Congrés Internacional d'ICOS sobre Ciències Onomàstiques/Names in daily life. Proceedings of the XXIV ICOS international congress of onomastic sciences, vol 2. Barcelona, Generalitat de Catalunya, pp 2642–2649. http://llengua.gencat.cat/web/.content/documents/publicacions/btpl/arxius/11_actes_icos.pdf

Toponimia: Normas para el MTN25. Conceptos básicos y terminología. Madrid, Instituto Geográfico Nacional, 2005. http://www.ign.es/web/resources/docs/IGNCnig/NormasToponimiaparaMTN25.pdf

Tort Donada J (2019) Hacia un *Atles integral de l'onomàstica catalana*? Consideraciones a propósito del proceso de inventario y sistematización de la toponimia en Cataluña. Perspectiva histórica, alcance actual y previsiones de futuro. In: Actes de la XII Jornada d'Onomàstica Onomàstica Hispànica. Normalització i Investigació. València, Academia Valenciana de la Llengua, pp 223–246

Urban Toponymy in Turkey

Alpaslan Aliağaoğlu and Abdullah Uğur

Abstract "Man [or object] does not exist unless he has a name. It must be named to exist" (Tesone, 2013: 73). Names are studied by different sciences. The act of naming is the bearer of meanings, and naming is the subject of urban toponymy. Cities consist of several levels of names. Urban toponymy addresses place names, such as street, square, park, school, and bridge names, at all levels in urban spaces. This research aims to present an evaluation of urban toponymy studies in Turkey. Hence 25 published articles are examined by using content analysis. As a result, research reveals an increase in the number of studies being done on the origin of urban place names. These are especially related to the classification of street and neighborhood names. In these classifications, the effect of Turkey's recent history on naming is elucidated by using 'republic' streets, referring to the government and those who founded the Turkish republic in 1922. Human characteristics of the geographical setting are extremely important in the analysis. Due to the upsurge in interest in cultural geography, urban toponymy is appraised here with a critical approach. For instance, local administrators with different political perspectives have tried to project their ideologies onto spaces by place naming. Also although not frequently, urban names are used to create more prestigious urban spaces for economic gain.

Keywords Urban toponymy · Street/neighborhood origin names · Critical toponymy · Turkey

Objectives

By the end of this chapter readers should be able to:

- Appraise the nature of urban toponymy in Turkey or the factors affecting the shaping of the urban toponymy there.

A. Aliağaoğlu (✉)
Balıkesir University, Balıkesir, Turkey
e-mail: alpaslan38ster@gmail.com

A. Uğur
Pamukkale University, Denizli, Turkey
e-mail: ugurabdullah@gmail.com

© The Author(s), under exclusive license to Springer Nature Switzerland AG 2023

G. O'Reilly (ed.), *Place Naming, Identities and Geography*,
Key Challenges in Geography, https://doi.org/10.1007/978-3-031-21510-0_14

- Become familiar with the most frequently used names in urban toponyms in the early period of nationhood in the modern state of Turkey.
- Be cognizant of how the Ottoman heritage continues to have an effect on the urban toponymy in Turkey.
- Gain insights into how political parties make use of toponyms in 'owning' spaces.
- Evaluate past and present urban toponymic classification studies in Turkey, and assess the current emerging critical approaches.

Introduction

People have always given different names to all animate and inanimate (beings) phenomena according to their ideas, thoughts, and situations. Nomenclature is one of humanity's oldest acts. The act of naming places or objects gains a personality. The personality of a place emerges as a result of its natural features and the arrangements that people have made in that place over time (Tuan 2005). The primary arrangement made by people in a place is the naming of that place. The nomenclature of places develops spontaneously in the natural and cultural environment of the place where it initially emerged with all its connotations. An activity that changes with the language over time has been preserved and transferred as a word. Thus, the continuity of the memory of that place is preserved and carried to the future by means of place names (Özkan and Yoloğlu 2005).

Studies on place names in Turkey are generally carried out by linguists, historians, and geographers. Linguists mostly focus on the properties of words and sentence structures. In this sense, a recent study that gives the theoretical background of the subject has been conducted by the linguist Şahin (2019) under the name "Onomasiology". Historians give information about the socio-economic characteristics, origin, ethnic structure, and settlement history in the past, based on place names. The article "An Essay on the Names of Neighborhoods and Districts of Istanbul" is one such study and was carried out by Semavi Eyice (1964). In this study, the origin of place names is discussed. In addition, historians dwell on the classification of place names. Geographers, alternatively, deal with place names in order to reveal the geography of social memory.

Although the studies of place names in Turkey started at the end of the nineteenth century (Gülbetekin 2017), they began to increase after 1960. The fact that Turkey has a rich historical past has provided much material on this subject. To date, there have been 284 studies on place names in Turkey. Some 152 of them are articles and books, 69 are papers, and 63 are undergraduate, graduate, and doctoral theses (Şahin 2010). The earliest work dates from 1928. This publication is an article: "Place Names of Turks in Anatolia" written by Hüseyin Nihal Atsız and Ahmet Naci. The "Turkey Place Names Symposium" held on 13–14 September 1984 became an organization where scientists related to the subject came together and different studies were exhibited (Şahin 2010). The study named "Turkish Onomasiology" by Saim Sakaoğlu (2001) was the first book written in Turkey on this topic. The work entitled

"Toponymy Techniques and Caucasus-Borcali Place Names" by Recep Albayrak in 2004 is another milestone in this subject area (Cited by Şahin 2019). The studies mentioned above were mostly dealing with rural settlement names. However, in Turkey, in recent times, urban place names have started to take on more importance. Among these studies, those that are handled from a geographical perspective hold a significant place.

Although studies on urban place names in the field of geography started in the late 1990s in Turkey, the real development took place after 2010. In this process, many articles have been published on the origin, formation, meaning, and classification of place names in cities. The most important study on this subject is Aliağaoğlu and Uzun's work on the taxonomy of street names in Turkey, in 2011. In the research, sorting is made by considering 4,825 street names from 24 settlements, which were the provincial centers in Turkey. This method of grouping was later used by Baysan and Kara (2014) for the city of Aydın, by Aliağaoğlu and Uğur (2018) for the city of Erzurum, by Asımgil (2018) for the city of İzmir, and by Çildam (2018) for the city of Bandırma. Also, Kahraman (2019) partially used this system for Istanbul. A similar classification was used by Özüpekçe (2021) in a study for the city of Kilis. However, the Aliağaoğlu and Uzun (2011) grouping is not used in all of the studies on urban place names that were carried out from a geographical perspective in Turkey. Günal, Şahinalp, and Güzel (2011) took into account physical, human and economic factors in their article for the city of Şanlıurfa. Some of the studies on urban place names in Turkey are related to the critical approach; however, there are not many.

Standardization of geographical names in Turkey started in the late 1930s under the Ministry of Interior. Today, studies on geographical names in Turkey are carried out by various national institutions. In order to standardize geographical names in Turkey, the work of these national institutions has been gathered under one roof which is the Ministry of Interior of Turkey.

Box 1 Turkey's activities within the scope of UNGEGN

Turkey has been involved in the work of the Geographical Names Experts Group (UNGEGN) that was established under the UN in 1949, with the committee formed under the Ministry of Interior. This committee consists of members from various institutions under the chairmanship of the General Director of Provincial Administration of the Ministry of Interior or the deputy general director to be appointed. These are members who are experts in different domains such as bureaucrats, soldiers, engineers, historians, linguists and geographers. The Geographical Names Experts Group consists of different sections, taking into consideration the languages spoken in the countries and their locations. Turkey is a member of Southwest Asia, Central East, Southeast Europe, and Roman Hellenic Sections. The main duties of the committee are as following: To establish the forms and principles regarding the determination of geographical place names. To standardize official and private names in Turkey and Turkish-speaking countries; to prepare and publish relevant guides; to represent and execute these guides under the supervision of institutions in the country, and abroad. To create and publish a geographical name index (gazetteer) in line with the rules and recommendations determined by UNGEGN. To keep up-to-date and publish names of the countries and their capitals with the spelling used in Turkish. To compile and publish the geographical place names written in Turkish that are different from the official language of the relevant country. To carry out studies on the standardization of geographical names with institutions corresponding to other Turkish-speaking countries and communities. (https://www.icisleri.gov.tr/illeridaresi/cografi-adlar-uzmanlar-kurulu-alisma-esas-ve-usullerine-iliskin-y-nerge).

The research presented here concerns the content analysis of urban place-name studies in the field of geography in Turkey. It was conducted with the reviewing method. Recent work on urban place names in the field in Turkey mainly consist of the classification of street and neighborhood names and the critical approach to urban place names. In the grouping studies of street and neighborhood nomenclature, Aliağoğlu and Uzun (2011) and Aliağaoğlu (2013) classifications were mostly used. For this reason, the present study is discussed under three headings of these groupings and approaches.

Urban toponymic studies in Turkey started quite recently. These were mostly done in the form of classification of urban place names and were mostly related to the grouping of street and neighborhood names. The primary contribution in this regard belongs to Aliağaoğlu and Uzun (2011). Many studies conducted in Turkey were based on the classifications made by Aliağaoğlu and Uzun. In some studies, neighborhood names were studied separately, while in other studies they were handled together with streets and avenues. Recently, research approaches have been more critical. The present work is based on a review of all these studies (Fig. 1).

Urban Toponymy in Turkey: On-Street Names

Street, boulevard, and road names are the most meaningful text fragments of cities. These remains connected to the main text are also memory spaces that store past experiences. Of course, urban toponymy does not consist only of streets and avenues. Squares and neighborhood names must also be included in the subject. Massey defines urban nomenclature as "domesticating the space by transforming the spatial into textual" (cited in Çetin and Şentürk 2019: 134). In a study conducted on the names of streets, boulevards, and roads in Turkish cities, the names of avenues and roads were classified under seven headings (Aliağaoğlu and Uzun 2011). These are the streets relating to:

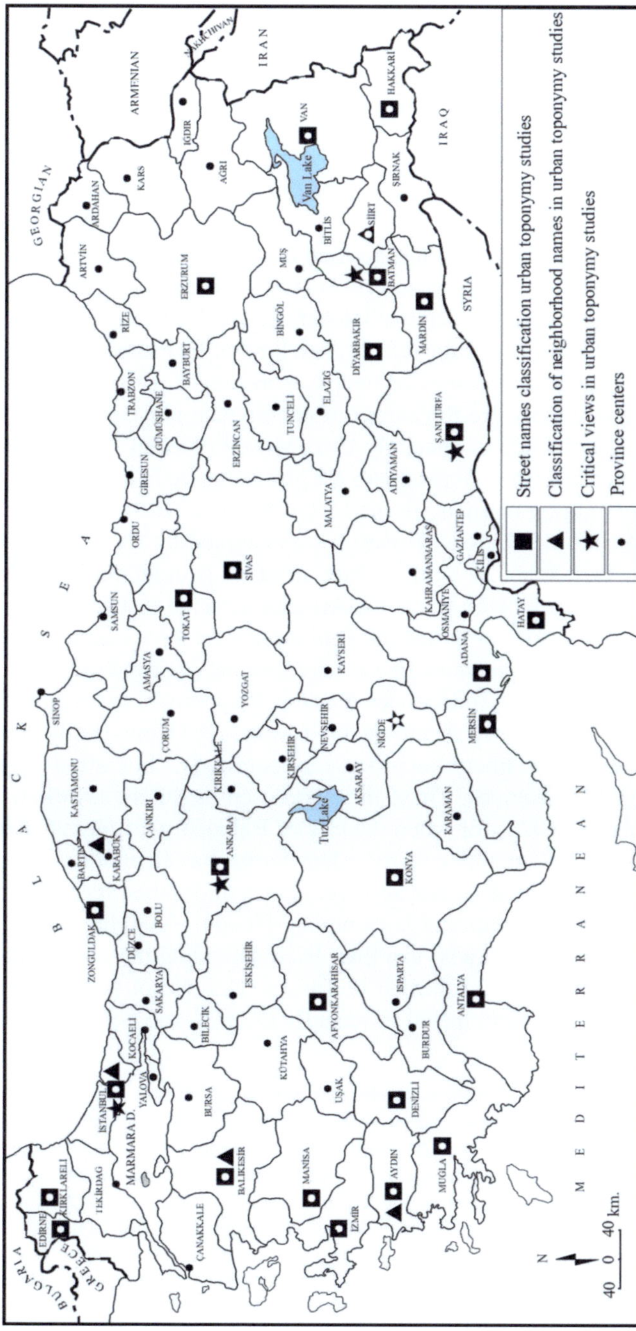

Fig. 1 Toponymic studies in Turkey

- the Republic,
- martyrs (heroes),
- friendship and brotherhood with other countries,
- place and direction,
- numerical named thoroughfares,
- important people,
- other streets/miscellaneous.

Republic street names

Republic street names in Turkish cities are used for the main boulevards and streets of the cities. These streets are the result of Turkey's nation-state creation process. The Republic of Turkey was founded on October 29, 1923. The republic was chosen as the form of government. Republic streets are a kind of historical map and include the events, dates, places, and names of people who played a role in the founding of the country. Among these are the names of people who played a role in the War of Independence (1919–1923), especially Gazi Mustafa Kemal Atatürk and İsmet İnönü. Street names related to independence are also included in this group. Independence was won by fighting on many fronts. After the First World War, the Ottoman Empire collapsed, and the lands of Turkey were occupied by the French, Russian, and Greek forces. Turkey had to fight on three fronts to gain its independence. War was fought with the French in the south, the Armenians in the east, and the Greeks in the west. Battles were fought in İnönü and Dumlupınar on the western front. Therefore, we can talk about the avenues of victory. Sakarya expresses different fronts during the Dumlupınar War of Independence and is frequently used as a place name. Every settlement in the occupied Anatolian lands was saved from enemy occupation at different times during the liberation process. Accordingly, "liberation day streets" are among the republic avenues. These dates are used as the names of streets, for example, September 6, 1922, for when the city of Balıkesir was liberated from the enemy occupation and September 9, for when İzmir was liberated. Other words related to independence are also used as street names under this title. Among these are the concepts of Liberty, National Sovereignty, Freedom and Homeland.

Streets named after martyrs who lost their lives defending the homeland against enemies and terrorists are also common in Turkey. This is not only related to the wars that the country has participated in recently (Korea[1] and Cyprus), but also to

[1] After World War II, Korea was divided in two by the 38th parallel. The north of the 38th parallel was occupied by Soviet Russia, and the south was occupied by the USA. With the division of Korea in two, the Korean problem also emerged. The expansionist attitude of the USSR led the USA and European countries to cooperate economically, politically and militarily. Economic organizations were formed and defense agreements were made. The most important of these was the North Atlantic Alliance, which was established for defensive purposes. Searching for security against the threat of the USSR, Turkey turned to the West and tried to take its place in Western institutions. Although it wanted to join NATO to address security concerns, it was not accepted for various reasons. In this period, Communist expansionism manifested itself with the attack of North Korea on South Korea. This time, the free world took action to stand against communist expansionism with the authority given by the United Nations under the leadership of the USA. By taking sides with the UN and deciding to send troops to Korea, Turkey saw this war against communism as an opportunity to

the members of the military and police force who died in the fight against separatist terrorist organizations. Captain Cengiz Topel()), who was martyred in the Cyprus peace operation (1974) and whose name is used in many settlements of the country, is an example in this regard. The number of martyrs who lost their lives against the separatist movements is high. They are used as urban place names depending on where they are from, like Birol Koç in the city of Balıkesir. In addition, names of those who lost their lives due to a recent coup attempt (July 15, 2016) were used for naming streets, avenues and squares (Aliağaoğlu and Çetinkaya 2021). Among the dead in the last incident, there are many civilians as well as soldiers or policemen. Accordingly, the name of someone like Ömer Halisdemir, who played a key role in the success of preventing the coup attempt, and names such as National Will and 15 July were used as urban place names. Thus, urban place names in Turkey experienced a rejuvenation.

Friendship and Brotherhood Streets

One of the important determinants of urban identity in Turkey is Friendship and Brotherhood Streets. These street names, which bear the names of people and places from outside Turkey, show that the impact of distant places or people are felt in Turkish cities. Friendship-Brotherhood Streets give information about the geography of the larger areas (the Balkans, especially Greece and Bulgaria) where Turks settled in the Ottoman period and express their lineage ties. Undoubtedly, the emergence of such avenues is not just about ancestry. Cultural and religious affinity emerges as another factor (Aliağaoğlu and Uzun 2011). In this context, there are Plovdiv, Thessaloniki, Tırnova, Drama, Rhodope and Delıorman streets; Bukhara and Tashkent streets; Kirkuk, Jerusalem, Bosnia and Herzegovina streets in Turkey. Such are the names of the ancestral areas of the Turks who migrated to Turkey from the areas within the borders of the Ottoman Empire during that era. It would be correct to describe them as immigrant or parallel place names. Some streets in the capital Ankara can be added to these and include the names of the countries or important people of countries with good neighborhood relations. For instance, streets with name Willy Brandt, Kennedy, and Argentina in Ankara are salient examples (Ayataç and Araz 2016).

A study on the same subject was carried out by Eşim and Zaman (2021). In this study, they revealed the spatial reflection of the Balkan migration in the urban place names in the Orhangazi district of Bursa province. Interestingly, 25 different names used in urban toponymy in the city and its surroundings are of Bulgarian origin. The settlement of Kardzhali, which is close to the Greek border in Bulgaria, was used most frequently. Plovdiv (3), Delıorman (3), and Tundzha (3) are other important settlement names. In the naming process, the names of the people who pioneered this migration from Bulgaria to Turkey were also used. Mümin Gençoğlu, Rusçuklu Mustafa and Koca Yusuf are among these names. Immigration was not only from

---

enter NATO. The Korean War began when North Korea attacked South Korea on June 25, 1950. Thereupon, the UN took a decision to keep the peace. The UN had asked member states to assist South Korea in restoring peace. Turkey complied with this decision and sent a land force of 5000 people to Korea (Haşimi 2014).

Bulgaria. Other Balkan countries also affected the naming in Orhangazi. For example, the Danube River is frequently used as a place name.

Place/direction and occupation street names

The street names indicating place/direction and occupation depend on memory and have functional, or non-ideological meanings. These streets are mostly inspired by physical features, landmarks, or material symbols adjacent to or near them. These are especially related to transportation (Aliağaoğlu and Uzun 2011). Station and port streets are among them. Areas pertaining to any kind of activity also become street names. School, university, prison, government house and museum streets are common examples.

The streets which indicate occupation or express economic activity are related to the dominant activity in the area in the past. They mostly reflect the Ottoman heritage. Old profession names such as coppersmith, blacksmith and felt maker are used as toponymic indicators. However, in Şanlıurfa and its surroundings, the agricultural activities and products are used for designations. As a result, names related to agricultural activities, especially vineyard and horticultural farming (especially orchards and pistachios) were used (Günal et al. 2011). The same feature applies to the city of Aydın. There the tailor, honey seller, and bag maker were used as the names of neighborhoods (Baysan and Kara 2014).

Fossil streets are those that remain in the city and indicate direction. For example, Muğla Road in the city of Aydın represents the way leading to the city of Muğla. Similarly, the Old Acıpayam Road, which remains in the city of Denizli, defines the highway leading to the Acıpayam district.

Numbers as street names

Numbers are also used as street names. Such streets that are classified as mathematically named thoroughfares, are mostly encountered in metropolitan cities such as Istanbul, Izmir and Ankara and in newly developing areas of these cities. However, recently (since 2006), street names are given numerical names or replaced with these names in order to keep the information processing easy.

Streets with personal names

Another category identified in street and boulevard naming is personal names except those used on republic streets. The names of the people (statesman, such as governor or prime minister), artists, sportsmen, soldiers, poets, writers, clergymen, etc.) who have served society in history or who have made a reputation in their fields on a national scale have been given to the streets and boulevards of Turkish cities. The name Adnan Menderes, who served as a prime minister in Turkey, is frequently used in this respect. Similarly, the names of local people who lived in that city and contributed to it in various ways (service, donation, etc.) were often included in the street and boulevard nomenclature. In total, 52 personal names were used in the city of Erzurum. While 30 of these are related to local persons, 22 of them are connected to nationally important persons. Among the local people, religious figures such as Habip Baba, Hacı Mevlüt Baba, Alvarlı Muhammed Lütfi Efendi come to the fore. The names of pashas, governors and mayors who served in the city in the past were also used in naming. In addition, Ottoman sultan names were used as nationally important street names (Aliağaoğlu and Uğur 2018).

"Names expressing cultural elements in the street, boulevard and road naming, names related to beliefs, religions, and discourses, names indicating weather conditions, names related to symbols and things, names related to surrounding entities (vegetation, flowers, and animals), names related to colors and names originating from human names are also frequently used. The names mentioned above are given under the title of "Other Streets" in the typology created in the study" (Aliağaoğlu and Uzun 2011: 131).

Urban Toponymy in Turkey: On Neighborhood Names

Neighborhood (in Turkish, Mahalle) is a word derived from the Arabic root, hall (halal and hulul), which means "to descend, land, settle down". In the Turkish dictionary, the neighborhood is defined as "each of the parts which a city, a town, or a large village are divided into" (Türkçe Sözlük 2011: 1603). "Place names, which are the reflection of cultural accumulation of cities, naming experiences and thus making spaces meaningful. The city, which is the golden cube of civilization, sometimes hides its past in neighborhood names. Neighborhoods are units in which social, cultural, natural elements and human will are shaped" (Çildam 2018: 1443). Neighborhoods have different names. These can be classified in different ways. Aliağaoğlu (2013) discussed and classified the neighborhood names in Balıkesir province. It is possible to list these categories as "neighborhoods connected to religious structure", "occupational districts", "court districts", "immigrant districts", "republican districts", "political districts", "urbanization districts", "neighborhoods depending on the physical and human characteristics or location of places", "neighborhoods named after a person" and "other neighborhoods" (Aliağaoğlu 2013). Çildam (2018) used the same classification for the province of Siirt.

Borrowing from geology, neighborhood names are like patches of land of different ages that appear on the surface in different parts of the city. In this context, many neighborhoods in Turkey keep the Ottoman period neighborhood heritage alive, at least in their names. In the Ottoman city, the neighborhood is a social unit rather than a physical unit (Aliağaoğlu 2013). "The neighborhood in the Ottoman city is a place where a community of people who know each other, are responsible for each other's behavior and are in social solidarity live. It is the place where the community worshiping in the same mosque lived with their families, as defined in the Ottoman era" (Ergenç 1994). In the Ottoman city, the neighborhood is physically shaped around a mosque or square and takes its name from the mosque. Grand Mosque neighborhood exemplifies this. Sometimes the name of the neighborhood is named after the person who built the mosque. There are many examples in the city of Erzurum in this regard (Kayserili 2015).

Another contribution of the Ottoman period to the neighborhood names of cities in Turkey was the names of professions. In the city of Balıkesir; Butchers, Millers, and Old Jewelers neighborhoods got their names from the professions. The Court neighborhood describes the Ottoman Empire in a different way. The multifunctionality of the judge's (in old Turkish, Kadı) house, that is, its use as a court, has led to the emergence of court neighborhoods. While the places named after a person mostly remain from the Ottoman period, they can sometimes relate to statesmen (such as Adnan Menderes) who lived in the near present period, and sometimes to people whose religious characteristics come to the fore. This situation is especially common in Erzurum, Siirt, and Şanlıurfa settlements, and it also implies religious devotion.

Republic neighborhoods are used for the reasons previously stated for streets and avenues. This explains the foundation of the Republic of Turkey and the decision process of the government regime and those who contributed to this process.

Local governments have reflected the ideologies they stand behind in different ways. This situation is the result of their efforts to show their existence, ensure their place and continuity, and reconstruct history in the local region (Özkan and Yoloğlu 2005). In light of this information, it seems possible to talk about neighborhoods with 'political content'. Labor Mahallesi (Bigadiç) is a good example in this context. The name of the neighborhood reflects the importance that social democrats in the present attach to labor. The same neighborhood name is also found in the city of Safranbolu (Deniz and Ersöz 2020).

Since the foundation of the modern Turkish state (1923), the phenomenon of urbanization developed slowly until the 1950s. Due to this situation, cities could not find many development opportunities horizontally. It can be said that "Cities in Turkey grew in the form of oil stains until the mid-1970s" (Tekeli 2009: 123). Beginning from the second half of the 1970s, "The form of growth of cities with the addition of individual buildings has changed to a form of growth with the addition of large parts to the city." (Tekeli 2009:124). All this process of change has been in the form of reflection of mass housing areas to neighborhood names. The districts of 1200 Houses (Denizli), 150 Houses (Ayvalık), and 600 Houses (Bandırma) exemplify the situation described which is related to the migration of rural populations to cities, which is not specific to Turkey.

Neighborhoods also take their names from the location or direction, as indicated by these regarding the sun or moon, as well as the location and human and physical characteristics of the site. In this respect, Karşıyaka, as the name of the neighborhood, is frequently used with the meaning on the opposite shore of a valley, a lake, or a sea. Karşıyaka neighborhood in Denizli is a salient example of this situation.

However, in the province of Siirt, unlike the Aliağaoğlu (2013) classification of neighborhood names, there are also "Tribal Neighborhoods" and "Displaced (Disaster) neighborhoods" specific to the area. The tribal system is very effective in eastern societies. It has been observed that this system, which has had a shaping or morphological effect on the social structure in the region, is also effective on nomenclature (Çildam 2018: 1453). In the province of Siirt, there are neighborhoods (3 units) revealed by this social phenomenon. There were Adiyan and Mallo Şakir tribes in the town of Pervari. Adian tribe in Aydın Neighborhood; Şakiran tribe lived in Şakiran District. The two oldest districts of Pervari are Aydın and Şakiran where these names have been used since 1927, are named after these tribes. The Garzan tribe, who gave its name to this place, lived in the Garzan district. There are also displaced (Disaster) neighborhoods in Siirt (1 unit). Many villagers were harmed as a result of rock-fall on houses in Demirkaya village within the borders of the central district of Siirt. The area in the central city was expropriated and it was decided to settle the victims' families here on June 20, 1975. Thus, the village was moved to the city center. This neighborhood was called Disaster Houses neighborhood in Siirt City (Çildam 2018).

Urban Toponymy in Turkey: On Critical Perspective

One of the approaches used in toponymic studies is the critical toponymic method. This approach, which gained importance in the 1980s and 1990s, is often referenced by the work of Maoz Azaryahu. He accepts that there are the reminder effect of streets and naming on populations or events; however, he emphasizes the importance of social processes in creating a common identity for the society and political struggle that stands behind the naming (Azaryahu 2009). "Urban names are used as an active tool to increase or consolidate the power of government in colonization and independence processes, nationalization and globalization periods, civil wars, regime changes, in short, in almost every political event" (Çetin and Şentürk 2019: 133).

The critical perspective on street naming in Turkey has remained limited in comparison to many other places. If Şanlıurfa (Günal et al. 2011) is not taken into account, the cities of Ankara, Niğde, and Batman are examined by emphasizing the local political struggle in the background of nation-building and place-naming. In the study which deals with the political capital city of Turkey, Ankara is analyzed by dividing it into three different periods. These epochs are respectively pre-1950, 1950-1960s and post-1970s. Respect for Gazi Mustafa Kemal Atatürk, modernization through westernization and Turkish-Islamic tendencies were dominant in these periods, respectively (Erdentuğ and Burçak 1998).

Ozkan and Yoloğlu (2005) also research Ankara. To them, before the Republic period, place names in the city of Ankara were formed spontaneously. There are traces of place names from the Ottoman period. After 1950, place names changed more ideologically, but a very significant change took place in 1994. While such subtle changes were evident to the elites of the national body politic and to the intelligencia in Turkey, it was less so for the mass of the citizens, and foreign observers.

After 1950, it represents the ideology of the Democratic party. In this period, the name of revolution was replaced by 'revolution'. Ziya Gökalp's name was used instead of Kazım Özalp Pasha's name. Veteran Mustafa Kemal was used instead of the name of Mustafa Necati, one of the former ministers. After 1994, it is seen that there have been important changes in the names of streets and avenues, as in the nomenclature of many places. Highly significantly, Ankara Metropolitan Municipality changed 500 street names between 1994–2005. Suggestively, the naming was not random, some neighborhoods were chosen deliberately. Naming forms can be considered under four headings according to their characteristics:

1. The names that evoke modern bourgeois names have been replaced with names that evoke (traditional) Islam and the Ottoman Empire.
2. Memory and accepted names have been changed. For example, the names of seven streets were changed to Ashgabat. (Possibly, referencing the capital of the central Asian republic of Turkmenistan and all the ethnic and historical connotations with the associated linguistic family.
3. The infrastructural works of the municipality were advertised. It is possible to see this in the names of the junctions on the streets, such as the 71st Day junction.
4. Names that prevent remembering are given. This situation has mostly been related to the current local government's (Justice and Development Party-AK Party) entering a structuring process to guarantee its dominance and existence over the city by marking the definition of national identity through Islam (Özkan and Yoloğlu 2005).

The effect of the geographical environment in the naming of streets and neighborhoods has also been investigated in the city of Şanlıurfa. The issue is similarly partially related to political tendencies. In Şanlıurfa, where personal names come to the fore, economic activities appear to be the second important reason for naming. The political tendencies of local governments are a significant factor that are effective in naming. However, local religious names were given priority due to conservative-oriented politicians in the city (Günal et al. 2011). Once again the subtleties between political and religious 'naming' processes in Turkey are not always evident to observes.

Street and avenue names in Turkey are given by local authorities, namely the city council (Box 2). In essence, due to this, election results determine the urban place names. The political background of efforts to dominate urban space in the city of Batman has changed periodically. During the 1989–1994 period, there was no change in the urban naming due to the fact that the city did not grow physically and conservative parties were in power. The situation did not change in 1994, and the ethnic origin party (DEP) did not participate in the elections, resulting in the dominance of conservative parties. The years between 1994–1999 refer to the period when the city grew and the conservative party won the election and religiously oriented naming was in vogue. 1999 witnessed the beginning of the years when names changed according to political identity. After 2000, when the party whose ethnic identity came to the fore (People's Democracy Party-HADEP) won control of the local government, the use of ethnic and political origin names, which experienced toponymic cleansing, increased. The new administration made 100 'new' namings in 2000 and it can be witnessed that the purpose of this was to erase the traces and memories of the old municipality administration and so to create a memory in line with the new administration's own thinking (Günal 2012).

Box 2 Legal regulation on place naming activities in Turkish cities

The first legal regulation on street and alley names in the Republic of Turkey is the Street Naming and Numbering Law, which was released in the Official Gazette on 20 April 1927. The main purpose of the enactment of this law at that time was to facilitate the general population census to be held on 28 October 1927. In the following process, a decree was issued on 7 October 1939 on how the streets and alleys would be named. However, there were more detailed rules on street naming with the Numbering Regulation released on 21 March 1963. In this regulation, there were some details to prevent giving obscene and ridiculous names in the naming of streets and alleys, and according to the regulation the names given should not be too long and difficult to pronounce. Again, in the same regulation, it is stated that the names of national heroes, statesmen, important individuals who grew up and served in that settlement, artists (such as writers or musicians) and natural elements such as mountains and rivers can be given in the naming of streets and alleys. Finally, in the Address and Numbering Regulation issued on 31 July 2006, it is stated that the same name and numbering cannot be made in different areas of a settlement. (Ayatac and Araz 2016). Today, the boulevard-street naming of cities in Turkey is given or changed by the decisions of the city councils within the framework of the legal regulations mentioned above.

The act of establishing a bond with the electorate of the political structure, which is the local power, is also valid for the city of Niğde. In two different periods, the space is perceived differently by the MHP (1999–2009) and the (Erdoğan's) AK Party (after 2009); it becomes partisan (Kara 2012).

Recently, depending on the trend of economic recession and deindustrialization, cities have increasingly adopted an entrepreneurial agenda and urban names have become commodified; an important economic tool and critical approach has gained a different content (Light and Young 2014). Çetin and Şentürk sets an example in this regard. The study is evaluated in the example of Fikirtepe (Istanbul). The area, which was a shantytown in the past, has turned into a place where land values are at their peak due to its geographical location close to the Bosphorus of Istanbul. In Fikirtepe, the gentrification process has been experienced through urban transformation. So much so that the low-income people were displaced, and the area was made to serve the upper class. In order to create a distinctive place with a high reputation, place names were changed with new and strong(er) names and presented to the market considering the preferences of the target class. For example, the more reputable Kadıköy or Göztepe names were used instead of Fikirtepe. This feature is seen both in housing projects and in street names.

Conclusion

Urban place names are part of the cultural landscape. They arise for many reasons. In other words, they are toponymic witnesses of experiences and memories. People create commemorations in various ways. This can happen in the form of museum exhibits, monuments, historical sites, and gatherings (rallies), as well as by naming streets, buildings, or parks (Rose-Redwood 2008). Cities are settlements where naming is concentrated. Urban place names consist of different names, starting from the streets in the city, to apartments, commercial workplaces, plazas, stadiums and public institutions. Cities are settlements far from 'nature'. This feature is often reflected in the nature of naming. Thus, it is inevitable for place naming in these areas to be of human origin. While giving names to places, the multitude of human reasons draws attention. Urban names may have historical content as well as be related to the horizontal development of cities, direction, security conditions, urban land use, and lineage ties depending on urbanization. Evidently, cities are settlements where interests are in conflict. One of the reasons behind the clash is the existent political power's desire to dominate the space. This desire can often be related to economic reasons also. In other words, street or avenue names or 'labels' have now been commodified or turned into commercial goods. All these reasons are active(ly) in the background of toponymic studies in Turkey.

Urban toponymic studies have been carried out in Turkey over the decades, although they are few in number. These have been more concerned with meaning, origin, and classification in an old-style academic, ethno-nationalistic-linguistic manner. In this respect, they can perhaps be described as traditional, which is nothing unique to Turkey, but experienced in several countries especially after the 1950s e.g. Britain, France, Israel, Ireland and so forth. However, it is against the nature of cities to consider every place-name in cities as part of the political struggle. Nevertheless, the critical perspective is hidden in the archaeology of classifications, albeit partially. As an example, classification of Republic streets is taken from the semiotic standpoint expressed by Rose-Redwood et al. (2010). These streets reflect the process of becoming a nation-state. Street naming is part of the demarcation process of nations (Azaryahu and Kook 2002). "In the establishment processes of nation-states, in nation-building, history must be reconstructed, community consciousness and the belief that they have a common past must be established. For this reason, while history and language are being reconstructed, new names of this language take their place on the streets" (Özkan and Yoloğlu 2005: 55). The urbanization of Turkey went through the urbanization process of the nation-state between 1923–1950. "It would be appropriate to perceive the Kemalist Project as an introduction to modernity and the nation-state project from the environment. In respect, Kemalism, as a bourgeois revolution led by the middle class, has undertaken the task of creating a nation-state, which is a part of the modernity project" (Şengül 2009: 112). Significantly, the same situation can be seen in the example of martyrs' streets; who and what defines the martyr (hero) is all important. Some observers see the 2016 coup d'état attempt movement as the result of struggles against an elite group that has hijacked genuine ideals from religion appealing to populism, in order to promote its own oligarchic agenda in disguise. Once again, this is not unique to Turkey, but evident in other functioning democracies as witnessed in the USA (2016–2021); India under the Modi regime promoting aggressive Hindu nationalism in a multi-ethnic federal state, not to mention the activities seriously questioned by the EU regarding the regimes of Victor Orbán in Hungary, and currently that of Poland. As this chapter is being written (2022) in Turkey, Russian forces under the control of the Putin regime have invaded Ukraine. Nonetheless, the regimes in Poland and possibly Hungary have backed the democracy project for Europe and adjacent neighboring countries. As real life takes place on the streets, often with 'hand to mouth' realities, geographical centrifugal and centripetal balances have to assessed in the light of the common good for Turkey's future generations. Traces of the struggle for Turkish and other peoples' existence can be traced on the streets of friendship and brotherhood. Personal names also emerge from the backyard of political actions, though not entirely. Each local power brings to the fore people who reflect their own political views. These individuals may be from the left- or right-wing or other, religiously inclined or nationalist or other, or have more innovative approaches. In the meantime, remember that I, the author lives here in Turkey, and not in EU or UCLA California, some 70 km. or 54 min from Disneyland., according to google!

Questions

1. What are the reasons that shape the urban toponymy in Turkey?
2. Which methods were used in urban toponymic studies in Turkey?
3. How did the Ottoman/Turkish culture influence the emergence of many toponyms in modern Turkey?
4. How did the process of nation-stateization affect the urban toponym in Turkey in the early period?
5. What is the influence of political parties on naming?

References

Aliağaoğlu A, Uzun A (2011) Şehirsel Toponimi (Hodonimi): Türkiye Için Bir Tipoloji Denemesi. Coğrafi Bilimler Dergisi 9(2):123–133

Aliağaoğlu A (2013) Şehirsel Toponimi: Balıkesir İlinde Mahalle Adları, Bir Sınıflandırma Denemesi. Balıkesir Üniversitesi Sosyal Bilimler Enstitüsü Dergisi 16(29):45–62

Aliağaoğlu A, Uğur A (2018) Şehirsel Toponimi: Erzurum'da Cadde Adları. Balıkesir Üniversitesi Sosyal Bilimler Enstitüsü Dergisi 21(40):1–26

Aliağaoğlu A, Çetinkaya S (2021) 15 Temmuz Darbe Girişimi Deneyiminden Sonra Türkiye'de Şehirsel Toponimi. SDÜ Fen-Edebiyat Fakültesi Sosyal Bilimler Dergisi, Sayı 53, s, 180–192

Asımgil B (2018) İzmir'de Kentsel Kimlik Bileşenlerinden Cadde ve Sokakların Toponimisi ve Milliyetçi, Etnik İzlerin Sosyal Yapı Oluşumuna Katkıları. Trak Univ J Eng Sci 19(1):21–36

Ayataç H, Araz S (2016a) Influence of renaming streets on urban memory: the case of Turkey. In: 17th international planning history society conference, Delft, Netherlands, pp 37–46

Azaryahu M, Kook R (2002) Mapping the nation: street names and Arab-Palestinian identity: three case studies. Nations Natl 8(2):195–213
Azaryahu M (2009) Naming The past: the significance of commemorative street names, İçinde. In: Berg LD, Vuolteenaho J (eds) Critical toponymies, the contested politics of place naming. Ashgate Publishing Company, Burlington, USA, pp 53–70
Baysan S, Kara A (2014) Aydın'da Mahalle, Bulvar, Cadde ve Sokak Adları: Şehirsel toponimik özellikleri. Coğrafi Bilimler Dergisi 12(1):23–48
Çetin R, Şentürk A (2019) Eleştirel Kentsel Toponimi ve Prestijli Mekânlar Yaratmada Kentsel İsimlerin Kullanılması: Fikirtepe Örneği. Megaron 14(1):133–144
Çildam SY (2018) Şehirsel Toponimi: Cadde Adları, Bandırma Örneği. Atatürk Üniversitesi Edeb Fakültesi Derg 61:1–25
Deniz T, Ersöz D (2020) Safranbolu city place names: space, identity and culture. Int J Geogr Geogr Educ (IGGE) 42:527–542
Erdentuğ A, Burçak B (1998) Political tuning in Ankara, a capital, as reflected in its urban symbols and ımages. Int J Urban Regional Res C 22(4):589–601
Ergenç Ö (1994). Osmanlı Şehrindeki 'Mahallenin' İşlev ve Nitelikleri Üzerine. In: Osmanlı Araştırmaları IV. Enderun Kitabevi, İstanbul
Eyice S (1964). İstanbul'un Mahalle ve Semt Adları hakkında Bir Deneme. Türkiyet Mecmuası, XIV, pp 199–216
Gülbetekin M (2017) Mekanın Hafızası, Yer Adları. Hitabevi Yayınları, Ankara
Günal V (2012) Batman'da Cadde Adlandırmaları: Toplumsal Bellek Oluşturma ve Şehirsel Mekâna Hâkim Olma Çabaları. Marmara Coğrafya Dergisi 25:171–197
Günal V, Şahinalp MS, ve Güzel A (2011) Coğrafi Ortamın Şehirsel Mekân Adlarına Etkisi: Şanlıurfa Şehri Örneği. Gaziantep Üniversitesi Sosyal Bilimler Dergisi 10(1):463–508
Kahraman M (2019) İstanbul'un Değişen Sokak İsimleri. Uluslararası Sosyal Araştırmalar Dergisi 12(65):255–263
Kara B (2012) Kentler Açısından Mekanın Toplumsallığı ve Yer İsimlerinin Siyasal Yapıya Göre Değişimi: Niğde Örneği. Zeitschrift für die Welt der Türken, Journal of World of Turks 4(1):149–163
Kayserili A (2015) Erzurum Şehri'nde Yer Adları Üzerine Bir Deneme. Zaman S, Coşkun O (eds) Coğrafyaya adanmış bir ömür: Prof. Dr. Hayati Doğanay içinde. Ankara: Pegem Akademi Yayıncılık, pp 653–667
Light D, Young C (2014). Toponymy as commodity: exploring the economic dimensions of urban place names. Internatıonal J Urban Regıonal Res 435–450
Özüpekçe S (2021) Bir Şehirsel Toponomi Araştırması: Kilis'te Cadde ve Sokak Sınıflandırması. Süleyman Demirel Üniversitesi Sosyal Bilimler Enstitüsü Dergisi 40:1–23
Özkan M, Yoloğlu AC (2005) Bir Bellek Projesi Olarak Sokak İsimlendirmesi: Ankara Örneği. Planlama, vol 4, pp 54–61
Rose-Redwood R, Alderman D, Azaryahu M (2010) Geographies of toponymic inscription: new directions in critical place-name studies. Prog Hum Geogr 34(4):453–470
Şahin G (2010) Türkiye'de Yapılmış Toponimi Çalışmaları. Adıyaman Üniversitesi Sosyal Bilimler Enstitüsü Dergisi 4(3):134–156
Şahin İ (2019) Adbilim. Pegem Akademi, Ankara
Şengül HT (2009) Kentsel Çelişki ve Siyaset. Ankara: İmge
Tekeli İ (2009) Modernizm, Modernite ve Türkiye'nin Kent Planlama Tarihi. Tarih Vakfı Yurt Yayınları, İstanbul
Türk Dil Kurumu (2011) Türkçe Sözlük, Ankara

Cultural Crossroads in Toponymy: Case Study of Bosnia and Herzegovina

Boris Avdić, Ranko Mirić, Nusret Drešković, and Marko Krevs

Abstract Bosnia and Herzegovina is a country that for centuries has been in an area where the influences of large cultural and civilizational spheres are intertwined. This type of influence has left a significant mark on local geographical names. Since the Slavic population has played a dominant cultural role in this area after the Great Migration Period, it is not surprising that most of the local toponyms have Slavic origins. However, the influence of other cultures or languages, such as Illyrian, Latin, Italian, Hungarian, German, and especially Turkish (in combination with Persian and Arabic), is also evident. The Ottoman Empire ruled this area from the 15th to the nineteenth century, which is formation period for most of today's towns in Bosnia and Herzegovina, so many toponyms emerged during that period. Sarajevo as capital city is a well known example. The confessional differences that arose then among the domestic population have persisted to this day, and certain implications of these differences are partly visible through toponymy. This became especially emphasized in some cases in the contemporary context when toponyms gained their ethnopolitical and nationalistic dimension.

Keywords Bosnia and Herzegovina · Balkans · Cultural heritage · Geographical names · Multiethnicity · Spatial identity

B. Avdić (✉) · R. Mirić · N. Drešković
University of Sarajevo, Sarajevo, Bosnia and Herzegovina
e-mail: borisavdic@pmf.unsa.ba

R. Mirić
e-mail: rankom@pmf.unsa.ba

N. Drešković
e-mail: nusretd@pmf.unsa.ba

M. Krevs
University of Ljubljana, Ljubljana, Slovenia
e-mail: marko.krevs@ff.uni-lj.si

© The Author(s), under exclusive license to Springer Nature Switzerland AG 2023
G. O'Reilly (ed.), *Place Naming, Identities and Geography*,
Key Challenges in Geography, https://doi.org/10.1007/978-3-031-21510-0_15

Objectives

By the end of this chapter readers should be able to:

- distinguish between different toponymic layers in Bosnia and Herzegovina
- recognize some of the most frequent appelatives in South Slavic toponymy
- compare various historic periods in relation to toponymic heritage of Bosnia and Herzegovina
- discuss (about) the socio-political significance of toponymic heritage

Introduction

Toponymy is a field in which geography, as a study of space and place, meets the humanities, such as linguistics, history, ethnology, culturology, and even political science. In this regard, it is necessary to apply an interdisciplinary approach to the study of toponymic phenomena of a particular region so that the conclusions can be placed in the appropriate context. In Bosnia and Herzegovina, a country with a multiethnic population, a pronounced intercultural dimension of the past, and a complex political present, the contextualization of public discussion of geographical names is a sensitive social issue. Given the lack of comprehensive literature on this topic and the frequent occurrence of tendentious interpretations of many toponyms and choronyms in Bosnia and Herzegovina, this (paper) chapter aims to contribute to an impartial view of the facts about the toponymic heritage of this country. From that aspect, it can have great use-value at different levels and spheres of education. The applied methodology is based on the analysis of different 'historical layers' of toponymic and identity heritage, which is why the introductory segment gives a brief overview of relevant historical processes and events in today's territory of Bosnia and Herzegovina.

Bosnia and Herzegovina is a country in Southeast Europe, located in the Western Balkans. It borders three neighboring countries—Croatia, Serbia, and Montenegro (all of which, including Bosnia and Herzegovina, became independent with the break-up of SFR Yugoslavia). It has a narrow coastline on the Adriatic Sea. It is primarily a mountainous area (Dinaric Alps) intersected by the valleys of numerous rivers. The lower terrain characterizes only the extreme north (the edge of the Pannonian Plain) and the south (Adriatic basin). The climate is mostly temperate, with a more pronounced influence of the Mediterranean in the south (Herzegovina region) and altitude modification in the mountains. Most of the local rivers belong to the Black Sea basin, where it flows through the Sava and the Danube, while a smaller part of the national territory belongs to the Adriatic basin, where the Neretva River is the most significant. Among the natural geographical features, it is necessary to mention the predominantly karst terrain in the southern and western parts of Bosnia and Herzegovina, as well as the fact that more than half of national territory is covered with forests (in local toponymy some species of trees, such as oak, beech, fir, etc. are extensively represented). In 2013 (according to the last census), the country had a population of 3.53 million people (BHAS 2016). The most densely populated is the valley of the Bosna River and its tributaries, representing this country's historical, population, transport and economic core. The majority of the population of Bosnia and Herzegovina are members of the three constituent peoples (ethnic groups)—Bosniaks (50%), Serbs (31%), and Croats (15%). The official languages are Bosnian, Serbian and Croatian—three mutually intelligible languages from the South Slavic group. Based on the last census conducted in 2013, there are 6,141 officially registered settlements or populated places (BHAS 2013a, b).

This area was inhabited in the prehistoric period—particularly important were the Butmir culture from the Neolithic and the Glasinac culture from the Iron Age. However, the first inhabitants of these areas about which there are historical records were the Illyrians, a common name for a group of tribes of Indo-European origin, which inhabited the western part of the Balkan Peninsula during the Ancient history. Among others, the most pronounced Illyrian peoples are Delmatae, Pannonians, Daesitiates, Iapydes, Autariatae, Daorsi, Mazaei, Ardiaei, etc. They remained historically best known for the tremendous resistance they offered to the Roman invaders, and most information about them was obtained from Roman sources. After this region was finally incorporated into the Roman Empire, the province of Illyricum was founded, which was then divided into two new ones—Dalmatia and Pannonia, which were named after (partly) Illyrian peoples. The Roman influence is visible primarily through the construction of roads and cities, mainly formed near mines, thermal springs, and strategically important locations. The Illyrian ethnolinguistic component, however, remained present until the Great Migration.

At the beginning of the Middle Ages, different peoples migrated to the Western Balkans, due to which the autochthonous Illyrian population retreated deeper into the mountainous areas. With the arrival of the Slavs in the sixth and seventh centuries, their complete assimilation with the newly arrived population took place in this area. In the following centuries, various Slavic states emerged, which were, however, mainly in the sphere of influence of two great regional powers—Byzantium and the Kingdom of Hungary. One of these states was Bosnia, first mentioned in the tenth century. It was formed in the upper part of the Bosna River basin (today's Sarajevo-Zenica Valley) and was, under the leadership of successful bans such as Kulin, Stephen II and Tvrtko, expanded to the surrounding regions, including the Adriatic coast. Tvrtko I Kotromanić was proclaimed king in 1377, and Bosnia, as a kingdom headed by him, experienced the peak of its territorial extension (parts of modern Croatia, Montenegro and Serbia). However, during the reign of his successors on the throne, Bosnia gradually lost its military and political significance, and its territory shrank. In religious terms, the population of medieval Bosnia was under the influence of Roman Catholicism from the West and Orthodoxy from the East, while, at the same time, there was an indigenous variant of Christianity—the Bosnian Church.

In the middle of the fifteenth century, a new military force spread across the Balkans—the Ottoman Empire. The Ottomans tore down the Kingdom of Bosnia in 1463, but they will gradually conquer the territory of present-day Bosnia and Herzegovina until the end of the sixteenth century. This was a significant turning point in the socio-political development of this land, which manifested itself primarily through the Islamization of a large part of the population. Although the Muslim population had a privileged position within this new empire, this area retained a multi-confessional character, given the continued presence of a number of Roman Catholics and the migration of Orthodox and later Jewish population. This religious identities of the local population will become the basis for the formation of modern ethnic nations in the nineteenth and twentieth centuries. Ottoman Bosnia was the first territorially organized as a sanjak and then as an eyalet—the highest administrative level in the Empire. During this historical period, numerous oriental-style urban settlements were established, including Banja Luka, Travnik and Sarajevo, which served as the main administrative centers of the eyalet. The western Bosnian borders (toward modern Croatia) were roughly defined by the Treaty of Karlowitz in 1699, followed by a century of frequent border wars with the Habsburg Monarchy and Venice, which did not result in significant and lasting territorial changes, at least in this region. In the nineteenth century, border conflicts calmed down, but internal instability, i.e. frequent uprisings of Muslim and Christian populations, came to the fore. Herzegovina eyalet (separate from the Bosnian eyalet) was also temporarily formed, which helped create a separate regional identity among the inhabitants of that (southern) region.

Since the militarily and economically weakened Ottoman Empire could not adequately respond to the increasing revolts on the Balkan Peninsula, at the Congress of Berlin in 1878, Austria-Hungary gained the right to occupy the territory of Bosnia and Herzegovina, while neighboring Serbia and Montenegro gained independence. In that way, the eastern Bosnian border was defined, which is still valid today. Bosnia and Herzegovina was granted a special status in this monarchy, as it was not classified in either the Austrian or Hungarian territories within the Monarchy. The Austro-Hungarian authorities introduced modernization into local social, infrastructural and economic factors. However, it was felt far more in urban than rural areas, where almost 90% of the population of Bosnia and Herzegovina still lived. This was also a period of consolidation of national movements when the Orthodox population developed a strong Serbian ethno-national identity, while Catholics started to clearly affiliate with the Croatian national cause. Bosnian Muslims (modern Bosniaks) were taken aback by the new social and political reality, which is why the process of developing their national identity will continue for most of the twentieth century, and many will migrate permanently to Turkey. The formal annexation of Bosnia and Herzegovina took place in 1908, causing an international geopolitical crisis since this unilateral act of Austria-Hungary conflicted with the interests of Serbia and its allies. The peak of political tensions on this route was reached with assassination of the Austro-Hungarian archduke Franz Ferdinand in Sarajevo in 1914, which can be considered as introduction to the First World War.

The defeat of the Central Powers in this war led to the disintegration of the Austro-Hungarian Empire, which created the opportunity to form (ation of) the first joint state of South Slavs. Thus, as early as 1918, the Kingdom of Serbs, Croats and Slovenes was established, headed by the Serbian Karađorđević dynasty. This newly formed state also included the whole territory of Bosnia and Herzegovina, with the initially unchanged borders of districts from the Austro-Hungarian period. Previously initiated economic development slowed down, and the social focus shifted to agrarian reform, which (has) caused multiple controversies due to the new redistribution of land. At the same time, interethnic tensions intensified at the level of the Kingdom, primarily between Serbian and Croatian political representatives, which culminated in 1928. The following year, the King dissolved the National Assembly and suspended the previous constitution, imposing a new one, according to which the state was renamed to the Kingdom of Yugoslavia. The new administrative structure with nine regions called banovinas eliminated historical borders, and the territory of Bosnia and Herzegovina was divided between four of them: Vrbas, Drina, Littoral and Zeta. This only temporarily suppressed the crisis but did not ultimately overcome it, as evidenced by the assassination of King Alexander in 1933. A new solution to the national question in Yugoslavia was offered by creating the Banovina of Croatia in 1939, which included a significant part of Bosnia and Herzegovina. The Kingdom of Yugoslavia was attacked by Nazi Germany in 1941, after which the Independent State of Croatia was proclaimed in 1941 as a puppet state of the Axis powers. In addition to most of present-day Croatia, it territorially encompassed the whole of Bosnia and Herzegovina, with a border on the Drina River.

However, due to the ethnic heterogeneity of the population, inaccessibility of the terrain, and poor transport connections, Independent State of Croatia failed to establish complete control over the entire territory, even with the help of the German and Italian armies. Under the leadership of the Communist Party of Yugoslavia, a movement of resistance to the fascist occupation was organized, as well as so-called free territories. (The majority of) Most Yugoslav partisan combat operations took place in Bosnia and Herzegovina, and in November 1943, key sessions of the National Anti-Fascist Council of People's Liberation of Bosnia and Herzegovina and the Anti-Fascist Council of People's Liberation of Yugoslavia were held, according to whose provisions, after the end of the war, the new socialist Yugoslavia was organized, consisting of six federal republics, one of which was Bosnia and Herzegovina within the borders from the Congress of Berlin. The new state authorities tried to overcome interethnic tensions, which were further inflamed in World War II, by actively promoting the idea of 'brotherhood and unity', within which Bosnia and Herzegovina has proclaimed a republic belonging equally to Muslims, Serbs and Croats. At the same time, there was an intensive socio-economic development—Bosnia and Herzegovina became an industrial republic from a predominantly agricultural society, education of the entire population became one of the national priorities, and effort was in full swing to build adequate urban and transport infrastructure. Sarajevo organized the Winter Olympic Games in 1984, which is an event that symbolically marked all the achieved social progress in the previous period. However, the economic problems that Yugoslavia will feel along with other socialist states in Europe will awaken repressed nationalism in the coming years.

The first democratic elections after the Second World War in Yugoslavia were held in 1990. In Bosnia and Herzegovina, a landslide victory was won by three so-called national parties—Party of Democratic Action, Serbian Democratic Party and Croatian Democratic Union. Their disagreements over the political future of Bosnia and Herzegovina will become very evident with the beginning of the turbulent process of dissolution of Yugoslavia, which began with the declaration of independence of Slovenia, followed by Croatia in 1991. After the majority of Bosnia and Herzegovina's citizens also voted for independence in a referendum the following year, the war escalated. The central government and its army were primarily supported by the largest Muslim population, for whom the ethnonym Bosniak has been used since 1993. The majority of the Serb population, which had previously boycotted the referendum, was politically organized through the self-proclaimed Republika Srpska, whose civilian and military authorities are charged with most war crimes committed during the conflict in Bosnia and Herzegovina, including genocide against the Bosniaks in Srebrenica. Areas predominantly inhabited by Croats were controlled during the war by the authorities of the self-proclaimed Croatian Republic of Herzeg Bosnia. The Washington Agreement integrated these areas with parts of the territory controlled by the central authorities into the Federation of Bosnia and Herzegovina. The Dayton Peace Accord ended the war in 1995, and since then Bosnia and Herzegovina has been defined as a complex country comprised of two entities (Federation of Bosnia and Herzegovina and Republika Srpska), where constituent peoples (ethnic groups) control powerful mechanisms for protection of its vital national interests. Although significant progress has been made in the peace process, political tensions have remained and frequent political crises have seriously jeopardized the functional dimension of the state. A particular social problem is a complete inconsistency in interpreting historical events from the past, which often tend to adapt to a particular identity context and narrative; (.) (T) toponymy (also) plays an important role in this.

Name of Bosnia and Herzegovina

The oldest documented mention of the name Bosnia is in the work of the Byzantine emperor Constantine VII Porphyrogenitus entitled *De Administrando Imperio* (On the Governance of the Empire) from 949. Namely, the last sentence in Chapter 32, titled *About Serbs and the country in which they now live*, reads "In baptized Serbia are the inhabited cities of Destinikon, Tzernabouskei, Megyretous, Dresneik, Lesnik, Salines, and in the small land of Bosnia (chorion Bosona) Katera and Desnek." Based on other historical sources, it is evident that the choronym Bosnia (Bosna) originally referred to a small area around modern Visoko (parish of Bosna), but later gradually expanded during the Middle Ages, including regions such as Usora and Soli, Donji Kraji, Podrinje, Hum, Zapadne Strane, Primorje, etc. In this regard, it is worth of mention one of the oldest documents in Slavic languages, the *Charter of*

Kulin Ban, by which this ruler of medieval Bosnia from the end of the twelfth century guarantees Dubrovnik merchants security and free business on his territory. Among the first words in it are: "I, Bosnian ban Kulin …" This is also the oldest documented domestic mention of the name Bosnia.

A river of the same name (Bosna) flows through the mentioned historical core of Bosnia, but the dilemma of whether the country is named after a river or vice versa has never been fully clarified. In fact, the etymology of this choronym/hydronym is generally unknown, although there are several hypotheses about its origin. Most of them contain an assumption about the pre-Slavic origin of this name. The opinion of philologist A. Meyer that the hydronym Bosna has origin from an Illyrian word containing Proto-Indo-European root *bos*, meaning *running water*, is most often cited. Some identify the present-day Bosna River with the ancient river Bathinus, which separated the Illyrian tribes Breuci and Osseriates. However, many authors reject such a conclusion, pointing to certain geographical discrepancies. The ancient Roman settlements of Ad Basante (noticeable on the Tabula Peutingeriana) and Bistua Nova are also associated with the origin of the name Bosnia. As alternative explanations, analogies also appear with the name of the Thracian tribe Besa, the Albanian word *bos* (salt), and the Latin term *bosina*, which means border. Vego (1982), on the other hand, advocates the thesis of the Slavic origin of this name and argues in detail the possibility that the Slavs may have brought it with them from Central or Eastern Europe during the Great Migration at the beginning of the Middle Ages. This claim is supported, among other things, by a large number of similar toponyms in other Slavic countries. As there is no explicit material evidence on any of these assumptions, the question of the very origin of the name Bosnia remains open.

The etymology of the choronym Herzegovina is much clearer. It originates from the title of an influential feudal lord, Stjepan Vukčić Kosača from the fifteenth century. For a significant part of his time, he managed to almost independently control the southern part of medieval Bosnia, that is the region until then known as Hum. It is important to note that this region is geographically significantly different from the rest of the country, regarding primarily (to) the terrain and climate. Stjepan Kosača was often in open conflict with the Bosnian king, and he needed a ruling title that would more clearly emphasize his power and importance than was the case with his inherited title of duke. It is assumed that, for that reason, he chose the title of *herzeg* (of Saint Sava), which was unusual for this region due to its Germanic origin. Namely, the Germanic word *Herzog* also means military leader, i.e. duke. Since then, the land he effectively ruled has been called Herzeg's land or Herzegovina. The Ottoman conquerors made a crucial contribution to the survival of this choronym, as in the second half of fifteenth century, they put under their rule almost the entire territory of medieval Bosnia and, in the area of the former Hum, they established a sanjak named Herzegovina. However, it should be emphasized that during most of the period of Ottoman rule, this sanjak was part of Eyalet of Bosnia. That means it was considered a subdued spatial (regional) determinant in relation to Bosnia.

Midžić (2010) states that the oldest mention of the dual name Bosnia and Herzegovina can be found on the *Delimitation of the Christian States and Turks* map by Pierre du Val from 1663, as a cartographic connection between Upper Bosnia and

Herzegovina. However, this cartographic document can be considered an isolated case since the choronmy Bosnia in almost all other available sources had an unequivocal primacy over Herzegovina. This changed in the nineteenth century. After the unsuccessful Great Bosnian Uprising in 1833, the Herzegovina Eyalet was formed and separated from the Bosnia Eyalet for the first time in several centuries. Although Herzegovina was returned to the Eyalet of Bosnia after approximately two decades, that period was clearly sufficient to develop an awareness of its regional separateness from Bosnia. The Herzegovinian uprising (1875–77), in which the Orthodox Christian population rebelled against the Ottoman authorities, also contributed to the separation of this choronym in relation to Bosnia. As a result, at the Congress of Berlin in 1878, the dual name Bosnia and Herzegovina was officially introduced for the first time for the territory that the Austro-Hungarian Monarchy will occupy lat(t)er in the same year. Since then, this dual name has been in continuous use. As an integral part of the FPR/SFR Yugoslavia, Bosnia and Herzegovina firstly had the status of People's Republic and then Socialist Republic. It declared independence in 1992, under the official name of Republic of Bosnia and Herzegovina. After the Dayton Peace Agreement in 1995, the prefix Republic was also abolished.

Pre-slavic Toponymic Layer

Given the lack of relevant historical sources, as well as the fact that generally little is known about the languages spoken in the territory of present-day Bosnia and Herzegovina before the Slavic settlement in sixth and seventh centuries (except for Latin), the analysis of the pre-Slavic toponymic layer has numerous challenges and limitations. It seems that hydronyms, i.e. the names of major rivers, actually have the longest continuity: Sava, Drina, Vrbas, Neretva, probably Una and Sana, and perhaps Bosnia. There is a possibility that they belong to the so-called Old European hydronyms, which explains the similarity of many hydronyms throughout Europe and partly in Asia. At the same time, they cannot be connected to the well-known local languages spoken in a given area. This theory implies the existence of common Indo-European roots in the names of rivers from this language group (Brozović 1995).

The ancient name Sava (Savus) is associated with the Indo-European root *sowos*, which refers to flooding. Drina River (ancient Drinus) also probably originates from the Indo-European root *dri*, which refers to the flow. Neretva River (originally Naron and later Narenta) comes from the Indo-European root *ner-*, which is identified with the verbs to immerse and submerge. The old etymological root of the Vrbas River (Urpanus/Urbanus) indicates a bend, and a type of tree—willow in the newer Slavic variant. The name of Adriatic Sea, on which Bosnia and Herzegovina has a short cderivative Mosorje isoastline, also belongs to the group of old hydronyms, but it is quite possible that it also has even pre-Indo-European roots (Ivšić 2013). It is (also) interesting that the only coastal town in this country—Neum, certainly has an ancient etymology. This may be the Municipium Neuense, first documented in the first half of sixth century.

In modern Bosnia and Herzegovina territory, it is possible to find more toponyms (oeconyms and oronyms) that originate from the pre-Slavic period. The most interesting examples are found in the inhospitable high karst area in the western part of the country. It is an area inhabited by the semi-Illyrian tribe of Delmatae, after whom the Romans named the entire province that included the westernmost part of the Balkan Peninsula—Dalmatia, a name that has remained to this day for the largest littoral segment of neighboring Croatia. It is evident that the influence of this tribe was significant since many ancient toponyms in this region have remained in some variant to this day, which is a phenomenon that is much less pronounced in other parts of the country. This may be explained by the relatively sparse population of these areas, because they were less attractive to the Slavic population than other regions.

It is known that the ancient political center of Delmatae, and later the nearby Roman municipium was called Delminium, whose etymology, together with the mentioned ethnonym, is based on the Illyrian word *delma*, meaning sheep. It indicates the dominant activity of this population—pasture cattle breeding. From this etymological root, two modern toponyms eventually emerged, Duvno (Dlmno—Dumno—Duvno) and Glamoč (Delmoč—Dlamoč—Glamoč). Glamoč is the name of the town and karst field where it is located, while Duvno today refers only to the field where the ancient Delminium used to be. Many prominent oronyms in this region are also associated with the pre-Slavic heritage. There is a thesis about the striking border mountain Dinara, after which the entire mountain system of the Dinaric Alps is named (or vice versa?), according to which this oronym originates from Dindarii, an Illyrian-Celtic tribe. However, they did not inhabit this area, so there are disagreements on this issue. The second border mountain Ujilica is mentioned in ancient sources as Ulcirus, in which Mayer (1959) recognized the Indo-European root *ulk* (=wolf). The highest mountain in western Bosnia is Cincar, which according to Nikić (1980), bears its name from the Illyrian-Roman period, and literally means five (it. cinque) heads, denoting five prominent peaks that exceed 1700 m.

In Bosnia and Herzegovina and the neighboring countries there are also an oronym Mosor (of the Illyrian words *mol*—hill and *sor*—source). Mosor is name of the mountain and the village in Central Bosnia (near Travnik), the village in Montenegro, the mountain peak on Suva Planina in Serbia, and derivative Mosorje is the name of the hamlet in Herzegovina (Ivšić 2012). The mountain massif above Split (Croatia) also bears this name.

Slavic Toponymic Layer

Despite the fact that Slavic tribes migrated to the Balkan Peninsula (way back) in the sixth and seventh centuries, means that we can speak about the existence of the first documented toponyms of Slavic origin only a few centuries later, considering the general lack of written sources from the early Middle Ages. However, by the end of the Medieval Period, Slavic toponyms became absolutely dominant within this region (Šabić, 2016). Those pre-Slavic terms that managed to remain part of the local toponymy in a specific form, usually got the characteristic Slavic prefixes or suffixes—they became corrupted to a certain extent. It is also worth noting that in this area the Slavs found a romanophone population (often called the Vlachs), and that some toponyms came into existence also under such hybrid influences. This holds true particularly in terms of the oronymic heritage, whose origins are linked predominantly to the Middle Ages. As far as hydronyms are concerned, it has already been highlighted that the root names of the major rivers of this area go back to antiquity, while the Slavs were left to name most of their numerous tributaries. Due to the extensive material, our analysis of Slavic toponyms is still primarily focused on the names of populated places (oeconyms), while taking into account only those that are clearly statistically defined (about 6000 of them), and leaving out of consideration numerous hamlets and other toponyms related only to parts of populated areas. An analysis of the official list of settlements in Bosnia and Herzegovina shows that most of them have a patronymic character, i.e. that they are named after the surnames of certain families, and in most cases have a typical suffix *-ići*, which in linguistic terms indicates the plural family names. Some of them originated in the Middle Ages, and others date from later periods (mostly the Ottoman era), since this suffix form of family names became permanently present among the local population, regardless of the ethno-religious background. In the text below, however, we single out some frequent examples of Slavic appellatives.

The noun *grad*, which in the same or similar variants (*gorod*, *hrad*, etc.) can also be found in all Slavic countries, represents the most common appellative in Bosnia and Herzegovina (Fig. 1). Its modern meaning suggests the presence of an urban type of settlement (town or city), but in its original form it mainly represented a kind of fortification, which is why the populated places that include this root word are usually situated near the former forts that originate from the Middle Ages or earlier periods. It is in this context that the form *gradina* also comes up very often. Other variants of these oeconyms with this characteristic root word include: Zagrađe, Podgrađe, Podgradina, Gradac, Gradišće, etc. In this group, the most prominent are modern-day toponyms that refer to certain urban centres (towns): Gradačac, Gračanica, Gradiška, Novi Grad, Mrkonjić Grad, Tomislavgrad and Višegrad. The Slavic appellative *most* (=bridge) is also specific to urban toponymy, referring to places where traditional roads were intersected with rivers. The toponym of Mostar, the main urban centre of Herzegovina, was created in this way. The same goes for Sanski Most, a town on the Sana River. Another similar appellative is *brod* (lit. ship), denoting the shallower parts of the river bed, where in the past it was much easier to cross over to the other side or river bank. One of the northernmost cities in Bosnia and Herzegovina is called Brod. It is situated on the Sava River and it was often referred to as 'Gate of Bosnia' in the past.

Many Slavic oeconyms are derived from different types of terrain. Among the most widespread is the term *polje* (=field), which refers to flatter parts of the terrain, which were most often used for agricultural purposes, and according to Skok (1971–1974), it come from the pan-Slavic and proto-Slavic *pol-jo*. A specific type of polje is the karst field, which in this original South Slavic variant came into use as part of the technical

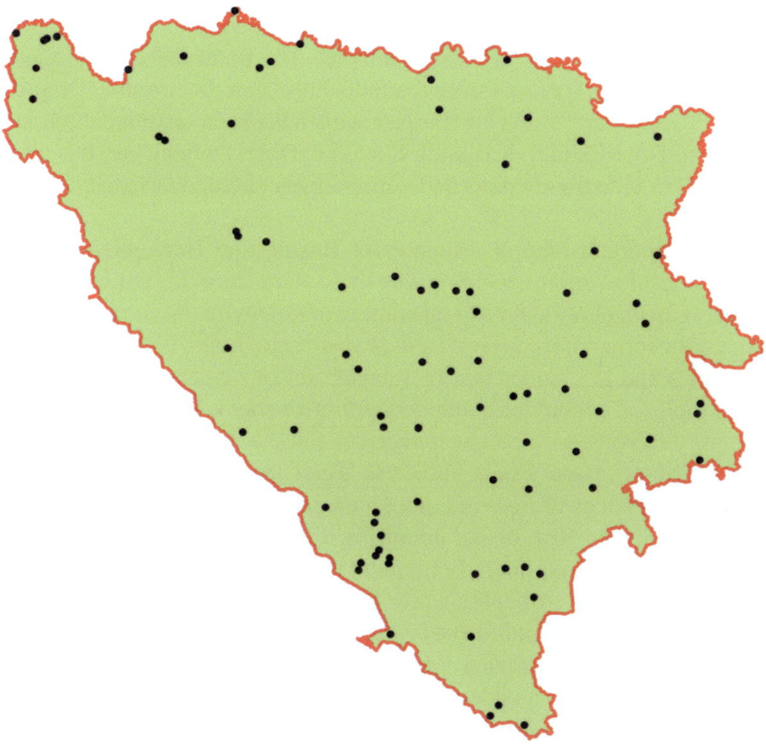

Fig. 1 Distribution of populated places with word root of 'grad'

terminology associated with karstology. This term has found its place in more than 80 formal oeconyms, among which the most notable are: Željezno Polje, Topčić Polje, Knešpolje, Roško Polje, Nević Polje, Poljice, etc. The next common terrain-related appellative is *brdo* (=hill), which due to the geomorphological characteristics of Bosnia and Herzegovina is also a very common relief form, present in a similar quantity as the fields. As a toponym it can largely stand alone, or in combination with some descriptive or possessive adjective, such as Čusto Brdo, Debelo Brdo, Radića Brdo, Golo Brdo, Tursunovo Brdo, Podastinjsko Brdo, etc. The older Slavic term for the hill is *hum*, which was nevertheless used much more in the Middle Ages than in any later times. Namely, the medieval name of Herzegovina itself was Hum or Humska Zemlja (lit. Hum Land) due to the typical landscape of karst hills (*humine*). Some toponyms marked with this root word still remain to this day, such as Hum near Foča (a place at the border crossing to Montenegro), Humac near Čelić, Humilišani near Mostar or Zahum near Prozor. In the modern context, the appellative *luka* can also be confusing, as it refers today to a sea or river port (harbour), while its original meaning in the South Slavic languages was a meadow situated along a river. This very explanation is equally applicable to the name of the second largest city in Bosnia and Herzegovina, Banja Luka (Banova Luka), as well as to many other smaller places (e.g. Olovske Luke and Oštra Luka). The term *dolina* or *do* (=valley) is also a very common term in local toponyms. The noun *dol* is confirmed in the Proto-Slavic language. It comes from the Indo-European root *dholos* (Dragić 2013). More than 60 oeconyms made with this root word have been recorded, such as Dolac, Dolovi, Stupni Do, Mrkodol, Kasindol, Krivodol, Dole, Doljani, etc. It is also worth noting that many villages are preceded with prefixes classified as *gornji* (=upper) or *donji* (=lower).

The most important natural resources of Bosnia and Herzegovina are waters, forests and mineral resources, which is why these elements of the natural environment have served as an inspiration for a great number of toponyms. Among the hydronymic appellatives, the terms *rijeka* (=river), *potok* (=stream), *jezero* (=lake), *bara* (=pond), *vrelo* (=spring) and of course *voda* (=water) stand out in terms of their frequency. Among examples of oeconyms combined with such root words are: Varoška Rijeka, Babin Potok, Jezero near Jajce (since recently famous for crater of the same name on the planet Mars), Bare, Bijela Voda, etc. The names of the places derived from the characteristic names of trees are quite numerous, which is understandable since the forests cover over 50% of the country's territory. As the root words of such toponyms, those that are particularly frequent include: *bukva* (=beech), *hrast* (=oak), *grab* (=hornbeam), *javor* (=maple), *jela* (=fir), *bor* (=pine), as well as fruit trees, such as *jabuka* (=apple), *šljiva* (=plum) and *orah* (=walnut). That is why the toponyms such as Bukovica, Bukovlje, Orahovica, Orašje, Hrasnica, Hrasno, Grabovica, Grabovik, Borovac, Borak, etc. are in a widespread use in multiple places. Animal appellatives are also widespread, though to a somewhat lesser extent compared with those named after plants. Motifs like *jelen* (=deer), *medvjed* (=bear), *vuk* (=wolf), *ovca* (=sheep), *koza* (=goat), *svinja* (=pig), *govedo* (=bovine), *lisica* (=fox), *žaba* (=frog) and *konj* (=horse) are particularly common. Finally, various mining appellatives, such as *ugalj* (=coal), *olovo* (=lead) and *srebro* (=silver) should also be mentioned. Based on them, the names of the mining towns like Ugljevik, Olovo and Srebrenica were established. Close to Srebrenica, there is a village with very unusual toponym Sase, which preserves the memory of the Germanic Saxon miners who lived here in the Middle Ages.

Ottoman Toponymic Layer

Most toponyms from the previous period survived under Ottoman rule, including some broader choronyms, such as Bosnia itself. However, at the same time, several new toponyms have emerged, whose origin comes mostly from Turkish and indirectly from Arabic and Persian languages. This is understandable given that a large number of new settlements were formed, both urban and rural. Particularly important was introduction of new oriental urbanization concepts, which was a complete novelty in this area. Until then, urban settlements were reflected almost exclusively in topographically elevated fortifications of a military or administrative character, in which only a small number of people could live. The new administration established new towns, often at the foot of important fortifications from the medieval period, and gave them a predominantly commercial and craft function. These cities were generally classified into three groups according to size and importance: *šeher* (cities with about 10,000 or more inhabitants), *kasaba* (medium-sized towns) and *varoš* (smaller urban centers, usually up to 2,000 inhabitants). This oriental terminology has entered the domestic toponymy, most often in the names of neighbourhoods in towns, as well as of some independent settlements (e.g. Novi Šeher, Kotor Varoš or Nova Kasaba). In urban toponymy, terms such as *čaršija* (central part of the city, with craft and trade shops), *mahala* (grouped section of town with a residential function) and *sokak* (oriental street) often appear.

Among modern large urban centers of Bosnia and Herzegovina, two cities stand out with a clear Turkish origin of the name—Sarajevo and Tuzla. Sarajevo was founded in 1462 (a year before the official fall of the medieval Bosnian state) by the Ottoman governor Isa-beg Ishaković, and this city, surrounded by mountains in the upper segment of the Bosna River basin, gained the status of a šeher for a relatively short period and ultimately became the administrative center of the Bosnia Eyalet. The city is probably named after the newly built court (tur. Saray—palace), while the origin of suffix in the city's modern name is a matter of dispute. According to one interpretation, the full original name of the town was Saray-Ovasi (tur. plain around the palace), while according to another, it is emphasized that *-evo* is a typical Slavic suffix added to many toponyms, even in cases where the root is of foreign origin, as in this case. Unlike Sarajevo, which originated in the medieval parish of Vrhbosna (lit. Summit of Bosnia), the toponym Tuzla in etymological terms is actually a reflection of centuries of continuity on the mineral wealth of rock salt, which has been exploited in this location since ancient times. This third-largest city in the country and the primary urban center of Northeastern Bosnia is first mentioned in historical documents under the Latin name Salinas, which was then Slavicized in the form of Soli, which was used as such a choronym for the surrounding parish in the Middle Ages. After the Ottoman conquests, the basis for the origin of this toponym is the Turkish word *tuz*, which, like the previous variants, means salt. Salt exploitation is still an important feature of local economy and one of the critical elements of identity of the city and its inhabitants.

New social circumstances also introduced certain changes in rural toponymy. More than 90% of the native population still lived in the villages and were dependant on self-sufficient agriculture. Their social life took place within the rigid framework of the Ottoman feudal system, in which the landowners were mostly Muslims, while the Christian population had subordinate serf status. The social reputation of feudal landowners was reflected, among other things, in their titles—*pasha*, bey/*beg* and *aga*. Many oeconyms were derived from such titles, especially in rural areas, which have survived to this day. Among the more prominent examples are Pašić Kula, Begov Han, Begovo Selo, Begluci, Agino Selo, Aginci, etc. Another very common toponym that directly originated from these feudal relations is *čifluk*, i.e. its corrupt variant *čitluk* (the consonant F did not exist in the language of the local population before the Ottoman period). Čifluk (tur. Çiftlik) were land estates in the Ottoman Empire, which were initially state-owned, only to be privatized by the said feudal lords from the sixteenth century, which worsened the position of the peasants. Today, there is the town of Čitluk in western Herzegovina, that was urbanized much later in the twentieth century. However, several villages with this toponym also exist throughout the country, such as in the municipalities of Goražde, Posušje, Kozarska Dubica, Šipovo, Tešanj, Visoko and Travnik. The feudal specificity of the Eyalet of Bosnia was the so-called *odžakluk-timar*, which represented estates that could be inherited by the family. The Turkish appellative *odžak* (lit. chimney) figuratively referred to the family and is also one of the most common toponyms of this period, primarily due to this type of property.

The influence of Islamic culture is visible through many other toponyms in Bosnia and Herzegovina too. Among the most interesting are those containing the word *vakuf*, which is the Bosnian version of the Arabic term waqf, whose meaning is defined by Sharia law. It is an Islamic endowment that a person voluntarily sets aside from his property for the public good. Thus, in Central Bosnia, in the Vrbas River valley, there are towns of Gornji (Upper) Vakuf and Donji (Lower) Vakuf, while in the west of the country, it is possible to find settlements of Kulen Vakuf and Skucani Vakuf, and in the recent past, there were also Skender Vakuf and Varcar Vakuf. Interestingly, there are two villages with the simple name Vakuf (near Gradiška and Lopare) in which almost exclusively Orthodox Serbs live. These are just two cases from many others that indicate the multicultural intertwining of Bosnia and Herzegovina. Among other examples of typical Muslim oeconyms that denote predominantly Christian settlements, the following stand out: Hamzići near Čitluk, Alibegovci in Usora municipality, Šehovina near Nevesinje, Hasanbegovci near Glamoč, Eminovo Selo near Tomislavgrad, Čengići near Bijeljina, Alagići near Kreševo etc. On the other hand, there are numerous reverse examples too, i.e. cases of predominantly Muslim (Bosniak) villages with toponymy typical of Christian background, such as Jurjevići near Zenica, Križevići near Olovo, Svinjarevo near Kiseljak, Krstac near Jablanica or Todorovo near Velika Kladuša. Practically no attention has been paid to these unusual phenomena in earlier toponymic research.

We conclude the analytical presentation of the Ottoman layer in the toponymy of Bosnia and Herzegovina by noting that the heritage of this historical period is difficult for full comprenesion, given the wide range of appelatives of Turkish origin that found their place in the toponyms of this territory. However, it is worth (of) mentioning at least these oriental terms: *han* (lodging), *kula* (fort/tower), *ilidža* (spa) or *majdan* (mine). They are also included in numerous local toponyms: Han Pijesak, Han Bila, Simin Han, Han Ploča, Fazlagića Kula, Kula Banjer, Pješivac Kula, Kula Grad, Ilidža near Sarajevo, Ilidža near Gradačac, Tomina Ilidža, Bronzani Majdan, Vareš Majdan, Stari Majdan, etc. The practice of merging Islamic roots (usually proper names) with characteristic South Slavic suffixes for obtaining a family name (most often *-ić*) among the Muslim/Bosniak population is also interesting, which was also reflected in toponymical phenomena. It has already been mentioned that patronymics played a significant role in naming numerous oeconyms in the Middle Ages, especially at the village and hamlet level, which continued throughout the Ottoman period. In this way, external influences permeated domestic practice once again.

Toponymic Changes in the Nineteenth and Twentieth Century

By the nineteenth century, the network of settlements in Bosnia and Herzegovina had been considerably consolidated, and a relatively small number of new ones were established in the period that followed. The case of infrastructural upgrading of the existing ones and their expansion was more pronounced. In that way, some toponyms that has previously designated rural settlements gradually became urban. Between 1878 and 1918, the Austro-Hungarian administration did not significantly change the acquired toponymic basis, which was already common among the local population. Instead, it even contributed to its preservation since it made the first detailed topographic maps of this area through extensive field and cartographic work, without which a certain number of toponyms (especially oronyms) would probably be forgotten in time. The most significant change that the Austro-Hungarians made in naming local settlements was the addition of the prefix *Bosanski/-a/-o* (=Bosnian) to the names of cities on the northern border of Bosnia and Herzegovina, i.e. mainly along the rivers of Sava and Una. This was done to make a distinction between identical toponyms on the other side of the river. Thus were formed Bosanski Brod (opposed to modern Slavonski Brod), Bosanski Šamac (opposed to Slavonski Šamac), Bosanska Gradiška (opposed to Stara Gradiška), Bosanska Dubica (opposed to Hrvatska Dubica) and Bosanska Kostajnica (opposed to Hrvatska Kostajnica). The same prefix was given to some places in the wider border area, such as Bosanska Krupa, Bosanski Petrovac and Bosansko Grahovo. Other aspects of Austro-Hungarian influence are only visible in some towns/cities, naming parts of urban units (e.g. neighbourhood of Marijin Dvor in Sarajevo) and individual streets. The original German names were used in a very small number of cases, such as the village of Windthorst (present-day Nova Topola) in Lijevče Field in the north of country, where only immigrant Germans lived at the time.

In the Kingdom of Serbs, Croats and Slovenes (since 1929 Kingdom of Yugoslavia), there has been a radical turn in naming hodonyms, i.e. city street names, since the entire state-building narrative has changed. As this was the first joint state of the South Slavs, new national motives and domestic historical figures came into focus, mainly from the Serbian ethnocultural circle. However, oeconyms remain generally unchanged, with three significant exceptions. The town in the western part of the country, which from the Ottoman period was called Varcar Vakuf, was given a new name—Mrkonjić Grad. Namely, the majority of the city's population were ethnic Serbs, and a local initiative was adopted to pay tribute to Petar I Karađorđević, King of Serbia and later the first King of the Kingdom of Serbs, Croats and Slovenes, whose nom de guerre during the anti-Ottoman uprising in Bosnia and Herzegovina in the 1870s was Petar Mrkonjić. Another example refers to a small town at the confluence of Vrbas and Sava rivers, which until 1933 was called Svinjar (named after pigs or possibly wild boars). Then, a new name was chosen—Srbac, where again the critical role was played by the ethnic structure of the local population (Serbian majority).

The third case refers to renaming of Duvno (etymology explained within the analysis of pre-Slavic toponymy) to Tomislavgrad. In this case, the 1000th anniversary of the coronation of the first Croatian King Tomislav was marked, which, according to unconfirmed sources, was performed on Duvanjsko polje (Duvno Field).

During the era of the Socialist Federal Republic of Yugoslavia, there was an ideological reorientation of general socio-political narrative, which was now based on socialist values, the concept of 'brotherhood and unity' and the victory of the People's Liberation Movement in the World War II. Again, the new social reality was most easily reflected through changes in hodonyms, especially because they were once again thoroughly changed during the fascist occupation in the period 1941–45. The names of city streets, but also the names of local communities (the smallest administrative and functional units in the republics of former socialist Yugoslavia) were named at a large scale after the names of national heroes, or those who were the most deserving individuals (often killed in battle) well-known for resisting the occupying forces and their collaborators. At the same time, there was a great expansion of urban settlements, as a result of which some entirely new neighbourhoods were created. They most often assume the names of local toponyms, which previously referred to rural areas (e.g. Grbavica and Alipašino Polje in Sarajevo). However, almost no major toponymic changes ever occurred in the names of larger settlements. Then again, there are a few exceptions that can be noted to this general rule. One of them was the custom of renaming a town in each of Yugoslav republics in honour of Josip Broz Tito, the first and lifelong president of socialist Yugoslavia. In Bosnia and Herzegovina, Drvar thus became Titov Drvar (lit. Tito's Drvar). Among other changes, it is worth noting that Tomislavgrad was renamed again as Duvno, then that the newly established town of Novi Travnik in Central Bosnia was once called Pucarevo (in honour of Đuro Pucar Stari, a prominent national hero and post-war political figure), and that there were some changes also in the names of rural areas, such as Kraljevo Polje (lit. King's Field) near Han Pijesak, which was renamed as Partizansko Polje (lit. Partisan Field), or Hrvatske Živinice (lit. Croat Živinice) and Muslimanske Živinice (lit. Muslim Živinice), which were renamed as Donje Živinice (lit. Lower Živinice) and Gornje Živinice (lit. Upper Živinice).

Toponymy and Contemporary Socio-Political Contexts

The breakup of Yugoslavia, the declaration of Bosnia and Herzegovina's independence and the war in the period 1992–95 are the processes that have created a completely new political reality, within which many toponyms have acquired new connotations. Over the centuries, the choronyms of Bosnia and Herzegovina have played a generally neutral role in relation to ethno-religious identification of the local population. Despite the support from imperial administration, attempts to create a single Bosnian nation in the Austro-Hungarian period failed due to the lack of interest among the local population, which was already roughly divided along religious lines. On the other hand, there were no distinctively negative sentiments against these choronyms among (neither) none of the three largest domestic ethno-religious groups. However, after Bosnia and Herzegovina became the official name of the independent and internationally recognized country, there followed a rapid polarization in the collectivist perception of what it actually represents. Thus, Bosnian Muslims, as the largest people in the country (43.5% of the total population according to the 1991 census), began to perceive Bosnia and Herzegovina more clearly as their only homeland and as a safeguard of their own survival. Thus, in 1993, the leading political and intellectual authorities within this ethnic group made a definitive decision to assume the ethnonym Bosniak and call their language Bosnian, instead of using the term Serbo-Croat language, which had been in use until the dissolution of Yugoslavia. Although both Bosniaks and the Bosnian language have been used as terms at earlier times in history, this is the first time they have begun to be used as almost the exclusive determinant for a people which was simply referred to as Muslims in the past.

At the same time, the dominant Serb perception of the name Bosnia (and Herzegovina) became radically different. The majority of Serbs (31.2% of total population in 1991) boycotted the referendum for the independence of Bosnia and Herzegovina. As a result of their resistance to this process, which was mounted with the aim of rounding up a Serb territory that would remain part of Yugoslavia, a parastate was created that was originally called the Republic of Serb People of Bosnia and Herzegovina, and thereafter the Serb Republic of Bosnia and Herzegovina, whereas eventually in August 1992 only the name Republika Srpska (lit. Serb Republic) remained, which has symbolically marked the complete rejection of Bosnian name and identity, even in a regional sense of the word. Strong resistance was also put up against the term 'Bosnian Serbs', which was often used for example by a number of international media journalists during the war. This was followed by the systematic removal of the adjectives *Bosanski/-a/-o* (=Bosnian) from all toponyms of cities that were under control of the Republika Srpska authorities (Bosanska Gradiška—Gradiška, Bosanska Dubica—Kozarska Dubica, Bosanski Brod—Srpski Brod, Bosanski Šamac—Šamac, Bosanska Kostajnica—Srpska Kostajnica, Bosanski Novi—Novi Grad, Bosanska Krupa—Krupa na Uni, Bosanski Petrovac—Petrovac, Bosansko Grahovo—Grahovo, and Bosansko Petrovo Selo—Petrovo). With the Dayton Peace Agreement, the towns of Bosanska Krupa, Bosanski Petrovac and Bosansko Grahovo were nevertheless passed to the Federation of Bosnia and Herzegovina and kept their pre-war oeconyms. Other towns, which remained part of the Republika Srpska territory, keep their modified names even today, but due to a decision rendered by the Constitutional Court of Bosnia and Herzegovina back in 2004, the ethnic adjective *Srpski/-a/-o* (=Serb) also had to be removed (Feldman, 2005). Apart from Brod and Kostajnica, the same was also done with regard to the names of municipalities and towns divided by the inter-entity boundary line, parts of which were called Srpsko Sarajevo, Srpsko Orašje, Srpski Mostar, Srpski Kupres, or Srpsko Goražde during the war. Some of them were then assigned another geographical determinant, such as in the case of Istočno (=East) Sarajevo, while in other examples there have been paradoxical situations in which two neighbouring municipalities bear the same name (e.g. two municipalities of Trnovo and two municipalities of Kupres).

During the war, the political leadership of the Republika Srpska went even steps further in the practice of changing toponyms. As part of the ethnic cleansing campaign, several prominent urban toponyms of Turkish or oriental origin were changed in line with the Serb nationalistic agenda (Carmichael, 2002). A notorious example is the renaming of Donji Vakuf to Srbobran (lit. Serb Defender) during the period between 1992 and 1995. The mountain town of Skender Vakuf, named after its Ottoman founder, was renamed as Kneževo (derived from *knez*—a South Slavic noble title, prince or duke), which still remains as its official name to this (very) day. Brozović (2016) also points out a bizarre example of an ideologically motivated change of toponyms, which resulted in the mistaken annulment of their own historical identity by the wartime authorities of the Republika Srpska. They probably came to a false conclusion that the name of town of Foča on Drina River is of Turkish origin, which is why they replaced it with an entirely new oeconym Srbinje. However, Foča (originally shaped as Hoča) is a medieval toponym definitely much older than the Ottoman conquest of Bosnia, and it originates from an Old Slavic (most likely Serb) name Hotko or Hotimir/Hodimir. Ultimately, by virtue of the said decision, the Constitutional Court has restored to the town of Foča its original historical name. In Banja Luka, the largest city in Republika Srpska, names of settlements that refer to the Ottoman part of its history have been formally changed: Gornji Šeher (now Srpske Toplice), Mejdan (now Obilićevo), Budžak (now Lazarevo), Šehitluci (now Banj Brdo) and Hiseta (now Kočićev Vijenac). In the hodonyms of almost all cities of the Republika Srpska, there remains hardly any reference to non-Serb personalities. The policy of renaming the streets and squares is indeed similar in the majority Bosniak areas as well, although with a somewhat more visible ethnic heterogeneity of represented personalities and motives. On the other hand, the local authorities in such areas in general did not engage in changing the toponymy of higher ranks. There are only known micro-examples of nationalistically motivated toponymic changes, such as Vasin Han near the City of Sarajevo being renamed as Gazin Han or Čekrčići near the town of Visoko being renamed as Čekrekčije.

The identity relationship to the choronym Herzegovina is also quite interesting. Regardless of their particular ethnicity, the inhabitants of this southern part of the country still have a strong regional identity, but it has not been exploited in terms of the affirmation of political regionalism. The perception of many Bosniaks is that this horonym is still regionally subordinated to Bosnia. Some of them advocate even its omission from the name of the state, but such an idea has never been seriously considered in socio-political discourse, due to an immense number of other problems that impose themselves as a higher priority. Since it originates etymologically from the medieval nobleman title herzog of Saint Sava, which indirectly refers to the founder of the Serb Orthodox Church, there are no distinctive negative sentiments among the Serb population towards the choronym Herzegovina, like in the case of Bosnia. However, this horonym is still not seen in the national, but instead mostly in the regional perspective. The Republika Srpska entity includes the eastern and very sparsely populated part of this region, which is why this issue is of peripheral importance for the contemporary identity of ethnic Serbs in Bosnia and Herzegovina.

Croats, who represent the largest ethnic group in Herzegovina and who are particularly homogeneously concentrated in its western part, cherish the most distinctive form of identity toward this region. The political-territorial entity that assembled the largest number of Croats from Bosnia and Herzegovina during the past war was called Herzeg Bosnia, which is why it is quite clear to which part of national territory the primary emphasis was placed. Extensive hodonymic changes have also been made in predominantly Croat communities, with a focus on a number of different characters from Croat history (Palmberger, 2012). Interestingly enough, much attention was paid to the names associated with medieval Bosnia, which sought to highlight the message that it is the Croats who largely inherit the medieval (pre-Ottoman) legacy of this land (see Korać and Beus 2021). For this purpose, the name of the local medieval parish of Uskoplje was added to the multi-ethnic town of Gornji Vakuf, and a in a similar way (was done) with the neighbouring municipality of Prozor, whose name is now Prozor Rama. At the same time, once again a change was made with regard to yet another oeconym—the name Duvno was reverted to Tomislavgrad.

Conclusion

The heterogeneity of toponymic heritage of Bosnia and Herzegovina testifies to its complex history, but at the same time points to the common Slavic roots of its current inhabitants, or three ethno-religious groups that make up the majority of the local population and dominate its socio-political development. This makes this country a very interesting case study in research of multidimensional identity links and differences within the population of a unique area, and its toponymy is one of the most evident elements of cultural perception of space. The image of Bosnia and Herzegovina as a cultural crossroads between East and West, as well as between Mediterranean and continent is further strengthened by a unique mosaic of geographical names of different origins. Nevertheless, the dominant Slavic linguistic basis gives even the 'imported' terms used as part of the toponymic heritage a certain autochthonous character. On the other hand, recent socio-political processes have influenced a partial change in the perception and treatment of certain toponyms, primarily for the purpose of strengthening ethnic identities, but also for achieving certain nationalist goals. From the aspect of academic work, it is extremely important to have an unbiased approach to research in this study field, and not to act in accordance with certain political agendas, which is unfortunately often the case. We hope that this text has created an adequate basis for further analysis of more specific topics related to toponymy in Bosnia and Herzegovina.

Questions

1. Which historical layers can be distinguished in the toponymy of Bosnia and Herzegovina?
2. Should the official name of Bosnia and Herzegovina be simplified or shortened? Why?
3. Compare multiple cultural influences (on) the toponymic heritage of Bosnia and Herzegovina with some other country.
4. How did the Ottomans/Turks influence on the emergence of many toponyms in Bosnia and Herzegovina, as well as on modern identity issues?
5. Why did certain toponyms get recently changed, and what are the social and political the implications of such decisions?

References

Agency for Statistics of Bosnia and Herzegovina – BHAS (2013a) Census of population, households and dwellings in Bosnia and Herzegovina in 2013

Agency for Statistics of Bosnia and Herzegovina – BHAS (2013b) Systematic list of municipalities and settlements of Bosnia and Herzegovina

Brozović D (2016) Changing names: a way to cope with identity issues in times of crisis – the ottoman legacy in Croatia & Bosnia and Herzegovina. Ethnol Balk 19:239–255

Brozović RD (1995) 'Staroeuropska' hidronimija. Filologija 24–25:81–86

Carmichael C (2002) Ethnic cleansing in the Balkans – nationalism and the destruction of tradition. Routledge, Taylor & Francis Group

Dragić H (2013) Appeals in the toponymy of the Stolac region. Školski Vjesnik 62(1):5–26

Feldman D (2005) Renaming cities in Bosnia and Herzegovina. Int J Const Law 3(4):649–662

Ivšić D (2012) On oronyms in ancient sources (Examples from the Croatian pre-Slavic toponymy). Folia Onomast Croat 21:97–119

Ivšić D (2013) Pre-Slavic stratum in Croatian toponimy – Doctoral thesis. University of Rijeka, Faculty of Humanities and Social Sciences

Korać D, Beus M (2021) Medieval Bosnian-Hum rulers and nobles in the hodonyms of Herzegovina. Hercegovina 7:71–88

Mayer A (1959) Die Sprache der alter Illyrier, Band II: Etymologisches Wörterbuch das Illyrischen – Grammatik der illyrischen Sprache, Schriften der Balkankomission. Linguistische Abteilung XVI

Midžić F (2010) Borders and territory of Bosnia and Herzegovina from the Peace of Karlovac to the end of the first world war on old geographical maps (1699–1918). Association of Archival Workers of Una-Sana Canton 2010

Nikić D (1980) The oronyms Slovini and Cincar – Two characteristic oronyms from southwestern Bosnia. Geogr Rev 24:89–99

Palmberger M (2012) Renaming of public space: a policy of exclusion in Bosnia and Herzegovina. MMG Working Paper 12–02. Max Planck institute for the study of religious and ethnic diversity 2012

Skok P (1971–1974) Etymological dictionary of the Croatian or Serbian language I–IV. JAZU 1971–1974

Šabić I (2016) Typological-structural analysis of toponyms in the medieval Bosnia. Bosan Jez 13:35–54
Vego M (1982) Postanak srednjovjekovne bosanske države. Svjetlost

Geo-Histories, Legacies, and Toponymy Transitions

Giving Identity to Space Through (Re)Naming: Practice of Village Renaming in the Period of the Republic of Turkey

Yılmaz Arı

Abstract The Ottoman Empire, lasting until the end of WWI (b), included multi-ethnic groups that often lived nearby simultaneously or in geographical areas consecutively. The Ottomans often continued to use the place names prior to their incorporation into the Empire. This situation began to change when the Union and Progress Party promoted renaming the villages in the 1910s. While some 'official' decisions were taken in this regard, there was no significant development within the Empire that witnessed a period of unstable administration, chaos, and wars. The founders of the new Republic of Turkey attempted to build a nation-state and considered name changes as part of this effort. Several studies have been done on the Turkishization of toponymy, focusing on different aspects of renaming implemented by the republicans. The renaming efforts included the decisions in the 2010s, allowing the use of some previously banned names. This work attempts to evaluate multiple dimensions of these initiatives: main motivations to change place names; criteria used; connection or continuity between new and previous names; and how ordinary people perceived name changes. Research is based on the laws, regulations, guidelines, previous studies on renaming, and the author's first-hand observations.

Keywords Toponymy · Renaming · Politics of renaming · Republic of Turkey · Critical toponymy

Objectives

- The aim of this chapter is to summarize the place name changes in Turkey in the last 100 years and the motivations behind this.
- By the end of this chapter readers should be able to learn the history, chronology and motivation for renaming in Turkey.
- Readers will learn to recognise the sources of the rich and diverse toponyms in Turkey.

Y. Arı (✉)
Department of Geography, Bandırma Onyedi Eylül University, Bandırma, Balıkesir, Turkey
e-mail: yari@bandirma.edu.tr

© The Author(s), under exclusive license to Springer Nature Switzerland AG 2023

G. O'Reilly (ed.), *Place Naming, Identities and Geography*,
Key Challenges in Geography, https://doi.org/10.1007/978-3-031-21510-0_16

Introduction

In Van, one of the easternmost provinces of Turkey, the old names of some settlements were surprisingly given back to places with a decision taken by the Metropolitan Municipality Council in November 2014. With the work of a local commission, the original names of 704 villages in Kurdish, Armenian, or other languages were returned (Sehrivan 2021). Accordingly, the old names of these settlements were (started) now to be used in official correspondence, and all road and map signs were changed. The interesting thing about this seemingly normal situation is that while the old names of these settlements were changed with Turkish names during the last century after a fierce debate with the central government's decision, the authority to return to the old names was transferred to the local municipal councils. Similar changes were made at many other settlements in the eastern part of Turkey in the past few years. Thus, the names of the settlements were subjected to a change again. The late Ottoman rulers initiated the previous period of change almost 100 years ago (Özdemir and Gül 2013).

The Ottomans often continued to use the place names before their incorporation into the Empire. This situation began to change when the Union and Progress Party gained some power in the first decades of the twentieth century and promoted the renaming of the villages. These names were changed at the end of a process that started in the first decades of the twentieth century and continued throughout the century. During this Turkishization process, many names were changed from Arabic, Kurdish, Bulgarian, Georgian, Armenian, Greek, Laz, Zazaki, and Syriac, even in languages that are no longer alive today. Even among these, Turkish names were changed unknowingly (Eröz 1984). However, among the people living in the Anatolian peninsula for centuries, there was a tradition of using place names by adapting them to the local language without changing them. Although various civilizations lived in Anatolia throughout recorded history, place names were not changed as such; they were only adapted to the local languages. This is more common in city names. For example, Smyrna to Izmir; Hadrianapolis to Edirne; Ankyra to Ankara; Mobola to Muğla; Prusa to Bursa; Iconion was transformed into Konya (Kütükoğlu 2012). This is also the case after place names were officially changed, and the public rarely used the new names entirely. Whether it was changed or not, the old names of the villages are also used by people in daily life (Alkan 2017). Ordinary people often do not know or adopt the new names given to the old settlements (Fields 2013).

While it was thought that the process of Turkishization of place names would continue forever, this situation was suddenly reversed in the 2010s, and the old names of many settlements began to be returned. In this context, the names of thousands of places were changed again with the old names reinstated. The names of these settlements in other languages began to be used instead of Turkish names. Sometimes it is difficult/impossible to determine in which language the name is since the place names in Anatolia have been used with minor changes by different ethnic and cultural groups without much change throughout the long history of settlements. Due to this confusion and difficulty in correctly identifying the origins, some Turkish names were also changed over the last century.

So why is naming essential, and why were place names changed? Naming is giving identity and personality to the space and place. The people living in a place, name it according to their culture, tradition, and lifestyle. Naming also indicates what the namers would like to establish in their imaginative geography. However, such a country is almost non-existent because there is hardly a region where the settlement names are entirely in the imagined language. Thus, it is seen that the processes of constructing identity with ideological motivations do not match with reality, most of the time. Place names, part of intangible cultural heritage, represent essential values in a society that constitute an intellectual wealth for national understanding and international public opinion. In general, it can be said that in the cultural context, they reveal to us an entire history of past, recent and ongoing processes, in the relationship between people, culture, and nature (Hausner 2013). It is possible to gather information about the historic and ethnic character of the rural settlements by examining the village names. Therefore, changing the village names makes it difficult to reveal the old situation and characteristics of the village; in a way, it breaks the connection between the village and its past with the name (İbret 2003). For this reason, the continuity of place names is crucial, and the loss of this continuity will mean the loss of the elements of intangible cultural heritage that the place name brings with it. This study will deal with the continuity and change in toponyms in Turkish villages.

Several studies have been done on the Turkishization of toponymy, focussing on different aspects, terms, and cultures of renaming implemented by the republicans. This work evaluates multiple dimensions of these initiatives: main motivations to change place names; criteria used; connection or continuity between new and previous names; how ordinary people received name changes; and local reactions. This chapter explores answers to such questions. Research is based on the laws, regulations, and guidelines on renaming, previous studies on the subject, and the author's first-hand observations in different parts of Turkey.

Toponymy in Anatolia Before the Republican Period

Settlements and Practice of Toponymy in Anatolia

The Anatolian peninsula is one of the oldest settlement places in human history. There are centres in every corner of Anatolia, where steps such as agriculture and domestication of animals, which enabled people to move from hunter-gatherer societies to settled order, can be followed. These religious, commercial, and culturally significant settlements date to 12,000 BC. After these first settlements, the Hattis, Hittites, Phrygians, Urartians, Ionians, Lydians, Assyrian Colonies, Trojans, Byzantines, Greeks, Seljuks and Ottomans, and many more ruled in Anatolia, respectively.

Each of these peoples migrated to Anatolia or became a continuation of the civilizations that still existed there; hence syncretic cultures developed. Throughout this long history of settlement, Anatolia was also invaded by various tribes temporarily; Persians, Macedonians, Crusaders, and Mongols are but some of them. The settlement history and place names of Anatolia bear traces of these different civilizations, states, and invasions.

Anatolian peoples used the names given in the past by adapting them to their language without much change (Kütükoğlu 2012). These place names can be in the language of any of the communities that lived in Anatolia. These communities can be in one of the languages that are no longer alive such as Hittite, Phrygian, Luwian, and those whose remnants or specters continue today. Until the beginning of the twentieth century, the language of the place names was not really questioned. The names of the settlements that persist in the official records are used precisely or similarly. The names of the abandoned settlements, which were turned into ruins, were sometimes forgotten and could only be discovered by examining the ancient records and archaeological excavations. In this sense, old writings such as the book—Geographica by Strabo of Amasya gain a reputation. For example, a find unearthed during archaeological excavations proved to be the place location of the ancient city of Daskyleion (Arı 2021). From the beginning of the twentieth century, place names in Anatolia began to be systematically converted into Turkish. In this case, how long does the history of Turkish place names in Anatolia extend, and how many of the place names have been changed or Turkified.

The existence of Turks in Anatolia is much older than is often thought. It has been determined that Sakas, one of the early civilizations on Earth, were of Turkish descent and in Anatolia in the fourth century B. C. (Kırzıoğlu 1984). During the Roman and Byzantine periods, Christian Turkish tribes such as Pecheneg, Cuman-Kipchak settled in parts of Eastern and Central Anatolia, starting from Bartın on the Black Sea coast in Anatolia. The settlement of Saka, Part, and Cimmerian Turkish tribes, which is also seen in some regions of Eastern Anatolia, is also quite old (Gülensoy 1996). These expeditions were carried out only in the form of military reconnaissance operations, far from the aim of establishing a homeland. The raids after the eleventh century, on the other hand, were planned and aimed at establishing a homeland in Anatolia. By then, Turks began to settle in Anatolia systematically and permanently with this westward migration. From research, the year 1071 is accepted as the turning point in this settlement movement.

After the Battle of Manzikert in the eastern part of Anatolia in 1071, Turkish communities who migrated to the west and groups that came with the Mongol invasion established new settlements in Anatolia. During the era of Great Seljuk and Anatolian Seljuk states, Turks were brought to Anatolia in masses and settled. As a result of this migration movement, some of which caused violent conflicts and wars, a significant part of the local people fled to the islands in the north and west, and Anatolia was mostly Turkified, except for some large cities (Kütükoğlu 2012). The number of people who came from the east and settled in Anatolia has increased rapidly since the eleventh century. These newcomers, who were nomadic in the past, settled down in time and established new settlements in Anatolia. They gave new names to these settlements according to their own tradition and naming culture (Yediyıldız 1984). Did they have a system of naming places? The next section explores this question.

Naming and Renaming as Identity Construction

There are basic tendencies that the Turks follow in the naming of places. Like geographers, nomadic, semi-nomadic, and settled Turks have been careful in naming spaces and places from ancient times. They gave meaningful names to both places and settlements, whether in planned or randomly established settlements, and this attention and care in naming (are) is always remarkable. Official authorities have also registered these names. It is possible to see the settlements named with this approach dispersed from Central Asia to the Balkans (Eröz 1984).

According to Eröz (1984), Turks firstly give the name of the tribe or community they belong to, to the newly established settlements. Today, many settlement names in Anatolia are the names of tribes of Turks when they were in Central Asia. Common names such as Kınık, Bayat, Kayı, Salur are such names. Religious influences are also seen in place names. Religiously referenced words such as Sheikh, Hodja, Molla, Haci, Tekke, Pir, Madrasa reflect this practice. In addition, the Turks consider the geographical elements around the settlement while giving their place names. Place properties such as mountainous, hilly, plain, slope, forest, rocky, barren, wet, arid, windy, etc., can be reflected in place names. At the same time, names such as white, gray, red, black can be given according to the colour of the surrounding soil and water. Names can also be given according to the landforms. In this case, names such as throat, nose, collar, cliff, and plateau draw attention. Natural woodlands that grow in a region, one of the physical geography elements, can also be the name of the settlements. Names such as pine, elm, oak, and poplar are the first to come to mind in this sense. Some settlements take their names from the agricultural products grown. Some of them reference wheat, rice, oats, pears, apples, olives, plums, and almonds. These product names turn into settlement names with some suffixes. The crafts that the people living in a place mostly do, can be reflected in the place names there also. Occupational names are also reflected in place names. Place names such as *Kazancı, Kovancı, Demirci, Okçular* serve as examples for crafts and professions. Thus, people demonstrate their culture by naming it in the landscape; on the other hand, they reflect the landscape's natural elements in their culture by giving their names to settlements.

On the other hand, history plays a vital role in naming places. The historical names are given by the civilizations that lived in Anatolia before the Turkish arrival are usually not changed but adapted to Turkish. In addition, the names of people who demonstrated heroism in history are given to the settlements. For example, there are many settlements in Turkey today named after the founding figure of the modern Turkish state, Mustafa Kemal Atatürk, or the Ottoman Sultans. In addition, some places where important historical events took place can be given names that remind

Table 1 Colours and directions in old Turkish

Colour	Direction
beyaz-ak (white)	batı (west)
kara-siyah (black-dark)	kuzey (north)
kızıl-al (red)	güney (south)
mavi-yeşil (blue-green)	doğu (east)

Source Arı (2012)

us of that historical event. It is often possible to identify locations inhabited in a period in history but are now abandoned by examining their names. These settlements are usually called ören, harabe or viran, indicating their ancient character. These names are preserved in contemporary settlements established at, or very close to their original places (Yediyıldız 1984). In areas with historically significant castles, fortresses, and mounds, settlements with these names were established (Başkan 1970).

Interestingly, a different way is followed in naming the seas. In old Turkish, each direction is paired with a colour, as shown in Table 1. According to the person who gave the name, the direction of the sea determines the name given to that sea (Arı 2012).

The seas around Anatolia were named according to their direction. The Red Sea in the south of Anatolia is called Kızıl Deniz, and the Aegean Sea, which coincides with the west of Anatolia, is called Ak Deniz, as it was once thought to be a part of the Mediterranean, and the Black Sea in the north is called Kara Deniz. Thus, the terminology is compatible with the direction colours. The Marmara Sea, surrounded by land on all sides except for the Bosporus and the Dardanelles, has taken this name because of the marble mostly extracted (in) from the islands inside it.

According to historians, the Turkish tribes came to Anatolia for reconnaissance purposes without an organized power much earlier than the Battle of Manzikert in 1071. But that war was the beginning of mass migrations and a turning point in the Turkifying of Anatolia. After the Seljuks, migrations from the east continued during the Ottoman period, and new tribes and groups came. While some of these newcomers lived a nomadic life, some settled down and established new villages. There are thousands of villages spread over the Anatolian geography whose names have not been changed in any way for centuries. These communities, which have settled down, have not given up on the nomadic life so to speak. They have established smaller settlements near the villages, moving in the summer and returning to the village in the winter. There are thousands of Turkish words in the names of these settlements.

These newcomers sometimes established new settlements; however, they mostly lived in the same villages with different groups and ethnicities living in Anatolia. The old names of these communal villages were never changed until the beginning of the twentieth century. In most of these settlements, people lived together despite different religions, ethnic origins, and cultures. As stated, the emergence of settlements was largely related to migrations and so Anatolia is a place where immigrants from the East and West often settled.

During the time of the Seljuk Turkish Empire (1037–1308), which ruled in Anatolia before the Ottomans, 681 place names with the Turkish parlance inn/caravan/palace/caravanserai were identified (Gümüşçü and Yiğit 2020, Fig. 1). In other words, there was a Turkish settlement system in Anatolia about 1000 years ago. Before establishing the Turkish Republic in 1923, there were thousands of settlements with Turkish names in Anatolia. The rate of settlements whose names were changed is around 35%, which means the rest had modern Turkish names. However, studies have always focused on the changed place names, and the unchanged place names have been neglected.

In the following periods, non-Turkish people settled in some settlements, which were transferred from the Seljuks to the Ottomans, and whose names were in Turkish in the Ottoman archival records. This issue, which is not mentioned much in the international literature, is the subject of many academic studies in Turkey. Accordingly, with the settling of non-Turks, the names were sometimes changed or adapted to the local language. This situation occurred due to the migration of the Turks who established the settlement in other places. For example, some Turkish groups who migrated to Anatolia earlier went back to Iran by reverse migration in the sixteenth century (Erpolat 2004). Others settled in their settlements and adapted the place names to their language. For example, a study examining Ottoman archival records revealed not many Kurdish place names in Diyarbakır province at the beginning of the sixteenth century. Kurdish names started to emerge (merely) alongside the localization of Turkish names. It has been determined that the word Amid (usually expressed as Amed in Kurdish), the old name of the city of Diyarbakır, as claimed by some, is not Kurdish but a word belonging to one of the previous civilizations (Erpolat 2019). Similarly, in a micro-scale study on the population and settlements of a town centre in the Black Sea Region in the northeastern part of Turkey in the sixteenth century, Demir (2011) extracted the statistics of the population living in the region from the Ottoman archival records. According to the course of historical events, there should have been a large number of Turkish-Muslim populations. However, archival records determined the proportion of the Muslim population to be 2.82% in 1518, 15.27% in 1530, and 23% in 1591. This situation did not coincide with the perceived historical realities, and there was a situation that needed to be explained. Demir analyzed the names of the village settlement to figure out who the first settlers were. The archival records indicated a significant part of the population was non-Muslim, yet the names of the villages were primarily Turkish. According to archive records, 94% of the names of the settlement in 1518; 90% in 1530, and 83% in 1591 were Turkish. So how was such a thing possible? Demir answers this by examining the political structure of the time. In the 1500 's, this region was the scene of first Akkoyunlu-Karakoyunlu and then the Ottoman-Safavid struggles and military conflicts. Due to the balance of power in these conflicts, the settled Turks migrated to other places. In other words, the Turks who established the settlements here migrated, but the villages continued to live with the same name. After the Ottomans completely dominated the region, the Turks returned, and according to the detailed Avarız registers in 1642, the Turkish population became the majority again. Undoubtedly, this situation cannot be generalized to the whole of Anatolia. Still, it

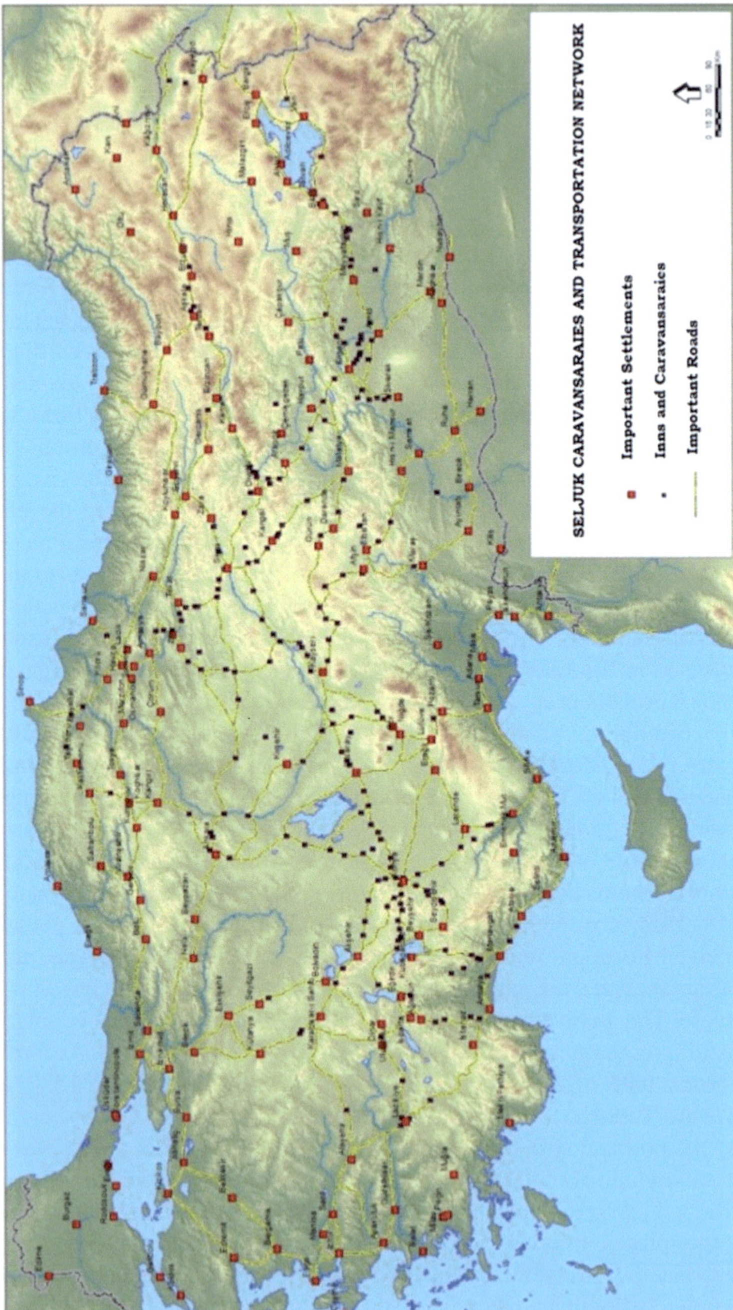

Fig. 1 The Seljuk Empire period (1077–1308 AD): inns and caravanserais in Anatolia. *Source* Gümüşçü and Yiğit (2020)

reveals that the names of settlements can tell a very different story than they first appear.

A reason for abandoning Turkish names is that some Turkish tribes adopted the identity and language of other groups living around them (Özdemir 2013). There are many studies in Turkish about this situation, which are not really encountered in the international literature. These studies have revealed through archival documents and fieldwork that some Turkish tribes lost their identity and became Kurdish under the influence of the Kurdish tribes living around them (Özdemir 2013).

It is helpful to remember (REMEMBER) here that being Turkish was not cared about or encouraged by (BY) the Ottomans as they were empire- builders encompassing many races, ethnicities and cultures (AS THEY WERE EMPIRE BUILDERS ENCOMPASSING MANY RACES, ETHNICITIES & CULTURES), and that it was not seen as different from other ethnic groups under (UNDER) the Ottoman administration. The essential elements of the Ottoman Millet system were Muslim and non-Muslim communities. There was no advantage to being Turkish. The issue of Turkishness and, accordingly, Turfication (TURKIFICATION) of place names came to the fore with the independence of some ethnic groups in the Balkans after the rise of nationalism during the nineteenth century (Batsashi 2019). They also adapted place names to their new languages. Therefore, the Anatolian settlement history is quite old and complex, and it is not always possible to accurately determine the origin of place names. Therefore, studies on place names can be highly controversial and political.

The practice of changing place names came to the fore only before the First World War, after the Empire lost large areas of land under the influence of nationalist movements. The first attempt was implemented after the "Regulation of Settlement of Muhajir" was issued on May 13, 1913 (Kocacık and Yalçin 2008). In the instruction of Enver Pasha, a leading figure of the Union and Progress Party, dated January 5, 1915, is stated: "the names of provinces, districts, villages, mountains and rivers in Armenian, Greek and Bulgarian languages in the Ottoman lands should be converted into Turkish by taking advantage of the positive opportunity offered by the wartime." In this instruction, how the name change should be made was also explained with the examples given. Some place names were translated into Turkish until this instruction was abolished on June 15, 1916. On this date, the Ottomans' practice of changing place names was abandoned. This is because place name changes were thought to cause confusion and misunderstanding during the war.

Aiming at building a nation-state around Turkish identity and led by Mustafa Kemal Atatürk, the Ankara government made some attempts to change place names from the beginning. On December 20, 1920, before establishing the new Republic, an Izmir deputy brought the issue to the agenda, complaining that the place names remained "non-national" in the newly established "national state." A deputy from Isparta, supporting these efforts, submitted a proposal to change the names of villages bearing foreign names. After these discussions, the names of some places were changed in 1922 on the condition of being limited. However, concrete steps to change place names were taken after the establishment of the Republic.

Practice of Place Name Change in the Republican Era (1924–2013)

During the first years of the establishment of the Republic, the practice of name change continued to be made without a political focus and a systematic approach. In this sense, some city names (for example, Kırkkilise-Kırklareli) were changed in 1924. In 1925, in the province of Artvin, located in the north-eastern corner of Turkey, the names of villages, the majority of which were in Georgian, were replaced with Turkish names by a decision of the province's council. In 1926, some city names were changed for practical reasons. Until that date, some provinces in Turkey had two names, and these names were confusing daily usage. Therefore, in this context, Canik was changed to Samsun; Saruhan to Manisa, Kocaeli to Izmit, Hinge to Muğla, Bozok to Yozgat, İçel to Silifke, Görle to Görele and Enderin to Andırın.

Another attempt was made to change the place names in Hatay. With the circular numbered 8589 prepared by the Ministry of Interior in 1940, it started the practice of "replacing the names of settlements and natural places with Turkish names that come from foreign languages and roots and cause great confusion in their use." The purpose of this circular was to change the Arabic place names into Turkish in the state of Hatay, which decided to join Turkey in 1939. However, as the Second World War continued, such issues were ignored in those years, and there was (**LITTLE**) little development in this direction (**AFTERWARDS**) afterwards. (NONETHELESS, W) Nonetheless, when WWII was over, some Arabic place names were changed into Turkish in 1947 in Hatay.

The legal regulation, which constitutes the infrastructure for the mass change of village names in Turkey, was enacted in 1949. With the Provincial Administration Law, No. 5442 passed that year, the process of changing place names was given a legal basis. However, when a significant change in government took place in Turkey in 1950, no new initiatives were made in this regard. In order to implement this law, eight years after its publication, the "Special Commission on Name Change" was formed and started to work. This Commission was composed of representatives from the General Staff, the Ministries of Interior, Defence, Education, and Ankara University. As a result of intensive studies, the Commission examined approximately 75,000 place names and changed about 28,000. The number of villages whose names were changed was around 12,000 (Tunçel 2000). The Commission also changed the names of approximately 2,000 natural places between 1965 and 1976 (Fig. 2). However, as expected, many erroneous practices were made during this name change. In the end, the Commission was abolished in 1978 on the grounds that it "changed the names of places with historical value." As a result of the work of the Commission, approximately one-third of the total number of villages in Turkey has been changed. The regions with the most remarkable change are the Black Sea Region and Eastern and South-eastern Anatolia Regions. The Georgian names were first changed in the Black Sea region, and then the Greek names were changed. In the Eastern Anatolia Region, non-Turkish names, predominantly Kurdish, Zazaki, and Armenian were changed. There were also hundreds of Turkish names among them (Eröz 1984). Apart

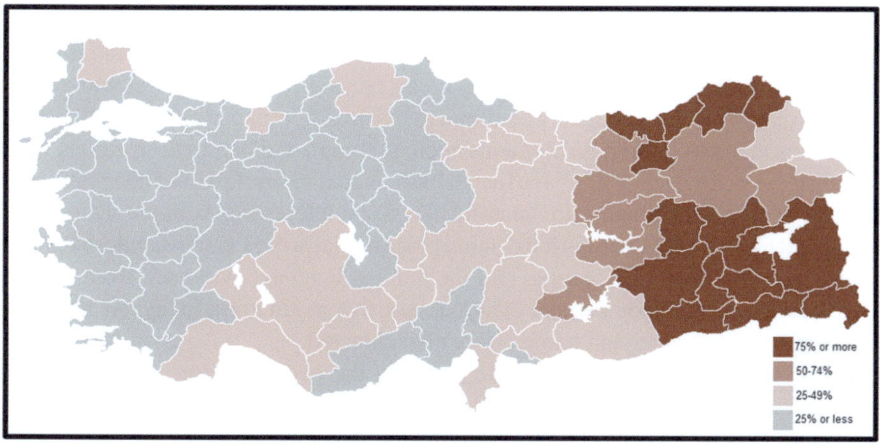

Fig. 2 Map showing the ratio of names changed after 1913 to total names in Turkey. *Source* Nişanyan (2011)

from this, the names of villages containing the words from Kurdish, Georgian, Tatar, Circassian, Laz, Arab, and Muhacir have also been changed, in order to avoid their being used to promote separatism by individuals or extremists in their environment (Tunçel 2000).

Although the commission's work was stopped, changing place names continued after the 1980 military coup. It seems that some place names were changed in this period depending on the sensitivity of the soldiers who took over the administration. In this period, the names of 280 more villages were changed (Tunçel 2000). These recent changes mainly reflected the political views of the 'specific' military administration. Some place names have been changed because they were thought to have fundamentalist connotations and contradicted the principles of the secular republic, while others have been changed because they were thought to evoke the ideology of communism. For example, place names with the word "sheikh" in them evoke radicalism, place names with the word "red" have been changed because they evoke communism. So why were the place names changed, and why were they transformed at that time? There were specific arguments used by the administrators when changing place names. The first is that each nation names its place to give (**s**) it an identity and reflects its character there. Every group holding political power uses (**does**) similar practices. When many places that used to be Ottoman Empire territory fell out of the hands of the Ottoman control, the place names were changed by new political forces (Hocaoğlu 2016). Even people were forced to change their names until 30–40 years ago, as in Bulgaria. Therefore, this practice is quite common.

The second reason is that some groups who gave names to the settlements no longer inhabit this geography. Just as millions of people came and settled in Anatolia from the lands left by the Ottoman Empire (Kocacık and Yalçin 2008), the people living here migrated to other places over time, settled in other locations with population change, or were established themselves in other areas by being subjected to forced migration. While migrating to and from Anatolia, people experienced great difficulties due to epidemics, banditry, extreme weather conditions, hunger, and famine, and some of them lost their lives. However, it can be said that whatever ethnic groups lived in Anatolia at that time, they continue to live in different parts of Anatolia today. The places that have Christian elements such as church and bell in their names were also changed because no Christians live in these places today. It would be appropriate to mention the Persian names in Anatolia here as an interesting example. During the Seljuk and Ottoman Empire periods, some place names were Persian because Persian culture was dominant in Anatolia, and the Persian language was used from time to time. However, people of Persian origin do not live in Anatolia today. As a natural consequence of this, Persian names were also changed.

However, the changed names were not only the perceived foreign ones. Some Turkish names have also been changed for different reasons. These changes largely depend on the ideological motivations of the period in which the change was made. Some Turkish names were changed, assuming they came from foreign roots since there were not enough cognitive studies on them. Names that do not fit the dialect of Istanbul in Turkish have also been changed. Turkish names reminding the traces of Shamanist Turks are also among the ones that have been changed. Some other names have been changed because they have a negative meaning. Village names such as *Atkafası*, *Çakal*, *Çürük*, *Deliler*, *Domuzağı*, *Haraççı*, *Hırsızpınar*, *Hıyar*, *Kaltaklı*, *Kıllı*, *Kötüköy*, *Kuduzlar*, *Sinir*, *Zurna* are of this type. Although some of them do not have slang or negative meaning initially, they need to be changed because they have shifted in meaning over time (Özdemir and Gül 2013).

A Special Commission on Name Change consisted of representatives of various institutions as required by law and had a bureaucratic working order rather than a scientific one. The establishment and structure of the board were determined with a top-down approach and did not allow much involvement of subject experts in the process. As a natural consequence, serious unwanted changes were made. Some of these are also contrary to the primary motivation of the name change attempt, which was to decrypt the names into Turkish, but the Commission changed the many village names in Turkish. In addition to this, many place names that did not include a 'Turkish word' in their names but are understood to be related to Turkish tribes have also been changed. However, the fact that these new names were thought of independently of historical processes and were artificial, caused them not to be adopted. It was later understood that some of the changed names were actually of Turkish origin, and some were restored. However, this has only increased confusion about renaming.

In order to evaluate all these studies on place names, a Turkish Place Names Symposium was held in Ankara between 11 and 13 September 1984, and the practices of changing place names in Turkey were critically evaluated. As a result of this, it was revealed that there is now a policy concerning changing place names, but interesting mistakes were made in practice. For example, Eröz (1984: 43–44) made the following conclusions:

> In the last 40–50 years, many Yörük (Turkish) tribes took place and established villages with their own efforts. They gave the names of the villages they founded according to the old tradition: they either took the names of their tribes or adopted the name of a well-known

> person. In the meantime, it was also seen that the names of animals bearing ancient totemic traces and place names in Central Asia were used. Despite this conscious act of our people, which can be praised and admired, some officials of the Ministry of Interior did not pass the test very well. While non-Turkish village names were changed, many Turkish names were also sacrificed. We want to believe that this is due to a lack of historical, sociological, and ethnographic knowledge but not a purposeful act. Many want to snatch Eastern Anatolia from us. On the other hand, we have destroyed old Turkish words that will prove the Turkishness of the place and have replaced them with rootless and baseless names. Although these new words are also Turkish, they bear the brunt of the destruction of the old ones and do not have a meaningful past.

In other words, the Turkish identity has become less visible due to the works that are expected to inscribe the Turkish name and identity on the landscape. This practice shows how complex the renaming issue was, and the work was interrupted. In the 2000s, the renaming was handled with a different dimension, and a reverse process began.

The Process Reverses: Democratization and the Return of Old Names (2013 on)

Discussions on the need to return old names started in Turkey after the 2000s. On these dates, Kurdish politicians especially attempted to return the settlement and place names. A deputy of the predominantly Kurdish Democratic Society Party (DTP) submitted a law proposal to use place and settlement names with 'old' names (Hurriyet 2008). Although this proposal was widely discussed in public, it was not enacted. However, Erdoğan, who was the prime minister of the Republic of Turkey at that time, said on 30.09.2013:

> We are making it possible for villages to retake their historical names, which they used until the 1980s, by removing the imposition clause in Article 2 of the Provincial Administration Law of 1949. As stated in the current law, the change of village names will be with the approval of the Ministry of Interior. In order to change the names of provinces and districts, legal regulation is required according to the current law. Such requests will be resorted to in this way. As the Government, we will consider the requests to change the names of provinces and districts.

This statement was made just before a period called the "Solution Process," which aimed for the members of the separatist PKK 'terrorist' organization to lay down their arms and engage in politics on a legal basis. In this process, which is a kind of peace process, the representatives of the Turkish government met with PKK representatives through official means for the first time and agreed that the PKK should lay down its arms in response to democratization steps. The process was enacted by being published in the Official Gazette on 16 July 2014 under the Law on Ending Terrorism and Strengthening Social Integration. According to this, the PKK armed forces would withdraw from Turkish territory, and the armed struggle would be ended. On the other hand, it has been determined by this law that the government will carry out the necessary work on the following issues within the scope of the resolution process:

- Making new reforms in political, legal, socio-economic, psychological, cultural, human rights, security, and disarmament aimed at ending terrorism and strengthening social cohesion.
- Deciding to make contacts, dialogues, interviews, and similar studies with people, institutions, and organizations in the country and abroad, if deemed necessary, and assigning the persons, institutions, or organizations that will carry out these studies.
- Taking the necessary measures to ensure the return of the members of the PKK who laid down their arms, their participation in social life, and their integration into the society.
- Providing accurate and timely information to the public.

The Kurdish Initiative required a paradigmatic change in the official view on the Kurdish Question. For the first time, official actors acknowledged that the Kurdish Question is not just a matter of terrorism or violence and that other approaches rather than security-oriented measures should be applied. Discussions worked well at the official and elite levels but failed to achieve a similar change at the social base (Köse 2017).

This process, which continued between 2013 and 2015, did not last long, and the democratization steps failed. In this process, the parties' expectations from each other were not met, the PKK saw the armed threat as a way, and the government, which could not find support from the popular base, preferred not to take any further steps. Returning the old names of the villages was discussed as part of the larger negotiations. Consequently, no systematic study has been conducted on place renaming. However, in 2014, many municipal councils decided to use the old names of the villages, and official correspondence, road signs, and signboards were changed accordingly. Although the Kurdish Initiative has ended, the village names have not been changed again, and the old names continue to be used now. Since the change of city names requires the central authority's decision, there has been no development in this direction.

However, since the central government did not have a systematic approach to the issue, the names of villages were changed by local decisions, and their area of influence remained limited. Thus, we do not have precise data on how many villages were renamed in the region during the process. Scientists who do toponymic research on the region indicate that some names that are thought to be in Kurdish, Armenian, or Turkish are in fact in one of the languages of the earlier civilizations such as Hittite, Hurrian, Urartu, Assyrian, Babylonian, Scythian, Parthian (Persian), Selefkos, Helen, Sassanid Roman, Byzantine, Hebrew, Syriac, Arab that was dominant or settled in Anatolia and Mesopotamia at certain periods.

These views, which are defended on the thesis that local cultures and ethnic diversity should be kept alive with a pluralistic approach, have caused an important debate in Turkey. Accordingly, some argue that place names should be returned to their original form. In contrast, others say this is not a case of place names but is part of a larger plan. They think this plan is a pillar of a political project that aims to divide Turkey into provinces and autonomous governments and then into independent states.

According to this view, this project, managed from a common center, continues not only in Turkey but also in Iran, Syria, and Iraq. The aim is to support building a new nation by consensus building and parity of esteem by changing place names in these countries. According to them, a Kurdish state is planned in the region, including parts of other countries (Özdemir and Gül 2013). Hence the eternal issue of who exactly is 'Us Vs Them', and 'who' or what elites, are enunciating these socially-constructed differences, in the centrifugal—centrifugal balances in identity politics and nation-state building.

Conclusion

Anatolia hosted many empires and/or states throughout its thousands of years of settlement history. Due to its strategic location between the East and West, rich mines, fertile agricultural lands, diverse land structure, rich water resources, and other reasons, it was invaded by many others. Throughout history, communities that spoke Hittite, Hurrian, Urartian, Assyrian, Babylonian, Scythian, Parthian, Persian, Seleucid, Hellenic, Sasanian, Roman, Byzantine, Hebrew, Armenian, Georgian, Syriac, Kurdish, Arabic, Turkish, and some other languages occupied the land for a certain period. Today, place names in Anatolia are like the collective memory of all these different communities, including parts of intangible cultural heritage from previous civilizations. If place names are taken as clues, significant information about important events, ethnic structure, history, geography, and living groups of people can be acquired. It is known that in Anatolia, toponyms were used only by adapting to the dominant local language, without undergoing a significant change until the beginning of the twentieth century. This is true not only for non-Turkish place names but also for Turkish ones. Undoubtedly, dual naming also occurred in some places.

At the beginning of the twentieth century, the project of changing place names into Turkish, which was initiated in the last decade of the Ottoman Empire, continued intermittently until the end of the century in various ways. The aim was to inscribe the Turkish cultural identity on the land. At the end of this process, more than 10,000 village names were changed to Turkish names. These names were mainly Kurdish, Greek, Armenian, and Georgian. However, in a place like Anatolia which has been inhabited for more than 10,000 years, it is impossible to determine the origin of place names with complete assurance without sufficient expertise and detailed scientific studies. A specialist group of people consisting of linguists, historians, geographers, archaeologists, sociologists, and political scientists is needed to analyse place names accurately. Since there was no such expertise, hundreds of Turkish names were cut from their historical roots during the twentieth century in Turkey and replaced with other meaningless Turkish names determined by bureaucrats at the desk. When the experts noticed this, some of these Turkish names were returned again, and the new Turkish names were canceled. The people have not accepted the proposed Turkish names in most places, and the old, non-Turkish names are still used in daily language.

At the beginning of the twenty-first century, place name changes in Turkey started to be discussed again as a part of Kurdish cultural identity. Some Kurdish political actors demanded the return of Kurdish place names, which were changed during the twentieth century, within the framework of the cultural rights and identity of the Kurds. During this request, it was suggested to use the historical names of many places where Kurds live today. Accordingly, after 2013, thousands of village names were changed to names thought to be Kurdish. However, previous erroneous practices while the Turkish names were proposed were repeated here. In other words, the names in other languages that are not actually Kurdish were changed as if they were Kurdish names. For example, the original name of Diyarbakır was proposed like Amed in Kurdish. Yet, the original name was Amid, belonging to one of the earliest civilizations in Anatolia. Therefore, just as identity construction through place names is not seen as a very persuasive practice for Turks, the situation is the same for the Kurds. These initiatives are intended to be used as a tool for some political demands and impositions.

Finally, it should be noted that while toponymy studies focus on identity construction through changed place names, unchanged place names are hardly mentioned. In many places where place names have been changed, there are as many or even more unchanged village names as changed ones. In this case, if it is claimed that an identity has been given to a place by renaming, aren't the unchanged place names also a part of this identity? Reading the identities over the changed place names may lead to misleading inferences about the history of those places. It does not seem very realistic to construct an identity based on place names only in an area like Turkey that has hosted countless civilizations and carries traces of each in the landscape. However, an analysis based on the unchanged village names, by reverse reading, reveals that the Turkish presence in Anatolia goes back about 2000 years, contrary to what is believed.

Questions

1. Place names in Anatolia emerged as the common memory of which civilizations?
2. Why did the Ottomans not consider changing place names until the last decade of the Empire?
3. What are the main motivations for place name changes during the era of the Turkish Republic since 1923?
4. What is the role of place names in creating identity and showing it through the landscape and naming?
5. Place names are part of intangible cultural heritage items belonging to the past. What are the linguistic, historical, and political problems that make it problematic to change place names? Discuss these in reference to Turkey.
6. Do you believe that the question of place naming to Turkey is unique to that country? What about countries or states neighbouring Turkey, in the MENA and European region e.g. Balkan region, German-speaking areas, and Ireland.

References

Alkan A (2017) Siirt ilinin eski köy adları üzerine coğrafi bir inceleme. Türk Coğrafya Dergisi 68:63–76

Arı Y (2012) Geography in Turkish higher education: sea names for textbooks and maps. In: Proceedings of the 19th international seminar on sea names: sea, sea names and mediterranean peace. Korea, pp 197–206

Arı Y (2021) Manyas Kuş Cenneti Milli Parkı: Kültürel, tarihi ve politik ekoloji. Çizgi, Konya

Başkan Ö (1970) Türkiye köy adları üzerine bir deneme. Türk Dili Araştırmaları Yıllığı-Belleten 18:237–251

Batsashi T (2019) Historical toponyms, hydronims and geographical name changes in Turkish Ottoman Empire and in the Republic of Turkey. Free Univ J Asian Stud 1:1–4

Demir A (2011) XVI. Yüzyılda Koğans nahiyesi: Nüfus ve yerleşme. History Stud (13094688), 3(2):125–245.

Eröz M (1984) Sosyolojik yönden Türk yer adları. Türk Yer Adları Sempozyumu Bildirileri 11(13):43–53

Erpolat MS (2004) XVI. Yüzyıl başlarında Diyarbekir eyaletinde viran köyler meselesi. In: I. Uluslararası Oğuzlardan Osmanlıya Diyarbakır Sempozyumu. Diyarbakır, pp 20–22

Erpolat MS (2019) Osmanlı arşiv belgelerinde Amid yerine Diyarbekir adının kullanılmasının tarihi seyri. In: Tanzimat'tan Günümüze Diyarbakır Bozan O, Asan H, Yıldız H, Erpolat MS (eds). Manas, pp 7–20

Fields DS (2013) State imposed place name change in Turkey and the response of Giresun residents, Doctoral Dissertation, Sabancı University, Istanbul

Gülensoy T (1996) Türk Dünyası ve Anadolu'da ortak yer adları. Erdem 9(26):693–710

Gümüşçü O, Yiğit İ (2020) Toponimiye mührünü vuran Selçuklular: 1/200.000 ölçekli Türkiye topografya aaritalarındaki han ve kervansaray adlı/ekli yer ve yerleşme isimleri üzerinden bir değerlendirme. Selçuklu Medeniyeti Araştırmaları Dergisi 5:97–109

Hausner I (2013) Priciples and rules of 'dual names/dual naming' (multiple names/multiple naming) systems in and outside Europe: an exemplary approach to the definition of two (?) new terms in the UNGEGN glossary. In Proceedings of the 19th international seminar on sea names: sea, sea names and mediterranean peace. Korea, pp 7–22

Hürriyet (2008) https://www.hurriyet.com.tr/gundem/dtpli-kaplan-yerlesim-yerlerinin-eski-isimleri-de-anilsin-8777632. Accessed 20 October 2021

Hocaoğlu B (2016) Varna ilinde ismi değiştirilen Türkçe yerleşme adları. Cihannüma: Tarih ve Coğrafya Araştırmaları Dergisi 2(1):73–92

İbret Ü (2003) Çankırı'daki köy adları üzerine coğrafi açıdan bir inceleme. Marmara Coğrafya Dergisi Sayı 7:53–80

Kırzıoğlu M F (1984) Selçuklu fetihlerinden (1064-1071) önce Doğu-Anadolu'da Türk boy ve oymaklarından kalma dağ ve su adları. Türk Yer Adları Sempozyumu Bildirileri: 75–96

Kocacık F, Yalcin C (2008) Forced migration from the Balkans to Anatolia: 1878–1900. Manas Üniversitesi Sosyal Bilimler Dergisi 10(20):251–262

Köse T (2017) Çözüm sürecinin yükseliş ve düşüşü. Türkiye Ortadoğu Çalışmaları Dergisi 4(1):13–40

Kütükoğlu MS (2012) Osmanlıdan günümüze yer adları. Belleten 76(275):147–166

Özdemir AR (2013) Kayıp Türkler: etnik coğrafya bakımından Kürtleşen Türkmen aşiretleri. Kripto, Ankara

Özdemir AR, Gül A (2013) Yer adlarını iade etmek mi, "Kürdistan"a sınır çizmek mi? Private Report. Ankara: 21st Century Türkiye Institute

Nişanyan S (2011) Hayali coğrafyalar: cumhuriyet döneminde Türkiye'de değiştirilen yeradları. TESEV, Istanbul

Sehrivan (2021) https://www.sehrivangazetesi.com/guncel/704-koy-ismi-iade-edildi-h14741.html. Accessed 21 Oct 2021

Tunçel H (2000) Türkiye'de ismi değiştirilen köyler. Fırat Üniversitesi Sosyal Bilimler Dergisi 10(2):23–34
Yediyıldız B (1984) Türkiye'de yer adı verme usulleri. Türk Yer Adları Sempozyumu Bildirileri. Başbakanlık Basımevi, Ankara, ss. 25–41

Geo-History of the Toponymy of Mohács Plain, SW Hungary

Norbert Pap, Máté Kitanics, Marianna Ács, and Péter Reményi

Abstract The Mohács plain had strategic and symbolic importance both during the Ottoman occupation and the seventeenth century liberation wars due to the decisive battles fought on, and military routes crossing the plain, transforming it into a military and memorial landscape. By the twentieth century it became one of the most conflicted Cold War border-scapes, where the Hungarian 'Maginot-line' was constructed against the Yugoslav army in the late 1940-ies. The Homeland War of Croatia was fought just south of the plain in the 1990-ies and the anti-migration fence of Hungary was also installed here in 2015. The region, which was owned for centuries by the bishops of Pécs and the abbots of Szekszárd but also played an outstanding role in the Hungarian Reformation, is inhabited by Serbs, Croats, Germans and Gipsies besides the majority Hungarians. The geographical names of the plain is well documented from the early Middle Ages, they reflect the (different) various migration waves, the military operations, and (different) various state policies. It is an ideal research area, which as a case study, well represents the different origins and bottom up changes of Hungarian place names as well as top down policies influencing them.

Keywords Mohács plain · Toponymy · Military landscape · Border · Battle of Mohács · Religion · Ethnicity

---

N. Pap (✉) · M. Kitanics · M. Ács · P. Reményi
University of Pécs, Pécs, Hungary
e-mail: pnorbert@gamma.ttk.pte.hu

M. Kitanics
e-mail: kitanics1@gmail.com

M. Ács
e-mail: acs.marianna@pte.hu

P. Reményi
e-mail: remko@gamma.ttk.pte.hu

© The Author(s), under exclusive license to Springer Nature Switzerland AG 2023
G. O'Reilly (ed.), *Place Naming, Identities and Geography*,
Key Challenges in Geography, https://doi.org/10.1007/978-3-031-21510-0_17

Objectives

- To describe the changes of place names in a multicultural Central European area during early modern and modern history.
- To explain how population changes as well as state-driven ethno-linguistic homogenization impacted on uniforming place names.
- To describe how different social groups are represented in local place names at different historical periods.
- To present the toponymic consequences of great historical events such as the Battle of (Mohacs) Mohács (1526).
- To show how a new state border drawn in the twentieth century affects place names in Central Europe.

Introduction

There are seven settlements in the Mohács plain, which lies along the southern border of Hungary and the Danube (Fig. 1). They are situated in a place of great strategic importance, inhabited since ancient times, where the Roman Limes extended and the medieval via *militaris*, the military road used by the Hungarian kings and the Ottoman sultans, passed through. This explains the rich conflict history of the area. Wars have been a heavy burden on the local population. From time to time, the population was replaced: they were forced to flee from the armies, and then some of the previous population returned, while other places were replaced by new inhabitants.

The multicultural character of the region is based on these waves of settlements. On the Danube war route from the Balkans, Serbs and Vlachs arrived, Croats from Bosnia and Slavonia settled, and Germans from Germanic regions sailed down the Danube and landed in Mohács. The later arrivals partly adopted the earlier geographic names, often reformulating them, but in all cases creating their own. The Jews and Gypsies who arrived in the last waves settled in concentrations and had only a minor impact. This is partly because their arrival coincides with the beginnings of the Hungarian ethnic homogenisation drive.

The various military confrontations in the border area also left their mark on the geographic pattern. These include the First and Second World Wars, the Cold War (the Yugoslav-Hungarian confrontation of the 1950s), but in this respect the building of the fence of the 2015 migration crisis has also had an impact on local conditions. In what follows, we attempt to examine the typical forms and types of geographical naming in the Mohács plain.

Multicultural Society on the Mohács Plain

The Mohács Plain is a multicultural microregion, where a small multi-ethnic town (Mohács) and villages of different ethnic backgrounds are located along the Danube. Kölked has been Hungarian since the Middle Ages and Calvinist since the Turkish

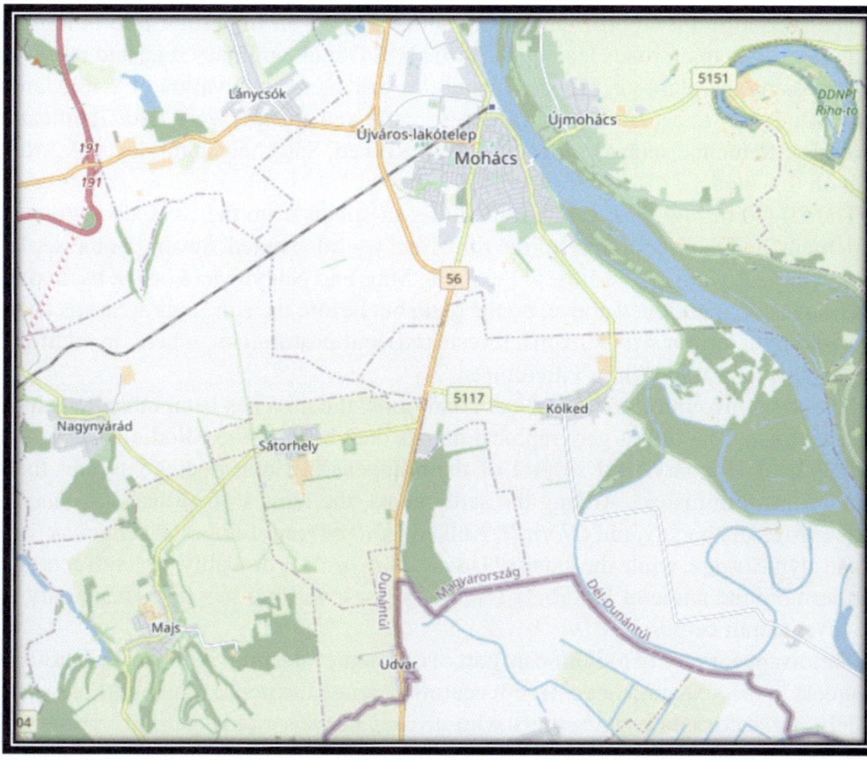

Fig. 1 Mohacs Plain, 2021. *Source* www.openstreetmap.hu

conquest (sixteenth century), Udvar and Nagynyárád are predominantly Catholic German, Majs is predominantly Catholic German, but until the early twentieth century partly Serbian Orthodox, Lánycsók is Catholic German and Hungarian, and to a lesser extent Catholic Croatian and Orthodox Serbian. Sátorhely was created later than the others from the central manor of a large livestock ranch, so its population is of mixed origin and its community was established late, only in the twentieth century. The battle of 1526, when the Ottoman army of Suleiman the Magnificent defeated the troops of King Louis (Jagiellon) II (.) of Hungary and Bohemia, gave the plain its particular historical importance. The longer-term consequence of this was that the central part of Hungary, including the area around Mohács, came under Ottoman-Turkish occupation for a century and a half.

Our study area consists of three parts, the physical geography of which was shaped by the Danube river. The first is the former pre-regulated floodplain of the river, where, due to the regular annual flooding, it was not practical to settle permanently until recently, and only temporary dwellings ('szállás') were traditionally built. However, its economic significance (fishing, cattle grazing, reed cutting, etc.) was very important.

The second is the flood-free terrace, which played an important role in human settlement. The main roads leading to it (the N-S Danube military road and the E-W road crossing the plain via Mohács Island), as well as the cultivation of arable land, horticulture and livestock farming (horses, cattle and sheep), were also significant. Several settlements, such as Mohács, Udvar, Kölked, Sátorhely and Lánycsók, were built on this land.

The 20–40 m high basin rim surrounding the plain from the west also attracted settlement, and the importance of the rim effect was also noted. Several of the settlements (settled) were on the edge of the plain: Majs and Nagynyárád on the basin rim, and Lánycsók, mentioned above, on the plain but before the rim. This area was characterised by a terrain with streams, forests (ed) and pastoralism, where, in addition, there was a long tradition of viticulture.[1]

Due to its favourable geographical conditions, the area has been inhabited since ancient times. The oldest geographical names date back to the Middle Ages of the Kingdom of Hungary, and several of them appear in Latin documents, but their number has diminished. Among the settlements, the names of Mohács (*Mohach*), Majs (*Moysa*), Nagynyárád (*Narad*), Kölked (*Kulked*) and Lánycsók (*Lanchuk*) are Árpád-dynasty age, while the name of Udvar (*Wdwarth*) is probably fifteenth century. Furthermore the name of Sátorhely (*Satoristye*), as we shall see, dates back only to the seventeenth century.

The disappearance of a significant part of the medieval names is due to the Ottoman conquest in the sixteenth-seventeenth centuries, when the population of the area was largely replaced, while the settlers who arrived in waves only partially preserved the earlier names, in many cases changing or giving new names to the geographical places.

Settlement in the region took place in several stages. The oldest layer is the Hungarian, of which Kölked is an example. Its population was a fishing community living along the Danube floodplain. The river and its marshes provided not only a livelihood but also shelter for the inhabitants, and it is the only Hungarian village to have survived the Turkish conquest. The settlement was surrounded on three sides by marshland and water, and could only be approached by land from the west, across a narrow landbridge. In case of danger, the inhabitants took refuge on the islands in the floodplain.

Already in Ottoman-Turkish times, the influx of Southern Slavs began, and for a time they played a dominant role in the life of the region. The Balkan Vlachs, the Orthodox Serbs and Catholic Croats from Bosnia and Slavonia arrived in several waves between the sixteenth-eighteenth centuries. A considerable number of land-use names and names of periodic hydrographic phenomena can be attributed to them.

The year 1704 was a critical period in the region. The previous year had seen the outbreak of the Rákóczi uprising, to which the Habsburg court responded in part by arming the Serbian communities loyal to it that had settled in the southern part of the country. Serbian irregulars attacked the Calvinist Hungarian and Catholic Croatian

[1] Pap et al. (2018).

populations. The survivors fled to the marshes, forests or further afield, and some of them never returned. Later that year, Kuruc troops arrived and began an ethnic cleansing similar to that of the Serbs, which left the Mohács area almost deserted for a number of years.

After the Kuruc uprising (1711), the local landowners (the Bishop of Pécs in the north and Prince Eugene of Savoy in the south) had to face a severe shortage of labour, together with the destruction of the villages. The only solution to this problem was mass settling. Therefore, in the eighteenth-nineteenth centuries, several waves of settlers arrived in the seven settlements.

The Serbs who had fled from the Kuruc troops to Slavonia (now Croatia) returned to the study area only at the end of the War of Independence. They settled in separate quarters in Mohács, and resettled in Lánycsók and Majs (and briefly in Nagynyárád). Similarly, some of the Catholic Croats (Sokac) returned to Mohács and were joined by new settlers. They came to Lánycsók mainly from here, by secondary settlement.

The Germans arrived in several waves from the 1710s onwards, as a result of landlord settlements, with the largest numbers arriving during the reign of Maria Theresa, from the 1740s onwards. The settlers came from Germany by boat, often in communities, accompanied by their priests. In addition to Mohács, where they occupied a separate district, they settled in the villages of Nagynyárád, Majs, Udvar and Lánycsók and engaged in agriculture. They mainly spoke the German dialects of Rhineland Franconian and Fulda.

Gypsies were certainly present in the area at the time of the Turkish conquest, but there are no reliable records of this. They arrived in the area sporadically in the eighteenth century and in masses in several waves in the nineteenth–twentieth centuries from Wallachia, and in smaller numbers from Serbia. Their settlements were established in Mohács, on the southern outskirts of the town, on the banks of the Danube, and in Kölked and Lánycsók outside the village. All three linguistic groups of Gypsies living in Hungary settled in Mohács: the Hungarian-speaking Romungros, who originally spoke the Carpathian dialect of Romanian, the Romani-speaking Oláh Gypsies and the Beyas, who spoke an archaic version of Romanian. In Kölked, the latter group has been identified as speaking the Munchan dialect. Many of them were involved in wood and metalworking and itinerant trade. By the early twentieth century, their numbers may have reached a few hundred.

The last settler layer to arrive in the area was the Jews. A large number of them started to settle in Mohács in the 1850s, and by the end of the century there were nearly a thousand of them living in the town on the Danube. They came mainly from Galicia, Russia and, to a lesser extent, from Austrian territories. Their language was Yiddish. In Mohács, which was ethnically organised into neighbourhoods, their houses stood on both sides of the main street. Their community had a school, a synagogue and a cemetery. They supported themselves through trade, small business and intellectual occupations.

From the end of the nineteenth century and throughout the twentieth century, the ethnic homogenisation of the area took place, and assimilation processes and Hungarianisation intensified. The area was under Serbian military occupation 1918–1921. In the 1920s, a significant part of the local Serb population opted to join the

Kingdom of Serbs, Croats and Slovenes. During the Second World War, the Jewish community and part of the Roma were victims of the Shoah. Around 1000 people from Mohács and the surrounding area were transported by train to Auschwitz, only a few of whom returned. On the basis of the principle of collective guilt, Germans were deported to Germany in large numbers after the war, especially in 1947. At the same time, and subsequently, under the Czechoslovak–Hungarian population exchange agreement, Hungarians resettled in Hungary were in many cases housed in the homes of displaced Germans in our region.

Another intersection of the research on geographical names is gender. How many geographical names of male and female origin are there in the area? Hungarian geographical naming tends to be male-centric, and even more so in the study area. The proportion of names of female origin is low, and there are hardly any examples. This is mainly due to the fact that the indigenous Hungarian community (see the fishermen of Kölked) and the German, Serbian and Croatian settlers created patriarchal, predominantly peasant societies. Gypsies and Jews who immigrated from the Balkans also lived in patriarchal communities. The role of women in the process of settlement was thus marginalised, as is reflected in the naming process.

Literature Review and Methods

Toponymic research in Hungary has a long history, mainly carried out by linguists, cartographers and ethnographers. Hoffmann and his colleagues have provided a comprehensive overview of the history of Hungarian geographical naming in their monograph.[2] Faragó summarized the considerations related to the use of geographical names mainly from a cartographic point of view.[3] Among the regional studies, István Csernicskó's research on the linguistic landscape of the border region of Ukraine with Hungary (the area of the former Kingdom of Hungary called Ciscarpathia) stands out.[4] The analysis of Fancsaly and his colleagues on the Hungarian language in Croatia is also valuable for the present study, as the area they investigate, the Drávaszög, is a spatial continuation of the Mohács plain from the south, beyond the border.[5]

The first surviving map of Hungary by Lazarus secretarius (1527/1982) in 1527 already shows the area around Mohács in a prominent position, and also the famous battle of 1526 appears (Fig. 2).[6] However, a more detailed modern study of the toponymy and hydronymy of the Mohács plain is only possible from the second half of the eighteenth century onwards, with a few minor exceptions, due to the lack of sources. It is only after the wars of liberation from Ottoman rule, when the first

[2] Hoffmann et al. (2018).

[3] Faragó (2014).

[4] Csernicskó (2018).

[5] Fancsaly et al. (2016).

[6] Lazarus secretarius (1527/1982).

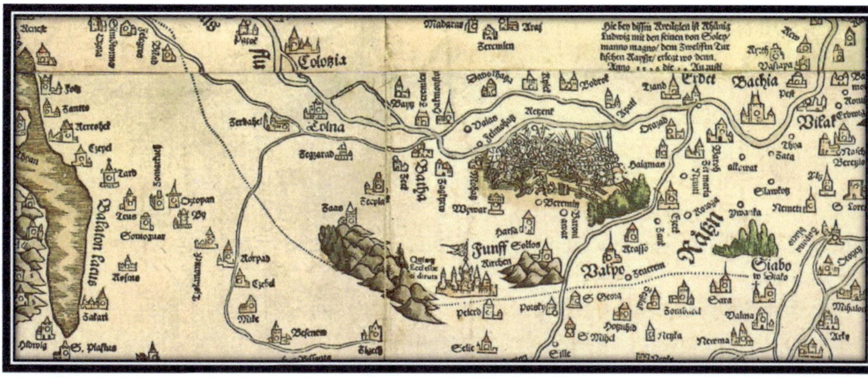

Fig. 2 (Mohacs) Mohács and its surroundings on the map of Lazarus secretarius (1527/1982)

primary sources and maps containing a significant amount of relevant information survive (from the 1760 and 1770s). These maps, which were produced by the Royal Hungarian Chamber in order to ensure the efficient management of the treasury's assets, not only show the land use, the hydrography, the roads, the boundaries of the settlements and the street network, with the boundaries of the plots and buildings, but also the names of the most important parts of the lands, hamlets, manors etc. belonging to each settlement.

The next important sources are from the second half of the nineteenth century, when cadastral maps were drawn up for the purpose of establishing the land tax for each of the settlements concerned. Of these, the 1865 surveys are particularly useful, as they provide a more detailed record of the lands of the settlements. It is also worth examining the military surveys, the first of which of 1783 and the second of 1858 contain less information, while the relevant map sections of the third military survey of 1880 are much richer in data on geographical names. The military survey of Hungary after the border changes of Trianon (1920), which was repeated by the Royal Hungarian Army Cartographic Institute in 1940–1944 with the Hungarian territorial growth recorded, also fits into this series.

In addition to the map sources, the description of Baranya County (1735/2017) from Mátyás Bél's *Notitia Hungariae novae historico geographica…*,[7] András Vályi's three-volume work *Magyar országnak leírása* (Description of Hungary) (1796–1799),[8] and the *Baranya vármegye helynévtára* (Local nomenclature of Baranya County) (1864–1865/2019) by Frigyes Pesty can be mentioned in connection with the examination of the relevant geographical names.[9] The Hungarian Central Statistical Office, partly based on the above works, has published historical gazetteers of some counties, including Baranya County, using a wide range of

[7] Bél (1735/2017).

[8] Vályi (1796–1799).

[9] Pesty (1864– 1865/2019).

primary and secondary sources, with different name variations of individual localities, hamlets, unincorporated inhabited places, lands, etc. with important demographic, religious, ethnic and economic characteristics.[10] The most important in this respect, however, is the two-volume work entitled *Baranya megye földrajzi nevei* (Geographical Names of Baranya County) (1982), edited by János Pesti, which, as a result of large-scale local collection work, recorded the geographical names of various spatial entities belonging to the individual settlements, often with their supposed or real origins, together with historical, ethnographic or archaeological information.[11]

Tent distinguished two main research approaches in toponymyc research.[12] After examining the studies published in the scientific journal 'Names' from 1952–2014, he found that one large group of studies focused on the origin and meaning of geographical names, while the other major group researched in a spatial context and focused on the spatial patterns of names. In his analysis, he also showed that in some cases, as in the present study, the two approaches can be mixed, but this is the case for only a few percent of the works examined.

Intensive place-name research looks to answer the wh-questions in the process of creating a 'biography' of a place name. Who named the place? When was it named? Why was it given that name? What does the name mean? What does it refer to? What is the linguistic and geographical origin? Where is the place in question? The list of questions is, of course, endless. Extensive place-name research, on the other hand, uses databases, a larger set of place-names, and does not focus on a specific name but on the naming characteristics of a narrower or wider area. From these investigations typologies can be created.

The dual function of geographical names is the same in our research area as everywhere else in the world. On the one hand, they help us to find our way around, they mark geographical locations, but they also store collective memory, they are cultural, historical and political representations and, as a result, they can also become political tools.[13] The state power exercises control over its territory by creating symbolic space through the choice of geographical names, among others, that reinforce and legitimise its own power and identity.

Physical or Geographical Features and Their Linguistic Representation

In the region, the smaller and larger elevated terrains (hillocks, rolling hills) have been named in a variety of ways. One mound (hillock) that barely rises out of the plain is the *Látó-hegy* (Watch hill) of Mohács, which emphasises the fact that the

[10] Kovacsics (2001), Lelkes (2011).

[11] Pesti (1982).

[12] Tent (2015).

[13] Baroch (2019).

whole field of Mohács can be seen from the 'hill' next to the road to Nagynyárád. Similarly, the *Planina-földek* (Planina fields) of Mohács were named because of their elevation from the surroundings, where the local inhabitants added the South Slavic 'planina' (mountain) to the Hungarian name of the cultivated area ('földek' means fields).

In many place names can one find the adjectives lower, upper, back, middle, small, big, short, long, wide etc. both in Hungarian, German or in the South Slavic languages. For example, in Nagynyárád you can find the German *Untere-*, *Obere-*, and *Hintern Hegwald* (lower, upper and back young forest), *Mittel Weingarten* (middle vineyard), *Kleine* and *Grosse Grund Wiesen* (small and large meadows), in Majs *Oberdorf* (upper village), in Mohács the Hungarian *Alsó-*, *Közép-* und *Felső-mező* (lower, middle and upper meadows),[14] and *Kurta-dűlő* (short fields),[15] in Lánycsók the meadow called *Siroko* (široko, meaning broad in South Slavic languages) and the German *Lange Wiesen* (long meadows), and on Mohács Island the area called *Sirina* (širina, meaning width in South Slavic languages).[16]

There are also geographical names with old or new adjectives, such as the German *Alte Dorf Schanze* (Old Village Schanze/Sconce) in the southern part of Mohács, the popular names of the *Rókus temető* (Rókus cemetery): *Öreg temető* (Old Cemetery) in Hungarian, Alter Friedhof in German, Staro groblje in the local South Slavic languages, or the manor in Sátorhely, called *Újistálló* (New Stables) in Hungarian and Neuer Stall in German.[17]

Among the names of the unincorporated lands, we can mention the *Sarok-dűlő* (Corner land) in Majs (*Csucsaj* in South Slavic languages),[18] which got its name from its location, as well as Sarok utca (Corner Street) and Szélső utca (Side Street) in Mohács,[19] or the land called Forgó (Turning) in Kölked. The latter was located on the Kölked branch of the Danube, where the river turned, creating a large loop, while the *Hajlok-part* (Bending-bank), also located here, takes its name from its curve.[20]

Also of particular interest in this group is the *Fekete-kapu* (Black Gate), unincorporated land of Sátorhely on the Osijek-Buda road, called *Schwarzes Thor* by the Germans and *Crna kapija* by the local South Slavs, which is also associated with the Battle of Mohács.[21] It is here—and also at the Iron Gate of Mohács–Lánycsók—that according to local tradition 300 students from Pécs sacrificed themselves in 1526 to resist the Ottomans, while another tradition says that until the end of the First World War, Bosnian Muslim pilgrims prayed here every year—besides the Turkish Hill—for the Ottoman martyrs who died in the Battle of Mohács. However the exact

[14] Cadastral maps (Nagynyárád, Majs, Mohács 1865).

[15] Pesti (1982).

[16] Cadastral maps (Lánycsók, Mohács 1865).

[17] Pesti (1982).

[18] Cadastral maps (Majs 1865).

[19] Pesti (1982).

[20] Cadastral maps (Kölked 1865).

[21] Pesti (1982).

meaning and origin of its name isn't known anymore just like that of the *Vaskapu-dűlő* (Iron Gate lands, or the German Eisenes Thor on the map section of the First Military Survey of 1783), although some recollections say that it was named after the black gate of a tavern that once stood here.[22] Finally, the physical characteristics of the *Büdös-árok* (Stinky ditch) in Mohács, named for its strong and unpleasant smell, are also noteworthy. This ditch, formerly a swampy stream, protected the town from the west during the Ottoman era, as can be seen on Henrik Ottendorf's map of 1663. Later, from the eighteenth century onwards, it separated the two parts of the town, *Ó-Mohács* (Old Mohács; in German Alt-Mohács) and *Új-Mohács* (New Mohács; in German Neu-Mohács).[23]

1700

The expulsion of the Turks from Hungary began in 1684, and after heavy fighting, the Habsburgs successfully ended it in 1699 with the Treaty of Karlowitz. The decisive battle for the liberation of this region took place in 1687 at nearby Nagyharsány (second battle of (Mohacs) Mohács). After the victorious battle, the reorganisation of the area began. The Islamised population was forced to leave the area, while new settlers arrived.

Initially, the Serbs played a decisive role in this process, moving into the country along the Danube after the Balkan campaign of the Habsburgs (1690), following their great migration. However, the extent to which their settlement was permanent in the 1690 s and early 1700s, in the midst of warfare along the main transport route, is questionable. In addition, in 1704 they fled to the other side of the Drava due to the attacks of the Kuruc troops.

The map below, dated 1700, is of great importance in the context of the foregoing for several reasons (Fig. 3). Firstly, it provides a visual record of the situation after the liberation from the Turks, including the settlements that were repopulated by the majority of Serbs (and Croats), according to the settlement hierarchy. On the other hand, it describes the moment when the Serb settlement had already taken place, while some settlements had only a few years left before the aforementioned depopulation in 1704.

In the year before the map was drawn (1699), Prince Eugene of Savoy, the victorious general of the Ottoman war, received the territory of the Bellye estate from the Habsburg emperor in recognition of his merits.[24] The map shows the change in local land tenure: among other things, it separates the bishop's estates around Mohács from the prince's estates and their villages further south. In recording the names of the settlements, we do not read standardised forms of names, but we see how the map-maker interpreted the geographical names and displayed them on the map. As

---

[22] Oral report by local resident, Szűcs József.

[23] Pap et al. (2020a, b).

[24] Pap et al. (2019).

Fig. 3 The (Mohacs) Mohács Plain in 1700[25]

far as the geographical names and the various inscriptions are concerned, it can be said with certainty that the map is in Hungarian, German and French, in addition to Latin. In some settlements, the network of settlements is also structured in a way by the marking of churches and houses: places with churches are higher in the hierarchy. The names of the settlements, as can be seen in Table 1, are mostly similar to their later names, with two notable exceptions: the cases of *Quilue/Kölked* and (F)*Utva*r, i.e. *Földvár*, are special.

The widespread 'ö' sound, or more precisely letter, in Hungarian, as well as consonant shifting, is a problem for Westerners when describing certain words. This is true for the spelling of both names. However, Kölked is well defined by its location on a marshland, by the fact that it is still a settlement in the same place and by the fact that its Hungarian population survived the Turkish conquest. The 'k' becomes 'g', while the 'ö' becomes 'u'… in transliteration.

The case of the village of *Utvar* on the map is different from that of Kölked, as there is no settlement in the area that can be clearly linked to it. The very similar-sounding village of Udvar was established only in the second half of the eighteenth century along the Mohács-Osijek road, in the western, wooded part of the Dályok village estate, with the name *Udvard*. However the identification of a certain settlement called Földvár and the question of how to spell its geographical name is part of one of the most important Hungarian historical debates that has been going on for about 100 years. Bishop István Brodarics, an eyewitness chronicler of the Battle of Mohács in 1526, mentions in his account that the Turkish artillery was deployed against the Christian army at a village called Földvár. The centre of the battle was therefore at the village and to the north of it.

Engineers who surveyed and mapped the area during the Turkish war (1684–1699) came across the name "Földvár" (Earth Castle) several times, because there

[25] Universum dominium Siklossiense…(1700).

Table 1 Explanation of the names on the 1700 map of Bellye estate

Names on the map	Today forms
Bata	Báta
Mohats	Mohács town
Insula St Brigita (latin)	The eastern side of the Mohacs Island
Duodecin navigia bellica imperialia (latin)	12 imperial warships (stationed on the Danube at Kiskőszeg/Batina)
Isip	Izsép village (today Topolje in Croatia)
Dalyok	Dályok village (today Duboševica in Croatia)
Donowitz	Danóc (former village in Croatia)
Quilue	Kölked village
(F)Utvar	Földvár (a destroyed village that no longer exists)
Mais	Majs village
Morse	Merse (destroyed former village, still living in the name of a dyke)
Orman	Ormány (destroyed former village, still living in the name of a dyke)
Beszedeg	Bezedek village
Ketu	Geta (in the Middle Ages there was a monastery and a village here and that lives on in the name of a periphery)
Lak	Lak (a devastated settlement that lives on in the name of a periphery)

were several settlements in the area of operations that bore this name. The most prominent are (Duna)Földvár and (Tisza)Földvár, but it seems that Földvár, situated on the Mohács plain and at the mentioned centre of the battle, also came into this circle.

In the works of Luigi Ferdinando Marsigli, we are confronted with the problem of name transcription on several occasions. During his mapping of the Danube, the (Duna)Földvár in Tolna County appeared on his maps and in his descriptions as *Feudvar* and *Fudvar*. And the name of the former Hungarian (Tisza, then Bács)Földvár along the Tisza, now Bačko Gradište in Serbia, was recorded exactly as it appears on the above map of 1700, i.e. as *Utvar*. In addition, the name of the water on the island of Mohács, *Udvar Palus* (Lake Udvar), was also given, and the term *Udvar* was used in its name (Fig. 4).[26] This body of water is *Lacus Földváriensis* (Mikoviny, 1720–1725), *Földvár-tó* (Lake Földvar), which was once part of the estate of which the village of Földvar on the plain of Mohács was the centre, and was therefore named after the village. On the basis of the above, it is likely that the mysterious *Utvar* is a distorted fixed form of the geographical name Földvár.

[26] Marsigli (1696/1726).

Fig. 4 Udvar Palus, Lake Földvár on (Mohacs) Mohács Island (1696/1726)

The geographical site in question has recently become the focus of landscape reconstruction and archaeological investigations.[27] The remains of a medieval settlement have been found along the Borza stream, visible on the map (Fig. 3), on the southern side of a crossing (and to a lesser extent on the northern side). The archaeological site in question is located two and a half kilometres west of the Osijek–Mohács road and three kilometres east of Majs. It is located where the village is marked on the 1700 map. This provides an additional basis for assuming that the location of Földvar, which marks the central area of the Battle of Mohács, has been identified.

Of the settlements in the Mohács plain that can be identified from the map, a (pretty) good number survived the period of destruction between 1704 and 1711 (e.g. Mohács, Majs, Kölked), while others were depopulated (Merse, Ormány, Földvár) and their territory became part of the settlements in the resettled area.

Linguistic Landscape Before Magyarization

The evolution of the geographical place names of the Mohács plain from the end of the seventeenth century to the end of the nineteenth century can be traced back to several origins. The most important of these is the change in the ethnic composition of the region, which had a fundamental impact on place names, their development and their changes.

---

[27] Pap et al. (2021).

As described above, by the second half of the eighteenth century, the ethnic composition of the settlements had been established and stablilized for a longer period of time.

If we examine the geographical names that have mainly survived on maps, we can see that the basic layer of names of the unincorporated lands and fields is formed by the names of the former Hungarian villages of medieval origin, which were destroyed in the Ottoman period or were not finally resettled after the Ottoman period. Although these villages no longer existed at that time, their names survived in the linguistically dominant South Slavic–Germanic environment: the parts of the territory once belonged to them together with their old names were merged into the unincorporated lands of the nearer settlement. This is how the former villages of Ormány, Merse (*Mersa*), Géth (*Getta*), Lak and Bátya (*Főlsi Battin, Also Batin*),[28] are shown as hamlets/preadiums belonging to the village of Majs on the maps of the 1760–1770s by the chamber cartographer Andreas Kneidinger, while the former settlement of Körtvélyes (*Körtviles*) became an inseparable part of the village of Nagynyárád (*Nyárád*).[29] In this respect, the map section of the First Military Survey of 1783, which is nearly the same age, preserves only the medieval name of Gáta in the German *Gátha Thal* (Gátha valley; spelled with the Hungarian letter 'á').[30] The cadastral maps of the 1850–1860s show the survival of the former village of Ösztyén (*Usztin, Ustine*) as an unincorporated land at Majs,[31] the aforementioned Gáta and Monyoród were listed as belonging to Lánycsók, while the former Bácsfalu with farm buildings is listed as a manorial centre of farming (*Bácsfalu, Puszta Bácsfalu*).[32] Mohács similarly preserves the names of the medieval villages of Jenő (*Jenyő*), Csele (*Cselle*) and Földvár.[33]

The names of these fields and unincorporated lands were markedly different from the surrounding South Slavic and German place names. However, it can also be observed that the original form used by the Hungarians was in some cases adopted to the language of other ethnicities. Thus, for example, the name of the medieval village of Bátya was first changed to *Bacsin* under the influence of the Slavs, only to be divided into *Unter* and *Ober Bacsin* (Lower- and Upper Bacsin in English) by the Germans who settled in the area alongside the Serbs. In the same line, we can also mention the aforementioned Ösztyén, whose former part, annexed to the outskirts of Majs, was called *Ustine* by the Serbs.

In addition to the above-mentioned problems of place names of medieval Hungarian origin, the cadastral maps of the 1850–1860s can be read as follows. Although in our study area these maps were produced in the period of imperial neo-absolutism following the Hungarian War of Independence of 1848–1849, when the German language dominated in Hungary under central pressure, this was not, or can

[28] Kneidinger (1772).

[29] Kneidinger (1767–1779).

[30] First Military Survey (Lánycsók 1783).

[31] Gemeinde Majs (1853), cadastral map; cadastral maps (1865).

[32] Cadastral maps (Lánycsók 1865).

[33] Gemeinde Mohács (1852), cadastral map.

hardly be detected in the use of geographical names on the maps of the Mohács plain. Basically, the geographic names, even if with minor or major disproportions, mostly reflected the ethnic composition of the settlements studied, in which, moreover, the German ethnic group was the dominant element in many places either way. Moreover, we can add that in the more intense years of neo-absolutism, i.e. the period of 1852–1853, some settlement cadastral maps not only show geographical names in the languages of the main ethnic groups, but also a greater number of geographical names in Hungarian. Even in settlements where the Hungarian ethnic group was certainly under-represented. Thus, for example, in 1853 in Majs, inhabited mainly by Germans and Serbs, place names like *Majsi rét* (Majs meadow), *Majsi föld* (Majs field) or *Alsó Batyin* (Lower Batyin) and *Felső Batyin* (Upper Batyin) was listed.[34] By 1865, however, the German equivalents of these same names were to be found in the form of *Maisser Äcker*, *Maisser Wiesen*, *Unter Bacsin* and *Ober Bacsin.*[35]

Similarly, in Lánycsók in 1853 the names *Parlag* (Fallow) or *Birkás* (Sheepy) are found,[36] which are replaced in 1865 by Parlog, adapted to the German language, and Schaflerwiesen in German.[37] We can also mention Nagynyárád, which in 1853 was still listed as Nagynyárád in Hungarian but in 1865 as *Gross-Nyárád* in German, while the name of one of the lands *Törökös* (Turkish) surrounding it, was translated into German as Türkei.[38] The reason for all this, however, can hardly be, in our opinion, a kind of Germanisation effort, as the 1860s were already just about to be marked by a freer atmosphere and a return to constitutionalism. Rather, the identity, ethnicity and language skills of the local respondents, and even more so of the surveyors, were of great importance in the unregulated work carried out in a multiethnic environment. Thus, the above may be explained by the fact that the mapping of 1852–1853 in all the settlements of the Mohács plain was carried out by a young Hungarian officer and military engineer, Ágoston Vucskics, who was a veteran of the War of Independence. The fact that in the case of geographical names there could hardly have been any centrally controlled Germanisation is confirmed by the example of Mohács, which, although it had a German population, still had almost entirely Hungarian (*Alsó-*, *Közép-*, *Felső-mező*, in English Lower, Middle and Upper meadow, *Előhegy*, in English Forehill) and South Slavic place names on its cadastral maps. Within this, there is also the spatial difference that the geographical names of South Slavic origin (*Sirina*, from the South Slavic širina, meaning width; *Topolovac*, meaning warm place, warm back, etc.), which can be traced back to the Turkish period, are concentrated more in the island part of Mohács.[39]

As to whether the geographical names of individual settlements were indeed adapted to the ethnicity of the inhabitants, the village of Majs is a case in point, where the Serbian and German names on the cadastral map of 1865 were also largely

[34] Gemeinde Majs (1853), cadastral map.

[35] Cadastral maps (Majs 1865).

[36] Gemeinde Lancsuk (1853), cadastral map.

[37] Cadastral maps (1865).

[38] Gemeinde Nyárád/nagy/ (1853), cadastral map; cadastral maps (Nagynyárád 1865).

[39] Gemeinde Mohács (1852), cadastral map; cadastral maps (Mohács 1865).

adapted to the settlement structure. Accordingly, Serbian names were concentrated in the broader area around the street of the Serbs, and German names around the street of the Germans. Thus, close to the street of the Serbs, one can see here, among others, the spatially more extensive *Vukobaba* area (vuk, or wolf, baba, or grandmother; later called *Farkas-tanya* (Wolf-farm by the Hungarians), *Travnik* (pasture, meadow), *Dolina* (valley), *Salašina* (residential buildings suitable for secondary economic activity), *Dubravica* (small oak), and also the charcoal-burning sites called *Uglenica*, which continued in the Dályok lands surrounding the village of Udvar. Likewise, the *Hofstellen* (the area behind the village courtyard), the *Maisser Äcker* (Maisser fields), the *Maisser Wiesen* (Maisser meadows), the *Udvarer Wiesen* (Udvarer meadows) and the *Hanf Thaeler* (hemp valleys) can be seen along the German Street. In the southern part of the settlement's area, next to the *Mersa Acker* (Merse field), other, mainly German names are shown as part of the *Maisser Wald* (Maisser forest), such as the *Akazienwald* (acacia forest), the *Salzlecken* (salt licker), named after the feeding of the game, the *Rehstand* (deer stand), the *Eichelacker* (acorn field) or the *Hütten* (huts). As mentioned above, according to the 1865 map, the German-speaking geographical names (*Mühl Wiesen*, or mill meadows; *Szaiker Hügel*, or Szaiker hill; *Dorf Weingebirge*, or Village vineyard hill; *Dorf Wiesen*, or village meadows; *Türkei Acker*, or Turkish field; *Schwarzes Thor*, or Black gate, etc.) dominated the German-speaking part of the German-inhabited Nagynyárád, including Sátoristye, and the only reference to the former Serbian inhabitants was a single German-language name for a vineyard (Raczen Gipfel, i.e. Serb summit). The same was true of Udvar, which had only a small area and whose unincorporated lands were entirely consisted of the so-called *Hofäcker* (also known as Hofstellen), and which, as a village split from the ethnically southern Slavic village Dályok, was surrounded exclusively by southern Slavic land names (*Veliki udvar*, i.e. Great yard; *Uglenica*; *Lisicse jame*, i.e. Foxholes; *Orlovo Gnizdo*, i.e. Eagle's Nest, *Duga megya*, i.e. Long field boundary). Also included in this line was Kölked, whose Hungarian land names (*Vásártető*, i.e. Market summit; *Várdomb*, i.e. Castle hill; *Csárda*, i.e. Tavern; *Hajlok-part*, i.e. Bending bank; *Kenderföld*, i.e. Hemp fields; *Uradalmi mező*, i.e. Manorial meadows, etc.) clearly indicated its Hungarian ethnic majority.[40] In the case of Lánycsók, however, even in the mid-nineteenth century, when ethnic relations changed and German names were already present (*Gata Wiesen*, i.e. Gata-meadows; *Mühl Bach*, i.e. Mill creek; *Bertoldsmühle*, i.e. Bertold's mill; *Rasen*, i.e. Lawn; *Bienen Garten*, i.e. Bee garden; *Schaflerwiesen*, i.e. Sheep meadows, etc.) there are strikingly few Serbian and Croatian names. Here, only *Siroko* (broad), a broad meadow near the mill creek, appears on the 1865 cadastral map,[41] while on the 1853 map the area called *Táboristye* has a South Slavic meaning.[42] The latter, however, was formed from the Hungarian word for 'tábor' (camp), the meaning of which was thus camp site, camp field.

A special mention should be made of Sátorhely/*Satoristye* (today's Croatian transliteration Šatorišće), where the origin of the geographical name can be traced

[40] Cadastral maps (Majs, Nagynyárád, Udvar, Kölked 1865).

[41] Cadastral maps (Lánycsók, 1865).

[42] Gemeinde Lancsuk (1853), cadastral map.

back to 1630/1631. It was at this time that Sultan Suleiman had a memorial site erected by Hasan, the beylerbey of Buda to commemorate his victory over King Louis II of Hungary and Bohemia at the Battle of Mohács in 1526. A few kilometres south of Mohács, along the Osijek-Buda road, he had a wooden pavilion built on a few metres high small mound, which had originally been constructed by the Romans, and had a well dug next to it. The site, known in Ottoman as *Hünkjar tepesi* (Emperor's Hill), was a favourite with pilgrims and was visited by many well-known Christian and Ottoman travellers and soldiers. In 1626 Athanasio Georgiceo, before the pavilion was built, named the hill as the site of Suleiman's camp in 1526, while Henrik Ottendorf in 1663 described it as the site of the Sultan's tent. In 1687, Tobias von Hesslingen drew attention to the 'wooden tent' here, which he said was built to commemorate the tent of the victorious 'Turkish emperor'. The site was then also marked on Samuel Mikoviny's map of 1720–1725 as the place marked by the entrenchments of Suleiman's tent (*Solman Tentorii Shanzense*), while Joannes Vötter placed the inscription *Shaturix* (still called *Schatriz* by those of German origin) on the hill on his map of 1766.

This inscription, which is a linguistic variant of the South Slavic *Satoristye*, meaning "place of the tent", clearly shows that this memorial site was originally called *Satoristye.* Hesslingen's note above is perhaps the most helpful in understanding the origin of the geographical name. According to which, the southern Slavic population of the Mohács plain in the seventeenth and eighteenth centuries believed that Sultan Suleiman's tent stood on the hill at the time of the Battle of Mohács. In this context, it was believed that the Ottomans had erected the wooden tent on the hill to symbolize the former sultan's tent, which they called *Satoristye* in Croatian and Serbian. However, as we have seen, the memorial site and its name survived the reconquest and the expulsion of the Ottomans. The reason for this was that in 1687, after the reconquest of the area, the Ottoman pavilion was converted into a Christian chapel, which, according to maps, was only destroyed in the 1770s. However, it was precisely in parallel with the destruction of the building here that, from the 1760s onwards, the manorial centre of *Satoristye*, which was to form the nucleus of the later village of *Satoristye*, began to develop under the authority of the village of Nagynyárád.

One of the first written records of the geographical name, already in its new place, in the form of '*Földvár alias Satanyistya*', dates back to 1780, in a document in which Queen Maria Theresa of Hungary donates the Bellye estate to her daughter, Archduchess Maria Christina.[43] But how did the name Földvár come to be added to the name *Satoristye*? Because the area, of which the first Ottoman and Christian memorial site and the newly established hamlet were all part, belonged to the former settlement of Földvár. Although this settlement of medieval origin was destroyed after the Ottoman period, as mentioned above, its name as *Földvár puszta* (*Földvár praedium*; Földvár'depopulated village'), not infrequently together with the name *Satoristye*, survived into the eighteenth and nineteenth centuries. In connection with the above, the name of the small mound has also undergone an interesting evolution.

[43] Pap et al. (2018).

After the local name *Satoristye* was transferred to the nearby hamlet, the mound was referred to in the South Slavic, German and Hungarian language versions of the local ethnic groups simply as *Törökdomb* (*Turski brig*, *Türkenhügel*, *Törökdomb*).

The geographical names of the various groups of Gypsies (Romanis) who arrived later than the above-mentioned ethnic groups are mostly found in Mohács, where the Gypsy neighborhood and its surroundings bear their names. This is how we can read on the maps, among others, about the *Czigány zátony* (Gipsy Shoal) and *Czigány utca* (Gipsy street), which refer to their presence.[44] The names of Gypsy street and Gypsy neighborhood appear variously in the languages of other nationalities: *Zigeuner Gasse*, *Zigeuner Viertel* (German), *Ceganergas* (Yiddish), *Kod cigana* and *Ciganszki Kraj* (in the local South Slavic languages). The various Romani-speaking groups of Gypsies refer to their street or neighborhood as *Romani vulica* (Gypsy Street), *Romani utca* (Gypsy Street) or *Stepan Ulica Hálung* (Stephen Long Street), or *Kumpányö dö cigánusztá* (Gypsy Company). They lived here for decades until the 1965 Danube floods destroyed their neighborhood of tiny houses. The ethnic character here was exceptionally reflected in the official naming, because the street name Dankó Pista referred to its ethnic character (Pista Dankó was a famous and nationally renowned gypsy musician in the nineteenth century). Gypsy neighborhoods existed outside Mohács in Kölked and Lánycsók. The authorities relocated the Gypsy population of Mohács to Lánycsók after the aforementioned 1965 Danube floods, but many of them migrated back to the city. The names of the places and institutions that play an important role in the life of their communities are mainly mirror translations of the geographical names of the ethnic groups that had arrived earlier.

Homes of Jews were concentrated along the main street in the centre of Mohács. The local geographic nomenclature was only minimally influenced by their names. The geographical names of earlier arrivals were adopted, possibly translated. They had a separate street where their synagogue, charity home and school were located. Here, in the place called *Judengasse* (Jewish Street) in Yiddish, Jews were almost the only inhabitants, while the surrounding area was called *Jidese Fertl* (Jewish Quarter). Their burial took place in the *Zsidó temető* (Jewish Cemetery; in Yiddish, *Beszëkvórësz*; in German, *Jüdischer Kirchhof*; in the local South Slavic languages, Židovsko groblje), separate from the other denominations. Outside Mohács, there are hardly any Jewish names on the Mohács plain, but in Kölked there is a place called *Zsidó rét* (Jewish meadow), which refers to its former tenant.

There are very few examples of geographical names with female names in the area. Most of them commemorate women of royal and aristocratic origin. In Mohács, a street is named after Saint Elisabeth, the canonised princess of the Árpád dynasty. In Mohács, a street and in Sátorhely, a housing estate were named after Dorottya Kanizsai, the second wife of the Hungarian kingdom's palatine, Imre Perényi. She buried the fallen of the battle of Mohács. The southern Slavic name of the garden of the Bishop's Palace of Mohács, '*Marijina basta*' (Mary's garden), may refer to Maria Theresa, who reigned during the office of Bishop Zsigmond Berényi of Pécs, who was the initiator of the construction of the palace. The name 'Izabella-föld'

[44] Cadastral maps (Mohács 1865).

(Isabella lands) refers to the name of the daughter-in-law of the largest landowner in the region, Archduke Albrecht von Habsburg. However, some women of civil origin also appear in the toponymy. It was not uncommon for family-run shops, pubs and factories to be named after the female owner or female family members. The *Schmidt Károlyné Téglagyár* (Brickworks of Mrs Károly Schmidt), whose popular name was *Schmitt banya gödre* (the pit of witch Schmitt), can be mentioned here. Also in this category is the steam mill in Mohács called *Erzsébet* or *Erzsi malom* (Elisabeth mill), named after the owner's wife. György Pécsi's restaurant in Mohács was called *Laci Mári kocsmája* (Laci Mari's tavern) after his wife, László Mária. But there is also example of sexist name giving: the name of a pub in Mohács called *Hatcsöcsű* (Sixtits), referred to the fact that at one time three unmarried, but large-breasted women worked there.[45]

The Border and Its Influence on Toponymy

Hungary's new borders following the lost First World War (1920), one of the most important traumas of Hungarian identity, also cut the Mohácsi plain in two, which resulted in the appearance of a political state border in the space, now "officially". The precise demarcation of the border brought with it the associated geographical names in the borderside settlements, such as the *Figyelő* (Watcher), the *Alsóbedai őrház* (Alsóbéda guardhouse), the *Határsorompó* (Border boom gate), the *Bekerített terület* (Enclosured lands), the *Határ-fok* and *Határgödör* (Border cape, Border pit—places close to the state border), the *Senki szigete* (No man's island an uninhabited place because of its border location) in Kölked or the *Magasfigyelő* (High watchtower), *Határkút* (Border well) and *Határ állomás* (Borderstation) in the village of Udvar.[46] This interwar state border could still be experienced, lived and named by the civilian population.

After the Second World War, however, Hungary's western border was closed by the iron curtain, and the southern border was even more militarised. After the Soviet Union rejected Yugoslavia's independent policy, the 630 km-long Southern Defence System was established along the Yugoslav–Hungarian border. To slow down the expected attack from the south, the Hungarian side built a multi-layered defence system reinforced with wire barriers, concrete works and mine barrages.[47] Elements of this system also appeared on the Mohács plain (Fig. 5), but by the late 1970s only one object in Majs had been given a name (the somewhat simplistic Tank Trap), despite the fact that structures were also built in Nagynyárád and Sátorhely. The reason for this lies first and foremost in the limited publicity given to the project. What is left of the objects was only taken over by the municipality in the 1990s and is colloquially known as bunkers.

---

45 Pesti (1982).

46 Pesti (1982).

47 Kitanics and Hegedüs (2021).

Fig. 5 Cold war fortress at Majs, 1951–53

The southern border has received renewed attention since 2015 due to increased migration to Europe. The Hungarian government has built a double fence along the Hungarian-Serbian border and a single fence along (Fig. 6) the Hungarian–Croatian border to restrict movement. The temporary security fence for border surveillance purposes, is known locally simply as a fence or border fence. Alongside them, high surveillance posts (*magasfigyelő*) have again been set up in the area, for which the abbreviated word *mafi* is used. However, in order for more geographical names to be associated with the border fence and for them to become permanent, the southern border fence and its elements need to be maintained for a long period of time.

Magyarization, Remembrance Politics, Battle of Mohacs

In Hungary, there is a long tradition of naming public spaces (from) after Hungarian national heroes, or historical figures expressing national unity. Although examples from almost every period can be found, a certain clustering can be observed. The main groups of historical figures are as follows:

- the Hungarian conquest of the Carpathian basin in the ninth and tenth centuries and the names of the Árpád Dynasty;
- Heroes of the anti-Turkish struggles of the 15th–seventeenth centuries;
- the national-liberal figures of the 19th-century "reform era" and the anti-Hapsburg liberation struggle of 1848–1849;
- writers, poets, artists and politicians of the first half of the twentieth century.

Fig. 6 Temporary security barrier for border surveillance purposes (2015). *Source* own photo

There is no balance between the above groups. The local authorities, seeking to conform to the ideological background of the political establishment, have repeatedly made changes in the areas under their jurisdiction. In the last century, there have been several renaming periods in Hungary affecting geographical names:

- First of all, the period between 1898 and 1912, when the official names of settlements were being determined, with a significant degree of Magyarization.
- During the 133-day rule of the Communist Republic of Councils in 1919, there were also name changes in several places (this did not affect the Mohács region, which was under Serbian occupation at the time).
- Renamings took place between the two world wars, but mainly in the late 1930s and early 1940s. As a result, names with German and Italian political connotations appeared on public squares (*Mussolini* and *Hitler tér* (Mussolini and Hitler squares) in Budapest, for example).
- In the first half of the 1950s, names with a Soviet-Russian influence and public squares named after communist martyrs/heroes appeared.
- In the early 1990s and after 2012, (an) other waves of renaming swept Hungary. In two phases, most of the names with a communist connotation were either changed to the former name or the names were changed to politically neutral ones.

In the examined region—as in other parts of Hungary—the state geographic naming policy began in 1898, when the National Municipal Register Committee was established (in) by the Austro-Hungarian Monarchy. In the course of its work until 1912, it reviewed the country's municipal names and intervened in many cases in local naming to eliminate nomenclatural duplications and to clarify geographical place names in line with modern life and the needs of the state. Its statutory task was to ensure that each municipality had only one official name and a clear spelling. In addition, it had a decisive role not only for the names of the old settlements but also for the names of the newly founded ones. In the Kingdom of Hungary at the end of the nineteenth century, there were about a hundred thousand names for the twelve and a half thousand settlements. The changes were not only practical, however, but also reflected a choice of values: the committee made efforts to promote Magyarisation throughout its work. All it needed was a foreign pronunciation to translate the names in multi-ethnic areas—such as the Mohács plain—and give them a Hungarian form and meaning. Thus, for example in our region, the medieval name *Lanchuk* became Lánycsók ('girl kiss') in 1903, or the South Slavic name *Satoristye* became Sátorhely ('place of the tent'), while the name *Udvard* was simplified to Udvar ('yard'). The name of *Gross Nyárád*, inhabited by Germans, became Nagynyárád in the twentieth century, when the adjective was changed to "nagy" (gross = nagy = big).

Since 1964, the Geographical Names Committee has taken over the role of the OKTB, with the task of establishing the official geographical names. It has been re-regulated several times and its current remit is "*to establish and give opinions on certain official geographical names and to register geographical names.*"[48]

Below we present the characteristics of the public names of the seven settlements of the Mohács plain and the changes that have taken place in the nomenclature.

The historical interest of the local society is demonstrated by the fact that the name of the Roman settlement of the area, Altinum, appears among the names of the public spaces in Mohács (Altinum street), together with the names of other nearby medieval

---

48 303/2007. (XI. 4.).

Table 2 Names of the heroes of the Battle of Mohács in the public spaces of Mohács, Sátorhely, Lánycsók and Kölked

Mohács	Sátorhely	Lánycsók	Kölked
II. Lajos street	II. Lajos street	II. Lajos street	II. Lajos street
Tomori Pál street	Tomori Pál street	Tomori Pál street	
Drágffy János street	Brodarics István street		
Brodarics István square	Kanizsai Dorottya housing estate		
Kanizsai Dorottya street	Perényi Péter street		
Perényi street			

settlements that no longer exist (there are streets named after the medieval Hungarian Földvár and Körtvélyes in Mohács). In addition, there are also street names related to the mythological past of the Hungarian people and to the conquest of Hungary (*Nimród*, *Megyer*, *Hunor*, *Géza fejedelem* [Géza Prince], *Árpád* and *Szent István* streets).

At the same time, the name of St Stephen appears in other settlements, including Nagynyárád. The names of such great figures of the anti-Turkish struggles as János Hunyadi, Mátyás Hunyadi or Miklós Zrínyi can appear as public names in any settlement in the country, so we can observe them in the seven settlements under study. At the same time, there is a peculiarity of the local nomenclature that is much less common elsewhere, namely the use of the Hungarian names of the heroes of the Battle of Mohács (Table 2), fought here in 1526 against the troops of Suleiman I. These persons are:

- Dorottya of Kanizsai, the noblewoman who, according to tradition, buried the dead of the battle with her 400 servants. Not only streets but also the local museum are named after her.
- István Brodarics, bishop of Serem, royal chancellor, humanist, chief chronicler of the battle. Not only a public place, but also the primary school of Mohács bears his name.
- According to tradition, János Drágffy, a magistrate, carried the banner in the battle and was killed in action.
- Pál Tomori, Archbishop of Kalocsa, commander-in-chief of the Hungarian troops, was killed in battle.
- Lajos II, King of Hungary and Bohemia, fled the battlefield after the defeat and drowned in the Danube.
- Péter Perényi, nobleman, took part in the Battle of Mohács and survived.

It is interesting to note, however, that in the villages inhabited by people of German origin, there is no public space named after any of the participants in the Battle of Mohács. In Udvar there is only one street, *Fő utca* (Main Street), and in Majs and Nagynyárád there are several Hungarian names, but no names of heroes from Mohács (Table 3). It is also striking that neither Majs nor Nagynyárád—despite their

Table 3 Street names of Majs and Nagynyárád

Majs	Nagynyárád
Kossuth Lajos street	Kossuth Lajos street
Táncsics Mihály street	Rákóczi Ferenc street
Petőfi Sándor street	Petőfi Sándor street
Károlyi Mihály street	Táncsics Mihály street
Ady Endre street	Szent István utca
Dózsa György street	Dózsa György street
Ifjúság street	Arany János street
Virág street	Budai Nagy Antal street
Szegfű street	Pincesor street

nationality roots—have any official public names of German origin. On the other hand, several names referring to Hungarian national belonging can be read off the maps.

The prominent figures of the reform era and of the 1848–1849 Revolution and War of Independence play a prominent role in the use of public names throughout Hungary, including the Mohács plain. Streets named after Lajos Kossuth (politician, governor of Hungary in 1848–1849), István Széchenyi (Hungarian aristocrat, promoter of westernisation and modernisation), Sándor Petőfi (poet and military officer, martyr/hero of the War of Independence), János Arany (poet), Mihály Táncsics (writer, publicist, utopian socialist) are typical public names.

There are two other interesting phenomena that strike the observer. As in the rest of the country, the name of Prince Ferenc Rákóczi II (leader of the anti-Hapsburg Kuruc uprising of 1703–1711) appears on street signs (Mohács, Nagynyárád, Lánycsók). He is almost the only Hungarian historical figure who has enjoyed unbroken popularity over the past century and a half, regardless of the period or political regime. He has been invoked and his memory upheld in every political era, although different elements of his complex historical legacy have been emphasised. The Horthy, Rákosi, Kádár and Orbán governments have all used his memory.

Another interesting feature is that a good number of historical figures who are compatible with the Marxist view of history (who lived before 1945 and can be interpreted on the basis of Marxist doctrine) are also represented in the public spaces of the seven municipalities. Among the members of this group are Antal Nagy Budai (soldier, leading a peasant uprising in Transylvania in 1437), György Dózsa (soldier, leading a peasant uprising in the Great Plain in 1514), Mihály Károlyi (the "Red Count", leader of the 1918 Revolution) and Attila József (a very popular "proletarian poet" in Hungary in the first half of the twentieth century).

The explanation for the above naming anomalies can be traced back in part to the events of the 1940s. After 1945, the ethnic German population was forced to suffer severe measures because some of them supported the Nazi occupation. Regardless of the fact of collaboration with the German invaders, which was not massive in this region, the wealthy, conservative parts of the population were mainly deported to the German territories occupied by the Allies. At the same time, their property (houses, land and other assets) was taken over by Hungarians expelled from Czechoslovakia under the Beneš decrees or by poor peasants from the eastern part of the country. These new settlers were loyal to the new power that allowed them to prosper, and formed a new Hungarian-speaking layer of leadership in the originally German-dominated settlements. Over time, a small number of the displaced Germans migrated back to the villages, but they were no longer able to play the role in village life that they had previously. In the naming as a kind of symbolic occupation of space, the common names among the Germans were replaced by those preferred by the new ruling class. These were either the names of consensual Hungarian historical figures, or the names of people who occupied points of alignment in the Marxist historical narrative—peasant uprisings, leaders of revolutions, representatives of socialist/communist views.

In the post-1989–1990 period, the names of the Communist party-state era have largely disappeared from the local registers. Streets and squares named after the leaders of the 1919 Republic of Councils, the worker movement figures of the Horthy era, representatives of the anti-fascist resistance and the socialist establishment after the Second World War, and the canonised figures of the communists who died in 1956, began to be renamed. The process was made compulsory by a legal provision in 2012, although it has not been implemented everywhere or in all cases.

In Lánycsók, *Imre Mező Street* was to be renamed (Imre Mező was a member of the international communist movement, shot in Budapest in 1956, and thus remembered as a communist martyr/hero in Hungary), but here it was decided not to change the street name. In Nagynyárád, however, the street named after the communist Máté Zalka (general, fought in Spain in the ranks of the international brigades, died in 1937 fighting the Falangists) became *Arany János utca*.[49]

Twenty-one streets and squares in Mohács were renamed in 1991. In the course of the action, popular names of the socialist period were replaced either by the former street names or by others. Among the names considered 'obsolete' were, for example, Béla Kun (revolutionary, leader of the Republic of Councils), Máté Zalka mentioned above, Imre Sallai and Sándor Fürst (two communist activists executed on false charges in 1932), Endre Ságvári (anti-fascist activist shot dead in a gunfight with the police in 1944) and György Matuzsa (communist secret service officer killed in 1956). The most prominent names were therefore heroes and martyrs of the communist movement. Other names that were changed referred to the socialist era in a more indirect way, such as *Úttörőliget* ('Pioneer grove'), which recalled the pioneer movement as an important arena for ideological education, or Tanácsház utca (the name referred to the Councils, borrowed from the Soviet administrative system), or *Terv utca* (Plan Street) and *Terv tér* (Plan Square) (() recalling socialist plan economy()).

In other parts of the country, the names of heroes who are completely unknown, and only locally important, are currently found in the study area in a negligible number. Only in two settlements (Mohács and Lánycsók) were public places named

[49] However, the Germans in Nagynyárád continued to use the old, traditional street names Altgaß, Neigaß, Ratzkhipl, Majser Tal (oral report by local resident, historian János Hábel).

after locally connected, recognised people found. In Mohács, streets were named after Ede Brand, a central person in the commemoration of the 400th anniversary of the 1926 commemoration, a school principal, and János Ete, a local historian and archaeologist who played an important role in the research of the history of the Mohács area. One might assume that the naming of a public space named after Lajos Muck, who was born in Lánycsók, reflects German national identity, but that is not the case. Muck (1886–1944) played an active role in the anti-fascist resistance and the leftist/communist workers' movement, organising the anti-Nazi movement of German-speaking workers during World War II. The name evokes the practice of the party-state era and not the emergence of a symbolic use of space by the German minority.[50]

If we look at the 21 streets and squares renamed in Mohács in 1991, we can see that in four cases public squares were renamed after local communists, anti-fascists and labour activists (György Matuzsa, Károly Bizony, József Frech and Aladár Hajdenák). After the change, these streets were given politically neutral names instead of local heroes from other backgrounds. *Matuzsa utca* (Matuzsa Street) thus became *Indóház utca* (Indóház Street or'House of launch'), *Bizony utca* (Bizony Street) became *Kodály Zoltán utca* (Zoltán Kodály Street, Zoltán Kodály was a twentieth century Hungarian music teacher, folk song collector and composer), *Frech utca* (Frech Street) became *Duna utca* (Duna Street), and *Hajdenák utca* (Hajdenák Street) became *Dobozi Mihály utca* (Mihály Dobozi Street, Mihály Dobozi was a semi-mythical figure who died in 1526 near Esztergom during the anti-Turkish struggles). In summary, the last renaming period, which began in 1990, saw a significant reduction in the number of names associated with locally relevant persons, as names with a communist connection were selected out.

Conclusion

Until the second half of the nineteenth century and the beginning of the twentieth century, the study area was characterised by the use of parallel languages. Its geographical nomenclature reflects the cultural and naming characteristics of the indigenous inhabitants of the area (Hungarians) and the settlers of the seventeenth and nineteenth centuries (Southern Slavs, Germans, Gypsies, Jews), which often correspond well to the structure of their settlement area.

Ethnic/linguistic homogenisation started at the end of the nineteenth century and progressed significantly: at the same time, the use of national minority geographical names was strongly reduced, mainly due to the state action after 1898. The Germanisation processes of the eighteenth and nineteenth centuries are either not perceptible in the region or their impact was so insignificant as to be intangible.

[50] There was also a street named after Lajos Muck in the XIII district of Budapest, but it was renamed Thurzó Street in 1990, after the regime change.

The top-down renaming policies observed in Hungary—with the exception of the period of the Republic of Councils in Hungary (1919), when the territory was under Serbian occupation—can also be observed in the study area. As a result, an almost completely homogeneous Hungarian public domain nomenclature has been created. Within this, the communist naming legacy of the second half of the twentieth century is still dominant, from which practically only the names of communist, anti-fascist (martyrs) heroes have been removed, but not even that in its entirety. The figures of the class-struggle explanation of history, which played a dominant role in the narrative of Marxist historiography, have remained in the public square.

The use of women's names in the name material can be considered negligible, below even the otherwise conservative Hungarian average. Names associated with the process of female emancipation that began in Hungarian society at the end of the nineteenth century do not appear in the region. A woman was named after a place if she was of royal descent, if she was a saint, if she was of noble origin and/or a national hero. In the few cases where a woman of civilian origin was designated by a geographic name in the area, the nominee was the owner of the named business, an employee or a relative of the owner.

The number of names of local heroes is negligible when compared to the national names of national affiliation. The 'locality', or local attachment, is no longer formed by actual locally-born politicians and activists, but by the abstract heroes of the Battle of Mohács in 1526.

Questions

1. How are different social groups (ethnic, gender, political, religious etc.) represented in the local toponymy?
2. What are the driving factors for place naming changes over time?
3. How did ethno-linguistic homogenisation affect local place names?
4. What is the role of the migration settling waves in local toponymy?
5. What is the role of the Battle of (Mohacs) Mohács (1526) in the given street names?

References

Baroch Cs (2019) Szimbolikus politikai fordulatok és térbeli mintázataik: közterületek átnevezése Budapest I. és V. kerületében. Földrajzi Közlemények 143(2):144–157

Bél M (1735/2017) Notitia Hungariae novae historico geographica... Tomus IV: Fejér, Tolna, Somogy, Baranya vármegyék. Budapest

Cadastral maps (1865) Mohács, Lánycsók, Nagynyárád, Majs, Sátorhely, Udvar, Kölked. https://maps.arcanum.com/hu/map/cadastral/?layers=3%2C4&bbox=2045551.7738559484%2C5756618.46966983%2C2118281.6062692935%2C5784403.329451492

Csernicskó I (2018) Nyelvpolitika és nyelvi tájkép többnyelvű környezetben: Kárpátalja példája. II. Rákóczi Ferenc Főiskola, Beregszász

Fancsaly É, Gúti E, Kontra M, Molnár Ljubić M, Oszkó B, Siklósi B, Žagar Szentesi O (2016) A magyar nyelv Horvátországban. Gondolat Kiadó—Media Hungarica Művelődési és Tájékoztatási Intézet, Budapest—Eszék

Faragó I (2014) Földrajzi nevek. ELTE Informatikai Kar, Térképtudományi és Térinformatikai Tanszék, Budapest

First Military Survey (1783) Lánycsók. https://maps.arcanum.com/hu/map/firstsurvey-hungary/?layers=147&bbox=2065142.6381849954%2C5781364.008425402%2C2074233.8672366636%2C5784837.115898111

Gemeinde Lancsuk (1853, cadastral map): Magyar Nemzeti Levéltár, S Kataszteri-gyűjtemény (1786–1948), S_78_-_026._téka_-_Lánycsók_-_1–5. https://maps.hungaricana.hu/hu/MOLTerkeptar/15212/

Gemeinde Mohács (1852, cadastral map) Magyar Nemzeti Levéltár, S Kataszteri-gyűjtemény (1786–1948), S_78_-_031._téka_-_Mohács_-_1–9. https://maps.hungaricana.hu/hu/MOLTerkeptar/15303/?list=eyJxdWVyeSI6ICJtb2hcdTAwZTFjcyJ9

Gemeinde Majs (1853, cadastral map) Magyar Nemzeti Levéltár, S Kataszteri-gyűjtemény (1786–1948), S_78_-_028._téka_-_Majs_-_1–7

Hoffmann I, Rácz A, Tóth V (2017) History of Hungarian Toponym. Buske Verlag, Hamburg

https://maps.hungaricana.hu/hu/MegyeiTerkepek/5058/view/?bbox=-7566%2C-9704%2C14155%2C285

https://maps.hungaricana.hu/hu/MOLTerkeptar/15250/?list=eyJxdWVyeSI6ICJtYWpzIn0

Kitanics M, Hegedüs N (2021) The Latest 'Southern Protection System' and the revived 'Fortress of Europe' topos in Hungary. Politics Central Europe 17:729–760

Kneidinger A (1772) Charten von der ka[mmer]al Ortschaft Mays… Magyar Nemzeti Levéltár, Kamarai Térképek, S_11_-_No._830:43

Kneidinger A (1767–1779) Charten von Mayser District in der… Magyar Nemzeti Levéltár, Kamarai Térképek, S_11_-_No._830:41.

Kovacsics J (2001) Magyarország történeti statisztikai helységnévtára 18. Baranya megye, KSH, Budapest

Lelkes Gy (2011) Magyarország történeti helységnévtára (1773–1808). Baranya megye. KSH, Budapest

Lazarus secretarius (1527/1982) Tabula Hungariae. Vas Megyei Levéltár, VAML T 541

Marsigli LF (1696/1726) Danubius Pannonico-Mysicus… Hagae Comitum: Apud P. Gosse, R. CHR. ALBERTS, P. De HONDT; AMSTELODAMI: Apud HERM. UYTWERF & FRANÇ. CHANGUION

Mikoviny S (1720–1725?) Comitatus Bacsiensis Pars. Magyar Tudományos Akadémia Térképtár Mo. 1

Pap N, Fodor P, Kitanics M, Morva T, Szalai G, Gyenizse P (2018) A mohácsi Törökdomb. Történelmi Szemle 60(2):325–345

Pap N, Kitanics M, Gyenizse P, Szalai G, Polgár B (2019) Sátorhely vagy Majs? Földvár környezeti jellemzői - a mohácsi csata centrumtérségének lokalizálása. TÖRTÉNELMI SZEMLE 61(2):209–246

Pap N, Reményi P (2020) Encounters Between Islam and Christianity: Mohács and Kosovo Polje. In: O'Reilly G (ed) Places of memory and legacies in an age of insecurities and globalization. Springer International Publishing, Cham, pp 285–305

Pap N, Gyenizse P, Kitanics M, Szalai G (2020a) Az 1526. évi mohácsi csata helyszíneinek földrajzi jellemzői. In: Pap N (ed) Mordortól Mohácsig: A mohácsi csatatáj történeti földrajzi kutatása. MTA Bölcsészettudományi Kutatóközpont, Budapest, pp 123–168

Pap N, Gyenizse P, Kitanics M, Szalai G (2020b) II. Lajos halálának helye. TÖRTÉNELMI SZEMLE 62(1):73–109

Pap N, Kitanics M, Gyenizse P, Szalai G (2021) A mohácsi síkon fekvő földvár település. ÚJKOR.HU—A velünk élő történelem. 2021/07.08, pp 1–4

Pesti J (1982) Baranya megye földrajzi nevei I-II. köt. Pécs

Pesty F (1864–1865/2019) Baranya vármegye helynévtára. Csorba Győző Könyvtár, Pécs

Tent J (2015) Approaches to research in toponymy. Names 63(2):65–74
Universum dominium Siklossiense (1700) Országos Széchenyi Könyvtár, kéziratos térképek. https://maps.hungaricana.hu/hu/OSZKTerkeptar/2142/
Vályi A (1799) Magyar országnak leírása 1–3. köt. Buda, pp 702, 736, 688

Recreating the Future: Modern Residential Neighbourhood and Existing Toponyms in Sarajevo

Elša Turkušić Jurić and Velid Jerlagić

Abstract Sarajevo's two largest municipalities, Novo Sarajevo (New Sarajevo) and Novi Grad (New City), were built in the second half of the twentieth century, as a result of ambitious and numerous modern urban planning projects. Because these pragmatic and rational processes treated the existing rural landscape as a tabula rasa, the original toponyms were only formally retained. In the perceptions of present-day residents, the urban areas in these two municipalities are an integral and embedded part of their identities and memories. This chapter starts from the supposition that, after several decades, these formerly new residential areas have given new meanings to pre-existing toponyms. It illustrates both the spatial and cultural characteristics of these meanings, and recent instances of their "re-decomposition", as a consequence of globalization. The research is based on the comparison and analysis of six modern housing estates. An analysis of the typological toponyms of Sarajevo's historic districts gives greater insight into these urban and social transformations.

Keywords Globalization · Residential neighbourhoods · Modernism · Sarajevo · Urban-planning

Objectives

- Explanation of the history of establishment and transformation of the Sarajevo's toponyms; determination of the link between process of change of toponymic meanings and spatio-organisational changes within modern residential neighbourhoods; insight of the social character of modern urban-planning of Sarajevo (;).
- By the end of this chapter readers should be able to understand (former and ongoing) levels of transformations in the meaning of toponyms in Sarajevo's new parts (New Sarajevo and New City).

E. T. Jurić (✉)
Faculty of Architecture, University of Sarajevo, Sarajevo, Bosnia and Herzegovina
e-mail: elsat@af.unsa.ba

V. Jerlagić
Sarajevo Historical Archives, Sarajevo, Bosnia and Herzegovina

© The Author(s), under exclusive license to Springer Nature Switzerland AG 2023

G. O'Reilly (ed.), *Place Naming, Identities and Geography*,
Key Challenges in Geography, https://doi.org/10.1007/978-3-031-21510-0_18

- Readers will learn to recognise the nature of socio-economic and global influences that have effects on the toponymic meanings (;).

Introduction

Sarajevo and its environs have a centuries-long tradition of settlement, in various socio-spatial organisational forms. Like every city, Sarajevo has survived and modernised through the morphological changes of its urban subdivisions. These changes have often been complex, and required longer periods of time to assess their effectiveness, justification, and level of acceptance as elements of the urban morphology and everyday lives of residents. One of the biggest changes in Sarajevo's development (and influences on its current state), was the construction of new, modern neighbourhoods in the second half of the twentieth century, with which the city sought to respond to the social challenges of industrialisation, modernisation and socialist self-government. Architecture has continuously played an important role in these societal processes. Space, its fundamental medium, is transformable, and becomes existential only through architectural actions:

"In addition to practical purposes, architectural structures have a significant existential and mental task; they domesticate space for human occupation by turning anonymous, uniform, and limitless spaces into distinct places for human significance, and equally importantly they make endless time tolerable by giving duration its human measure." (Pallasmaa 2009, 17).

The significance of architecture is twofold: its actions can transform a space by inscribing in it the visionary marks of urban development, and it can imbue that same space with new meaning. The latter process is fuelled by both human ability and the need to give meaning to a space, because "a prerequisite for housing is to establish a meaningful relationship between a human being and a particular environment" (Norberg-Schultz 1990, 30).

The character, efficiency and directional changeability of these processes, observed through the spatial inscription of meaning and reflection on the city's potential future development, was the starting point of this research. The most appropriate examples for analysis were the completely planned residential neighbourhoods that are the fundamental fabric of the new urban municipalities of Novo Sarajevo (New Sarajevo) and Novi Grad (New City). *Topos* means place, and the name of a place carries the meaning of the space it creates. Although the toponymic meanings of the modern neighbourhoods analysed here have changed, their names have been retained. This research applies chronological analysis, and compares five neighbourhoods built from 1948 to 1991, and located between the urban centre of Marijin Dvor and the western edge of the city (Sarajevo International Airport). To understand the intensity and character of the toponymic transformations in these neighbourhoods, it is necessary to first explain the history of the toponyms creation; the chronological development of the spaces they represent; manner and social context in which they were developed.

Neighbourhoods are "specific spatio-social entities", with real and symbolic aspects. The former comprises their physical condition, organisation, dimensions and capacity. The latter is a process of psychological and social appropriation: the building of memories. We therefore start from the fact that these neighbourhoods differ from one another, and that although they are inspired by an original urban architectural concept, they are also in a state of constant change (Čaldarović 2011, 37).

The presentation in this chapter highlights the nature of the ongoing transformations in these residential neighbourhoods, and shows that their characteristics affect the meanings carried by toponyms. Elements of social cohesion in the space, such as the way vacant space is used, and signs of identification and realisation of privacy, are also presented.

Starting from the fact that space is transformative and dynamic, and can be "domesticated" through architecture, this chapter examines three levels of transformation that indicate a change in the meaning of the toponyms of certain neighbourhoods. The first level of transformation takes place with the construction of completely planned neighbourhoods, whose dimensions and architecture change the existing (rural or natural) landscape entirely. After this stage, the toponyms begin to take new meanings, directly related to the characteristics of newly-built suburbs: new parts of the city. The second level of transformation occurs as changes to existing neighbourhoods, and involves the analysis and detection of spatial elements that enable integrity and functionality in the everyday spaces of residents. The third level is based on the analysis of new built structures, and relates to changes within the neighbourhood that were not part of its original concept.

The aim of the chapter is to determine whether the intensive process of change to toponymic meanings is currently occurring again, and if so, whether this is linked to spatio-organisational changes within neighbourhoods. Or, have these processes become independent of each other?

Interpretation of the Topology of Sarajevo

Story of Sarajevo's Toponymy

Absence of the older Palaeolithic artefacts in the core area of the city does not imply the absence of Palaeolithic people, for geomorphological circumstances and climate of the period were favourable for human habitation. Wider city area, especially in karst relief regions (caves, lithic shelters, chasms and cave wellsprings), revealed significant archaeological data[1] pointing at an equal habitation scheme in today's core city area.

---

[1] Bijambare, see: "Arheološki leksikon BiH", Tom 3, str. 39., Reg. 15, lok. 15.11, Zemaljski muzej BiH, Sarajevo 1988.

Abundant material remains from the core city area revealing a major human presence in the Neolithic period—their representativeness being fascinating as a historically sudden emission of powerful creative energy, but also as a toponymy presence indication.

Localized on the confluence of Tilava and Željeznica, in the open field, Sarajevo's main Neolithic settlement was conditioned by natural protection of the watercourses: fertile soil and vast hunting grounds, along with the river that carries suitable type of stone used for tools and arms, were the most important conditions for its setting 5000 years ago.[2]

Sarajevo's Neolithic settlements disappeared suddenly, but certain localities of the late Neolithic and Endolithic (copper) show significant change in the settlement scheme: free life in the flatlands became substituted with habitation in protected locations with prominent features, ranging from tumps on basically flat alluvial deposits to highly inaccessible points (Benac 1954, 13–16).

With the arrival of peoples acquainted with bronze metallurgy, life in the flatlands became almost impossible: as only strategic locations with excellent views are now being used for habitation (Mesihović 2007, 154–164).

The ethnic or linguistic affiliation of the Bronze Age inhabitants remains unknown, as well as the question if their toponymy ever merged into the city's current place-names.

Only with the appearance of the Indo-European population—the Illyrians and Celts—the historical context for the city's modern toponymy arises. Although the matrix of habitation remains similar to the Bronze Age, bound to hilltop forts accompanied with appropriate seasonal and transhumant settlements, the first linguistic localization of the wider area begins at that time. As a result of the multi-century topological Romanization, and later Slavization of the area, the presence of the Illyrian or Celtic linguistic substratum in today's toponymy of Sarajevo is not indisputably determined.

However, the Illyria-Roman epoch is the first to have proven toponyms, preserved in historiography through few and fragmented archaeological but mostly lapidary artefacts, along with rare written sources. A turbulent past has made the Roman toponymy of the Sarajevo area, including the official 'Aquae S…' name, is virtually extinct. The exception is one toponym, which proved to be of the ultimate importance for the region: more precisely the hydronym, for the river Bosna—*Bathinus flumen,* (Mesihović 2011, 13) whose source lies in the south-western corner of Sarajevo field. In later times the whole river basin came to bear the name—Bosna—initially as a name of the region, medieval district, and finally as a political and administrative unit: Banate/Kingdom of Bosnia. The name of the river Bosna is the only surviving toponym of the antique linguistic topological substrate in the area of today's Sarajevo.

Establishment of the larger part of present-day toponymy results from Slavic conquers of the area in the early middle ages. The matrix of Slavic tribal settlement and its later transformation from clan organization into political subjects laid

[2] Butmir, see: "Arheološki leksikon BiH", Tom 3, str. 40., Reg. 15, lok. 15.38, Zemaljski muzej BiH, Sarajevo 1988.

a foundation for the establishment of the local toponymy as we know it today. It is a period of foundation of habitational forms characteristic for the modern era (city, village, hamlet) along with logical exception of purely medieval types of settlement toponymy (fortified city-burgs, walled city with suburbia, a marketplace) and some recent forms (metropolis) (Anđelić 1963).

Further development of toponymy continued in periods of strong influx with oriental languages: Turkish, Arabic and Persian, into the Slavic Bosnian language. The orientalization of toponymy in situ encompassed words from religious, administrative, economic or merchant vocabulary. With the arrival of the Ottomans (1450 s), the concept of urban spatial development, from existing village or empty space to *mahala* (neighbourhood), was supplemented by its merger with administrative facilities (residence of the governor) and *Čaršija* (city market), economic area of the city, which was the medieval communal market square.

Although toponymic exceptions from Germanic and Romance languages existed in the late 19th and early twentieth century, it's actually the Ottoman cultural heritage that gave birth to the prime toponym i.e. the city of Sarajevo. The term "*Saray-Bosna*" (governor's residence in Bosnia) and "*Saray-ovasi*" (governor's residence in the field) are place-names derived directly from the administrative and military-political organization of the Ottomans: the building of the Saray—court and administrative centre of the Ottoman governor in Bosnia (Kreševljaković 1956, 13–22).

As a legacy of the nineteenth-twentieth century, urbanization, construction and expansion of the city led to a new phenomenology of understanding the city's toponymy, primarily through political stances, but not without immense influence of the spontaneous linguistic and emotional expression embodied in everyday activities and reflection of the city's inhabitants themselves.

Topology of the Habitational Structure of Sarajevo's Historical Core

The strategic location of the historical core, situated in the narrow alluvial flats along Miljacka river's rocky gorge, an outlet in the amphitheatre surrounding the nearest terminal slopes of Mount Trebević and the Bukovik massif, is the main determinant for its topological attributes.

The agglomeration gravity centres of the Sarajevo field have shifted through history: in the Neolithic it was Butmir, Debelo brdo in the Bronze and Iron Ages, Aquae S… in Ilidža in the Roman era, Blažuj in early medieval times, until the late medieval period, when the far eastern tip of the field became the district centre—the historical core of Sarajevo.

Although heavily debated, topology of medieval settlement complexes in the area of modern Sarajevo are undoubtedly pointing at the significant economic and political hub of the medieval *Vrhbosna*. Localization of all toponyms is indefinite, but historical sources, archaeological artefacts, cartographical finds and terrain survey indicate the existence of typical medieval urbanization with significant status. Protected by the single, historiographical nameless fortress on Vratnik, urbanization was a natural confluence of the district area and adjacent villages. The fortress had its two market squares—separated by less than a mile—"*Staro*" and "*Novo*" *Trgovište*, both located on important road intersections.

All toponymy of medieval villages in *Vrhbosna* is of Slavic origin. Brodac, Kovači, Bistrik, Hrid, Radilovići, Budakovići, Koševo, Bjelave (Bilave), Kovačići and Gorica are the villages that would merge with the city's commercial, administrative and religious centre and become its neighbourhoods. With the establishment of the Ottoman administrative centre not far away from the oldest city core, the development of Sarajevo's toponymy would stay in close relation to the development of the *mahalas* and their gradual amalgamation with the economic core.

Mahalas are named by mosque founders, religious endowments (*waqf*), and other persons, but always had colloquial names along with the official ones, both often inherited from the medieval period. Continuous use of unofficial and historical names, enabled the expression of Sarajevo's modern toponymy.

The last decades of Ottoman rule were the period of the first formalization of the administrative conformation of the city: streets were officially named, and the city itself was organized as a township. Since then, the administrative policy would mainly follow three topological concepts: traditional, that retained the original place-names; organic (colloquial) in which the inhabitants themselves created and enrooted place-names that would eventually become official; and memorial, where the naming of streets, roads, squares, or planned urban settlements would include important persons or dates from political history of the social arrangement at the time concerned.

Sarajevo's Post-WWII Urban Development: A Socio-Economic Perspective

Sarajevo's current appearance dates from the three decades that followed WWII, which was a time of intensive urbanisation, industrialisation and de-agrarization.[3] The city's area increased fivefold during this period, and the number of households had increased 2.5 times by 1970.[4]

At the end of WWII, Bosnia and Herzegovina became one of six constituent republics of the newly-created Socialist Federal Republic of Yugoslavia (SFRY), with Sarajevo as its capital and social, cultural and commercial centre. The city expanded merging with its rural outskirts and forming, over time, the municipalities of Novo Sarajevo (New Sarajevo) and Novi Grad (New City).

[3] Before WWII, Bosnia and Herzegovina had been a predominantly agricultural nation. Rural migration had already begun at this time, and had created a new class of workers (especially in the mining districts of central B&H). This process accelerated during the socialism era, and led to a new urban working class.

[4] In 1948 there were 38,999 households; in 1970 there were 99,679. (Aganović 2014, 190).

The 1948 urban plan foresaw the city's development towards the wider Sarajevo plain, whose realisation began on the basis of guidelines detailed in the 1965 General Urbanistic Plan (GUP). This plan was based on the concept of functional spatial organisation and contained a fundamental principle of modern urbanism: clearly defined spaces for work, housing, recreation and leisure. The left bank of the Miljacka River was intended for housing. The right was to be for commerce and industry, partly because it contained infrastructure from earlier periods (including industrial plants, a railway station, a furnace and workshop, and freight terminals). Completely planned residential neighbourhoods were the basic organisational unit of a modern socialist city. In the new parts of Sarajevo, neighbourhoods were arranged along the city's axis- boulevards' Dragon of Bosnia and Meša Selimović (former 6th Proleterian Brigade) and, between busy transversal (north–south) and longitudinal (east–west) traffic streets. This formed groups of *microraions*, or residential communities, which gravitated towards *raions* or suburban centres.[5] In addition to the existing centres of Baščaršija and Marijin Dvor, new, evenly-populated ones were planned along the length of the city: Otoka, Čengić Vila and Mojmilo (Fig. 1).

The Spatio-Functional, Design and Cultural Characteristics of Modern Residential Neighbourhoods

Complete residential neigbourhoods were built according to the principles of modern urbanism: free-standing buildings were placed within extensive green spaces compensated for the smallness of the apartments; the absence of traditional streets; grande vistas from living areas; and the design of apartments was spatially organised to be functional.

"The organisational design of residential neighbourhoods should provide all the conditions for an easier, more comfortable and more cultured life, especially with regard to the development and education of children. [...] Individual apartments and the neighbourhood should be considered an inseparable whole"(Krstić 2014, 201).

The sociological structure of the population in most new neighbourhoods was diverse: high-ranking officials, intellectuals, and residents with lower incomes or levels of education lived together in one building; buildings for military personnel and buildings for civilians were in one neighbourdhood (Grbavica 1, Breka, Vojničko Polje, Dolac Malta).

Every neighbourhood was supposed to have its own (public) facilities and commercial, cultural, educational, health and recreational services, whose capacity was based on the number of residents. The same principle was applied to green spaces, which were generally at least the same size as built-up areas. Often, financial constraints prevented the total (or even partial) realisation of these facilities and services, some of which would have encouraged diverse activities among residents.

---

[5] These residential communities (microraions) had an area of ca. 20–30 ha, and 10,000–12,000 inhabitants. A raion had ca. 20,000–30,000 people. (Source: ARH, 1963).

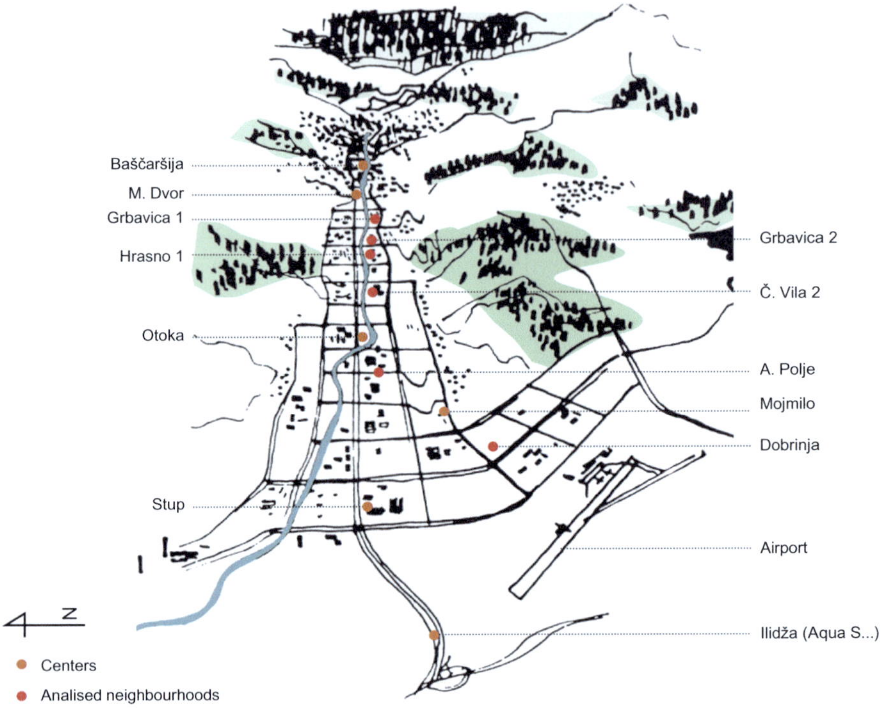

Fig. 1 The topography of Sarajevo is picturesque and interesting: spatio-historical sequence is followed by the transition from valley to plain. Following the linear urban structure from east to west, one has the impression of unwinding a chronological film strip, on which the city's various historical periods are outlined: from Ottoman to Austro-Hungarian to modern functional socialist periods. *Source* ARH 2–3; redesigned by Elsa Turkusic Juric; technical support: Amila Terzić

Their absence, therefore, contributes to the monotony and difficulty of everyday life. The unexpectedly slow development of new city centres has compounded the problem, and transformed these spaces into "dormitory neighbourhoods". The reasons for this lie partially in the uncontrolled influx of residents into these newly-built neighbourhoods, which happened in larger numbers than expected.

Also partly to blame is the sociological character of the neighbourhoods, which, although they provide the basic conditions for a healthy lifestyle, fail to take into account the habits, wishes and social needs of residents (Kulić et al. 2012, 121).

Inhabitants of the central and old parts of the city generally view the new neighbourhoods as unattractive; not only do they lack a wide range of facilites, but they were created ex nihilo, and treated existing localities as tabula rasa. This is compounded by their lack of visual and spatial cohesion, and the uniformity of their facades (which on some buildings do not have balconies or loggia), and the massiveness of the architectural structures.

Nevertheless, some buildings were given local (of) or unofficial names: “Crvena/Red building” and “Teheranka/Tehran building”(according to its prominent façade’colour), “Kifla”, “Kula” and “Šibica” (according to their shapes), “Metalka”, “Zvijezda”, “Loris”, “Palma”, “Turbo”, (according to the shop (in) on its ground-floor), “Bolnica” or “Pancirka” (according to its “their role/function” during the 1992–1995 War Siege of Sarajevo), “Karingtonka/ Carrington building” (according to high-ranking officilas residents), etc.

For a long time during and after their construction, these fully planned neighbourhoods were subject to criticism; urban-sociological and cultural-anthropological research; and expressions of dissatisfaction from members of the public and architecture and urbanism experts.

They have been described as soul-less, a sentiment that could be applied to most post-WWII residential neighbourhoods in Europe (Kulić et al. 2012, 121). The social perception of these modernist “dormitory” neighbourhoods changed, however, in the final decades of the twentieth century. This is because the accumulated lived experience of residents and development of multi-structured social relations (such as moving in, spending childhoods and establishing social ties) have given these spaces native characteristics.[6] Cultural anthropology and urban sociology have contributed most to this change in social perception, because they address the phenomenon of “the urban”, starting with the fact that architecture and urban environment strongly influence how people’s (residents’) lives are shaped. Daily experiences and social narratives are inscribed in the fabric of the neighbourhood, which records the space’s collective and individual memories, and the changes that take place within it. This becomes symbolic and meaningful in ways that are recognised and accepted by new residents.[7]

Housing Policy: Rational and Continuous Mass Construction

A socially-oriented accommodation policy was seen as the best response to the high demand for housing, regardless of the social category of the population. The residents became the owners of the apartments; apartments could be inherited on a “right to occupancy” basis, but not individually sold.

Such an efficient, rational and standardised construction of housing could only be implemented with a steady source of funding (a self-contribution fund for housing construction).[8] A number of conditions was set that influenced design solutions,

---

[6] At the end of the 1970s, a tendency had emerged that revalued urban life and developed arch-urban projects that also considered sociological aspects. An example of this is the Ciglane neighbourhood (former Đuro Đaković) in Sarajevo.

[7] Significant migrations took place after the 1991–1995 war, as a consequence of the break-up of the SFRY, which changed the structure of the population in neighbourhoods.

[8] The SFRY Law on Contributions was passed in 1955, and ensured lasting finance for the construction of housing (and thereby the development of cities) in Yugoslavia. The law states that all labour

such as the typification of buildings structures, and the use of normative, modular and prefabricated constructions (Box 1).

Housing policy (1956–1965)[a]: *Fixed price for an entire neighbourhood*	Housing policy (1966–1991): *Apartments as marketable products*
Implemented by the organised cooperation of all stakeholders (designer, contractor and investor); and city's coordinator- the Sarajevo Directorate of Housing (founded in 1957, and renamed the Sarajevo Fund for Housing Construction in 1962) Neighbourhoods: Grbavica 2 (former Slobodan Princip), Koševo (former Zuko Džumhur), Čengić Vila 1 and 2, Svrakino selo (former Pavle Goranin), etc.	Socially-oriented housing adapted to social and market changes, to become "self-governing housing communities of interest, through which housing construction was directed towards a market of known buyers" (Aganović 2014, 216) Neighbourhoods: Hrasno 1 and 2, Dolac Malta (former Novo Sarajevo), Otoka, Ciglane, Mojmilo, Alipašino Polje, Breka, Dobrinja, etc.

[a]The number of apartments built quadrupled compared to the previous decade - 19,866 and 4,394 respectively(Aganović 2014, 203–355). The construction of Čengić Vila 2 brought with it the lowest cost for an apartment in Yugoslavia, mostly as a result of the cooperation of all stakeholders. A daily newspaper ran the headline "An Agreement that Binds Everyone": evidence of the urgency to resolve the housing issue, which in turn affected the quality of the housing construction culture.

Examples of Completely Planned Modern Residential Neighbourhoods

Grbavica: From Landscaped Units into the Pavilion-Based Neighbourhood

History

Grbavica is a former rural landscape situated between Kovačići and Hrasno, previously longitudinally intersected by Kovačići stream. The origin of the place-name is rather indistinctive, possibly related to: uneven terrain (*grba*), hornbeam groves (*grab, grabovica*) or a common toad (*gubavica*). Thanks to its abundantly fertile lands, Grbavica was presumably related to pre-historic, antique and late antique settlements on Debelo brdo and Soukbunar (refugium, open-type settlement) together with one in Kovačići (Roman brickyard). The population there started before 1555, when it was mentioned for the first time in recorded history (Bejtić 1973, 43). Another mention of Grbavica from 1682 indicates its status as a minor village or hamlet.

Since the last decades of Ottoman rule and up to World War II, Grbavica was inhabited sporadically, with summer houses, isolated farms, or individual houses

and other organisations are obliged to allocate 4 percent of personal income contributions to housing construction.

grouped in family clusters. Only in the first half of the twentieth century did a physical annexation of Grbavica to the city of Sarajevo truly begin. With the construction of Sarajevo's first brewery in 1866, it becomes the city's 'first' industrial area; development in the following decades was marked only by a communal slaughterhouse (1881), cattle market and some buildings belonging to the military.

Due to the proximity of the main train station (1882), the true industrial point of gravity was transferred to areas of the newly formed *Neusarajevo* (New Sarajevo) (Kreševljaković 1969, 201). Albeit administratively included into the city of Sarajevo together with Neusarajevo and Pofalići after 1895, settlement started to modestly appear only in the period between two World wars (Kreševljaković 1969, 197).

At first, 20 working-class houses were raised, only to be replaced by construction of planned settlements Grbavica 1 and 2 (former "Slobodan Princip Seljo"). Between 1948 and 1968, the territory of Grbavica fully changed its appearance and function, becoming totally integrated into the city.

Grbavica 1 (1948–1958)[9]

Grbavica 1 is composed of small residential pavilion units (square or rectangular in shape with greenery in between) (Fig. 2).

Several types of residential building are present, all of which have balconies or loggias, and most of which have flat, accessible roof terraces. Some of them are unofficially named as Vojni Paviljoni/Military pavilions, and "Metalka" buildings. The buildings are positioned in either an east–west or north–south orientation. According to the original project, the ground floors of all residential buildings were intended to be activated spaces for public use, but a lack of available apartments meant that they were used instead for residential purposes, a role they have maintained to this day.

The public facilities are concentrated between two areas: the covered and open market "Grbavica pijaca" (in the S-W part of the neighbourhood); and the three high-rise building near the Miljacka River, to the north-east, unofficially named "Zvijezda neboderi"/Star towers. Both areas offer a range of on-street shopping and hospitality facilities. The "Grbavička" main traffic streets within the estate are one-way.

The open pavilion spaces with their various features (defined entrances, garages, parks and playgrounds) give the neighbourhood a harmonic, diverse atmosphere. The way these areas have been transformed, either by residents or the municipal/communal administration, into spaces for leisure shows an effort to raise the quality of housing in the neighbourhood (Fig. 3a, b).

Grbavica 1's connection to surrounding areas is especially pronounced, due to the proximity of the Marijin Dvor urban centre. There is an emphasis on pedestrian traffic, mainly to and from the museum complex, along attractive, tree-lined Wilson's Promenade, via the newly-built Ars Aevi footbridge (architect: R. Piano, 2002).

[9] The western microdistrict has an area of 6.8 ha; 10 objects; and 656 apartments. Landscape architecture: S. Klaića; urban and building designers: Z. Kovačević, M.Peterčić, B. Kalajdžić, V. Zarahović, Ž. Janković, Z. Likić. (Source: Aganović 2014).

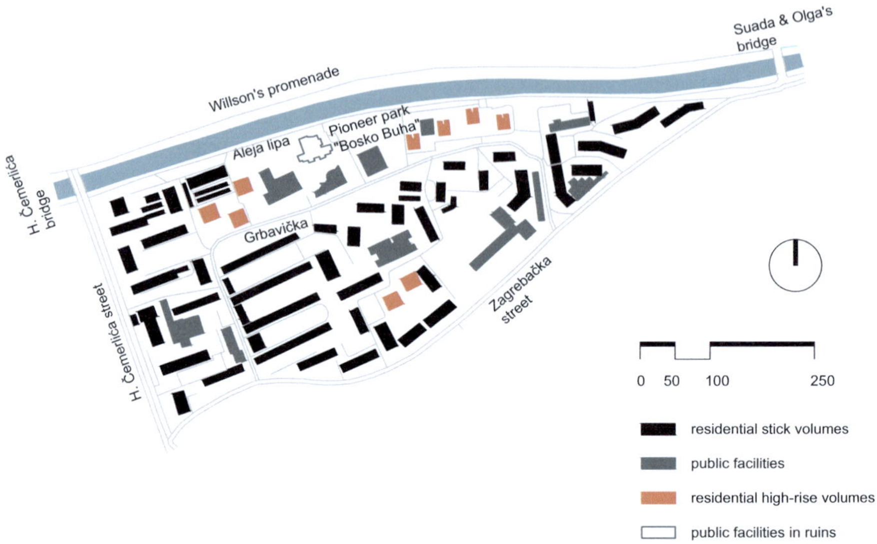

Fig. 2 The neighbourhood Grbavica 1 comprises residential stick volumes (4, 6, 8 and 12 storeys high) with green spaces, organized in pavilions; three groups of 16-storey residential buildings, as spatio-visual verticals; public buildings consist of: two primary schools "Kovačići" and "Grbavica 1"; open and covered markets "Pijaca Grbavica"; a youth cultural centre "Children's House" with a landscaped "Pioneer Park -Boško Buha" with stone sculptures; a kindergarten (today in ruins); a community centre for high school students (now a commercial building); the Forestry Faculty of the University of Sarajevo (part of which has been returned to the Catholic seminary); two administration buildings (the smaller is in ruin today); and a Jordanian mosque (built after 1995.). *Source* Elsa Turkusic Juric; technical support: Amila Terzić

Grbavica 2 (1958–1962)[10]

The neighbourhood consists of two parts: a peripheral residential zone (pavilions) and a central public space that is bordered by a tree-lined one-way vehicular roads (B. Mutevelića Street and K. Kapetanovića Street) (Fig. 4).

The uniform height of the buildings, the significant green spaces, and pavilion atmosphere define the spatial and design characteristics of the neighbourhood. The buildings are positioned in either an east–west or north–south orientation.

User-led transformations within the neighbourhood are sporadic, and mostly involve the areas around building entrances. Colourful facades attempt to break the monotony of the neighbourhood, which is characterised by a lack of active ground floors and public facilities along the street as a whole (Fig. 5a, b).

The "Bristol" footbridge amplifies the connection between Grbavica 2 with the other side of the river (Wilson's Promenade, the hotel "Bristol", administrative and

[10] Area 21.5 ha; 52 objects; 2,612 residents. Sources: Z. Kovačević, M. Peterčić, A.Paljaga, H. Muhasilović, V. Zarahović, H. Salihović (buildings along the Miljacka River), I.Štraus (residential buildings and shopping centre). (Source: Aganović 2014).

Fig. 3 Military pavilions with intensive greenery; Pavilions with playgrounds. *Source* Elsa Turkusic Juric

Fig. 4 The central part of the neighbourhood Grbavica 2 comprises: three 16-storey residential buildings "Kula/Tower 1,2 and 3"; a 3-storey shopping centre "Šoping"; a primary school "Grbavica 2" with a sport field; primary musical and balet school "Novo/New Sarajevo"; a nursery (today in ruin); a park with a small open-air amphitheatre; a centre for children and youth and a mosque (built after 1995). The residential zone (along the eastern, western and southern boundaries) comprises 6- and 8-storey residential buildings, and central open areas (green spaces and carparks). The Strojorad former industrial complex occupies the north-western part of the neighbourhood. The northern zone, on the banks of the Miljacka River, comprises four 12-storey residential buildings and a kindergarten "Mašnica". *Source* Elsa Turkusic Juric; technical support by Amila Terzić

educational facilities). The neighbourhood's northern attractive tree-lined footpath (Aleja lipa), with rhythmically placed buildings, allow an open view of the neighbourhood, and from it to the river and footpath. While, the neighbourhood's southern face borders busy traffic streets, and the sports and recreation facilities around the Grbavica Stadium.

Fig. 5 Residential pavilions with user-led transformations; Central core with residential towers “Kula” and “Šoping”. *Source* Elsa Turkusic Juric

Čengić Vila: An Historic Building Name, Important to the "Scattered" Residential Neighbourhood

History

Traditionally, this area fell into the municipal domain of Dolac village, and was largely uninhabited, until it become part of the city in 1945. The place name was derived from a building whose age is not reliably determined—a private summer house complex (villa) that belonged to the Ottoman governor of Bosnia and Herzegovina Topal Osman-paša (1861–1869), who sold it to Dedaga Čengić, son of the famous Smail-aga from the Ivan Mažuranić's epic. The Čengić family used the building until 1933 (Bejtić 1973, 37–38). In the late 19th and early twentieth century, the area hosted one inn on the main road to Mostar (Kreševljaković 1957, 113). After WWII a planned development begins, and until the 1960s housing complexes would completely fill in this previously sparsely industrialized area and annex it to the city. Since the place name was deeply rooted in colloquial communication, it was introduced into to official usage. Today, the area is divided into two neighbourhoods: Čengić Vila 1 and 2.

Čengić Vila 2 (1963–1966)[11]

Čengić Vila 2's spatial planning and design are not derived from the topography, which, with its proximity to green slopes (Hrasno Hill) and the Miljacka River, is more pronounced than in the previous neighbourhoods (Fig. 6).

Čengić Vila 2's composition is based on the principle of free-standing volumes, but without clear spatial, functional or programmatic organisation, which makes it difficult for pedestrians to navigate, especially in its southern part. This makes it almost impossible to create diverse housing environments or smaller spatial units, and as a result, instances of activation of public spaces by residents are rare. The design of the buildings is modest and uniform. Many do not have balconies, and in some cases basements are used as apartments (Fig. 7).

A busy two-way traffic street (Gradačačka Street) passes through the neighbourhood, and is lined with commercial and hospitality facilities. The proximity of the footpath (Aleja lipa) along the Miljacka River to the buildings makes it unattractive, as does its lack of trees. Čengić Vila 2 is closely connected to the Otoka nearby urban centre on the other side of a river.

User-led transformations within the neighbourhood are sporadic.

---

[11] Area: 13.87 ha; 25 objects; 1,752 residents; (Source: Aganović 2014).

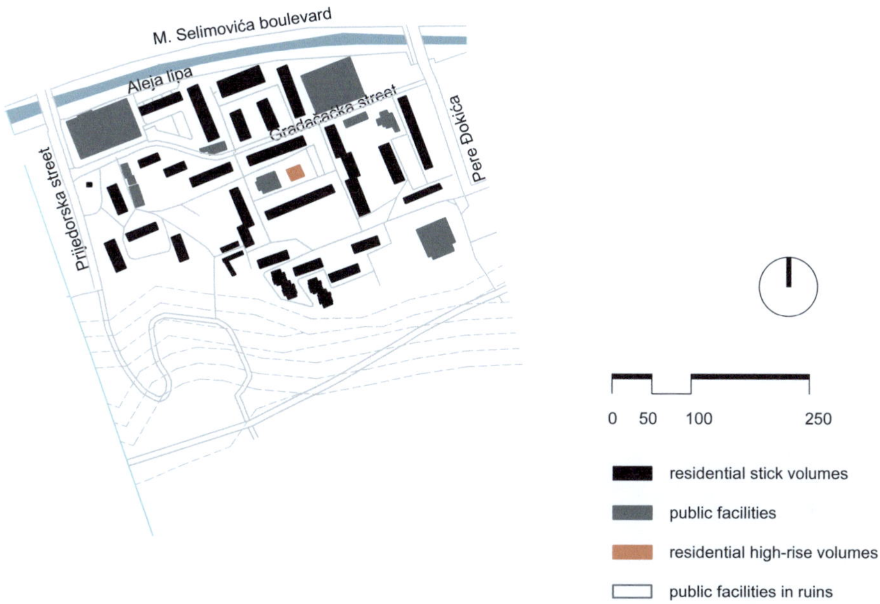

Fig. 6 The N-E part of Čengić Vila 2 contains the "Lastavica/The Swallow" kindergarten, and the "Osman Nakaš" primary school with a sports field. In the N-W part is a trolleybus terminal, and the "Otoka" shopping centre and market. This part of the neighbourhood is the most frequented, and attracts residents from surrounding areas. In the central part, one representative 12-storey residential building stands alone, without any surrounding public facilities. The rest of the neighbourhood comprises 6-storey residential buildings with a dominant E-W orientation, and large green spaces, especially in its southern part. *Source* Elsa Turkusic Juric; technical support by Amila Terzić

Hrasno: From Village to Pavilion-Based Residential Neighbourhood with a Public Core

History

Hrasno is an old village, situated in the continuation of the river Miljacka's alluvial plateau, just west of Grbavica, where the confluence of the Kovačići stream was located before it was covered. The origin of the place name is derived from an oak meadow that once existed in the area. Mentioned in written sources under the current name in 1604 (Group of authors 2000), and 1682, it is possible that the toponym *Hrastište* from the 1469 and 1555 censuses is an older form. Some 63 recorded households in 1606 tell a story of a sizeable village next to the old road to Herzegovina, easily reached by traversing the meandering Miljacka (river crossing, later a wooden bridge), although it was part of the communal area of the much larger Dolac. Throughout Ottoman and Austro-Hungarian rule, Hrasno kept its rural form, but was significantly upgraded 1929–1941, when cooperative settlements of railway workers were built. The village of Hrasno, with its rural characteristics,

Fig. 7 Modest residential buildings surrounded by large green spaces. *Source* Elsa Turkusic Juric

will completely vanish as a separate rural settlement and be replaced with modern housing blocks. That was its physical integration into Sarajevo, even though the village was previously attached to the city limits. The only remains of the previous rural settlement are, the village mosque (built 1940, R. Kadic), and *Hrasno brdo* hill settlement—situated on the mt. Trebević's lower slopes, where old agricultural estates and vast orchards were replaced by a large number of illegally built individual houses from 1970 to this day.

Hrasno 1 (1967–1969)[12]

The specificity of this neighbourhood lies in the uninterrupted sequence of green areas and pedestrian zones along its central axis. The aim of this was to create a continuity of public facilities, and increase the dynamism of the community's everyday life (Fig. 8).

The main one-way traffic streets (Alejla Lipa and Porodice Ribar) forms a ring inside the neighbourhood, which in the northern zone passes a footpath along the river. The neighbourhood's access to the river is less direct than Grbavica's, and its

[12] Area 21.17 ha; cca 37 objects; cca 3,919 apartments. Sources: Z.Likić, H. Muhasilović, A.Hadžiosmanović, Ž.Janković; (Source: Aganović 2014).

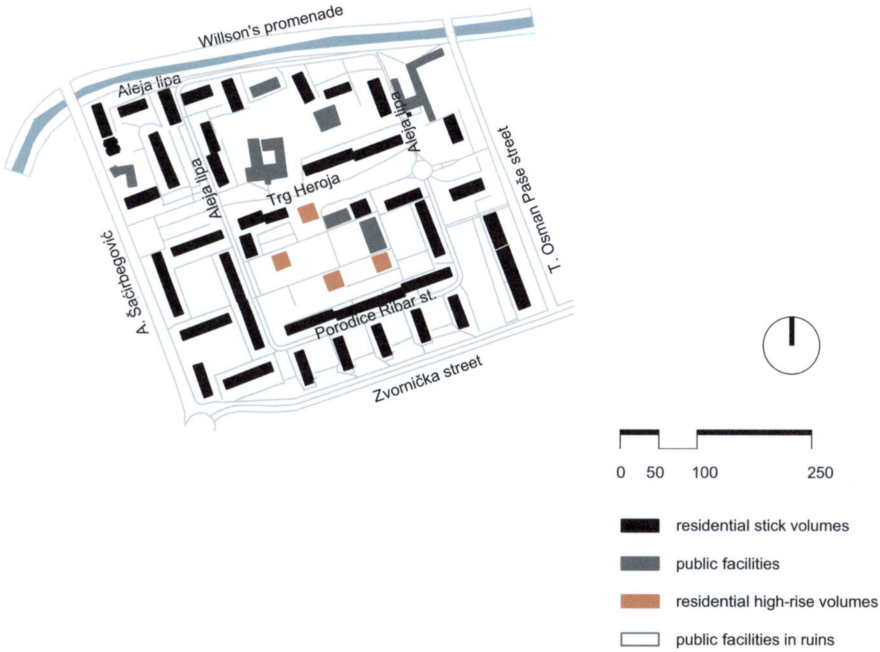

Fig. 8 In the centre of Hrasno 1 are four (vertically) spatially and visually representative 21-storey residential buildings; the small "Squares of Heroes" surrounded by 5-storey residential buildings; two 2-storey administrative buildings; and a green belt, alongside which are the "Košuta/Doe" kindergarten and the "Hrasno" primary school with a sports field. The surrounding residential zones are spatially organised in different ways. The southern zone, through which a busy road (Zvornička Street) provides an exit to surrounding areas, is defined by 8-storey residential buildings in a line, at regular intervals with spacious green areas between them. That height of the buildings continues in the eastern zone as well, but with a different modern design of the buildings, unofficially named the "Loris" and "Palma/Palm Tree". *Source* Elsa Turkusic Juric; technical support by Amila Terzić

footpath lacks natural surroundings. Unlike those of the neighbourhoods around it, most of the ground floors of Hrasno 1's buildings are public rather than residential. The facades have a variety of shapes, and include balconies, loggias, French doors and accentuated entrances with gardens. Residential initiatives that aim to improve the use of open public areas include a children's playground and landscaped green belts in front of building entrances. Despite this, Hrasno 1's lack of public facilities and the nature of the surrounding areas mean its connection with the outside is less pronounced (Fig. 9a, b).

Fig. 9 Central axis with both green pedestrian zones and vertical core (now transformed to traffic zone). *Source* Elsa Turkusic Juric

Alipašin Most and Alipašino Polje: From Arable Land and a Bridge into Residential Neighbourhood of Mega-Blocks and Green Squares

History

The area where Miljacka valley starts to widen towards the confluence with the Bosna, and where the slopes of the mt. Trebević massif terminate, was always sparsely populated. The Miljacka river meandered there heavily (Otoka), and with its sharp bend towards the north-west, cluttered with amounts of sand deposits, interrupted by the traditional road link from Sarajevo to Herzegovina and other parts of Bosnia. Such a place required either a river crossing to cross on horsebacks and in carriages during low water levels, or a bridge. So, the first known wooden bridge was built as a personal endowment of Ali-paša (fifteenth century). And the name of the complete area from Čengić-vila to Stup bears the name of Ali-paša, the first bridge builder.

The torrential river Miljacka used to destroy bridge((e)s) repeatedly so it often needed renovation in the following centuries (Čelić and Mujezinović 1969, 103–105).

With the completion of a narrow railroad, and building of the train station here in 1882 (Bejtić 1973, 39) this was supposed to serve dozen villages on the slopes of Žuč massif, eventually giving an industrial character to the area. That would climax after WWII, when industrial plants filled up the area's northern part.

The southern part, a flatland situated on elevated alluvial terraces connected to the northern slopes of the Mojmilo hill, was surely once in the ownership of Ali-paša. Scarce and disjointed ancestral settlements in the area were probably conditioned by agricultural landlord-peasant relations, but all surveys and cadastral censuses mention this locality under the same name: *Alipašino polje* (*Ali-paša's field*). Until the 1970's, it remained sparsely inhabited, and the city authorities retained its colloquial place-name, even in the cadastre. Today, Alipašino polje is a large residential neighbourhood, built in three "phases": A, B and C.

Alipašino Polje (1974–1980)[13]

Alipašino Polje's spatial and design organisation did not arise from the area's topography, but from the need to avoid the creation of a "dormitory" neighbourhood. This consideration differentiates it from the neighbourhoods previously discussed, as its concept was to create better social dynamics and cohesion. Alipašino Polje is a ring-shaped (B and C zones) and U-form-shaped (A zone) groups of stepped residential volumes (mega-blocks), which create open internal space—a square that is available to residents for various activities (Fig. 10).

[13] Area 65.0 ha; 8,200 apartments. Sources: M.Medić, J.Milić, N.Muftić, Z.Likić, S.Stojanović, Š.Omerović, (Source: Aganović 2014).

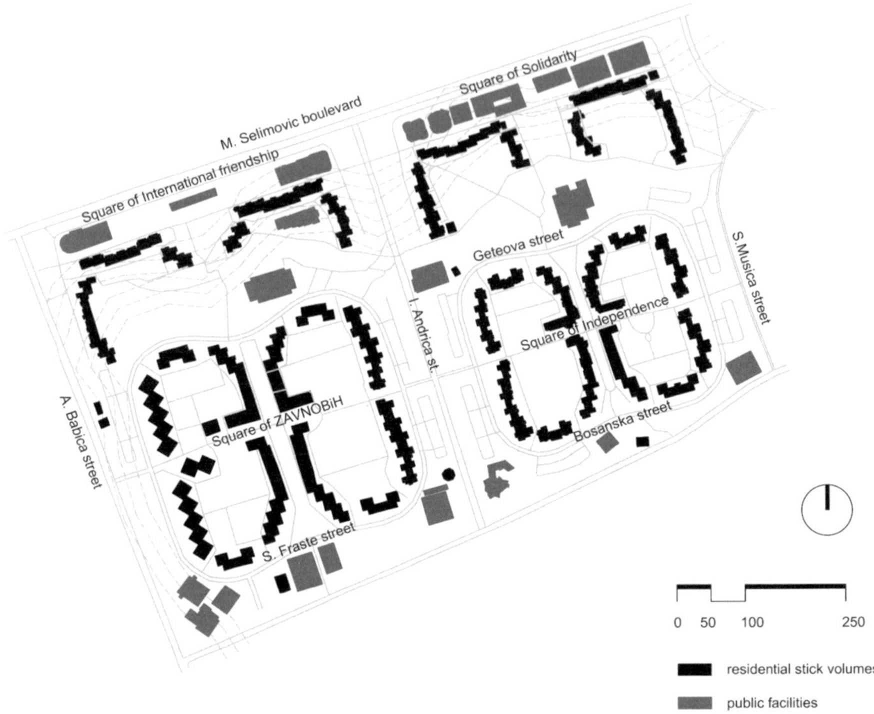

Fig. 10 The buildings are 5 storeys to 18 storeys, and are applied as several repeating types (officially named as C1, C2,C3 and C4 only in the C zone). The public buildings consist of two primary schools "Meša Selimović" and "Fatima Gušić"; the "Vjeverica" kindergarten; the "Ramiz Salčin" sports hall; a market; and the "King Fahd" mosque with religious centre, the "Bosnian" mosque, the "St Luka" church (all built post-1995 on the south neighbourhood's outskirts); commercial and administrative buildings (all built post-1995 on the north neighbourhood's outskirts- former pedestrian zone). (in the.) *Source* Elsa Turkusic Juric; technical support by Amila Terzić

Vehicular traffic is kept outside the neighbourhood, so public areas within it are intended exclusively for pedestrians, further enabling cohesion. This design and the programmatic concept have resulted in the creation of an artificial topography or "vertical facades", which block the view of the neighbourhood from outside and reinforce the impression of closeness.

Transformations of public areas by residents are rare and consist of inactive green spaces and small playgrounds for children (Fig. 11a, b).

Fig. 11 Inner pedestrian zone (the Square of ZAVNOBiH); Ring-shaped mega-blocks in zone C

Dobrinja: From Village into "New Type" Combined Residentia Neighbourhood

History

The region of Dobrinja, south of the Mojlmilo hill massif in the lowlands along the Dobrinja stream, fed by Lukavica and Tilava streams, according to known historical sources was inhabited no earlier than medieval times. It was recorded as a village in *Župa Tilava* with ten households in an Ottoman census from 1455 (Šabanović 1964, 16). Later in 1468/1469 it was mentioned as "*Dobrino*", property of Kemal, castellan of the Hodidjed fortress, with only 8 households (Aličić 2008, 196), while in 1604 it counted 23 households (Group of authors 2000, 213–214). Throughout the era, Dobrinja existed as a scattered-type village, organized in family clusters, even though situated along the frequented road that led from north to south of the region. The beginning of Dobrinja's merger with Sarajevo started with the construction of the Butmirska and Lukavička roads (Bejtić 1973, 125–235) in the first half of the twentieth century. That process is continued through preparations for the 14th Olympic Winter Games, held in the city in 1984. The first built Dobrinja's residential blocks housed visiting journalists and other international guests. Dobrinja was built in several phases (1–5), as the last (but biggest) of Sarajevo's neighbourhoods.

Dobrinja 1982–1991[14]

Unlike the other planned residential neighbourhoods analysed, Dobrinja is characterised by its undisturbed pedestrian zones, dominant green spaces (with no tall vegetation), and a footpath along the regulated course of the small Dobrinja River. Dobrinja was created as a self-contained neighbourhood, positioned adjacent to Sarajevo's international airport, with impressive views towards the massif of Mount Igman, and near the Mojmilo Forest Park recreational zone (Fig. 12).

Along the neighbourhood's central axis, parallel to the river, are two main corridors: a vehicular thoroughfare with a trolleybus line (Mimar Sinan Boulevard); and a green pedestrian zone (Branilaca Dobrinje Boulevard). The latter forms the heart of the neighbourhood, and the ground floors of its buildings house public services (Fig. 13a, b).

Dobrinja's positioned at the end of the city is partly responsible for its characteristic local identity, which developed through an expressed need for more indoor and outdoor public facilities. Some of its open spaces still require better cultural and recreational content, however, especially those along the river.

---

[14] Area 265 ha; 15,100 apartments; 40,000 inhabitants. Sources: M. Medić and B. Povlakić; Z. Likić, S. Stojanović, M. Jašarević, A. Hadžiosmanović and D. Karanović.

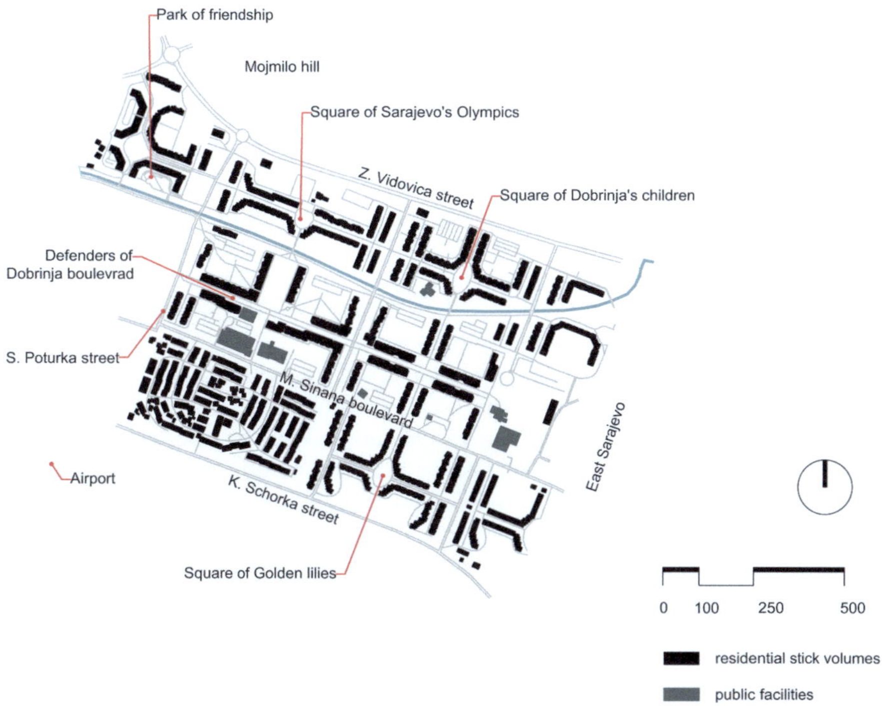

Fig. 12 Dobrinja has a range of commercial, hospitality and cultural facilities; three primary schools “Skender Kulenović”, “Osman Nuri Hadžić” and “Mak Dizdar”; the 5th Gymnasium; a sports hall; the “Mercator” shopping centre; the two “Dobrinja” mosques, the “Francis of Assisi” church; and several landscaped parks. A cinema, swimming pool, theatre and tram line are planned, and await implementation. Dobrinja’s residential zones were conceived as “semi-atrium” pavilions. They are composed of rows of multi-storey buildings (4, 6 and 8 storeys) in a “U” shape, surrounded by large green spaces. Pairs of pavilions are connected by tree-lined streets, and the diverse atmospheres suggest better spatial orientation and encourage a range of activities. The housing in the part of the neighbourhood adjacent to the airport (quadrant 5C) has a more pronounced individual character, because it is less-densly populated and has lower buildings (3 storeys) and smaller green spaces. *Source* Elsa Turkusic Juric; technical support by Amila Terzić

Recomposing Modern Residential Neighbourhoods: The Third Level of Transformation

The completely planned neighbourhoods presented in this research started as unfinished ones, with plans to integrate public and ancillary facilities later. This, along with the social changes that took place at the beginning of the twenty-first century, has meant that the transformation of these neighbourhoods has taken different paths. These processes can be seen through various phenomena within the neighbourhoods, and are not guided by improved design or programmatic integrity. They have been followed by changes in the offical names of streets, squares or public buildings, while the most of the names of the neighbourhoods (are) have remained unchanged (except the neighbourhoods that had the names of communists: Grbavica 2—former Slobodan Princip Seljo, Aneks-former Pavle Goranin, Ciglane—former Djuro Djaković).

Personalised and sporadic interventions on facades (such as the closing of loggias, and the enlargement or replacement of openings) are one such phenomenon. Although this has been part of these neighbourhoods since the first decades after their construction, it has intensified in recent years. There are no examples of the uniform treatment of facades, although this would be preferable. Additionally, several buildings in Grbavica 1 and 2 and Hrasno 1 have been upgraded, which has created a design discrepancy within the neighbourhoods.

Further, the trend of "filling cities" has led to the conversion of green spaces into construction plots. Buildings in Hrasno 1, Čengić Vila 2 and Dobrinja were built on such "vacant" plots. The parcellation of neighbourhoods has also been implemented, not only in these parts of the city but also in its historic centre, and is manifested in the construction of freestanding buildings, representative only of themselves, with no social or public significance. As a result, it is impossible to maintain the continuous, logical and organic development of neighbourhoods, and therefore cities, and the quality of housing inevitably declines. Urban sociologists call this "Disneyfication",

Fig. 13 Green pedestrian zone Bulevar Branilaca Dobrinje; Facades of Dobrinja's buildings are dynamically designed, with large balconies, various kinds of opening, and more elaborate volumes. *Source* Elsa Turkusic Juric

Fig. 13 (continued)

or the creation of "cities as attractions", with individual "project-planned" buildings. These new buildings within neighbourhoods do not have either local or attractive names, unlike new buildings in the central and historic parts of the city ("Hrđa/Rusty" building, BBI Centre–former Sarajka, etc.). Also, changes in names can be seen through the processes of the literal takeover of foreign names in the historic part of the city (the "Europe hotel"-former "Evropa hotel", etc.) or in newly developed attractive residential suburbia (Poljine Hills, etc.).

Examples of this can be seen in Zone A of Alipašino Polje, where buildings have been constructed in a former pedestrian zone, alongside the main road. This has made it impossible to create a "longitudinal pedestrian centre", and has interrupted views and communication within the neighbourhood. More importantly, the insufficient distance of the new buildings from existing residential structures has affected the quality of life of residents, by reducing privacy and natural light in their apartments, and replacing their previously open views with vertical walls (Fig. 14).

The motorisation of the neighbourhood is also taking place, through an increased number of parking spaces and the conversion of pedestrian zones to vehicular ones. This limits the movement of pedestrians and interrupts the continuity of public facilities. These transformations are completely at odds with the intentions of architects when they design residential neighbourhoods and are especially pronounced in Alipašino Polje and Hrasno 1. Part of the fault lies in insufficient city public transport links and the citizents' (anti)culture of car use. The expansion of road networks in Sarajevo at the expense of green spaces has worsened the position of residential buildings in Grbavica 2, where a green buffer zone was removed between them and the (now even busier) traffic street.

Such problems do not arise solely from issues within the neighbourhoods but are necessarily related to processes occurring in the city as a whole, and to global trends (such as Disneyfication, fragmentation and the concept of the spectacle in urban landscapes). The processes analysed here often contradict the very concept of residential neighbourhoods. This creates inhumane living conditions, which in turn can cause social segregation and criminal activity. Residents are becoming increasingly conscious of this, which indicates their developing relationships and sense of identification with their living spaces. At a public debate in 2019, citizens and professional organisations (Association of Architects in B&H) voted against the planned Strojorad project in Grbavica 2, because they opposed the redevelopment of the site.[15] In December 2021, residents of Dobrinja launched the #spasimoDobrinju (#saveDobrinja) campaign, in order to prevent the construction of a petrol station and kerosene storage facility immediately adjacent to their neighbourhood, and in a Level 3 water protection zone.[16] Similarly, residents of Čengić Vila 2 (in 2018) and Dobrinja (in June 2021) took action to prevent the conversion of large green spaces into car parks.[17]

[15] https://aabh.ba/odrzana-javna-rasprava-grbavica-ii-strojorad/.

[16] https://www.klix.ba/vijesti/gradjani-ne-odustaju-pozivaju-na-nastavak-borbe-protiv-gradnje-naftnog-terminala-na-dobrinji/211217033.

[17] https://radiosarajevo.ba/vijesti/bosna-I-hercegovina/Gradani-dobrinje-protiv/419376.

Fig. 14 Capitalist utopia vs architectural dystopia and the quality of life for residents-Buildings have replaced the former pedestrian zones in Alipašino Polje (the Square of Solidarity and the Square of International Friendship). *Source* Elsa Turkusic Juric

Conclusion

The toponyms of the areas occupied (and changed by) these planned modernist neighbourhoods in Sarajevo's post-WWII suburbs originated from the names of the bridges, buildings, fields, regions and villages that previously occupied them. The residential settlements, with their various spatial and design manifestations, fit within a scheme of functionalist urban structural development and imbue the areas they occupy with new meaning: pavilion neighbourhoods of the Grbavica 1 and 2, pavilion neighbourhoods with a public core (of) as in Hrasno 1; scattered neighbourhoods with markets of the Čengić Vila 2; mega-blocks with green squares of the Alipašino Polje; and the Dobrinja as self-contained neighbourhood. These concepts arose from a need to find the most rational and efficient housing solutions. Efforts by the profession to offer alternative concepts that would correct the mistakes identified in this (paper) analysis and improve these "soul-less neighbourhoods" are not inconsiderable. In each modernist residential area, architects designed solutions to improve social ties and encourage inhabitants to inscribe their own meanings. This is particularly visible in Alipašino Polje, Dobrinja and Hrasno 1, although the concepts have not been fully justified, especially in the former. It is, however, difficult to properly assess the conceptual effectiveness of these neighbourhoods, because they remain unfinished: the public facilities that should enable recreational and social activities have developed too slowly, or been only partially realised (as the reasons have been previously discussed through the socio-economic perspective of Sarajevo's post-WWII development).

The first and second levels of transformation began to develop in parallel, through the (institutional) building and filling-in of neighbourhoods on one side, and the (symbolic) inscription of meaning within them by residents on the other. For example, there were no official names for the buildings, but some of them gained and retained local names over time. The transformation and creation of new toponymic meanings are mostly the results of "social adaptation", in which residents become more satisfied with their living conditions the longer they inhabit a place (user-led transformations in the neighbourhoods are sporadic but evident). The design and programmatic characteristics of neighbourhoods also add meaning in an area's specificities and position within the greater city structure allowing different processes of adaptation to develop in each accordingly. For example, the Grbavica 1 is completely different from the Dobrinja neighbourhood, by spatial urban structure, appearance in the city's panorama and connections to surrounding areas (as previously specified through maps).

The third level of transformation has the same starting point as the first two, but its manifestations occur outside neighbourhoods' conceptual and semantic frameworks. These transformations are partly determined by global phenomena (such as Disneyfication and fragmentation), which change the accepted meaning of toponyms. More importantly, they disrupt the continuity and functionality of neighbourhoods and hinder their ongoing processes of adaptation.

The arrival of local activism on the part of residents, who demand more transparent management of their immediate living space, is evidence that the processes of identification and social cohesion within neighbourhoods have begun. The ways in which the consequences of third-level transformations are acknowledged and "repaired", and the redirection of them towards the conceptual and semantic interiors of neighbourhoods, will indicate the future characters of these residential areas. Further transformations are inevitable and may allow some neighbourhoods to reobtain the inscribed meanings of their toponyms.

Questions

1. How did the Slavic and Ottoman cultures influence the emergence and transformations of many of Sarajevo's toponyms in modern identity issues?
2. Do the residential buildings in WWII neighbourhoods have official or unofficial names, and how are those names established?
3. What are the main differences between the three levels of transformation of the modern neighbourhoods in Sarajevo?
4. Which modern neighbourhood and its toponym, are mostly exposed to transformation in the period of globalization, and how?
5. Why did certain toponyms change, while some others remained unchanged in the post-socialist period?

References

Aganović M (2014) Sarajevo–savremene urbane, stambene i komunalne dileme i problemi razvoja grada, Štamparija Fojnica, Fojnica

Aličić AS (2008) Sumarni popis sandžaka Bosna iz 1468/69. godine, Islamski kulturni centar Mostar, Mostar

Anđelić P (1963) Trgovište, varoš i grad u srednjovjekovnoj Bosni–Prilog tipologiji naselja, GZM "Arheologija", Zemaljski muzej BiH, Sarajevo

Arheološki leksikon BiH (1988) Tom 3, Reg. 15, lok. 15.38, Zemaljski muzej BiH, Sarajevo

Bejtić A (1973) Ulice i trgovi Sarajeva–Topografija, geneza i toponimija, Muzej grada Sarajeva, Sarajevo

Benac A (1954) Sarajevo kroz arheološke spomenike at "Sarajevo od najstarijih vremena do danas", Muzej grada Sarajeva, Sarajevo

Čelić Dž, Mujezinović M (1969) Stari mostovi u BiH, Veselin Masleša, Sarajevo

Čaldarović O (2011) Urbano društvo na početku 21. stoljeća, Naklada Jesenski i Turk, Zagreb

Kreševljaković H (1956) Saraji ili dvori bosanskih namjesnika (1463–1878), Naše starine, Zavod za zaštitu spomenika kulture, prirodnih znamenitosti i rijetkosti Bosne i Hercegovine. Sarajevo III:13–22

Kreševljaković H (1957) Hanovi i karavansaraji u Bosni i Hercegovini, Izabrana djela VIII, Odjeljenje istorisko-filoloških nauka knj. 7, Naučno društvo NR Bosne i Hercegovine

Kreševljaković H (1969) Sarajevo za vrijeme austrougarske uprave, Izabrana djela IV, Sarajevo

Krstić B (2014) Atinska povelja i misao arhitekata i urbanista FNRJ 1950-ih, Beograd

Kulić V, Mrduljaš M, Thaler W (2012) Modernism In-between, Jovis

Mesihović S (2007) Prapovijesne i protohistorijske gradine Sarajeva, Znakovi vremena (38), Institut "Ibn Sina" Sarajevo, pp 154–164

Mesihović S (2011) Antiqui homines Bosnae, Filozofski fakultet Sarajevo, Sarajevo; CIL III, 3198b (p 2275, 2328,19) = CIL III, 10156b = ILJug I, 263
Norberg-Schultz Ch (1990) Stanovanje, stanište, urbani prostor, kuća, Građevinska knjiga, Beograd
Pallasmaa J (2009) Space, place, memory and imagination. Spatial Recall: memory in architecture and landscape. Routledge, New York and London, pp 16–41
Šabanović H (1964) Krajište Isa-bega Ishakovića: Zbirni katastarski popis iz 1455. godine, Book 1, Volume II, Monumenta turcica historiam Slavorum Meridionalium Ilustrantia: Defteri, Orijentalni institut u Sarajevu, Sarajevo

Suggested Reading

ARH časopis društva arhitekata Sarajeva, 2–3(I), 1963, Sarajevo
ARH časopis društva arhitekata Sarajeva, 5 (II), 1964, Sarajevo
Arhitektura i urbanizam, 14(3), 1962, Beograd
Arhitektura i urbanizam, 94–95 (24), 1984, Beograd
Low M (2006) Promišljanje grada, Naklada Jesenski i Turk, Zagreb
Štraus I (2013) Arhitektura Jugoslavije/Architecture of Yugoslavia 1945–1990, Sarajevo Green Design Foundation, Sarajevo

Street Naming in Malta as a Geo-Cultural and Political Exercise as Seen from Local Sources

John A. Schembri and Ritienne Gauci

Abstract Street naming in Malta is essentially a reflection of the colonial and post-colonial influences. This is seen within a number of spatial scales: the national (country), regional (towns and villages), and local (streets). A further factor that influences the naming of streets is the language used in identifying their location with a spray of vernacular Maltese (Semitic), Italian (Romance) and English. Although the Maltese Islands were occupied by a succession of rulers who were intent on exerting their influence in the central Mediterranean, it was the Arabs (870–1090), Knights of St. John (1530–1798) and the British (1800–1964) who left their mark on the written and spoken word. This chapter presents a description and critique of the initiation, evolution, change and purposes of street naming in Malta using three main sources. The first are maps and sketches drawn by the official designate such as the Public Works Department (Malta) and the Ordnance Survey (UK); secondly, the minutes and proceedings of the street-naming committees in both Malta and Gozo; and, thirdly, a ground-truth exercise at selected sites. It is expected that the geo-political influences at local, regional and international levels, together with the transition from socio-religious to secular attitudes are reflected in street naming.

Keywords Street naming · Geo-cultural · Socio-religious · Transition · Ground truth · Malta

Objectives

By the end of this chapter readers should be able to:

1. distinguish between different toponymic layers in Malta
2. recognize the changing socio-cultural influences
3. value the polyglot attributes of the Maltese and their language
4. compare various historic periods in relation to the toponymic heritage of Malta

---

J. A. Schembri (✉) · R. Gauci
University of Malta, Msida, Malta
e-mail: john.a.schembri@um.edu.mt

R. Gauci
e-mail: ritienne.gauci@um.edu.mt

© The Author(s), under exclusive license to Springer Nature Switzerland AG 2023

G. O'Reilly (ed.), *Place Naming, Identities and Geography*,
Key Challenges in Geography, https://doi.org/10.1007/978-3-031-21510-0_19

5. identify with past personalities and their socio-cultural contributions
6. discuss the socio-political significance of toponymic heritage
7. appreciate the importance of a name in identifying a street
8. connect geographic attributes to a street name
9. figure out the importance of street networks in association with names.

Introduction: "What's in a Name?"

The use of names has for long been the focus of numerous empirical theory and practices in various fields of studies, such as in political geography with toponyms (e.g. Radding and Western 2010) and in biology with taxonomy (e.g. Härlin and Sundberg 1998). Toponyms and taxonomy are in many instances largely based on the traditional ideas that names are abbreviated definite descriptions or representations of landscape semantics. In geography, place names are not assigned through arbitrary decisions but rather are bestowed in order to impart a meaning which conveys familiarity and spatial identity through societal associations. In the philosophy of language, this is known as the descriptivist theory of proper names, i.e., the view that the meaning or content of a proper name fits a description associated with it by speakers, while their referents (i.e., the named objects) have characteristics that satisfy or fulfill such a description (Textor and Rami 2015).

Conversely, over the years, non-descriptivist scholars have gained traction in response to the seminal work by Kripke (1972) who argued that naming is 'picture' of reference (rather than fitting a meaning or a description) and that such a name reference is determined by a simple causal connection with the object created by the speaker and transmitted through communities of speakers. A vivid example of this non-descriptivist mindset is best illustrated by writers and poets who playfully and artistically challenged the descriptivist use of names and the meaning of association behind names. Memorable passages have been created in various works of literature, such as the famous oration by Juliet in Shakespeare's Romeo and Juliet:

> 'Tis but thy name that is my enemy;
> Thou art thyself, though not a Montague.
> What's Montague? it is nor hand, nor foot,
> Nor arm, nor face, nor any other part.
> Belonging to a man. O, be some other name!
> What's in a name? that which we call a rose.
> By any other name would smell as sweet;
> So Romeo would, were he not Romeo call'd,
> Retain that dear perfection which he owes.
> Without that title. Romeo, doff thy name,
> And for that name which is no part of thee.
> Take all myself.
> Romeo and Juliet (2.2.38–49) by William Shakespeare.

In these lines, Shakespeare conveys the idea that the naming of things is irrelevant and arbitrary. Juliet dismisses the meaning of Romeo's last name, Montague, in being associated with her family's sworn enemies and orates how much his last name does not describe the essence of Romeo and what his persona represents to her. This passage anticipates by a number of centuries the evolution of Kripke's theory, by implying that a last name is therefore a causal connection created by a family history and should not hold any truth-value to the description of the stand-alone identity of its bearer.

These on-going debates in the philosophy of language, though still unresolved, surely serve to enrich empirical discussions about the use of toponyms in geography. Tichelaar (2002) points out how the use of toponyms is based on the relationship between numbers of speakers and the importance of a language in a community. Two aspects underpin this relationship: (i) the level of geographic (or regional) attachment of a language and (ii) the presence of 'historic rights' that speakers of a language can enjoy in the land they occupy. In fact, it is within the realms of language, that De Saussure (1916) considers place names as 'signs' that are "situated socially and chronologically by a reference to a certain community and a certain period of time" (1916: 110). The aim of this work is therefore to situate the evolution of street naming within the geo-cultural and political context of the Maltese Islands as a manifestation of the language development and historical rights in the use of language and as a response to the islands' colonial and post-colonial history.

Street Names in Onomastic Research

Street names are referred to as ononyms and are identified as important components in onomastic research (Rusu 2021). They have been linked to various applications, such as those being considered as "political weapons and produce memories linked to past events, individuals and attach an intimacy to an urban environment" (Bobin 2021, p. 1); however, their use in urban geography can distort history (Bobin 2021). Street names can also be considered as the "cumulative commemorative decisions of municipalities" (Oto-Peralías 2017, p. 1) and as "participants in the cultural production of a shared past…however in a fragmented manner" (Azaryahu 1996, p. 311). They can also encapsulate tension between administrators and consumers as both street and square names can be used as onomastic show-off between the authors of the names and the individuals' attribute to such names (Meital 2007). Reiger (2010) furthers the importance of street names where two streets meet at corners and he eulogizes Walter Benjamin who considered language and urban space as important elements in a cultural set-up. Onomastics, as defined by the Concise Oxford Dictionary, refers to the study of the history and origin of proper names (Soanes and Stevenson 2004). Extended within the spatial and geographic milieu, as a derivative of the term, the word 'toponomastic' belongs or is related to place names. Applied to numerical values, the term 'streetonomics' is used for the quantitative analysis of street names.

This chapter examines street naming in Malta as a reflection of the cultural base of the islands through a description and critique of the initiation, evolution, change and purposes of street naming in Malta. The main themes are religious, secular, historical, linguistic, and colonial/post-colonial influences. The first sources are maps and sketches drawn by the official designate, such as the Public Works Department (Malta) and the Ordnance Survey (UK); secondly, the minutes and proceedings of the street-naming committees in both Malta and Gozo; and, thirdly, a ground-truth exercise at selected sites. It is expected that the geo-political influences at local, regional and international levels are reflected, together with the transition from socio-religious to secular attitudes.

A Palimpsest of Tongues in Malta

The chequered history of Malta saw the archipelagic group, in the centre of the Mediterranean, passing on from one occupier to another over millennia (Schembri 2019). The need to control the narrow Sicilian Channel that divides the semi-enclosed sea saw successive maritime powers vying for mastery of the islands. Table 1 lists in chronological order these powers and the time-span during which they controlled the Islands. Within the broader Mediterranean the three islands are practically equidistant from the eastern and western shores and also from the northern and southern shores of the Mediterranean. In addition to their location, the Islands are situated in the Sicilian Channel, which at 145 km in width, links the eastern and western halves of the Mediterranean. The coastal geography, of the island of Malta especially, consists of a very irregular shape giving rise to a number of bays, inlets, and harbours some of which became ports following human settlement and the introduction of maritime-related activities (Gauci and Schembri 2019).

As a result, Malta became an attractive pole for powers competing for superiority in the Mediterranean and subsequently had geopolitical importance far greater than its size. As for Gozo, its smaller size and overall rectilinear coast subjected it to a secondary role to the main island of Malta, although its linguistic roots and culture are very similar. This geopolitical importance is further enhanced by the almost all-weather ports situated on the eastern coast of Malta where Port Benwarrad (Burmarrad), known also as Salini Bay, was an important maritime station for the Romans. The Grand Harbour and Marsamxett Harbour were vital ports of call for the Knights and the British, whereas Marsaxlokk Harbour provided the main anchorage for attacks on Malta by the Ottomans in 1565 and Napoleon in 1798. As a result of the location and geography of the Maltese Islands, foreign occupation influenced their socio-cultural landscape. This is manifested in a number of vestiges including the language with the spoken and written word: a manifest in the combination of the Semitic, Romance, and English.

Considering that the Mediterranean Sea has an area of 2.5 million km^2 and the central area is approximately 583,000 km^2, the Maltese Islands form a minute part both within the territorial expanse of the lands (8 million km^2) enclosing the sea and the marine areas. Although the Mediterranean region itself has more than 2,000

Table 1 Malta: a chronological list of occupiers and important selected events

Year/s	Occupier/Event
5,200 BC	Stone Age farmers
3,600–2,500 BC	Temple building period
800 BC	Phoenicians
400 BC	Carthaginians
218 BC	Romans
Ca 60 AD	Shipwreck of St Paul
395 AD	Byzantines
870	Arabs
1090	Normans
1194	Swabians
1266	Angevins
1283	Aragonese
1412[a]	Castillians
1530	The Spanish King fiefs Malta to the Knights of St John
1565	The Great Siege
1566	De Valette starts building Valletta
1798	French occupation starts
1800[b]	French surrender and British takeover
1914–1918[c]	World War I
1939–1945[d]	World War II
1964	Malta gains independence
1974	Malta becomes a republic
2004	Malta joins the European Union

Adapted from https://www.visitmalta.com/en/a/timeline/
[a]Aragon and Castille unite and Malta becomes part of the Spanish Empire.
[b]Malta becomes British protectorate in 1800; a de facto colony in 1813; and, confirmed as such by the Treaty of Paris in 1814.
[c]World War I: Malta named "The Nurse of the Mediterranean" for its role in the war.
[d]World War II: Malta bombed heavily by the Italians and the Germans and in 1942 King George VI awards the people of Malta the George Cross.

islands dotting mainly the northern areas, it is the strategic importance of Malta that conditioned its geopolitical history. The islands have a population of about 520,000 and the length of the road and street network is over 2,400 km (Attard 2020) giving a 7.6 km length of road per square km of terrain.

The links with the powers that governed Malta are reflected in a number of geographies pertaining to place names designated to rural, coastal and urban localities throughout the islands. Wettinger (2000) and Salafia (2020) are two exponents who detailed names linked to the land and shores as a result of extensive research in archives, ground truth and as personal communications by residents, farmers and shepherds. In addition, the semitic links of the islands to North Africa were studied also by Mifsud Bonnici (1960) and Serracino Inglott (1981). The overall result is that most localities, zones and individual fields bear names linked to the Arab occupation of Malta of a millennium ago. However, street names are a more recent introduction with the vestiges dating back to the building of Valletta in the mid-sixteenth century.

Valletta: The Changes in Street Names of the Capital

Name-changing of streets has traditionally been one of the hallmarks of whoever administered the islands. An excellent case in point are the changes made to the streets and piazzas of the capital city of Malta, Valletta. The first stone of the capital city of Malta was laid by the French Grand Master Jean Parisot de Valette on the 28th March 1566 following the success of the Great Siege of May to September 1565. Whilst most of the village streets had hardly any recorded names, those in the capital were identified, initially by the Knights of St. John, with names dedicated to Catholic saints. The grid-iron pattern of streets was the final outcome of the capital (Fig. 1) and an example of one of the streets in Fig. 2 is seen today as it was originally planned. The city was completed by 1571 and the Knights moved from their original residence at Birgu to Valletta. As seen previously in Table 1, Malta was successively occupied by various powers especially those who vied for military control of the central Mediterranean. The names also reflect the linguistic base of the island. Whilst the overall population used Maltese (a Semitic-based language with Romance intonations introduced as the sixteenth-nineteenth centuries unfolded) the official communications were in Italian and Latin, especially for administrative and ecclesiastical purposes. This is reflected in Table 2 which highlights this flow of name-changing in the capital over five centuries.

The arrival of the French following the capitulation of the Knights to Napoleon marked a succession of rules and regulations to turn Malta into a French dependency, also through the name-changing process to give a more secular identity with names reflecting the spirit of the French Revolution of 1789 and its immediate aftermath. Although the British displaced the French after two years at the turn of the century the names given back to the principal streets in Valletta showed Malta's links to Italian with 'Porta' (trans. door), 'Piazza' (trans. open space) and 'Strada' (trans. street) for over a century. Visitors to Valletta were generally impressed by the layout of the town, its buildings, and amenities. However, not all travellers favoured the architecture of the streets, such as Lord Byron.

It was only in the 1920s that the British changed the names to English versions. This reflected a British policy towards Malta following the interest and overtures being shown by the fascist Italian dictator Mussolini to increase his control over the Mediterranean. Following independence from Britain, Malta's successive administrations changed the names into Maltese with connotations of political events. Cases in point demonstrate that the principal entrance to Valletta was changed from Kingsgate to Bieb il-Belt, (lit. trans.: Door of Town) whilst retaining the commonly corrupted Italian version of Porta Reale (lit. trans.: Royal Door) to the local 'Putirjal'.

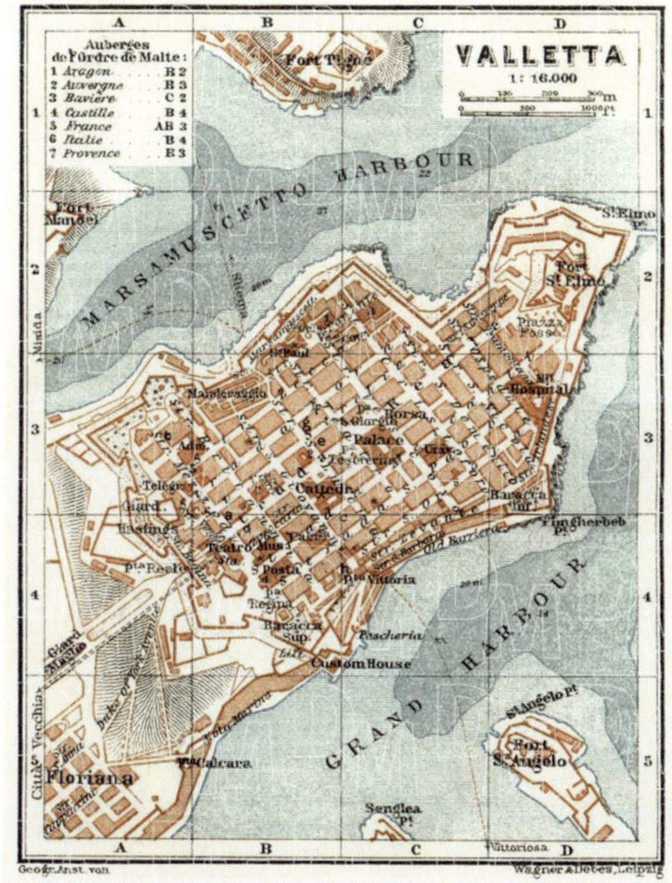

Fig. 1 Old Map of Malta and Gozo 1888. *Source* Gif-map.com, under CC-BY-3.0

A study on the street names of Valletta shows that out of 75 streets, 25 have a religious link (Table 2). Generally, the street names deal with the important buildings such as the treasury, mint, hospital, theatre, bakery and mill. Geographic orientation is also included with the four main compass directions represented. The list also illuminates structural features such as steps and spur, as well as hydrological attributes to fountains and wells. The list provided in English as the main language also gives the Maltese version of each street together with the Italian version for some. As an example, Old Theatre Street is given as the Maltese version and also in vernacular, and as the popular Strada Teatro in Italian with the older generations.

Further details on the names of Valletta streets can be examined in Table 2 and Box 1 where the list provided shows that the bi- and, on occasions, the tri-lingual versions are used. Concluding the section on Valletta three significant quotations are presented, one by the English poet and peer Lord Byron (1788–1824) who on a visit

Fig. 2 St Ursula Street, Valletta. *Source* Gif-map.com, under CC-BY-3.0 and by permission from Leslie Vella

to Malta in the early nineteenth century seemed to have been glad that his sojourn on the island was over as represented from his verse where he cursed the stairs lining some streets. The second by the two-time Prime Minister of the United Kingdom, Benjamin Disraeli (1804–1881) who visited Malta in the 1830s and was awed by the capital city of Malta describing Valletta as a city of palaces built by gentlemen for gentlemen. Thirdly, in 2004, an academic paper published by landscape architect Trude Jaeger, emphasized how the street plan of Valletta was a model of modern urban design.

Table 2 Evolution of historical names of Valletta's streets and squares

Original 1571	Informal	French 1798–1800	British 1800/14–1964	British post-1926	Maltese post-1964
Porta di San Giorgio		Porte Nationale	Porta Reale	Kingsgate	Putirjal/Bieb il-Belt
Piazza San Giorgio	Piazza dei Cavallieri	Place de la Liberté		Palace Square	Pjazza San Ġorġ
Piazza dei Cavallieri	Piazza della Città	Place de l'Egalité	Piazza Regina	Victoria Square	Pjazza Repubblika
Strada San Giorgio	Strada delle Corse	Rue Nationale	Strada Reale	Kingsway	Triq ir-Repubblika
Strada San Giacomo	Strada di Castiglia	Rue de Marchands	Strada Mercanti	Merchants Street	Triq il-Merkanti
Strada San Luigi	Strada Sant'Aloisio	Rue de la Constitution	Strada Levante	East Street	Triq il-Lvant
Strada San Paolo	–	Rue de la Constitution	Strada San Paolo	St Paul Street	Triq San Pawl
Strada San Pietro	Strada della Chiesa di San Rocco	Rue de la Barraque	Strada St Ursula	St Ursula Street	Triq Sant'Ursula
Strada San Giovanni Battista	–	Rue des Fours	Strada Forni	Old Bakery Street	Triq l-Ifran
Strada San Sebastiano	Strada Toro	Rue de la Monnaie	Strada Zecca	Old Mint Street	Triq Żekka
Strada Stretta	Strada Vanella	Rue Etroite	Strada Stretta	Strait Street	Triq id-Dejqa
Strada Pia	Strada del Gran Falconiere	Rue de la Félicité Publique	Strada Britannica	Britannia Street	Triq Melita
Strada di Monte	Strada dei Carcerati	Rue de Peuple	Strada San Giovanni	St John Street	Triq San Ġwann/In-Niżla tal-Ganċ
Strada del Popolo	Strada di Aragona/Strada dei Greci	Rue des Libérateurs	Strada Vescovo	Archbishop Street	Triq l-Arċisqof

Source Adapted from Chetcuti (2015)

Box 1 List of street names in Valletta

- Archbishop Street (Maltese: Triq l-Arċisqof; alt. Strada Vescovo)
- Battery Street (Maltese: Triq il-Batterija)
- Bishop Lane (Sqaq l-Isqof)
- Carmelo Street (Triq Tal-Karmnu)
- Carts Street (Triq il-Karrijiet)
- Castile Place (Maltese: Pjazza Kastilja)
- St Elmo Place (Maltese: Misraħ Sant'Iermu)
- Old Bakery Street (Maltese: Triq l-Ifran; alt. Strada Forni)
- St. Biagio Street (Triq San Bjaġju)
- Boat Street (Maltese: Triq il-Lanċa)
- Bounty Street (Triq l-Għajnuna)
- Bull Street (Triq il-Gendus)
- Eagle Street (Maltese: Triq l-Ajkla; alt. Strada Aquila)
- East Street (Maltese: Triq il-Lvant; alt. Strada Levante)
- Engineers Lane (Sqaq l-Inġinieri)
- Felix Street (Triq Feliċ)
- G. Cassar Road (Triq Ġirolomu Cassar)
- West Street (Maltese: Triq il-Punent; alt. Strada Ponente)
- Knight Street (Triq il-Kavallier)

- Nelson Avenue (Vjal Nelson)
- Nix Mangiari Steps (Taraġ Nix Mangiari)
- North Street (Triq it-Tramuntana)
- Pope Pius V Street (Triq il-Papa Piju V)
- Scots Street (M.A.Vassalli; alt. Strada Scozzese)
- South Street (Triq Nofsinhar; alt. Strada Mezzodì)
- St Andrew's Street (Triq Sant'Andrija)
- St Anne's Street (Triq Sant' Anna; alt. Strada Sant'Anna)
- St Anthony's Street (Triq Sant'Antnin)
- St Barbara's Bastion (Is-Sur ta' Santa Barbara)
- St Charles Street (Triq San Karlu; alt. Strada San Carlo)
- St Dominic's Street (Triq San Duminku; alt. Strada San Domenico)
- St Fredrick's Street (Triq San Federiku; alt. Strada San Federico)
- St George's Street (Triq San Ġorġ)
- St John's Street (Triq San Ġwann; alt. Strada San Giovanni)
- St John's Cavalier Street (Triq il-Kavallier ta' San Ġwann)

- M.A. Vassalli Street (Triq Mikiel Anton Vassalli)
- Mediterranean Street (Maltese: Triq il-Mediterran)
- New Street (Triq il-Ġdida)
- Republic Street (Maltese: Triq ir-Repubblika; alt. Strada Rjali, Kingsway)
- Old Hospital Street (Triq l-Isptar)
- Old Mint Street (Triq Żekka; alt. Strada Zecca)
- Old Treasury Street (Triq it-Teżorerija)
- Old Theater Lane (Sqaq tat-Teatru l-Antik)
- Old Theater Street (Maltese: Triq it-Teatru l-Antik; Strada Teatro)
- Ordinance Street (Triq l-Ordinanza)
- Great Siege Road (Maltese: Triq l-Assedju l-Kbir)
- King's Garden Street (Maltese: Daħlet Ġnien is-Sultan)
- Lascaris Wharf (Xatt Lascaris)
- Liesse Hill (Telgħet Liesse)
- Lower Barrakka Lane (Sqaq il-Barrakka t'Isfel)
- Marsamxett Street (Maltese: Triq Marsamxett)
- Melita Street (Maltese: Triq Melita)
- Merchants Street (Triq il-Merkanti; alt. Strada Mercanti)
- Mill Street (Triq il-Mitħna)

- St Joseph Street (Triq San Ġużepp)
- St Lucy Street (Triq Santa Luċija)
- St Michael's Street (Triq San Mikiel)
- St Nicholas Street (Triq San Nikola)
- St Paul Street (Triq San Pawl; Strada San Paolo)
- St Patrick's Street (Triq San Patrizju)
- St Sebastian Street (Maltese: Triq San Bastjan)
- St Ursula Street (Triq Sant' Orsola; alt. Strada Sant'Orsola)
- Zachary Street (Triq Żakkarija; Strada Zaccaria)
- St Christopher Street (Maltese: Triq San Kristofru)
- St Mark Street (Maltese: Triq San Mark)
- Spur Street (Maltese: Triq l-Ixpruna)
- Fountain Street (Triq l-Għajn)
- Steps Street (Triq it-Turġien)
- Strait Street (Triq id-Dejqa; alt. Strada Stretta)
- Quarry Wharf (Maltese: Xatt il-Barriera)
- St Dominic Street (Maltese: Triq San Duminku)
- Toni Bajada Lane (Sqaq Toni Bajada)
- Victory Street (Triq il-Vitorja)
- Wells Street (Triq l-Ibjar)

Source National Archives Malta

A Historical Geography of Secular and Religious Street Names

Aggregated records of local street names can be traced only to mid-nineteenth century census counts, where notwithstanding the enumeration of the population every decade the records published included the individual streets only for 1861 and 1871. For the purposes of this study the names of the streets were taken as those enumerated in the official census lists. All other previous censuses and those of the twentieth and twenty-first centuries only give data by towns and villages thus foregoing the details by street.

The Mid-Nineteenth Century

An interesting case in point relevant to this study is the proportion of streets having a secular and a religious name. In the mid-nineteenth century, the Maltese Islands were still predominantly Catholic with the church exerting influence at all levels of society. Street names within the core areas reflected this situation and the results from the official lists of street names published in the 1861 and 1871 censuses testify to this. However, the aggregated results giving all the names of the streets in all localities as shown in Table 3 with the 1888 map in Fig. 3, identifying the localities, show that only one-third of the streets had a religious name. The others were either linked to local trades, geographical features, or prominent personalities of the era. In addition, the geography of this secular/religious feature that can be observed is the fact that although over two-fifths of the names in the highly urbanized towns of the walled cities around Grand Harbour bear religious names, the proportion declines in districts I and III, the localities nearest to the harbours. As for the rural localities in the south of Malta, the proportions were about two-thirds religious. However, the lowest figure was for Gozo with barely one-fifth having religious names.

The Early Twentieth Century

In addition, in the earlier part of the twentieth century a series of sketch maps were drawn by employees of the local Public Works Department and published in 1907. This included 49 sketch maps of villages in Malta and Gozo identifying names with selected streets. The evolution of street naming in Malta can be gleaned from this series through three tangible parameters. The first is the identification of the actual street name as written on the sketches within the space provided by the representation of the street width. The second is the identification by an arrow of the direction towards the nearest village. The third is the use of, in the absence of an arrow, the term "to"/"Road to"/ name of village. As a corollary, in another case no names are

Table 3 Distribution of religious and secular street names in Malta and Gozo

Location	Number of streets in district	Number of streets with religious names	Per cent of streets with secular names	Per cent of streets with religious names
Valletta, Floriana, Cospicua, Vittoriosa and Senglea	185	75	59.5	40.5
District I Notabile, Rabato and Dingli	67	18	71.2	28.8
District II Zebbug and Siggiewi	59	12	79.7	20.3
District III Birkirkara, Msida,Sliema,St. Julians, Lia, Balzan and Attard	95	27	71.6	28.4
District IV Mellieha, Mosta and Gargur	66	14	78.8	21.2
District V Curmi, Luqa, Tarxien and Paola	63	24	61.9	38.1
District VI Zurrieq, Qrendi, Safi, Mqabba and Kirkop	73	44	39.7	60.3
District VII Zejtun, Zabbar, Gudia and Axiak	85	48	43.5	56.5
TOTAL MALTA	693	262	62.2	37.8
District VIII Gozo	223	43	81.7	19.3
GRAND TOTAL	916	305	66.7	33.3

Source Censuses (1861, 1871)

identified with the street shape and pattern drawn. The sketches also identify selected buildings such as churches, schools and chapels.

The changing aspect in the use of the written and spoken language is also seen through the street names where both English and Italian were used. Clear examples of this are the use of both English and Italian versions of the terms "road" in English and "via/Sta./Strada" in Italian. The Maltese word of Semitic origin "Triq" was not yet used but came into fashion later on in the last quarter of the century. In addition, the names of the villages in question were written without the prefix "Ħal/Ħ'" which is the diminutive version of "Raħal", the Semitic term for village. It is to be noted that in practically all the maps drawn during the British period the prefix was not used, although in the local vernacular, villages were always prefixed with "Ħal/Ħ'. Further cultural influences can be identified with the Zabbar locality where a street named "Prince of Wales" and another as "Strada Chiesa and Church" are identified.

Geographies can also be identified within the street names in two categories. The first is giving names to streets that serve as indications of direction to nearby villages. Attard is a typical example where the locality's position in central Malta allows for the

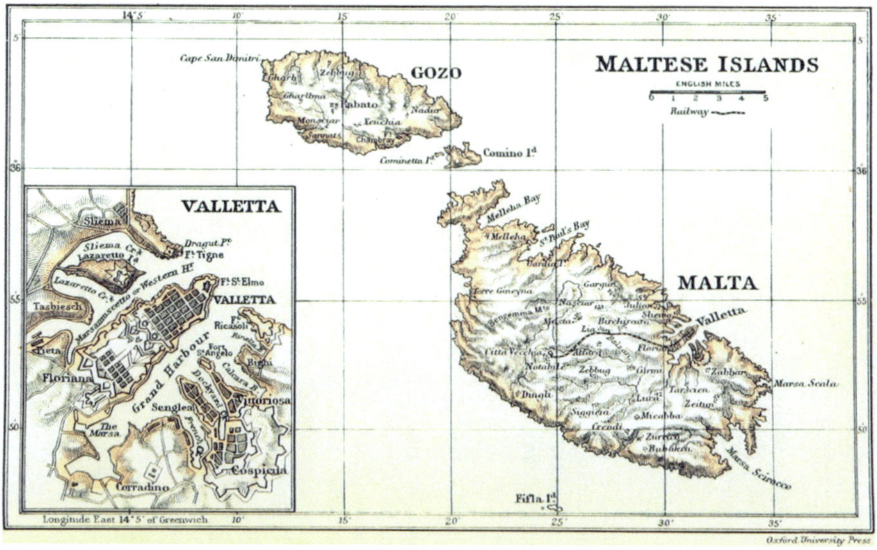

Fig. 3 Map of Malta, 1888. *Source:* Gif-map.com, under CC-BY-3.0

pattern of streets that radiate from the centre towards different directions. *Strada San Giorgio a mare* at Safi indicates a street towards St George's Bay limits of Birzebbugia whilst the same locality has a street showing the direction "To Birzebbugia". The second category refers to either a natural or human-made feature. For Qormi (Curmi) Strada Vallone (vallone meaning valley) is a case in point. The term "Molo" for jetty is also used to identify coastal streets along Msida, Pieta and Calcara (Kalkara).

Street names in Malta have also been used as geographical indicators to guide people in the proper direction from one locality to another. This is especially evident in the villages where names of adjacent towns and other contiguous localities are indicated as a street name with a compass bearing towards the adjacent destination. Table 4 lists the names of the streets in each locality. The list was compiled from a series of sketches drawn by the local Public Works Department and issued on 7th June 1907.

Prominent among these are the street names in village cores that are identified with geographic direction. A case in point is Ħ'Attard (Fig. 4) a village in central Malta where the list includes seven streets: Qormi, Valletta, Birkirkara, Mdina, Mosta, Żebbuġ and Lija.

Street-Naming Regulations

Regulations regarding street naming is under the supervision of the Electoral Commission and involve a number of steps. These include, as a first instance, an application that has to be endorsed by the local council; secondly, the street-naming committee accepts or refutes the name proposed which should be up to a maximum

Table 4 A street-name inventory in 1907, compiled by authors from a sketch-map series of villages in Malta and Gozo published on 7th June 1907 by the Public Works department

Mellieha S. Cimitero S. Tramuntana S. Parroco Road to Marfa - > To St. Paul's Bay (ESE) Road to Qasam Barrani (SW)	Dingli S. Rdum Depiru S. Dietro La Chiesa Private Road Road to Rabato (NW) Via Boschetto (SE)	Lija To C. Balzan (SE) To Musta & Naxaro Strada dietro La Chiesa
St. Paul's Bay	Siggiewi To mellieha Strada Molino To Ca Zebbug - > (NNE) To Curmi - > (NE) To Rabato (W)	Balzan Strada dietro La Chiesa Via Birkirkara - > (ENE) Via Birkirkara To Lia - >
Mgiarro To Gneina To Mosta (E) - > Bengemma	Naxaro > To Musta (W) - > Notabile	Curmi Strada Piazzetta Road to Valletta - > (ESE) Strada Vallone Strada Reale San Giorgio (near St. Sebastian) Strada Reale (also) To Zebbug - > (W) Road Tal Handaq (NW) To C. Attard - > (NW)

(continued)

Table 4 (continued)

Musta R to Naxaru (E) R to Wied il-Ghasel (N) R to Notabile (NW) Stra Molino via Valletta (SW)	Rabato	B'Kara Sta. Naxro (NW) Sta Ganu Via C. Balzan (W) Road to Gharghur (n) Strada Misida (E) Strada Fleur de Lys via S. Giuseppe (this is arrowed or road arrowed eg Road to Gharghur S. Giuseppe *See notes in original*
Attard Cas? Qormi (E) Valletta (E) Birkirkara (E) Strada Musta and separately Road to Musta NW & N Zebbug (Strada Casal Zebbug) (S) Mdina Via Notabile X2 Piazza Chiesa Lija (NNE) IMP crossing the railway line bridge 3 bridges/4bridges	St Julians To Birkirkara X2 (SW) To Pembroke X2 ? Strada Birkirkara Strada Stretta	Sliema To Msida - >
Luca Road to C. Curmi - > To Valletta - > Strada Nuova to Gudja - > (SE) To Tarxien - > (ESE) Via casal Zurrieq (S) Sta. Hal Farrug (although no roads are here)	Msida Molo Ta Xbiex Molo Pieta Strada Guardamangia (SE) Strada Sliema	Pieta Sta Guardamangia Msida Creek Pieta Creek (put accent on Pieta)

(continued)

Table 4 (continued)

Zebbug Strada Infetti Via Valletta - > (E) Sta Casal Attard - > (N) Road to Notabile (NW) Bishop Caruana Street Via Wied is Sewda	Hamrun Road to Curmi & Attard - > To Valletta To Msida Road to Casal Curmi - > (SSW)	Calcara Molo Calcara > To Villa Portelli
Zabbar Strada Chiesa & Church Avenue Piazza?? Prince of Wales Strada Capuccini - > Xgheira - > (NNE) Sta Marsascala To C Zeitun - > (S) Piazza Giacomo	Tarxien Piazza Brittanica Via Casal Luca - > (S) To Cospicua - > (NE) To Zeitun - > (ESE) To Casal Axiak - > (SE)	Chircop Strada Maggiore To c Gudja (E) *no name of street* Via Valletta (NNW) *no name of Street)* Piazza Maggiore San Leonardo San Andrea To Safi & Zurrieq - > (S)
Casal Zejtun Piazza Brittanica *1* To Cospicua - > (N) To Ca Tarxien - > (NW) Piazza Maggiore 2 *Nos 1 and 2 are near church on either side* To Axiak - > (W) Strada Marsascirocco (S)	Micabba To Chircop - > (E) To Zurrieq - > (E) Strada Santa Caterina To Siggiewi - > (W) To Siggiewi - > (NW) Via Valletta - > (E) To Hal Farrug - > To Qrendi - > (S) STR Torre (check capitals)	Safi To Casal Zurrieq - > (SW) To Chircop & Micabba (NW) Strada San Giorgio a mare To Birzebbugia (E) To Casal Gudja - > (N) Strada Hlantun S. san Tommaso

(continued)

Table 4 (continued)

Axiak Via Zeitun - > (E) To- > cospicua - > (N) Sta Maria S. Giuseppe S. Filippo S. Paolo	Gudja To Birzebbugia - > (S) To C. Kirkop - > (SW) To Luca (NNW) S. Stretta *All other names of Santo* *Santa Maria ??*	Zurrieq Strada Brittanica Sta. S. Andrea (&) Via Casal Safi (SE) *same one* Strada Reale Via Valletta - > *(near Govt School)* Via Qrendi (WNW) To Wied iz- Zurrieq - > (SW) Road to Nigret (SW)
Qrendi Strada Reale Strada Parrocchiale Strada Misericordie Strada Zurrico - > (ESE) Strada Stretta Strada Mercato	San Tumas Cala San Tumas To san Tumas Tower - > (NE) To Marsascala - > (W) To casal Zeitun - > (W) Rod tal Gzira	M'Xlokk
S. Giorgio & Birzebbugia C San Giorgio no names Strada San Giorgio Via San Giorgio Strada Birzebbugia		

Source National Archives Malta

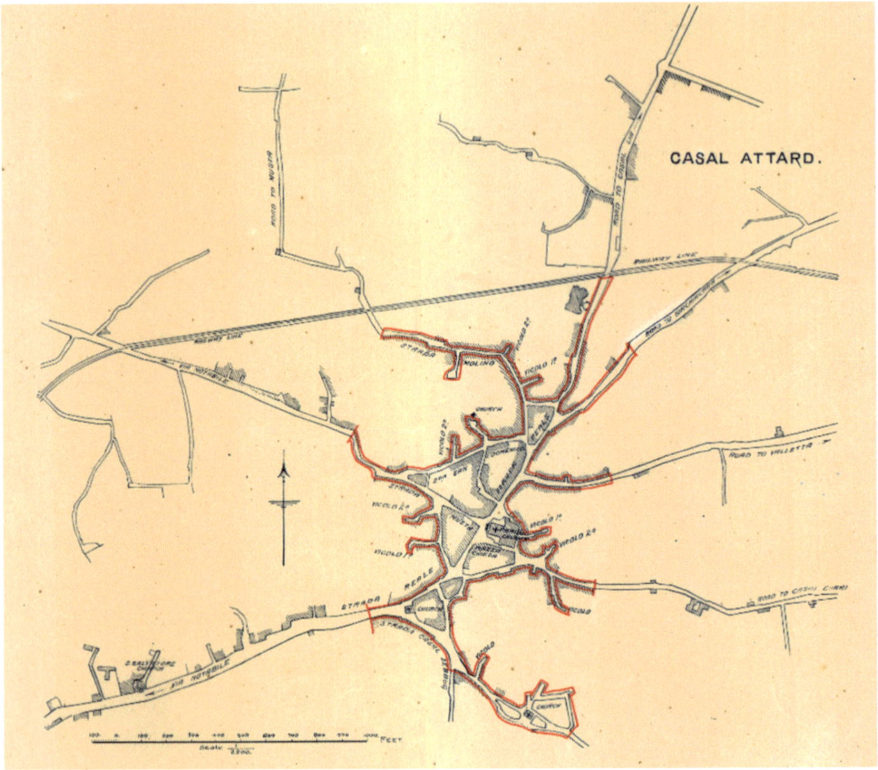

Fig. 4 Sketch map of Casal Attard, 1907. *Source* National Archives Malta

of 40 letters and can be pronounced by the locals; and thirdly, the street is physically identified with a name. The proposed name cannot be of any living person. A number of geographies are involved in this process at a series of scales. Firstly, details are requested about the area in the locality, these include, whether the area is urban, rural or coastal; and, partly built-up, in the process of development or an open space including a garden. The geographical bearings of the proposal must also be given as would be the form of the street whether it is rectilinear in shape, and having access at least at two ends, an alley (one entry point, cul-de-sac), or a piazza (an open space). A plan of the locality with details drawn must also be submitted. An officially-appointed committee vets all applications.

Contemporary Geo-Cultural Aspects

The geo-cultural dimension of street naming in Malta is very evident throughout the islands and certain localities bear witness to it. Typical examples can be found in

Ħ'Attard, a locality in central Malta, that spreads from the traditional compact core to the outskirts experiencing a six-fold increase in its population to 12,000 and a built-up area from 1/6 km^2 to 1 km^2 from the mid-1970s to date. This urban spread gave scope for the street-naming committees to consider appropriate proposals put forward by individuals, NGOs, local councils and central government. Two cases in point illustrate this. The first concerns a cluster of streets developed after 1975 having as a central theme the cultivation of grapes in vineyards and subsequent traditional production of wine. Triq id-Dielja (Vine Street), Triq l-Għenba (Grape Street), Triq il-Ġellewża s-Sewda (a local grape variety) and Triq l-Inbid (Wine Street) and Triq Bacchus, commemorating the god of wine. Furthermore, the area was also famous for the cultivation of oranges in gardens and therefore Triq iċ-Ċitru (Citrus Street) and Triq il-Ġonna (Gardens Street) are clear samples. Other street names are linked to the cultivation of vines. The citation in the application reads: "These names are inspired from the methods used in the growing of vines, starting from the zargun, type of grapes and ending up with the final product which is the wine" (The Malta Government Gazzette, Tuesday 17 December 1985, No 14,518 pp 3849–3850).

The locality also commemorates a prominent lexicographer and researcher, Annibale Preca. The citation in the street-naming committee's reports in justifying the proposal was as follows: "he did an important study about the connection between the Maltese language and that of Kanaan [Lebanon], which study was published under the name of 'Malta Cananea'". Further naming of streets after prominent personalities included one at Marsa after Giovanni Francesco Abela through a request dated 26/05/1955 (NAM 724/1955).

Following the construction of major roads that were built to alleviate traffic congestion names associated with members of national and international importance were introduced: Dun Karm for the national poet; Triq Mikiel Anton Vassalli recognized as the Father of the Maltese language; and also, for Aldo Moro, the former Italian Prime Minister assassinated by the Brigate Rosse in 1978. However, a major road running through the island connecting various localities: Tul il-Kosta formerly Coast Road, St Paul's Bay Bypass, and Dawret il-Mellieħa, the latter skirting the northern locality.

A number of streets also experienced a name change. George Borg Olivier (Prime Minister during the period of Malta being granted independence by Britain) has names in a number of localities: St Julian's and St Paul's Bay. In Sliema, the street where the former Prime Minister lived also bears his name having replaced the former colonially-influenced name of Victoria Terrace. However, the local council has promoted this transition by identifying the street as: "This is a Historical Street name". On close observation of the said image part of the original street name can also be seen painted on the brickwork. Furthermore, a hotel having an entrance from the street honours Queen Victoria in being named "The Victoria". Other former prime ministers also have multiple streets named after them on both islands, these include Nerik Mizzi, Dom Mintoff and Gerald Strickland.

Other interesting street names in a newly developed area at L-Ibraġ (l/o Swieqi) and identified with age-old weights and measures as listed in Table 5 A letter/petition to the Prime Minister was sent on 6th May 1985 with suggestions. It was signed by 354 residents.

Table 5 Streets named after traditional weights and measures

Name of street—measures	Meaning and measure
Triq il-Kejla	Kejla stands for 18.74 m^2 It is also used as a volumetric measure at ca 127.9 cc, or as a measure of dry units at 303.1 cc (eg capers)
Triq is-Siegħ	Siegħ stands for 187.4 m^2 = 1/6 tomna
Triq it-Tomna	Tomna stands for 1,124m^2. A standard measure for field areas
Triq il-Wejba	Wejba stands for X4 of tomna = 4,497m^2
Triq il-Modd	Modd stands for X16 of tomna = 17,988 m^2 = 1,799 ha
Measures (human and urban) terminology	
Triq il-Pulzier	One inch (2.183cms)
Triq ix-Xiber	The Maltese foot measure (26.19 cm)
Triq il-Qasba	$6'10^{1/2''}$ (2.096 m) The local fathom, but used also as a terrestrial measure, generally in the building industry to measure spatial extent of house facades
Weights	Equivalences
Triq l-Uqija	26.46 grms 14/15 oz The Maltese ounce
Triq ir-Ratal	793.8 grms
Triq il-Wiżna	3.969 grms = 5 ratal (rotolo)
Triq il-Qantar	100 ratal 79.38 kgs

Source National Archives of Malta (NAM)

Politically the proposed name changes were a very sensitive issue and made the round of cabinet ministries. The secretary of the Street-naming committee in proposing a number of new street names for various localities to the Prime Minister included endorsement of the Minister of Justice, Lands, Housing and Parliamentary Affairs, the Minister of Works and Sport, and the Minister of Development, Energy, Ports and Telecommunications. A case in point was a letter signed by E. F. Mallia, the Secretary of the Street Names Committee dated 19 December and registered at the Office of the Prime Minister on 24 December 1980.

Naming of new streets was also seen in the changing of the party in government. In 1957, two years after a Labour government was elected in Malta, a spate of new street names were announced. 'First of May Street' in Zabbar (NAM/427/1957), proposed by the MLP district committee of the village with proposal dated 25/04/1957; 'Labour Avenue' at Naxxar (NAM/737/1957), proposed by Calcidon Agius; 'Labour Street' as part of Irish Street at Vittoriosa (NAM/1039/1957), proposed by the Minister of Education through letter dated on 01/10/1957.

Politically charged situations were also identified. A case in point was a letter (4th December 1948) where a correspondent wrote to the Prime Minister and also to the Minister of Reconstruction and Works suggesting that "all Maltese streets should be named to the many different British cities, towns and villages, and, putting forward the idea that "little Malta transferred to little England". Attaching a list of 115 localities he was promoting this idea because of the part played by Malta in the Second World War.

As a reflection of the linguistic issue in Malta between the use of English and Maltese languages it was proposed on 31/10/1955 to have all the streets renamed in Maltese and Italian (1228/1955) by Cassar Torregiani. These two instances reflected the linguistic problematic between English and Italian.

Again…. "What's in a Name?"

Another example of a town that commemorates elements in Maltese colonial and military history linked with the Mediterranean and the rest of the world is Pembroke. Pembroke evolved as a direct result of the island's military importance, as well as to accommodate the needs of the British forces in Malta. Although the locality has been populated only over the last three decades and inclusion of its demographic and housing data was only included into the 1995 census it has an association with military history since the late nineteenth century when a fort was built to enhance the fortifications of eastern Malta. The locality bears street names of important battles of World War II, in the Mediterranean, northern France (Normandy) and the Far East (Table 6).

Pembroke was the first major expansion project targeted to cater for the infantryman—the fighting soldier. The earliest construction works for a first stone encampment date back to 1859, when Secretary of War, Sidney Herbert, son of 12th Earl of Pembroke, sanctioned the construction of Pembroke Camp. From 1860, the War Office authorised for the lease of land in many parts of Pembroke for the construction of infantry battalion barracks such as St George's Barrack Guard Room (1860), overlooking St Julian's Bay, St Andrew's Barracks (1901) with its own Guard Room and an imposing clock tower in 1903 and St Patrick's Barracks in 1938. Fort Pembroke was built in 1875–1878 to guard the northward seaward approach to the Grand Harbour and then strengthened with Pembroke Battery in 1897–1899.

Existing rifles ranges along the coast were extended and new sport facilities were also built for the troops to practice both live weapon firing but also tennis, cricket, rowing, football and athletics. A number of married quarters blocks were built for sergeants and senior non-commissioned officers, whilst warrant or regimental officers resided in private small dwellings within the town such as in Juno House. Along with the barracks expansion, a military cemetery, Pembroke Military Cemetery was built in 1908 for the soldiers and their dependants stationed at the camp. Today this cemetery is administered by the Commonwealth War Graves Commission. In November 1915, Australia Hall was built as a social centre for the many wounded ANZAC soldiers brought to Malta from Gallipoli. During the Second World War, a part of Pembroke was converted into a Prisoner of War camp for captured German servicemen. It also suffered heavy air raid bombings during the war especially St Patrick's and St Andrew's despite being marked as hospitals. The last housing to be constructed by the British in Pembroke was in 1960s with the White Rocks officers' married quarters. Following the closure of the military facilities with the British withdrawal in 1979, most of the barracks and former military residences were rented as social housing and several housing schemes were developed to accommodate a growing influx of new residents.

Table 6 Pembroke—military links to street names

Street name	Commemoration
Misraħ il-Forti Pembroke	Fort Pembroke built between 1875–1878
Misraħ Il-Paċi	Peace
Trejqet Ir-Reġimenti Maltin	Maltese Regiments
Triq Alamein	Commemorating North Africa Campaign in WWII
Triq A.N.Z.A.C	To commemorate Australian, New Zealand and Canadian Forces who were nursed in Malta during WWI
Triq Burma	Commemorating Far East Campaign in WWII
Triq Anzio	Anzio was a landing site in Italy of the Allied Forces used in Operation Shingle during WWII
Triq il-Battalja Ta' Malta	The Battle of Malta (1940–43)
Triq il-Mediterran	Commemorating Mediterranean WWII operations
Triq Juno	WWII invasion of Europe on Juno beach in Northern France
Triq Mandalay	Commemorating Far East Campaign in WWII
Triq Normandy	Commemorating Normandy landings 6th June 1944
Triq Suffolk	Either named after the Suffolk Regiment, or a cruiser in WWII
Triq Arnhem	To commemorate Operation Market Garden when the British and the Polish were tasked to secure a bridge at Arnhem in 1944
Triq Medjez	Place of a final offensive of Allied Forces in Tunisia, in which the war in N Africa came to an end with a defeat of the Axis Powers in 1943
Triq Tobruk	Town in Libya, scene of WWII Battles
Triq Tunis	Commemorating the Battle of Gabes in S Tunisia in 1943 and the Fall of Tunis on the same year
Wesgħet Ix-Xelter	Shelter
Triq Falaise	The Falaise Pocket was an area in France, in which the Allied Forces tried to encircle and destroy the German armies
Triq Cassino	A Benedictine monsatery on Monte Cassino in Rome that was destroyed in 1944 due to air raids

Source Darmanin (2010)

Toponomy and Contemporary Socio-Political Context

The advent of tourism has further intensified the use of English in practically all of the hospitality establishments with even European nationals expected to communicate in English whilst working in Malta. Commercial establishments, shops, restaurants and offices convey their services through English whether it is through their shop sign, menu, letterhead or printed bill. Practically all technical and engineering manual instructions are in English. The extensive use of electronic technology over the last two decades has intensified the use of English to grass-root levels and relegated further the use of other languages including Maltese.

The effects of the tourist industries also had their influence on names pertaining to coastal areas. Drury (2019) considers these alternative names to British Services' using it and using the area as recreational zones and Mallia (2021) provides a table listing some of the most popular ones. These include Il-Bajja tal-Mixquqa (Trans: The Bay of the Cracked [Faulted] Promontory) known as Golden Bay identified by the colour of the sandy beach but missing out on the spectacular scree slopes surrounding it. Another popular tourist spot lies between the island of Comino (3.5 km^2) and the islet of Cominotto (0.25 km^2) called Blue Lagoon but otherwise named as Bejn il-Kmiemen (Trans: Between the Cominos). A number of sites had their local name modified to the English version, some of which pertain to colonial times but which have now been used in tourist promotional material. Nine of the sites listed hereunder are located to the north of the Maltese Islands.

A well-developed coastal tourism on the islands has over the years actively featured the images and names of bays, creeks, sea arches, circular sinkholes, sea channels on local and international media and promotional material (Fig. 5). Much less has been done for names of streets and roads that equally represent a strong cultural and historical identity and which could visibly benefit from a similar promotional approach to heritage. Mallia (2021) highlighted the combined use of names in Maltese and popular touristic English names (with traditionally the largest share of incoming tourists being in fact British). Examples of such combined use of names for these coastal features has been compiled as follows:

(i) Beaches: Ir-Ramla tal-Mixquqa (Maltese); The Bay of the Cracked One (English translation of Maltese name); Golden Bay (Popular touristic name); Il-Mellieħa (Locality in Maltese); Mellieħa (Locality in English).
(ii) Sea Arches: Il-Ħnejja (Maltese); The Arches (English translation of Maltese name); Blue Grotto (Popular touristic name); Il-Qrendi (Locality in Maltese); Qrendi (Locality in English).
(iii) Sinkholes: Il-Qawra (Maltese); The Circle (English translation of Maltese name); Inland Sea (Popular touristic name); Id-Dwejra, Gèawdex (Locality in Maltese); Dwejra, Gozo (Locality in English).
(iv) Fault-generated Creeks: Il-Prajjiet/Ix-Xquq (Maltese); The faults (English translation of Maltese name); Anchor Bay (Popular touristic name); Il-Mellieħa (Locality in Maltese); Mellieħa (Locality in English).
(v) Sea channels: Bejn il-Kmiemen (Maltese); Between the Cominos (English translation of Maltese name); Blue Lagoon (Popular touristic name); Kemmuna (Locality in Maltese); Comino (Locality in English).

Conclusion

With increasing urbanization in Malta, the number of streets is also increasing and the process of new names being introduced is becoming more efficient and also involving more the local population's wishes to include late individuals' names from the

Fig. 5 Examples of tourist promotional material (hand-held fans for summer tourists), rebranding coastal heritage and which could be similarly be applied to re-imagining street heritage. *Photo* R. Gauci, author

locality. Prominent among these would be politicians and locals who contributed to the locality's socio-cultural development. However, the environment and its geographies are also being included. Recently-published names for the northern locality of Dingli (Ħad-Dingli) include a mixture of religious, secular and also floristic and geomorphological features such as Triq l-Irdumijiet (trans: Cliffs' Street), Trejqet il-Ħarrub (Carob Street [trejqet, the diminutive for triq, a short street]), Trejqet is-Sagħtar (Thyme Street [short]),.and Triq Widnet il-Baħar, the Maltese Centaury, *Cheirolophus crassifolius,* Malta's national plant. In addition, the Dingli rural environment is popular for its rugged topography, mainly karst terrain with some still-surviving shepherds tending to their flocks. In this regard, Triq il-Merħla (trans: Flock Street) is also included in the Government Gazzette No 20,734 published 12th November 2021 and Triq il-Felu (trans: a colt or the donkey's offspring) honour the animals that traditionally assisted in the cultivation of the rural environment.

Regarding colonial influences, it seems that even less street names have been dedicated to the British period over the last decades and even less street names were devoted to British political personalities which is normal as a post-colonial measure by authorities with an exception to Sqaq Ernle Bradford (1922–1986) at Kalkara, an alley in the locality where he lived. Bradford was an internationally acclaimed writer who wrote about Malta and the Knights of St. John, However, the use of English as a medium of written communication is still extensively used by state and private entities. In addition, personal names, having morphed into English from Maltese and Italian over the last century, are now extensively in fashion. Furthermore, a pass at ordinary level English is still the mark applied to one's success at secondary school certification, progression to tertiary education, university, and, as prospects for employment opportunities.

Street names from the pre-Christian era have also been introduced and these are geographically associated with Marsaxlokk, the popular fishing village to the southeast of Malta visited by tourists and locals alike. As the locality is near to the tas-Silġ temple, that was successively inhabited by pre-Christian societies, a number of streets have been named accordingly. Examples include: Triq Melqart, the protector and patron of the Phoenicians and a deity in Phoenician and Punic pantheons; Triq Axtart (trans: Astarte) 1750 BC, and Triq Cippi (Cippus Street). Two Cippi, originating in the second century B.C. were unearthed in the seventeenth century and they assisted in deciphering the ancient Greek and Phoenician alphabets. The latter are foremost among other socio-cultural contributions that Malta bestowed to the Mediterranean and Middle Eastern historiography and historical geography.

It is encouraging to note that in village cores, street names having a geographic bearing and dedicated to catholic saints have been retained together with the numbering of alleys and cul-de-sacs emanating out of the main square towards the outer parts of the core. The main reason is that the street and alley naming retains the original designation congruent with retaining the built-up fabric of the core. Actually few streets had their name changed in the old core areas except for the the post-colonial attitudes of replacing the "Brittania Street" with "Main Street" in a number of localities.

In fact, as indicated above, street naming is very organized now and most of the work and results done are published in a bi-lingual format in Maltese and in English online and can easily be accessed through the Electoral Commission's website. As a result, it is expected that more locals seek information on the street-naming mechanisms and following encouragement from the authorities more information be communicated as an ongoing reaching-out exercise to inform the public, school children and foreigners alike of the presence of the cultural wealth that the islands possess as demonstrated by the street names.

The evolution of street naming in Malta flowed from naming streets within what are now the village cores principally to show direction to contiguous villages and to Valletta, through religious and British-colonial based names, followed by a surge of secular designations, to over the last two decades, in honour of voluntary organizations such as Triq il-Volontarjat at Qormi, and Triq id-Donaturi tad-Demm (Blood Donors' Street) at Rabat in Gozo. Additionally, sporting personalities are also making it through with Triq Tony Nicholl (1916–1999) at Sliema, honouring a legend of Maltese football. Further details of the local urban environment are seen in the introduction of names to previously nameless steep streets made up of flights of stairs instead of the normal smooth surface. Taraġ Sant'Elija (St Elija Stairs) at Cospicua (Bormla) is a case in point where besides showing that religious names are marginally coming to the fore again the new designation of stairs is a recent inclusion to the street-naming nomenclature. Further additions can be expected honouring individual voluntary organizations and non-governmental organizations attentive to the environment and social causes.

It is proposed, and also expected, that a street named after the present epidemic, Covid-19, will be shortly added to the list of streets named following the example set by former street-naming committees in commemorating past epidemics and their victims that ravaged the island and upset its demographic profile. It is also valid that international organizations and especially Maltese individuals in particular who contributed to various sectors at global dimensions and had a world-wide impact should have a local street named after them.

Questions

1. Which historical layers can be distinguished in the toponymy of Malta?
2. Should old street names be changed? Why?
3. Compare multiple cultural influences on the toponymic heritage of Malta with another country.
4. How did the British culture influence the emergence of many toponyms in Malta, as well as impacting on certain modern identity issues?
5. Why did certain toponyms get recently changed in Malta, and what are the social and political implications of such decisions?
6. Can more geographical features be included in street names in Malta?
7. What geographies of street-naming patterns can be identified in Malta?
8. Does street naming in Malta conform to other former British dependencies who now form part of the Commonwealth of Nations?
9. Is it justified to use street names to rekindle old terms of equipment that are no longer used?
10. Can it be considered as politically divisive when streets are named after past political personalities with opposing views to those of present-day residents?
11. Should more information be included in the street name dedicated to personalities in addition to the years of their life-span?
12. Does having clusters of streets named after a particular trade confuse pedestrians and motorists in walking and moving around an area? Why?

Acknowledgments The authors would like to thank the following: Professor Maria Attard (HOD Geography at University of Malta); Dr Charles Farrugia, Malta's National Archivist and CEO of the National Archives of Malta; Mr. Melvin Caruana, Public Services Co-ordinator, from the National Archives of Malta; Mr Leslie Agius; Dr Dean Hili, Mayor and Mr Kevin Borg, Executive Secretary from the Pembroke Local Council; Mr Herman Grech, Editor-in-Chief, Times of Malta; the staff of University of Malta Melitensia Library and GuideMeMalta.com. A final thanks goes to volume editor Professor Gerry O'Reilly for his invitation to contribute this work.

References

Attard M (2020) Mobility justice in urban transport—the case of Malta. Transp Res Procedia 45:352–359

Azaryahu M (1996) The power of commemorative street names. Environ Plan D: Soc Space 14(3):311–330

Bobin F (2021) The politics of street names. The Elephant, p 2

Chetcuti K (2015) Where streets have four names, Times of Malta, 1st June 2015. Available on: https://timesofmalta.com/articles/view/where-the-streets-have-four-names.570511. Last accessed 07 Feb 2022

Darmanin DA (2010) Pembroke, from a British Garrison to a Modern Civlian Town. Pembroke, Malta

De Saussure F (1916) Nature of the linguistic sign. Course Gen Linguist 1:65–70

Drury M (2019) The exciting camping sites in Malta will make you go wild. Available at: https://www.guidememalta.com/en/these-exciting-camping-sites-in-malta-will-make-you-go-wild?fb_comment_id=1958062160933688_5679103062162894

Gauci R, Schembri JA (eds) (2019) Landscapes and landforms of the Maltese Islands. Springer, Cham

Härlin M, Sundberg P (1998) Taxonomy and philosophy of names. Biol Philos 13(2):233–244

Jager T (2004) The art of orthogonal planning: Laparelli's trigonometric design of Valletta. J Soc Arch Hist 63(1):4–31

Kripke SA (1972) Naming and necessity. In: Semantics of natural language. Springer, Dordrecht, pp 253–355

Mallia J (2021) Bejn il-Kmiemen and il-Bejta tal-Fenek? Did you know the Maltese versions of these popular bays? GuideMeMalta.com

Meital Y (2007) Central cairo: street naming and the struggle over historical representation. Middle East Stud 43(6):857–878

Mifsud Bonnici R (1960) Dizzjunarju bijo-bibljografiku nazzjonali. Malta, Dipartiment ta' l-Informazzjoni

Oto-Peralías D (2017) What do street names tell us? The 'city-text' as socio-cultural data. J Econ Geography lbx030. https://doi.org/10.1093/jeg/lbx030

Public Works Department (1907) Sketch maps of villages of Malta and Gozo. Public Works Department, Valletta, Malta

Radding L, Western J (2010) What's in a name? Linguistics, geography, and toponyms. Geogr Rev 100(3):394–412

Regier A (2010) The magic of the corner: Walter Benjamin and street names. Ger Rev 85(3):189–204. https://doi.org/10.1068/d140311

Rusu MS (2021) Onoma: Journal of the International Council of Onomastic Sciences Street naming practices: A systematic review of urban toponymic scholarship. Onoma 56:269–292

Salafia S (2020) The place-names of Gozo, vol 1, Part 1. Gutenberg Press, Malta

Serracino Inglott E (1981) Il-Miklem Malta. Klabb Kotba Maltin, Malta

Schembri JA (2019) The geographical context of the Maltese Islands. In: Landscapes and landforms of the Maltese Islands. Springer, Cham, pp 9–17

Soanes C, Stevenson A (eds) (2004) Concise Oxford English dictionary, vol 11. Oxford University Press, Oxford

Textor M, Rami D (2015) Proper names: philosophical and linguistic perspectives. Erkenntnis 80(2):191–194

Tichelaar T (2002) Toponymy and language. DGSD-UNGEGN Toponymy Course Enschede/Frankfurt am Main August, 12–23

Wettinger G (2000) Place-names of the Maltese Islands, ca 1300–1800. P.E.G. Malta

Toponymic Study of the Map of New Lusitania: A Portuguese Cartographic Monument from the Eighteenth Century

Paulo Márcio Leal de Menezes and Manoel do Couto Fernandes

Abstract A set of four versions of the map "Carta da Projecçaõ Geographica Espherica Orthogonal da Nova Luzitania or America Portugueza and Estado do Brazil", is a monument of Portuguese cartography in the eighteenth century. The wealth of details about Portuguese America, Spanish America, and other European domains is impressive. For instance, toponymic content, presented in colonial and indigenous languages illustrating a hitherto unique set of the location of indigenous groups and is considered one of the first ethnographic maps of the Americas and perhaps worldwide. The research focus is to carry out comparative studies of the toponymy used in each version, as well as documents that allowed its creation. From this, a database is being created for the identification of historical toponymy, and extraction of cultural elements, areas, and behavior at a spatial level. Extraction of toponyms from the 1798 copy revealed identification and classification of 4,750 toponyms, according to geographic groups, language, linguistic and geographic motivations.

Keywords Portuguese cartography · Historical-cartography · Brazil · Toponymy · Database creation · South America

Objectives

By the end of this chapter the reader will be able to:

- To appreciate the unique importance of the toponymic study of the map of New Lusitania
- a Portuguese cartographic product from the eighteenth century.
- To observe the different skills in comparing and contrasting different versions of historical maps.

P. M. L. de Menezes (✉) · M. do Couto Fernandes
Federal University of Rio de Janeiro (UFRJ) - Department of Geography–Cartography Laboratory (GeoCart), Rio de Janeiro, Brazil
e-mail: pmenezes@acd.ufrj.br

M. do Couto Fernandes
e-mail: manoel.fernandes@urfj.br

© The Author(s), under exclusive license to Springer Nature Switzerland AG 2023

G. O'Reilly (ed.), *Place Naming, Identities and Geography*,
Key Challenges in Geography, https://doi.org/10.1007/978-3-031-21510-0_20

- To be familiar with Toponymic Extraction Methodology and its applications.
- To be able to define the different stages in the Toponymic Extraction Methodology
- To appraise the importance & value of GIS in furthering research and knowledge in historical cartography.
- To discuss the viewpoint that the map of New Lusitania is important not only for Brazil and neighbouring countries but also for world heritage

Introduction

The map called "Carta da Projecçaõ Geographica Espherica Orthogonal da Nova Luzitania ou America Portugueza e Estado do Brazil" ("Map of the Orthogonal Spherical Geographical Projection of New Lusitania or Portuguese America and the State of Brazil"), from 1798, together with the other versions, is undoubtedly one of the cartographic monuments developed by Portuguese mapping at the end of the eighteenth century. The toponymy (of which) is rich, in relation to the Portuguese domain but also in areas under Spanish, French, and Dutch influence. Because it presents a unique set of indigenous groups throughout South America, it can be considered one of the first ethnographic maps covering this region. The four versions of which, under the coordination of Antônio Pires da Silva Pontes Leme, designers José Joaquim Freire and Manoel Tavares da Fonseca supported by a plethora of geographers, astronomers, and engineers, despite their prominence, have not had their toponymy studied more deeply yet. These versions are currently spread around the world. The main version of this study, the New Lusitania Map of 1798 is in the Army's Historical Archive (Rio de Janeiro—Brazil), and the other versions are located in the Biblioteque Nationale de France (BNF) (Paris—France), Office of Archaeological Studies of Military Engineering (Lisbon—Portugal) and at the Astronomical Observatory of the University of Coimbra (Coimbra—Portugal). The four versions present clear differences, sometimes marked, regarding the spelling of the names, as well as different densifications, changes in position and even additions, and sometimes decreases in the placement of toponyms.

The purpose of this chapter is to present as its main purpose an analysis of the toponymy existing in the 1798 copy of the New Lusitania Map, as well as to raise questions about the differences found in the other versions, especially with regard to orthographic differences, number of toponyms, toponymic positioning, and changes. However, the study will not be complete, due to the number of 4,750 extracted and identified toponyms, whose analysis and associated attributes, idiom, language, and motivations will still be studied in continuity. It is assumed, according to the study planning, that all analyzes and conclusions on all versions will be completed in approximately three years, including master's dissertations and doctoral theses. In this chapter, preliminary examples of the main toponymic differences between the four versions will be shown.

The Versions of the New Lusitania Map

Several authors initially indicated the existence of two versions of the New Lusitania map, one in Rio de Janeiro, in the Army Geographical Service and the other in Portugal (Complementary) (Coelho 1950; Furtado 1969). Corrêa-Martins (2011) explicitly presents the different versions of the New Lusitania map. It also shows that, according to Faria (2001), the "Resumo dos Mappas Chartas Geographicas, e Plantas, que se tem Copiado, e Reduzido, no Real Jardim Botanico, por Ordem do Ill.mo e Ex.mo Senhor D. Rodrigo de Souza Coutinho, desde 26 de março, de 1797, até o presente anno de 1803"; there were 3 complete versions built from New Lusitania, the first from 1797, the second from 1798, and the third from 1803 (Corrêa-Martins 2011).

The 1797 copy, belonging to the Astronomical Observatory of the University of Coimbra, is entitled "Carta Geografica de Projeção Espherica da Nova Lusitania ou America Portuguesa e Estado do Brazil" (Geographical Map of Spherical Projection of New Lusitania or Portuguese America and State of Brazil—Astronomical Observatory of the University of Coimbra) and its dimensions are 142 cm wide by 128 cm high. Figure 1 presents a view of the map of Coimbra. This copy has three inserts, Rio Grande de S. Pedro, Salvador and Rio de Janeiro.

The 1798 copy, belonging to the Brazilian Army Geographical Service, currently in the Army's Historical Archive, has several differences in relation to the Coimbra copy, including the word Orthogonal in its title, its dimensions, design and toponymic content. The dimensions of which are also distinct, having 148 cm wide by 133 cm high. The names of 14 personalities are listed, including geographers, astronomers, engineers, among others, who helped in its preparation (Fig. 2). This map is the basis for all work in this article. This copy has four inserts, showing the Baia de Todos os Santos, Rio Grande de São Pedro, Rio de Janeiro and Barra do Pará.

The 1803 version is referenced by the designers Jozé Joaquim Freire and Manoel Tavares da Fonceca, when applying for the promotion to 1st Lieutenant in 1803 (Faria 2001). According to Corrêa-Martins (2011), this map was produced for the Viscount of Anadia, João Francisco de Sá e Melo. Also according to Corrêa-Martins (2011), this copy is believed to be the one in the BNF (Biblioteque Nationale de France), entitled *Carte de l'Amérique équinoxiale et du Brésil*. The map is incomplete, showing only its top. Authorship, according to the associated metadata, is given to José Lopes Santo (sic) and Antonio Pires da Silva Pontes, with the edition date of 1798. This version has dimensions of 156 cm by 68 cm, and features at least three inserts, from Barra do Pará, Recife and Olinda and Baia de Todos os Santos (Fig. 3).

The copy from Lisbon, Portugal is 202 cm wide by 199 high, the largest of all versions. It is at the Office of Archaeological Studies of Military Engineering, Department of Army Infrastructures, in Lisbon (Corrêa-Martins 2011). This version was recognized in 1903 with one designed by Silva Pontes and its main designer, José Joaquim Freire. The copy is incomplete, showing gaps in the northwestern part of South America and in the part of Patagonia Argentina. The area for its caption is still blank. It has five inserts, Baia de Todos os Santos, Rio Grande de São Pedro,

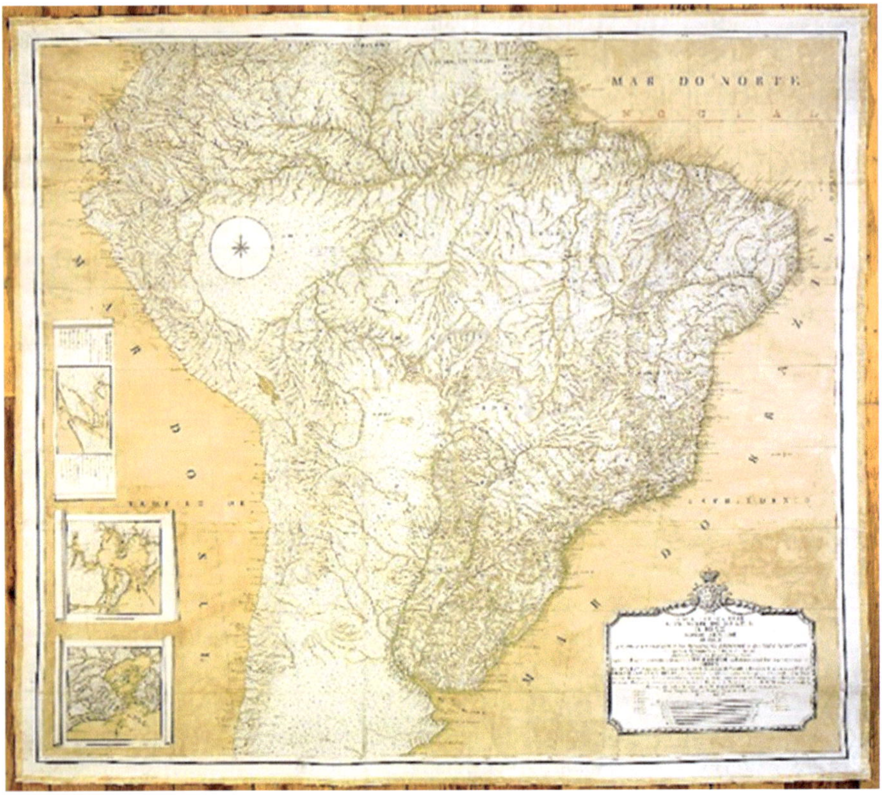

Fig. 1 Geographical Map of Spherical Projection of New Lusitania or Portuguese America and State of Brazil—Astronomical Observatory of the University of Coimbra

Rio de Janeiro, Barra do Pará and Cabo Frio, part of which are presented in the other versions. It is assumed that this copy is the model for the other versions, prior to 1795, the year in which the preparation of the 1797 and 1798 copies began (Corrêa-Martins 2011). However, a question arises, as there is no proof of authorship and the date attributed. There is an assumption, also without any proof so far, that this map is after 1820, due to differences in relation to the other versions (Menezes et al. 2021). Figure 4 shows the map of Lisbon.

The copies existing in the Mapoteca do Itamaraty, made over the copy of the Lisbon version in 1866 and two heliographic copies of the 1798 version, existing in the National Library of Rio de Janeiro, should also be mentioned.

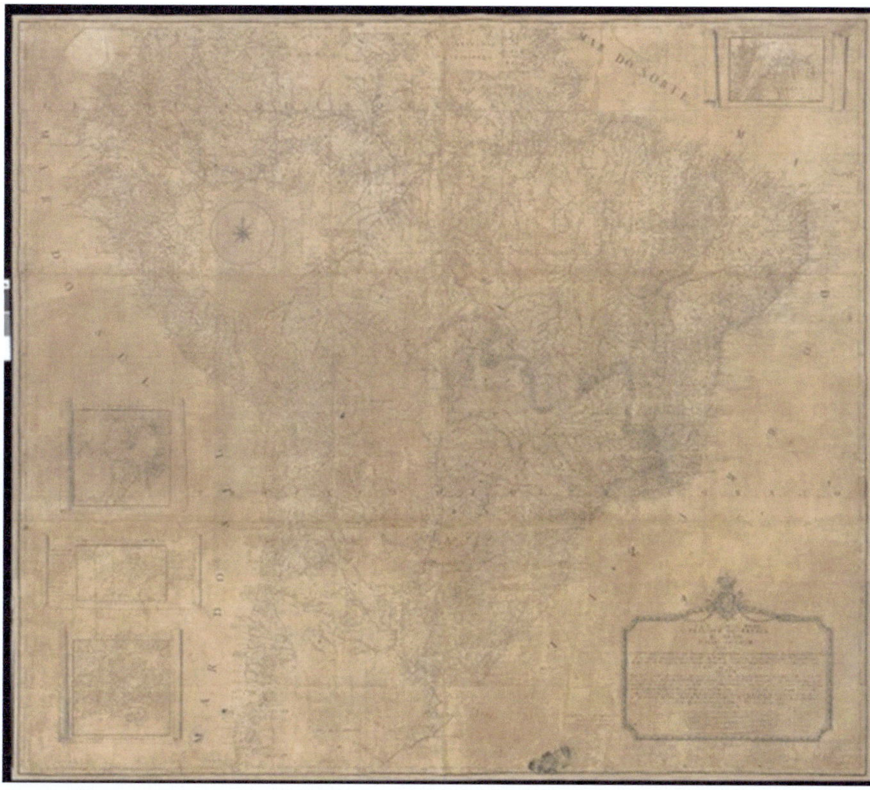

Fig. 2 Geographical Map of Spherical Projection of New Lusitania or Portuguese America and State of Brazil—AHEx—Brazil

Fig. 3 Carte de l'Amérique équinoxiale et du Brésil—BNF

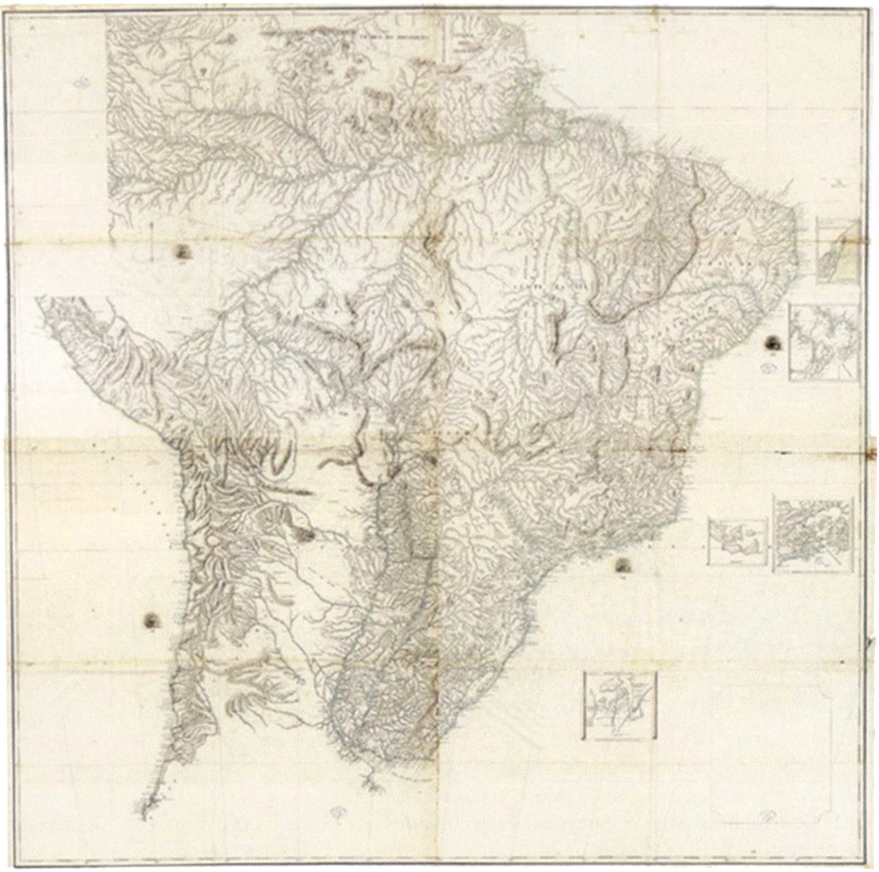

Fig. 4 Lisbon Version—GGEAM

The Toponymy of New Lusitania

The toponymy of New Lusitania will be referenced to each of its versions, since there are significant differences, in terms of spelling, geographic positioning and volume of toponyms, between each version. The complete extraction was performed in the 1798 version, still requiring revisions due to the map's degradation state. Despite being a relevant topic, the complete research on the toponymy of the other versions has not been fully developed yet, but the various common characteristics between the different versions of New Lusitânia are under study. The decision to start this toponymic study on the 1798 version was due to the availability of a digital copy in 300 DPI, as well as the ease of access to the original, if necessary, in the Army's Historical Archive, in Rio de Janeiro. On the other hand, the large number and wealth of toponyms existing in the various versions should be highlighted.

From the toponymic analysis of this version, it is proposed to extract the toponymy of the other versions, enabling a complete comparative analysis, both in terms of positioning and linguistic aspects, toponymic abundance and scarcity, lack, presence, and location errors in relation to each one of the versions. It is well known how long a survey of this magnitude will take.

Toponymic Extraction Methodology Applied to the 1798 Version

ArcGis®10.2 software was chosen for the extraction and creation of the initial table. Thus, the following methodology was prepared (Menezes et al. 2019):

- Study and division of the map into 6 (six) sheets, following the sheets of its drawing. Each sheet was in its turn divided into 6 parts, numbered from top to bottom, according to Fig. 5. This division allowed to optimize and sectorize the extraction and analysis of toponyms, as well as serving as a comparative basis for the toponyms of the other versions.
- Extraction of toponyms manually and inclusion in the data table in the ArcGis®10.2 software. Association of identification and geographic group.
- Linguistic identification relating to the source language and associated idiom.
- Definition of geographic motivation and linguistic motivation. The identification of linguistic motivation is a geographic interpretation of the name, through the determination of geographic associations for its creation. Linguistic motivation was defined by the application of physical and social characteristics developed by Dick (1990).

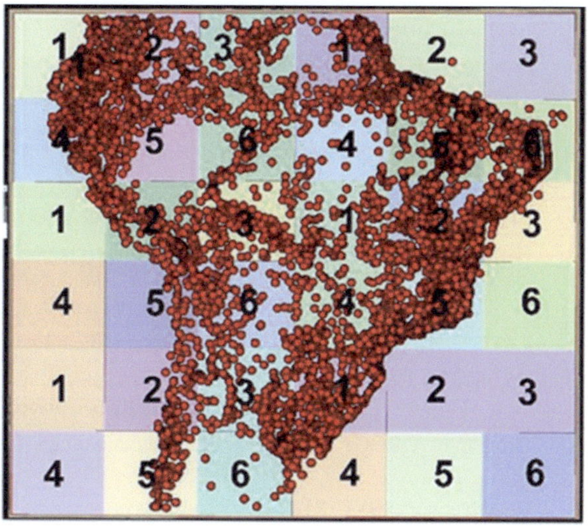

Fig. 5 Map of 1798 and its divisions for toponymy extraction

Extraction and Identification of Toponyms

The extracted toponyms were included in a table with the following elements: Toponym, identifying the specific term of the geographic name; Id_Geo, identifying the general term or geographic feature of the toponym; Grupo_Geo, defining the grouping of features with similar characteristics; Polygon, identifying the area of the map where the toponyms are located; Motiv_Geo, characterizing the toponymic motivation with a geographic view of each toponym, when identified; Motiv_Ling, identifying linguistic motivation, which was classified according to Dick (1990); Idiom; identifying the idiom of the toponym; Language, identifying the Language of the idiom; Pais and Capit, respectively indicate the country according to the current borders and the captaincy, according to the limits defined on the map. Table 1 presents the geographic groups and the geographic features or indicators that compose them.

A total of 4,750 toponyms were extracted, most of them perfectly identified. Several toponyms could not be identified yet, due to poor definition problems, mainly caused by being in places that had folds, degradation or even in restored areas that lost part of the document. These toponyms, however, will not be abandoned, as many were identified by comparison with the other versions, and this process will be applied in the study of the other versions, benefiting the final toponymic identification, relating to all versions.

Table 2 shows the total number of toponyms extracted by geographic group and toponyms without identification, with very low percentages in relation to the number of toponyms involved. It is important to add that after extraction, the percentage of unidentified was considerably higher, around 2.5% of the total, but all toponyms were

Table 1 Geo groups and geographic features or indicators

Nr	GRUPO_GEO	Geographic feature or indicator (Id_Geo)
1	Administrative (ADM)	Captaincy, country
2	Descriptive (DESCRIPTIVE)	All features that do not fit in with the others and that describe a place: place, landmark, forest, wood, coppice, camp,
3	Ethnic (ETHNICITY)	Indigenous groups, tribes, village, nation, family, descriptive,
4	Extractivism (EXTRACT)	Mine, gold mine, iron mine, saltpeter, saline, crystal
5	Hydrography (HDR)	rivers, streams, channels, waterfalls, lagoons, lakes, shoal, beach, port, island, cape, tip, bay, bar, hole, bathymetry, inlet, Igarapé, parcel, fishing, reef and ford
6	LATITUDE	Places with indication of their latitudes on the continent
7	Locations (LOCATION)	Site, register, parish, fort, fortress, town, village, tower, church, barracks, place, farm
8	Orography (OROG)	Hill, mountain range, canyon, volcano, mount, peak, desert,
9	Road Network (RVIARIA)	Roads, ways

Table 2 Number of toponyms by geographic group and percentage of unidentified names

Nr	GRUPO_GEO	Extracted	Percentage extracted (%)	Not identified	Not identified percentage (%)
1	Administrative (ADM)	8	0.17	0	0
2	Descriptive (DESCRIPTIVE)	158	3.33	15	9
3	Ethnic (ETHNICITY)	103	2.17	3	3
4	Extractivism (EXTRACT)	23	0.48	0	0
5	Hydrography (HDR)	2,205	46.42	49	0.45
6	LATITUDE	33	0.69	0	0
7	LIMITS	5	0.11	0	0
8	Locations (LOCATION)	2,122	44.67	48	0.44
9	Orography (OROG)	86	1.81	0	0
10	Road Network (RVIARIA)	7	0.15	0	0
	Total	4,750	100	115	0.41

checked and many of them confronted with the Coimbra, Lisbon or Paris versions, allowing to resolve the non-identification.

The number of toponyms distributed by current countries can be seen in Table 3.

Linguistic Identification

The next step was the identification of the idiom and language associated with each toponym, in a first phase analyzing only the toponyms located in Brazil. Subsequently, the other toponyms will be analyzed. This task was not a problem in relation to those classified in Portuguese, Spanish or French languages, as the language and idiom are the same. However, in relation to toponyms in the indigenous language, there were doubts about the different linguistic variations of each indigenous group, family or branch expressed on the map.

The Brazil of 1500 presented itself as an immense territory, in a framework of total multilingualism, where approximately 1175 languages existed, without any associated writing. It was assumed that there were 6 million indigenous people from different linguistic groups. There was, however, a relative linguistic unit along the coast, where indigenous people from the Tupi trunk, belonging to the Tupi-Guarani

Table 3 Spatial distribution of toponyms by current countries and their percentage

No	Countries	Extracted	%
1	Brazil	3,177	66.87
2	Argentina	208	4.37
3	Bolivia	222	4.67
4	Chile	183	3.85
5	Colombia	193	4.06
6	Ecuador	132	2.77
7	Suriname	6	0.13
8	French Guiana	44	0.92
9	Guyana	7	0.15
10	Paraguay	57	1.2
11	Peru	412	8.67
12	Uruguay	80	1.68
13	Venezuela	29	0.61
	Total	4,750	100

family, had been settled. Even those who were historically enemies, such as the Tupiniquins and the Tupinambás, had a very similar spoken language.

In the interior and in the Amazon, it is possible to verify the existence of groups whose linguistic trunk was quite adverse, such as the Arawak, Karib, Je, Pano, Tukano, Bororo, Kariri and even hundreds of other smaller groups, showing the linguistic diversity existing in earth. Figure 6 shows the distribution of the Tupi trunk along the coast and Fig. 7 shows the distribution of the main ethnic groups in the Brazilian territory.

The map features 103 names related to ethnicities. For the area of Brazil, there are 32 indigenous groups or families, some of which have different languages, which it is intended to identify, if possible, most of them, as can be seen in Fig. 8a, b.

The linguistic classification analyzes each toponym identifying the different languages and idioms. In this way, a classification was defined for each toponym according to the language of origin, Portuguese, Spanish, French, English, Dutch, indigenous and hybrid languages, when linguistic elements of two or more languages were joined. For this stage of the work, the focus of linguistic classification was defined for toponyms within the limits of Brazil and the results can be seen in Table 4, which shows the quantities classified according to the established criteria.

The percentage values were close, according to the work by Menezes et al. (2014), where the percentages between indigenous and Portuguese toponyms are similar to the research carried out on cartography to the millionth of Brazil.

On the other hand, today it is considered that the vast majority of toponyms are of Tupi-Guarani origin. However, the identification of a considerable number of toponyms from other ethnicities and linguistic groups is expected. So far, toponyms of Aruak, Bororo, Jê, Kariri, Pano, Karib, Kaikang, Puri and Guarani origins have

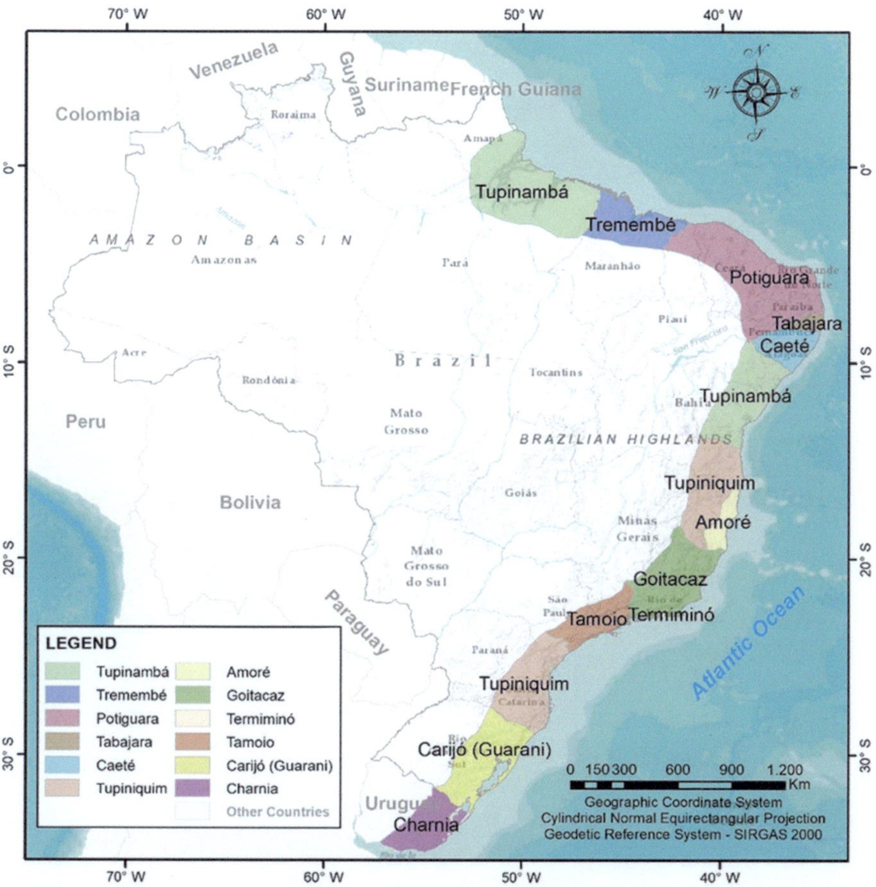

Fig. 6 Distribution of the Tupi trunk along the coast. *Source* modified from Arruda (2001)

been identified. This identification will still be the object of study, and it is estimated that it will be ready in 2022.

Comparison of the Toponym with the Other Versions

Preliminarily, some comparisons were made between the versions, in relation to toponymy, based on the names of the 1798 version; however, according to the need to clarify doubts about the legibility of toponyms, given the degree of degradation of the 1798 version, the comparison with the other versions showed a series of divergences, in relation to the following aspects:

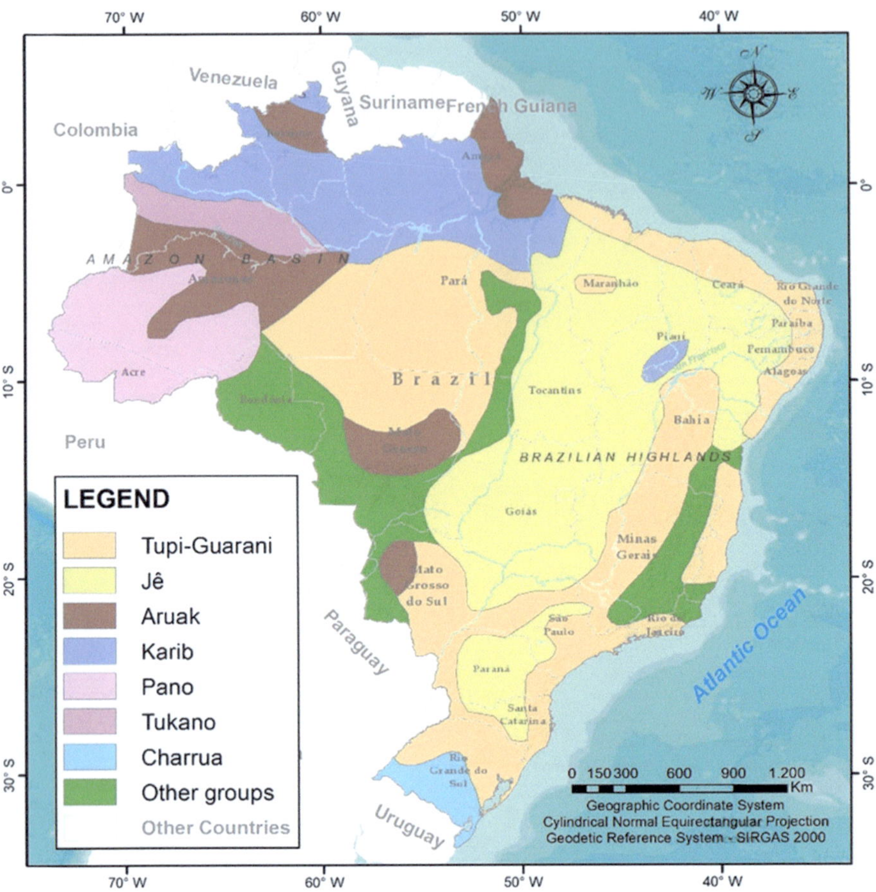

Fig. 7 Distribution of the main ethnic groups in the territory of the then colony. *Source* modified from Arruda (2001)

- Spelling divergence between toponyms;
- Divergence in name;
- Identical toponyms at different points identifying different features;
- Number of existing toponyms in each version.

Spelling Divergence Between Toponyms

The greatest divergences in spelling among the toponyms were found in the 1797 version. As an example, Table 5 presents some toponymic variants in spelling.

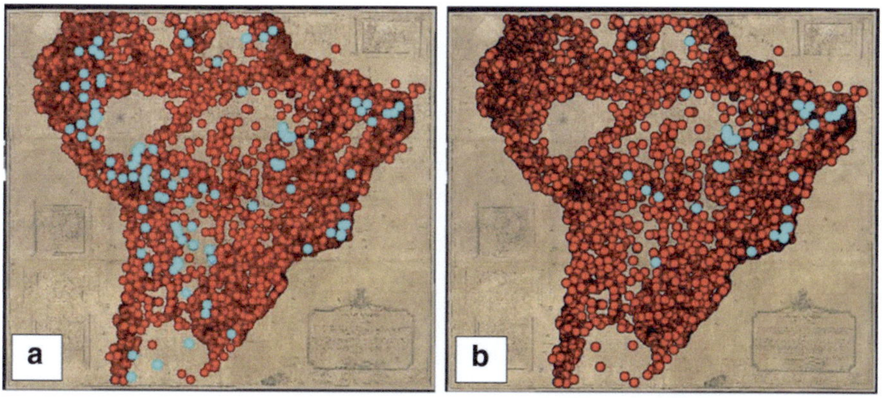

Fig. 8 Distribution of the Ethnic Group across the 1798 map (**a**) and in Brazil (**b**)

Table 4 Number of toponyms extracted by language about Brazil

Toponyms	Quantity	Percentage (%)
Portuguese	1,859	39.13
Spanish	929	19.56
Dutch	7	0.15
Indigenous	1,763	37.12
Afro	7	0.15
French	43	0.91
Hybrid	22	0.46
Not identified	45	0.92
No References	75	1.59
Total	4,750	100

Table 5 Examples of spelling variation in toponyms in different versions

Nr	1798	1795 (Lisbon)	1797 (Coimbra)	1803 (BNF)
1	Parayba	Parayba	Parayba	–
2	Rio Tiethe	Tiethé	Tieté	–
3	Rio Tapicurú	Rio Itapicó	Rio Itapicurú	Rio Itapicurú
4	Parnayba	Parnayba	Parnaiba	Parnayba
5	Seará	Ciará	Siará	Siará
6	Rio Tucantins	Rio Tocantins	Tocantins	Tocantins
7	Rio Jacuhipe	Jacuype	–	–
8	R. Gurupy	R. Gurupi	R. Gurupi	R. Gurupi
9	Guariuvacus	Guariuvacus	Guaricuvacus	Guarusvacus

Divergence in Name and Identical Toponyms in Different Positions

The next two items can be very well analyzed in the Santa Anna Island area, or the current Bananal Island, in the Araguaia River.

The following divergences and different toponym positions were found:

- Left branch affluents of the Araguaia River (Table 6);
- Toponyms existing in the 1798 version and not included in one or more versions (Table 7).

Figure 9 shows these toponymic differences that occurred in the mentioned area.

Table 6 Toponymic variation in the affluents of the Araguaia River (Left arm—Santa Anna Island)

Nr	1798	1795 (Lisbon)	1797 (Coimbra)	1803 (BNF)
1	Rio Casca	Rio da Casca	Rio da Casca	Rio da Casca
2	R. Farto—R. Farto	R de S. João—R. Farto	R de S. João—R. Farto	R de S. João—R. Farto
3	Rio Tapirapés	Rio das Vertentes	Rio das Vertentes	Rio das Vertentes
4	–	R. Cuxuru	R. Cuxurú	R. Cuxuru
5	R. Cuxuru	R. Tapirapés	R. Tapirapés	R. Tapirapés
6	R. Aquyquy	R. de Ponta	R. da Ponta	R. de Ponta
7	R. Tapirapés	–	Rio dos Tapirapés	Rio dos Tapirapés
8	Lamasaes Carajás	Carajá	Lamasay	Carajá
9	Ladário	Codajal	Ladário	Carajal
10	Aldeya do Gentio Ximbiuá	Gentio Ximbiva	Gentio Ximbiva	Gentio Ximbiva

Table 7 Toponyms existing in the 1798 version and not included in one or more versions

No	1798	1795 (Lisbon)	1797 (Coimbra)	1803 (BNF)
1	Xingu River	–	–	–
2	NOVA BEIRA	–	–	–
3	–	Gentio Xavante	Gentio Xavante	Gentio Xavante
4	Rio Barubó	–	–	–
5	Rio dos Bois	–	–	–
6	Morrinhos	–	–	–
7	R. Taboca	–	–	–

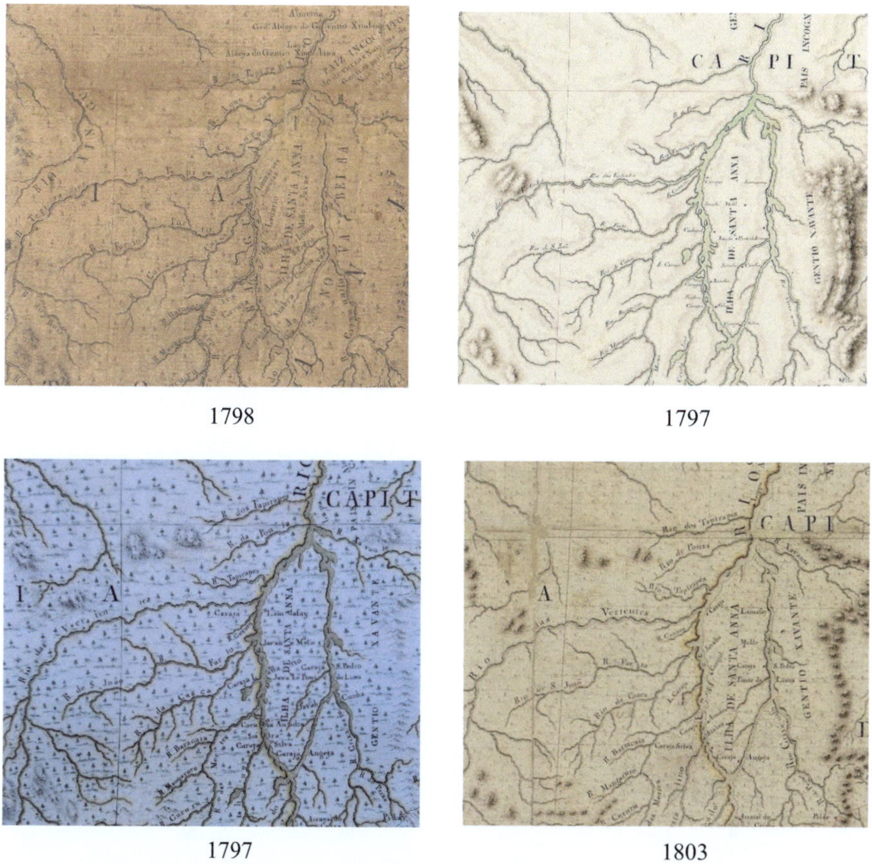

Fig. 9 Toponymic variation in affluents of the Araguaia River

Number of Existing Toponyms in Each Version

In relation to this topic, apparently, the 1798 version is the one that presents the largest number of toponyms in relation to the others. In some areas, such as the Northeast, around Piauí and Maranhão, the appearance is similar, while in others, such as Rio de Janeiro and Minas Gerais, it is more prominent. However, this issue can only be fully elucidated when toponymic extraction from the other versions is completed. It is evident that the 1803 version will suffer from this comparison. Figure 10 shows an example of toponymic density in part of Rio de Janeiro, Minas Gerais and Espírito Santo.

Fig. 10 Toponymic density in Rio de Janeiro, Minas Gerais and Espírito Santo, in the 1798, 1795 and 1797 versions

Conclusions

The complete study of the toponymy of the four versions of New Lusitania will not be an easy job to complete. The linguistic classification of each of the identified indigenous languages will certainly require the concurrence of specialists in ethnography, anthropology and linguists, professionals in dead indigenous languages.

Regarding languages that have been considered extinct, there is little information available, but it is still possible that some kind of linguistic fossil can be detected on the map.

The next phase, in relation to the 1798 version, is the extension of linguistic identification to other toponyms. Once this version is ready and serving as a comparative basis for the others, each of the others will be integrated into the developed toponymic process.

At least three years of research are expected, given what is being worked on in this first map.

Questions

1. What is the unique importance today of the toponymic study of the map of New Lusitania, a Portuguese cartographic product from the eighteenth century?
2. Why is it important to compare and contrast different versions of historical maps and also more modern ones right up to the present day?
3. What is Toponymic Extraction Methodology and how can it be applied.
4. Define the different stages in the Toponymic Extraction Methodology
5. Appraise the importance of GIS in furthering research and knowledge in historical cartography as illustrated in the chapter presented.
6. Evaluate the viewpoint that the map of New Lusitania is important not only for Brazil and the Americas but also for world heritage.

References

Arruda JJA (2001) Atlas Histórico Básico. 17a ed. Ática, São Paulo

Coelho PDA (1950) Obra Cartográfica dos Militares no Passado e no Presente do Brasil. Palestra apresentada na Escola Superior de Guerra, mimeografada, Rio de Janeiro

Corrêa-Martins FJ (2011) As várias faces da Nova Lusitania, de Antonio Pires da Silva Pontes Leme. Simpósio Luso Brasileiro de Cartografia Histórica; IV. Available at: <http://eventos.letras.up.pt/ivslbch/comunicacoes/79.pdf>. Accessed on 30 April 2021

Dick MVPA (1990) Motivação toponímica e a realidade brasileira. São Paulo: Arquivo do Estado

Faria MFA (2001) Imagem útil–José Joaquim Freire (1760–1847) desenhador topográfico e de história natural: arte, ciência e razão de estado no final do Antigo Regime. Universidade Autónoma Editora, Lisboa

Furtado SS (1969) A Cartografia Portuguesa, Roteiro de Glórias. Apêndice 1 de Cartografia

Menezes PML, Santos CJB, Resende ACB (2014) From 1500 to 2000, the Evolution of Toponymy in Brazilian Maps. Symposium on Place Names Changes, Joint IGU/ICA Commission/ Working Group on Toponymy, Academia Nazionale dei Lincei, Rome

Menezes PML, Fernandes MC, Corrêa-Martins FJ, Santos KS, Antunes FS, Oliveira PL, Santos JG (2019) Análise Toponímica da Carta da Nova Lusitânia. Simpósio Luso Brasileiro de Cartografia Histórica; VIII. Available at: <https://ler.letras.up.pt/uploads/ficheiros/18291.pdf>. Accessed 22 June 2021

Menezes PML, Santos KS, Brito TLF, Corrêa-Martins FJ, Fernandes MC, Santos JG, Lapaine M (2021) Map projection analysis of the Nova Lusitania map. Kartografija i geoinformacije 20:48–69

Names and Naming of Collective Farms in (the) Soviet Estonia

Taavi Pae, Ats Remmelg, and Jussi S. Jauhiainen

Abstract Place naming is part of national and local identity-making. Names of important territorial objects are often changed when state ideologies fundamentally change. Names are ideologized for the inhabitants' suitable education. However, some older names might remain in use despite being dubious after the changes in political power. Some people refuse to use new names or create nicknames based on these. This chapter analyzes farm names and their naming process in Estonia during the Soviet occupation era after World War II. The focus here is both on names and naming processes: the principles and practices of farm names; their connections to ideologies; and curious international, national, and local features and contradictions in these. Following World War II, Estonia became part of the Soviet Union. The ideological organization of politics and the economy also included agriculture. Former privately owned small farms were collectivized into large units that were important locations in the everyday lives of people in rural Estonia. Thousands of *kolkhozes* (cooperative collective farms) and *sovkhozes* (state-owned farms) needed names. Most names reflected directly the new political and economic ideology. These names were meant to honor major communist leaders, such as Lenin, Stalin, Kirov, Kalinin, and so forth; or the communist system with such farm names as *Uus tee* (New road), *Ühistöö* (Collective work). Nonetheless, names deriving from Estonia origin were also permitted if they outwardly fitted in with the state ideology. However, the authorities overlooked some interpretations of national romantic names, such as those from the epic poem *Kalevipoeg*. In the early stages, politically oriented individuals involved in the farms independently suggested names. Later, an administrative regulation was introduced for approving names. Throughout the decades, the share of place-connected names increased and the number of names inspired by the cult referring to a person, Soviet patriotism and national (political) romanticism declined.

T. Pae (✉) · A. Remmelg
Department of Geography, University of Tartu, Tartu, Estonia
e-mail: taavi.pae@ut.ee

J. S. Jauhiainen
Department of Geography, University of Turku, Tartu, Estonia
e-mail: jusaja@utu.fi

Institute of Ecology and Earth Sciences, University of Tartu, Tartu, Estonia

© The Author(s), under exclusive license to Springer Nature Switzerland AG 2023
G. O'Reilly (ed.), *Place Naming, Identities and Geography*,
Key Challenges in Geography, https://doi.org/10.1007/978-3-031-21510-0_21

Following the restoration of Estonia's independence in 1991, farmlands were privatized again, and large farm units were broken up. At the same time, almost all of the Soviet-era names vanished—but not entirely.

Keywords Estonia · Collective farm · Place naming · Soviet Union · Stalin

Objectives

By the end of this chapter readers should be able to:

- (know) see how place naming and agricultural farms connect to each other
- (know) understand how place naming took place generally (place) in Estonia before and during the Soviet occupation era
- (know) comprehend how political ideologies supported wiping out former place names and allowing new names to emerge in the Soviet Estonia
- (know about) witness the evolution of collective farm (s') names in the Soviet Estonia

Introduction

Place naming is a cultural and political practice aimed at fostering people's collective territorial identity and sense of connectedness and continuity in time. Governing territories includes influencing place naming that suits the ruling political powers' current ideology. Toponymic inscription aims to create politically charged linguistic landscapes to communicate intended political agendas (Mamvura 2020), to affirm their legitimacy; control their territory, and physical and social space; and promote their ideological norms (Gill 2005; Horsman 2006).

Place names tell the stories of both the past and present as they fuse history with the contemporary. Place names are worthy of traditional etymological and archaeological analysis. However, Giraut and Hoyssay-Holzschuch (2016) argue that the place-naming processes should also be the focus of inquiry, rather than only place names. The names of places change over time. Sometimes place names are changed following the top-down governmental ideological practices. Such place names are then imposed in wider use. However, at other times, people resist and refuse to use the new or existing place names. Instead, they create colloquial nicknames for their everyday uses and can even publicly request that unsuitable place names are changed as the recent postcolonial, anti-racist, and feminist movement actions illustrate over the globe.

A place name represents a geographical object on the ground and is a landscape marker. The study of place names falls within the field of onomastics in linguistics. In traditional onomastics, all place names undergo continual development as language changes. A phonetic type of change consists of primarily sound shifts in a language. Semantic change occurs when some words are lost from common usage along with cultural transformation (Saparov 2003). Place names have close links with the academic discipline of geography because they are often presented cartographically. Political geography addresses relationships between power, identity and place names. In the early twenty-first century, critical toponymy studies emerged (see, for example, Berg and Vuolteenaho 2009; Rose-Redwood et al. 2010) in response to the atheoretical mapping and cataloguing of place names or conceptually narrow linguistic studies regarding toponymy that mostly did not take into account the social and political contexts in which the place names emerged and were used.

Numerous transition societies and various authoritarian states collapsing are relevant objects of research. For example, numerous articles on place name changes in the (former) Soviet realm have appeared in the Anglo-American scientific literature since the collapse of the Soviet Union (Gill 2005; Kaşikçi 2019; Saparov 2017), and even before that time period (Bursa 1985; Peterson 1977). Much of the contemporary research on place names regards the urban environment because it is a priority for the dominant power of toponyms (Light and Young 2014). Research has been conducted on changes in street names, i.e. what was before, what is now, why such changes took place, and who was content and who was discontent with these changes (Capdepón 2020; Forrest 2018; Light 2004). The scientific literature on toponymics has devoted less attention to rural areas.

This chapter studies the names and naming processes of collective farms in Estonia. These emerged in the 1940s with the Soviet Union's occupation of Estonia. Then, they disappeared following the restoration of Estonia's independence in 1991 and consequent land privatization and land reform. However, names and naming processes link to various scales, from local to national, in the Soviet Union and beyond it, as well as to spatial and urban technologies.

The empirical material consists of maps and administrative materials indicating the evolution of collective farms and their names in Estonia. Furthermore, additional material also consisted of local and other Estonian newspapers containing information about collective farms, their names and naming processes. The data on collective farm names were compiled for three periods: the 1950s, 1960s and 1980s. For the 1950s, the main source was *Eesti NSV skeem maakasutuse piiridega* (Scheme of the Estonian SSR with land use boundaries) from 1957 (1:100,000). The data of the 1960s derives from the *Eesti NSV skeem maakasutuse piiridega* (Scheme of the Estonian SSR with land use boundaries) from 1967 (1:200,000). The data for the 1980s were from the map series *Eesti NSV Põllumajandusettevõtted* (Farms of the Estonian SSR) from 1986 (1:300,000).

The collective farm names were identified from these comprehensive sources. With the material-driven content analysis, five categories of farm names were identified:

(1) a person's name,
(2) a Soviet patriotic name,
(3) a place name,
(4) a national romantic name, and
(5) an abstract name.

Farm names based on an individual person's name formed a clear category because they pointed to an existing person, such as Lenin or Stalin. However, there were also names derived from a person's name but were used as slogans, such as *Leninlik Tee* (Leninist Road) or *Leninlik Võit* (Leninist Victory). Such names were classified in the category of "Soviet patriotic names." In general, such names usually had a communist undertone. These often reflected the victory of war, communist symbolism, socialist slogans, or referred to something common or new, for example, *Kommunismi Tee* (The Path of Communism), *Viisnurk* (Pentagon), *Uus Elu* (New Life), etc. Most collective farms associated with the place name received it directly from the name of the larger settlement near to the farm. There were also farms whose names referred to a larger geographical area, such as *Põhjarannik* (North Coast). Another group of farms had names with a national romantic undertone, such as *Kalevipoeg* (National hero) or *Hiie* (sacred natural site). Other names not belonging to any of these groups were classified as "abstract," such as *Edasi* (forward) and *Jõud* (power).

Three particularities relate to the analysis of the collective farms in Estonia and their names. First, this research studied places and organizations that no longer exist, at least not by their name or their former function. In rural Estonia, thousands of names appeared along the major reorganization in agriculture and rural people's everyday lives. Then, both this phenomenon and the related names disappeared. Nevertheless, it is important to analyze such significant changes as they are gradually disappearing from the population's direct memories and the landscape. Some material ruins of this period remain in the physical landscape of rural Estonia, but this research questions whether some names still remain, and if so, which ones.

In the Soviet system, there did not exist privately owned land; therefore, former private agricultural farms in the Soviet Union merged into large collective farms that also needed to have names. The collective farms there, were already were created in the early 1930s (Belov 1956). *Sovkhozes* (*советское хозяйство*, sovetskoye khozyaystvo, Soviet farm) were large, state-owned farms in the USSR and *kolkhozes* (коллективное хозяйство, *kollektívnoye khozyáystvo*, collective farm) were farming collectives in the Soviet Union. In the USSR, the average size of a *kolkhoz* (*kolhoos* in Estonian) was around 6,000–7,000 hectares and their number gradually declined from 44,000 in 1960 to 29,100 in 1990. A *sovkhoz* (*sovhoos* in Estonian) was a substantially larger unit and they were fewer. In Estonia, during the Soviet period, the initial 3,000 collective farms were merged into about 300 units.

Second, *sovkhozes* and *kolkhozes* were more than simple collectivized agricultural units in rural Estonia. From one perspective, the Soviet-era farms could be considered state-led or cooperative "enterprises," which are often not counted as place names. However, in the Soviet system, these farms had a large administrative power and were a type of functional area in which people conducted many living practices. Besides work, there were schools, cultural events and other social and political activities were organized. In numerous ways, they were more important for people than the municipality or county in which the collective farm existed. There was a village council in the state administration system, but it was primarily a statistical unit for population management. As will be discussed later, the collective farms were large agricultural production cooperatives or other state units that possessed substantial economic, political, social and technological infrastructure and power in the area in which they were located. They were the key everyday references for people living in rural areas.

Third, their naming was significant in the economic, social and political organization of these collectivized farms. New names were needed to convey that the former capitalist agricultural system based on privately owned plots of land, was modernized into the technocratically rationalized Soviet-style agricultural system. This highlights the top-down technocratic-administrative character of place naming and how place names function as social facts (Berg and Vuolteenaho 2009, 4, 9). From one standpoint, giving them proper names differentiated each individual and unique unit of the collective system. From another perspective, the general ideology in the place naming bound these units together because names held a connection to the Soviet ideology—i.e. relevant persons, goals or achievements. As there were many mergers during the decades, some 'invented' names of farms remained while others disappeared and completely new ones were created again.

Place Naming in Estonia and the Collective Farms

Historically, most areas of Estonia have used place names in two languages. The Baltic Germans used the German-speaking names and Estonians had their own place names in Estonian (e.g., *Reval/Tallinn*). In addition, Russian and Swedish place names also existed on the outskirts of Estonia. Historically, for the most part, there were parallel names in Estonia, where the name in one's own language was used according to the situation.

After Estonia's independence in 1918, the Estonianization of place names began. While changing street names was important at the beginning of the independence period, parish names were the primary target of changes in the 1930s. The whole process connects to the Estonianization of personal names and rise of nationalism in society. In 1930, *Eesti Rahvusluse Ühing* (the Estonian Nationalism Association) launched a campaign to change all foreign parish names. In particular, the aim was to change the names that came from the name of the former manor; for instance, Schilling to Kiling, Wrangell to Prangli (Pae et al. 2016).

Due to the challenging past with the Baltic Germans, a substantial part of the Estonian society was relatively anti-German in the 1920s and 1930s. Foreign language was supposed to disappear from place names. In 1935, preparations for the administrative territorial reform in Estonia began and it entered in full force in 1938. In the same year, the law on organizing the names of places and land units were issued. According to this, all municipalities and settlements had to bear Estonian names, and the foreign, vulgar, or bad-sounding names in use were subject to change. Many Estonian place names were also changed if they had a connection with the families of the German manors, such as Taagepera, which came from the surname Stackleberg (Pae et al. 2016).

World War II put the Estonianization of names on hold. Place name changes became frequent during the occupation periods that accompanied World War II. Already in the first Soviet period (1940–1941), several street names were changed to highlight the new political regime. During the German occupation period of 1941–1944, several names were changed because of the new power. However, as Germans withdrew from Estonia in 1944 and Estonia again became part of the Soviet Union, place names again started to change (Pae et al. 2016). Names were ideologized for the purpose of the citizens' communist education (Raag 2015, 180). Common examples were new street names in urban environments—even a new naming of the Saaremaa county main town of Kuressaare to Kingissepa (in 1952), which happened because V. Kingissepa was a communist revolutionary originally from Saaremaa. Further, the first *kolkhoz* created in Estonia was named after him. However, for a decade, many elderly people used the old place names of streets and other places in informal settings. For many, new Russian/Soviet names evoked unpleasant feelings of the foreign regime (Raag 2015, 181).

While many place names were changed in the urban areas, this was less so in rural areas. However, this started to change with the collectivization of rural farms. The planned economy in the Soviet Union transferred agricultural production to the collective farming system. Nevertheless, Estonia did not progress much in this direction in the 1940s, despite the collective farms being created there. The process only gained greater momentum after 1947 when the Central Committee of the All-Union Communist Party decided to create *kolkhozes* in the Lithuanian, Latvian and Estonian SSRs (Sirendi 2007). In principle, the legislation suggested that one did not need to join the collective farms in rural areas. However, in reality, merging rural farms into collective farms was a coercive process. The accelerated collectivization in Estonia was initiated after approximately 21,000 rural inhabitants were arrested and deported to Siberia and Central Asia within a couple of days in March 1949. According to Rausing (2004, 9), deportations were designed to weaken resistance to collectivization (see also Taagepera 1980). As a result, by October 1949, there were already 3,000 collective farms and 78% of the farms had joined them (Taagepera 1979). By the early 1950s, practically all former farms belonged to *sovkhozes* or *kolkhozes* (Maandi 2009).

The new farms needed names both for administrative purposes and also to indicate to the rural population that the regime had changed (Kasepalu 1974; Neve 2008). The initial place-naming process was a bottom-up initiated practice. Usually, at the founding meeting of a *kolkhoz* or *sovkhoz*, the participants submitted one or more names for consideration. Then, the general board meeting selected an appropriate one. Often, devoted Estonian communists put forward the proposals, which others with similar political views then supported. There was undoubtedly self-censorship in the face of terror, and only authorized names came out. Most likely, the district's central authority would have intervened if the farm's name was deemed inappropriate for the new Soviet regime.

However, in the early 1950s, many collectivized farms needed to be merged into larger ones because of the initial limited size; they were not economically viable nor suitable for the Soviet agricultural system. Due to mergers, many farms had to have name changes again, even during the first year. In Estonia, of the original approximately 3,000 collective farms (both *kolkhozes* and *sovkhozes*) formed in the 1940s and early 1950s, only 1,000 farms remained (in) 1952. In other words, 2,000 farms (and thus their farm names) had disappeared (Mertelsmann 2009). By the 1970s, their number had been reduced to less than 500. From the end of the 1970s until the collapse of the Soviet Union in 1991, the farm system was relatively permanent and acquired the meaning of a clear administrative and economic unit. By the end of the Soviet (time) epoch, there were around 300 collective farms, thus 300 names for these units (Fig. 1). For example, in the country of Saaremaa, 157 collective farms existed in the early 1950s and only 14 of them remained in 1985.

As territorial units, there were particularities in collective farms. Each collective farm's territorial borders were precisely defined and demarcated on maps. Gradually, rural areas became comprised of one collective farm next to another. As collective farms were jointly managed and sustained units, they had substantial effects on the everyday lives of the people associated with the *kolkhoz*. Besides employment, a successful *kolkhoz* would provide cultural activities, schooling, housing construction, landscape planning and so forth. Therefore, to a certain extent, people in rural communities "belonged" to their nearby *kolkhoz*. Rausing (2004, 1–4) indicated how the everyday lives and identities changed in rural Estonia when large societal and ideological transformations took place, including implementing different systems for social structures, and breaking and creating connections and traditions. Collectivization informed the worldview of the collective farm workers and their families—often interiorizing fragments of the Soviet (Marxist-Leninist) ideology and related political discourses. As an example, the daily newspapers contained relevant and irrelevant information and rankings about the practical achievements of the best and

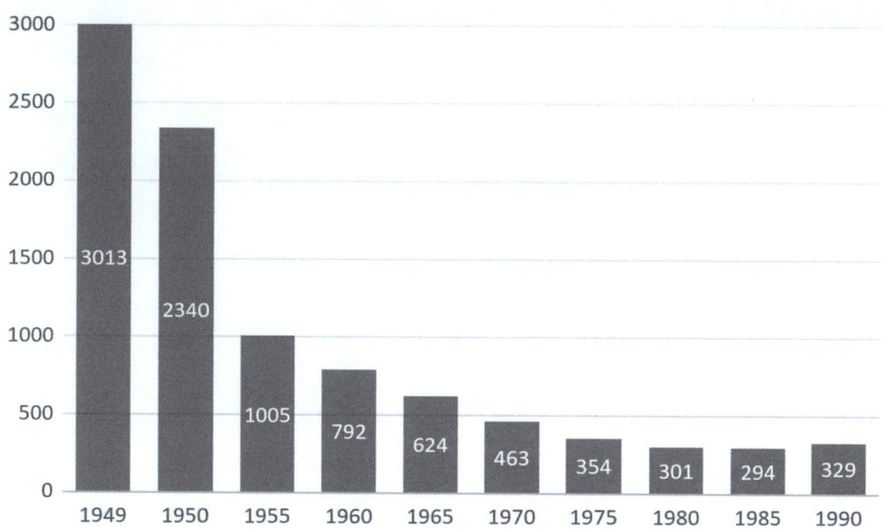

Fig. 1 Number of collective farms in Estonia, 1949–1989. *Data source* Sirendi (2007): 196

worst performing collective farms. Successful farms were cited as an example for the others, with tables published in the press about the development of various seasonal works e.g. harvesting, hay storage. Thus, the farms and their names were frequently visible in the everyday lives for the broader audiences elsewhere in Estonia. In addition, Estonian agriculture was highly developed compared to the rest of the Soviet Union. The agricultural sector was a priority, and in the 1980s, the standard of living achieved on farms was on average better than in the cities (Kannike 2008).

Nonetheless, despite a clear territorial cartographic demarcation of each collective farm, such borders were rarely demarcated into the physical territory. However, the collective farms' presence was evident with each particular name and the symbols that were visible on signposts, tractors, buses and other equipment that moved in and around the collective farm. The machinery and the fleet of farms bore individual farm's symbols and signs. This increased the visibility of the farm and helped to associate the place-based character of farms and their activities.

Most machinery associated with agriculture belonged to these farms and the farms' names and symbols were inscribed on the machines. (see Figs. 2 and 3). The space of a collective farm was simultaneously absolute (a fixed container), relative (in comparison with other collective farms) as well as relational (changing networks of production and consumption of its materiality and symbolic meaning locally, in (the) Soviet Estonia and the Soviet Union).

Fig. 2 The name of one collective farm, Ühistöö (Collective Work), inscribed on its truck. *Photo courtesy* Agnes Joala, private collection

Fig. 3 Signpost to *Rahva Võit* (People's Victory) *kolkhoz* in Estonian and Russian. *Photo courtesy* Estonian History Museum, AM N 16,194

Evolution of Collective Farms' Names in the Soviet Estonia

Collective farm names changed substantially from the 1950s to the 1980s (Table 1). The general trend in the collective farm names from 1957 to 1987 i.e. from the immediate post-Stalinist era to the emerging Estonian national consciousness, was a significant increase in the proportion of farm names based on place names. At the same time, there was a substantial decline of politically charged names that referred to Soviet ideology through patriotic and communist persons' names.

In 1957, there were 1,024 collective farms in Estonia. A third (34%) of the farms had a Soviet patriotic name, followed by a rather neutral place name (25%) and a

Table 1 Collective farm names in Estonia, 1957–1987

Year	1957		1967		1987	
Name category	Farms	Share (%)	Farms	Share (%)	Farms	Share (%)
Person	179	17	129	15	28	9
Soviet patriotic	349	34	250	29	22	7
Place	259	25	306	35	234	75
National romantic	90	9	77	9	12	4
Abstract	147	14	110	13	17	5
Total	1024	100	871	100	313	100

person's name (17%) (Table 2). The latter often referred to communist leaders of the Soviet Union, as will be discussed later (see Table 2).

From 1957 to 1967, the collective farms declined by 153 units (−15%). The only name category that grew absolutely (to 306 units, 35% of all farms) was the collective farms based on place name (47 such farm names more, i.e. a growth of 18%). Soviet patriotic farm names declined by 99 units (−28%) as did collective farms with a person's name, which declined by 50 units (−28%). The absolute number and share of national romantic names and abstract names of collective farms also decreased slightly.

From 1967 to 1987, there was a rapid decrease of collective farms from 871 to 313 units (−64%). Then, three out of four collective farms (75%) had names based on a place—but there was a decrease of 72 collective farm names (−26%). However, there was a dramatic decline of the Soviet patriotic names (decreased by 228 units, −91%). Further, the number of collective farms with national romantic names diminished (by 65 units, −84%), farms with abstract names (declined by 93 units, −85%) and farms with names referring to a person (decreased by 101 units, −78%). In 1985, the agricultural sector in Estonia consisted of 142 *kolkhozes* and 142 *sovkhozes* (Maandi 2009).

Person's Names in Collective Farms' Names in (the) Soviet Estonia

The Soviet-era also started a new naming formula practice in Estonia. A person's name was attached to many types of organizations and places, including the collective farms (David 2011; Raag 2015). Almost 200 collective farms in Estonia had, at least at one moment, a name referring to an actual person. As seen in Table 2, all of these persons were suitable for the Soviet ideology. Most of them were communist leaders, war heroes, political (communist) martyrs, and revolutionaries, as well as scientists, writers, and artists approved of, in the Soviet context.

Table 2 Collective farms in Estonia with a person's name and the frequency of names in 1957, 1967 and 1987

Person's name		Year of birth and death	1957		1967		1987	
			Number	Share (%)	Number	Share (%)	Number	Share (%)
M. I. Kalinin	SSSR State	1875–1946	20	11	15	12	–	-
V. I. Lenin	SSSR State	1870–1924	19	11	17	13	8	29
S. M. Kirov	SSSR State	1886–1934	16	9	13	10	4	14
J. V. Stalin	SSSR State	1878–1953	16	9	5	4	–	–
V. Kingissepp	EST revolutionary	1888–1922	16	9	15	12	2	7
L. Koidula	EST artist	1843–1886	12	7	9	7	1	4
I. V. Michurin	SSSR scientist	1855–1935	11	6	8	6	1	4
K. Marx	RUS revolutionary	1818–1883	6	3	4	3	–	–
A. A. Zhdanov	SSSR WWII	1896–1948	6	3	4	3	–	–
M. M. Gorky	SSSR writer	1868–1936	5	3	6	5	–	–
V. M. Molotov	SSSR State	1890–1986	5	3	–	–	–	–
V. R. Viljams	EST scientist	1886–1944	5	3	4	3	–	–
J. Lauristin	EST revolutionary	1899–1941	4	2	3	2	1	4
E. Vilde	EST writer	1865–1933	3	2	2	2	2	7
C. R. Jakobson	EST writer	1841–1882	3	2	1	1	–	–
K. J. Voroshilov	SSSR State	1881–1969	3	2	1	1	–	–
A. H. Tammsaare	EST writer	1878–1940	2	1	2	2	1	4
A. V. Suvorov	RUS war	1730–1800	2	1	1	1	–	–
A. Matrosov	SSSR WWII	1924–1943	2	1	–	–	–	–
T. D. Lysenko	SSSR scientist	1898–1976	2	1	–	–	–	–
M. Härma	EST musician	1864–1941	1	1	1	1	1	4
F. R. Kreutzwald	EST writer	1803–1882	1	1	2	2	–	–
C. R. Darwin	Scientist	1809–1882	1	1	1	1	–	–
A. S. Pushkin	RUS writer	1799–1837	1	1	1	1	–	–

(continued)

Table 2 (continued)

Person's name		Year of birth and death	1957		1967		1987	
			Number	Share (%)	Number	Share (%)	Number	Share (%)
F. E. Dhzerzhinsky	SSSR State	1877–1926	1	1	1	1	–	–
J. Tomp	EST revolutionary	1894–1924	1	1	1	1	–	–
P. S. Nakhimov	RUS war	1802–1855	1	1	1	1	–	–
A. Sommerling	EST revolutionary	1898–1924	1	1	1	1	1	4
V. I. Chapayev	RUS war	1887–1919	1	1	1	1	–	–
G. M. Malenkov	SSSR State	1902–1988	1	1	–	–	–	–
H. Heidemann	EST revolutionary	1896–1925	1	1	1	1	–	–
M. Lillevere	ESSR party martyr	1903–1949	1	1	1	1	1	4
K. Umal	ESSR party martyr	1893–1941	1	1	–	–	–	–
V. Sassi	ESSR party martyr	1899–1941	1	1	1	1	–	–
J. J. Lyakhov	SSSR WWII	1922–1944	1	1	1	1	1	4
J. Vares	SSSR State	1890–1946	1	1	1	1	–	–
M. Sibul	ESSR party martyr	?–1947	1	1	1	1	–	–
H. Vall	ESSR party martyr	1930–1950	1	1	–	–	–	–
J. M. Sverdlov	SSSR State	1885–1919	1	1	1	1	1	4
R. Pälson	EST revolutionary	1900–1924	1	1	1	1	1	4
Y. Gagarin	SSSR cosmonaut	1934–1968	–	–	1	1	1	4
L. Pärn	SSSR WWII	1903–1974	–	–	–	–	1	4
			178		129		28	

Collective farms in Estonia received names from 42 different persons when looking at the years 1957, 1967, and 1987. Of these, 18 came from the Soviet Union (USSR), i.e. not from Estonia, the Estonian Soviet Socialistic Republic (ESSR), Russia, or elsewhere; 13 from Estonia before its annexation to the Soviet Union; 5 from Russia before establishing the USSR; 5 from the ESSR; and 1 from elsewhere.

The latter is a unique exception of a person not involved in communist, military or pro-working-class activities, i.e. a collective farm named after the world-known scientist Charles Darwin.

The most commonly used person's name for collective farms in Estonia was derived from V. I. Lenin. From 1953 to 1955, there were 25 Lenin-named farms in Estonia as well as 7 farms called *Leninlik tee* (Leninist road), *Leninlik tahe* (Leninist will), or *Leninlik võit* (Leninist victory)—associated with Leninist ideology. By 1981, ten of these were removed, and seven still remained in 1987. The name Lenin, originally part of a small farm, persisted in the land mergers. Apparently, it was not appropriate to dismiss Lenin's name in terms of mergers. For several farms, the anniversary of Lenin's birth played a role. In 1960, on the 90th anniversary of Lenin's birth, two farms (Allvee 1980) received Lenin's name. On Lenin's 110th birthday in 1980, on one collective farm, the Estonian national romantic female name *Linda* was changed to *Lenin* (Kasepalu 1981).

One of the most commonly used names in the initial period of collective farms in Estonia was Stalin. Several farms received this name already during his lifetime. Stalin's death in 1953 had an immediate impact in that several farms were re-named to pay tribute to him. A contemporary of that time, Mr. K. Oskin described the situation in 1953, in which the *kolkhozes Ühistöö* (Collective work) and *Kungla* merged a few weeks after Stalin's death. At the *kolkhoz* meeting, the name of the recently deceased leader was unanimously taken as the new name of the merged collective farm (Oskin 1953). However, after a while, Stalin's personality cult disappeared and his name thus disappeared from farms (Jones 2013). For example, the former kolkhoz *Stalin*, located in the Jõgeva district, was re-named *kolkhoz Estonia* by a decision of the *kolkhoz* general meeting in 1960.

In addition to the farms named after Lenin and Stalin, several other farm names related to the highest management of the Soviet State. For example, in the late 1950s, the most common person-related name of collective farms in Estonia was Kalinin, who was the chairperson of the Presidium of the Supreme Council of the USSR from 1919 to 1946. Kalinin was popular in Estonia because prior to the Soviet revolution of 1917, he had been employed in Tallinn and his wife was from Estonia. Furthermore, during the Soviet period, one of Tallinn's districts and many streets in several Estonian towns were named after him.

In addition to political leaders, the Soviet military leaders e.g. A. Zdanov; K. Vorošilov, and the names of several historical Russian and World War II soldiers e.g. A. Suvorov, V. Chapajev, were also used in farm names, including those utilized for propaganda purposes such as A. Matrossov. Several farms were named after a suitable person having lived near the farm. For example, these included a Soviet army fighter as well as communist party martyrs killed by Estonian pro-independence underground fighters, the so-called Forest Brothers.

In addition, some farms were named after Estonian communists. Some of them had been sentenced to death in the 1920s for their actions against the Republic of Estonia. A well-known communist political activist and leader in Estonia, A. Sommerling, was killed by the Estonian police in 1924. A collective farm bearing his name was located just outside Tallinn (Fig. 4). Here, the name of the *kolkhoz* changed from

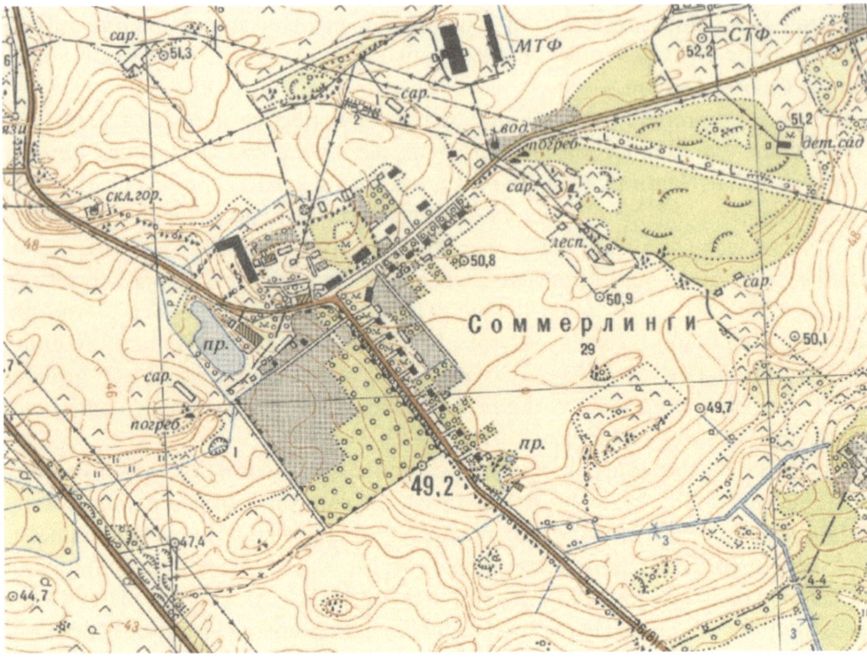

Fig. 4 A collective farm and a village named after an Estonian communist A. Sommerling (USSR O-42 topographical map 1:10 000; 1959)

a place name to a person's name. As discussed earlier, from the 1960s onward, collective farms were not named after persons. One exception to this occurred in 1961 after Y. Gagarin's journey to space, the *sovhoos Viiratsi* (a name based on a nearby place) took the name of the *sovhoos Gagarin.*

In several cases, the farm name was addressed to scientists, in particular agricultural scientists. In the Soviet system, the most important agricultural scientist was I. Michurin, one of the founding fathers of scientific agricultural selection and plant hybridization. In Estonia, several farms bore his name. The names associated directly with the Soviet rule prevailed, but several other names of Estonians, such as writers, appeared. These included authors F. R. Kreutzwald and C. Jakobson who played an important role in forming Estonia as a nation. Such names could also have an undertone of the Estonian nation and Estonians instead of the Soviets. The most popular name among these was from the mid-nineteenth century poet L. Koidula, after whom 12 farms were named in the late 1950s. She had been active in the national awakening of Estonia (against the Germans) in the nineteenth century. However, at the same time, her pro-Estonian father (J. V. Jannsen) cooperated with the Germans during the national awakening, thus his name was fully absent from the collective farms. Other accepted names of writers were those who had cooperated with the Tsarist State of Russia and had seen the future of Estonians in the Russian Empire. Names of suitable popular writers included A. H. Tammsaare or E. Vilde, the latter was famous for social democratic ideas describing class controversies in nineteenth century Estonian society. The Russian author M. Gorky appeared because he was a popular representative of the classic socialist realism. Local people's names were more permanent because the names of internationally renowned communists decreased or disappeared over the years. For example, the most used name from 1957 to 1960, Kalinin, disappeared from the collective farm names by the 1980s. By 1987, with the exception of Lenin, the collective farm names originating from Estonians were more common compared to those originating from elsewhere in the Soviet Union.

The Place Name Changes in Post-Soviet Estonia

When the Soviet power began to collapse by the end of the 1980s, many Estonian towns began to change their Soviet-related place names. Street names were most commonly changed, generally returning to the name they had before World War II. The desire to change names also appeared in the politicized names of collective farms. However, it was not often possible to turn to the historical name because in many cases, the initial farm no longer existed. In the early 1990s, land reform occurred and it was privatized in Estonia. In general, the collective farms fell apart; therefore, the need to change the names for that clearly territorially definable unit disappeared. In addition, new farms lost their role in the rural areas' administrative organization. The municipal authorities took over this task. Laansalu (2019) argues that a special layer of commemorative names of the *kolkhozes* and *sovkhozes* names vanished from the Estonian toponymy, and hardly any noticeable trace remained.

However, some collective farm names are still somehow present in rural areas in Estonia. In total, there are a dozen bus stops in Estonia (these are official place names) associated with the name of a former collective farm. These names are derived from the national romanticism era, such as *Kungla*, *Kalevipoeg*, and *Hiie*; therefore, their name does not directly refer to the period of the Soviet regime, despite these names being used in collective farms. One particular case regards the name of the Vambola farm in the village of Soe in Viljandi County. In 2008, the head of the Soviet-era *Vambola* collective farm proposed to designate Soe village as Vambola because there were several villages called Soe in Estonia (Kukk 2008). Many people were against this name change because it was reminiscent of the period of Soviet rule and collective farms. At the same time, the name Vambola has existed since the late nineteenth century. In the book by the Estonian writer A. Saal, Vambola was the main hero who struggled for the freedom of the Estonians at the beginning of the thirteenth century when the Germans Christianized people in the areas of current Estonia. In the end, the village name was not changed, but in Soe village, there is a bus stop called *Vambola*, there is *Vambola kindergarten,* and a private Vambola farm (*Vambola OÜ*).

At least one former collective farm name that was based on a communist person has been preserved in Estonia. The *Lillevere kolhoos* (*Lillevere kolkhoz*), located in

Põltsamaa village, was named after the pro-Soviet activist M. Lillevere, who the Forest Brothers killed in 1949. The farm with this name remained there until the end of the Soviet period. Today, the name *Lillevere* appears as one district name of Põltsamaa town. However, only a few people associate this name with the activist's name because many place names in the Põltsamaa area end with "vere," and the word *Lillevere* means "flower's side."

Conclusions

In this chapter the authors analyzed the names and naming of farms in Estonia during the Soviet occupation era (1940–1941, 1944–1991) and what remained of these names after the restoration of independence in 1991. During the Soviet era, private farms were collectivized into large *sovkhozes* (state-owned farms) and *kolkhozes* (cooperative collective farms). There were 3,000 of these farms by the end of the 1940s, but decreased to slightly over 300 by the end of the Soviet rule due to the necessity of merging the farms together to form larger units. Thousands of names had to be invented for these collective farms. They were more than functional agricultural units—they were important and meaningful places for people in rural Soviet Estonia and beyond it.

Place naming is part of identity making locally, nationally and beyond. Collective farms in Estonia were agricultural organizations that followed the Soviet regulations on agricultural production for the Soviet Union's purposes. Nonetheless, these farms were significant in terms of defining and designing rural people's everyday lives. Regarding this, the farms were more important than the territorial administrative units in which they were located. The farms had proper names, symbols and possessed substantial infrastructure beyond agricultural matters. Collective farms were places with various simultaneous spatial dimensions—absolute (as a fixed container of agricultural production), relative (in comparison with other collective farms) as well as relational (changing networks of production and consumption of its materiality and symbolic meaning locally, in the Soviet Estonia and Soviet Union).

In the early period of the Soviet regime in Estonia, the Soviet patriotic names referring to the new communist rule were the most common for the farms. The most frequently used names of farms based on people were those with Lenin, Stalin and Kalinin—the latter because he had been in Estonia and his wife was from Estonia. Place naming was initiated often as a bottom-up process by individuals in the collective farm. However, the names needed to resonate with the upper-level political ideology. Over the years, the names referring to the political regime and individuals diminished substantially and the names taken from the nearby environment, usually from the largest settlement, became the most common. In the 1980s, three out of four collective farms used names derived from nearby settlements or the environment.

Names of territorial objects are often changed when state ideologies fundamentally change. During the Soviet era, all names of collective farms in Estonia needed to be suitable for the prevailing political ideology. However, names of Estonian writers and artists were used, as were Estonian names from national epic poems that could be interpreted in many ways, such as those from the epic poem *Kalevipoeg*. The authoritarian control often overlooked these potential double meanings. When the Soviet regime ended, the once collectivized farms were dismantled as the land was privatized and restituted to the former owners. The names that were present for a couple of generations disappeared from maps and the local environment. Today, some physical ruins of collective farms remain in Estonia, but almost all names related to them have disappeared. However, there are a few exceptions, as with the names of bus stops near to former collective farms.

This analysis illustrated how the central authority-level political ideologies supported wiping out former place names and allowing new names to emerge. However, these were erased again when the political regime changed once more. It is important these names be remembered and recorded as part of Estonia's heritage because they tell much about the society and how the people, places and ideologies were (dis)connected over space and time during a very eventful twentieth century.

Questions

1. How do place naming of collective farms connect (ed) to political ideologies in the Soviet Estonia?
2. What were the main types of collective farm names in the Soviet Estonia?
3. What were the ethnic and political origins of people according to (who) which farms were named in the Soviet Estonia?
4. What remains of the names of collective farms in the post-Soviet Estonia?
5. How did collective farms appear (ed) as spaces and places during the Soviet occupation era?

References

Allvee Ü (1980) Kui kolhoosile anti Lenini nimi. Ühistöö, 19 IV

Belov F (1956) The history of a soviet collective farm. Routledge, London

Berg L, Vuolteenaho J (2009) Toward critical toponymies. In: Berg L, Vuolteenaho J (eds) Critical toponymies. The contested politics of place naming, pp 1–18. Ashgate, Aldershot

Bursa GRF (1985) Political changes of names of Soviet towns. Slavonic East Eur Rev 63(3):161–193

Capdepón U (2020) Challenging the symbolic representation of the Franco dictatorship: the street name controversy in Madrid. Hist Mem 32(1):100. https://doi.org/10.2979/histmemo.32.1.06

David J (2011) Commemorative place names—their specificity and problems. Names 59(4):214–228. https://doi.org/10.1179/002777311X13082331190074

Forrest C (2018) What's in a name? A feminist reflection on street name changes in Durban. Agenda 32(2):53–61

Gill G (2005) Changing symbols: the renovation of Moscow place names. Russ Rev 64:480–503. https://doi.org/10.1111/j.1467-9434.2005.00371.x

Giraut F, Hoyssay-Holzschuch M (2016) Place naming as dispositive: toward a theoretical framework. Geopolitics 21(1):1–21

Horsman S (2006) The politics of toponyms in the Pamir mountains. Area 38(3):279–291. https://doi.org/10.1111/j.1475-4762.2006.00697.x

Jones P (2013) Myth, memory, trauma: rethinking the Stalinist past in the Soviet Union, 1953–70. Yale University Press, New Haven
Kannike A (2008) Argikultuur Nõukogude ajal. In: Viires A, Vunder E (eds) Eesti rahvakultuur, pp 541–548. Eesti entsüklopeediakirjastus, Tallinn
Kasepalu A (1974) Eesti NSV sovhooside (1944–1950) ja kolhooside (1947–1950) nimed. Tallinn
Kasepalu A (1981) Majandid, mis kannavad Lenini nime. Ühistöö, 21 April
Kaşikçi MV (2019) The Soviet and the post-Soviet: Street names and national discourse in Almaty. Eur Asia Stud 71(1):1–22
Kukk JE (2008) Külarahvas valib kahe nime vahel. Sakala, 2 October
Laansalu T (2019) Transferred place names in Central Harjumaa. Linguistica Uralica 2:109–125
Light D (2004) Street names in Bucharest, 1990–1997: exploring the modern historical geographies of post-socialist change. J Hist Geogr 30:154–172
Light D, Young C (2014) Habit, memory, and the persistence of socialist-era street names in postsocialist Bucharest, Romania. Ann Assoc Am Geogr 104(3):668–685
Maandi P (2009) The silent articulation of private land rights in Soviet Estonia: a geographical perspective. Geoforum 40(3):454–464
Mamvura Z (2020) Reconstituting the cultural geography in Zimbabwe: place renaming in Zimbabwe's 'New Dispensation.' Geopolitics. https://doi.org/10.1080/14650045.2020.1850443
Mertelsmann O (2009) Ühismajandid. Estonica. Entsüklopeedia Eestist. http://www.estonica.org/et/%C3%9Chismajandid/
Neve A (2008) Eesti Stalini-aegsed kolhoosinimed—ideoloogilised konstruktid. In: Pärli Ü, Lepik E (eds) Nimetamise strateegiatest eesti kultuuris, pp 108–124. Tartu Ülikooli Kirjastus, Tartu
Oskin K (1953) Kolhoosile suure juhi—Stalini—nimi. Kolhoosi Küla, 24 March
Pae T, Maidla M-J, Tammiksaar E (2016) Vallanimede küsimus 1930. aastate vallareformis. Keel Ja Kirjandus 10:755–769
Peterson C (1977) The nature of Soviet place-names. Names J Onomast 25:15–24
Raag V (2015) Ploščad Proletarskoj Diktatury: On ideologized names and the reestablishment of pre-Soviet names. In: Eckhoff H, Rosén T (eds) Essays in Honour of Irina Lysén. Uppsala Universitetet, Uppsala, pp 167–183
Rausing S (2004) History, memory and identity in post-Soviet Estonia. The end of a collective farm. Oxford University Press, Oxford
Rose-Redwood R, Alderman D, Azaryahu M (2010) Geographies of toponymic inscription: new directions in critical place-name studies. Prog Hum Geogr 34(4):453–470
Saparov A (2003) The alteration of place names and construction of national identity in Soviet Armenia. Cahiers Du Monde Russe 44(1):179–198
Saparov A (2017) Contested spaces: the use of place-names and symbolic landscape in the politics of identity and legitimacy in Azerbaijan. Central Asian Survey 36(4):534–554
Sirendi A (2007) Põllumajandusettvõtted. In: Sirendi A (ed) Eesti Põllumajandus XX sajandil: I osa Inimene ja Ühiskond. Tallinn, pp 182–208
Taagepera R (1979) Soviet collectivization of Estonian agriculture: the taxation phase. J Baltic Stud 10(3):263–282
Taagepera R (1980) Soviet collectivization of Estonian agriculture: the deportation phase. Sov Stud 32(3):379–397

Toponymy: Narratives, Languages, Culture, and Education

Reading Ireland's Colonial and Postcolonial Toponymic Landscapes

Jonathan Cherry, Brian Ó Raghallaigh, and Úna Bhreathnach

Abstract One of the most basic, yet powerful and symbolic acts of geographical appropriation is the naming of places. With a focus on the island of Ireland, this chapter proffers a reading of Ireland's rich and diverse toponymic landscapes as integral components of the cultural landscape capable of providing unique insights into the social, political and cultural attitudes and perceptions of those who have both named and renamed places through time. In illustrating the themes of the politics of naming, appropriation, conflict and identity in both historical and contemporary urban and rural settings, a range of examples, from both the Republic of Ireland and Northern Ireland, and a case study from Dublin have been deployed. An overview of the various bodies involved in managing and promoting Ireland's place names today and the legislative framework within which it operates is also given.

Keywords Place names · Street names · Identity · Cultural landscapes · Legislative framework · Bilingualism · Ireland · Irish language

Objectives:

By the end of this chapter readers should be able to:

- Identify the colonial imprint on place names in Ireland between 1600 and 1900
- Recognise the agency and impact of the Ordnance Survey on Ireland's place name history in the nineteenth century
- Account for the significance of renaming as part of decolonisation and commemorative processes in Ireland during the twentieth century

J. Cherry (✉)
School of History and Geography, DCU, Dublin, Ireland
e-mail: jonathan.cherry@dcu.ie

B. Ó Raghallaigh · Ú. Bhreathnach
Fiontar & Scoil na Gaeilge, DCU, Dublin, Ireland
e-mail: brian.oraghallaigh@dcu.ie

Ú. Bhreathnach
e-mail: una.bhreathnach@dcu.ie

© The Author(s), under exclusive license to Springer Nature Switzerland AG 2023
G. O'Reilly (ed.), *Place Naming, Identities and Geography*,
Key Challenges in Geography, https://doi.org/10.1007/978-3-031-21510-0_22

- Identify trends in naming of streets and urban developments since the 1960s through the use of a detailed case study
- Detail the legislative framework of place names in the Republic of Ireland and Northern Ireland.

Introduction

Place names as central components of cultural landscapes are reflections of the varied social, political and cultural attitudes, aspirations and identities that have sought to appropriate, control, 'civilise' and 'make' places through time (Alderman 2008). In colonial and post-colonial contexts, as experienced in Ireland, the conflicting naming and renaming of places points to the centrality of place names as expressions of hegemony over space and ideologies (Nash 1999; Berg and Kearns 2009). Taking a broadly chronological approach, this chapter initially considers the significance of naming and renaming places and streets as a key component in the colonist's cultural arsenal, with particular emphasis on the period circa 1600 to circa 1920. The use of English as the state language resulted in the gradual decline of the Irish language over the centuries. The naming and renaming of places and streets that occurred during the 1920s, a symbolic part of the ritual of revolution as the country gained independence, is explored briefly; it illustrates the value placed on symbolically erasing the colonial imprint and inscribing the toponymic landscape with a new nationalist identity. Naming was an important tool in exercising hegemony, and during the 1930s up to the 1960s, the pervasive influence of the Roman Catholic Church, in conjunction with the state, found particular expression in the naming of new local authority urban residential developments as detailed. The naming of new residential estates and the associated roads in suburbs since the 1970s and to the present day is explored for one discrete case study area in Dublin, illustrating how naming was and is employed in creating and augmenting 'a sense of place' and 'image' of a particular area or development. The chapter concludes by setting out the legislative framework for establishing, translating and using place names across Ireland today and how these names are managed and recorded as integral parts of our cultural heritage.

Place Names in Ireland: An Overview

The legacy of past generations naming of places has resulted in Ireland being 'described as one of the most densely named countries in Europe' (Mac Giolla Easpaig 2008, 164). Names have been ascribed to the various administrative divisions of the country both civic—province (of which there are 4), county (32), barony (300+), district electoral division (4,500+), townland (61,000+)—and ecclesiastical, parish and diocese. In addition, physical features (e.g. mountains, rivers), human-made features (e.g. churches, castles) and centres of population (towns, villages) all the way down to individual fields have also been uniquely identified through this naming process. At an urban level, some 45,000 street names have also been identified.

Most Irish place names have been influenced by either dominant features of the local physical landscape, water bodies such as rivers, lakes, inlets and coastlines or settlement features such as castles or churches, with over 90% of administrative place names identified as being of Irish-language origin, some which were in use by the seventh century (Mac Giolla Easpaig 2009, 82). As the Irish writer Manchán Magan has noted place names 'offer tantalising hints and encoded reflections of our culture, psyche and past practices,' providing 'glimpses into the historical, geographical and anecdotal qualities of our past' and 'our ancestors' lives, their knowledge of environments and their folk beliefs' (2020, 139). Such names were not merely indicators or markers in differentiating one place from another, but rather as the historical geographer Patrick Duffy has noted, 'Naming places is a primary act of geographical appropriation, a demonstration of control over nature, the landscape and everything in it' (2007, 63). It is to this aspect of naming as a hegemonic tool that our focus now turns.

The Colonial Toponymic Imprint

As a country located in Western Europe, Ireland, unlike many other European countries which acted as colonising or imperial powers, had the distinction of being colonised (Smyth 2006, 9–14). By the sixteenth century ownership, England's colonial policy for Ireland hinged on dismantling the existing indigenous land structures and the transfer of ownership of land with all of the attendant social, economic and political advantages it bestowed to a new landowning elite loyal to the English government and Crown. This process gained momentum in Ireland during the sixteenth and seventeenth centuries as a sustained series of colonial schemes of land confiscation and plantation were initiated to bring the entire country under English control, a position achieved by the later seventeenth century. The planning and implementation of these projects was underpinned by extensive surveys and mapping of the existing landscape, settlement features, land ownership boundaries and place names.

The recording and mapping of toponyms as Vuolteenaho and Berg have observed was 'an ancillary form of knowledge-production in the service of a wider scientific-geopolitical project of knowing the world as accurately as possible as part of the process of controlling its spaces' (2016, 4). The large-scale mapping of the confiscated lands of Ulster in the early seventeenth century exemplifies the significance colonisers placed on recording place names, as an integral part of their desire to control space. In providing legal basis to this subjugating colonial enterprise, government officials and administrators based in London employed maps and surveys in delimiting boundaries and extent of the landed estates. The names of the small land divisions, now referred to as townlands in Ireland, were listed in composing the legal title for these land grants (Margey 2009).

Maps and recording of place names remained important throughout the seventeenth century. In suppressing the 1641 Gaelic rebellion, the English administration confiscated 11 million acres of land transferring ownership to those 'loyal' to the English Crown. Once again the legal basis of this rested on the mapping and recording of names, with the resulting Book of Survey and Distribution and the monumental Down Survey maps,[1] illustrating again how 'Maps were used to legitimise the reality of conquest and empire' (Harley 1988, 282).

One of the crudest tactics employed by colonial powers in the subjugation of a people was through the naming and/or renaming of places. In Ireland, while many of the original names were retained by the colonisers, others were erased, resulting in the loss of indigenous names alongside the imposition of names in the coloniser's language reflecting their ideologies and values. This process of renaming was part of a suite of acts—including construction of urban settlements and dismantling or repurposing of previously important or symbolic sites—that sought to simultaneously disorientate, inspire awe and cultivate loyalty amongst the colonised. One early example of the power of this naming process was the formation and naming in 1556 of King's County and Queen's County, from the core of the Gaelic lordships of O'Connor of Offaly and O'Moore of Laois as a precursor to plantation with English settlers. In a further act of homage to the English monarch Mary I of England and her husband Philip, the primary urban settlements in each of the new counties were named *Philipstown* and *Maryborough*.

As the historical geographer William J. Smyth has remarked, these towns in the colonial context were 'frontier outposts created to dominate and reorganise hitherto hostile territories' (2006, 427), and as important 'civilising' centres they were 'the central instruments of imperial expansion and control' (2006, 219). The names selected for some of these new towns were integral to promoting and sustaining the ideological values that underpinned the broader colonial projects and in the achievement of hegemony, as 'part of the broader process of colonial, cultural and political subordination' (Nash 1999, 461). The naming of a new urban settlement established as part of the Ulster plantation on a greenfield site in south county Cavan as *Virginia* was undoubtedly influenced by the earlier use of the same name by the British for one of their newly established colonies in North America. Evidence that such names rankled the Gaelic people emerges from depositions gathered from the colonial settlers in the aftermath of the 1641 Rising when the remaining Gaelic families attempted to forcibly upend the colonial project in Ireland. In these depositions the English settlers recounted their experiences and losses as a result of the rebellion. The deposition of a settler based at Virginia, County Cavan detailed how the local Gaelic family, the O'Reillys, had said that they would enact laws preventing the use of English and 'that all the names given to Lands or places should be abolished, and the ancient names restored'. In specific reference to Virginia, one of the insurgents, the Earl of Fingal, '…asked the deponent [the settler] what was the ancient name of

[1] To view the Down Survey maps see http://downsurvey.tcd.ie/ [last accessed 11.2.22].

Virginia: He answered Aghanure (as he remembreth): The Earle then said that must be the name againe'.[2] The desire amongst the Gaelic population to revert from *Virginia* to the original name *Achadh an Iúir*—the field of the yew tree—signals the levels of unease felt at, and resistance offered to, the imposition of colonial toponyms.[3]

Further insight into the importance placed on imposing place names in English as part of the colonial project 'to secure the political authority and to preserve the cultural identity of the New English settlers against the cultural and material threat of the "native Irish"' (Nash 1999, 461) is exemplified in a statute passed by King Charles II in 1665.

> His Majestie taking notice of the barbarous and uncouth names, by which most of the towns and places in this kingdom of Ireland are called, which hath occasioned much damage to diverse of his good subjects, and are very troublesome in the use thereof, and much retards the reformation of that kingdom, for remedy thereof is pleased that it be enacted, and be it enacted by the authority aforesaid, that the lord lieutenant and council shall and may advise of, settle, and direct in the passing of all letters patents in that kingdom for the future, how new and proper names more suitable to the English tongue may be inserted with an alias for all towns, lands and places in that kingdom, that shall be granted by letters patents; which new names shall thenceforth be onely names to be used, any law, statute, custom, or usage to the contrary notwithstanding. (Quoted in Nash 1999, 461).

This attempt to initiate a reformation in naming does not appear to have materialized to any great extent, but the desire to 'civilize' Ireland's toponymic landscape did find expression in Crown grants of land to families during the 1660s. In County Wicklow for example, Robert Kennedy of Balligarny was granted lands on the basis that 'The manor shall be called Mount Kennedy'.[4] The usage of the family surname as a component of a new place name became popular amongst the landowning elite, under whose aegis most of the naming and renaming of places in Ireland during the eighteenth and nineteenth centuries occurred. Such naming processes reflect the sentiments of Psalm 49, verse 11 that 'Their inward thought is, that their houses shall continue for ever, and their dwelling places to all generations; they call their lands after their own names'. These names were usually associated with the large country houses and ornate demesne landscapes constructed by landowning families during the period. Historian Roy Foster observed that 'the Ascendancy desire to build and to plan… may indicate an obsession with putting their mark on a landscape only recently won and insecurely held' (1988, 192) and he concludes that they 'built in order to convince themselves not only that they had arrived but that they would remain' (1988, 194). A cursory examination through the first pages of *Burke's Landed Gentry of Ireland*—which details the familial lineages of landowning families and records their 'seats'—illustrates the variety of ways in which the naming process worked.

> Ancketill of Ancketill's Grove, Emyvale, Co. Monaghan
>
> Archdale of Castle Archdale, Irvinestown, Co. Fermanagh

---

[2] See https://1641.tcd.ie/index.php/deposition/?depID=833227r167 [last accessed 5.2.22].

[3] See https://www.logainm.ie/en/4307?s=Virginia [last accessed 5.2.22].

[4] See https://www.logainm.ie/en/55643?s=mount+kennedy+demesne [last accessed 5.2.22].

Heaton-Armstrong of Mount Heaton, Co. Tipperary

Beresford-Ash of Ashbrook, Co. Derry

Bagot of Bagotstown Castle and Bagotstown House, Co. Limerick

Barron of Carrick Barron and Barron Court, Co. Waterford

Blakeney of Castle Gallagh (Castle Blakeney), Co. Galway and Mount Blakeney, Co. Limerick

Bowen of Bowen's Court, Co. Cork

Burton of Burton Hall, Co. Carlow

Other landowners in a romantic vein sought to honour their wives, by naming their lands and home after them. In 1623, the Montgomery lands at Comber in County Down were renamed *Mount Alexander* in honour of the owner's new wife Lady Jean Alexander.[5] In County Fermanagh, Sir John Cole named his new country mansion (constructed in 1719) *Florencecourt* after his wife Florence Wrey, displacing the earlier Irish name for the area—*Mullach na Seangán*—translated as the hilltop of ants'.[6]

For other landowners, inspiration in renaming their places came from 'idealised' and 'exotic' names (Flanagan and Flanagan 1994, 84). The lands of Droim Caiside—translated as the ridge of Caiside (a family name)—in County Cavan which were granted to the Scottish Sanderson family in the mid-seventeenth century, had by the early nineteenth century been renamed *Clover Hill* (See Fig. 1). The origins for this name remains unknown, but the view across some clover clad hills in the drumlin landscape may have provided the inspiration.[7] A number of 'exotic' French and Italian names inspired by the Grand Tours of the eighteenth century were also introduced onto the Irish landscape with Belvedere, County Westmeath and Marino, Dublin as representative examples.

Family names were also employed in naming urban settlements and the use of either, *Newtown-* followed by the family name, or the suffix *-town* preceded by it indicates the influential role played by individual families who arrived as part of colonial projects in Ireland in establishing and developing towns and villages from the seventeenth century onwards. By 1682, the settlement at Mostrim, County Longford had been renamed *Edgeworth Town* after the Edgeworth family.[8] Newtown, County Fermanagh was renamed *Newtownbutler* in 1715,[9] while in County Longford *Newtown Forbes* replaced the Irish place name for the village (An Lios Breac) in 1750[10] and by the early nineteenth century, Bunclody, County Wexford had been renamed *Newtownbarry*. Each of these examples illustrative how local landowning

[5] See http://www.placenamesni.org/resultdetails.php?entry=11969 [last accessed 5.2.22].

[6] See http://www.placenamesni.org/resultdetails.php?entry=13047 [last accessed 5.2.22].

[7] See https://www.logainm.ie/en/1393740 [last accessed 5.2.22].

[8] See https://www.logainm.ie/en/1412094 [last accessed 5.2.22].

[9] See http://www.placenamesni.org/resultdetails.php?entry=13712 [last accessed 5.2.22].

[10] See https://www.logainm.ie/en/1166313?s=newtownforbes [last accessed 5.2.22].

Fig. 1 Place name signage for Droim Caiside, Cloverhill, County Cavan. *Source* Jonathan Cherry

elites used place names to elevate and accentuate their dominant positions locally, further inscribing their presence and agency in maintaining hegemony.

Colonial Urban Naming

The naming and renaming of streets and bridges in urban areas, in conjunction with the construction of a suite of other features such as statues, monuments and a range of civic and ecclesiastical structures were strategic components of the colonial arsenal deployed in exercising hegemony in Ireland. As dominant forces in the governance of Dublin during the eighteenth century, the colonial landowning elite commemorated themselves: individual lord lieutenants (the most important representative of the English administration in Ireland), notable military figures and English monarchs, gave their names to streets and bridges throughout the city (Whelan 2003, 95). Such processes reflect how street names 'are embedded into the structures of power and authority' and how 'political regimes and elites utilize history to legitimate and consolidate their dominance and reinforce their authority' (Azaryahu 1997, 480). During the eighteenth century the leading landowning families in Dublin, used their family names in naming the streets, squares and lanes in the fashionable residential quarters they developed. In the south of the city, developments by the Fitzwilliam family of Merrion resulted in *Fitzwilliam Square*, *Fitzwilliam Street*, *Fitzwilliam Lane*, *Merrion Square* and *Merrion Street*. On the north side of the city the Gardiner family were the dominant developers and in addition to naming several streets after family names and titles, a new street that they developed in the mid-eighteenth century was named *Cavendish Street* after William Cavendish, lord lieutenant of Ireland between 1737 and 1745. Such a move proved loyalty to the English administration in Ireland. While such processes found their greatest expression in Dublin they filtered down to smaller urban settlements across the country. At Maynooth in county Kildare, the proprietors of the town, the dukes of Leinster, lent their family name to the main thoroughfare, Leinster Street, while the Farnham family in Cavan named the new street that they laid down there in the early nineteenth century *Farnham Street*.

The usage of general and specific names relating to the royal family in naming the numerous quays along, and bridges across the River Liffey which bisects Dublin resulted in the naming of Queen's Bridge in 1776 after Queen Charlotte wife of George III, King's Bridge, named after the visit of George IV in 1821 and later Victoria and Albert Bridge commemorating its use by the royal couple during a visit in 1861. Several quays also displayed royal connections including George (George I) and Victoria (1871), while the commemoration of the naval figure the Duke of Wellington, victor at the battle of Waterloo (1815), resulted in the naming of both a bridge in 1816 and a quay in 1817 in his honour (Whelan 2003, 96–99). Later in the 1840s two new residential streets in south Dublin were named *Wellington Road* and *Waterloo Road*.

Commemorations of significant events also provided naming opportunities. Aughrim Street in Dublin, was named in 1792 as a centennial commemoration of the Battle of Aughrim, county Galway (1691), which proved decisive in King William III securing Ireland under English control (McCready 1892, 4). Such naming practices were thus hugely significant, 'in particular, when used for commemorative purposes… [to] inscribe an official version of history onto the cityscape and introduce this version of history into myriad networks of social communication that involve ordinary urban experiences that seem to be separated from the realm of political ideology.' (Azaryahu 2009, 54). A new row of houses constructed in Clones, county Monaghan at the end of the nineteenth century was named *Jubilee Terrace* to celebrate Queen Victoria's sixtieth year as monarch, complementing the town's existing Whitehall Street, bringing a 'heart of the Empire' feeling to this small Irish county town.

Despite the dominance of the colonial imprint in Dublin's nomenclature, opportunities for naming and renaming streets emerged in the last decades of the nineteenth century as the local urban authority, known as Dublin Corporation came to be dominated by nationalist councillors. The significance of using naming as an act of resistance in supporting the growing nationalist political agenda and identity as an alternative to the colonial regime is reflected in the numerous proposals received, some of which were acted upon. The renaming of Carlisle Bridge to *O'Connell Bridge* (after Daniel O'Connell, the 'Liberator' of Irish Catholics) in 1880 was symbolic of the shifting balance of local power in Dublin (Whelan 2003, 101).

Not all renaming in Dublin city was politically motivated. The renaming of Lower Temple Street to *Hill Street* in 1885 was an attempt to retrieve the street's reputation 'owing to the fact that certain houses … have been for some time past occupied by immoral characters' (Quoted in Whelan, 2003, 104). At the same time 'members of the respectable working classes' residing in Upper Mecklenburgh Street petitioned for the renaming of their street as 'many of the houses in Lower Mecklenburgh Street are used for improper purposes, and inhabited by persons of the worse character' (Quoted in Whelan 2003, 104). The campaign proved successful, with the renaming as *Tyrone Street*.

Standardising and Mapping Place Names: The Role of the Ordnance Survey

The most dramatic and widespread colonial impact on the place names of Ireland occurred—1820s-1830s as the entire country was surveyed and mapped under the agency of the Ordnance Survey (Mac Giolla Easpaig 2008, 166–68). The resulting archive—alongside the voluminous body of surveys and reports generated during the nineteenth century—was, 'an important strategic resource in increasing official knowledge and control over landscape and society in Ireland… part of a colonial project to regulate and re-order a society and economy perceived as inherently backward and impoverished' (Duffy 2011, 562). While the Ordnance Survey brought standardisation to place names in the English language it is important to acknowledge that the process of anglicisation of place names was well established by the early decades of the nineteenth century, and generally the use of such place names in English when established went unchallenged (Ó Muraíle 2014).

As John Andrews has detailed, existing Irish-language place names were subjected to processes of substitution, translation, transcription from Irish-language documents, dictation and restoration, in arriving at standardised names in English (Andrews 1992). The majority of place names were arrived at through dictation, 'in which a non-Irish speaker recorded in English orthography a place name spoken by an Irish-speaker and turned to English words that partially matched the sound of the Irish name-elements, but obviously not the meaning' (Nash 1999, 465). Other 'standardised' names were arrived at through substituting unrelated English words for approximate Irish sounds for example, drum for *droim* (a ridge), knock for *cnoc* (a hill), kill for *cill* (a church) and more for *mór* (great or big), resulting in *Cill Mhór* emerging as *Kilmore* in English, and *Na hUamhanna* ('the caves') becoming *Ovens*. This distortion and anglicisation coincided with a mass shift in language from Irish to English in the nineteenth century, accelerating the loss of historic, cultural and geographic knowledge inherent in place names (Doyle 2015, 129) (See Table 1).

Table 1 Examples of transliteration and translation in place names. Note the variety of approaches to some names. *Source* Compiled by authors

Transliteration	Bántír (Co. Cork) > Banteer
	Bearna (Co. Galway) > Barna
	Na hUamhanna (Co. Cork) > Ovens
	Muiceanach idir Dhá Sháile (Co. Galway) > Muckanaghederdauhaulia
	Tamhlacht > Tallaght (Co. Dublin), Tamlaght (Counties Antrim, Fermanagh, Tyrone), Tawlaght (Co. Kerry), Tamlat (Co. Monaghan)
	Clais an Ghainimh > Clashaganny (Co. Roscommon), Clashaganniv/Clashnaganiff/Clashganniv/Classes (Co. Cork)
Translation	Cluain Eocrach (Co. Roscommon) > Keyfield
	Cuan na gCaorach (Co. Donegal) > Sheephaven
	Clais an Ghainimh > Sandholes (Co. Tyrone), Sandpit (Co. Louth), Sandville (Co. Cork)
Transliteration *and* translation	Ceann an Dúna (Co. Kerry) > Doon Point

Erasing the Colonial Imprint: Post-Colonial Place Naming

By the 1920s, part of Ireland known as the Irish Free State—had gained independence and a new native government was established in Dublin. As others have noted the, 'renaming of streets figures prominently in periods of regime change and revolutionary transformations' (Azaryahu, 2009, p.59). The centrality of this process as part of the ritual of revolution reflects the significance and power of street names in building and augmenting new narratives and identities. As Azaryahu has observed, 'street names conflate history and geography and merge the past they commemorate into ordinary settings of human life' (1997, 481).

While the process found its greatest expression in the new capital, Dublin, similar nomenclature cleansings were recorded in many towns and villages as both the new state and local authorities sought to erase and replace colonial narratives with nationalist ones. The renaming process varied depending on each local urban authority. In some instances, ratepayers, residents and business owners were asked to vote on the change and if a certain proportion were in favour the renaming could proceed. In other circumstances, unilateral decisions regarding renaming were taken by individual local urban authorities.

In Dublin, the principal streets and thoroughfares renamed after independence included Great Brunswick Street to *Pearse Street* in 1922; Queen's Square to *Pearse Square*; Great Clarence Street to *Macken Street*; Sackville Street to *O'Connell Street*; Wentworth Place to *Hogan Place*; Denzille Street to *Fenian Street* all in 1924; Rutland Square to *Parnell Square*; Gloucester Street Upper to *Cathal Brugha Street* and Gloucester Street Lower to *Sean MacDermott Street* in 1933 and Stafford Street to *Wolfe Tone Street* in 1943 (Whelan 2003, 224). The authorities in selecting names found inspiration in commemorating high profile individuals. Those who had died in the quest for independence included Theobald Wolfe Tone in 1798; Pádraig Pearse, Peadar Macken and Seán McDermott during the 1916 Rising and Cathal Brugha who had died in the early days of the Civil War in 1922. Others had local connections to a particular street, such as John Hogan, a prolific sculptor, who was commemorated through the renaming of Wentworth Place where he had once resided, to *Hogan Place*.

While the renaming of Sackville Street to *O'Connell Street* was hugely significant as it was the most prestigious thoroughfare in the capital, what is most striking is the small number of streets renamed and the protracted nature of the process. Ireland's experience of purging colonial street names was not as radical or dramatic as might be expected (particularly compared to countries that transitioned from Communist to post-Communist governance (Azaryahu 2009, 59)) and as a result, the origins of many of the contemporary street names in Dublin may be traced to the eighteenth-nineteenth centuries, reflecting the difficulties of erasing the deeply embedded impacts of colonialism. A cursory examination of a contemporary map of Dublin's streets will show Nelson Street, Victoria Street, Waterloo Road and Wellington Road, while Erne Street and Terrace and Creighton Street, in Dublin's south inner city, commemorate the former owners of this part of the city, the Creighton family, the earls of Erne.

The changing of a street name was a relatively cheap process in terms of replacing existing name plates and signage, but gathering sufficient support from those who resided on or operated businesses from these streets proved harder to garner. Some residents may have found that the suggested new name did not reflect their own personal politics, while others may have harboured concerns over the appropriateness of the name to the image of the area, wondering perhaps if it would be 'a good fit' or detract from the area. Many businesses did not want to endure the expense of changing an address on their headed notepaper, invoices and receipts, shop signs and branded goods, while anxious to avoid the hassle and impact of delayed deliveries or disorientated customers. Some of these concerns were voiced during the early 1940s by residents and ratepayers as an unsuccessful attempt to rename Talbot Street to *Seán Treacy Street* was made.[11] As a result of such challenges in Dublin, it was the existing bridge network spanning the river Liffey that facilitated the greatest expression and fervour in nationalist renaming in the immediate aftermath of Independence. In 1922, amongst others, Queen's Bridge (constructed in 1764) was renamed Queen Maeve Bridge; Sarah Bridge (1791) renamed Islandbridge; Richmond Bridge (1813) renamed O Donovan Rossa Bridge; Wellington Bridge (1816) renamed Liffey Bridge and Victoria & Albert Bridge (1858) renamed Rory O'Moore Bridge (Whelan 2003, 225).

Besides erasing and replacement of street names, attempts to provide Irish forms for existing English street names had been ongoing from the early twentieth century. In Dublin, Nassau Street, named in 1749 by Lord Molesworth in marking the birth

[11] See https://comeheretome.com/2012/04/11/ailtiri-na-haiseirghe-sean-treacy-and-talbot-street/ [last accessed 15.11.21].

of his son Richard Nassau, was originally part of St Patrick's Well Lane. In 1905, it was suggested that the street's name in Irish was *Sráid Thobar Phádraig*. In 1921 a motion that Nassau Street be renamed *Tubber Patrick Street* was rejected by Dublin Corporation and today the street name in Irish remains as Sráid Nassau.[12] Contemporary street signs for Beresford Street in Dublin, named in the late eighteenth century, show both the contemporary Irish translation *Sráid Beresford*, which keeps the proper name, and an earlier *Sráid Dúinsméara* which means, roughly 'Berry Fort Street'.[13]

Away from Dublin, the renaming of towns and administrative areas post-Independence reflects this broader theme of erasing place names and replacing them with names that could be employed in augmenting the nascent state's nationalist independence narrative. The renaming of Queenstown, county Cork, named in honour of the visit of Queen Victoria in 1849 to *Cóbh* in 1921 and the designation of Queen's County as county *Laois* and King's County as county *Offaly* are amongst some of the best known examples where references to monarchy was removed. In County Wexford, a vote by the rate payers of the town of Newtownbarry resulted in the name reverting to *Bunclody* in 1950.[14]

Renaming as part of the 50th anniversary commemoration of the Easter Rising in 1966 reflects the longevity of attempts in cleansing Ireland's toponymic landscape. In 1966, a number of train stations across Ireland were selected to be renamed in commemoration of the main leaders who had led the rising. In Dublin Kingsbridge Station was renamed *Heuston Station* and Amiens Street Station became *Connolly Station*. The main train station in Cork originally named Glanmire Road Station was renamed *Kent Station*, Sligo Station was renamed *Mac Diarmada Station* and Galway station renamed *Ceannt Station*. During 1966, Jubilee Road in Clones, county Monaghan named in 1897 to mark the 60th anniversary of the coronation of Victoria, was renamed *'98 Avenue* in marking the 1798 Rebellion, while the town's Erne Square was renamed *Fitzpatrick Square* in memory of Matt Fitzpatrick, the officer commanding of the 5th Northern Division of the IRA who had been killed in 1922.[15]

While opportunities to change existing names in Dublin proved quite challenging, reconstruction of existing parts of the inner city and construction of new developments in the suburbs from the 1930s onwards, aimed at addressing the city's housing crisis (Brady 2014), provided a clear canvas where the new power brokers in the form of the state and Church could inscribe their identity, status and hegemony (Whelan 2003, 214). The road names in the vicinity of Kimmage in south Dublin, 1930s-1960s were inspired by names associated with some of Ireland's most famous historic monastic settlements including Bangor, Cashel, Clogher, Clonmacnoise, Downpatrick, Ferns, Kildare, Kells, Leighlin and Saul. Saints' names were frequently employed in naming local authority housing schemes such as St Teresa's Gardens and St Laurence's Mansions in Dublin (Brady 2016, 170). Most Irish country

---

[12] See https://www.logainm.ie/ga/1383555) [last accessed 18.11.21].

[13] See https://tinyurl.com/3pcyxcwm for a Google Street view of Beresford Street name signs.

[14] See http://www.irishstatutebook.ie/eli/1950/si/281/made/en/print [last accessed 11.2.22].

[15] See https://www.rte.ie/archives/2017/0425/870227-what-should-happen-to-queen-victoria-statue/ [last accessed 16.2.22].

towns had at least one development, normally local authority housing, that was named after one of the two most popular saints in Ireland, Brigid or Patrick. The designation of 1954 as a Marian Year—a year for particular devotion to Mary, the mother of Jesus—by the Roman Catholic Church resulted in some streets and housing developments completed during that year including the name *Marian*. Such a phenomenon speaks to the level of influence enjoyed by the church in Ireland at this time. Besides socio-religious and political factors, Brady has examined how other motivations in seeking name changes centred on how the naming of new developments 'might impact adversely on … property values' (2016, 172), a theme that emerges in the next section of this chapter.

Naming in Modern Ireland

With a view to capturing a snapshot of naming practices in Modern Ireland (*c.*1960–2020), we carried out a case study on the modern and contemporary north-east suburbs of Dublin where we looked at residential suburban hodonyms (i.e. road names). The case study area equates approximately to the postcode districts of Dublin 5, Dublin 13 and Dublin 17, but also includes part of County Dublin. More specifically, it equates to the historic civil parishes of Artaine|Ard Aidhin, Balgriffin|Baile Ghrífín, Baldoyle|Baile Dúill, Coolock|An Chúlóg, Howth|Binn Éadair, Kilbarrack|Cill Bharróg, Killester|Cill Easra, Portmarnock|Port Mearnóg, Raheny|Ráth Eanaigh. Our analysis revolves around the historic townlands contained within these civil parishes. We chose this area as it comprises suburban developments built from the 1960s right up to today, including areas currently under development.

Situating our case study within the historic civil parishes was somewhat contrived, but it was convenient since defining the boundaries of the contemporary neighbourhoods (=areas) of Dublin is difficult. Muiris de Buitléir made the most definitive attempt at this so far in his mapping of what he termed Dublin's *City Districts* in *A Portrait of Dublin in Maps* (de Buitléir 2013, 128–129). These are the toponyms Dubliners use to describe what *area* or *part* of Dublin they are from when asked, or perhaps what they put down on the optional second line of their postal address. District boundaries can be fluid and there are many examples of road name changes designed to reassign a road from one district to another, usually for social reasons. For example, *Brookwood Rise* in Artaine, formerly the south west end of *Harmonstown Road*, was renamed (Brady 2014, 174) to distinguish the privately owned houses on that end of the road from the social housing further up the road, effectively relocating that part of the road from Harmonstown into Artaine. De Buitléir's districts which overlap with our case study area (excluding large parkland and industrial zones) are *Beaumont, Donnycarney, Killester, Artane (sic), Harmonstown, Raheny, Kilbarrack, Kilmore, Coolock, Darndale, Clare Hall, Donaghmede, Baldoyle, Sutton, Howth.* Portmarnock was outside the scope of his map.

Since de Builtéir's map was published, however, new districts have emerged within the area in question, namely, *Belmayne* and *Clongriffin*. Using the historic civil parishes allowed us to have a well-defined boundary that captured all of the residential developments in north-east Dublin since 1960 right up to today, including these new ones. In addition, the civil parishes subdivide neatly into townlands. With the green belt rapidly narrowing between the city limits at Baldoyle (Dublin 13) and the first north-east satellite suburb of Portmarnock (County Dublin), we have included Portmarnock to cover this area of development as well as to capture the use of townland names in residential road names outside the city limits.

Focusing on groups of residential roads within townlands allowed us to look at how much naming practices in modern and contemporary road developments relate to local history. Historic townland names, about one third of which in County Dublin are of Irish-language origin (de hÓir 1972–1973), are not generally recognised in urban and suburban Dublin City, but they are widely used in County Dublin, outside the city limits. Within the city and suburbs, however, townland names are sometimes incorporated into the naming of modern roads and housing developments, e.g. *Brookville Park* (Artane, Dublin 5). The case study area contains 62 townlands.

The method for our case study was as follows. Firstly, we compared the residential road names within each civil parish with the historic townland names therein. Secondly, we categorised the names occurring within the research area. Thirdly and finally, we described the place name elements we saw occurring in the dataset. The toponymic data was retrieved from OpenStreetMap (OSM) via the Overpass API on a townland by townland basis, using OSM's own townland boundary data. The analysis and categorisation of road names in this case study was not based on extensive local knowledge or contact with housing developers, but rather on the surface forms of the road names supplemented by some local knowledge.

Obtaining a list of all residential ways (i.e. roads) in OSM in each townland allowed us to compare the townland name with the roads contained within the townland boundary in each case. This is important because we wish to show whether or not developers, public or private, are looking to the historic names for inspiration when naming roads in new residential developments. As highlighted by the Placenames Commission in their 1992 booklet entitled *Streetnames: Guidelines*, Government and local councils have in the recent past encouraged this, but it is not mandatory. Nonetheless, you would expect many new roads to be named after the places they traverse. What we see in the data we looked at (i.e. *c.*1k roads) is that only 5.64% of roads (61/1,081) had names that were based in whole or in part on the name of a townland they traverse. Examples of these names include *Grange Abbey Crescent* (Grange), *Newtown Court* (Newtown), *Tonlegee Road* (Tonlegee) and *Burrow Court* (Burrow). It is likely, however, that some of these are coincidental with the village or parish name, e.g. *Killester Park* (Killester North < Killester) and *Raheny Wood* (Raheny North < Raheny). Roads named after neighbouring townlands were not included, but nonetheless, the practice of naming new residential roads using historic townland names appears to be quite rare.

The second part of our analysis required us to broadly categorise the road names that occur within the case study area. We established the categories by reading through an alphabetic list of all the road names grouped by parish and townland, looking out for (recurring) themes. We assumed that some names would follow common types (e.g. famous people, landmarks, destinations), but we were on the lookout for other themes as well. What emerged were the following overlapping categories: Irish language (i.e. names in Irish only albeit with acute accents removed, e.g. *Dal Riada*, and names with Irish-language-based elements, e.g. *Carrick-*), flora and fauna (i.e. trees, animals, flowers, shrubs), built environment (e.g. buildings, parks, enclosures), physical landscape (e.g. fields, rivers, rocks, bays), eponymic places (e.g. Irish baronies, towns, townlands, lakes, foreign places, destinations), historic (i.e. townland and some other historic names), people (e.g. Irish politicians, Irish saints, biblical figures), ecclesiastical (e.g. church buildings, church lands) and colonial (e.g. English place name elements such as *wood*, *dale* and *brook*). Once we had established our categories, we assigned every road to one or more categories. Overlap happened most often between the *Irish-language*, *historic* and *ecclesiastical* categories, but also between other categories, e.g. *built environment* and *colonial*. Results are shown in Table 2 with examples. An interesting theme, which was not captured by the broad categorisation, is present in the townlands of Artaine (Domville) and Bonnybrook where a group of seven roads built in 1969 are named after the moon landing of that year (e.g. Aldrin Walk).

The third and final part of our analysis involved extracting a frequency list of toponymic elements in the dataset (*c.*1k residential roads). This allowed us to further analyse the factors influencing the naming of residential roads in modern Dublin. The 12 most common disambiguation elements found in the dataset were *Road* (172), *Avenue* (145), *Park* (120), *Drive* (96), *Court* (64), *Grove* (63), *Crescent* (44), *Close* (37), *Green* (24), *Walk* (24), *Lawn* (24) and *Lane* (21). The 12 most frequent place name elements (including compound elements) found were *saint* (150), *-wood*

Table 2 Case study road name categories showing prevalence. Overlapping examples are shown in bold. *Source* Compiled by authors

	Category	Examples
23%	Irish-language	**Adare Park**, **Clanmahon Road**, Coolrua Drive, **Donaghmede Road**, Dromawling Road, **Edenmore Park**, **Kilbarrack Road**, **Mask Green**, **Moyclare Avenue**, **Tonlegee Road**
17%	Flora and fauna	Ashgrove, Buttercup Drive, Foxfield Crescent, **Thornville Avenue**, Verbena Park
15%	Built environment	Castle Court, **Martello Court**, **Millbrook Drive**, Watermill Avenue
15%	Physical landscape	Beaumont Grove, **Edenmore Park**, Seacliff Drive, Wheatfield Road
14%	Eponymic places	**Adare Park**, **Clanmahon Road**, **Mask Green**, **Moyclare Avenue**
13%	Historic	**Donaghmede Road**, **Edenmore Park**, **Grange Abbey Crescent**, **Harmonstown Road**, **Kilbarrack Road**, **Martello Court**, **Tonlegee Road**
11%	People	Ardilaun, Collins Park, Grace O'Malley Drive, **Harmonstown Road**, McAuley Park
10%	Ecclesiastical	**Donaghmede Road**, **Grange Abbey Crescent**, **Kilbarrack Road**, Temple View Close, Churchwell Mews
10%	Colonial	Newbury Wood, **Millbrook Drive**, **Thornville Avenue**

(82), *-field* (61), *-more* (45), *grange-* (42), *ard-* (40), *-side* (33), *brook-* (30), *castle-* (30), *-ton* (30), *-hill* (27) and *kil(l)-* (27). Irish-language native elements emerged prominently, e.g. *-more* (45), *ard-* (40), *kil(l)-* (27), *clon-* (25), *ros-* (22), *carrick-* (16), *carrig-* (12), *clan-* (10), *bal-* (9) and *moy-* (9). The physical environment was prevalent within the native elements. English-language colonial elements emerged slightly more prominently than the Irish-language elements, e.g. *-wood* (82), *brook-* (30), *castle-* (30), *-ton* (30), *-hill* (27), *-mont* (16), *mill-* (15), *-mount* (15), *-town* (13) and *-ville* (11). The built environment was prevalent within the colonial elements.

The results of the case study show us clearly that the colonial names that characterise Dublin City (McCready 1892) are not prevalent in the modern and contemporary suburbs of Dublin. Interestingly, names commemorating independence from Britain are not prevalent either. What we see instead is an overall tendency towards more neutral naming involving things like flora and fauna, the built environment and the physical landscape, coupled with homogenisation through the frequent use of common descriptive disambiguators (e.g. Crescent). Nonetheless, colonial name elements (e.g. *Wood*) are still quite common and outnumber Irish-language-based elements (e.g. *-more*), perhaps because the colonial-sounding elements are used by some developers in an attempt to assign prestige to new roads.

While it would have been interesting to accurately capture the naming tendencies over time from 1960 to 2020, it was outside the scope of this case study to assign a year to each residential road, and this data was not readily available. However, we posit that while the preference that existed for Irish-language names prior to the modern era may have waned in the subsequent decades, the use of Irish-language names and name elements persists to this day. In addition, through the use of elements derived from local townland and settlement names, as well as the names of baronies, parishes, towns, townlands, lakes and other places outside of Dublin, while obviously contrived in certain cases, Irish history and Irish language retain prominence in the names being chosen.

Protecting, Recording and Promoting Place Names as Integral Parts of Cultural Landscape and Heritage

Legal framework for naming in the Republic of Ireland

Under the Irish Constitution, the Irish language is the first official language, and the English language is recognised as a second official language. Place names in both languages are equally valid, except for Gaeltacht, or Irish-speaking, areas, where only the Irish version can be used on road signs.

There is no central authority for the creation of new English-language place names, administrative or otherwise, in Ireland. Local authorities are responsible under the Local Government Act, 1946, for keeping a record of new names in their area.

An Brainse Logainmneacha/The Placenames Branch, situated in the Government department with responsibility for the Irish language, undertakes research into the place names of Ireland to provide authoritative Irish language versions of place names for official and public use. The Irish language versions of administrative and certain non-administrative names determined by the Placenames Branch are given legal status by means of a place names order made by the Minister of State for Gaeltacht Affairs. The primary responsibility of the Placenames Committee is to advise the Minister in relation to the place names of Ireland as defined in Sect. 31 of the Official Languages Act, 2003: '"placename" includes the name of any province, county, city, town, village, barony, parish or townland, or of any territorial feature (whether natural or artificial), district, region or place, as shown in the maps of Ordnance Survey Ireland'. Once an order has been made in respect of a Gaeltacht place name, the English version of the place name ceases to have any legal effect.

The Placenames Branch has, since its establishment in 1956, worked to determine the original Irish-language form of anglicised administrative place names, and to suggest suitable Irish-language forms for the minority which are not of Irish origin (Mac Giolla Easpaig 2008). The Placenames Committee (formerly the Placenames Commission) meets regularly to consider the Branch's recommendations. The work of the Placenames Branch on Irish-language place names is available to the public through the *Logainm* ('place name') database, developed by Fiontar & Scoil na Gaeilge in Dublin City University since 2007 with funding from the Government department with responsibility for the Irish language. The database contains English and/or Irish equivalents for over 120,000 places.

In addition to this basic information, all the digital records of the Placenames Branch, and most of their paper records, have been incorporated in the Logainm database so that much of the research trail of the Branch is available. A full bibliography and database of some 7,000 sources is also shared. The research trail stretches back to the Surveys and maps mentioned above as well as written records and audio recordings of fieldwork. This open approach has, ironically, led to an increase in correspondence to the Placenames Branch, often owing to a misinterpretation of the raw research, with local residents sometimes questioning or challenging decisions about both Irish and English place names. Naming can be controversial, and occasionally members of the local community can object to official naming policies, most notably in the case of Dingle, County Kerry. This Gaeltacht town and tourist hub was officially designated *An Daingean* in 2004, reflecting the long-established local Irish version of the name. As the Official Languages Act 2003 specified that this Irish version would have been the only name used on maps and official signage from then on (see information box above), residents of the town, fearing loss of tourist revenue, campaigned for a plebiscite on the name. On foot of the results, in 2011 the

Minister of the Environment and Local Government changed the name of the town to *Dingle* in English and altered the official Irish version to *Daingean Uí Chúis*.

In recent years, Logainm editors have worked to link Logainm place entity identifiers to Ordnance Survey Ireland records, and this has led to corrections in both datasets of incorrectly located or spelled place names. Logainm data is openly available and Irish-language names are supplied to An Post GeoDirectory, and fed through to the Eircode (post code) system and to Google Maps. OpenStreetMap, Apple Maps and GeoNames have all ingested Logainm Irish-language toponymic data. Open online availability of correct Irish-language forms has, anecdotally, led to a reduction in misspelled or mistranslated forms in signage and in public usage. Although the Placenames Branch is not the official source for the provision of Irish language forms of street names (this responsibility is delegated to the relevant local authorities), the Logainm database holds *c*.38,000 street names, derived from Ordnance Survey Ireland records and research carried out by the Placenames Branch in conjunction with local councils, and gaps in Irish-language provision (where no extant Irish-language name can be found) are gradually being filled.

The other major category of place names in Ireland is non-administrative minor place names, such as physical features and human-made features. Field names are an example of minor place names; they have no official status or administrative function, but they can be extremely important as local sources of information regarding settlement patterns, agriculture and industry, amongst other things. Knowledge of field names is not as widespread as it once was due to demographic and occupational shifts. Minor place names are a popular topic with local historical and research groups, and field-name recording projects have been successfully undertaken in several counties, often with funding from city and county councils. An example of this is the Kilkenny Field Names Recording Project, which was funded by the Heritage Council and resulted in the collection of over 11,000 field names, a book (*Meitheal na bPáirceanna: The Kilkenny Fieldnames Recording Project*, 2016), and a short film. Minor names are generally more recent than administrative names and are more likely to be of English origin in non-Gaeltacht areas. In County Kilkenny, where Irish has not been spoken for several generations, 14% of all field names were of Irish origin or in Irish, and this percentage was higher in areas where Irish had been spoken somewhat more recently (Counihan 2016, 29). The majority of field names collected were utilitarian and related to ownership, to location, to size and to topography or shape (Counihan 2016, 12–17). The vast number of minor place names in Ireland is reflected in the *c*.300,000 references collected by the Cork and Kerry Place Names Survey since 1976. Meitheal Logainm.ie, a tool for recording and preserving minor names, was established in 2016 to help such projects and ensure long-term preservation of the information collected.

Legal framework for naming in Northern Ireland

Northern Ireland addresses are stored in the Pointer address database, maintained by Land & Property Services (LPS) with help from local councils and Royal Mail. The authority for naming is the local councils, and names can also, rarely, be designated by Letters Patent: this happened in 2021 when Hillsborough was renamed Royal Hillsborough on foot of an application by the local city council.

English is the official language in Northern Ireland and Irish has no official status. Local councils may however choose to pursue a bilingual approach. The Local Government (Miscellaneous Provisions) (Northern Ireland) Order 1995 bestows the power on councils to erect street name signs in a language other than English but the name must always be present in the official English form also.

Unlike the Republic of Ireland, there is no statutory imperative in Northern Ireland to use Irish-language forms of place names, although that may change in the near future with the enactment of a Languages Act. In the meantime, the Northern Ireland Place-Name Project works to elucidate the origin and meaning of the names of Northern Ireland, irrespective of their linguistic origin. The Project provides Irish-language forms to those local councils that choose to take a bilingual approach to signage and publications. While signage is politically contentious owing to the politicised nature of the Irish language in a Northern Ireland context,[16] place names research projects can be a way of creating a shared, non-sectarian pride of place, and the Northern Ireland Place-Name Project regards names as a 'manifestation of shared languages and shared space' (Ó Mainnín, 59). 'There can be fierce attachment to a name such as Drumnahuncheon (a name of Irish-language origin, Droim na hUinseann 'ridge of the ash-tree'), and pride on the part of the person who received it from their forebears… At the same time, there was resistance from the same person to the thought of the name appearing in its original Irish form in bilingual signage in the locality.' (Ó Mainnín, 76).

The introduction of the postcode and street name address system by Royal Mail in Northern Ireland in the 1970s meant that townland names were no longer always used in addresses, leading to concern that they would be forgotten.[17] Street names are generally much newer, and townland names are more likely to be of Irish-language origin. An address such as

[16] See, for example, 'Irish language sign damage a "hate crime" police say', *The Irish News* 29 October 2021: https://www.irishnews.com/news/northernirelandnews/2021/10/29/news/irish-language-sign-damage-a-hate-crime-police-say-2492719/.

[17] See, for example, a debate in the Northern Ireland Assembly on 1 October 2001: https://www.theyworkforyou.com/ni/?id=2001-10-01.9.0.

Lambeg

Lisburn

County Antrim

changed to:

Ballyskeagh Rd

Lisburn

BT27 5TE

Although similar concerns were voiced prior to the introduction of the Eircode postcode system in the Republic of Ireland in 2015, it is too early to measure any effect on knowledge and use of townland names.

Conclusion

This chapter has sought to offer a reading of Ireland's toponymic landscape over the past four hundred years in charting how the agency of the state and influential groups and individuals, from the colonial to postcolonial era periods through to the present day have sought to name and rename places and record these names in either strengthening or challenging dominant ideologies and values. In looking at more recent influences on naming in the period 1960 to the present day, the case study of parts of north Dublin is hugely insightful in illustrating the breadth and range of influences that have a bearing on the choice of street name. The impact and significance of such inscribing, erasing and recording on the cultural landscape is a useful indicator of broader changing socio-economic and political structures that has to date, in the Irish context, not received the attention it merits. In concluding the chapter, consideration is given to how the rich toponymic heritage of both administrative and non-administrative minor place names in the Republic of Ireland and Northern Ireland are protected, recorded and promoted, but also contested for a variety of reasons.

Questions

1. With a particular focus on Ireland, detail the varying motivations in naming and renaming streets?
2. Use the www.logainm.ie website to identify five places in Ireland whose Irish name and English name are not direct translations of each other and discuss.
3. For an administrative region/urban settlement that you are familiar with, consider how the place names reflect those who have influenced and controlled it over time.
4. How are place names recorded and protected in a country that you are familiar with?

Acknowledgements The authors would like to thank Aindí Mac Giolla Chomhghaill (Placenames Branch) and Frances Kane (Queen's University Belfast) for their advice on some points in the chapter.

References

Alderman DH (2008) Place, Naming and the Interpretation of Cultural Landscapes. In: Graham B, Howard P (eds) The Ashgate research companion to heritage and identity. Ashgate, Aldershot, pp 195–214

Andrews J (1992) More suitable to the English tongue: the cartography of Celtic placenames. Ulster Local Stud 14:7–21

Azaryahu M (1997) German reunification and the politics of street names: the case of East Berlin. Polit Geogr 16(6):479–493

Azaryahu M (2009) Naming the past: the significance of commemorative street names. In: Berg LD, Voulteenaho J (eds) Critical Toponymies: the contested politics of place naming. Ashgate, Farnham, pp 53–70

Berg LD, Kearns RA (2009) Naming as norming: 'race', gender and the identity politics of naming places in Aotearoa/New Zealand. In: Berg LD, Voulteenaho J (eds) Critical Toponymies: the contested politics of place naming. Ashgate, Farnham, pp 19–51

Brady J (2014) Dublin, 1930–1950: the emergence of the modern city. Four Courts Press, Dublin

Brady J (2016) Dublin, 1950–1970 houses, flats and high rise. Four Courts Press, Dublin

Counihan A (2016) Harvest time: at work in the fields. In Alan Counihan and Dearbhala Ledwidge (eds) Meitheal na bPáirceanna: The Kilkenny Field Name Recording Project. Kilkenny pp 8–33

de Buitléir M (2013) A Portrait of Dublin in Maps. Gill & Macmillan, Dublin

de hÓir, Éamonn (1972–1973) The anglicisation of Irish placenames. Onoma XVII, pp 192–204

Doyle A (2015) A History of the Irish language from the Norman invasion to independence. Oxford Linguistics, Oxford

Duffy PJ (2007) Exploring the history and heritage of Irish landscapes. Four Courts Press, Dublin

Duffy PJ (2011) Ordnance survey maps and official reports. In: Murphy J (ed) The oxford history of the Irish book, volume IV: the Irish book in English, 1800–1891. Oxford University Press, Oxford, pp 553–562

Dunlevy DA (2021) Place names and the complexity of language recognition in Northern Ireland. In: Blackwood R, Dunlevy DA (eds) Multilingualism in public spaces: empowering and transforming communities. Bloomsbury, New York, pp 137–157

Flanagan D, Flanagan L (1994) Irish place names. Gill & Macmillan, Dublin

Foster RF (1988) Modern Ireland 1600–1972. Allen Lane, London

Harley JB (1988) Maps, knowledge, and power. In: Cosgrove D, Daniels S (eds) The iconography of landscape. Cambridge University Press, Cambridge, pp 277–312

Mac Giolla Easpaig D (2008) Placenames policy and its implementation. In: Pháidín CN, Ó Cearnaigh S (eds) A new view of the Irish language. Cois Life, Dublin, pp 164–177. https://www.logainm.ie/ga/res/26. Accessed 9 Nov 2021

Mac Giolla Easpaig D (2009) Ireland's heritage of geographical names. In: Geographical names as a part of the cultural heritage, Wiener Schriften zur Geographie und Kartographie, vol 18, pp 79–85. https://www.logainm.ie/en/res/92. Accessed 9 Nov 2021

Magan M (2020) Thirty-two words for field. Gill, Dublin

Mainnín Ó, Mícheál B (2021) Empowering multilingualism? Provisions for place names in Northern Ireland and the political and legislative context. In: Blackwood R, Dunlevy DA (eds) Multilingualism in public spaces: empowering and transforming communities. Bloomsbury, New York, pp 59–87

Margey A (2009) Representing plantation landscapes: the mapping of Ulster, c.1560–1640. In: Lyttleton J, Rynne C (eds) Plantation Ireland settlement and material culture c.1550–c.1700. Four Courts Press, Dublin, pp 140–164
McCready CT (1892) Dublin street names, dated and explained. Hodges, Figgis, and Co, Dublin
Montgomery-Massingberd H (1976) Burke's landed gentry of Ireland. Burke's Peerage, London
Nash C (1999) Irish placenames: post-colonial locations. Trans Inst Br Geogr 24(4):457–480
Ó Muraíle N (2014) Translations: the placenames work of the OS in Ireland. https://www.ria.ie/mapping-city-town-and-country-1824-ordnance-survey-ireland-0. Accessed 9 Nov 2021
Smyth WJ (2006) Map-making, landscapes and memory a geography of colonial and early modern Ireland c.1530–1750. Cork University Press, Cork
Whelan Y (2003) Reinventing modern dublin streetscape, iconography and the politics of identity. UCD Press, Dublin

Translating Topographies: Brian Friel's Approach to Language, Landscape, and Toponymy in Ireland

Charles Travis

Abstract Brian Friel's play *Translations* takes place in 1833 and is set in a "Hedge School" near the village of Baile Beag, in Donegal, in Ulster during the period Ireland was being mapped by the British Ordnance Survey. Tension between Gaelic *Dinnseanchas* (Lore of Place Names) perceptions of place and 'Anglo-Cartesian' appellations of space suffuse the drama. Yolland, the English soldier/topographical engineer is befriended by Owen, the urbane son of Hugh the Hedge Schoolmaster as he seeks to translate Gaelic place names into English words and labels for the survey's new maps. This chapter discusses *Translations* in reference to past debates between Friel's 'fictive character' and geographer John Andrew's 'historic' perspectives on the impact of the British Ordnance Survey not only in Ulster, but on a wider linguistical landscape. It also contextualizes *Translations* within other literary geographies detailing global nineteenth century industrialization processes in addition to the current post-Brexit milieu of Northern Ireland as a means to address the complexity of landscapes, identities, and senses of place at play in twenty-first century Ulster.

Keywords Brian Friel · *Translations* · Drama and language · Gaelic/Irish toponymy · Literary geography · Irish Troubles

Objectives

The objectives of this chapter are:

- To appraise Brian Friel's play *Translations* (1980) set in 1833 Ulster, when Ireland's toponymy was being (re) named and 'cartographically standardized' by the British colonial administration.
- To explore the nexus between toponymy, identity and culture regarding the local, native population and colonial administration.

C. Travis (✉)
Department of History, University of Texas, Arlington, USA
e-mail: ctravis@tcd.ie

Trinity Centre for Environmental Humanities, Trinity College Dublin, Dublin, Ireland

© The Author(s), under exclusive license to Springer Nature Switzerland AG 2023

G. O'Reilly (ed.), *Place Naming, Identities and Geography*,
Key Challenges in Geography, https://doi.org/10.1007/978-3-031-21510-0_23

- To review the play *Translations,* referencing past debates between Friel's 'fictive characters' and geographer John Andrew's 'historic' perspectives on the impact of the British Ordnance Survey in Ulster, and on the wider linguistic colonial landscape.
- To contextualize *Translations* within other literary geographies detailing industrialization processes and the post-Brexit milieu of Northern Ireland as a means to address the complexity of toponymy, landscapes, identities and senses of place at play in twenty-first century Ulster.

Introduction

... remember that words are signals, counters. They are not immortal. And it can happen -to use an image you'll understand -it can happen that a civilization can be imprisoned in a linguistic contour which no longer matches the landscape of ... fact.

Brian Friel, *Translations* (1981)

The whole landscape, a manuscript / We have lost the skill to read.

John Montague, *A Lost Tradition* (1972)

Overview

This chapter will discuss the tensions and transgressions of toponymical practice in Irish playwright Brian Friel's 1980 drama *Translations.* The British Ordnance Survey first mapped the island of Ireland between 1824 and 1842 and produced baseline six inch to one-mile maps (1:10,560) charting the locations of structures, townlands, roads, fields, rivers, streams, bogs and other features of the physical landscape. Part of this process involved the Anglicization of Gaelic/Irish place names. Whilst topographically accurate, the first ordnance survey in the nineteenth century was a political project of empire that rewrote cultural landscapes of Ireland with an intent, -to create a scale of appraisal and establish the basis for Griffith's Valuation of land and property. The drama *Translations* takes place in August 1831 and is set in a "disused barn or hay-shed or byre," [1] that serves as a hedge-school for the townland of *Baile-Beag / Ballybeg (Baile* for "Town", *Beag* for "Little") a Gaelic/Irish speaking community in north County Donegal during one of the numerous cartographical surveys conducted by British map-makers and orthographers. Much has been made of the linguistic colonization of Ireland by these nineteenth century Ordnance surveys, but fundamentally, it was the institution of the national school system's curriculum that pressed the English language on native Gaelic speakers.

Hedge schools emerged in the wake of Penal Laws established after the seventeenth century Cromwellian, conquest of Ireland when Catholic masses and Gaelic language instruction were prohibited by law. Patrick John Dowling's *The Hedge*

Schools of Ireland (1968) estimates that by 1824 hedge schools made up the great majority of the 9,352 "pay schools" then in existence in Ireland.[2] Schoolmasters tutored students in mathematics and the classics, but in the eighteenth century "the medium of instruction was largely Irish," and traditions of Gaelic literacy in Irish were preserved "by copying and preserving Irish language texts in manuscript form, compiling dictionaries and grammars" and by writing poetry in the circumscribed language.[3]

In *Translations*, Hugh, a poteen fueled school master, is assisted by his disabled son Manus. For a small fee they provide the denizens of *Baile-Beag* curricula in Gaelic, Latin, Greek and logic. The hedge school roster includes Sarah John Sally, a woman struggling to gain speech; Jimmy Jack Cassie, a bachelor in his sixties known as the "Infant Prodigy," who is fluent in Latin and Greek; Maire Catach, a young woman in her twenties, -the love interest of Manus-, who wishes to learn English so she can emigrate to America; Doalty Dan Doalty "an open-minded, open hearted, generous and slightly thick young man" and Bridget "a plump, fresh young girl, ready to laugh, vain and with a countrywoman's instinctive cunning."[4]

Other off-stage characters include George Alexander, the local Justice of the Peace, who has invited Hugh to take over as headmaster of the new national school; Nelly Ruadh and her baby (who sadly dies, providing *Translations* with a christening and a wake); Anna na mBreag ('*Anna of the Lies*') pub owner and poteen distiller and the Donnelly twins- fishermen -parsed as ciphers for violent Republican resistance to British colonization. In the play, Owen Hugh More, urbane son of Hugh the school-master, returns from Dublin to assist George Yolland an English orthographer, attached to the British Ordnance Survey in Donegal which is led by Corporal Lancey. Their aim is:

> *… to take each of the Gaelic names -every hill, stream, rock, even every patch of ground which possessed its own distinctive Irish name -and Anglicise it, either by changing it into its approximate English sound or by translating it into English words.*[5]

Within *Translations*, language and geography are rendered by Friel on a continuum over which perceptions and experiences of landscape, identity and sense of place oscillate between phenomenological impressions and Euclidian framings of space and orthography. Gaelic toponymy populates a wide geographical swath that runs from the west and north of the island of Ireland across the western islands of Scotland to its Highlands. Gaelic place names originated with the early bardic traditions of *dinnsheanchaís* ('knowledge of the lore of places.') This practice of place-naming constitutes an intimate and deep "geography based on *seanchaís*, in which there is no clear distinction between the general principles of topography or direction-finding and the intimate knowledge of particular places."[6] Twentieth century Monaghan poet Patrick Kavanagh's renderings of the Ulster townlands, drumlins and fields of his native county reflect such a geo-topographical sensibility:

> To know fully even one field or lane is a lifetime's experience. In the poetic world it is depth that counts and not width. A gap in a hedge, a smooth rock surfacing a narrow lane, a view of a woody meadow, the stream at the juncture of four small fields -these are as much as man can fully experience.[7]

In *Translations*, the "password of the tribe" is tied to *seanchaís* and a "communal dialect which identifies its members at birth according to their native origins -the name of their parents and local birthplace."[8]

Writing *Translations*

Set over a few days in late August 1833, Translation is written and performed in the English language, though all of the characters except for the British Surveyors Lancey and Yolland in the play speak either Gaelic, Latin or Greek (for instance, Lancey mistakes Hugh's Latin greeting for Gaelic.) The English employed for 'Gaelic' dialogue with the people of Baile Beg is a 'Hiberno' inflected English. Friel claims "Ritual is part of all drama [...] Drama is a RITE, and always religious in the purest sense,"[9] and contends that "there are bigger truths beyond that of the literal translation."[10] During the period of drafting *Translations*, Friel was engaged in translating Anton Chekov's play *Three Sisters.* As he was considering the philological smithing of a Hiberno-English, Friel noted that versions of Chekov "read in this country always seem to be redolent of either Edwardian England or the Bloomsbury set. Somehow the rhythms of these versions do not match the rhythms of our own speech patterns. [...] This is something about which I feel strongly –in some way we are constantly overshadowed by the sound of the English language."[11] He observed that in Chekov's 19th nineteenth-century work, characters behaved "as if their old certainties were as sustaining as ever -even though they know in their hearts that their society is in melt-down and the future has neither a welcome nor even accommodation for them. Maybe a bit like people of my own generation in Ireland today."[12] Reflecting on the process of writing drama Friel stated:

> … At any given time every playwright has half a dozen ideas that drift in and out of his awareness. For about five years before I wrote *Translations* there were various nebulous notions that kept visiting me and leaving me: a play set in the nineteenth century, somewhere between the Act of Union and the Great Famine; a play about Daniel O'Connell and Catholic emancipation; a play about colonialism; and the one constant - a play about the death of the Irish language and the acquisition of English and the profound effects that that change-over would have on a people.[13]

The play emerged from a hermeneutic crucible that included Dowling's The *Hedge Schools of Ireland*, Thomas Colby and T.A. Larcom's, *Ordnance Survey of the County of Londonderry* vol. 1: *A Memoir of the City and the North-West Liberties of Londonderry. Parish of Templemore* (1837) and the work of historian John H. Andrews (1927–2019): "in 1976 I came across *A Paper Landscape* and suddenly here was the confluence—the aggregate—of all those notions that had been visiting me over the previous years: the first half of the nineteenth century; an aspect of colonialism; the death of the Irish language and the acquisition of English. Here was the perfect metaphor, accommodate and realize all these shadowy notions - 'map making'."[14]

Whilst the practices of cartography and orthography were figurative metaphors in *Translations,* Friel's poetic license in regard to history and Andrews' historical empiricism met a cross-purpose. The latter stated that "Ordnance maps did not circulate widely in nineteenth-century rural Ireland, and the inhabitants of a real-life Ballybeg may never have known what Colby's officers called their village. Even today many spellings promulgated by the Ordnance Survey of Ireland are ignored in local newspapers, directories, and public notices."[15] The transposition of violence between Irish Republicans and the British Army in 1970s Northern Ireland into the landscape of early nineteenth-century Donegal was a sub-text to Friel's drama. A dialogue was facilitated between the historian and dramatist, and Andrews despite his concern that members of the British Ordnance were depicted as militants, rather than surveyors conceded: "Despite having myself written a factual account of the Ordnance Survey, I found Friel's inventions effective not just dramatically but in expressing a legitimate attitude to modern Irish history as a whole."[16]

Placing Baile Beg / Ballybeg

Humanist geographers have observed that 'literature is the product of perception, or, more simply, *is* perception,'[17] and long maintained a 'view of literature as a valuable storehouse of vivid depictions of landscapes and life.'[18] Writing about places names in Ulster, the geographer E. Estyn Evans noted:

> Here are a few which I have come across [...] Ballywillwill, Ballymunterhiggin, Aghayeevoge, Treantaghmucklagh. In all Ireland there are no less than 5,000 townlands beginning with 'Bally,' forty-five of them named Ballybeg (little town).[19]

In Friel's literary imagination, the intersecting linguistical spaces of Baile-Beag (*Gaelic*) and Ballybeg (*Anglo*) seemingly a ubiquitous feature on the Irish landscape, have been mapped to the Donegal town of Glenties, the birth-place of his mother, Mary McLoone, a native of the parish who had worked for a time in the local Post Office. Friel himself was born in Omagh, County Tyrone, Northern Ireland in 1929 and relocated with his mother and father Paddy to Derry in 1939 where he attended St. Columb's College (as did his Field Day colleagues poets, Seamus Heaney and Seamus Deane.) Located in the south Donegal *Gaeltacht* (Gaelic speaking community) Glenties, nestled between the shoulders of Meenalargan, Shalloganbeg, and Crockmore Hills, became Friel's boyhood summer home and is known as a region where traditional music incorporating fiddles, spoons, tin-whistles, *bodhrans* (goatskin hand drums) blends with storytelling and *sean-nós* singing sessions in local pubs and homes.

Its milieu and landscapes strongly shaped his perceptions and representations of place, language, and toponymy. In 1963 Friel composed an essay, *A Fine Day at Glenties* describing a street-fair where animals and goods were sold. Fair-Day provided Friel a portal from which to view the idiosyncrasies and vagaries of the human condition as the bartering and trading commenced: "Glenties is the stage," he noted, "and the fairgoers are the players."[20] Sitting at the heart of a regional (the) hand-knitting industry and host of the MacGill Summer School where Ireland's cultural and political *intelligentsia* are annual *habitués*, Glenties, like other rural towns on Ireland's Atlantic fringe is at the mercy of seasonal fishing, agricultural, pastoral, cultural tourism and knitwear labor markets. Despite the perennial flux of emigration, Donegal's Gaelic communities survive and cling to the surrounding mountains, glens, boglands, and along the seacoasts, ports, and cliffs of the remote, rugged, and rocky county.

Glenties has oft been identified as the inspiration for Friel's imaginary *Baile-Beag/Ballybeg*, the town that serves as the setting for his successive plays *Philadelphia, Here I Come!* (1964), *Living Quarters* (1978), *Aristocrats* (1979), *Faith Healer* (1979), *Translations* (1980), *The Communication Cord* (1982), *Dancing at Lughna*sa (1990), *Wonderful Tennessee* (1993), and *The Home Place* (2005). However, its location and sense of place shift according to the dictates of the playwright's dramatic perspective and perhaps because of what Friel describes as his own "sense of rootlessness and impermanence. It may well be the inheritance of being a member of the Northern minority [...] where you are certainly at home but in some sense, exile is imposed upon you."[21] Chronologically, *Baile-Beag / Ballybeg* has been transposed to the times of the Battle of Kinsale (1602), the 1831 Ordnance Survey of Ireland, to 1878 during the Land War and the dawn of Irish Home Rule, the 1930s, and into the mid-twentieth century.

Baile-Beag / Ballybeg's first appearance was in *Philadelphia, Here I Come!*, (1964) and the town was depicted as a "stagnant, rural backwater, full of people lost in their own delusions because of the meanness of their lives."[22] In *Living Quarters* (1978), *Aristocrats* (1979), *Faith Healer* (1979), *Making History* (1988), *Wonderful Tennessee* (1993) and *Dancing at Lughnasa* (1990), *Molly Sweeney* (1994), and *The Home Place* (2005) the environs of *Baile-Beag / Ballybeg* are variously situated and described as a "remote" and "run-down village" situated "in the wilds" of "north-west Donegal," but paradoxically "not far from Donegal town," and possessing a "pier". Ballybeg is also described as "a named destination—democracy, ... heaven", as well as "off its head" with a "fever" anticipating a Celtic harvest dance.[23] Over the dramatic canon of Friel's plays, *Baile-Beag / Ballybeg* is inhabited by the seventeenth century McDevitt clan, a nineteenth-century hedge school-master and English anthropologist, a retired African missionary priest, a shopkeeper, and a university lecturer.[24]

Mapping *Translations*

It can be surmised that *Baile-Beag / Ballybeg* in *Translations* was imagined and pieced together from a composite geography formed by Donegal's glens, hills, townlands, seascapes and conjured from a constellation of place names signifying landscape features such as the *Sliabh Liag* sea-cliffs in the south, to the Bloody Foreland on the west coast, and the Inishowen Peninsula, which juts out into the Atlantic Ocean in the county's north-east. Friel's portrayal of the imaginary town is impressionistic, illustrating Gaston Bachelard's proposition that: "space that has been seized upon by the imagination cannot remain indifferent space subject to the measures and estimates of the surveyor. It has been lived in, not in its positivity, but with all the partiality of the imagination."[26] However, one can approximate where Friel cartographically situated *Baile-Beag* / *Ballybeg* in *Translations*. In the play, the Irish Rebellion of 1798, a major rising against British rule in Ireland, inspired by the 1776 and 1789 revolutions in America and France is recalled by Hugh. The rebellion, organized by the Society of United Irishmen, formed by Presbyterian radicals angry at being excluded from power by the Anglican establishment, was joined by members of the Irish Catholic population.[27] In 1833, as Hugh's Gaelic world seems to be collapsing around him, he recalls:

> The road to Sligo. A spring morning. 1798. Going to battle. Do you remember, James? Two young gallants with pikes across their shoulders and the *Aeneid* in their pockets. Everything seemed to find definition that spring – a congruence, a miraculous matching of hope and past and present and possibility. Striding across the fresh green land. The rhythms of perception heightened. The whole enterprise of consciousness accelerated. We were gods that morning […] We marched as far as -where was it? - . . . Glenties! All twenty-three miles in one day. And it was there, in Phelan's pub that we got homesick for Athens, just like Ulysses. The *desiderium nostrorum* – the need for our own. Our *pietas* James, was for older, quieter things. And that was the longest twenty-three miles back I ever made (Toasts JIMMY.) My friend confusion is not an ignoble condition.[28]

The speculative route for Hugh and Jimmy Jack's march from a probable location of the imaginary Baile-Beg can be seen in Fig. 1, as can the locations of other Donegal Gaelic place names listed in the play's text. In addition, the location of the island of *Inish Meadhon*, where Manus is offered a position with a hedge school is described by Owen as about fifty miles south of *Baile-Beag* / *Ballybeg*—which would place the village perhaps on the Inishowen peninsula- but nowhere near only 23 miles from

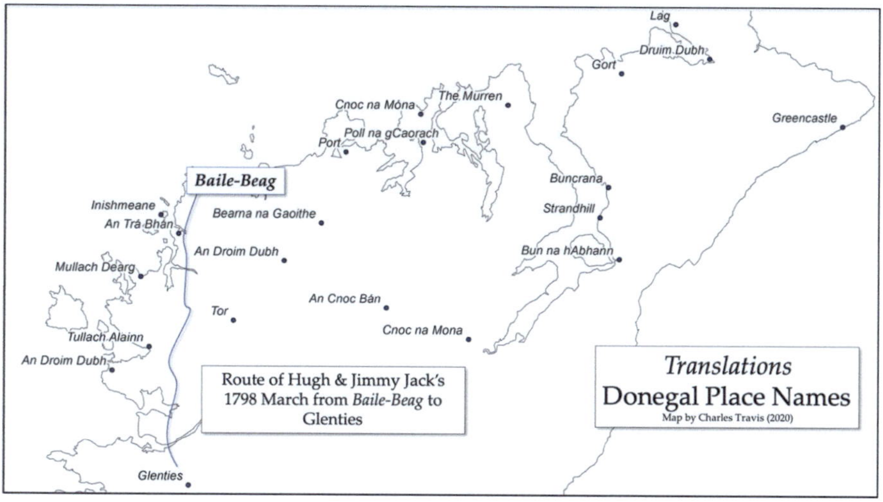

Fig. 1 Speculative site of Baile-Beag and Location of Donegal Place Names in Brian Friel's *Translations* (1980). *Source* Map hand drawn by Charles Travis[25]

Glenties. In one of *Translations*' more comical scenes, the British Ordnance Survey sappers are described at work. Jimmy Jack's parsing of the name of their mapping tools invokes Donna Haraway's 'God-Trick' critique of twentieth-century geospatial technology.[29] In turn, Doalty's mischievous interference with the sapper's mapping process underscores how a cunning, cultural landscape often eludes the snares of Euclidean space:

> **DOALTY:** Up in the bog with Bridget and her aul fella, and the Red Coats were just across at the foot of Cnoc na Mona, dragging them aul chains and peeping through that big machine they lug about everywhere with them -you know the name of it, Manus.
>
> **MAIRE:** Theodolite.
>
> [. . .]
>
> **JIMMY:** Theodolite -what's the etymology of that word, Manus?
>
> **MANUS:** No idea.
>
> **BRIDGET:** Get on with the story.
>
> **JIMMY:** *Theo - theos* -something to do with a god. Maybe *thea* -a goddess! What shape's the yoke?
>
> **DOALTY:** 'Shape!' Will you shut up, you aul eejit you! Anyway, every time they'd stick one of these poles into the ground and move across the bog, I'd creep up and shift it twenty or thirty paces to the side.
>
> **BRIDGET:** God!
>
> **DOALTY:** Then they'd come back and stare at it and look at Their calculations and stare at it again and scratch their Heads. And Cripes, d'you know what they ended up doing? [. . .] They took the bloody machine apart! (*And Immediately he speaks in gibberish -an imitation of two very agitated and confused sappers in rapid conversation* . . .)
>
> [. . .]
>
> **MAIRE:** That was a clever piece of work.

MANUS: It was a gesture [. . ..] Just to indicate a . . . presence.[30]

In a subsequent scene, Owen and Yolland attempt to determine the best translation for a placename, -a task that is proving to be equally elusive:

YOLLAND: Let's see . . . Banowen

OWEN: That's wrong. (*Consults text.*) The list of freeholders

calls it Owenmore -that's completely wrong: Owenmore's

the big river at the west end of the parish. (*another text*)

And in the grand jury lists it's called ~ God! -Binhone! -

wherever they got that. I suppose we could Anglicise it to

Bunowen. but somehow that's neither fish nor flesh.[31]

Friel's set pieces underscore that "in the case of *Translations* we were talking about the function of a fractured language, an acquired language and a lost language."[32] The drama plays with the indeterminacy and subjectivities of mapping and place-naming -and the slippages and malapropisms that can occur in the process. Noted Friel:

> I think that the political problem of this island is going to be solved by language [...] Not only the language of negotiation across the table, but the recognition of what language means for us on this island [...] Because we are in fact talking about the marrying of two cultures here, which are ostensibly speaking the same language but which in fact are not.[33]

Over the course of the play, Owen and Yolland, assisted at times by poteen from Anna n mBreag's pub, Anglicize the Gaelic place names of *Baile-Beag*. During the process, Yolland begins to 'go native' despite Owen's advice that the townland is 'no Eden.' In spite of this Maire Catach, who yearns for America, and the young sapper falls in love after a local dance, despite being barely able to communicate across their Gaelic-Anglo language divide. This earns the rage of Manus who suspicious of Owen's assistance to the Ordnance Survey becomes the "lame scholar turned violent," [34] by impotently throwing a stone that misses the Englishman. The next morning Yolland has vanished, supposedly killed by the Donnelly twins. This in turn causes Corporal Lancey to threaten to shoot the livestock of *Baile-Beag*, start evictions, and level surrounding townlands if the missing orthographer is not found. In the midst of the trouble that erupts, Jimmy Jack, lonely in his hillside abode at *Bean na Gaoithe*, and so deeply immersed in Homer, and the myths of Ancient Greece, plans to marry the goddess Athene. But in spite of his delusional condition, his dilemma brings a clarity to the violence that Maire's tryst with Yolland seems to have instigated in the parish. Advising Maire, Jimmy-Jack asks her:

> Do you know the Greek word *endogamein*? It means to marry within the tribe. And the word *exogamein* means to marry outside the tribe. And you don't cross those borders casually -both sides get very angry.[35]

In the wake of Yolland's disappearance, Manus is hired to run a hedge-school on the island of *Inis Meadhhon*, but defers his appointment by going on the run to the Errigal peninsula in County Mayo. Then Bartley Timmins, a schoolmaster from Munster is selected as the head of *Baile-Beag's* national school instead of Hugh, prompting the following exchange:

HUGH: Ha-ha-ha-ha-ha! The Cork bacon-curer! *Barbarus hic ego sum*

quia non intelligor ulli'- James?

JIMMY: Ovid.

HUGH: *Procede.*

JIMMY: 'I am a barbarian in this place because I am not understood by

anyone.'[36]

As the play closes, Hugh pragmatically sensing the arrival of a new cultural landscape, promises to teach English to Maire Catach so she can emigrate to Brooklyn. Pondering the *Name Book* that Owen had been assembling with Yolland, Hugh informs his despondent son:

> We must learn those new names [...] We must learn where we live. We must learn to make them our new home [...] it is not the literal past, the 'facts' of history, that shape us, but images of the past embodied in language [...] we must never cease renewing those images; because once we do we fossilize.[37]

References to other locations and place names (English, French, Gaelic, Latin, and Greek) in *Translations* are supranational and wider in scope and go beyond Donegal and the shores of Ireland. Illustrating the geographical horizons of modernity and empire, the list of place names dots the globe from Brooklyn and Nova Scotia in North America in the west, to the imperial brain of London, and the revolutionary Bastille in France, and reaching as far to the east as Bombay in British ruled India.

Gaelic Language Diaspora

On a wider geographical scale, the rationalization of Gaelic topographic names into an Anglo-Euclidean system in *Translations* speaks to global processes attached to nineteenth-century industrialization and the dispersion and loss of Gaelic language dialects, cultures, and communities. In *Translations*, Doalty, concerned about the cattle stampeding into a lake because of the summer heat declares "The bloody beasts" end up in Loch an Iubhair."[38] By sharing a toponym with *Loch Iubhair*, located in the Scottish Highlands, Friel's drama alludes to the Gaelic fringes of Western Europe. Specifically, themes in Friel's drama resonate with the Scottish Highland Clearances instigated in the eighteenth century. Indeed a corollary to *Translations* is historical geographer Charles Withers' exploration of cultural identity and place naming in Gaelic Scotland by English-speaking officials of the Ordnance

Survey in the nineteenth-century Scottish Highlands.[39] Regarding the historical diffusion and sharing of Gaelic culture between Ireland and Scotland, M.W. Heslinga in his work *The Irish Border as a Cultural Divide* (1971) contended,

> … It is frequently said that it was the Scots who made modern Ulster. There is as much reason for saying that it was 'Ulstermen' who made ancient Scotland. For the colonists from east Ulster gave Scotland her name, her first kings, her Gaelic language, and her faith. It is often suggested that modern Ulster is an extension or projection of Scotland, but the first concept of Scotland was really an extension or projection of 'Gaelic Ulster'.[40]

The Jacobite loss at the Battle of Culloden (1746) and eviction of Gaelic-speaking inhabitants of the Highlands and western islands of Scotland commenced in the mid-to-late eighteenth century and continued sporadically into the middle of the nineteenth century in order to allow landlords to introduce modern methods of sheep farming. This contributed to the disintegration of the traditional Scottish clan society, and a process of rural de-population leading to emigration to Ulster and beyond. The collapse of Gaelic language traditions highlighted in *Translations* also chimes with Ulster poet John Hewitt's study of the eighteenth and nineteenth-century "Rhyming Weavers" of counties Antrim and Down whose Ulster-Scots dialect poetry faded as the craft of hand-looming declined with the rise of Northern Ireland's steam-powered and mechanized linen mills during the industrial revolution.[41] Hewitt, from a Methodist background, was interested in Gaelic toponymy. In the poem *Ulster Names,* he declared:

> *They move on the tongue like the lilt of a song.*
>
> *You say the name and I see the place-*
>
> *Drumbo, Dungannon, Annalong.*[42]

Between the seventeenth and nineteenth centuries, successive waves of 'Scots-Irish', 'Ulster-Scots' and 'Gaelic Irish' communities crossed the Atlantic into North America. As new arrivals, these diaspora populations clustered in the mill and factory towns and cities of New England, the Great Lakes in the Midwest before following the spines of the Appalachian and Piedmont Mountain ranges into the Ohio and Tennessee Valleys and migrating westward across the Mississippi River to the Great Plains and Rocky Mountain West. The wider contextual themes of the historical processes of industrialization and cultural displacement in *Translations* find echoes in English writer George Eliot's *Middlemarch* novels (1871–1872) and Texas author Larry McMurtry's *Lonesome Dove Chronicles* (2010). It has been noted that Eliot's *Middlemarch,* as a piece of historical literature set in the English Midlands countryside in 1831 during a period of agricultural and parliamentary 'reform' in Britain, intersects with the period in which Friel's *Translations* takes place. [43] Eliot's novel depicts rural England being colonized "by the dominant technological and empiricist practices of an expanding capitalism." [44] In *Middlemarch,* railway surveyors attempt to map rural fields with spirit levels and measuring chains, but local field-hands offer violent resistance to the new technological order, by running the surveyors off the land.[45]

Likewise, Texas author Larry McMurtry's *Lonesome Dove Chronicles* (2010) a saga on the closing of the American west, intersects with the period in which the Industrial Revolution in England catastrophically unsettled the agrarian and labor market economies of the 'British Isles,' -the prime example being the 1848 Famine in Ireland- causing the "despised races of the Celtic fringe- Wales, Ireland and Scotland," to migrate to America- many of whom headed West, beyond the boundaries of the Anglo-centric east coast of the United States.[46] McMurtry, like Friel's blending of Gaelic storytelling tropes in *Translations* weaves "folklore and American frontier history" to depict the transgression of "physical, psychic, and geographical borders" symbolized by artefacts of place such as a "a simple adobe wall, an otherworldly dimension, or a mountain chain."[47] Appropriating the character tropes of the Errant Knight and Sancho Panza from Miguel de Cervantes' *Don Quixote* (1605–1615), McMurtry's *Lonesome Dove Chronicles* recounts the lives of the filibusters, Texas Rangers and cattlemen Augustus 'Gus' McRae, a raconteur from Tennessee, and his stolid partner Woodrow Call, who is reminded by Gus "You was born in Scotland [...] I know they brought you over when you were still draggin' on the tit, but that don't make you no less a Scot." [48] The *Lonesome Dove* saga spans the 1840s to the 1890s, a time when the statistical cartography of U.S. Census Bureau enumerated the lower 48 states and the indigenous and colonial *conquistador* trails of the West transformed into the industrial webbing of telegraph wires and railroad tracks to remap the natural landscapes of the once North American 'frontier.'[49] One of Gus' rueful reflections in the *Lonesome Dove Chronicles* on the 'closing' of the American West's frontier echoes Yolland's identification with the fading Gaelic culture of Baile-Beag:

> In the north, the Army had finally taken the fight against the Comanches away from the Rangers and had nearly finished it ... After the [American civil) war, the cattle market came into existence and the big landowners in south Texas began to make up herds and trail them north to the Kansas railheads. Once the cattle became the game ... he and Call quit rangering. It was no trouble for them to cross the river and bring back a few hundred head to sell ... they had roved too long, Augustus concluded ... They were people of the horse, not of the town; in that they were more like the Comanches than Call would ever have admitted.[50]

In turn, Call's aim is to drive a herd of cattle from south Texas up north to see the unsettled landscape of Montana before bankers, lawyers and railroads arrive in the territory. McMurtry compares *Lonesome Dove* to Eliot's *Middlemarch* and its historical contextualization of emerging English railways, the accession of King William IV and the passage of the Whig sponsored 1832 Great Reform Act by the British Parliament.[51] Just as Eliot's novel depicted a period during which the map of parliamentary constituencies was redrawn to make it more representative of the English population, the *Lonesome Dove Chronicles,* re-plots the historiographical map of the settling of the American West bringing in a more spatially democratic representation of different ethnic composition and classes of settlers from store, saloon and bordello keepers, with cattle-men and illiterate working class hands, in addition to endowing heroic and villainous dimensions to Anglo-Euro-Americans, indigenous and Mexican characters. Like Cervantes before him, Friel in *Translations* "does not disentangle the story from the history but points its telescope at the ill-defined frontier itself."[52] In a program note for his drama *Making History*, Friel stated that:

> … When there was tension between historical 'fact' and the imperative of the fiction, I'm glad to say I kept faith with the narrative […] But then I remind myself that history and fiction are related and comparable forms of discourse and that an historical text is a kind of literary artifact.[53]

Along with McMurtry's *Lonesome Dove Chronicles* and Eliot's *Middlemarch*, Friel's *Translations* maps out a topography and geography of nineteenth century cultural loss that blurs "the boundary between history and story."[54]

Conclusion

Translations was the first project of The Field Day Theatre Company, founded in 1980 by Friel and the actor Stephen Rea (who played Owen in the play's first performance run). The company actively engaged in the cultural politics of the Northern Ireland 'Troubles,' (1968–1998) publishing pamphlets in addition to plays with an aim to contribute to "the solution of the present crisis by producing analyses of the established opinions, myths and stereotypes which had become both a symptom and a cause of the current situation."[55] Friel's claims that *Translations* "has to do with language and only language,"[56] and in his drama, the oscillation between the toponyms of *Baile-Beag* and *Ballybeg*, creates an incantation of place formed from a bricolage of Gaelic, English, Latin and Greek language traditions to interrogate notions that sustain "the old prejudices and myths handed down through the republican tradition and from the myths of official British history."[57] Turning to the myths of Irish nationalism, Friel questioned "the wholeness, the integrity of that Gaelic past. Maybe because I don't believe in it."[58] Though lauded and equally critiqued in various quarters for its purported 'idyllic' and 'Edenic' representations of a prelapsarian (i.e. before the Fall of Man) Irish Gaelic past -the fact is that the characters in *Translations* remain largely intoxicated (Hugh and Jimmy Jack) (&) and are involved in hard, agrarian manual labor (Maire, Brigid, and Doalty) and are assumed murderers and victims (the Donnelly Twins and Yolland). *Translations* allude to, as the historical-geographer Kevin Whelan, describes "the harshness, the poverty and the degradation of the life of many" in nineteenth-century Ireland including "the darker side of life—drudgery, drunkenness, violence, ill health, exploitation."[59] In addition, Sarah John Sally -Manus' mute pupil perhaps serves as a prescient symbol for the elision of women's voices that would mark the first three volumes of *The Field Day Anthology of Irish Literature* edited by Seamus Deane.

Following the plantation of Ulster by Protestant English and Scottish settlers during the sixteenth and seventeenth centuries, the largely Gaelic and Catholic population of Ireland was relocated to the Atlantic hinterland of the island. The Gaelic

"West" of Ireland where the imaginary *Baile-Beag / Ballybeg* is situated was "effectively constructed between 1650 and 1840, when the limits of cultivation... rose from c.500 to c.900 feet" and "the 'spade and the spud' conquered its contours and frontiers."[60] In 1833, the year *Translations* was set in, Donegal located in northwest Ireland was an Ulster county on the British governed island of Ireland. With the rise of cultural nationalism in the late nineteenth century after the great potato Famine of 1845–1848, isolated Gaeltacht communities, that had suffered feudal landlordism, poverty, and emigration, took on totemic significance in nationalist and republican mythology. During the Irish War of Independence, (1919–1921) six counties of Ulster were partitioned under the 1920 Government of Ireland Act, and a border was established, Northern Ireland, a British statelet, separate from the twenty-six counties of southern Ireland, including Donegal. The southern Irish Free State was founded in 1922 followed by its declaration as a Republic in 1949.

Partition of the island in 1920 secured a Protestant Unionist majority for the six-county British province in the north, in which Catholic and Irish nationalist communities were marginalized to the status of second-class citizens. There are other more detailed sources and insufficient space in this chapter's conclusion to describe the failure of Northern Ireland's polity and the eruption of the war in the 1960s, '70s, '80s and '90s between the Provisional IRA (Irish Republican Army), the British State and Army, and Loyalist paramilitaries. This conflict set all warring parties in the North at odds with an ambivalent Irish society in the southern Republic -whose official language is Gaelic. The 1994 PIRA Ceasefire and the 1998 Good Friday Agreement brokered a cold peace between Unionist and Nationalist communities in Northern Ireland. In recent years, *Translations'* themes of conflict, cartography, and language have come to the fore again. The United Kingdom's withdrawal from the European Union and the remapping of a customs border in the Irish Sea under the Northern Ireland Protocol between the United Kingdom and Northern Ireland has infuriated Unionists and Loyalists as it is perceived as a threat to their sense of British identity. Another political division in the power-sharing government of Northern Ireland between the Republican party Sinn Fein and Unionists stems over an Irish language education bill proposed by the former but reject by the latter-perhaps because of the 'weaponization' of the Irish language during the Troubles by a minority of nationalist and republican groups.

The growth of interest in Gaelic particularly in parts of Loyalist East Belfast is exhibited by initiatives such as the *Turas* language project (defined as a journey or pilgrimage in both *Irish Gaelige* and *Scots Gàidhlig* to connect Protestant communities to their own history with the Irish language), and *Naíscoil na Seolta* a religiously and ethnically integrated school for toddlers to access Gaelic through the medium of play.[61] However, Ireland's linguistical landscape which shares a heritage in part with Scotland's Highland and western island fringes remains a contentious issue in regards to identity and sense of place in Ulster. *Translations,* was written in the late 1970s when the violent conflict was reverberating nihilistically outside of the borders of a province with a contested political and cultural identity. Reflecting on the divided statelet of Northern Ireland, Friel observed: "if England were to go tomorrow morning, that wouldn't solve it. We still have got to find a *modus vivendi* for ourselves within the country."[62] Friel recognized that "one can never go back to the old culture, but it could extend to the present day,"[63] and hoped his drama would contribute a road map that would "lead to a cultural state, not a political one," and claimed that "out of that cultural state, the possibility of a political state follows."[64]

Box 4: Derry/Londonderry.

"They are the people, too, who keep calling Derry 'Londonderry.' That's a name that's not used by any indigenous Northerner, not even the Protestants, for they'd be too self-conscious. The right name for the city is Derry from the Irish Doire Cholm Chille—meaning the oak-grove of Colmkille. It got the name Londonderry from a company of swindlers that were founded in London, in the seventeenth century, to drive the native Irish off the land and to settle the place with English and Scots." Brendan Behan, (*Brendan Behan's Island (*()1962).

Derry vs. Londonderry: The city lies astride the ancient historic boundaries between the Ulster counties of Donegal (in the Republic) and (London) Derry in Northern Ireland (United Kingdom) and is Northern Ireland's second city, and fifth largest on the island. It is labelled on official maps and signage in Northern Ireland, as "Londonderry" or "L'Derry." It is not uncommon to see signage with the word "London" defaced in an attempt to remove it. While the vast majority of people who live in the region call it Derry; in some circumstances (e.g. on official occasions), the pro-Unionist-UK community identifies the city with the toponym "Londonderry," though the municipality is officially administered by the "Derry City Council." In Donegal, in the city's westwards natural hinterlands, and rest of the Republic, "Derry" appears on signage followed by the Irish word "Doire" referring to the pre-colonial name of the region—"Oak Grove." The name Derry is reflected in media and government statements in the Republic of Ireland, while Londonderry is the appellation employed by British media and government sources; such usages are often mirrored within the nationalist, republican, unionist and loyalist communities within Northern Ireland. The place-name "Doire" was changed from the Anglicised "Derry" in 1613, during the colonial land confiscations and Plantation of Ulster, with English and Scottish settlers, to "Londonderry," so as to reflect the establishment of the city and (re)settlement of the surrounding lands by the commercial London guilds as part of the English / British imperial project. This laid the foundations for the "us and them" division reflected in tying ethnic identities to the practice of (re)naming and (re)settling places. Such "Divide and Rule" tactics were nothing new, being practiced by imperial polities ranging from ancient Rome to the modern European colonial powers as employed by British rulers, creating cleavages between Ireland's native Roman Catholic Gaelic-speaking population, and the "planted" English and Scottish newcomers. Such strategies were also witnessed in Southern Africa, the Indian subcontinent including Ceylon, the Northern, Meso and Southern Americas, Indonesia, and elsewhere. According to Royal Charter (1662), the city's official name is Londonderry that was reaffirmed in a High Court decision in 2007, due to petitions for its "official name" to be changed to Derry. Attachment to the name Londonderry for many unionists and loyalists, reflects emotive connections to their histories and foundation mythology—the colonial "settlement" experience and particularly the role that the city played from the 17th to the twentieth centuries, while managing to just remain geographically and (geo) politically within the boundaries of the six counties partitioned from the rest of Ireland's twenty-six counties in 1921 with the creation of two states, representing just under 18% of the territorial area of the island, but remaining as a province of the United Kingdom. During WWI and WWII, Londonderry / Derry, as well as providing recruits for the UK forces, also offered an excellent harbour, port and industrial facilities that played important roles for the combined British, American and Canadian forces. Such provisions were not available in Éire after the Free State's declaration, like Switzerland and Sweden of neutrality in 1939. In Unionist narratives, Londonderry like Belfast proved its loyalty to the UK, especially being bombed by the Nazi Luftwaffe during WWII. In nationalist mythologies and collective memory, putting London before the name Derry, recalls the negative aspects of the British imperial project, including the attempted undermining or attacks on the identity of a people and their sense of place. In more recent history, particularly since the resurgence of the conflict known as the "Troubles" 1968–1998, the place-name Londonderry was perceived to denote state facilitated oppression of nationalist and Catholic Northern Ireland citizens, collectively a demographic minority in the province's political and commercial institutions. In particular, the city and it's disputed toponymy came to international attention after 1968 due to the communication revolutions in media. Televised marches led by the Northern Ireland Civil Rights Association brought global attention to discrimination against Catholics and nationalist enforced by the Unionist administration's system of gerrymandering or rigging electoral boundaries after the partition of the (Irish) island and foundation of the "state" of Northern Ireland in 1921. The Association's activism mirrored civil rights marches taking place elsewhere during that period—ranging from Montgomery and Selma in the USA to Sharpeville and Soweto in South Africa.

In 1969, the "Battle of the Bogside" in Derry / Londonderry took place, with the establishment of "Free Derry," a no-go area for British state authorities that lasted only a few days, but was reminiscent of the bloody Paris Commune events in 1871. What did last, however was an enduring legacy of associated place names, monuments and murals, expressing socio-political, economic and cultural grievances from the disenfranchised nationalist and Catholic communities. Also in 1969, the paramilitary Provisional IRA (Irish Republication Army) was created, as had Loyalist paramilitary groups including the Ulster

Volunteer Force (UVF) in 1966 and the Ulster Defence Association (UDA) in 1971, both in collusion with elements within the Royal Ulster Constabulary (Northern Ireland police force) and allegedly the British military, intelligence and political establishments which lead to a 30 year "dirty-war" euphemistically known as the "Troubles." It could be argued that the evolution in democracy and application of the rule of law in many areas was different for Northern Ireland in contrast to the rest of the UK. On 30 January 1972 known as "Bloody Sunday" 13 unarmed civil rights marchers were shot dead by British paratroopers in the Bogside area of Derry, with another 13 civilians wounded. After several disputed official enquiries, including those embedded in the internationally legally binding Belfast / Good Friday Agreement (1998), it has now been "officially" established that the civilians were unarmed. However, the victims' families, supporters and Irish government are still calling for the principals responsible for the killings to be brought to trial for murder. The 1998 Agreement was greatly facilitated by British and Irish European Union membership, collaboration, institutions and norms that established open borders and free movement between both jurisdictions. The 1998 Agreement was much aided by US President Clinton's administration (1993–2001).

English remains the only official language in Northern Ireland and Irish has no official status, unlike in the Republic. Local councils may however choose to follow a bilingual approach. The Local Government Miscellaneous Provisions, Northern Ireland Order of 1995 bestows the power on councils to erect street-name signage in a language other than English, but the name must always be present in official English also. Unlike the Republic, there is no statutory imperative in Northern Ireland to use Irish-language (Gaelic) forms of place-names; nonetheless, this may change in the future with the creation and passage of a Language Act. Efforts and projects continue to facilitate, research and present the shared community heritages of toponymy in Northern Ireland.

Many disputed physical monuments and statues (reminiscent of the "Lost Cause" Confederate memorials that dotted the southern USA, after the American Civil War) find confluence in the eyes and mind of the perceiver and their experience of the landscape. Likewise, place-names utilized in "official spheres" may be seen as offensive to large sections of a population, due to the systemic subliminal power messages and attitudes being conveyed, embedded in commemorative statues, monuments and landscapes. The name "Derry" has been used in the names of the local government district and council since 1984, when the council changed its name from "Londonderry City Council" to "Derry City Council." This also changed the name of the district, which had been created in 1973 including both the city and surrounding rural areas. Reforms of local government in 2015 merged the district and council with the Strabane district to form the Derry City and Strabane district and council. In the first decades of the 21th first century, "Derry / Londonderry" was once again a peaceful and vibrant border and regional city with a thriving cultural and hospitality industry, but the issue of place names remains and lurks just beneath the surface of civil society. Such a disputed heritage can be exploited negatively by dissident republican and loyalist factions alike stoking up populism and especially in the context of the unfolding realities being experienced on the ground due to the implementation of the 2016 Brexit Referendum results, in which the UK electorate voted to leave the European Union by 52% to 48%, while in Northern Ireland, 56% of the electorate voted to remain in the EU as did 62% of the electorate in Scotland. Sources: Curl, James Stevens (2001). "The City of London and the Plantation of Ulster". BBC1 History Online. Archived from the original on 13 September 2011. Mícheál B. Ó Mainnín, 'Empowering multilingualism? Provisions for place names in Northern Ireland and the political and legislative context' in Robert Blackwood and Deirdre A. Dunlevy (eds) Multilingualism in Public Spaces: Empowering and Transforming Communities, (Bloomsbury 2021), 59–87. Gerry O'Reilly 2022.

Questions

- Appraise the statement that the Arts, including theater, can contribute to toponymic and hydronymic research?
- Brian Friel's play *Translations* (written in 1980) is set in 1833 in Donegal, Ulster, when Ireland was being mapped by the British Ordnance Survey. What cultural and language tensions do you observe between the local people and the foreign ordnance surveyor; and why?
- When *Translations* was produced in 1980, civil and military strife was ongoing (1968–1998) in six of the nine counties of Ulster under UK rule, and adjacent to Donegal in Ulster, the most northerly county in the Republic of Ireland. Appraise the role of toponymy in the colonial project and post-colonial discourses, in order to enhance peace-building, as now promoted in the Good Friday Peace Agreement and also SDG-16.
- 'This chapter discusses *Translations* in reference to past debates between Friel's 'fictive character' and geographer John Andrew's 'historic' perspectives on the impact of the British Ordnance Survey not only in Ulster, but on a wider linguistical landscape.' Discuss.
- How does Friel's *Translations* contextualize within other literary geographies detailing global nineteenth century industrialization processes in addition to the current post-Brexit milieu of Northern Ireland as a means to address the complexity of toponymy, landscapes, identities, and senses of place at play in twenty-first century Ulster?
- Reflect on place naming in your home area, or region, or country, and how this may be presented in the Arts.

Recommended Reading

Doyle A (2015) A History of the Irish language from the Norman invasion to independence. Oxford Linguistics, Oxford

Duffy PJ (2007) Exploring the history and heritage of Irish landscapes. Four Courts Press, Dublin

Duffy PJ (2011) Ordnance Survey maps and official reports. In: Murphy J (ed) The oxford history of the Irish book, volume IV: the Irish book in English, 1800–1891. Oxford University Press, Oxford, pp 553–562

Dunlevy DA (2021) Place names and the complexity of language recognition in Northern Ireland. In: Blackwood R, Dunlevy DA (eds) Multilingualism in public spaces: empowering and transforming communities. Bloomsbury, New York, pp 137–157

Giraut F & M Houssay-Holzschuch (2022) The Politics of Place Naming: Naming the World. Wiley

Mac Giolla Easpaig D (2009) Ireland's heritage of geographical names. In: Geographical names as a part of the cultural heritage, Wiener Schriften zur Geographie und Kartographie, vol 18, pp 79–85. https://www.logainm.ie/en/res/92

Mainnín Ó, Mícheál B (2021) Empowering multilingualism? Provisions for place names in Northern Ireland and the political and legislative context. In: Blackwood R, Dunlevy DA (eds) Multilingualism in public spaces: empowering and transforming communities. Bloomsbury, New York, pp 59–87

Smyth WJ (2006) Map-making, landscapes and memory a geography of colonial and early modern Ireland c.1530–1750. Cork University Press, Cork

Notes

1. Friel, B. *Translations*, London: Faber and Faber, 1981, pg.11.
2. Grillo, R. D. *Dominant Languages: Language and Hierarchy in Britain and France.* Cambridge: Cambridge University Press,1989. Pg. 100.
3. Ibid.
4. Friel, *Translations*, 11; 15; 17; 10.
5. Ibid., 34.
6. Bowen, C. A Historical Inventory of the *Dindshenchas, Studia Celtica* 10–11, (1975/76). Pg. 115.
7. Kavanagh, P. *Patrick Kavanagh: Man and Poet,* (ed.) Peter Kavanagh, Maine: National Poetry Foundation. University of Maine at Orono, 1986, p. 15.
8. Kearney, R. "Language Play: Brian Friel and Ireland's Verbal Theatre." *Studies: An Irish Quarterly Review* 72, no. 285 (1983): 20–56. http://www.jstor.org/stable/30090491. Pg. 36.
9. Quoted in Dantanus, U. *Brian Friel: A Study*. London and Boston: Faber and Faber, 1988.Pg. 87.
10. Quoted in *Brian Friel: Essays, Diaries, Interviews: 1964–1999*, edited by Christopher Murray London: Faber and Faber, 1999. Pg. 100.
11. Ibid., p. 84.
12. Quoted in *The Cambridge Companion to Brian Friel*, edited by Anthony Roche, Anthony, Cambridge: Cambridge University Press, 2006. Pgs. 104–116.

13. Friel, B., Andrews, J. and Barry, K. Translations and a paper landscape: Between fiction and history. *The Crane Bag* 7, no. 2 (1983): 118–124. Pg. 122.
14. Ibid.,123.
15. Andrews, J.H. Notes for a Future Edition of Brian Friel's "Translations". *The Irish Review* (Cork), Winter, 1992/1993, pp. 93–106. Pg. 103.
16. Ibid., 93.
17. Pocock, D.C., *Humanistic Geography and Literature: Essays on the Experience of Place*, New Jersey: Barnes & Noble Books, 1981, p. 15.
18. Meinig, D.W., Geography as an art. *Trans. Inst. Br. Geogr. N.S.* (1983) 316.
19. Evans, E.E., *Irish Folk Ways*, London: Routledge 1957, p.28.
20. Quoted in, Mulholland, J. 'Glenties is the stage' Brian Friel's Donegal, *The Irish Times*, 3 October 2015 < https://www.irishtimes.com/culture/brian-friel/glenties-is-the-stage-brian-friel-s-donegal-1.2376262 >
21. Quoted in, O'Toole, F. 'Brian Friel's interview with Fintan O'Toole: 'I'm not really very good at this kind of question.' *Irish Times* 2 October 2015. < https://www.irishtimes.com/culture/brian-friel/brian-friel-s-interview-with-fintan-o-toole-i-m-not-really-very-good-at-this-kind-of-question-1.2372705>.
22. Friel, B. *Philadelphia, Here I Come!* London: Faber and Faber. 1964.
23. Corbett, T. *Brian Friel: Decoding the Language of the Tribe.* Dublin: The Liffey Press. 71–72; Friel, B., *Dancing at Lughnasa*, New York: Dramatists Play Service, Inc., 1993. Pg. 19–20.
24. Ibid.
25. Place-name spellings referenced in instances from *logainm.ie* < https://www.logainm.ie/en/>.
26. Bachelard, G. *The Poetics of Space.* London: Penguin. Pg. xxxii.
27. Dickson, D., Keogh, D. and Whelan, K., *The United Irishmen: Republicanism, Radicalism and Rebellion.* Dublin: The Lilliput Press, 1993.
28. Friel, *Translations*, 67.
29. Haraway, D. *Simians, cyborgs, and women: The reinvention of nature.* New York: Routledge 1999.
30. Friel, *Translations*, 17–18.
31. Ibid., 35.
32. Quoted in McGrath, Francis C. Language, myth, and history in the later plays of Brian Friel. *Contemporary Literature* 30/4. 1989, 534–545. Pg. 543.
33. Friel, quoted in Kearney, *Language Play*, 26.
34. Friel, *Translations*, 55.
35. Ibid., 68.
36. Ibid., 64.
37. Ibid., 66.
38. Ibid., 39.
39. Withers, C. Authorizing landscape 'authority', naming and the Ordnance Survey's mapping of the Scottish Highlands in the nineteenth century. *Journal of Historical Geography* 26 /4. 2000: 532–554.
40. Heslinga, M.W. *The Irish Border as a Cultural Divide.* Assen, 1971.
41. Hewitt, John Harold. *Ancestral Voices: The Selected Prose of John Hewitt.* Blackstaff Press, 1987.
42. Hewitt, J. H., 'Ulster Names' in Ed. Sean Dunne, *The Ireland Anthology*, New York: St. Martin's Press, pg. 187.
43. Breen, R. *Place and Displacement in the Works of Brian Friel and Seamus Heaney*, Unpublished Ph.D. Dissertation, University of Warwick, 1993, pg. 48.
44. Ibid.
45. Ibid.
46. Quintelli-Neary, M., Establishing Boundaries in the Irish American West. *Études Irlandaises* 29, no. 2 (2004): 22.
47. Ibid., 44.
48. McMurtry, L. *The Lonesome Dove Chronicles.* N.Y.: Simon & Schuster, pg. 1144.

49. The *Lonesome Dove Chronicles* is an omnibus containing the novels *Dead Man's Walk* (1995), *Comanche Moon* (1997), Lonesome *Dove* (1985) and *The Streets of Laredo* (1993).
50. McMurtry, *The Lonesome Dove Chronicles*, 1203.
51. Walton, R.J., *Larry McMurtry and the Victorian Novel.* College Station: TAMU P, 1994.
52. Wardropper, B. W. " 'Don Quixote': Story or History?' *Modern Philology* 63:1 (1965), 1–11. Pg. 5.
53. Brian Friel, Programme Note, *Making History*. Derry: Field Day Theatre Co. 1988. Pg. 7.
54. Wardropper, *Don Quixote*, 11.
55. Friel, B., Rea, S. *et. al., Ireland's Field Day Theatre Company*. 1985/1986. Pg. vii.
56. Friel, B., "Extracts from a Sporadic Diary (1979): Translations," in Murray, *Brian Friel*, p. 75.
57. McGrath, *Language, myth, and history in the later plays of Brian Friel*, 535.
58. Friel, Extracts *from a Sporadic Diary*,74.
59. Whelan, K., A *Geography of Society and Culture in Ireland since 1800*, Unpublished PhD dissertation. Dublin: University College Dublin, 1981. Pgs. 185;188.
60. Pierce, D., Light, *Freedom and Song: A Cultural History of Modern Irish Writing*. New Haven & London: Yale University Press, 2005. Pg. 194.
61. McKay, S. The Irish language can give us all a sense of home- if we save it from sectarianism, *The Guardian*, 10 August 2011; *Turas*, East Belfast Mission, The Methodist Church in Ireland < https://www.ebm.org.uk/turas/>.
62. Friel, "In Interview with Mel Gibson (1991)," in Murray, *Brian Friel*, 142.
63. Friel, "In Interview with Desmond Rushe (1970)," Ibid., 27.
64. Friel, Quoted in Kearney, *Language Play*, 27.

The Overlaid Past: The Politics of Space and Memory in Gibraltar's 'Doubling' Street Naming Principle

Jennifer Ballantine Perera

> *When he arrives back home and confronts the radically remodelled city, its reconstructed parts rush at him 'de sopetón'—como un tortazo en la cara. The huge office blocks, the multi-storey car parks, the expanded and refitted stadiums, the luxury penthouses for tax dodging millionaires. It all confuses him, disorients him, increases the sense of strangeness he already feels being back in his hometown after an absence of decades. At times like these you could say that he is aware of two Gibraltars – the one that lies before him and the Gibraltar that exists in his mind, the two of them fusing together like a pair of overlaid plastic transparencies.*
> M.G. Sanchez, 'Ruina', 2021 (157)

Abstract This chapter engages with the resilience of traditional streets names at Gibraltar together with the resistance they posed to the implementation of official street names. The permanence of traditional names, in use pre-1704 until the mid-nineteen seventies, raises questions regarding Edward Said's theory of imaginative geographies regarding the colonially informed mapping of a territory. Whilst this (did) level of mapping was exercised at Gibraltar post-1704, I suggest that this process proved to be incomplete when it came to the territory's urban streetscape, a space which had already been imaginatively mapped by the inhabitants though a linguistically and culturally codified street naming principle. These traditional street names existed only in their imagination as they did not appear on street signs, yet they were extremely effective in disrupting access to a military based settler community. As newcomers to the territory, this sector was linguistically and culturally ill equipped to decipher the norms underpinning these names, rendering this group quite unable to easily find locations in the territory's city centre. Testament to this was the need to publish lists of street names even after official street names were placed on street corners in the 1870s. However, rather than leading to greater access or effective mapping, these publicly displayed official names created a doubling through the

J. B. Perera (✉)
Gibraltar Garrison Library, Gibraltar, Gibraltar
e-mail: j.ballantine@gibraltargarrisonlibrary.gi

University of Gibraltar, Gibraltar, Gibraltar

© The Author(s), under exclusive license to Springer Nature Switzerland AG 2023

G. O'Reilly (ed.), *Place Naming, Identities and Geography*,
Key Challenges in Geography, https://doi.org/10.1007/978-3-031-21510-0_24

co-existence of two parallel albeit very different sets of nomenclature. This (piece) chapter aims at unpacking the terms under which traditional street names were generated and sustained, together with the challenge they posed to a colonially informed mapping necessary for fixity and control over the territory. Attention will be drawn towards initiatives to publish lists bearing both official and traditional street names; the aim was to decodify if not decentre these traditional names, but at a practicable level, these lists were deemed as necessary tools to be able to navigate Gibraltar's streets. Whilst difficult to fully measure their success, the need to publish lists over a prolonged period of time offers insight into the limitations of a colonially informed mapping process when faced with a far more powerful mapping principle that existed only in the minds of Gibraltarians.

Keywords Topography · Imaginative geographies · Naming principles · Doubling · Colonial mapping

Objectives

- To explore the resilience of traditional street names in Gibraltar.
- To view how Gibraltar's streetscape had already been imaginatively mapped by the inhabitants through a linguistically and culturally codified street naming principle before official names were attributed.
- To see how street names existed orally but did not appear on street signs, so disrupting access to a military-based settler community that was linguistically and culturally ill-equipped to decipher the norms underpinning these names, rendering this group quite unable to easily find locations in the territory's city centre.
- To understand the doubling of street names through the co-existence of two parallel different sets of nomenclature.
- To unpack the terms under which traditional street names were generated and sustained, together with the challenge they posed to a colonially informed mapping necessary for fixity and control over the territory.
- To evaluate initiatives to publish lists bearing both official and traditional street names with the aim to decodify them.

Introduction

In a short piece entitled 'Ruina' (2021),[1] Gibraltarian author M G Sanchez writes of the disconcerting reality faced by his protagonist when he returns to Gibraltar after many years away. Sanchez describes this rather accurately as a doubling that comes about between the prevailing memory of a Gibraltar of the past, and the one his protagonist faces upon his return, which is very different to that of his youth and which he finds disconcerting. Equipped with a navigational tool, his memory, he sets about to

[1] M.G. Sanchez, 'Ruina', *SureS: Gibraltar en SureS*, Tangiers: Litograf, 2021, p. 157.

revisit the old haunts of his childhood which he remembers by their traditional albeit unofficial names, rather than through the official street signs situated on buildings and street corners. This, together with the unfamiliarity of a Gibraltar topographically transformed though waves of land reclamations and newbuilds, leads Sanchez to acknowledge the presence of two Gibraltars, the physically changed one, and a historical one retained only in memory. These two versions of the same place come alive if only in his imagination 'like a pair of overlaid plastic 'transparencies'. The co-existence of both versions invariably creates a sense of loss, but they also denote a messiness as the past resists yielding to the present. Memory and nostalgia are at play here, but in referring to street names as a central motif of his short story, Sanchez enters into a very recent drive at Gibraltar to revive traditional street names. These now have limited currency, but they are being reclaimed as a means to recover an aspect of Gibraltarian social history that aims at retelling a history that had been pushed to the margins to give way to the primacy held by the military and colonial histories that have so dominated the telling of Gibraltar's story. The present-day recovery of these names, however, seems to me to be devoid of the significance of a traditional street naming principle that ultimately decentred, during a prolonged period of time, the official, colonial naming principle for these very streets. Whilst it is the case that many of these traditional names are no longer in use as before, they represented a counter-mapping of the urban centre which, as discussed below, served to disrupt a British colonial naming principle within this densely populated space.

The traditional naming of Gibraltar's streets originated from a pre-1704 British Gibraltar. Street signs were not in existence at that stage and not until the 1870s, and so what we see is an imaginative mapping of Gibraltar by its inhabitants, with names of places memorialised and passed down orally through generations thus keeping the imaginative process going. There was also a high degree of subjectivity in how names where generated but as a general rule, these traditional names were based upon fixing the location of streets against landmarks in that vicinity, historical points of reference, the names of prominent citizens or revered Roman Catholic saints, with this naming principle reinforced on a daily basis though passage through these streets. These names did not have a physical presence as they only existed in the inhabitant's memory. In reality, Gibraltar's streetscape had effectively been imaginatively mapped and named by its inhabitants, with this plan of the city existing only in their minds, but nevertheless functioning as an efficient memory driven GPS.

It would be too easy to say that all this changed once Gibraltar is taken by Anglo-Dutch forces in 1704, as that was not to be the case, because what occurs instead is a grappling by the mainly British garrison to decodify and dismantle a rather nebulous naming principle already in place and resistant to the imposition of a paradigm shift. Gibraltar's urban streetscape had in fact become quite inaccessible to the Garrison and their personnel; not because physical access had been blocked but because Garrison personnel were unable to penetrate the highly codified street naming principle necessary to find locations. Whilst this unknowability factor is commonplace during the early stages of a colonial venture, as was the case for the newly arrived Garrison at Gibraltar, what we see instead is a continued uncertainty when it came to finding locations in the city centre. The fact that this state of play remained for over 250 years, also leads us to suppose that traditional street naming had primacy over the official. During this time service personnel and settlers were ill equipped to navigate these streets. This was not meant to happen, and the resilience of these names seems quite remarkable when we consider the efforts made by the garrison to decentre them. In fact, once official street signs are put up in the 1870s, a doubling effect is created with both sets of names coming into use, leading to confusion and importantly, exacerbating the matter further. This last raises questions as to why these colonially informed mapping initiatives failed.

To this end, I shall be drawing upon a number of lists produced in response to the persistence of these street names and the difficulty these posed when finding ones way through Gibraltar's streets. The fact that some of these lists were produced after official street names were established is suggestive of an incomplete colonial mapping of Gibraltar, and a nod towards the primacy held by traditional street names. Unpacking the contexts for these lists is important when it comes to understanding their historical timeliness, their impact and indeed, their limitations. My aim is to offer an insight into the terms under which traditional street names are generated and sustained as a means to understand the strong foothold they had. Given the disruptiveness of these street names, the question begs as to whether they functioned at a deliberately subversive level manner centred around maintaining a civilian stronghold within the urban centre.

The Mapping of a British Gibraltar: 18th Century

> For the first time, with Baudelaire, Paris becomes the subject of lyric poetry. This poetry is no hymn to the homeland, rather the gaze of the allegorist, as it falls on the city, is the gaze of the alienated man.[2]
>
> Walter Benjamin, 'Baudelaire and the Streets of Paris' *Arcades Project*.

My point of departure in respect to naming lists is determined by the 1704 taking of Gibraltar by Anglo-Dutch forces and the beginning of a British Gibraltar. This span of over three hundred years is significant as it brings in to play a colonially informed mapping of the territory with the aim to transform Gibraltar into a garrison and fortress. Whilst the rationalisation of the territory is an imperative, a list compiled in c. 1704–1711 comes to mind when considering the terms under which the British set about establishing their imprint over a small piece of land of which they knew little about, except for the strategic aims underpinning the taking of Gibraltar.

By way of background, when Gibraltar is taken in 1704 as part of the Spanish War of Succession (1701–1714), a topographical rupture takes place. Gibraltar, as

[2] Walter Benjamin, *Arcades Project*, trans. Howard Eiland & Kevin McLaughlin, Cambridge, Massachusetts & London: Harvard University Press, 1999, p. 10.

a British Fortress and later a colony (1830),[3] is born from the territorial boundaries established in 1704, later to be ratified by the 1713 Treaty of Utrecht. Prior to this, Gibraltar and El Campo had been one and the same area. The town of Gibraltar, within the city walls, was the dormer town whereas the economy and sustainability were derived from its 'campo', that is, its cultivations fields that spread for some considerable miles beyond the hermitage of San Roque.[4] The splitting of this region in two brought about a loss of territorial integrity; the drawing of a previously non-existent line demarcating a British Gibraltar from mainland Spain had immediate impact on the pressures placed upon the garrison. The culture informing the streetscape, however, remained, and whilst most of the Spanish inhabitants left upon capitulation, the linguistic predominance of the Spanish language crept back-in through a porous border. Ongoing migrations from the Mediterranean and North Africa also brought new cultural influences and a high degree of polyglotism amongst the civilian population.

Since her taking, and during the years of some ambivalence between 1704 and 1713, Gibraltar was de facto under British rule. Testament to this is the 1706 declaration by Queen Anne of Gibraltar's port as Open, together with the steps taken by the British forces to map-out the town for defence and revenue raising purposes. This process was physically (through the building of fortifications) and discursively articulated as a means to rationalise, measure, and indeed, to understand Gibraltar, the city and the territory. Lieutenant Colonel Thomas James was one such recorder, whose efforts led to the discursive mapping of Gibraltar in his *History of the Herculean Straits, now called the Straits of Gibraltar* (1771).[5] James offers descriptions of the territory and town through the use of prominent landmarks, illustrating some of these features with plans of fortifications and boundaries. Interestingly, whilst he offers some descriptive passages on the city, his naming process is underdeveloped in that he only refers to established landmarks or topographical features from Moorish or Spanish time to underpin his mission. James reverts to classics and academic references to understand Gibraltar's strategic positionality and its history. His is a scholarly attempt to 'fix' Gibraltar at what were the early enough stages of her taking, but James does not get down to the messiness and uncertainty of this venture. We do, however, see a high degree of confusion in an earlier attempt at mapping

[3] This takes place with the passing of the Fifth Charter of Justice in 1830 which created a Supreme Court and civil jurisdiction in a territory that was until then under martial law. The introduction of this jurisdiction transforms Gibraltar into a colonially run territory although the term colony is not mentioned in the document.

[4] Francisco Maria Montero *Historia de Gibraltar y Su Campo*, Cádiz: Imprenta de la Revista Médica, 1860, and *Documentos Del Archivo Municipal de San Roque (1502-1704)*, Edición de Adriana Pérez Paredes, Algeciras: Ilustre Ayuntamiento de San Roque, Delegación de Archivos, 2006. See also Jennifer Ballantine Perera, 'El desarrollo de la población de Gibraltar durante los siglos XVIII y XIX dentro del sistema dual fortaleza-colonia', *Gibraltar y los Gibraltareños: Los originéis la situación de un enclave estratégico en las puertas del Mediterráneo* (pp. 65–90), Eds., Enrique Ojeda Vila & Rafael Sánchez Mantero, Sevilla: Fundación Tres Culturas, Colección Ánfora 3, 2008.

[5] Lieutenant Colonel Thomas James' *History of the Herculean Straits, now called the Straits of Gibraltar*, Vol II, London: Charles Rivington, 1771, p. 92.

Gibraltar's streetscape. Written by an unnamed soldier, 'A List of the Houses which are possessed by the Inhabitants of Gibraltar, excluding those of the Old Inhabitants, on the Capitulation with the Prince of Hesse with what rents are Monthly paid to the Governor', c.1704–1713, this document offers some insight into a most remarkable if not tentative process. As seen in Fig. 1, it soon becomes evident that the premise upon which this list is taken is problematic, as the unnamed list taker seemed to be new in town, and it becomes apparent that he struggled to find streets and people in a place he knew little about. His lack of local knowledge is a major stumbling-block, as we enter into a process as long-winded as the title suggests, in which we view these streets though his eyes. Our soldier can only but be defined as an amateur detective on a mission of discovery, with this endeavour placing him as the heart of Gibraltar's cityscape. Although an identifiable figure as he walks down these streets, his alienation within these surroundings shows. I would suggest that there is some affinity between this soldier and Walter Benjamin's articulation of the flâneur figure. Our soldier and the flâneur are both features in their respective environment because they stand-out as different from everybody else, with both occupying a liminal space that is neither inside nor outside of this experience. Their lack of embeddness defines them as individuals who are alienated from their surroundings, although there is a difference here in that the flâneur's is a native Parisian, and his aloofness is a choice. On the other hand, the alienation experienced by our soldier is not a choice, but rather one that presses hard upon him. All he is faced with is the unfamiliarity of a place he neither knows nor belongs to. Our solider is an out of place feature in the streets he now occupies. He is unable to communicate in a meaningful manner as he lacks the language skills to engage with many of the inhabitants he encounters and for whom English was not the *lingua franca.*

The list taker is evidently hampered by this language barrier, but this does not deter him from attempting to identify people and their location, all this in the absence of a map or street signs that could assist. It must be noted that his list does not include a line drawing of the city streets he navigates, which he could have populated with data as he progressed. Instead, we see a discursive practice in which he identifies those that are known to him by name, if only to situate others located close-by. A sample of the list reveals his haphazard methodology:

> A Gardener near Watergate.
>
> Michael the Gardener (besides all greens[6] for the Governor's House).
>
> Another Gardener next to him.
>
> A Geonese near Machada's house.
>
> The lame shoemaker.
>
> The Gardener in the Moats at South Gate.
>
> The Confectioner near the Great Church.

---

[6] The 'greens for the Governors house' could be the current location of the Gibraltar Garrison Library, which was built on the site of the Governor's Garden, a large plot of cultivation ground where life stock was also kept. We can see a small corner of this on Fig. 1 although most of the ground is covered by the plan's legend. See also Dorothy Ellicott, *Place Names in Gibraltar: Memories of Long Ago*, Gibraltar Tourist Board, n/d, c.1970s, p.14.

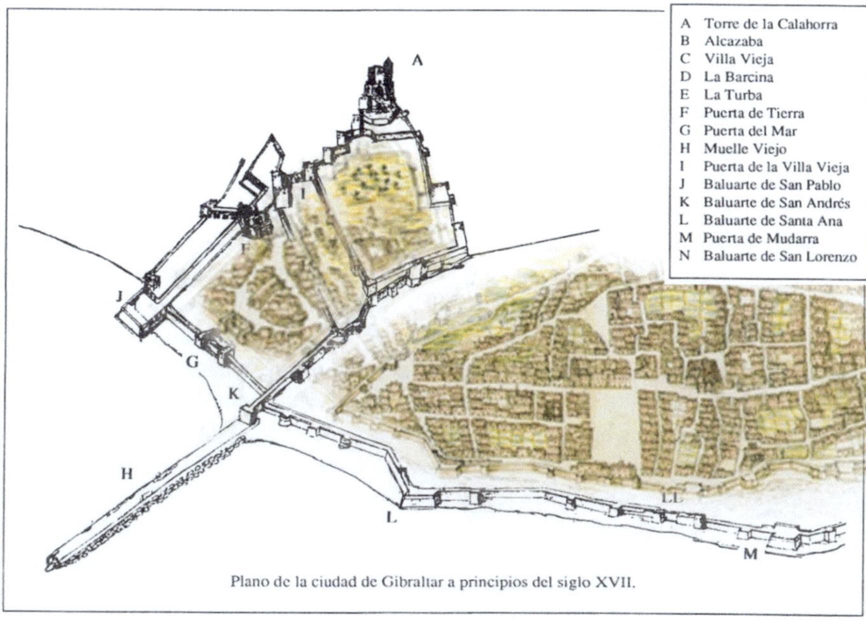

Fig. 1 Plan of the Town of Gibraltar by Portillo. In Portillo's plan *La Turba* seems like an empty space, but the overlay of the 1627 Bravo de Acuña map offers a far more accurate interpretation of the density of *La Turba* as seen the main area centre right of this plan

> 28 Jews Shops in the Great Street.
>
> All other Jews cannot be known but need to pay.
>
> 2 English Taverns over against La Merce.
>
> An English tavern near Haines.
>
> A Spanish barber next to him.
>
> An English tavern near Peter the Baker.
>
> Sergeant Stokes in the rabble.
>
> The Dutch Musick House Another Dutch house next to it.[7]

And so, it continues, referring to the landmarks he is aware of, such as churches, La Merce, and the Great Church, whenever possible. He mentions the Great Street, which is clearly identifiable in Fig. 1 as the long street that cuts across *La Turba*, which is referred to in the list as 'the rabble', which is where Sergeant Stokes lives.

[7] 'A List of the Houses which are possessed by the Inhabitants of Gibraltar, excluding those of the Old Inhabitants, on the Capitulation with the Prince of Hesse with what rents are Monthly paid to the Governor', n/d c.1704–1711. Gibraltar Garrison Library, 314/4689. Whilst the list is not dated the fact that a Dutch presence remains suggests that the list was compiled in the years following 1704 and certainly before 1713.

Antonio Torremocha Silva in his notes on Portillo's *Historia de Gibraltar* (1620–1622), suggests that the name *La Turba*, comes from the Latin, which denotes multitudes of people, noise and riotous behaviour. This is a description which corresponded with the activities taking place in the most highly populated area of Gibraltar where the high street and taverns were located. It seems likely that the more recent military presence exacerbated the already tumultuous culture of the zone, leading to it being referred to as 'the rabble'.[8] Tito Benady, however, posits that the name comes from the Arabic, *Turba al-Hamra* (red hill) on which *La Turba* was built.[9] As such, the naming of this area is already loaded with an historically informed naming backdrop, none of which was accessible to our soldier. Streets here were narrow, with alleys leading to many different areas as seen in Fig. 1, which could account for the soldier's trepidation in delving deeper into a space that seemed unruly and prohibitive. Instead, he holds on to the names of those known to him, but perhaps not so for others reading this list, with these names becoming his signposts to determine the approximate location of those living close by. The use of physical characteristics also become markers both of identity and location, such is the case for the 'lame shoemaker', a disability that would be visible. Language clearly served as a barrier given the demographic of the residents, and this, together with our soldier's lack of local knowledge, shows. The streets of Gibraltar seem impenetrable to him to the extent that he caveats the unknowability factor with comments such as 'all other Jews cannot be known but need to pay' and 'besides which many Magazines and other Place which I cannot find out'.[10]

Admittedly, our soldier was not a surveyor or a cartographer, but this also draws attention to the messiness of what was for Gibraltar a shift away from Spanish rule and the commencement of a British colonial venture.

Whilst these initial steps are seemingly ad hoc, they are also revealing of a process described by Edward Said as imaginative geography, a practice that sets about to transform something that is unknown into something that is through the discursive representation of a space. The outcomes become overlaid onto existing physical and social formations, transforming these in the process. We could perhaps consider the endeavours of our list taker as a first move towards creating a narrative informed by these aims and power relations—that between the authority directing the discourse/mapping, and the power it seeks to have over both territory and people.[11] Still, as I suggest in my opening paragraphs, I am not entirely convinced that this particular effort at mapping out the urban centre was fully achieved, especially if

---

[8] Antonio Torremocha Silva, 'Introduction' of Alonso Hernández Del Portillo, *Historia de Gibraltar (1610–1622)*, Algeciras: Centro Ascociado de le UNED, n/d, p. 53.

[9] Tito Benady, The Streets of Gibraltar: A Short History, illustrations George Felipes, Grendon, Northants: Gibraltar Books, 1996, p.10.

[10] 'A List of the Houses which are possessed by the Inhabitants of Gibraltar, excluding those of the Old Inhabitants, on the Capitulation with the Prince of Hesse with what rents are Monthly paid to the Governor'.

[11] Edward Said, *Orientalism,* Penguin Modern Classics, 2019. See also A. Dawson, (2013). Edward Said's Imaginative Geographies and the Struggle for Climate Justice', *College Literature*, 40(4), 33–51. http://www.jstor.org/stable/24543185.

we consider that this zone had already been imaginatively named and mapped by its inhabitants. What is clear is that Gibraltar's streets seemed quite impenetrable, and whilst this may have been because of the soldier's inexperience and his recent arrival, his inability to gain deeper and more relevant knowledge was hampered by the fact that this was already a highly codified space that pre-dated his posting to Gibraltar. Street names were based on any number of references, cultural colloquial, historical, linguistic, the names of prominent inhabitants and so it would have been difficult to find connecting strands as most if not all of these names pre-dated the arrival of the British. Local knowledge and a certain degree of cultural embeddedness would have been essential to find one's way around, and he unfortunately lacked both.[12]

An important aspect of 'A List of the Houses which are possessed by the Inhabitants of Gibraltar' pivots on the balancing act that takes place between what could be discovered, which was quite limited, and the unknowable. A similar problem is expressed in a report by Col. Bennett in regards to the state of Gibraltar since 1704. Bennett reported that he was experiencing difficulty in knowing how people continued to enter Gibraltar, and once embedded, the same difficulty arose when attempting to find them.[13] Anxiety over this is evident in Bennett's correspondence to the Secretary of State for the Colonies, as the Garrison was having great difficulties in housing and provisioning civilians once embedded. These could only be discovered if entered in the agent victualler's list.[14] Even this was a problem because once entered, these provisions for food were paid from garrison expenditure. Over a period of 6 years the cost of feeding inhabitants far exceeded that of the defence of the town and the repair to the fortifications.[15] The situation was, therefore, messy and remained so in spite of attempts to map out and name, imaginatively or otherwise, Gibraltar's urban centre. Instead, we see a garrison town that had cultivated a captive population of civilian inhabitants who lacked self-sufficiency as there was no agricultural land on which to cultivate or keep animals. This land, 'El Campo', was now on the Spanish side, which meant that all food was imported from Morocco and further afield. The on-going stream of migrations by individuals seeking work in what was considered a thriving economy under the British, placed enormous stresses upon provisions and indeed, housing. But we also see the resilience of these migrants in securing a foothold at Gibraltar, and this led to a level of subversion of the rules the British were attempting to implement. As such, the codified fabric of traditional names also brought protection as they could secrete themselves in places few could find.

[12] See Manolo Galliano, *A Rocky Labyrinth: The History & Memories of the Ramps, Lanes, Roads and Streets of the City of Gibraltar*, as yet unpublished but is due under the Gibraltar Heritage Trust imprint in 2022. References below to Manolo Galliano's work refer to the as yet unpublished manuscript.

[13] TNA/Colonial Office Original Correspondence Secretary of State 1705/1731—Letters from the Inspectors of the Colony to the Earl of Dartmouth with enclosures dated Lisbon 14th March 1713.

[14] *Ibid*.

[15] *Ibid*.

19th–20th Century Gibraltar

Traditional street naming remained in the hands of the civilian population long after these initial years, but the remarkable aspect here is that Gibraltar was otherwise managed by the Garrison and the Colonial Office through the implementation of legislation enforcing permits and ongoing list taking to determine who was entering Gibraltar.[16] The same could not be said about the urban centre, where services personnel and colonial agents remained perturbed by the lack of ready access to and through Gibraltar's streets. Gibraltarians had no issues in finding their way around as they were the generators of these traditional names which they understood and continued to transmit within their community. Whilst there is no evidence to suggest that this knowledge was withheld from the wider settler community, it is doubtful that translating these names to English would have resolved matters given their codification.

Street names did not appear in public spaces until around 1876. Still, even when official names are finally put up, little changes. These official street plaques were not English translations of traditional names but rather new names placed on buildings and street corners, and physically present for all to see. Yet this was not enough to break the dominance of the pre-existing names. In fact, attempts to supplant the traditional with official names was received with mixed responses. Tito Benady notes that the implementation of official names caused confusion amongst services personnel, who were now confronted with a doubling of names, with the overlaid official names creating some tension between both as Gibraltarians continued to use traditional names.[17] Manolo Galliano, however, has a different take on this, as he argues that the level of confusion was far more extensive, with Galliano suggesting that the display of signage posed a challenge to the Gibraltarian dependence on local knowledge as a navigational tool.[18] Whilst it seems to be the case that official names were intended as a strategy to supplant traditional names, this did not occur, at least not to the extent that was intended. I would further suggest that the cultural points of reference underpinning traditional street names remained a very powerful pull and primarily because they had been generated over hundreds of years of passage though these streets. These names were in essence embedded in the Gibraltarian imagination. They existed only in their imagination as they had no physical representation such as signage, but neither did they require signage as traditional names functioned at a very different level from the official given their cultural underpinning. Their reinforcement continued to be passed verbally across the community and across generations. It

[16] See T. J. Finlayson, *The Fortress Came First*, Grendon, Northants: Gibraltar Books Ltd, 1991; Stacie D. A. Burke and Lawrence A. Sawchuk, 'Alien encounters: The *jus soli* and reproductive politics in the nineteenth century fortress and colony of Gibraltar', *History of the Family* 6 (2001); and Gibraltar National Archives (GNA)/History of The Permit System 1704–1871.

[17] Tito Benady, The Streets of Gibraltar: A Short History, Illustrations George Felipes, Grendon, Northants: Gibraltar Books, 1996, pp. 1–2.

[18] Manolo Galliano, *A Rocky Labyrinth: The History & Memories of the Ramps, Lanes, Roads and Streets of the City of Gibraltar*, p.9 of manuscript.

seems evident that this was enough to keep this imaginative mapping alive. The fact that these names were not written anywhere also added to their subversive potency. Indeed, there was nothing to pull down as would have been the case if written on some form of signage. Gibraltarian culture was also very different to that of the British and so decentring traditional street names would not have been enough to eradicate them, as at the core of these names lay a hybrid Gibraltarian cultural identity that was different from the British settler's expectation.[19]

Still, traditional street names were eventually exposed, as suggested by Galliano in his reference to an 1884 publication, *Notes and Queries*, which carried an article on the streets of Gibraltar by the then Chaplain to the Garrison, the Reverend Robert Stewart Patterson. This piece was necessary because official street names had proved to be ineffective, and this compelled Patterson to compile a list in which he juxtaposed traditional street names alongside the official. His aim was to provide a tool for greater access through these streets, and perhaps also to decodify them by rendering traditional names visible on the page. In doing so he also drew attention to the doubling effect created by these competing sets of naming principles.

> The streets of this city for the most part bear English names which are painted in the usual manner on the corners, and by which they are known by the Garrison and English residents. Many of these date from the period anterior to the capture of the fortress by the British, and may be considered to be preserved in the columns of N. & Q. I give the English first, followed by the Spanish with the translation of the later.[20]

As part of the decodifying exercise Paterson offers a translation, but he also makes an important distinction between two sets of communities at Gibraltar; members of the Garrison and English residents, to whom the list is directed at, and Gibraltarians, denoted by the reference to the Spanish language and for whom the list was not necessary. This move directs our attention towards the primacy held by a local sector by virtue of their local knowledge of these streets. Patterson is in effect drawing attention to the presence not only of two naming principles, but also of two Gibraltars, now defined though a list that included both sets of street names to offer a cross-referencing navigational guide. A doubling nevertheless prevailed as in placing both names alongside each other brought visibility to a set of street names which had otherwise existed only in the imagination of Gibraltarians. Importantly, these did not need visibility to be reinforced, a point made by David Lowenthal who refers to De Quincy's view that the brain is a 'natural and mighty palimpsest' that '[piles] up everlasting layers of ideas, images and feelings'.[21] As such, the mind can be

[19] See for example George Borrow, who describes Gibraltar and its inhabitants as, 'A strange city […] inhabited by men of all nations and tongues, its batteries and excavations, all of them miracles of art…'. Imbued with an orientalist eye, Borrow found Gibraltar to be elusive and full of contradictions, but also hybrid to the extent that this challenged his conception of Britishness. George Borrow, *The Bible in Spain*, London: John Murray, Albemarle Street, 1849, Ch LI, p. 294.

[20] The Reverend Robert Stewart Patterson, 'Gibraltar Street—Nomenclature', *Notes and Queries*, Sixth Series, Volume Tenth, London: John C. Francis, July-December 1884, p. 84. Cited in *Manolo Galliano. A Rocky Labyrinth: The History & Memories of the Ramps, Lanes, Roads and Streets of the City of Gibraltar.*

[21] David Lowenthal, *The Past is a Foreign Country*, Cambridge University Press, (1985) 2003, p. 16.

considered as a far more efficient instrument for the perpetuation of a traditional naming principle than a street sign.

As it happens, this was a question of mind over matter as Patterson's list failed to have the desired effect. A new list was produced by Rodolfo Bandury, the Deputy Librarian at the Garrison Library, in 1890. As seen in Fig. 2, Bandury's list was similar to Patterson's in the juxtaposition of official English names set against and their equivalent traditional names in Spanish. Published in the 1913 *Gibraltar Directory*,[22] the suggestion was that finding places in Gibraltar remained difficult, with this list was seen as yet another guide primarily for services personnel and English residents.

In producing a far more exhaustive list than Patterson, Bandury continued with the doubling effect that so resonates with MG Sanchez's 'Ruina'. Whilst 'Ruina's' protagonist navigates streets through memorialisation and his familiarity of these locations, those new to Gibraltar would not have this recall, nor direct access to the cultural points of signification underpinning the street naming process. It is difficult to know how successful these lists were in practice, but the prevailing need to produce them suggests that the official naming of streets had become part of the problem rather than the solution. No longer was it only a case of traditional names having primacy, but of a doubling that led to further confusion when attempting to pinpoint a location. It seems that official nomenclature became disrupted by the continued use of these 'other' names. Rendering the traditional names visible on lists did not change this dynamic but rather added to it, perhaps because their listing also served to not only acknowledge but also legitimise the traditional.

Official names had become overlaid by the traditional rather than vice versa, but so was the past overlaid onto the present given the permeance of a local population versus the transience of an armed-forces based settler community. Gibraltarians enjoyed a culture that had grown from their shared experience, with these values, and indeed sense of belonging, continually being reinforced across generations and through daily interactions within what essentially was (and remains) a very small place. The close proximity in which people lived together with the density of the urban space would have served to further consolidate the cohesiveness of this community. We are in essence dealing with a very small urban footprint but size, together with density, supported the reinforcement of traditional street names. Alongside this cohesiveness we find a transient English settler community posted to Gibraltar for short spurts of time, usually only three or four years. Once their term was over, they would leave, taking with them whatever local knowledge they acquired over that time. It seems very likely that even if they did acquire knowledge of the cultural norms informing the street naming principle, their leaving would rupture any effective level of transferral to the new intake, who would need to start this process all over again.

These two Gibraltars, each culturally and linguistically different, occupied the same space but the permeance of a Gibraltarian population clearly held them at an advantage as they were free and able to continue their imaginative mapping in an unfettered manner. Although visible only on the lists produced, these traditional

[22] *The Gibraltar Directory and Guide Book 1913*, compiled by Sir Cavendish Boyle, KCMG & R. Bandury, Gibraltar: Garrison Library Printing Establishment, 1913, p. 41.

1913 GIBRALTAR DIRECTORY. 41

English Names of Streets in Gibraltar and their local equivalents.

Abecasis' Passage	... Callejon de la Paciega.
Arengo's Lane	... Callejon de Chiappi.
Benzimbra's Lane	... Callejon del Moro.
Boschetti's Ramp	... Callejon del Tio Pepe.
Crutchett's, or Portuguese Town	... La Calera.
Cloister Ramp	... Los Baños de Scotto.
College Lane	... Callejon de Risso.
Cornwall's Parade	... Plaza de la Verdura.
City Mill Lane	... Calle las Siete Revueltas.
Commercial Square...	... Plazuela del Martillo.
Castle Street	... Calle Comedias.
Church Street	... Calle de la Iglesia.
Castle Ramp	... Calle del Castillo.
Civil Hospital Ramp	... Cuesta del Hospital.
Cannon, and Gunner's Lane	... Detrás de la Iglesia.
Casemates Square	... La Esplanada.
Devil's Gap	... Escalera del Monte.
Engineer Lane	... Calle Ingenieros.
Fountain Ramp	... Callejon de la Fuente.
Fraser's Ramp	... Escalera de Benoliel.
Flat Bastion Road	... Cuesta de Mr. Bourne.
Governor's Street	... Calle Cordoneros.
Governor's Parade	... Plaza de Artilleros.
George's Lane	... Calle Vicario el Viejo.
Gowland's Ramp	... Callejon del Hospicio.
Hargrave's Parade	... Plazuela de los Artificers.
King's Yard Lane	... Callejon de la Paloma.
Lime Kiln Road	... Callejon de Dolores Corbe.
Lime Kiln Gulley	... Callejon de Segui.
Library Garden	... Huerta Riera.
Library Ramp	... Cuesta del Ball-Alley.
Line Wall	... Las Murallas.
Market Lane	... Callejon de la Carnicería.
Market Street	... Calle de la Policía.
Parliament Lane	... Callejon de los Masones
Prince Edward's Ramp	... Cuesta de Cárlos Maria.
Prince Edward's Road	... Cuesta de Sandunga.
Paradise Ramp	... Escalera de Cardona.
Rodger's Ramp	... Los Espinillos.
Serruya's Lane	... Calle Peligro.
Serfaty's Passage	... Callejon de Bobadilla.
Serruya's Ramp	... Escalera de Maqui.
Southport Street	... Calle Puerta Nueva.
Turnbull's Lane	... Detrás de los Cuartos.
Tuckey's Lane	... Callejon del Jarro.
Town Range	... Calle Cuarteles.
Victualling Office Lane	... Callejon del Perejil.
Waterport Street	... Calle Real.
Willis' Road	... Buena Vista.

Fig. 2 Rodolfo Bandury's 'List of English Names of Streets in Gibraltar and their local equivalents', from the *Gibraltar Directory and Guide Book* 1913

names had become incribed onto the landscape. All this brought about a resilience strong enough to challenge widespread usage of official names, but ultimately, the use of a codified naming principle also created an invisible barrier when it came to easy passage access to these streets. It is not as if Gibraltarians had far greater agency than the Garrison, but rather that each sector had divergent points of cultural signification, one stronger and more permanent than the other, leading to civilian primacy over this street naming aspect. Whilst none of the lists indicate that the use of traditional names had become weaponised as vehicles for subversion, there is a subversive undertone here, if only at a practical navigational level.

Despite the attempts at de-codification offered by both Patterson and Bandury, a further list became necessary in 1939, leading to the publication of a *City Council of Gibraltar, Road Book for General use, but Particularly Intended for the use of Civil Defence Organization.*[23] Notably, the Foreword explains that the contents were for the benefit of a specific user, namely, the large numbers of services personnel arriving at Gibraltar upon the outbreak of the Second World War. Understandably, the aim was to offer a comprehensive guide of Gibraltar's streets at a time when fast responses were necessary. Additionally, most of the civilian population of Gibraltar, excepting men in essential services, had been evacuated to the UK, Madeira and Jamaica, with many of the private properties in the city centre being requisitioned by the War Office.[24] A number of those arriving would be billeted in these and concern remained that the doubling of street names would hinder access in and around the territory.

As a municipal body comprised of local councillors, Colonial Office and joint-services representatives, the City Council was the best appointed to produce the book as Gibraltarian input was essential to inform the cultural points of reference and the colloquial nuances informing the unofficial naming of streets. Their objectives were to devise a book (interestingly no map was included in this venture) to 'readily [find] the name, situation, limits and traditional or colloquial name of the public highways and other localities in Gibraltar'.[25] 'Readily' is presented as a key term.

The *Road Book* consists of an alphabetically arranged list of official names of 'the public highways, including also naval and military ways, private passages and other well-known localities, which cannot be classified as public highways and etc.'[26] In a column alongside these names, as seen in Fig. 3, we find their 'situation', suggesting location, followed by their corresponding traditional names. By way of being as conclusive as possible, a second list comprised of alphabetically organised traditional names, their 'situation' and corresponding official names was added. This second list was for the purpose of finding the official name when only the vernacular had been given as reference.

---

[23] *City Council of Gibraltar, Road Book for General use, but Particularly Intended for the use of Civil Defence Organization*, Gibraltar: Beanland Malin, 1939.

[24] T. J. Finlayson, *The Fortress Came First*.

[25] *City Council of Gibraltar, Road Book for General use, but Particularly Intended for the use of Civil Defence Organization*, p. 1.

[26] *Ibid.*

Official Name	Situation	Colloquial or Spanish Name
Abecasis' Passage	Between Hospital Ramp and Castle Road, north of the Colonial Hospital.	Callejón de la Paciega *
Arengo's Palace Lane	East of Castle Road, opposite Police Barracks.	Palacio de Arengo
Benoliel's Passage	A *cul-de-sac* on the North Side of Fraser's Ramp	
Boschetti's Steps	Between Governor's Street and New Passage.	Callejón del Tío Pepe
Casemates Square	Between the north end of Main Street and Waterport	La Explanada
Castle Ramp	Between Castle Road and the Road to the Lines	Calle del Castillo
Castle Steps	Between Hospital Ramp and Willis's Road, passing through Castle Road and being almost a continuation of Castle Street	
Commercial Square	An open area between Main Street and Line Wall Road, wherein the Exchange Building and the City Hall stand.	El Martillo**
Engineer Road	Extending from a point on the east side of Europa Road, about 50 yards to the south of Europa Road Reservoir, to the Jews' Cemetery in Windmill Hill.	
Library Ramp	Between Governor's Parade and Prince Edward's Road, to the North of the Garrison Library	Cuesta de Ball Alley***

Fig. 3 A sample taken from the 1939 *City Council of Gibraltar, Road Book for General use, but Particularly Intended for the use of Civil Defence Organization*

From the above we see that a number of these traditional names are not Spanish translations of the official. Abecasis Passage is one such example as it was known traditionally as *Callejón de la Paciega,** which refers to a traditional Spanish breed of red dairy cattle called paciega. It is difficult to know how this naming came about but it clearly has no direct reference to Mr Abecasis, after whom the passage was named. Commercial Square, was known as '*El Martillo*',** the hammer, as the Square had been the site for public auctions with the association based on the auctioneers hammer. In the case of Library Ramp, traditionally known as '*Cuesta de Ball Alley*',*** the reference is far more oblique in its nod towards a fives court in that vicinity. As such there is no rule to how these names came into being as they are not exclusively based on translations from, or corruptions of, the English, but based on historical or personal points of reference and usage. This additional information is not contained in the *Road Book*, but the detailed descriptions in the 'situation column' reveal a considered attempt to offer directions to all these places. Attention is nevertheless drawn to the complexity in describing the situation of a location, as this is not the same as pinpointing a location on a map. As such, the 'situation' of any place on the list is based on identifying their relative proximity to other landmarks. These are expressed in narrative form punctuated with terms such as 'between' this and that street, or 'extending from a point' and 'about 50 yards to the south'. A considerable degree of cross referencing across other streets on the list would have been necessary, which somehow defeats the 'readily' aspect of the *Road Book*. Some level of local knowledge would have also been necessary as well as, possibly, a compass and a tape measure. Notably, the placing of traditional street names alongside their official counterparts in an official publication serves to bring them on an equal enough footing, short of legitimising them as official. Implicit in the *Road Book* is the recognition of traditional names given the impossibility of rendering them obsolete. As with the Patterson and Bandury lists, these traditional street names continue to be given a visibility they never had. It is difficult to know how successful the *Road Book* was in achieving its aims, but it could be surmised that wartime Gibraltar, occupied almost entirely by a services contingent, could have been easier to navigate. The evacuation of civilians would have also served to collapse, if only temporarily, the perception of two Gibraltars.

The Recovery and 'Immortalisation' of Traditional Street Names

Whilst the level of visibility of these traditional street names in the *Road Book* was primarily for the benefit of services personnel as the sector that had the most difficulty in navigating Gibraltar's streetscape, there is also evidence that the use of these names started to give way to the official ones in English. This was down to the increased use of English in a post-war Gibraltar following the establishment of a state-run education system with a national curriculum in English. Prior to this, education had primarily been in the hands of the Roman Catholic Church, and generally delivered in Spanish.[27] These post-war generations may have continued to hold on to a Spanish maternal tongue, but the increasing predominance of English should be considered as a factor in the gradual erosion of the oral transferral of traditional street names and the increased use of the official.

Indicative of this shift is a booklet written in the 1970s by Dorothy Ellicott, a local historian who dedicated much of her time towards recovering an unrecorded history of Gibraltar. Her *Place Names in Gibraltar: Memories of Long Ago*,[28] is written in this spirit. She is also driven by a Tourist Board initiative to attract tourism to Gibraltar following the closure of the border with Spain in 1969, and as such, her publication would also appeal to new visitors to Gibraltar as well as for local consumption. Ellicott's opening sentence explains that 'Military writers have chronicled the history of Gibraltar, recording the names of its Batteries, Walls and Bastion in great detail, but

[27] See Antonio de Bethencourt Massieu, *El Catholicismo en Gibraltar Durante el Siglo XVIII*, Estudios y Documentos No. 25, Universidad de Valladolid, 1967; and EG Archer & AA Traverso, *Education in Gibraltar 1704–2004*, Great Britain: Gibraltar Books, 2004.

[28] Dorothy Ellicott, Her *Place Names in Gibraltar: Memories of Long Ago*, Gibraltar Tourist Board, n/d, c.1970s.

hitherto no-one has delved deeply into our civilian place names'.[29] Interestingly, she recalls the presence of several 'so-called' towns within the city, such as Portuguese Town, Hayne's Town, Spanish Town, Irish Town and Black Town, but she is not able to offer historical details of why these towns came into being, Still, the existence of these towns within the city centre also suggests the demarcation of areas if not quarters within an already dense city and of walls within walls. Ellicott relies heavily on official sources and streets named in English, such as Lynch's Lane, so named after Henry Lynch, a 'Protestant landowner whose property suffered heavily in the Siege [1779–1783]'.[30] Whilst not entirely focussed on traditional names, her booklet offers an insight into the meaning of official street names, an area that had not been delved into before, as all previous lists were attempts at decodifying traditional names as a means to have ready access to these streets. Ellicott was concerned with recovery and a historical context. She mentions in her booklet some of the colloquially named streets, such as *El Ball Alley* for Library Ramp, a location which resonates in the lists discussed above. In a final section of her booklet Ellicott incudes a section entitled 'Hurried Notes', where she lists both English and Spanish colloquial names together their corresponding location or meaning, such as 'Crooked Billet', which she sites as 'probably', Convent Place, and Mr Fraser's Lane, which was most likely sited at Parliament Lane.[31] Ellicott's list is far from exhaustive but rather the start of a recovery.

A far more exhaustive account can be found in Tito Benady's *The Streets of Gibraltar*, in which he goes back to a mediaeval Moorish Gibraltar through to a Spanish and a post-1704 British Gibraltar to get to the root of the naming principle of these streets, as does Manolo Galliano in his soon to be published manuscript 'A Rocky Labyrinth'. Benady relies on historical contexts for the naming principle as one which comes about from the multiple occupancies informing this space. He also focusses on the cultural practices of the population, and on a military naming principle based on the strategic construction of fortifications or the billeting of men. An important feature of Benady's book is how his narrative structure mimics a walking tour of Gibraltar, the point of departure of which is historic, commencing at Casemates, where the 'engineers sent by the Sultan Abd-al-Mummin of Morocco landed in May 1160, with instructions to lay the foundations of the new and noble City of Victory (Medinat-al-Fath) at the foot of the Mountain of Tarik'.[32] Thus, he underpins the origins of the establishment of the City of Gibraltar before embarking on his tour, pointing out streets and landmarks which are accompanied with descriptive passages and illustrations. These illustrations are essential as identifiers, as the reader can visualise their location. Whilst a plan of these streets is not provided, the tour follows, on the whole, a sequential passage with each street leading to the next. This level of visualisation was missing from all the lists discussed above, as none included a map or drawings of these places. Benady's guide is also intended as a record of these streets

[29] *Ibid*., p. 1.

[30] *Ibid*, p. 8.

[31] *Ibid*, p. 17.

[32] Tito Benady, *The Streets of Gibraltar*, p. 3.

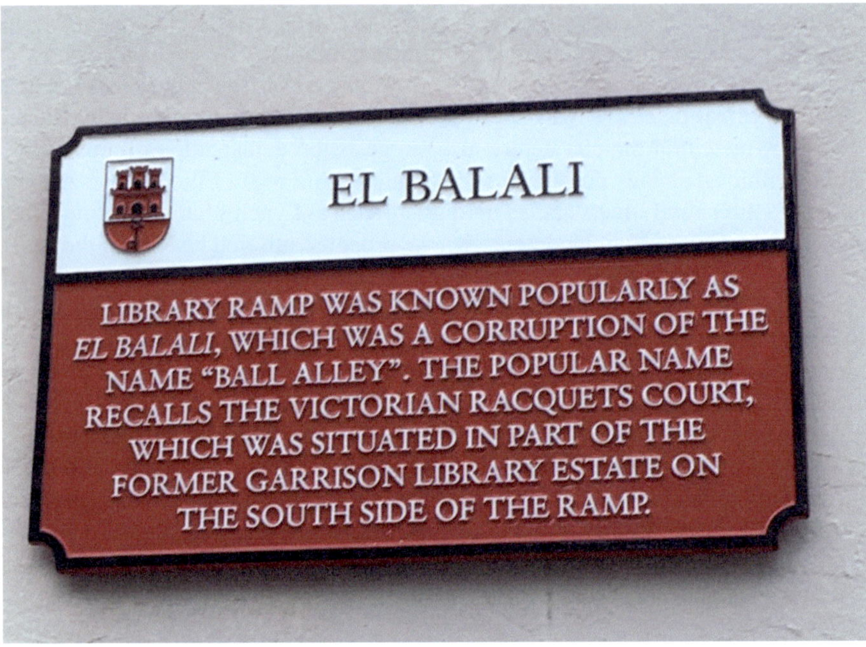

Fig. 4 The plaque for '*Cuesta de Ball Alley*', is now transformed into *El Balali*, a further corruption from the original articulation of the name and an indication that traditional names were not stable and remain subject to change. *El Balali* refers to the phonetical spelling of 'the ball alley' as it would have been pronounced in Spanish or *Llanito*, a code switching register that incorporates, amongst other languages, English and Spanish. Installed at Library Ramp in February 2020

and landmarks as he was writing at a time of rapid change, one which commenced with the opening of the border with Spain in 1982/5, and continued during the term of the GSLP Government between 1988 and 1996, which brought an important impetus of land reclamations and a wave of new developments. Benady's work, and 'Ruina' for that matter, have echoes of David Lowenthal's suggestion that the 'familiarity evoked by the past is far more comforting than the uncertainty of change and a future which now seems quite unfamiliar'.[33] Both Benady and Sanchez approach the future with some trepidation given the speed at which Gibraltar is becoming transformed.

Through their revival, these traditional street names have become part of a wider drive to recall the past to generate some level of stability in times of change, which rather perversely turns on its head the disruptive presence these street names created in the past. Importantly, the naming principle of traditional street names has become symbolic of a 'Gibraltarianess' and perhaps even of resistance. Their lack of visibility contributed to their staying power, but even this has now changed following a governmental initiative to have plaques bearing traditional names placed alongside their corresponding official name plaques. Whilst there is a heritage driven side to

[33] David Lowenthal, *The Past is a Foreign Country*, p. 69.

this enterprise, as well as touristic one, these name plaques are also concerned with the cultural recovery of Gibraltar's social history. In February 2015 the *Gibraltar Chronicle* covered the story of the installation of the first five plaques of traditional street names. In an interview given by the minister for Transport, Paul Balban, he referred to his grandparents speaking of 'many old names'. He would listen in awe learning about these places, but the transferral of this local knowledge proved to be incomplete when he recalled that during his time as a taxi driver, a taxi radio operator (of a certain generation) 'would confuse other drivers by using old street names'.[34] In fact, such was the case that these street names posed a navigational issue because this younger generation of taxi drivers could only find locations though their official names. Balban here reveals a tension between the recovery of the past in the present. Balban continued that his aim was the 'immortalisation' of these names and in doing so the past becomes overlaid onto the present. A further five plaques were installed in February 2020 with more to follow, but I wonder how this bodes for an unofficial naming process that existed only in people's imagination. Their placing alongside the official in plain sight is significant when we consider that these names were never meant to be visually present, and certainly not on any list discussed above, as their power lay in their opaqueness and codification. Their inclusion on these lists invariably fixed them, especially when we consider the descriptive sentences found in the City Council *Road Book* informing upon the 'situation' of these streets. These may have added to the confusion and the doubling, but those descriptions created a layer of knowledge, of meta-data, that did not originate from the imagination of those Gibraltarians who created the naming principle. As seen in Fig. 4, the recovery of these names in the 21st Century will no doubt continue with this layering given the inclusion of additional information on the new plaques. Although part of a recovery process, it seems to me that these moves fail to convey how traditional names were born from an imaginative process. Ultimately, they were resilient because they occupied an imaginative space; one which invariable also generated subversive effects that disrupted an official form of mapping and naming.

Conclusions

Prior to the installation of official street signs in the 1870s, street names had been constructed mainly by Gibraltarians using their own cultural points of reference, and language, which suggests that there was limited scope for the transferral of these values to an English speaking Garrison sector. The longevity of these traditional names clearly posed difficulties when it came to understanding the streetscape for access purposes. The fact that these street names were constructed in a non-discursive manner and held only in the imagination of Gibraltarians also rendered this naming

---

[34] Mark Viales, 'Gibraltar's old street names 'immortalised'', *The Gibraltar Chronicle,* 20th February 2015.

process resistant to decentring moves. Although we could consider that the installation of official street names had a hand to play in diluting the primacy held by the traditional, the reality was that it took around 100 years from the installation of the official names in the 1870s, to the point at which we see the publication of Dorothy Ellicott's c1970s booklet, aimed as a vehicle to recover street names which at that stage were being forgotten. It is nevertheless difficult to determine when the use of traditional street names starts fading into a historic past. The advent of a post-Second World War education system in English in 1944, would have, by the 1970s, yielded a new generation of Gibraltarians far more fluent in English than ever before. Further, as Edward Picardo suggests, this new education system also functioned as a vehicle to inculcate 'British *imperial* values on the population'.[35] The breakdown, therefore, of a differentiated Gibraltarian cultural identity, one which they had been free to cultivate and express, would have led to transformations with Gibraltarians becoming culturally and linguistically closer to Britain.

This naming principle was primarily linguistically articulated in Spanish, but the cultural mores that informed the naming belonged to Gibraltarians. As discussed above, the traditional naming principle was in based on rather nuanced points of reference based on how Gibraltarians not only imagined their community but also their streets between the Eighteenth to mid-Twentieth century. Gibraltar is now a very different space, and it seems very unlikely that traditional street names will be fully recovered for daily use. As minister Balban stated in his press release, it would be impractical, but it remains remarkable that Gibraltarians successfully challenged a colonially informed imaginative geography of their city through the deployment of an imaginative mapping of their own. In doing so attention was drawn to the schism between the Gibraltar they had domain over and the Gibraltar held by the Garrison.

Questions

1. Explain what is meant by the resilience of traditional street names in Gibraltar.
2. How did street names that existed orally but not on street signs in Gibraltar disrupt access to a military-based settler community?
3. Unpack the terms under which traditional street names were generated and sustained in Gibraltar, together with the challenge they posed to a colonially informed mapping necessary for fixity and control over the territory.
4. Evaluate the initiatives to publish lists bearing both official and traditional street names in Gibraltar.

---

[35] Edward Picardo, 'The War and the Siege: Language Policy and Practice in Gibraltar, 1940–1985', unpublished PhD, University of Birmingham, 2012, p. 127.

From Historical to New Place Names. The Case of Italy

Laura Cassi

Abstract Place names, which were created by the need to identify their setting and, as a result, to proceed to their organization, are a basic component of historical memory of a territory. Thus they are a valuable part of the cultural heritage and expression of perception by generations throughout the centuries of their environment. This is all the more important in today's world which is subject to enormous processes of certification and standardization. They deserve to be protected and foregrounded by way of careful data collection and specific enquiry into the complex links between place names and individual landscape and territorial categories (waterscapes, vegetation, shapes and characteristics of the land, settlements, historical road systems and human activities) as well as the complex relationship between the whole place name corpus of a territory and its landscape, and the birth of new place names, originating, for example, in the tourist industry. After the general geographical lines of development of place name research in Italy, an outline will be presented of a number of examples both nationally and regionally relevant to the above mentioned themes.

Keywords Cultural heritage · Geo-toponomastic research · Historical place names · New place names · Landscape · Italy

Objectives

By the end of this chapter, the reader should:

- To discuss the concept of the 'historical memory of a territory'
- To define what the 'cultural heritage and expression of perception by generations throughout the centuries of their environment' means.
- To make arguments why place names are one of the most important components of the intangible cultural heritage.
- To explain the reasons why the processes of certification and standardization is so important today.

L. Cassi (✉)
University of Florence, Florence, Italy
e-mail: laura.cassi@unifi.it

© The Author(s), under exclusive license to Springer Nature Switzerland AG 2023
G. O'Reilly (ed.), *Place Naming, Identities and Geography*,
Key Challenges in Geography, https://doi.org/10.1007/978-3-031-21510-0_25

- To enquire into the complex links between place names, individual landscape and territorial categories.
- To be familiar with the narratives of place naming in Italy, from the examples given both nationally and regionally.

Introduction

Place names are a special kind of nomenclature, the initial aim of which is to establish a system of spatial reference for the identification of individual geographical features, such as areas of settlement, isolated dwellings, waterways, landforms, territorial sections and so on. Thus their role appears to be basically similar to that of geographical coordinates i.e. to identify a place exactly. So place names, inasmuch as they are aids for identification of places, which, owing to their names are foregrounded in relation to their surroundings, are of fundamental importance in meeting human needs, such as direction and locationing, that are as old as human existence.

By identifying and delimiting, they contribute to spatial compartmentalisation processes, their role in communication linked functions being clear. The important role of place names in the past should not be forgotten, up to the point when the introduction of the geometric and numerical land registry facilitated rationalisation of spatial positioning, thus avoiding the need for identifying a geographical object by way of its qualitative and topographical features. Concerning the frequency of place names in a given territory, it is obvious that apart from revealing natural diversity it also depends, above all, on the intensity and detail of human occupation of space.

Place name coverage is complex, as it is the result of centuries of stratification, in line with other reconfiguration processes linked with population movements, socio-economic factors, environmental modifications and the wear and tear of time. The result is the crucial role of settlement continuity in the survival of place names, otherwise undermined by demographic substitution or reinstatement by way of different ethnic input. Since their origin as a practical purpose linked by a direct relationship with space, place names are a kind of representation, which is the result of the perception of environmental characteristics by the groups of people who created them. Therefore, they are the expression of collective knowledge, by means of which local identity and the sense of belonging to a specific area emerge.

As a kind of identity label symbolically applied by the various cultural groups following one another in a specific area, place names contribute to recognition of linguistic variation characterising the various phases of conquest and reoccupation of land. Thus, while the role of naming in the initial stages of spatial settlement is undeniable, this is also the case with the fact that it represents a kind of "disciplinary intersection", to quote Giovan Battista Pellegrini, who argued that the analysis of the relationship with a territory both in the present and the past cannot do without the ability to 'read' names on a linguistic basis. This is basic to the geographical, historical, and naturalistic investigation, especially in ancient settlements like those in Italy, where language has been subjected to substantial development, and inter-lingual contacts have been frequent.

That investigation of place names is fascinating as well as the complex is undeniable, even obvious. Naturally one is curious to know why a specific site has a particular name, why that name and not another one, and, furthermore, who thought of that name and why it changed over time. Many questions come to mind, to which there is often no plausible answer. At the same time, it should be pointed out that most place names, even in ancient densely populated settlements, as is the case in Italy, are "mere identifications" i.e. of common standard linguistic or dialect terms promoted to proper noun status, as pointed out by Giulio Cornelio Desinan and before him by Olinto Marinelli. Even apparently obscure names like *Arno*, *Bisenzio*, and *Serchio* are nothing more than the result of ancient descriptive terms ("flowing water") which, over time, have lost semantic clarity.

Though this clarification can appear to diminish the fascinating character of place names, they remain attractive, since the fact that they mostly consist of common nouns turns them into useful markers of present or past environmental conditions, as is the case with place names reflecting wooded areas which no longer exist. Furthermore, since naming is the result of mental processes of registration of specific environmental and territorial aspects, the result is an interpretation of a landscape filtered through the gaze of generations.

Interest in Geo-toponomastic Studies in Italy

After these preliminary observations, since the antiquity of names can even suggest a reference to ancient environmental conditions, place name research can undoubtedly contribute to our knowledge of a territory and the gradual process of human occupation. Reference to ancient hydrographic and woodland characteristics as well as the nature of the soil or of former economic activities and previous settlement is abundant and significant, as can be seen from research concerning Italy with specific surveys of various regions.[1]

Despite these potential applications, place name research has not received adequate attention from geographers. This was less the case when, in the late 19th and early twentieth centuries, geographical research in Italy covering place names aimed mostly at the elaboration of appropriate terminology for the description of

[1] A systematic survey of Italian place name research was published by linguist Fiorenza Granucci, *Prontuario bibliografico di toponomastica italiana, con prefazione di Carlo Alberto Mastrelli*, Firenze, Dipartimento di Linguistica dell'Università, 1988. The appearance in 1995 of the *Rivista Italiana di Onomastica* (RIOn), edited by Enzo Caffarelli, the purpose of which is "to communicate and enhance Italian research on onomastics with special attention to links between linguistics, literary and philological studies and other disciplines: history, psychology, sociology, demography, statistics, law, geography, literature, philology and archaeology" provides bibliographical updating of place name research. For a geo-toponomastic bibliography see: M. Fuschi and G. Massimi, eds, Toponomastica italiana. L'eredità storica e le nuove tendenze, Atti della giornata di studio, Pescara, 13 dicembre 2007, in *Memorie della società Geografica Italiana*, vol. LXXXV, 2008; V. Aversano and L. Cassi, Geografia e nomi di luogo, *Geotema* 34, 2010 and L. Cassi, *Nomi e carte*, Pisa, Pacini, 2015.

environmental aspects and characteristics; this was to such an extent, that a special geographical congresses session took place dealing with toponymics at the time. The leading geographical journals, such as *Bollettino della Società Geografica Italiana* and *Rivista Geografica Italiana*, devoted a special section to this research field.

Nevertheless, after a period of considerable lack of interest, the second half of the twentieth century witnessed a revival of geo-toponomastic research. This was a reflection of general interest in the cultural heritage and new attention to questions of specific local matters. There is no doubt that the rise of a new 'esprit des lieux', promoted by a kind of reaction to standardization and uniformity processes in recent industrialized societies favoured the recognition of the value of place name research as an expression of a sense of belonging to places felt by a local population.

Place Name Collections and Revision of Place Names on Maps

Since place names had been awarded the status of part of the cultural heritage, widespread interest in their preservation increased. They have been endangered by a decline in the traditional economy. Therefore, the need arose to produce collections, with special attention to micro-toponyms, which face the risk of extinction, owing to a fall in careful farming of the land, a good example is the traditional farming set up in Tuscany known as 'mezzadria' aka 'métayage'.

Creating a collection of place names initially requires indexing, which calls for knowledge of the latest terminology required to define objects and phenomena which even experts have difficulty in labelling in Italian, seeing that geographical terminology is often of foreign origin. It should also be remembered that in place name research much of the terminology is old fashioned, on occasion obscure and difficult to interpret with its links to an economy that no longer exists in a situation of great difficulty for anyone used to living in the superficial surroundings in our modern urban environment.

Following repeated requests for rationalization of the terminology in use in place name collections, mostly lacking even in those that are scientifically sound, linguistically speaking, for a correct, unambiguous terminology concerning the various aspects of the landscape to which proper nouns have been applied, rationalization of this terminology was attempted.[2]

In the conviction that place name collections could take advantage of geographical experience in the case of two fundamental aspects: recognition and correct naming of the geographical object to which the proper noun is attributed and a synthetic description of the territorial context in which the object is situated, a special tool was created. This was a kind of glossary, the purpose of which was the acquisition of the necessary terminology for the compilation of files of place names, also including a

[2] C. A. Mastrelli, (1992), Geonimi e indicatori geografici, *Geografia*, 2–3, pp. 87–91; L. Cassi - P. Marcaccini (1992), Gli «indicatori geografici» per la schedatura toponomastica. Criteri e norme per la loro definizione, *Geografia*, 2–3, pp. 92–102.

cross-reference to ancient traditional geographical terms in the standard language or local dialects to present-day terminology.

For the purpose of correct identification of geographical objects, for example registering the hyperonym "waterway" instead of "stream", "river" or "ditch" – a terminological list was set up in a relational database following a logical connection in line with geographical space characteristics where place names are inserted, incorporating geographical terms with coherent logic from which the majority of place names derive. Figure 1 illustrates the geographical space framework followed for placing terms starting from the lithosphere (rocks), hydrosphere (waterways) and atmosphere (meteors), ground morphology, hydrographic elements, vegetation, the presence of human activities, and their types (settlements, spatial arrangements, activities) and on to the geographical position (mathematical and topographical: peaks, slopes, valley floor). Geographical space is produced by the above grouping and its interactions.

The framework—on the increase with respect to the original arrangement (1,700 terms)—includes 1. A suggestion for classification for the various types of objects that can be subject to specific labelling, such as rural dwellings, built-up areas, plots of land, waterways, highlands etc., and their definitions; 2. A large number of specific items concerning characteristics, objects and phenomena of geographical space which can potentially 'enter' the toponomastic repertoire of a territory as conceptual place name matrices; 3. A series of useful entries for the description of the environmental context containing the place names.[3]

With reference to the task of publishing large scale microtoponymy collections, the autonomous Italian province of Trento has played a leading role. As of 1980 a place name collection project has been ongoing, and in 1987 the Dizionario Toponomastico Trentino was set up with the purpose of spreading information on the pronunciation, use, meaning, tradition and origin of place names, by way of a repertoire for each municipality. The project is very close to completion, 21 volumes having been published so far.

Another theoretical methodological aspect is the criteria followed for place name cataloguing and their review on large scale maps. Following enquiry by the Tuscan Regional Administration, and advice from linguists the compilation of a project was addressed as being of general interest, albeit based on examples from Tuscany. The review was based on a feature of the 1:5000 technical regional map, using information from maps dating from different periods, written records e.g. old guide books, chorographic dictionaries, tithes from the eighteenth century Tuscan Grand Duchy and so forth, precise inspections and interviews.[4] Figure 2 shows a typical file from

[3] L. Cassi - P. Marcaccini (1998), *Toponomastica, beni culturali e ambientali: gli "indicatori geografici" per un loro censimento*, Memorie della Società Geografica Italiana, LVI, Roma.

[4] L. Cassi - P. Marcaccini (1991), Appunti per la revisione della toponomastica nella cartografia a grande scala. Saggio di correzione ed integrazione di un elemento della carta tecnica regionale 1:5000 della Toscana, *Geografia*, n. 2–3, pp. 100–110. It should be pointed out that the Tuscan researchers' theoretical, methodological approach not only originated with a tradition of geo-topomomastic studies by the geographer from the Friuli region Giovanni Marinelli and his pupil from Trento Cesare Battisti – but also research by another leading Italian geographer by the name of

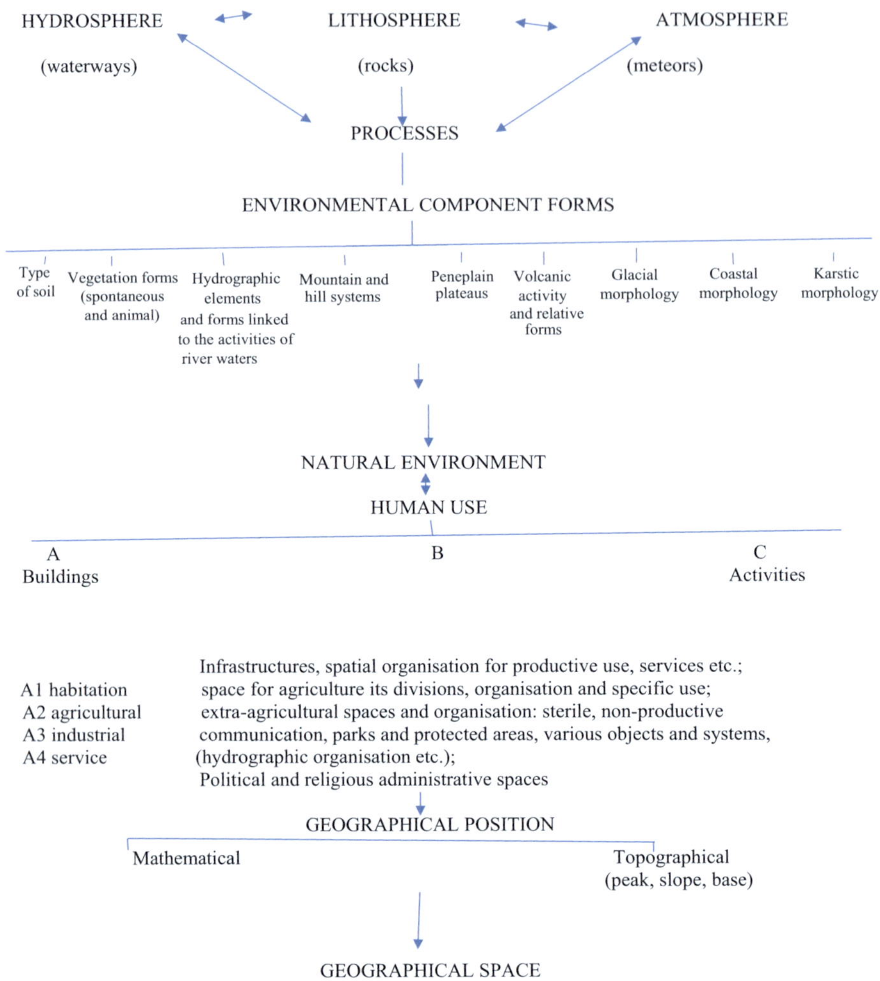

Fig. 1 Key to general reading of the relationships between the physical spheres and between them and human activity

the period. Since there is not enough space to provide further information, attention is directed to the original text, which is now thirty years old. Nevertheless, some reflex can be found in a recent regional place name repertoire containing geographically referenced Tuscan place names, gathered from the main large-scale historical map sources, such as 19th-century land registers, maps from the Italian Istituto Geografico Militare, and the present-day regional ones. RE.TO.RE can be accessed at www.regione.toscana.it.

Aldo Sestini, who, although he was a naturalist showed interest in the field, and encouragement by Florentine linguist Carlo Alberto Mastrelli, who was particularly interested in the systemic aspect of place name research.

CATEG. B	Monte Oriuolo		
Fonte toponimo	*Nome*	*Posizione*	*Oggetto*
CTR	Mezzoso	1683895–4843590	edificio
CT	Mezzosso		edificio
IGM 1952	Mezzosso		"
IGM 1900–4	Mezzosso		"
CDF	Il Mezzosso		"
CGT 1832	Mezzosso		"
CGT altra ed.	Mezzosso		"
altra fonte			
cartografica:			
fonte scritta:			
Carocci			
altra fonte:			
sopralluogo	Il Mezzosso		edificio
proposta	Il Mezzosso		"
notazioni			

Fig. 2 File for place name revision in large scale mapping CTR Carta Tecnica Regionale (Technical Regional Map) 1:5000; CT Property Register 1:5000; IGM Istituto Geografico Militare: maps 1:25.000 editions; CDF Map of the environs of Florence 1:10.000, 1900; CGT Catasto Generale Toscano (General Tuscan Property Map)

Metaphor and Landscape in Place Names

Being characterized by considerable fascination, place names reflect, at the same time, objective conditions and perceptions, as shown, among other things, by the frequent use of metaphors and expressions opposite to the normal meaning of a word. The interest in place names also extends to seemingly marginal, but, at the same time, attractive aspects, such as the frequency of metaphors to be found from the Alps to Sicily, where the landscape often communicates by means of allegory. We only need to think of names like *Il Ramo del Diavolo* (*the Devil's Horn*), *Il Monte Scavezagenoci* (*Mount Nutcracker*), *L'Omo Morto* (*the Dead Man*), *Il Poggio Stancalasino* (*Weary Donkey Hill*), *Cefalù* (*High Headland*), to mention only some of these expressions, which, in many cases, are portraits of an environment.

The numerous examples from Tuscany include *L'Oncino* (*the Hook*), a winding river with a particularly striking contour, *La Piazza di Siena* (*Siena's main square*), describing a wide, levelled ridge just to the south of *Monte San Michele* in the Chianti area, *Monte Libro Aperto* (Mount Open Book) in the Apennines of the Province of Pistoia, *Il Podere Ruzzolapaioli* (a smallholding whose name recalls a pot rolling slope), *Il Podere Bramasole* (a smallholding longing for the sun), *Casa Mezzòsso* (*Half-bone House*) (names like this, originating from a rather poor, previous share-cropper [métayage] context, provide significant information on their history and meaning). Furthermore, the interest created by naming strategies expressed by many of the names should not be forgotten, such as those inspired by political history and territorial marketing. Photos 1, 2, 3.

So, are we allowed to state that the landscape communicates by means of place names? The answer is a definite 'yes'. This takes place by means of the creation of new descriptive terms over centuries by human communities, the result of the way in which their territory was perceived, living there, exploiting it, and planning it. It is, however, necessary to pay attention to what the names 'say', openly or indirectly, or even, occasionally, ironically, so as to point out natural or human characteristics, which might even be 'denied', as is the case with names like *Montaùto* (*Sharp Mountain*), to the south of Florence, which is neither a mountain nor sharp.

If we consider a number of significant categories of landscape, such as the vegetation cover, terrain shapes, water, human activities settlements, and roads, the terms of this 'dialogue' between names and landscape can easily be identified. A glance is enough at the distribution in Tuscany of place names evocative of beech woods to realise that it retraces the actual vegetation area, despite a number of exceptions that actually make this distribution even more interesting. The same can be said about place name reflections of hill and mountain areas, in certain cases sensitive to local perception, which is able to point out a slight rise in the terrain compared with its surroundings, by use of the label *mount*. Place names linked to water are just as interesting, both in hydrographic terms as well as concepts. In the case of the relationship between names conceptually referring to roads and the actual historical road network, place names fulfil their role as "fossils of human geography", as they

Photo 1 Cefalù (High Headland)

Photo 2 Omo Morto (Dead Man, Apuan Alps)

were called by Jean Brunhes. In this case however, we must again be cautious, since a place name may not remain linked to the original object forever.

Another interesting category is that of place names deriving from personal names, an indirect mirror of landscape components, though significant because it reflects individuals who had contributed to its shape over the centuries. We begin with very ancient, pre-Indo-European names followed by those of Latin, then the Germanic origin, and so on up to modern times.

Photo 3 Monte Libro Aperto (Mount Open Book)

Nonetheless, the first point to the foreground is that the study of place names, in the words of Giovan Battista Castiglioni, is a "disciplinary crossroads", and that analysis of the links between the present and past situations presupposes the ability to recognize names, starting out from linguistic study, which precedes geographical, historical and naturalist investigation, particularly in places inhabited for centuries, and in which there have been significant linguistic changes and contacts with other languages.

If the landscape 'speaks' by means of toponymy and is not lacking in wit, in the words of Franco Farinelli, place names are no less so, and feedback should be carefully examined. Do a *Casa al Faggio* (*Beech Wood House*) or a *Casa La Sughera* (*Cork Oak House*) refer to a large number of beech or cork oak trees, or else is it their exceptional presence that attracts attention? If we were to argue, somewhat deterministically, that toponymy is unconditionally linked to the landscape, we would be mistaken, because, the fact should not only be taken into consideration that a name given to a place is not the result of only one possible choice but also could have been suggested by something exceptional attracting attention. Some scholars have stated that place name research is more likely to 'deny' rather than assert: a *Bosco del Lupo* (*Wolf's Wood*) could refer to a wood populated by wolves or a wood where a wolf has exceptionally been seen; are we dealing with numbers or a single case?

It would, nevertheless, be just as imprecise to insist on such a negative stance. There are many examples of a positive approach. I can mention a simple, though explicit, one. A historical linguist, one of my university colleagues, saw nothing unusual when I had pointed out that the names linked to Saint Colombanus on large scale Italian maps are all in the north and never below the Tuscan-Emilian Apennine range, the reason being that the Saint in question was an Irish monk who had been sent to Bobbio by the Lombard King Agilulf and had never traveled further south.

Place Names and the Environment

The opportunities geography offers to the analysis of the relationship between place names and the environment have been explored under several headings with more qualitative than quantitative positive results.[5]

One of the most significant study areas, as I have already mentioned, is that of the relationship between landscape and place names linked with vegetation, a landscape-environmental component particularly representative of the naming process, as can readily be seen in a large number of relevant place names, when compared with other categories. Surface vegetation is undoubtedly one of the most eloquent, visually striking aspects of the appearance of a territory, also being of considerable importance from the viewpoint of a traditional economy. In Tuscany, for example, names linked with spontaneous vegetation are one of the richest sources for the naming of settlements and various geographical features, such as mountains, valleys, waterways, more or less extensive land areas linked to sensations like *Cerreto (Quercus cerris or Turkey oak Wood), Suvereto (Cork tree Wood), Faeta (Beech tree Wood)* as well as single examples like *Casa al Pino (Pine tree House), Podere la Sughera (Cork oak tree Smallholding),* where the choice could have been determined by an exception i.e. an isolated tree (not forgetting the fact, however, that names in the singular can also stand for a numerous group).

If many place names by association give us an idea of the species (*Carpineto i.e. Oakhornbeam trees, Pineta i.e. Pine Grove, Fargneto i.e. Quercus robor or English Oak Wood, Borro di Faètole i.e. Beech Gorge*), there are also many that suggest a wood in general terms, often accompanied by additional evocative adjectives (*Boscotondo i.e. Circular Wood, Selvabuia, Dark Wood, Selvagrande, Large Wood*), like the case of the place name *Selvamaggio* i.e. *Selvamaggiore* (*Main Wood*), which now designates an industrial area. The evocative and symbolic values communicated by woodland areas have probably contributed to the wealth of the related place name corpus. The forest has always expressed a clear contrast with cultivated, organized, and inhabited land. The forest is the symbol *par excellence* of wild nature, the opposite of culture in its disorderliness. It is the place of fear of the unknown and, in the popular imaginary, its inhabitants are characters symbolizing danger and risk for those who enter. It is also the place where one gets lost and a journey or adventure begins, with decisive encounters, as we learn from Dante's *Divine Comedy*, and which we can read about in poetic tales of chivalry and a host of fairy stories. The forest is impenetrable and inhospitable, but not for everybody: a hermit is not afraid, in fact, he makes it his home. Limited areas such as isolated wooded scrubs or groves, on the other hand, are seen as places for gatherings and meetings, in classical antiquity reserved for worshipping the gods, and they remained sacred spots for Celtic and Germanic peoples.

Apart from their fascination—arguably greater nowadays than in the past—names associated with vegetation are of interest for a variety of reasons. They can rightly be

[5] What follows concerning the situation in Italy mostly makes use of examples from Tuscany, which has been the object of a great deal of diversified geo-toponomastic research.

seen as useful for the reconstruction of the ancient extent of a wood. The geographical distribution of names such as *Bosco (Wood)*, *Selva (Forest)*, *Lecceto (Holm oak Wood)*, *Faggeto (Beech Wood)*, and so on has been a clear indication of the presence of woods where nowadays none can be seen. Information is also supplied on different climatic conditions from present-day ones and different set-ups in the past of the vegetation cover. Additional information is occasionally supplied by heterotypic synonym areas near place names in quite different positions to present-day areas of vegetation.

In the nineteen-thirties Alberto Chiarugi found some beech shrubs at a low level near the Tyrrhenian Sea coast. Near these beech heterotypic synonym areas, the large scale Istituto Geografico Militare map shows place names like *Faeta (Beech tree Wood)*, a clear indication of the presence of vegetation, proof of different climatic conditions than present-day ones, possibly dating back to the Little Ice Age of some centuries ago, when the climate was more humid and colder than nowadays, thus allowing beech trees to grow at a lower altitude closer to the coast. The cork oak also has some place name connections outside its present-day area, as is also the case with the holm oak, which occurs in place names at an altitude of 1,200 m. It is these exceptions that are a stimulus for research, since they offer information on different conditions as compared with those of today, and which by way of names have left traces of their existence.

Further confirmation of the relationship between names and spontaneous vegetation comes from the marked general nature of place name distribution linked to oak trees and oak forests, corresponding in our region to the general nature of distribution of a number of species of oak tree such as the downy oak. On the 1:25.000 map of the Istituto Geografico Militare Italiano (IGMI) limited to Tuscany there are more than 4,000 place names connected with vegetation.

Another interesting landscape-environmental component linked with place names is landform, especially in hilly areas, basically owing to the fact that morphology is a kind of landscape imprinting, an initial and clear characteristic locating and identifying it.

Continuing with the case of Tuscany, the widespread hilly character of the Region is frequently reflected in place names, with a sequence deriving from words such as *poggio* (*hillock*), *colle* (*hill*), *monte* (*mount*), in their simple or composite forms often qualified by adjectives specifying certain conditions: *Montebello (Beautiful Peak)*, *Poggio Secco (Dry Hillock)*, *Poggio Deserto (Desolate Hillock)*. There are also frequent references to the shape of quarries: *Vallecava*, *Vallicella*, *Vallone* and the appearance or nature of the land: *Pietra del Diavolo (the Devil's stone*, a bizarre, picturesque limestone rock sticking out from the marlstone at *Pania di Corfino* in the Apuan Alps in north western Tuscany; *Sassi Neri (Black Rocks)*, from the dark colour of the ophiolite rocks at Impruneta near Florence). I should like to recall here the names linked with the exposure of a slope such as *Bacìo (Shady)*, *Solatìo (Sunny)*, *Lubaco (In the Shade)* and those linked with the microclimate, like *Casa Spazzavento (House sweeping the wind away)*, *Podere Scaldagrilli (Warming crickets Smallholding)*, *Nebbiolo (Mist)*.

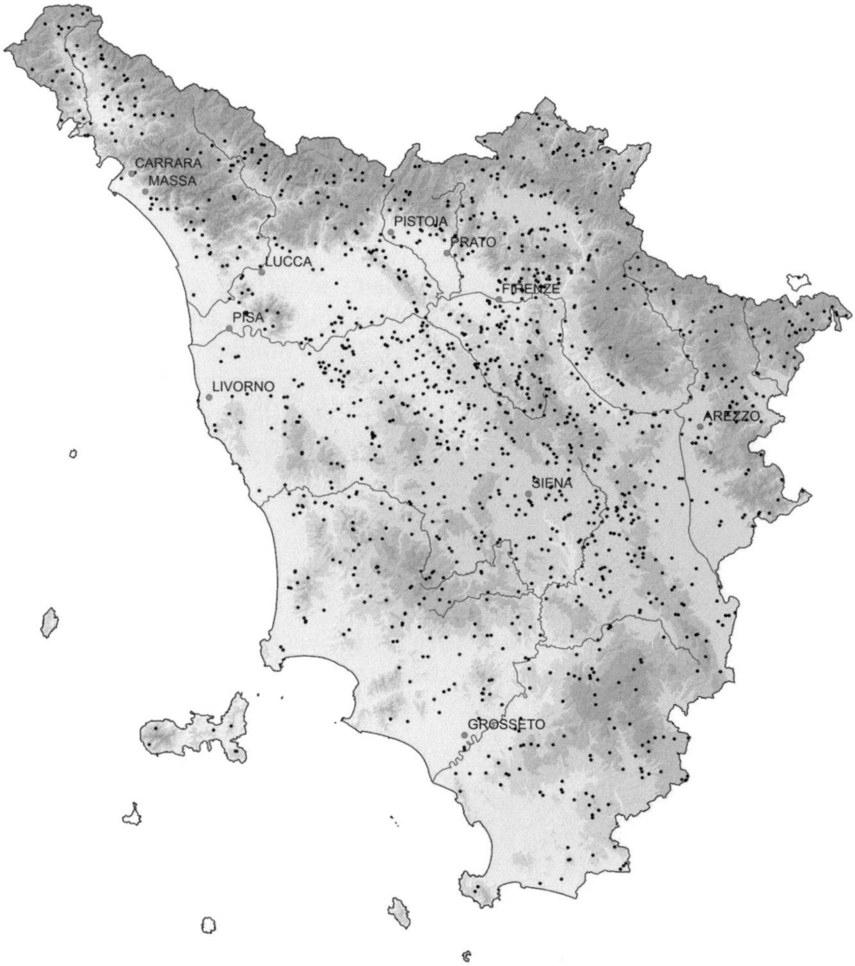

Fig. 3 Distribution of place names from the label *mount* in Tuscany (in large scale cartography 1:25.000)

In the hilly area around l'Impruneta, to the south of Florence, there are many place names such as *Monte (Mount)*, *Mezzomonte (Half Mount)*, *Montoriolo*, but there are no mountains near Impruneta! Here we have a case of the perception of the environment of the person who made up the name and who considered a modest rise in the land a 'mount' which evidently stood out in the surroundings (Olinto Marinelli had already pointed this out more than a century ago). It is no accident that there are as many as 1,400 place names containing the word *monte*, in its simple or composite form and often accompanied by a qualifying adjunct, as already pointed out all over the large scale regional maps, without a specific relationship to a genuine mountainous area. And if in the really mountainous areas, such as the Apuan Alps and Apennines, they are plentiful, they are mainly concentrated in central Tuscany, in the hilly area between Florence and Siena, the term *monte* (*mount*) being used for these uplands, which, however modest, are typical of the area without actually being true mountains (Fig. 3).

Concerning names like *Montebello (literally Beautiful Mountain)*, mention should be made of the fact that, in traditional place names, the aesthetic component is rare: 'bello (beautiful)' stands for 'useful' or, in the case of names such as *Belvedere* and *Bellavista*, which are quite frequent in Italy, reference is made to the position of the place and the view of it from higher ground. 'Bello' in the truly aesthetic sense dates from the emergence of new place names, such as those created in Sardinia in recent decades, i.e. *Cala degli Angeli (Angels' Bay)*, *Cala Paradiso (Paradise Bay)* and so on, inspired by the imaginary of the tourist industry (see further on). There are names like *Casa Paradiso (Paradise House)* in traditional toponymy, but they mostly reflect exposure, as in the use of *Inferno (Hell)* mostly symbolizing morphological characteristics.

Water and place names

Water is one of the most significant place name research categories, both because hydronyms are generally the most ancient ones in a territory, and owing to the intrinsic and symbolic ties with water. Many names of Italian rivers derive from very ancient linguistic strata, even Pre-Indo-European ones, gradually adapted to population and language shifts. It should also be remembered that, generally, hydronyms pre-date mountain names (with the exception of the names of mountain chains), since waterways have always represented a fundamental reference point for orientation and localization and they have been used for communication. Moreover, water, inasmuch as it is a fundamental resource for life and human activities, is a strong symbol, buried in the depths of human consciousness, at the same time being a bearer of love-hate relationships. Water is the expression of the source of all forms of life, as well as of dispersion and drowning: floods wipe out those who displease the Devine Being; rain is associated with energy and prosperity; placid rivers are a symbol of an ordered lifestyle, while whirlpools represent danger and difficulties. The links of water with the afterlife are as old as human beings, as is clear from the myth of the immersion of the sun to warm the world of the dead by night, water being a sacred purifying, and ritual entity, as recalled by the ancient religious cults linked with thermal springs. In the popular imaginary, lakes and springs have often contained natural spirits, water sirens and genies but also terrifying demons.

Water is a well-represented category in Italian toponymy, in relation to naming springs, running water, stagnant water, and solid state water, as well as place names conceptually linked to water regardless of the actual object. In

Tuscany, for example, the connections between the antiquity of a populated area, distributive typology and the density of the hydrographic pattern are so close that even the tiniest waterway has a name.

Most of the names of Italian rivers derive from terms or phrases meaning ‘running water’. So we have *Bisenzio*, *Reno* (the Italian translation of Rhine), *Arno*, *Serchio*, *Isonzo*, *Isarco*, *Serio*, *Sarca*, *Semenza* and *Sarno*. Clearly these are both hydronyms and place names conceptually linked to water. Many names of rivers derive from colours, geomorphology, instrumental metaphors, plants, animals, names of gods, or personifications of mythological or fairy tale characters. Categories referring to water include: *Acqua Calda* (*Hot Water*), *Acqua Longa* (*Long Water*), *Acqua Viva* (*Living Water*), *Acquapendente* (*Hanging Water*). Linked to ‘*rivus*’ (Latin for river) we have: *Risecco* (*Dry River*), *Rio Torbido* (*Murky River*), *Rio Tosto* (*Tough River*), *Riomaggio* (*Main River*), linked with ‘*fiume*’ (Italian for river): *Fibbio, Fiobbo, Fluvione* (*Big River*), and *Fimón*. Calabria has several river names linked to *pòtamon* (potamòs = Greek for river): *Xeropòtamon* (*Dry Riverbed*), *Serrapòtulu, Sciarapótamo, Sciarapóttolo, Zarapòtamo*. Reference to border hydronyms is linked to *Finale* (to be found in Piedmont as well as Liguria, Emilia, Venetia and Sicily). A river metaphor is ‘*Marteddu*’ = Sicilian dialect for ‘martello’ (hammer) which can also be found in Sardinia. This is probably suggestive of the hammering nature of rushing water. The same image is conjured up by *Lima* and *Lesina* (in the Lucca area) as is *Bacchiglione* from Venetian dialect ‘baccagliare’ = making a loud noise. From imaginary beings and ancient gods, we have *Drago* (Sicily) and *Dragone* (Apennines in Tuscany and Emilia) and *Mèrcure* and *Janare* from the category of ancient classical gods.

The Massaciuccoli Lake (Tuscany)

Acqua Cheta (Silent Water), Tuscan-Emilian Apennines

Coming to the human components of the landscape, roads are one of the most representative of the human presence. The road has always had a vitally important role as the framework of settlement *par excellence,* and, as a consequence, its reflection in place names is of particular interest. Settlements were set up along roads and so we find place names such as *Strada (Road), Via (Street or route),* and those linked with crossroads (*Croce, Crocicchio*), referring to divergent routes (*Trebbio, Trivio, Bivio)* or to distances along a road *Quarto (fourth), Quinto (Fifth), Sesto (Sixth),* (influenced by ancient Roman milestones), to services along the route as in *Tavarnelle (Little Taverns), Tavarnuzze* (refreshment stops for travellers), or *Spedaletto (Hospitality Service).*

The links between place names and roads are well known to linguists and scholars studying ancient road networks, concerning the advantages of etymological analysis of local names for historical topography, with its aim of identifying ancient roads and traces of Roman land measurement called centuriation, as applied by Giovanni

Uggeri. This scholar has reconstructed many ancient routes by way of, among other things, survivals, which are variously hidden, from the point of view of linguistic evolution, of the name of the road or names referring to constructions and services relevant to it.

There is no doubt that suggestions from place names can, in some cases be decisive. However, though etymological interpretation can provide correct linguistic solutions, we must be careful about accepting or suggesting hypotheses, since this method must always be compared with the data on the land and in the archives. The distribution of a place name category, as has been pointed out, though clearly symptomatic, is not in itself the only direction to follow in examining the concept.

Nevertheless, there is no doubt that interest in names linked with roads is particularly keen, when they preserve the memory of otherwise unknown infrastructures, for example, in cases when place name indications allow the identification of unknown or confusing routes and variants, possibly in the category of secondary routes. However, the fact that we are dealing with general names like *Strada* (road) or *Via* (street or route), and only a few of them are accompanied by qualifying elements like *Via Reggi i.e Via Regia (Royal Route), Via Maggio (Via Maggiore, Main Route)*—and they are even more significant when they no longer refer to main routes—this fact obscures the link between place name and hierarchy of road routes.

But if it is true that, from this category of names, albeit with the necessary caution, one can identify a kind of template of the historical road network, more or less accurately, according to individual cases, and if the present distribution of names generally agrees, it is not a binding rule that names linked with roads are actually evidence of previous routes, since names occasionally are shifted from the original site. As is well known, during the centuries following the fall of the Roman Empire, for example, the coastal areas were gradually abandoned and internal sites were preferred. The settlements favoured hilltops, while, in more recent times, the opposite has happened. In some cases, names may have followed the inhabitants of an abandoned site.

Undoubtedly it would be interesting to check—in the context of local names in a given territory—the percentage of names linked with roads out of the total and their relation to other name types, thus identifying the weight of the influence of the road system in the context of the components of the landscape reflected in toponymy. Routes are one of the main features, not only because they are visible, but also because they are fundamental, in consideration of their function as a basic infrastructure, for a series of movements, such as the circulation of people and goods (the latter linked to ideas) up to a very recent past, to the extent of taking on the role of an energetic factor in settlement and economy localisation. For example, the road was the most characteristic feature of the Roman rural landscape.

However, since it is, in most cases, the major roads (independently of the duration of their importance) that give rise to place names, it is clear that this category cannot play a leading role, when numerically compared, for example, with the names linked with locations, vegetation or the shapes of the land, these being more widely distributed and spread over the territory. It is true that the role of roads had been of considerable importance in the past, and, as such, was one of the territorial elements clearly envisaged by human beings when place names were created, that is when it was a duty and desire to identify objects on the landscape with a specific name.

Locations, another basic category in the appearance of a territory, are the certain source for a rich, varied place name corpus, as has been fully demonstrated by recent research. Hopefully, seeing that an overview, albeit not a truly comprehensive one, but at least illustrating the relationships between parts of the landscape and place names, has been presented a number of considerations emerging from a look at the entire place name corpus of a territory are called for.

What follows concerns an investigation of a countryside area with urban development belonging to the Florentine landscape, a hilly district densely inhabited since very ancient times characterised by scattered settlements and intense exploitation of the territory. The survey aimed, on the basis of the almost complete collection of names, to establish, with the aid of a quantitative calculation of the contribution of the various natural and human phenomena, how the geographical landscape, including historical and natural traits, was reflected in place names. What emerged clearly was that almost all of the names vividly expressed the process of the creation of the place name stock, reflecting the various natural and historical aspects of the area under examination and, despite the fact that they did not allow an organic reconstruction of the appearance of the territory, owing to changes over time in the historical stratification of names, there was a close relationship between local names and the physical and human characteristics of the territory. Besides, the conceptual categories of naming, although somewhat varied, turned out to be devoid of imagination and aesthetically banal, as can often be noticed in modern urban developments. This coherence was also confirmed by research in other parts of Tuscany. Hopefully, there will be further investigations of other territories.

From all the examples mentioned a number of mechanisms have emerged basic to the process of place naming, they are: the unconscious operations of registering specific ‘images’ provoked by a particular significant phenomenon or aspect and, complementary to this, the conscious, voluntary aim to clearly identify a place, foregrounding an aspect considered especially expressive. As already pointed out, place names are not banal, chance objects indifferent to surrounding reality. On the other hand, since they satisfy the need to identify an object, they aim to express a significant qualification in relation to the environment in question. In this way, they appear to be the expression of a mental operation aiming at highlighting specific features. In the case of a house in the country it could be the owner’s name or another evident characteristic such as a dovecote and so on. We should not forget that the same place can have different names given by people speaking a different language, just as in our towns and villages names, being mostly of ancient origin, reflect the way in which inhabitants of the past described them.

Landslides are also frequently present in Italian place names on the 1:25,000 Istituto Geografico Militare maps together with the actual presence of the phenomenon. Figure 4 shows the distribution of *lama-e* place names, one of the most frequent terms for the landslide.

Fig. 4 Distribution of place names from the label *lama, lame* (in large scale cartography 1:25.000)

Naming Strategies and New Place Names

There is no doubt that place name research is a diversified field, useful for investigation of previous territorial characteristics, as well as retrieval of historical memory of a territory aiming to inspire a feeling of belonging, as well as illustrating old and new naming strategies.

Seeing that place name research mostly addresses a spontaneous new naming, as a consequence of the need to acknowledge, point out, and circumscribe, in many cases place names are the product of a conscious strategy, which goes beyond the simple, albeit fundamental, need to locate and address the desire to claim possession, prestige, and feeling of belonging.

Here informed naming strategies followed by explorers, colonizers, dominant political regimes, and so forth come into play. It is enough to recall the naming strategy followed by Christopher Columbus in the Americas, which was quite different from that of Amerigo Vespucci, who, without being aware of it, gave his name to the new continent and another different case i.e. that of Giovanni da Verrazzano who, in his letter to the King of France and in the the map made by his brother, 'reconstructed', as it were, the Florentine landscape in America. And again, concerning Italy, one should recall the almost complete cancellation, after World War Two, of place names linked with the monarchy in many Italian towns, like avenues and squares such as *Corso Vittorio Emanuele* (after King Victor Emanuel the first King of united Italy), and *Piazza Torino* (the city where his dynasty originally ruled). Street names like *Via Roma*, (linked with Italian unification with Rome as its capital in 1871) remained. Reference can also be made to territorial marketing strategies mostly with tourism in mind (as will be treated further on).

Another interesting aspect of the development of place name coverage, as can be observed on the new 25,000 IGMI maps, concerns the appearance of new names in Italy.[6] The origin of these new names lies in the territorial organization—especially land reclamation and reform, housing (coastal resettlement, increase in built-up areas, new urban districts, etc.) productivity (with special attention to industrial areas), tertiary functions (names of large areas occupied by retailing and names linked with tourism), transport and communication infrastructures. Concerning territorial organization, the main repercussion of place names is the land reclamation which, as is well known, boasts a long history in Italy.

After 1950, a series of new place names—originally noted by Roberto Almagià—were given to houses and small villages built on expropriated large estates and land parcelled out into holdings, for example in the area between the Arno and Tiber rivers, where the Maremma Development Board (after the name of a desolate area in southern Tuscany) operated in areas which had been deserted for centuries.

As for the tertiary functions, the analysis of the recent 1:25,000 maps, together with help from the 1:10,000 technical regional maps, shows that the new names rely mostly on tourism. These names are often aesthetic in nature, linked with specific features. Note those assigned to tourist accommodation along the coasts all over Italy, including hotels, campsites, residences, as well as bays conjuring up peaceful, heavenly images, following advertising criteria ever more influential over recent decades, examples being *Costa degli Angeli* (*Coast of Angels*) and *Baia Paradiso* (*Paradise Bay*).

The character of these recent place names is certainly more general than the traditional one, which is so concrete and tied up with the local context. Nevertheless, it also has precise functions; first of all, that of attracting visitors' attention and affecting their perception, by signalling the fact that the place in question is ready to satisfy their

[6] L. Cassi, Nuovi toponimi, in *Italia. Atlante dei tipi geografici*, Firenze, Istituto Geografico Militare, 2004.

expectations. Tourism, besides being representative of the main economic driving force in the creation of new place names, makes use of strategies and evocative allusions which are totally different from traditional place naming processes.

The case of Sardinia is a good example, seeing that the island's coastline has been the object of what we could call a kind of 'place naming re-colonization, to the extent that the new place names are a long way from the traditional imprint, as a consequence of the impact of tourism on coastal areas. Some of these names are already official, since they appear on the new 25 000 sections, while others are only to be seen on tourist maps.

Even a quick look at some of these names foregrounds the desired image of the Sardinian coastline, in a combination of aesthetic emphasis and the satisfaction of the tourist's generic imaginary: *Costa Smeralda* (*Emerald Coast*), *Costa Dorata* (*Golden Coast*), *Costa Corallina* (*Coral Coast*), *Torre delle Stelle* (*Tower of the Stars*), *Villaggio Spiaggia Bianca* (*White Beach Village*), *Costa Rei* (*King's Coast*), *Spiaggia Due Mari* (*Two Seas Beach*), *Costa Paradiso* (*Paradise Coast*), *Valle della Luna* (*Moon Valley*), *Eden Beach*, *Costa Verde* (*Green Coast*), *Costa del Sud* (*Southern Coast*), *Cala Romantica* (*Romantic Cove*), Residences: *Villaggio Alga Bianca* (*White Seaweed Village*), *Villaggio Cala Paradiso* (*Paradise Cove*), *Villaggio Le Farfalle* (*Butterflies Village*).

In cases like Sardinia new place names paint a new picture of the territory in question, quite different from that of traditional place naming, and could even be seen as a kind of 'environmental aggression'; however, they have new territorial functions, in the face of a territory elevated to internationally famous tourism, the *raison d'être* of which is the island's undeniable beauty, which, at the same time, adheres to the consolidated logic of world-wide models.

The banality of 'tourist' place names is only superficial, seeing that place naming has proved to be sensitive to the value assigned to leasure activities, which have now become an essential part of society (Fig. 5).

Fig. 5 New place names along the coast of Sardinia

Geographical Information Systems and Place Names

The methodology and basic setup of work regarding 'geographical and indicators' for the indexing and collection of place names, apart from the revision of place names large scale mapping, are still valid in their basic assumptions. However, it cannot be denied that this kind of investigation can be greatly aided by geographical information systems. This is the case both at their minimum potentiality level, such as the representation of the distribution of a specific term and related place names,

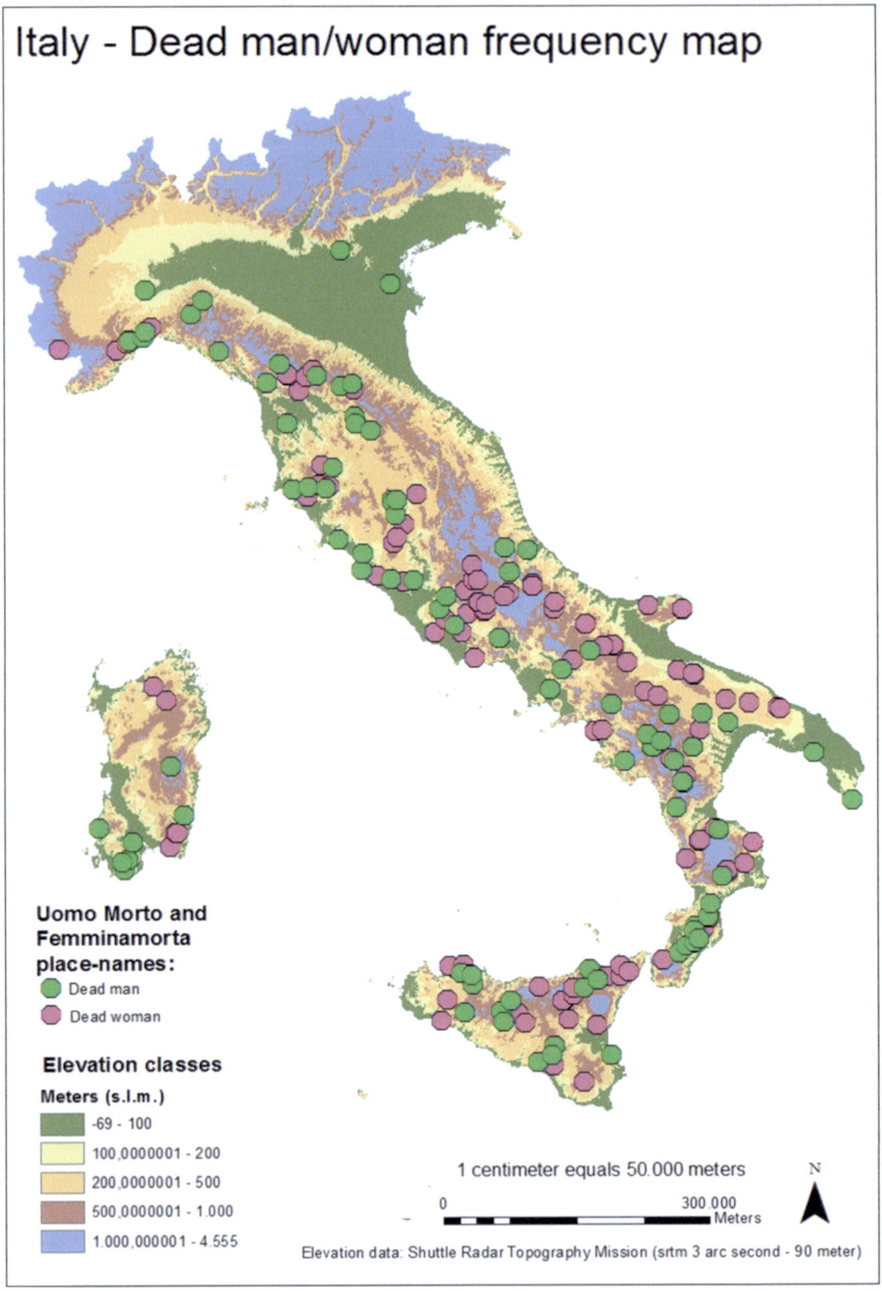

Fig. 6 *Uomo Morto* (*Dead Man*) and *Femmina Morta* (*Dead Woman*) place names (in large scale cartography 1:25,000)

as well as when they are more complex. They are a useful tool for the short term, simple production of geographically referenced data bases, which can be analysed on the basis of spatial relations and processes, which will allow the production of thematic maps for future use. Figure 6 illustrates the distribution of *Omo Morto* (*Dead Man*) and *Femmina Morta* (*Dead Woman*) place names in Italy taken from 1:25,000 mapping.

Concluding Remarks

From all the examples a number of basic mechanisms of the formation process of place names have been illustrated: unconscious registration of specific 'images' deriving from a significant phenomenon or aspect and complementary to this, the conscious, voluntary aim of identifying a place by foregrounding an aspect considered particularly expressive. Naming a place goes beyond the simple need to distinguish it or signal it topographically. It nearly seems to transmit the function and history of the named object, thus acquiring a specific cultural value. Place naming mirrors a direct relationship between human beings and the land, a reflection of the environment and landscape to which it intrinsically belongs. Place names thus represent one of the most complex and, at the same time, emblematic aspects of the process of becoming acquainted with and organizing a territory by human social groups and is one of the strongest identifying traits of local culture as well an evident manifestation.

Questions

1. Elucidate the concept of the 'historical memory of a territory' in Italy's regional landscapes.
2. Define what the 'cultural heritage and expression of perception by generations of their environment' means in a globalizing world.
3. Make at least three arguments why place names are one of the most important components of the intangible cultural heritage of a region and/or country.
4. Explain the reasons why the processes of certification and standardization of place names is so important today.
5. What is the complex links between place names, individual landscape and territorial categories?
6. What are the narratives of place naming in Italy, from the examples given both nationally and regionally?

Geographical Names as Indicators of the Environment: Case Study in Bandung Basin, West Java, Indonesia

Multamia R. M. T. Lauder, Titi Bachtiar, and Cece Sobarna

Abstract Indonesia is a vast country with 17,000 (thousand) islands covering three time zones and has a cultural diversity with 718 languages. One of the interesting areas to be discussed from a hydronymy point of view is (in the area of) the Bandung Basin. It is home to the Sundanese ethnic group (. The Bandung Basin) and is the base of the Ancient Bandung Lake which was formed due to the eruption of *Mount Sunda* 105,000 years ago. The Ancient Lake collapsed around 16,000 years ago, gradually receding, leaving behind a vast lake and wetlands. Archaeological findings around the lake suggest (ing) that the surrounding area once supported early human habitation that provided water and food. The Sundanese have a legend called *Sangkuriang* that mentions the existence of the lake, as well as the mythical origin on the lake's creation. This might suggest the collective memories about the lake's existence (,) was transmitted by oral tradition through the ages. In this Bandung Basin, we found an abundance variety of generic terms for hydronyms. Business and tourism interests appear to be less compliant with the UNGEGN resolution to use local names. Changing geographical names actually eliminates knowledge about natural conditions including earthquake mitigation.

Keywords Bandung basin · Collective memory · Generic terms of hydronym · Tourism · Earthquake mitigation

---

M. R. M. T. Lauder (✉)
Universitas Indonesia, Depok, Indonesia
e-mail: multamia@ui.ac.id

T. Bachtiar
Bandung Basin Research Group, Bandung, Indonesia

C. Sobarna
Universitas Padjajaran, Bandung, Indonesia

© The Author(s), under exclusive license to Springer Nature Switzerland AG 2023
G. O'Reilly (ed.), *Place Naming, Identities and Geography*,
Key Challenges in Geography, https://doi.org/10.1007/978-3-031-21510-0_26

Objectives

By the end of this chapter readers should be able to:

- (understand that) see how toponymy studies require a multidisciplinary approach
- understand that collective memory can preserve geomorphological information (in) about the past
- (understand) be cognisant that composition of generic and specific (for) hydronyms are cultural artifacts
- (discuss about) debate the consequences of changing geographical names related to disaster problems
- (discuss about) elucidate socio-political and economical pressures to preserve geographical names

Introduction

Indonesia is one of the world's largest archipelagic states. Indonesia has 17,000 islands (Badan Informasi Geospasial 2022) with the five largest islands: Java, Sumatra, Kalimantan, Sulawesi, and Papua. The terms archipelagic state and archipelago are defined in the United Nations Convention on the Law of the Sea (UNCLOS 1982) Article 46 (a) that "archipelagic State" means a State constituted wholly by one or more archipelagos and may include other islands and (b) "archipelago" means a group of islands, including parts of islands, interconnecting waters and other natural features which are so closely interrelated that such islands, waters and other natural features form an intrinsic geographical, economic and political entity, or which historically have been regarded as such (United Nations 2009, p. 23).

The territory of the Republic of Indonesia stretches from Rondo island off the northern tip of Sumatra (Latitude 6° 05′ North) to the southwest point of Roti island (Lat. 11° 15′ South), and from Longitude 94° 58′ to 141° 00′ East (Cribb and Ford 2009). With an overall distance of nearly 5,000 km from east to west Indonesia (covers) covering an area which is almost from UK to Greece which stretches over three time zones. Forbes (2014) states that nearly 80% of the area between Indonesia's geographical extremities consists of seas.

The vastness of the territory and also the very diversity of languages and cultures in Indonesia directly impact the geographical names. First, based on the numbers, there are geographical names which are abundant that should be managed in the National Gazeteer. Second, there are variations in the structure of different names because each ethnic group has a different language, and there are different naming structures including composition of generic and specific names.

The diversity of language and culture can be seen from language mapping throughout Indonesia based on research from 2560 villages it has documented 718 languages (Badan Pengembangan dan Pembinaan Bahasa 2019). In general, it can be seen that the number of local languages is bigger in the east, especially in the regions of Sulawesi, Maluku and Papua.

The presentation of the Distribution of Languages in Indonesia is the result of a collaboration between the Geospatial Information Agency (Badan Informasi Geospasial) and the Language Development and Fostering Agency (Badan Pengembangan dan Pembinaan Bahasa) (Fig. 1). In the past, the Geospatial Information Agency managed geographical names only to the extent that they were written on the map. Currently, the Geospatial Information Agency is the Indonesian National Names Authority (who) that is responsible for developing the National Gazeteer. All geographical names in the Gazetteer are to be the authoritative and sole reference for use in society. This includes names in all official documents, (the) mass media, and school text books. Also, importantly, it is the main reference source for map editors. Maps provide various views of the natural and manmade environment on or below the earth's surface. The Geospatial Information Agency creates a master map which is to be the sole basis for all other maps such as maps for mining, (for) tourism, (for) property development, or (for) natural disaster prediction. It is important that maps which are used as an important reference are reliable and standardized with reference to names and labels (Lauder 2014; Lauder and Lauder 2015).

In the recent developments, many ministries and agencies are starting to acknowledge the importance of the geographical name, the urge to have a firm standpoint for the standardization arises. The existing standardization regulation issued by Geospatial Information Agency considered needs reinforcement to gain a higher legal power and has a wider reach in society. However, further discussions revealed that the geographical name holds a vital role in many aspects of national interests, such as preserving the local language, national sovereignty and identity in international society. Therefore, the urgency to have an even more elevated legal base for geographical standardization is inevitable.

Before proposing a new regulation, actually Indonesia already (have) has several existing Regulations and Laws which is fragmented and beginning to be dated such as (1) Minister of Home Affairs Regulation Number 39/2008 on General Guidance in Standardization of Geographical Names; (2) Minister of Home Affairs Regulation Number 35/2009 on Guidance in the Formation of Committee Standardization of Geographical Names; (3) National Law Number 4/2011 on Geospatial Information; (4) National Law Number 23/2014 on Regional Government; (5) Presidential Regulation Number 116/2016 on The liquidation of National Team on Standardization of Geographical Names; (6) Geospatial Information Agency Regulation Number 6/2017 on Standardization of Geographical Names; (7) National Law Number 24/2019 on Flag, Language, State Symbol and Anthem; and (8) Presidential Regulation Number 63/2019 on Bahasa Indonesia. The level of regulations also varies from National Law down to the organizational level. The Geospatial Information Regulation Number 4/2011 noted that the geographical name is one of elements that forms a topographical map which the Geospatial Information Agency is obliged to provide, and also the National Law Number 24/2009 mentioned that a geographical name has to be in Indonesian or can use the local language if it has historical value and supports the local culture.

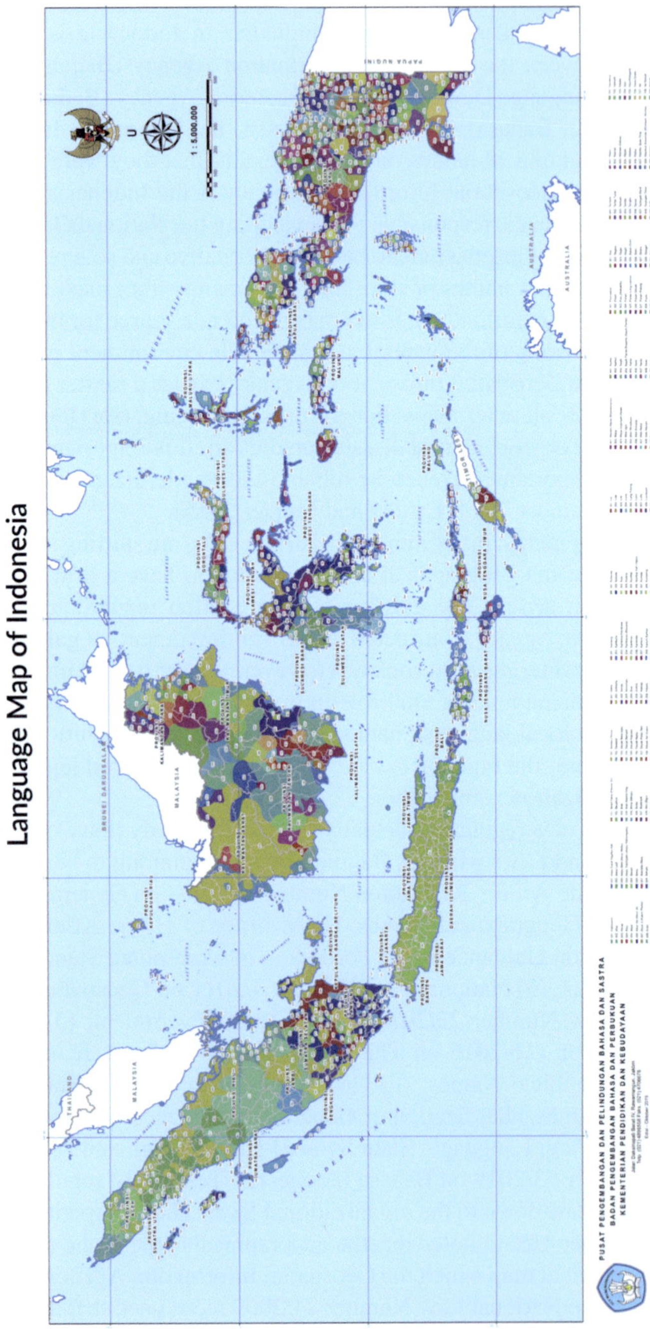

Fig. 1 The Distribution of Languages in Indonesia (map courtesy of Ministry of Education)

The Ministry of Home Affairs (Kementerian Dalam Negeri) and the Ministry of National Development Planning (Kementerian Perencanaan Pembangunan Nasional) agreed and fully supported the idea to uplift the regulation concerning the procedure of geographical name standardization to a presidential regulation level. The drafting of the presidential regulation took about two years, from mid-2018 to early 2020. It turns out that it is not easy to unify all regulations that can be accepted by various ministries and agencies. Each agency has a different focus and interest so that long discussions and debates cannot be avoided. Finally, after undergoing several revisions, the presidential regulation contains comprehensive standardization of the geographical name and was signed by President Djoko Widodo on January 7th, 2021 and released to the public on January 8th, 2021. The regulation encompasses all the standardization processes, started from data acquisition, data gathering, verification, and publication of the standardized name in the form of a national gazetteer.

National Law (Undang-Undang) Number 24/2019 on Flag, Language, State Symbol and Anthem, Section 36 Regulation for Geographical Names:

1. Indonesian must be used in geographical names in Indonesia
2. Geographical names have only one official name
3. Indonesian language must be used for names of buildings, roads, apartments or settlements, offices, trade complexes, trademarks, business institutions, educational institutions, organizations established or owned by Indonesian citizens or Indonesian legal entities.
4. Naming can use regional languages or foreign languages if they have historical, cultural, customary, and/or religious values.

The Presidential Regulation, Number 2/2021, on Standardization of Geographical Names, Section 3 Regulation for Geographical Names:

1. Use Indonesian Language
2. May use local language or foreign language if the geographical features has historical, cultural, customs, and/or religious values
3. Use the Roman script
4. Use one name for geographical features
5. Respect the existence of ethnicity, religion, race and class
6. Limit names to a maximum of three words
7. Avoid using proper personal names of people who are still alive or may use the name of the deceased person for minimum of five years after the death of the person
8. Avoid using the name of an agency or institution
9. Avoid using names that are contrary to national and/or regional interests
10. Fulfill the orthography principles of geographical names

In general, the standardization falls within the Geospatial Information Agency's lead, except for the administrative boundary's name, the Ministry of Home Affairs is responsible to provide official name from the provincial level down to the village. The regulation also highlights the significance of the national gazetteer as the authoritative document of Indonesia's geographical name. The national gazetteer will be regularly updated and reported to the President and become the sole reference for Indonesia's standardized geographical name. Last but not least, the regulation encompasses the technical part regarding the tool to perform the standardization. It states that the Government has to provide the official instrument to achieve the standardization.

Challenges Preserving Geographical Names

Data from the field shows that it is not easy to preserve and standardize topographical names because there are many economic and tourism incentives to use foreign names. Actually, changing the toponym is the same as changing the info of cultural artifacts about natural situations and migration of people. For this reason, a case study that focuses on hydronyms in West Java will be presented.

Toponymy is the study of place names (Kadmon 2000). Place names provide evidence that people in one country were aware of the existence of other countriesand this suggests the possibility of contacts between countries or civilizations. The methodology of toponymy can help reveal the original meaning of place names. It uses methodologies that combine information that may be in different languages in old texts with field work (Taylor 2016). The study of place names is of use in a number of fields, such as historical linguistics, archaeology, geography, history, and other subjects (Hough 2016).

This study stirs the lost memories of former villages that are gone. One of the evidences is the existence of development that requires a very large area of land, namely the construction of three dams (in) on the Citarum river. This development eliminates dozens of villages for the construction of the dams, namely the construction of Ir. H. Djuanda Dam, which was built in 1957 and inaugurated in 1967, uses an area of 8,300 ha, thus drowning 14 villages. Further upstream from this lake, Saguling Dam was built, which operated in 1986, covering an area of 5,600 ha. Between Lake Ir. H. Juanda Dam and Saguling Dam, another dam, namely Cirata Dam, was built in 1988 with an area of 7,111 ha. In addition, Jatigede Dam has an area of 3,035 ha which dams up the Cimanuk River in the Sumedang Regency. The dam was inaugurated in 2015, but only became fully operational in 2017 by (sinking) flooding 28 villages. Automatically, the names of the villages also disappeared. The condition of economic needs triggers the difficulty of preserving village names which are cultural artifacts and local historical records (Bachtiar and Syafriani 2016).

In addition to the missing villages being (sunk) flooded for development purposes, there are also sites in South Bandung that have gone unrecorded. To compare the condition of the Bandung Basin in the early nineteenth century with the Bandung Basin in the early twenty-first century, it can be examined through colonial maps, compared to maps made today by the Geospatial Information Agency. Bandung continues to grow. The appeal of Bandung as a dream city that you can climb the peaks of social and economic status continues to be the dream of many people. Residents began to arrive in Bandung, so areas that were originally in the form of wetlands such as swamps began to be stockpiled, so an area called *Situsaeur* 'the swamp that was stockpiled' was formed. Furthermore, there are parts of the area which were later named *Sukaleueur* 'soil heaps when exposed to rain becomes slippery' and *Soledat* 'slippery soil that can derail people'. Even the swamp that was stockpiled was used as land for rice fields, so that some locations were named *Bebedahan* 'opening new land', *Pasawahan* 'rice field area', and *Lembursawah* 'kampung rice fields'. There is also a fish pond and it is named *Balonggede* 'big pond'.

In the beginning around the year (of) 1800, when people started to come to Bandung, the natural condition around the Bandung Basin was still green, so that the meteoric water caught by the forest was directly absorbed through the tree roots into the ground, then the water appeared regularly as a spring which the locals called *seke*. Gradually, many newcomers came to Bandung and chose to live near water sources. Currently, there are many village names that use the generic term *seke*. Examples of the name of a spring or *seke* which later became builtup as a residential area exist, namely *Ciseke; Sekeawi*; *Sekebalingbing; Seke; Sekereundeu; Sekeandur; Sekeangkrih; Sekebunar; Sekebungur; Sekebulu; Sekecariu; Sekedangdeur; Sekeburuy; Sekegentong; Sekehonje; Sekegawir; Sekejengkol; Sekejolang; Sekejati; Sekekuda; Sekekukumbung; Sekekondang; Sekelimus; Sekeloa; Sekelapa; Sekemalaka; Sekemandung; Sekemala; Sekemerak; Sekemirung; Sekemeer; Sekepicung; Sekepondok; Sekepanjang; Sekesalam; Seketimbang.*

The Bandung Basin

In 1996 UNESCO designated Sangiran on the island of Java as a world heritage site. The island of Java is a historic island, 50% of homo erectus worldwide (are actually) inhabited (in) Sangiran for the past one and a half million years (Widianto and Noerwidi 2020). So, Java island is a very (an old) historic place which supports human living (way back) for millions of years. Bandung Basin located (in) on Java Island.

Inside the Bandung Basin there is Bandung City as the capital of the West Java Province as the third largest city in Indonesia with a population of 2,527,854 people (Biro Pusat Statistik 2022 (https://bandungkota.bps.go.id)). Bandung City (is) in (The) Bandung Basin (which) is the base of the Bandung Ancient Lake. The city of Bandung was inaugurated as the new capital of Bandung Regency with a decree from the Dutch East Indies Government on September 17th, 1810. Now, September 17th is considered to be the anniversary of the City of Bandung.

That ancient lake was formed due to the eruption of *Mount Sunda* 105,000 years ago. The ancient lake collapsed around 16,000 years ago, gradually receding, leaving behind a vast lake and wetlands. Archaeological findings around the lake discovered several obsidian stone tools. Obsidian is a stone naturally formed from quickly cooled lava (Binder et al. 2011). This findings suggesting that the surrounding area once supported early human habitation that provided water and food. The Bandung Basin has been inhabited by humans for millenia. (at least thousands of years.) At the end of 2003, four intact skeleton fossils of Homo Sapiens were found in Pawon Cave

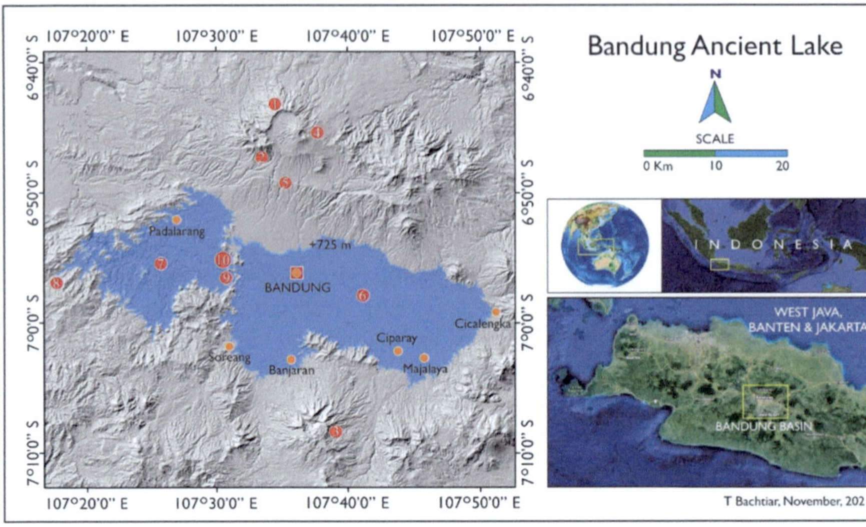

Fig. 2 Bandung Ancient Lake. *Source* Titi Bachtiar, November 2021

north of Padalarang near the Bandung area, which were estimated to be 9000 years old (Kartodiwirio 2006).

The height of Mount Sunda +4,000 m above sea level, (erupted) experienced several periods of eruption. According to Kartadinata (2005), the eruptions of Mount Sunda occurred between 128,000 and 105,000 years ago. The massive eruption of Mount Sunda in prehistoric times can be classified into the type of plinian eruption, an eruption with very high gas pressure, causing material from inside the mountain to be ejected into space, then fall, then spread to various areas in an area of 200 km^2. Eruption material that slid down the southern slope, immediately dammed Citarum river in the north of Padalarang. Due to the large amount of material from the body of the mountain that was scattered into the sky, there was a void in the body of the mountain, causing the upper part of the body of the mountain to no longer be strong enough to support the load, and then collapsed to form the caldera of Mount Sunda (Fig. 2).

On the map of the early twentieth century there are still remnants of Ancient Bandung Lake in the form of a very wide swamp. The location is south of Ujung-berung, the name is *Rawa Gegerhanjuang* 'Gegerhanjuang Swamp'. The area of this swamp is at least 75 km^2. Information on the existence of the *Rawa Gegerhanjuang* is recorded on the *Het District Map of Oedjoengbroeng Kidoel* which is a collection of the National Archives of the Republic of Indonesia, which was done by F. de Haan (1911). The location of this very large swamp is also recorded in the *Kaart van Het Eiland Java* (1845), in the *Etappe-Kaart van Java en Madoera* (1878), and in the *Geologische Beschrijving Map van Java en Madoera* (1894) but the name of the swamp is not written. *Rawa Gegerhanjuang* is the remnant of East Bandung Ancient Lake which has begun to recede, whose remains are still visible in the South

Bandung area. Now the rest of the ancient lake has been managed into ponds and rice fields. In line with the development of the city, the development of all the needs of its citizens has turned the landscape into a builtup área (Bachtiar and Syafriani 2016).

Cultural Collective Memories

Collective memory (ies) is used here to describe the knowledge within a community, often with only spoken language and without a writing system, passed down from generation to generation and used to be of benefit to them. People's collective memories play an important role in storing significant knowledge learned from past natural disasters. Some of those messages contain important lessons on how to cope with natural disasters if they should occur again (Tulius 2020). Based on evidence, reported at ITIC-IOC (2022) we know that the Indonesian island of Mentawai has experienced several tsunamis in 1797, 1833, 2007, and 2010. However, the series of tsunamis memories were not passed on and preserved in the collective memory, so that when recently a tsunami occurred, many died because they did not know what was happening or what had to be done to escape. However, there (There) are examples of when the collective memory works well. One of these is the Simeuleu Recency in Aceh, which nurtured and left behind a collective memory of the disaster which they called *seumong*. Thanks to that, many lives were saved when the tsunamis disaster came again (Gadeng et al. 2018). Of course, various collective memories of the disaster are stored in their respective local languages. Based on good practice in Aceh, efforts to use local languages have an important value and so need attention (Lauder 2021).

How about collective memory in Bandung? If the sky is clear, in the north of Bandung you can see an active volcano that looks like an overturned boat. That is Mount Tangkubanparahu, which is a marker of the earth, as a city landmark. The basic geomorphology of the flat Bandung Basin, with Mount Tangkubanparahu whose peak is cut flat, has given birth to the legend of Mount Tangkubanparahu with the main characters Sangkuriang and Dayang Sumbi. This legend is an answer in the past to the question, why is Mount Tangkubanparahu shaped like an overturned boat? The legend explains how the damming of the river occurred and how Mount Tangkubanparahu was formed. This is the answer to the natural events that occurred millenia ago. (at that time.)

This legend is very old, because it was known and very popular in the community in the fifteenth–sixteenth centuries. Bujangga Manik, a traveler from the Sunda Kingdom, recorded the legend while passing by on the outskirts of the Bandung Basin. (These days are) This is probably similar to a travel blogger today who travels to different places, writes about their experiences and publishes them onto an online blog. Bujangga Manik wrote his experiences (in) on palm leaves. The Bujangga Manik manuscript is one of the ancient Sundanese texts about his journey around the islands of Java and Bali. The manuscript describes the geography and topography at the time it was written. More than 450 names of places, mountains and rivers are mentioned. Most of these place names are still in use and/or recognized today. The manuscript in the form of a narrative poem has been kept in the Bodleian Library at the University of Oxford since 1627 (Noorduyn and Teeuw 2006).

The following is the final chapter of the legend of Mount Tangkubanparahu mentioned by Bujangga Manik. It is said that after Sangkuriang walked away from his mother, he continued to walk east and learned knowledges from various elders. After studying for a long time, Sangkuriang did not realize he had walked back to the west. Somewhere, he met a beautiful charming woman. Sangkuriang fell in love with her. However, in the course of time, Dayang Sumbi finally found out that the dashing young man was (his) her own son who had long gone away, so it was impossible to marry (his) her own child. But Sangkuriang did not believe it. He did not believe that the woman was Dayang Sumbi, his mother, so he remained in his position to ask for her hand. Dayang Sumbi finally proposed three terms to Sangkuriang if he was serious about asking for her hand. First, he has to stem Citarum river into a lake. Second, he has to build a boat to cruise on the lake. Third, the lake and the boat must be finished in one night, (finished) completed before dawn. If it is not completed, the engagement will be canceled. Sangkuriang agreed to the three conditions. First he dammed the river, and in a flash the lake was finished, and the water started to pool. After making a lake, he immediately cut down the best tree to build a boat. Seeing this situation Dayang Sumbi began to worry. She immediately asked the Almighty to cancel all the agreements, because it was impossible to marry (with his) her own child. After praying, Dayang Sumbi sprinkled the pieces of the shroud to the east, which turned an instant to dawn, a sign that the sun will soon rise. Dayang Sumbi was relieved, but it was very disappointing for Sangkuriang. With mounting anger, he kicked the almost finished boat upside down. Sangkuriang immediately chased Dayang Sumbi. On a small hill, Dayang Sumbi disappeared. On the hill where she disappeared then beautiful flowers bloomed. That hill was named *Gunung Putri* 'mount of princess'.

The upside down boat kicked by Sangkuriang is the shape of the mountain, that's why the Sundanese people named the mountain as *Tangkubanparahu. Tangkub/Nangkub* 'upside down' and *parahu* 'boat'. This collective memory (ies) about the ancient lake and mountain also stored as legend (and) is retained in a *pantun* 'Sundanese poetry'.

Generic Terms of Hydronym in Sundanese

One of the interesting areas to be discussed from a hydronymy point of view is found in the province of West Java, especially in the area of the Bandung Basin. It is home to the Sundanese ethnic group. The (L)language that they use is Sundanese. According to SIL International, Sunda [sun] has 27,000,000 speakers, Class: Austronesian, Malayo-Polynesian, Sundanese. Dialects: Banten, Bogor, Karawang, Priangan, Cirebon. Location: Western third of Java Island (SIL 2006: 78).

After Bandung Ancient Lake collapsed, the lake water receded, leaving vast wetlands in the sunken area, which in Sundanese are called *situ* (lake) and *ranca* (swamp). That is why, at the base of the Bandung Basin, from Cicalengka in the east to Rajamandala in the west, from Dago in the north to Majalaya, Baleendah, Banjaran, and Soreang in the south, many geographical names are found which have the characteristics of the earth as a watery area or have a very close relationship with the surrounding water, as can be seen from the use of the generic term *situ* 'a large natural pond or lake', *ranca* 'swamp, wetlands which is always flooded naturally', or *bojong* 'headland, land that juts into the water'.

We found 96 hydronyms that use the generic term *ranca* 'swamp', namely: *Ranca; Rancaendah; Rancakondang; Rancapateuh; Rancabadak; Rancaengan; Rancakuluwung; Rancapelem; Rancabali; Rancaeunteung; Rancakuya; Rancapeuteuy; Rancabango; Rancagarut; Rancalame; Rancaputung; Rancabaraya; Rancagede; Rancaloa; Rancarayud; Rancabatok; Rancagempol; Rancalonggong; Rancasaat; Rancabatuk; Rancagirang; Rancamalaka; Rancasabir; Rancabawang; Rancagoong; Rancamalang; Rancasadang; Rancabayawak; Rancaherang; Rancamanuk; Rancasagatan; Rancabelut; Rancaheulang; Rancamanyar; Rancasari; Rancabentang; Rancailat; Rancamekar; Rancasawo; Rancabeureum; Rancajigang; Rancamelong; Rancaselang; Rancabiuk; Rancakalong; Rancameong; Rancasenggang; Rancabogo; Rancakamuning; Rancamidin; Rancasepat; Rancabolang; Rancakandang; Rancamoyan; Rancatanjung; Rancabulan; Rancakasa; Rancamulya; Rancatawu; Rancabungur; Rancakasiat; Rancanangkub; Rancatengah; Rancabuntu; Rancakaso; Rancanata; Rancatiis; Rancacagak; Rancakasumba; Rancanilem; Rancatungku; Rancacangkuang; Rancakembang; Rancanumpang; Rancawaas; Rancacatang; Rancakemit; Rancanyiruan; Rancawali; Rancacenang; Rancakendal; Rancaoray; Rancawaliwis; Rancacili; Rancaketan; Rancapacing; Rancawangi; Rancaekek; Rancakihiang; Rancapanjang; Rancawaru.*

We found 17 hydronyms that use the generic term of *situ* 'lake', namely: *Babakansitu; Situciparay; Situsaeur; Bloksitu; Situdarma; Situsari; Cisitu; Situemuh; Situtarate; Situ; Situgede; Situtengah; Situaras; Situgunting; Situwangi; Situbolang; Situhiang.*

We found 79 hydronyms that use the generic term of *bojong* 'headland', such as: *Bobojong; Bojonggede; Bojongkunci; Bojongpeuteuy; Bojong; Bojonggempol; Bojonglaja; Bojongrangkas; Bojongasem; Bojonghaleuang; Bojonglandeuh; Bojongrengas; Bojongasih; Bojonghaur; Bojongloa; Bojongreungit; Bojongawi; Bojongjambu; Bojongmalaka; Bojongsalak; Bojongbabakan; Bojongjati; Bojongmalati; Bojongsalam; Bojongbadak; Bojongjaya; Bojongmanggu; Bojongsari; Bojongbaraya; Bojongjengkol; Bojongmanjah; Bojongsayang; Bojongbecik; Bojongkacer; Bojongmanyar; Bojongsempur; Bojongbolang; Bojongkalapa; Bojongmekar; Bojongsereh; Bojongbuah; Bojongkalong; Bojongmenje; Bojongsero; Bojongbubu; Bojongkasiat; Bojongmenje; Bojongseureuh; Bojongbungur; Bojongkasur; Bojongmonyet; Bojongsoang; Bojongcibodas; Bojongkawung; Bojongmulyasari; Bojongsuren; Bojongcijerah; Bojongkihiang; Bojongmuncang; Bojongtahu; Bojongcitepus; Bojongkoang; Bojongnangka; Bojongtanjung; Bojongemas; Bojongkondang; Bojongpacing;*

Bojongtepus; Bojongeureun; Bojongkoneng; Bojongpari; Bojongwaru; Bojong-galatik; Bojongkukun; Bojongpari; Cibojong; Bojonggaok; Bojongkulon; Bojong-peundeuy.

We also found 15 hydronyms that use the generic term *tanjung* 'cape', namely: *Babakantanjung; Rancatanjung; Tanjunglaya; Bojongtanjung; Tanjung; Tanjung-pura; Curugtanjung; Tanjungjaya; Tanjungraya; Karangtanjung; Tanjungkerta; Tanjungsari; Nanjung; Tanjunglaksana; Tanjungwangi.*

Hydronyms that (was) were used by the local people are cultural artifacts that store information about the natural conditions of the location. Naming from the ancestral side, is not arbitrary but really describes the conditions under which the name was made. Based on Nystrom's categorical meaning (2016), all hydronyms in Bandung generally refer to natural conditions such as *Bojongawi* 'bamboo trees in foreland basin' or about hope and expectations such as *Tanjungjaya* 'cape of victory'.

Beside dominant generic term of *ranca, bojong, situ, seke* and *tanjung* there are (few) some other generic terms which also convey the characteristics of the earth as a watery area or have a very close relationship with the surrounding water (S) such as *Ancol* 'foreland basin'; *Bantar, babantar,* or *parung* 'part of the river where the flow is heavier'; *Balong* or *Empang* 'a large pond where fish are kept'; *Beber* 'the river flow is slightly blocked, so it is a bit stagnant'; *Bugel* 'closure of the river mouth by sand because the river flow decreases during the dry season'; *Dano* or *Talaga* 'a vast natural body of water, surrounded by land'; *Lengkong* or *Bobojong* 'headland, land that juts into the water'; *Leuwi* 'part of the river whose bottom is deeper'; *Muara* 'where the river ends in the lake or in the main river'; *Nambo* 'the old course of a river now deserted, the former bed of a river'; *Parakan* 'a place for parading fish in a swamp'; *Rawa* 'swamp, wetlands which is always flooded naturally'; *Tanggeung* 'the part of the river whose bottom is slightly uphill, so the flow slows down'; and *Teluk* 'bay, body of water that juts into the land'. Considering that the Bandung Basin originates from an ancient lake, the related generic terms for hydronym are very rich and varied to store the collective memory of the local population.

The water stored in the *ranca* or *situ* there, has ecological and economic benefits. The lake water can cool the temperature of an area. The sun's rays can be softened, so the air becomes cool. With the disappearance of swamps and lakes in the Bandung Basin, the temperature in the area will heat up, moreover the built-up area is getting wider, the number of motorized vehicles is increasing, without being accompanied by reforestation of the area. In the past, the residents of the Bandung Basin in naming geography features were not arbitrary, and were closely related to the natural and cultural conditions of their people. For scientists and legislators alike, the environmental messages passed down should be clear.

Earthquake Mitigation for Bandung Basin

Today, most of the residents of the Bandung Basin build their houses on the former bottom of the Ancient Bandung Lake, or on lake sediments. On a flat surface with an altitude between +665 m (above sea level) and +725 m above sea level, it is now densely packed with settlements, schools, places of worship, factories, shops, markets, sports fields, office buildings, apartments, and roads. It was there that the residents of the Bandung Basin socialized, on the former sediment of the Ancient Bandung Lake which had been dried. Toponyms can help provide information about the local geographic situation and conditions of reducing the impact that will occur if and when a natural disaster occurs.

Another thing that must be considered in building at the base of the Bandung Basin is earthquakes. Earthquakes are a natural necessity, so they will keep repeating themselves. Naturally, the Bandung Basin is surrounded by fault lines, both in the north, west, east, and south of the region. However, what has been studied in detail is only the Lembang Fault. Daryono (2019) researched all known evidence of the movement of the earth's crust, such as what happened in the Lembang Fault over a span of 10,000 years, the fault is classified as an active fault but now has (are already) dense settlements. Naturally, (A) along this fault line, buildings will suffer damage in the event of an earthquake. Mitigation efforts are needed regarding materials, construction, and rescue methods before the earthquake actually occurs. The potential threat of danger originating from the Lembang Fault can be estimated at its magnitude, with full strength, it can trigger an earthquake up to a magnitude of 7.

Now, *Lembang* as a lake at the foot of the Lembang Fault is one of the favorite tourist destinations because of its beautiful scenery and cool mountain air. Distance from Bandung City to Lembang is only 8 miles. As a result, in the Lembang area there are many restaurants, tourist attractions, and hotels starting from 5 star hotels down to homestays, so the area is increasingly crowded. Lembang is located at one point in the stretch of the Lembang Fault, starting from Lembang in the east to Mount Palasari and in the west until the Cimeta river flows. In the past there was a large puddle of water, so that it became a distinctive feature of the earth, so that's why the area was named *Lembang* 'large puddle of water'.

By knowing the potential threat from the Lembang Fault movement, the best way is to be prepared. Earthquake mitigation is key in all actions. Both the mitigation carried out by the State Authorities in various regulations and their application in the community, as well as the mitigation carried out by the community. In fact, it is not only along the Lembang Fault that must be carried (y) out mitigation, but also among the entire population in the Bandung Basin. Because from one earthquake source, the waves will propagate in various directions. Moreover, the active fault that criss-crosses in the Bandung Basin is not only the Lembang Fault, but there are other faults that cut across the (Bandung) Basin. The potential threat is clear, but it is not known exactly when it will occur. Therefore, one must always be alert, where to build houses and buildings, and how to act if an earthquake does occur. Communities and local governments must know what to do before, during and after an earthquake.

Geographical names can give basic information about the area and the results of scientific research can be used as the basis for the State Authority that issues the building permit. It seems that the attractiveness of economic profit does not pay attention to geographical names which are valid records of natural conditions. That is why, one of UNGEGN's important resolutions is to use local names and (do) not change local names because it will confuse cultural artifacts and the collective memory of local residents.

Contemporary Geographical Names and Tourism

Geographical names given by the ancestors are not arbitrary. The names of places are mostly related to the character of the earth of that place, both the physical state of the earth, the hue of the earth, plants, animals, events, and cultural activity. In the development of the lives of residents who live or come from that place, place names will be attached to their mind, and will be written forever in various information for (various) multiple purposes, such as written on birth certificates, identity cards, marriage certificates, and also tombstones. Some public figures may even write the geographical name behind their own name, such as: *Nawawi Al Bantani* 'Nawawi from Banten' or *Syekh Siroj Garut* 'Sheikh Siroj from Garut'. Another phenomena, a local community changes the original name of their religious leader with the word *Ajengan* 'religious title in West Java' and then this is then followed with the village name, such as Kyai Haji Ilyas Ruhiyat became *Ajengan Cipasung* 'religious leader at Cipasung'. Same case for *Ajengan Cibedug, Ajengan Cianting, Ajengan Pasir Meong*. Place names are very important and can even replace (the) a person's original name to (and) become part of a person's identity.

In addition, from the point of view of linguistic landscape, quite a lot of changes occur in place (names) or street names due to socio-political or economic interests. An example of changes on odonyms, in toponymic terminology, names of streets and roads are referred to as odonyms. Erikha (2021) conducted research on changing street names in Bandung. Referring to the Dutch file *Gewijzigde Straatnamen van Bandung* (1950) there were 565 changes in street names from the Dutch colonial period to the Indonesian independence period. In addition, based on his observations, at present, there are also changes in street names in Bandung, such as *Jalan Cimandiri* 'Cimandiri street' which is a hydronym replaced with *Jalan Hayam Wuruk* 'Hayam Wuruk Street' which is the name of former king of the Javanese Hindu State of Majapahit Kingdom.

Geographical names, such as the names of villages which are so attached to the local people (Kostanski 2009), (recent days) have sometimes been changed. The development of housing estate that is growing (so fast and) rapidly in the Bandung Basin, has taken advantage of the very large village land, so that it has eliminated the geographical names in the area. The very large residential area was originally associated with beautiful places such as hills, valleys, springs, rivers, rice fields, and villages. All of those places actually have local geographical names, but it was often changed by the developer with so called modern names that they think may attract buyers.

To understand the recent situation in the Bandung Basin, we can compare an area formerly called the Kacakaca Wetan area, in Bandung City (at) as listed in the topographic map 1910 from the Dutch (time) period, with and an online satellite image map 2022 at the same location now called Burangrang Village, Lengkong District, Bandung City. The area which was originally green in the form of swamps and rice fields, has now turned into a dense urban area with housing and apartments. Generally, the name of the housing or apartments that they built is toponymically (completely) unrelated, both naturally and culturally (in) to the area. Even with some big developers, naming their buildings or housing area with the name of their company. (Including n) New names appear in foreign languages such as *Green Valley Residence* in Mandalajati District or *Dream Hill Residence* in Padalarang District. Even in Padalarang District there is a hill they called *Nemo Highland*. How many local people in Pangandaran are familiar with the Disney cartoon of *Finding Nemo*? This, in fact, violates the national law, presidential regulation, UNGEGN's Resolutions and also confuses people's memories of an area.

Suyatna et al. (2016) observed that the place naming of tourist destination in West Java can be classified into three types: firstly, can be translatable and literally descriptive; secondly, can be the significant marker of the intimate relationship between people and territory; thirdly, (are) have potential to locate and bind people both geographically and historically.

Reality on the ground, the names of tourist destinations are sometimes given meaningless labels (names), which are very foreign to the local people who have lived in the area for generations, for example, the location for river tourism in the Pangandaran area is *Cukang Taneuh* but they now called it (as) *Green Canyon*. Foreign tourists generally look for original and unique tourist destinations based on local culture. Light (2014) states that renaming places does not always enjoy local support. These foreign names trigger the curiosity of local tourist more than tourists from abroad. (This) These kinds of tourism strategies are not appropriate because they eliminate (eliminating) local names that (should not be changed because they) store the history of the location, including the migration of people who occupy the location.

Conclusion

People's collective memories play an important role in storing significant knowledge so as to learn (ed) from the past. Indonesia has an abundance of geographical name data due to its vast territory and diversity of language and culture. Laws as well as government regulations (have been) are in place to manage geographical names in Indonesia. However, the implementation of the regulation still requires coordination from various ministries and local governments. The renaming of places and the use of foreign names that mess with collective memory needs to be actively curtailed (reorganized).

Questions

1. Why is it necessary to understand toponyms, especially local geographical names?
2. Who else needs to understand toponymy rules and policy other than the Local Government and the National Names Authority?

3. To what extent can collective memories in the form of myths, legends, or oral traditions help to preserve geographical names as indicators of the environment?
4. What kind of earthquake mitigation measures can local residents take before, during and after a disaster?
5. What should be done so that local geographical names are preserved but business and tourism can thrive?

References

Bachtiar T, Syafriani D (2016) Bandung Purba Panduan Wisata Bumi. Masyarakat Geografi Indonesia, Bandung

Badan pengembangan dan Pembinaan Bahasa (The Language Development and Fostering Agency) (2019) Bahasa dan Peta Bahasa di Indonesia. Edisi Keenam. Kementerian Pendidikan dan Kebudayaan, Jakarta. https://petabahasa.kemdikbud.go.id

Badan Informasi Geospasial (The Geospatial Information Agency) (2022) Country Report of Republic of Indonesia. UNGEGN Asia South-East Division

Binder D et al New investigations of the Grolludag obsidian lava flows system: a multi-disciplinary approach. J Archaeol Sci 38(12):3174–3184

Biro Pusat Statistik (Statictical Central Bureau) (2022) Kota Bandung dalam Angka. https://bandungkota.bps.go.id

Cribb R, Ford M (2009) Indonesia as an archipelago: managing island, managing the seas. In: Indonesia beyond the water's edge: managing an archipelagic state. Institute of Southeast Asian Studies, Singapore

Daryono MR et al (2019) Earthquake geology of the Lembang Fault, West Java, Indonesia. Elsevier Tectonophysics 751:180–191

Erikha F (2021) Application of the concept of critical toponymy to street name changes in bandung, yogyakarta and surabaya. Paradigma, Jurnal Kajian Budaya 11(1):25–41

Forbes VL (2014) Indonesias delimited maritime boundaries. Springer-Verlag, Berlin

Gadeng AN et al (2018) The value of local wisdom *seumong* in tsunami disaster mitigation at Simeuleu Regency, Aceh Province. Earth Environ Sci 145:1–6

Hough C (ed) (2016) The Oxford handbook of names and naming. Oxford University Press, Great Britain. ISBN 978-0-19-965643-1

ITIC-IOC (2022) International Tsunami Information Centres – International Oceanographic Commision. https://itic-ioc-unesco.org

Kadmon N (2000) Toponymy: the lore, laws, and language of geographical names. Vantage Press, New York

Kartadinata MN (2005) Tephrochronological study on eruptive history of Sunda-Tangkuban Perahu volcanic complex, West Java, Indonesia. Doctoral dissertation, Kagoshima University, Japan

Kartodiwirio SK (2006) Bandung: Kilas Peristiwa di Mata Filatelis Sebuah Wisata Sejarah. PT Kiblat Buku Utama, Bandung

Kostanski L (2009) 'What's in a Name?': place and toponymyc attachment, identity and dependence. A case study of the grampians (Gariwerd) national park name restoration process. Doctoral dissertation, University of Ballarat, Australia

Lauder MRMT (2014) Memahami Manfaat Toponimi bagi Jatidiri Bangsa. In: Rapat Pendampingan Dekonsentrasi Pembinaan dan Pembakuan Nama Rupabumi Unsur Alami. BIG & Kemendagri, Jakarta

Lauder AF, Lauder MRMT (2015) Ubiquitous place names: standardization and study in Indonesia. Wacana J Humanit Indones 16(2):383–410

Lauder MRMT (2021) Merawat Memori Kolektif melalui Toponim. Seminar Nasional Merawat Budaya: Studi Dokumentasi dan Revitalisasi Bahasa-bahasa di Indonesia (Jawa Barat, NTT, dan Maluku). Kantor Bahasa Provinsi NTT, Kupang
Light D (2014) Tourism and toponymy: commodifying and consuming place names. Tour Geogr: Int J Tour Place, Space Environ 16(1):141–156
Noorduyn J, Teeuw AH (eds and trans) (2006) Three Old Sundanese poems. Royal Netherlands Institute of Southeast Asian and Caribbean Studies (KITLV) Press
Nyström S (2016) Names and meaning. In: The Oxford handbook of names and naming. Oxford University Press, Great Britain. ISBN 978-0-19-965643-1
SIL International (2006) Bahasa-Bahasa di Indonesia (Languages in Indonesia) 2nd Edn. Jakarta, SIL International Indonesia Branch
Suyatna ETS et al (2016) Place naming of tourism destinations in Jawa Barat: a toponymy study. Int E-J Adv Soc Sci II(5)
Taylor S (2016) Methodologies in place-name research. In: The Oxford handbook of names and naming. Oxford University Press, Great Britain. ISBN 978-0-19-965643-1
Tulius J (2020) Lesson from the past, knowledge for the future: roles of human memories in earthquake and tsunami narratives in Mentawai, Indonesia. Parad J Kaji Budaya 10(2). e-ISSN 2503-0868
United Nations (2009) United Nations Convention 1982 on the Law of the Sea (UNCLOS). Nova Science Publisher, New York
Widianto H, Noerwidi S (2020) It's time to look to the west: a new interpretation on Homo erectus findings distribution of Java. J Berk Arkeol 40(2):153–178

Suggested Reading

Hough C The Oxford handbook of names and naming. Oxford University Press, Great Britain
Langendonck Wv (2007) Theory and typology of proper names. Mouton de Gruyter, Berlin. ISBN 978-3-11-019086-1
Malpas J (2004) Place and experience: a philosophical topography. Cambridge University Press, Cambridge
Nash J (2015) Is toponymy necessary? Stud Lang 229–234

The Vital Question of Placenames and Naming of Places in Geographical Education: Concepts, Activities, and Questions for Reflection

Ruth McManus

Abstract This chapter provides resources to encourage both self-directed readers and educators to reflect on the themes of the volume. A wide range of sources and methodologies is presented, combined with both chapter-specific and broader questions, project exemplars and suggestions for extending learning.

Keywords Geography · Toponymy · Place · Identity · Case-studies · Education · T&L · Resources

Introduction: The Importance of Unpacking Placenames in Geographical Education

The chapters in this volume have emphasised the importance and complexity of understanding placenames, including their evolution, standardization and changes at different scales and within different local, national and international contexts. This chapter focuses specifically on the value and importance of exploring both placenames (toponyms) and the naming of places (toponymy) within the context of geo-education.

At a basic level, everyone—including students and their teachers—needs accurate placenames in order to actually locate themselves in their own environment. A toponym, the proper name given to a geographical feature, is essential in order to locate oneself, and place names can be considered to be the most useful and commonly-used geographical reference system in the world. Clearly there is a practical need for consistency and accuracy in referring to a place in order to prevent confusion. This necessity has been intensified with the advent of digital technologies and the requirement for geospatial data and the importance of standardised names in SDIs (spatial data infrastructure). Although placenames might be simply thought of as convenient labels, their use in everyday life including cartography and public signage can become highly charged and controversial.

R. McManus (✉)
School of History and Geography, Dublin City University, Dublin, Ireland
e-mail: Ruth.McManus@dcu.ie

© The Author(s), under exclusive license to Springer Nature Switzerland AG 2023

G. O'Reilly (ed.), *Place Naming, Identities and Geography*,
Key Challenges in Geography, https://doi.org/10.1007/978-3-031-21510-0_27

Placenames hold many connotations beyond their straightforward functionality as locational labels. Every student will have encountered the perceptual impact of certain toponyms. This frequently relates to the negative connotations of particular locations, or what might sometimes be termed 'the wrong side of the tracks'. Even the smallest human settlement will have locations which are considered more or less desirable than others, and their placenames often serve as short-hand for these positive or negative associations. An important element of geographical education is to make students aware of their learned prejudices and attitudes, which are frequently the outcome of exposure to lazy stereotyping in the media, where placenames are used as a short-cut to encompass a broad range of negative attributes. This can extend from the reporting of crime linked to poor socio-economic areas in a town or village, to coverage of war-torn cities and victims of conflict or environmental disaster. By contrast, certain placenames have highly positive associations as decidedly attractive locations which, in turn, might lead to their appropriation for economic purposes such as tourism or real estate.

At an existential level, toponyms hold a further layer of importance beyond their functional and perceptual associations. Yi-Fu Tuan considers the ways in which people feel and think about space and how they form attachments to home, neighbourhood, and nation. Topophilia, the affective bond with one's environment which includes a person's mental, emotional, and cognitive ties to a place, naturally encompasses placenames that may contribute to a sense of place and place experience. Placenames are part of the process whereby the abstract concept of *space* becomes invested by humans with cultural meaning to become *place*. This attachment of meaning to a specific site or locale is very powerful. Appreciating this fact helps us to understand why renaming, including the imposition of alternative placenames or erasure of specific names, can be so controversial and may elicit a visceral response in individuals.

As outlined above, placenames may be considered an element of intangible cultural heritage. Although initially created through the need to identify and subsequently organise a setting or locale, toponyms ultimately become a basic component of the historical memory of a territory, often reflecting past cultures, migration waves, and historical events associated with a particular location. Such is their cultural significance that placenames have a role to play in relation to SDG16 (peace, justice and strong institutions) as a means of enhancing peace-building and righting past wrongs. It can therefore be argued that education about placenames is an element in the development of citizenship. This is explored further later in this chapter.

The suggested activities and questions provided below build upon the state-of-the-art toponymic analyses of the preceding chapters. They are intended as starting points to facilitate the application of the concepts and case studies in this volume within an educational setting, and most offer suggestions as to how the activity in question might be tailored to specific groups. Activities range in complexity from simpler tasks designed to help students develop a basic awareness of placenames and toponymy, to more sophisticated analyses which use the study of placenames as a lens through which to examine deeper issues including contemporary geopolitics, treatment of indigenous peoples, colonialism, and post-colonialism, cultural heritage and more.

A broadly constructivist educational approach is taken in this chapter. Constructivism is an umbrella term, based on the idea that students construct their own meaning by building on their previous knowledge and experience. Among the characteristics of constructivist learning environments are the provision of multiple representations of reality and avoidance of oversimplification, an emphasis on authentic tasks in a meaningful context rather than abstract instruction out of context, and the use of real-world settings or case-based learning where possible (Jonassen 1999).[1] With an emphasis on knowledge construction instead of knowledge reproduction, constructivist approaches encourage thoughtful reflection on experience, enable context- and content-dependent knowledge construction, and support collaborative construction of knowledge through social negotiation rather than competition among learners for recognition. The following questions and activities build on these elements, encouraging inquiry-based learning in order to deepen understanding, with an emphasis on exploring the complexity of the real world and appreciating multiple perspectives. Suggested project work and other practical hands-on activities help to provide a meaningful context for learning. All of the activities are designed to support real engagement with relevant concepts while also encouraging the development of various transferrable skills relating to research, analysis, and presentation.

Introducing Placenames in the Classroom: Purpose, Meanings, Collection and Perceptions

For most students, placenames are a taken-for-granted aspect of their everyday world. As younger children their encounters with toponymy likely focused on lists of mountains, rivers, or administrative areas to be memorized. By their very ubiquity and commonplace existence, placenames are rendered invisible and unremarkable. Some introductory tasks are therefore appropriate, to encourage students to become aware of placenames and to begin to consider their importance at a functional, perceptual and cultural level. Reference is made to specific chapters which contain discussions or case studies of particular relevance to the activity topic. The initial activities outlined in the boxes below are as follows:

- Discovering placenames in your locality.
- What do our placenames tell us? Two related activities are linked to the meanings of placenames, linked to physical geography attributes, and to different naming forms.
- Gathering placenames: what is important to gather and why? Volunteer-based placename projects to preserve intangible heritage.

---

[1] Jonassen, D.H. (1999) Designing constructivist learning environments, in C.M. Reigeluth (ed), Instructional-design theories and models: a new paradigm of instructional theory, NJ: Lawrence Erlbaum Associates, pp. 215–239.

- Challenging perceptions—activity linked to both positive and negative perceptions associated with placenames.

The initial activities are intended to generate interest in the topic and an awareness of some of the complexity associated with placenames and their collection. They should also provoke an interest in the process of naming and a recognition of its links to power, identity, and heritage. Note that the activities can be applied at different spatial scales—the national (country), regional (towns and villages), local (streets, buildings), and so forth.

BOX 1: Discovering placenames in your local landscape: field trip (or virtual field trip)

Walk around your town (or use Google StreetView to explore another location) to find placenames (street signs, names on houses, directional signs, plaques on locations of interest, etc.). Choose at least five.

Take pictures using a camera or mobile device. Alternatively, download pictures from the internet.

Try to answer the following questions in relation to your selected placenames:

- Where did you find the greatest concentration of placenames? Why do you think that is?
- Who put up the signs? Are these 'official' or 'unofficial' placenames?
- What differences did you notice between placenames and the way in which they were indicated?
- Does anything surprise you about the placenames you recorded?
- What do the placenames mean? Do they tell us anything about the physical or human geography of the location?

Either

Use the Google Earth application to make an interactive map by inserting the pictures in the correct place. Write up a small report about each one of them.

- Are there any differences between the placenames which you recorded during fieldwork and the placenames used by Google Earth?

Or

Create inspiring stories in Arc GIS story maps (https://storymaps.arcgis.com/) by combining text, maps, and other multimedia content.

BOX 2: Exploring meanings of placenames

These two associated tasks were undertaken by a group of undergraduate geography students in Ireland as part of a research methods course where they undertook an in-depth geographical exploration of their home area.

In Ireland, the study of placenames is greatly facilitated by the availability of a user-friendly national placenames database (www.logainm.ie) which includes geographical locations and meanings of placenames in both Irish and English, as well as overall distribution maps showing the presence of specific placename elements across the country.

Task A: recognition of links between toponyms and physical geography

Students were asked to list four placenames from their home locality at a range of scales (i.e. townland, the smallest spatial unit used in Ireland; nearest village; town; and any important local landmark names). They then researched the meanings of these placenames using the logainm database described above, with a particular emphasis on those elements which might indicate physical geography features (e.g. placename elements linking to a hill (*cnoc*), ridge (*droim*), or bogland (*móin*) etc.). The students took photographs of the four locations to demonstrate the relationship between the placename and the physical geography of that location. They created a poster that superimposed these placename elements and their associated photographs onto a digital map illustrating the physical geography of the area, together with a short paragraph explaining the placenames and their relevance.

Task B: Toponyms as a record of both human and physical landscape

At the start of this simple activity, students were asked to select the local placename with which they identified most closely (whether a local farm or townland name, the name of their housing estate, or that of their nearest village). In most cases, this was a placename that they had previously examined in Task A, but if not, they followed the same process to identify the meaning of the toponym using the logainm database. The students were then asked to move around the classroom space, asking their colleagues about their placenames, in order to find partners with similar placename elements. For example, students were asked to identify whether their placename signified a human impact on the landscape (e.g. bridge, ford/crossing point on river, castle, town) or a physical feature of the landscape (e.g. hill, valley, lake). They physically clustered in the room, joining fellow students with similar placename elements. This was a way of highlighting the different patterns of naming across the home locations of the student cohort, and a way of encouraging them to think of the ways in which placenames might have originally been chosen or imposed.

BOX 3: Gathering placenames

The activities suggested here are intended to familiarize students with the process of collecting and standardizing placenames, drawing on the guidelines and information provided by the United Nations Group of Experts on Geographical Names (UNGEGN).

Task A: Learning about the work of toponymy

Students are tasked with using online sources (including the useful Wikipedia entry: https://en.wikipedia.org/wiki/Toponymy) to find definitions and answer specific questions, suggested below:

- What is a toponym?
- Define the terms toponymy and toponymist.
- How are placenames implemented, removed, or changed?
- On accessing the UNGEGN website, what publications can be found by the general public to learn about toponymic standardization?
- Why is the standardization of geographical names important for everyone?
- Drawing on the chapters of this book, and the UNGEGN and other similar websites, explain how a toponymist determines names with established local usage (hint: maps, local histories, interviews).
- What information should be gathered by toponymists during name surveys? (hint: application, specific language, pronunciation, origins, meaning).
- Who is responsible for standardizing geographical names at the global and national levels?
- Which experts should participate in standardizing geographical names?

Additional information in relation to responsibility for naming and how naming decisions are made can be found in a number of chapters, including Chaps. 7–9 (responsibility for naming) and Chaps. 10–13 and 22 (how decisions are made). At this point it might be considered useful to look at a specific example of the regulation of geographical names, using the case study of Slovenia following independence in 1991 (see Chap. 9). A careful review of this material will help students to answer the following questions: why is standardization important, how is it undertaken, who participates and what problems need to be addressed? Particular challenges for standardization in bilingual or multilingual areas could also be flagged.

Task B: Local level placename projects protecting cultural heritage

While national-level organisations undertake responsibility for the recording of placenames at particular scales, a variety of local projects have undertaken work (often crowdsourced) to record and preserve minor or extremely localized toponyms which are considered to be a valuable part of local intangible heritage.

For example, the Meath Field Names project was a volunteer-led project established in 2008 to record and publish the field names of County Meath along with their history, features, name origin, and folklore (see http://www.meathfieldnames.com/). It was recognized that these minor placenames, together with associated folklore and features, were at risk due to significant landscape change (e.g. motorway construction, an amalgamation of fields, modern farming practices) and alterations in population patterns (such as the dwindling population of farmers and older rural dwellers who carried these names in the folk memory).

Students are asked to visit the Meath Field Names project website and address the following questions:

- Why was the project undertaken?
- What interest groups were involved?
- What specific data was collected?
- What were the key learning points from the project?
- Are there any similar projects being undertaken in your home country?

BOX 4: Challenging perceptions

This activity is used to help students to identify the existence of both negative and positive associations which may be felt in relation to particular placenames. It helps to challenge preconceptions that may have been absorbed due to media images and portrayal of certain locations in their daily lives and encourages the students to pause and reconsider the degree to which these perceptions are fair or accurate representations of reality. While every student will have been exposed to negative portrayals of nearby locations (e.g. stigmatized or ghettoized areas of poor quality housing associated with poverty, crime, and anti-social behaviour) it is recommended that this activity is undertaken using more distant locations, to avoid further stigmatizing such local communities.

The activity begins with brainstorming, whereby the educator calls out a list of placenames and the students record their immediate response to that name, ticking the appropriate box on a sheet (i.e. positive or negative association) and including a word or number of words which immediately comes to mind when they hear that toponym (e.g. war, violent, beautiful, scenic, etc.). The selected list of placenames should include a range of locations globally and is dependent on what is topical and might have resonances for the student body (i.e. whereas students in the 1980s might have had particular negative perceptions of placenames such as Belfast or Lebanon, this is less likely to be the case in the present).

Once the initial brainstorming activity is completed, students are asked to compare their perceptions with those of their colleagues within small groups. Likely many will be very similar, although they should be encouraged to discuss any variations and the reasons behind them (e.g. a student who has visited a location or has relatives living there may have a more positive association with the placename than someone who is wholly reliant on media for their perceptions).

The educator will then present the students with a range of images that are unlabeled, asking them to match these photographs to the placename. These images should be deliberately selected to include some which are immediately recognizable media portrayals of a location (e.g. conflict, environmental disaster, images for tourism promotion with eternally sunny skies and an absence of crowds), and others that challenge perceptions by including very different experiences of places which would typically have strong negative or positive associations. For example, a photograph from Chernobyl, the site of a major nuclear accident, could focus on the local flora and fauna, while a photograph of Yellowstone National Park, generally considered to be a location of great natural beauty, could emphasize the impacts of over-tourism or littering.

The final phase of the task is to provide the students with the 'correct' answers, by showing which photographs are associated with which placenames, and for the students to discuss what they have learned about both positive and negative stereotyping in the context of placenames.

Toponyms as Clues to the Past

Human perceptions, layers of memories, and emotions may be associated with toponyms, thus linking to Tuan's concept of topophilia. Because the naming of places is linked to social, political, economic, and cultural processes, many current placenames reflect past cultures, borders, migration waves, and other historical events. For example, Chap. 19 shows how different rulers left their mark on toponyms in the Maltese islands. While at their most basic, placenames were created due to the need to uniquely identify a particular location, as Chap. 25 points out, over time these toponyms become a basic component of the historical memory of a territory and a valuable part of the cultural heritage.

In some cases, toponyms may have been altered many times (e.g. Chap. 11 re Zimbabwe) or may even serve the purpose of palimpsests with traces of past names bearing competing histories and discourses, reflecting position at a cultural crossroads (e.g. Chaps. 15 and 17 referring to Bosnia and Herzegovina and Hungary respectively).

In Chap. 20 the use of historical cartography to examine placenames is discussed. The value of comparing and contrasting different versions of historic and more modern maps is also raised. The activity below similarly draws on historic maps to develop an understanding of the evolution of toponyms.

Activity: Using historic maps and gazetteers to understand the evolution of your home area

This activity was undertaken with undergraduate students in Ireland but could be adapted for different cohorts and different national settings. Students were introduced to two historic sources, the colour first edition Ordnance Survey maps surveyed in the 1830s at 6 inches to one mile (1:10,560) available online and Lewis's Topographical Dictionary of Ireland (1837), also available and fully searchable online. They were also provided with a contemporary map of their local area and access to logainm that is, the national placenames database.

Students were asked to read the relevant entry from Lewis's Topographical Dictionary (a form of gazetteer) for their home area and to summarise the key points of the description in their own words. They were then asked to identify the toponyms mentioned by Lewis on the 1st edition Ordnance Survey map (i.e. both physical and human landscape elements, including individual named houses of the wealthy). Finally, they compared the 1st edition map with its contemporary equivalent, noting which placenames and tangible features had persisted. Where placenames had changed, they researched the timing and reason behind the change.

This activity gave students the opportunity to engage with a range of historic and contemporary sources, and to consider processes of naming, renaming, and loss of names over time.

Processes of Naming as a Reflection of Power Dynamics

As this volume makes clear, the process of naming the landscape is inherently political and forms a complex part of both national and local identity-making. How, why and what names are used can create/reflect a conflict between competing groups, beliefs, and ideas (see Chap. 10). Placenames are inscribed with meaning, which can be both overt and hidden. As observed in Chap. 21, names are often changed when state ideologies fundamentally change, a point which is also made in Chap. 2's consideration of the case of Stalingrad—Volgograd. Naming is a fundamental expression of power and control, which is evident with regard to both colonial and post-colonial placenames discussed across the volume. Indeed, as Chap. 22 asserts, the naming of places is one of the most basic, yet powerful and symbolic acts of geographical appropriation. Disputes regarding naming may reflect different perceptions, territoriality, historical legacies, and claims to sovereignty, as seen in the case of Japan and Korea discussed in Chap. 3. At times the role of placenames may be ambiguous. For example, where one group may feel that a specific toponym reflects certain political conflict or is negatively perceived within their community, they may also recognize the economic rationale for its retention, e.g. for the purposes of attracting tourists, particularly those engaged in 'dark tourism'.

The act of naming may serve a variety of purposes, including the promotion or imposition of specific identities, erasure, remembrance, commodification, respecting heritage and more. It is important to always consider who is responsible for naming and what are their motivations.

The first case study in this section asks students to consider the process of naming in what might be considered a *tabula rasa*: the Antarctic. This experience can then be linked with and/or contrasted with the ways in which colonisers renamed landscapes as part of processes of cultural appropriation. The second activity refers specifically to names associated with the colonial and post-colonial period in Africa, but a similar methodology could be applied to any location where names have been changed over time.

Case Study: geographic nomenclature in Antarctica

Antarctica is the only continent without an indigenous human population and therefore lacks a history of cultural ties to placenames. Most human presence has been temporary, but the continent has been visited and explored by individuals and teams from many different nationalities. Antarctica is not a sovereign territory and instead is managed under an international agreement, the Antarctic Treaty. This leaves the region in a unique situation with regard to geographic nomenclature. Although most major features of Antarctica have been discovered, named, and mapped, there are many secondary features that remain unnamed and only partially delineated.

Student task:

Select five toponyms from a map of Antarctica.

Using an encyclopedia and/or other online sources, research their origins. Who are they named after?

Find out how new names can be proposed for features in Antarctica, drawing on the following website or other sources (https://www.usgs.gov/us-board-on-geographic-names/antarctic-names).

In a group, discuss the patterns you have noted in relation to naming in Antarctica.

What parallels and contrasts do you think might exist between the Antarctic experience and other locations which had existing or indigenous populations at the point of 'exploration' and colonization (e.g. South America, Africa, Australia, New Zealand Aotearoa, etc.).

Activity: naming African countries—a reflection of colonial and decolonial processes

Students are broken into teams and asked to research one of the following country names: Gold Coast/Ghana, Dahomey/Bénin, Southern Rhodesia/Zimbabwe, Upper Volta/Burkina Faso. In each case, the first placename of the pair is the name inherited from the colonial period, while the second is the name chosen by the newly-independent state in order to break with their colonial history. For their allocated country, students address the following questions:

- What was the colonial name of this country?
- When was it colonized, and by which country?
- What was the meaning of the colonial name? (e.g. Gold Coast directly refers to the resource being exploited by colonial powers)
- When did this country gain independence?
- What is the meaning of the new country name? (e.g. Burkina Faso means 'the country of honest men', renamed following a coup against a compromised regime).
- Does the new country name draw on toponymic heritage or other aspects of cultural heritage?
- Why was this new name selected and what does it tell us about decolonization?

These examples help students to explore one aspect of colonialism and the activity could be used as part of a deeper study of the topic. In relation to placenames specifically, the answers to these questions should help students to understand the importance of appropriating the national name, distancing from colonial heritage, and legitimizing the new power (especially by drawing on or referring to pre-colonial historical entities). This case study emphasizes naming at the national scale, but see also the discussion of placenames within Zimbabwe and South Africa (Chap. 11). This chapter points to the coexistence of different toponyms, e.g. colonial/apartheid and indigenous toponyms in post-apartheid South Africa.

Activity: Placenaming and Identity

As many of the chapters demonstrate, renaming of toponyms involves the exercise of politics and power (e.g. Chap. 2 re Stalingrad/Volgograd, Chap. 13

re Turkish streetnames). Placenames are often intentionally used to create a certain identity. This may be the case in the act of naming new places which can be used to shape the perceptions of the country's citizens (see Chaps. 4 and 7). The intentional creation of identity may relate to attempts at building a nation-state (e.g. Turkey, discussed in Chap. 16) or state-driven ethno-linguistic homogenization—Magyarization (as in the case of the Mohács plain in SW Hungary discussed in Chap. 17). The following questions encourage students to explore aspects of placenames and identity. In each case, the relevant chapter is indicated in brackets at the end of the question.

- What are the successive place-naming experiments that France has investigated to name its administrative territories (departments, regions) and its communes? (Chap. 4)
- 'The social, political, cultural and economic dimensions of toponymy are expressed in the act of naming, that is, in the contradictory production of toponyms.' Do you agree? (Chap. 4)
- What are the overt and hidden meanings of sea names and water sources in Israel? (Chap. 7)
- How are religions, politics and nationalism involved in naming places in Israel? (Chap. 7)
- How did the process of nation-stateization affect the urban toponyms in Turkey in the early period of the foundation of the state in 1923? (Chap. 14)
- What is the influence of political parties on place naming in Turkey? (Chap. 14)
- Why did the Ottomans not consider changing place names until the last decade of the Empire? (Chap. 16)
- What are the main motivations for place name changes during the era of the Turkish Republic since 1923? (Chap. 16)
- How did the Mohács plain play major roles in toponymy from the Ottoman era on with transformations in the cultural landscape? (Chap. 17)
- 'The Mohács plain became one of the most conflicted Cold War borderscapes, during the Cold War (1947–1991) and again in the Yugoslav wars of the 1990s.' Discuss this statement with reference to place naming (Chap. 17).
- How could the 'anti-migration fence' of Hungary installed in 2015 to prevent the flow of migrants and asylum seekers impact place naming and cultural landscapes? (Chap. 17)
- How have religion, languages, and cultures left their toponymic prints on the Mohács and Hungarian landscapes (Chap. 17).
- 'The Mohács plain is an ideal research and case study area, representing different origins and bottom-up changes of Hungarian place names as well as top-down policies influencing them.' Evaluate the importance of this statement in the context of the European Construction Project (Chap. 17).

- What is the role of place names in creating identity and showing it through the landscape and naming? (ALL)

Changing Attitudes to Indigenous Names and Contested Names

While colonial naming practices are discussed above, this section looks specifically at indigenous placenames and their treatment, both as reflections of changing society (Box 5: Indigenous and derogatory names: Aotearoa New Zealand and the United States of America, and as embedding cultural memory which can be of tangible benefit (Box 6: Indonesia).

Box 5: Indigenous and derogatory names: Aotearoa New Zealand and the United States of America
Chapter 10 discusses changing thinking about naming practices in Aotearoa New Zealand. This is a reflection of social and political shifts since the 1970s, including changes in attitudes towards restoring indigenous Maori rights. Students are asked to read the chapter and answer the following:

- In what ways is a political shift towards reflecting Maori rights changing how place name decisions are made?

Students are then directed to the US Board on Geographic Names (https://www.usgs.gov/us-board-on-geographic-names) to find answers to the following questions:

- When and why was the US Board on Geographic Names established?
- What does it explain about the importance of standardization?
- What is the process of suggesting a new name? Why might it be necessary (see current example re 'squaw')?
- Who do you think should be involved in the process of renaming, and why?
- Now visit the section of the website dealing with derogatory names and read the following document relating to Order No. 3405 'Addressing Derogatory Geographic Names' (https://www.doi.gov/sites/doi.gov/files/elips/documents/so-3405-508.pdf). Which stakeholders are listed as being involved in the renaming process? Do you think that this achieves a good balance on the naming committee (explain your reasoning)?

In the final stage of the activity, students are asked to reflect on the experience of their own country:

- Reflect upon your own country's practices of using indigenous names. What is your country's approach to using indigenous names, and how has this changed over time?

Box 6: Case study of Indonesia

Because of their intangible heritage, there is clearly value in protecting place-names in a world which is subject to increasing homogenization. Toponyms provide links with the landscape and its characteristics, history, and human activities, but also clues as to collective memory including natural conditions. Chapter 26 shows that this collective memory can be valuable in relation to earthquake mitigation. Therefore, if names are changed this does not just represent a loss of intangible heritage, but could also potentially result in very real and tangible consequences relating to disaster problems if past geomorphological information is lost. In this activity, students are asked to study the chapter and respond to the following questions:

1. 'Indonesia is a vast country with 17,000 islands covering three time zones and has a cultural diversity with 718 languages.' (I) List and explore three major challenges that could be faced by the central government in trying to standardize place naming in Indonesia. (II) List and explore three major advantages that local and central government could gain due to the diversity of place naming in Indonesia.
2. What should be done so that local geographical names are preserved but business and tourism can thrive?
3. Why is the Bandung Basin area in the province of West Java One, of particular interest regarding toponymy?
4. Why is the Sangkuriang legend that has been transmitted orally through the ages by the Sundanese people in the Bandung Basin area of much importance regarding linguistics, hydronomy, and toponymy?
5. Explain the significance of the research that found hydronyms dating back over centuries in the Bandung Basin area.
6. Evaluate how archaeological findings, legends, oral traditions and collective memory, can contribute to generic forms of hydronyms in areas of the Bandung Basin inhabited by the Sundanese people of Indonesia.
7. Why is it necessary to understand toponyms, especially local geographical names?
8. To what extent can collective memories in the form of myths, legends, or oral traditions help to preserve geographical names as cultural artefacts?

Placenames and Resistance

Given that placenames can be used for political purposes by powerful elites, it is unsurprising that one form of resistance to those in power lies in the refusal by 'ordinary people' to use certain names. For example, the refusal to use newly imposed names of collective farms and the creation of nicknames was a form of resistance seen in Soviet-era Estonia (see Chap. 21). In Gibraltar (Chap. 24) the resilience of the use of traditional street names and resistance to the implementation of official street names shows the strength of imaginative geographies and is an expression of the presence of two Gibraltars, represented by the doubling of names. Students could be asked to identify local examples of resistance or reluctance to adopt name changes which can, occasionally, result in a reversion to the use of previous names. See also Chap. 16 for relevant examples in relation to Turkey.

Placename Studies as a Vehicle to Build Empathy and Citizenship

The activities in this section move beyond simply exploring toponyms and instead use contested placenames as a vehicle to explore issues of conflict, to develop deeper historical understanding, and to build empathy and citizenship as pathways to peace-building.

Box 7: Brainstorming, Discussion, and Organizing—A framework for analysing contested placenames

Discussions around placenames have been particularly prominent at points of historic and geopolitical change. Examples in this volume are drawn from the former Soviet Union and its satellites, Turkey, central Europe, and other territories which have experienced colonialism. Debates around naming of localities can provide opportunities for students to encounter and explore issues of conflict in the present or past, develop a deeper understanding of their own or others' histories, and consider issues of citizenship and peacemaking which align with the SDGs.

This framework is a starting point for potential group discussions around contested placenames that could be applied to a contemporary or historic example. The structure encourages students to see alternative perspectives and learn to debate issues in a reasoned and respectful manner.

- Example: specified controversial toponym
- What: description of toponym currently in use, together with other variants and/or previously-used toponyms
- Where: location map

- Meaning: elaboration with detail and the historic narrative(s)
- Who decided on the toponym and when (date) [if known]
- Why did 'they' decide on this toponym—Commemorative purposes? Identity-building? Propaganda? Territorialisation? Social or mind control?
- Is there past conflict(s) surrounding the toponym? Why?
- Is there current conflict(s) surrounding the toponym? Why?
- Are alternative placenames and/or nicknames in use and, if so, by who? Why?
- Arguments for and against retaining this placename (some typical responses are suggested below, but there may be different or specific aspects to the location which you are analysing):

Arguments for retaining this placename	**Arguments against retaining this placename**
• Enhancing the cultural landscape • Paying tribute to a person(s), event or place—in the local, national or international context, so that they will not be forgotten • Celebrating positive things e.g. contribution to 'humanity' or other • Commemorating events and the people involved • Commemorating tragic events and the people or victims involved, so that they will not be forgotten • To preserve the heritage and culture • To inform/educate people • Tourism: Visitors may be attracted to visit because of certain associations with the placename	• Out of place/inappropriate in the current cultural landscape • Paying tribute to a person(s), event or place—in the local, national or international context, so that they will not be forgotten—but that caused great harm to sections of humanity • Celebrating negative things and giving approbation to it, e.g. destructive to sections of 'humanity' or other • Destructive/negative events and the people involved should not be remembered • Glorifying tragic events and the people involved, whose messages continue to be used for negative purposes • To preserve the heritage/culture of destructive and negative forces that humiliates other people, can be used to stereotype them or to justify the negative legacies left • To (mis)inform/(mis)educate the public, propaganda • Tourism: disrespect for people and places, generating funds for purely commercial purposes

Note that SDGs are specifically mentioned in Chap. 5 (Argentina) and Chap. 23 (Good Friday Peace Agreement) which may be of relevance in this discussion.

Box 8: If I were you—The empathy game

This activity is designed to encourage an understanding of different perspectives. In this case, students are encouraged to put themselves in the position of one of the actors who uses a particular toponym, so as to appreciate the perspectives of different 'sides' in the creation or perpetuation of any contested toponym.

Steps

1. Read carefully the stories that lie behind each one of the placenames (each bullet point represents a selection of contested toponyms, or the educator may choose an alternative set of examples)
   - Lake Kinneret, Lake Tiberias, Sea of Galilee
   - Derry, Londonderry (Northern Ireland)
2. Consider why each group uses a different name for the same physical location
3. Choose one of their characters and write a text by answering the following questions (in each case 'you' refers to the character that you chose):
   - What is the placename you prefer?
   - What do you think of when you hear or use this placename (NB you refers to the character you chose)?
   - How do you feel?
   - How do you feel when a different placename is used for this location?

Box 9: Stakeholders in the naming of new places, a role-playing activity

This activity takes the form of a role-play in which the students consider the various perspectives of different stakeholders. The task assigned by the educator relates to an imaginary new housing estate in their local area (although other relevant imaginary or concrete examples might be used as appropriate). Each team is assigned the role of a different stakeholder within the community (e.g. community president or mayor, planner, real estate developer, toponymy expert, potential resident etc.). All participants have access to contemporary and historic maps as well as other research materials which can be used to learn about existing toponyms, historic events in the area, and so forth. Every team draws up a written proposal for the naming of the new housing estate, providing a rationale for their choice.

- What name are you suggesting? What does it mean?

- What will this placename symbolize? What would you like people to know, understand, or feel in the future?
- What is the key purpose of your placename choice (e.g. real estate development team may focus on commercial resale value and positive connotations of particular names, whereas the community president or mayor team may be more interested in the development of place attachment and links to topophilia for instance.
- Do you anticipate any controversy because of the 'message' or meaning associated with your placename choice?

Following the in-class discussion about the different choices, the activity might be expanded in a number of directions. For example, students could review online newspaper archives to find examples of housing estates with 'prestigious' names, or of attempts to rename localities to improve property values. Can they find other examples of the use of certain names to create more prestigious urban spaces for economic gain or to suit tourism (see Chap. 25) or conduct an online search to find out about the Irish tourism industry's branding of the 'Wild Atlantic Way'. Another avenue is to explore planning regulations to find out whether policies exist in relation to the use of historic or local names for new developments. Finally, students could examine a range of current advertisements for new housing estates to see the role played by placenames and the degree to which estate agents exploit naming practices in order to market the new development in particular ways.

Concluding Remark

This chapter aimed at encouraging reflection concerning place naming and its relevance on multiple scales drawing on the respective chapters in this volume. The exercises and activities suggested are based on the real-life experiences of educators in the classroom, infield, and virtual environments including the author's direct participation, observation, and individual and group mentoring sessions concerning geographical education. Educators, in this case, refer to teachers at all levels from primary to the third level, but not exclusively. It can also refer to trainers or facilitators for voluntary local heritage, history, and development groups, or similar as experienced by my work with such people outside the academic environment. Residents associations, County Council, and local library management personnel have been especially active and collaborative with me in organizing workshops, discussions, and lectures. Such experiences were enhanced by being commissioned by a private builder construction company to research and write up the history of the firm over the past one hundred years using data from their private archival material and from public institutions that included the naming processes involved in going from extensive 'green field' sites to built-up areas.

Therefore, the suggested exercises and activities given in the chapter can be adapted and reworked to the environment and language that the educator is working in whether in Ireland or elsewhere in Europe, Africa, the Americas, Asia, Australasia, and so forth.

The Relationship Between Geographical Naming and Cultural Politics

The Nexus Between Geographical Naming, Place, and the Politics of Power

Gerry O'Reilly

Abstract This chapter reviews the standpoints and conclusions of forty-eight multidisciplinary researchers in geography and associated disciplines working on geographical naming in different parts of the world. They explore and analyse from different perspectives the state of the art in place naming due to its multi-level significance in issues of identity, perceptions, culture, polity, sovereignty, geopolitics, and GIS database creation and management. This includes the technical geospatial perspectives alongside the imperative of name standardization in a globalizing world but also the dynamic aspects of intangible cultural heritages embedded in names and cultural politics. It is highly noteworthy in supporting the objectives and targets of the UN SDGs such as numbers 5 and 16. The researchers work in numerous universities, regional, national, and international geographical naming-related institutions, and organizations including the UNGEGN (UN Group of Experts on Geographical Names), IGU (International Geographical Union), and ICA (International Cartographic Association) among others.

Keywords Geographical place naming · Multidisciplinary · Cultural politics · GIS · Name standardization

Objectives

- To present a holistic perspective on geographical naming ranging from multiple perceptions to its standardization including technical GIS aspects, culture, and politics.
- By the end of this chapter, readers should be able to deconstruct the variables, and actors involved in place naming.
- Readers will learn to recognize the significance of geographical naming in relation to the SDGs, intangible cultural heritage, and geopolitics.

G. O'Reilly (✉)
School of Geography and History, Dublin City University, Dublin, Ireland
e-mail: Gerry.oreilly@dcu.ie

© The Author(s), under exclusive license to Springer Nature Switzerland AG 2023
G. O'Reilly (ed.), *Place Naming, Identities and Geography*,
Key Challenges in Geography, https://doi.org/10.1007/978-3-031-21510-0_28

Introduction

Forty-eight multidisciplinary researchers working on geographical naming in different parts of the world explore and analyze different perspectives, and state of the art in place naming due to its multilevel significance in issues of identity, heritage, perceptions, culture, polity, sovereignty, and GIS and GPS—Geographical Information Systems creation and management. The researchers work in numerous universities, regional, national, and international geographical naming-related institutions, and organizations including the UNGEGN (UN Group of Experts on Geographical Names), IGU (International Geographical Union), and ICA (International Cartographic Association).

The main objective was to investigate the dynamic of place naming from geographical and cultural political perspectives from the local scale to wider national and international levels. In the cultural and language dynamics, local and outside naming actors and forces have to interface and/or coexist to normalize the appellation(s) peacefully or else in a conflictual manner that may be resolved by working within democratic and or legal structures where the rule of law prevails, or else, by open conflict, violence, and war. We tragically witnessed this in 2014, and again in 2022 on a larger territorial scale with the Russian invasion of Ukraine. Hence, geopolitical processes are at play in geographical naming. Significant research exists on territoriality and territorialization processes, particularly since the latter part of the 19th century concerning the occupation of a place or territory, and the associated defense of land, seas, resources, and culture. Geopolitical research postulates that territorialization is due to: (i) innate behaviourism, or else (ii) generational cultural learning from social constructs such as state institutions and educational curricula, or else (iii) a combination of both of these.

In an individual habitus (to live and experience) or group place and space, there are always other places in the hinterlands and beyond the horizon. Naming gives a sense of distinction to a place, and types of real or perceived ownership or polity with the greatest expression of territorialization being the state construct itself and the name given to it, and names given by it to territories within and beyond its boundaries. Therefore, issues of power and naming exist at the level of the state, village, street, or rivers and seas.

Historically, mapping with place labels helped to provide or locate the individual and groups with an image of their place spatially in contrast to other places. While a cadaster or cadastration, especially from the 19th century onwards aimed at providing a comprehensive recording of the real estate or real property's measurements and the bounds of a state on the map. Such a blueprint combined with ordinance surveys and population census data provided the political, military, and commercial elites of empires and states to manage power relations in the territories for a range of reasons such as tax collection, policing, conscription, commercial activities, and other geostrategies. Within this, the state imperative bolstered by the bureaucracy continued to standardize and harmonize the geographical and naming data within the territories under its control. In the globalization processes, and due especially to the Digital

Revolution and strength of algorithms, this data can now be stored, manipulated, and used by state and governmental authorities, but also non-governmental actors including the commercial sectors for a vast range of purposes. The main meta-parameter is the GIS-GPS for place coordinates. However, there are physical and human geographies embedded in each place. Physical geographies include specific geology, geomorphology, topography, resources, and so forth, while human geographies are multi-layered with deep human geo-histories and political archaeologies including the names attributed to the place that forms part of its cultural heritage, which can be of an intangible nature. Cognizance of this is articulated in the UN SDGs and particularly in numbers 5 (Education) and 16 (Peace, justice, and strong institutions). A spectrum of organizations and institutions at national and international levels promotes research in place and geographical naming that can feed into resolving disputes and enhance peaceful development and sustainability.

While the authors recognize the need for standardization of geographical naming in a glocalizing and globalizing world, they emphasize the specificity of each place, the associated geo-histories, and cultural politics, as well as the dynamic nature of language and place naming. This is especially evident in the wide range of case studies presented.

Challenging Conceptual and Theoretical Approaches to Place Naming

In Part I the authors set out the aims and objectives concerning their analyses and discussions of geographical naming. Concluding arguments include the salient need to recognize the validity of critical toponymy perspectives in elucidating human perception of places. This standpoint is underscored by an analysis of the contrasting memories contained in the name of Stalingrad and addresses the political and power groups in Russia that exhibit contrasting views on the renaming of the city to Volgograd. Differing perspectives promote Stalin as a hero and others as an authoritarian villain, while others distinguish between the different periods in his life and the changing contexts. A certain modus operandus has been achieved, even if uneasy at times, in keeping both names functioning in different ways despite the range of emotive perspectives, and also practical decisions that have to be made regarding the name. This is enhanced with reference to the pluralism of memory and commemoration contained in a single place name and the resultant accumulation of different senses of place. In the Stalingrad/Volgograd study, geographical names in South Korea are references, where the author is writing from in his home culture. The outside-insider approach enriches the analysis and especially regarding the conceptual and theoretical aspects of critical theory, proving that they can be applied almost anywhere while respecting the unique aspects of a place such as Volgograd (Choo 2018, 2019).

This outside-insider approach complements an appraisal of the cartography-digitalization and name standardization nexus in globalization processes, and Google's potential power in place naming. Here conflicting appellations concerning Japan and Korea regarding both land and sea are appraised from an external perspective by a European-based researcher. The analysis includes an evaluation of the geopolitics of emotions regarding geographical names in Japan-Korea relations, by elucidating perceptions, territorialization(s), historical legacies, and claims to sovereignty in the post-Colonial contexts. The processes whereby toponyms are selected and prioritized at home and in the international community are reviewed, including the use of historical maps in disputed geographical naming. Both Japan and Korea continue to monitor the post-colonial and post-WWII legacies that they share regarding disputed geographical names. The case study shows the characteristics of geographical naming regarding geopolitics at different geographical scales, in South Korea and Japan. But by association, throughout the Northeast Asia region with its local, regional, extra-regional, and global balances of power, culture, emotions, sovereignty, and ideologies reflected in China's maritime (geo)policies on many levels and more specifically the creation of geostrategic islands on reefs that have different appellations given by the riparian states. Naming and renaming places enhance claims to cultural politics, and polity. This becomes evident over time with natural reefs and human-made island construction targeting sovereignty and the pursuant claims to jurisdiction over maritime zones such as territorial waters and particularly Exclusive Economic Zones (O'Reilly 2021).

The above perspectives when contrasted with an overview of the French experience of naming political territories—geographical administrative divisions and local government units by using the theoretical constructs of neotoponymy, political and critical toponymy alongside the hypothetical aspects of territorialization(s). On an applied level, comparisons are made concerning the experiences and initiatives of political place naming in different historical contexts in France and applications of neotoponymy in its different aspects. Once again, there is a continuous balancing of centripetal and centrifugal factors and forces in order to prevent violent conflict concerning the polity adapted for geographical names, and levels of power within a place, and from outside that affects it. The minute data analysis of geographical political scales in France and the dynamic socio-cultural constructs clearly shows the power of neotoponymy (Giraut and Hoyssay-Holzschuch 2016).

However, it is essential to recognize the impact that the technological revolution and new cultural paradigms have had on the global management of geospatial information. Evidently, the identification of the social actors involved in the management of geographical names as part of geospatial Big Data, and actions that have arisen have to be continuously monitored by democratic means. This is assessed in Argentina and its response to this new technical scenario. This necessitates distinguishing the core problems and challenges that management of geographical names encounters in the specific context of Argentina for example, considering its own characteristics but also those of the current global technological, cultural, and environmental context. Education and training in this sphere are vital in order to recognize the importance of geographical names in SDIs and in compliance with the SDGs (e.g. SDG 5 and 16) not only in Argentina but globally, hence the urgent need for name standardization. Application of concepts developed in the analyses of case studies in Argentina and some other countries can provide samples of evolving best practices (Vescovo 2019).

From this standpoint, but from a different perspective it is argued that clarification of the fundamental ideas about the concept of toponym and its theoretical dimension (as a geographical proper name, endowed with a meaning related to space), and its practical dimension (with regard to its application in cartography and in public signage) must be considered. That helps to recognize the problem of the cartographic treatment of toponymy, and especially the issues that affect the scale and the change of scale. This helps capture the diversity of interactions and correspondences that occur between 'macrotoponymy' and 'microtoponymy', taking into account that these are the two great levels or scales into which naming is conventionally divided. However, from a practical point of view, the problem of cartographic representation of the spatial meaning (or semantic content) of toponyms is a reality that must be dealt with, a vibrant example is provided by the phytotoponyms of the Spanish region of La Rioja as they are represented in the cartography and the more specifically in the cadaster (Tort-Donada 2020).

To recap, so far we have looked at geographical naming from the viewpoints of multiple perceptions, critical approaches, cultural politics, power in compound forms, changing technologies, and universalizing Googlization juxtaposed with case studies showing insider and outside perspectives on the essence in place appellations at various geographical scales. The authors present the advantages, but also the challenges regarding the application of geographical name standardization by cartographers and institutions, and especially the use of GIS and Big Data in Argentina and Spain. So what does this vast complex naming nexus mean to ordinary citizens on the ground, including teachers and students?

A perspective on this is given in the case study of the Land of Israel and its toponymy and hydronomy from historical, political, geopolitical, and economic perspectives. It is postulated that understanding the complexity of the names of places in Israel is due to the uniqueness of Israel in the Middle East, and the rich history of the region religiously and politically. This entails fostering recognition skills in order to see the ambiguous situation of local geopolitics, that are core to place but embedded in deep archaeological knowledge and wider multiple geo-cultural identities and emotions, with geopolitical scales. The author reviews the hydronomy of three water bodies. Depending on which language you speak—Hebrew, Arabic, or English, each of them has three very different names with distinctive etymological origins, connotations, and cultural-political attributes. This could provide much material for a teacher and class in Jerusalem, Tel Aviv, Eilat, or elsewhere. Interestingly when you Google, the name shown on the map corresponds to the Geocode location of your device. For instance, the Gulf of Aqaba or the Gulf of Eilat at the northern tip of the Red Sea. The Sea of Galilee is also known as Lake Tiberias, Kinneret, or Kinnereth in the respective languages in the region. The hydronym—the Dead Sea in English is named Yam hamMelaḥ in Hebrew and Āl-Baḥrū l-Maytū in Arabic. These three examples are of more than linguistic and cultural significance, due to the disputed nature of territoriality and sovereignty in the region. Once again, questions concerning place name standardization and languages become evident in a multi-layered geographical naming world. This may work on different coexistence levels on the ground, but now in the globalizing omnipresent GIS sphere, greater attention is imperative regarding the creation of the algorithms for geocoded locations and associated geographical names, along with their rapid diffusion.

The work of UNGEGN (United Nations Group of Experts on Geographical Names) supports the core aims and ideals of the UN and SDGs. Among its achievements, it provides an educational infrastructure and training that was developed and evolved over time in response to facilitating best practices concerning geographical naming and so in promoting peaceful resolutions of disputes. UNGEGN openly displays what it considers the necessary content of its teaching programs, and what expertise can be transferred through its training Courses (Ormeling 2017).

Approaches to Implementing Standardization of Place Names

In Part II the aims and objectives concerning geographical naming come from a range of interdisciplinary, multilingual academic, and applied perspectives with viewpoints from Slovenia and its internal management of place names. An inter-country team from Bosnia-Herzegovina and Slovenia analyzes the management of geographical naming in BiH. Contrasting case studies from New Zealand, Southern Africa, Spain, Catalonia, and Urban areas in Turkey debate the power of cultural politics in their contemporary environments.

Analyses underscore the importance of standardizing geographical names for each nation or language, how standardization takes place, who participates in it, and what problems exist. This necessitates being familiar with the most important and common terms associated with geographical names and their standardization. However, this depends on an awareness of the current level of standardization of the geographical names. Perspective is given by an applied research team in Slovenia, where the state started standardizing in 1995, four years after its independence from Yugoslavia and the Communist sphere. It is argued that understanding the basic differences between endonyms and exonyms from the standpoint of an individual nation is crucial for citizens and the state alike. For peaceful sustainability, this has to be with democratic awareness as to standardization in bilingual or multilingual areas, and that citizens understand how bilingual geographical names are to be standardized for practical coexistence purposes. In that way, citizens know, engage, and feel that they 'at home'. This place-name and subliminal associated cultural and language recognition can help attenuate real and/or imagined grievances of people's human geohistories in the area ranging from political-cultural domination to violence, war, and shifting boundaries and borders (Perko et al. 2020). These issues have come especially to the fore in Catalonia and Northern Ireland in recent decades. Multi and bilingual place name signage is particularly evident in France and particularly in the respective southern and eastern regions close to Spain, Italy, and Germany, Luxemburg, and Belgium.

Though the written history of New Zealand can be said to be quite recent in comparison to Slovenia, the state there is now attempting to assure that all its citizens feel at home by recognizing in research, official signage, documents, media, and place names the original Maori cultures. The New Zealand Geographic Board plays a crucial role in this by helping inform legislators and citizens alike about how geographical names are: created, positioned, removed, or changed. Here subliminal association issues of the past colonial project, racial and supremacist attitudes, culture, democracy, and human rights are being addressed in subtle ways. This has evolved in New Zealand, as its values have had to change in response to shifts in cultural politics since the mid-twentieth century abroad and at home (Kearns and Lewis 2019).

Due to the long dynamic and organic nature of languages (oral and written), including linguistics, dialects, and etymologies, as well as people, communities, societies, and histories, geographical names in southern Africa are multi-layered with a multiplicity that impacts citizens' daily lives. In understanding this, the concept of palimpsest is used i.e. a manuscript or writing material on which later writing has been superimposed on or effaced earlier, reused or altered but still bearing visible traces of its earlier form. On old maps and cultural landscapes including relic boundaries—often this is observable. This metaphorical framework aids to deconstruct names and settings in Zimbabwe (Mangena 2020). Close to the idea of palimpsest is pentimento in art i.e. change(s) made by the artist(s) during the process of painting. Even if a canvas is completely overlaid, with a new painting, with time and aging old images come through. The experts can use X-ray techniques to see this. These techniques are frequently used in cleaning or restoration work. Such metaphors are highly noteworthy in avoiding oversimplified binary black-and-white logic that may eventually result in bloody conflict and war. Nelson Mandela and his supporters transposed the palimpsest image into the 'rainbow nation' concept, with the ending of the state's racist Apartheid regime in 1991. Unlike many other countries that were colonized, their formal independence was not followed by a massive bloody exodus of ruling elites and refugees, nor a gory civil war pursued by the different factions in the liberation struggle and now trying to take absolute or authoritarian power. Understanding the co-existence of colonial/apartheid and indigenous geographical and place names in post-apartheid South Africa is central to progressing the peace and stability of the new Rainbow Nation with its transgenerational traumas. Obviously, place names dedicated to the most brutal architects of the colonial and apartheid project could not, and cannot remain an overarching shadow in the daily lives of the overwhelming majority of citizens and especially those who suffered under it, and many of its legacies still today. Having said that, numerous traces of past names bear competing histories and discourses in postcolonial countries that are being worked through in a non-extreme way in the RSA.

From a technical, applied, and democratic perspective, countries like Spain have been trying to work through geographical naming challenges and especially since the demise of the dictatorial Franco regime in 1975. Therefore, the importance of official and agreed normalized toponymic databases being able to cater to multiple place names is extremely significant. Recognizing and resolving contentious issues that may arise in the process of creating a new place name, or else making one official that already exists in the vernacular of the area, or else changing an existing one, and the role played by the toponymic authorities, advisory bodies, and broader society in their resolution needs to be part of the solution. The work undertaken by advisory bodies, and their role, and strategies are core to the creation of the appropriate appellations linked to the geocode in GIS databases. Even following democratic principles in richer countries such as Spain with the capacity in technology and educated personnel and experts, getting the appropriate linguistic policies and guidelines for geographical names is conceivably more challenging than creating the right GIS applications and algorithms when there is a diversity of languages. Due to the different types of legal autonomies catered for in Spanish legislation, legal and language place naming policies, have diverse interpretations as implemented on the ground. There is often a nexus between language and geographical and place naming, that can be used in political bargaining at various different levels and in the real or imagined quest for independence or establishing a counter-state movement e.g. Catalonia. Iterations of such territorial and language-cultural politics exist in Northern Ireland, Corsica, and the Berber regions in Algeria.

Few commentators would deny the existence of cultural heritage(s) as they continue to morph and inspire new ideas and cultural and economic developments whether in the arts, landscapes, or hospitality industries. Like language, culture is dynamic and cannot be constrained by closed parameters at any particular date in time. This axiom impacts geographical naming if we reflect on such place names as the Roman Latinized Londinium, and its disputed etymological origins possibly 'Lud' from Celtic, Liguria, Gaul, or other origins. Similarly, the Big Apple and New York, and Manhattan, with the latter word etymologically derived from Pre-Colombian people and languages, or New Amsterdam (17th Century) as named by Dutch adventuring colonists. While the placename York in England has its origins in the Viking word Jorvik. Here the point is that dynamic narratives exist around geographical naming with bottom-up grassroots and standardizing top-down name forces constantly at play. When attempts at over-standardization occur, then power agendas are obviously at play as witnessed in the case studies from Spain and elsewhere as in Brazil, Estonia, Zimbabwe, Indonesia, New Zealand, and Turkey.

By analyzing place naming at small geographical scales such as a street, village, or area of a town or city we see the different variables at play. This mirrors similar patterns to the naming processes at larger geographical scales, such as the region or state. This point is crucial in enhancing knowledge of place labels, the power dynamics of naming, cultural politics and the rounded skills that are necessary for geospatial data recording and processing. Thus the importance of this on the education trajectory from the primary sector upwards.

By appraising the nature of place naming in urban environments, we explore the grass-root smallest geographical scales in Turkey, in other words, the factors affecting the shaping of the urban historical and cultural politics there. This necessitates becoming familiar with the most frequently used names in urban environments in the early period of nationhood-building processes in the modern state of Turkey in the early twentieth century. Becoming cognizant of how the Ottoman heritage continues to have an effect on the urban odonymy and gaining insights into how political parties today make use of this in 'owning' spaces is insightful. Evaluating past and present urban odonymy classification studies in Turkey, and assessing the current emerging critical approaches enhances our comprehension of the significance of 'naming' in territorial, place, ideological, and power constructs. The place and power transitivities in place naming processes in Slovenia, Turkey, Zimbabwe, RSA, Ireland, New Zealand, Argentina, and Brazil attest to common experiences in different languages and historical political archaeologies; however, the constant factor remains power relations.

In this context, distinguishing between diverse toponymic layers in Bosnia and Herzegovina provides an imminent pressing imperative due to its recent conflict history in the 1990s, and the efforts of many agencies to bolster sustainable peace in the region. Attempting to erase long-term, or delete place names has had very limited success in the Balkans, like elsewhere. But due to the laws of nature and everyday realities, having to live in the present does not mean cultivated ignorance of the past and the dangers that this can run, with extremism as witnessed with genocide in its human and cultural forms. In this context, recognition of some of the most frequent geographical appellatives in South Slavic landscapes is imperative, as it compares various historic periods in relation to the cultural heritage of Bosnia and Herzegovina but that is open to discussing the socio-political significance of such heritages (Brozović 2016; Mikša and Zorn 2018).

Geo-Histories, Legacies, and Toponymy Transitions

In Part III conclusions are based on the interdisciplinary work of geographers, historians, architects, cartographers, GIS developers, and practitioners who carry out research in Turkey, Hungary, Bosnia—Herzegovina, Malta, Brazil, Estonia, and elsewhere.

Geographical name changes in Turkey during the past 100 years are critically analyzed. This gives sharp insights into the history, chronology, and incentive for renaming places there and cultural politics. For this, the researcher has to be able to recognize key sources of the rich diverse geographical names and their archaeologies in the country (Ari 2012). Juxtaposed with this are the changes of geographical names in multicultural Central European areas during the early modern and modern historical eras. Explanation of how population changes, as well as state-driven ethnolinguistic homogenization policies, affected attempts at 'uniforming' place names. The input of different social groups existed in local geographical names from diverse historical periods up to the present. The toponymic legacies of major historical events such as the Battle of Mohács (1526) in Hungary where allied Christian forces fought against the advancing Ottoman armies of the legendary Sultan Süleyman the Magnificent resurface and are, used by successive generations and powerful elites. This is evident in the Mohács landscapes right up to the present and reinforced by government perspectives appealing to a collective consciousness or memory of national and self-defense against the invaders. Bearing this in mind, the EU and international community witnessed the construction of a border fence in 2015 to prevent immigrants and asylum seekers from entering Hungary and consequently the EU Schengen area. This stands in contrast to an almost 'open door' policy in the majority of EU states concerning the flow of people from Ukraine after the Russian invasion in 2022. Therefore, the drawing of new state boundaries in the twentieth century continues to affect geographical names in Central and Eastern Europe and beyond.

At a different geographical scale, a team of architects and geographers explore the history of the establishment and transformation of Sarajevo's geographical names and the determination of the link between processes of change of place name meanings and spatial-organizational changes in modern residential neighbourhoods. That gives insight into the social character of urban planning in Sarajevo today. Deconstructing the former and ongoing levels of transformations in the meaning of names in Sarajevo's new residential areas—New Sarajevo and New City—illustrate embedded old name continuities that may seem anachronistic today in very different functional spaces and with the current socio-economic realities. This recognizes the nature of local, regional, and global influences that have effects on the meanings of geographical names. As in other case studies, the authors also reference the glocalization of names taken from other cultures and the ideological impacts of key words in names taken from the market economy and international advertising.

Comparative and contrasting experiences exist in Malta with its toponymic multilayers recognizing the changing socio-cultural influences. This is evident in the polyglot attributes of the Maltese and their language and place-signage due to the various historical periods regarding the compound and multi-level name and archaeological heritages (Gauci and Schembri 2019). In cultural politics, past and present selected personalities and their socio-cultural contributions exist in the geographical names. Critical skills are required in identifying and decoding place names and the odonymy, its geographic attributes, interconnections, and networks associated with the appellations. A wealth of historical cartographical, image and documentary data exists on geographical naming in Malta and outside it that can be enhanced by GIS.

Historical maps with their place labels can provide much information for Big Data and GIS-related research as proven by the unique importance of the toponymic study of the map of New Lusitania (Brazil and surrounding areas), a Portuguese cartographic product from the eighteenth century. Different skills in comparing and contrasting various versions of this historical map are employed including the Toponymic Extraction Methodology and its applications in furthering GIS research and knowledge in historical cartography and layering methods up to the present. Naturally, as well as the intangible cultural heritage contained in the geographical names, on land and water found on the maps, this can affect the current dynamic cultural politics. Therefore, such work on the map versions of New Lusitania is important not only for Brazil and the neighbouring countries but for world heritage and scientific approaches to place appellations on international scales. Highly advanced and innovative research on geographical naming involving GIS and associated physical and cultural naming data exists in a dynamic network of institutions in Brazil and Argentina as in the University of Sao Paulo (Menezes et al. 2021).

The research focus on specific geographical scales in naming goes right down to analyzing agricultural production and farms in Estonia before and during the Soviet era, and now. Core socio-economic variables exist in all historical periods; what changes are the political ideologies that influence the place names. Soviet policies attempted to erase former geographical names with the construction of large-scale collectivized kolkhoz farms, somewhat distinct from the Sovkhoz that was entirely state-managed and controlled. Therefore, new ideological place names emerged in Estonia. With independence in 1991 and EU membership in 1993, Estonia embarked on a different liberal economic ideological trajectory encouraging privatization and breakup of the collectivized kolkhoz system. While readopting some place names that existed prior to the communist period took place, new ones were created, while certain place appellations associated with the Soviet-era survived by stealth in the landscape.

Overall, the authors argue that top-down ideological cultural politics and processes drive geographical name standardization. Hence, the challenge is to ensure that democratic principles are embedded in the relevant governance systems and associated bureaucracies, where Big Data and GIS applications are monitored in a transparent manner, and algorithms are not left to the vagaries of globalizing competing market actors.

Toponymy: Narratives, Languages, Culture, and Education

In Part IV regarding geographical names, perspectives given in Parts I-III are enhanced by research analyses and discussions led by geographers, linguists, an expert in cultural studies, and another in post-colonialism research, an anthropologist, and educators that bring together theoretical aspects embedded in the case study material.

In the toponymic analyses of Ireland, colonial imprints are identifiable on place names from the 1600 to 1900 centuries and then reviewed revealing the structures, agency, and impact of the Ordnance Survey in the nineteenth century. The significance of renaming in decolonization and commemorative processes during the twentieth century is deconstructed at variable scales, revealing trends in the naming of streets and urban developments since the 1960s. Detailing the legislative toponymic framework in the Republic of Ireland and Northern Ireland reveals the multi-layered identities, and cultural politics, within and between the respective jurisdictions in their approaches to state construction after 1922. The www.logainm.ie database offers a range of insights into geographical naming in Ireland with its nexus of historical, cultural, language, and power complexity. The website is open to the public, schools, and state institutions at national and local governance levels. Its role in research and providing the government with data and reports have added a strong dimension to geographical naming studies and place name standardization in the Republic where there are two official languages Irish and English, with each being derived from very different linguistic families i.e. Celtic and Germanic respectively. This analysis is followed by an appraisal of the fictional theatre of Brian Friel's *Translations* (1980) set in 1833 Ulster when Ireland's toponymy was being (re)mapped and standardized by the British administration. The play now forms part of the secondary school curriculum in the Republic. Audiences experience the nexus between toponymy, language, identity, culture, and politics in an empathetic and often enjoyable manner. Psychologically, this literally reaches far beyond the island connecting with other peoples' narratives concerning cultural politics and decolonization e.g. Scotland and Wales, and more widely the independent states of the former British, French, Spanish and Portuguese empires. Such artistic narratives are embedded in (de) colonization and help successive generations to understand colonial and independence legacies, including the civil and military strife experienced in Northern Ireland (1968–1998) or the strongly polarized standpoints there regarding the anti-EU rhetoric, surrounding the 2015 Brexit campaign in the UK and the pursuant outfall for Northern Ireland now constitutionally outside the EU. Some 52% of voters in the UK as a whole voted to leave the EU in the 2016 Brexit Referendum. This is in sharp contrast to those areas that voted to remain in the EU: Northern Ireland (58%), Scotland (60%), and the British Overseas Territory of Gibraltar (96%).

Certain aspects and iterations of the socio-cultural experiences of geographical naming in Ireland exist in the British Overseas Territory of Gibraltar (BOTG) while the geo-historical narrative is different. However, in both Ireland and Gibraltar, the colonial administration manufactured a dual society with the associated cultural politics and pursuant legacies i.e. 'the locals and us'.

In 2007, the British authorities renamed the Crown Colony to the British Overseas Territory of Gibraltar (BOTG). In the colonial process there, one cultural power strata was exclusively Anglo-British, while the other was made up of economic migrants from the Mediterranean region. In the eighteenth century, the Gibraltar peninsula was occupied and the local inhabitants fled north. The naval and military forces and ship maintenance and supplies facilities had to be sustained with the help of imported labour. The British administration facilitated entrepreneurial traders to settle in the Colony i.e. civilians from throughout the Mediterranean region (O'Reilly 1999, 2015). Today, essentially these people are the Gibraltarians who want to be 'decolonized' in the administrative political and cultural context, but territorially to remain constitutionally with the UK.

The specificities of geographical naming in Gibraltar are often overshadowed for outsiders by the abundance of appellations associated with British naval, military, and names of the royal family, as well as battles and 'heroes'. Sometimes this is reminiscent of areas in Malta. However, there is the resilience of the use of traditional street names in Gibraltar together with the resistance civilians posed to the implementation of official top-down street names. The permanence of traditional names raises questions regarding Edward Said's theory of imaginative geographies, which whilst active in Gibraltar discursively and colonially mapping the territory, proved less effective when it came to the urban centre. Unpacking street nomenclature, the tensions generated through the (co)existence of native and colonial, using two very different sets of geographical and place naming invariably stem from a similar imaginative process in Gibraltar.

Significantly, the British and Gibraltar Government authorities officially name the Bay area to the west of the British Territory, the Bay of Gibraltar, while the Spanish official name is the Bay of Algeciras, the same appellation as the main Spanish port city on the Bay. As well as the Spanish sovereign claims to the entire BOTG territory and waters, there are also micro-territorial disputes within the bay area that include territorial waters and maritime rights. This often leads to spats or 'cat and mouse games' between the respective navies, port authority agencies, embassies, and governments, but sometimes with high levels of bellicose rhetoric from right-wing politicians taken up by populist media in Britain and Spain. The historical and collective memory and transgenerational experiences of the Gibraltarians and especially that of the Franco era (1936–1975) remain strong. On Google maps, the name given is the Bay of Gibraltar followed in brackets by The Bay of Algeciras.

Regarding the geo-historical experiences in Italy, the concept of the 'historical memory of a territory' underscores what the 'cultural heritage and expression of perception by different generations throughout the centuries' means for their environment. This justifies arguments why geographical names are one of the most important components of intangible cultural heritage. To protect this, the processes of certification and standardization are more important now than ever due to globalization because of the complex links between geographical names, individual landscapes, and the territorial scales and categories that exist. Therefore, narratives of place naming in Italy are core variables supported by legislation enhancing the rich contrasting landscapes that exist nationally, regionally, and locally. Geographical product labeling for food, wine, and tourism forms a key element in Italy's economy. Unquestionably, geographical nomenclature is all-important to issues of identity and is highly significant in local, regional, and national politics and the associated parties and dynamics.

The intangible heritage of place appellations for cultural and applied human security reasons are analyzed with infield anthropological and linguistic research in the Bandung Basin region in Java, Indonesia, and confirm that toponymic studies require a multidisciplinary approach. That is in order to appreciate that the collective memory can preserve geomorphological information from the past in recognizing both generic terms for hydronyms and toponyms, a case in point being the consequences of changing geographical names related to disaster problems. This substantiates the socio-political and economic pressures to preserve geographical names and especially local ones contrasted with their current standardized administrative names that warrant appropriate policies implemented by the Local Government, and the National Names Authority. Though not always evident to local populations or outsiders, place names in collective memories in the form of myths, legends, or oral traditions help preserve geographical names as also do cultural artifacts, but this is even more challenging when local languages change over the centuries.

In the physical environment of the Indonesian archipelago with such phenomena as volcanic eruptions and natural island creation processes, alongside intricate hydrological changes on land, witnesses to these events throughout human history, name these places and have passed on these appellations through oral traditions. The quest to find an explanation and to record histories can become a strand in the mythology, folklore, changing languages, and etymologies of the geographical name. In 21st-century Indonesia regarding specific time and place contextual realities, economic rights, and a strong legal framework for local geographical names must ensure local appellations be protected, alongside legitimate business and tourism interests (Lauder, Multamia RMT 2021).

Like other case studies in this tome, the research from the Bandung Basin region in Java illustrates the intricate nexus between the physical and human environment, economy, and social-cultural realities and heritages including geographical naming. Hence, state policy and legislation must embrace balanced long-lasting sustainable development and implement it on the ground as promoted in the SDGs. Throughout, territorial control and power dynamic processes of cultural politics are implicit. With the imperative of standardizing geographical names for functional reasons on the ground and in a globalizing world digisphere, technology and especially GIS plays a key role. Matching the geospatial data—GPS, with the appropriate name culture data is imperative and supported especially in SDGs 16 and 4.

Aligned with SDG4 is education that enhances a critical understanding of geographical names. This is a necessary task of research and its practical application within democratic legislative frameworks. Hence, there is a vital need for educators to develop appropriate relevant adaptable pedagogies and material that students can connect with for functional geographical reasons but also an appreciation of place names as a significant element of the intangible cultural heritage. Although initially created through the need to identify and subsequently organize a setting or locale, geographical naming ultimately becomes a basic component of the historical memory of a territory or sea area, often reflecting past cultures, migration waves, and historical events associated with a particular location. Their cultural importance has a role to play in enhancing peace-building and righting past wrongs. Therefore, education regarding place names is a significant element in the development of good citizenship, for instance in Northern Ireland, Scotland, Spain, Turkey, the Balkans, Asia, and elsewhere. The work of UNGEGN and especially its education and training staff attests to this.

In this tome, there is a range of targeted and thematic questions provided for readers in each chapter. Enhancing this, there is a full chapter on the activities and questions that build upon the state-of-the-art geographical naming analyses. This acts as starting points to facilitate the application of the concepts and case studies in the volume within an educational setting. Suggestions are offered as to how the activity in question can be adapted to specific age groups whether in Dublin, Derry, Durban, or Dallas; Belfast, Bristol, Belgrade, Bandung, or Buenos Aires; Christchurch, Capetown, or Chichester; Seoul, Shikoku or Sydney. The activities range in complexity from simpler tasks designed to help students develop a basic awareness of place and naming, to analyses that are more sophisticated. This uses the study of geographical names as a lens through which to examine deeper issues including contemporary geopolitics within and between countries, treatment of indigenous peoples, colonialism, and post-colonialism, cultural politics, heritage, and more.

Questions

1. Appraise the concept of discovery learning for individuals, groups, and researchers regarding the significance of the geographical names of areas and places.
2. Make a list of primary and secondary sources for geographical naming that exists in your home area, followed by a list of such sources outside your area.
3. Compare and contrast the conceptual and theoretical approaches to geographical naming and the realities on the ground in at least two places discussed in the book e.g. Turkey, Slovenia, Bosnia-Herzegovina, Ireland, UK, Hungry, Spain, France, Italy, Malta, Zimbabwe, Republic of South Africa, Korea, Japan, Brazil, Argentina, New Zealand or elsewhere.
4. What actors, forces, contexts, and circumstances are driving the standardization of place and geographical naming in the contemporary world?
5. Evaluate the challenges faced by GIS developers and database managers regarding geospatial data e.g. GPS, and the associated naming, cultural and political data.
6. Why is geographical naming, subtly embedded in the SDGs, and associated UN recommendations and policies?
7. Has place naming become a brand in the state and other sectoral contexts and agendas?

References

Arı Y (2012) Geography in Turkish higher education: Sea names for textbooks and maps. In: Proceedings of the 19th international seminar on sea names: sea, sea names and Mediterranean peace. Korea, pp 197–206

Brozović D (2016) Changing names: a way to cope with identity issues in times of crisis—The Ottoman legacy in Croatia & Bosnia and Herzegovina. Ethnologia Balkanica 19:239–255

Choo S (2018) The essence and practicality of dual naming: considerations toward the East Sea/Sea of Japan dual naming. J Korean Cartogr Assoc 18(3):33–43

Choo S (2019) Towards a multi-dimensional critical toponymic research. J Korean Geogr Soc 54(4):449–470

Gauci R, Schembri JA (eds) (2019) Landscapes and landforms of the Maltese islands. Springer, Cham

Giraut F, Hoyssay-Holzschuch M (2016) Place naming as dispositive: toward a theoretical framework. Geopolitics 21(1):1–21

Kearns R, Lewis N (2019) City renaming as brand promotion: exploring neoliberal projects and community resistance in New Zealand. Urban Geogr 40(6):870–887

Lauder Multamia RMT (2021) "Merawat Memori Kolektif melalui Toponim". Seminar Nasional "Merawat Budaya: Studi Dokumentasi dan Revitalisasi Bahasa-bahasa di Indonesia (Jawa Barat, NTT, dan Maluku). Kantor Bahasa Provinsi NTT. Kupang, 31 Desember 2021

Mangena T (2020) Rhodes, Mugabe and the politics of commemorative toponyms in Zimbabwe. Geopolitics 25(4):1015–1036

Menezes PML, Santos KS, Brito TLF, Corrêa-Martins FJ, Fernandes MC, Santos JG, Lapaine M (2021) Map projection analysis of the Nova Lusitania map. Kartografija i Geoinformacije 20:48–69

Mikša P, Zorn M (2018) The, "battle" for the mountains: Germans versus Slovenes in the South-Eastern Alps in the late 19th and early 20th century. Adv Environ Res 65:199–227

O'Reilly G (2021) Geography, geopolitics and naming places in the digisphere. In: The society for east sea (ed), geographical naming in the digital era: proceedings of the 26th International Seminar on Sea Names, Gangneung, Korea, 17–19 Sept 2020, Seojeon Printech, Seoul. ISBN 979-11-951305-5-9

O'Reilly G (1999) Gibraltar: sovereignty disputes & territorial waters 7(1):67–81. https://www.researchgate.net/publication/238724623_Gibraltar_Sovereignty_disputes_and_territorial_waters

O'Reilly G (2015) Remnants, relics and shadows of empires in the gibraltar region. Duluth, USA. https://middlegroundjournal.files.wordpress.com/2015/11/oreillygibraltararticlesfall2015themiddlegroundjournal-org.pdf

Perko D, Ciglič R, Zorn M (2020) Slovenia: a European landscape hotspot. In: Perko D, Ciglič R, Zorn M (eds) The geography of Slovenia: small but diverse. Springer, Cham, pp 1–20. https://doi.org/10.1007/978-3-030-14066-3_1

Tort-Donada, J (2020) On the connection between the physical environment and the Toponym. Geography's contribution to clarifying the problem. In: Balode L, Zschieschang C (eds). Onomastic Investigations II. Latvian Language Institute, Riga, pp 286–299

Vescovo A (2019) Geographical names in the era of big data. A necessary updating in the curriculum design of the geosciences. Micro, Macro & Mezzo Geo Information (Issues) 13, 16–28. ISSN: 1857-9000 (printed version) EISSN: 1857-9019 (electronic version) UDC: 528:004

Suggested Reading

Adebanwi W (2018) 'Coloring "rainbow" streets in South Africa: The struggle for toponymic multiracialism in urban post-apartheid South Africa. In: Rose-Redwood R, Alderman D, Azaryahu M (eds) The political life of urban streetscapes: naming, politics and place. Routledge, London and New York, pp 218–39

Azaryahu M (2018) Revisiting East Berlin and Haifa: A comparative perspective on renaming the past. In: Rose-Redwood R, Alderman D, Azaryahu M (eds) The political life of urban streetscapes: Naming, politics and place. Routledge, London and New York, pp 56–73

Batsashi T (2019) Historical toponyms, hydronyms and geographical name changes in Turkish Ottoman empire and in the Republic of Turkey. Free Univ J Asian Stud 1:1–4

Berg LD, Kearns RA (2009) Naming as norming: 'Race', gender and the identity politics of naming places in Aotearoa/New Zealand. In: Berg LD, Voulteenaho J (eds) Critical toponymies: the contested politics of place naming. Ashgate, Farnham, pp 19–51

Boorstin B (2009) Director, Google's public policy team, when sources disagree: borders and place names in Google Earth and Maps. Google Public Policy Team Blog, December 4

Capdepón U (2020) Challenging the symbolic representation of the Franco dictatorship: the street name controversy in Madrid. Hist Mem 32(1):100. https://doi.org/10.2979/histmemo.32.1.06

Chabata E, Mumpande I, Mashiri P (2017) Mangling of place names and identity corruption amongst the Tonga of Zimbabwe. In: Mashiri P, Chabata E, Mamvura Z (eds) Zimbabwe naming practices and patterns: a sociolinguistic approach. University of Zimbabwe Press, Harare, pp 109–127

Choo S, Jin S (2020) Pluralism of memories and commemoration and the accumulation of placeness contained in place names: the case of Volgograd/Stalingrad. J Korean Geogr Soc 55(4):409–426
Choo S (2018) People, places and place names. Hanul M Plus, Paju
Ciglič R, Perko D (2013) Europe's landscape hotspots. Acta Geogr Slov 53(1):117–139. https://doi.org/10.3986/AGS53106
Dempsey C (2012) The politics of Google's mapping. GIS Lounge, 18 May 2012. https://www.gislounge.com/the-politics-of-googles-mapping/
Deniz T, Ersöz D (2020) Safranbolu city place names: space, identity and culture. Int J Geogr Geogr Educ (IGGE) 42:527–542
Dillon S (2005) Reinscribing De Quincey's palimpsest: the significance of the palimpsest in contemporary literary and cultural studies. Textual Pract 19(3):243–263
Dodds KJ, Yusoff K (2005) Settlement and unsettlement in Aotearoa/New Zealand and Antarctica. Polar Rec 41(2):141–155
Dragić H (2013) Appeals in the toponymy of the Stolac region. Školski Vjesnik 62(1):5–26
Dunlevy DA (2021) Place names and the complexity of language recognition in northern Ireland. In: Blackwood R, Dunlevy DA (eds) Multilingualism in public spaces: empowering and transforming communities. Bloomsbury, New York, pp 137–157
Feldman D (2005) Renaming cities in Bosnia and Herzegovina. Int J Const Law 3(4):649–662
Forrest C (2018) What's in a name? A feminist reflection on street name changes in Durban. Agenda 32(2):53–61
Gadeng AN, Maryani E, Romat D (2018) The value of local wisdom Seumong in Tsunami disaster mitigation in Simeuleu Regency. Aceh Prov Earth Environ Sci 145:1–6
Geršič M, Kladnik D, Brnot M (2020) Standardization of geographical names in officially recognized bilingual areas of Slovenia. Onomàstica 6:145–167
Geršič M, Kladnik D, Repolusk P (2017) Geographical names in the languages of official minorities in Slovenia. Onomàstica 3:105–126
Giraut F (2020) Plural toponyms: when place names coexist. EchoGéo 53:1–8. https://doi.org/10.4000/echogeo.20760
Gregory D (2005) The colonial present. Blackwell Publishing, Oxford
Habib A (1997) South Africa-the rainbow nation and prospects for consolidating democracy. Afr J Polit Sci/Revue Afr Sci Polit 15–37
Hang'ombe K, Chabata E, Mamvura Z (2019) Syungu Namutitima or Victoria Falls? Contest for place and place naming. Nomina Africana: J African Onomast 33(1):19–31
Hausner I (2013) Principles and rules of 'dual names/dual naming' (multiple names/multiple naming) systems in and outside Europe: an exemplary approach to the definition of two (?) new terms in the UNGEGN glossary. In: Proceedings of the 19th international seminar on sea names: sea, sea names and mediterranean peace, Korea, pp 7–22
Hoffmann I, Rácz A, Tóth V (2017) History of Hungarian Toponym. Buske Verlag, Hamburg
INSPIRE (2010) Implementing Directive 2007/2/EC of the European Parliament and of the Council as regards interoperability of spatial data sets and services (COMMISSION REGULATION (EU) - No 1089/2010 of 23 November 2010). http://eur-lex.europa.eu/LexUriServ/LexUriServ.do?uri=OJ:L:2010:323:0011:0102:EN:PDF,
INSPIRE Thematic Working Group Geographical Names (2014) Data Specification for Geographical Names. European Commission Joint Research Centre. https://inspire.ec.europa.eu/Themes/132/2892
Kangaspuro M, Lassila J (2017) From the trauma of Stalinism to the triumph of Stalingrad over Volgograd: the toponymic dispute. In: Fedor J et al (eds) War and memory in Russia, Ukraine and Belarus. Palgrave Macmillan, Cham
Kitanics M, Hegedüs N (2021) The latest 'Southern Protection System' and the revived 'fortress of Europe' topos in Hungary. Polit Central Europe 17:729–760
Kladnik D, Geršič M, Perko D (2020) Slovenian geographical names. Acta Geogr Slov 61(3):7–190. https://doi.org/10.3986/AGS.9394

Kulić V, Mrduljaš M (2012) Unfinished modernisations—between utopia and pragmatism, UHA/CCA, Zagreb
Kumer P, Zorn M, Žorž G, Gašperič P (2020) From the Great War to interwar fortifications: changing narratives attached to the military landscape in Western Slovenia. In: O'Reilly G (ed) Places of memory and legacies in an age of insecurities and globalization. Springer, Cham, pp 247–263. https://doi.org/10.1007/978-3-030-60982-5_11
Leshem E (2021) Why does the same search word get different results in Google around the world? Haaretz, 12 July (Hebrew)
Maandi P (2009) The silent articulation of private land rights in Soviet Estonia: a geographical perspective. Geoforum 40(3):454–464
Mac Giolla Easpaig D (2009) Ireland's heritage of geographical names. In: Geographical names as a part of the cultural heritage, Wiener Schriften zur Geographie und Kartographie (18), pp 79–85. https://www.logainm.ie/en/res/92. Last Accessed 9 Nov 2021
Mamvura Z (2020) Reconstituting the cultural geography in Zimbabwe: place renaming in Zimbabwe's 'New Dispensation.' Geopolitics. https://doi.org/10.1080/14650045.2020.1850443
McLaughlin A (2008) How Google determines the names for bodies of water in Google Earth, Google Public Policy Blog. https://publicpolicy.googleblog.com/2008/04/how-google-determines-names-for-bodies.html
Medzini A (2017) The role of geographical maps in territorial disputes between japan and Korea. Eur J Geogr 8(1):44–60, February 2017. Association of European Geographers
Medzini A (2012) The war of the maps: the political use of maps and atlases to shape, national consciousness—Israel versus the Palestinian authority. Eur J Geogr (EJG) 3(1):23–40
Menezes PML, Santos CJB, Resende ACB (2014) From 1500 to 2000, the evolution of toponymy in Brazilian Maps. In: Symposium on place names changes, Joint IGU/ICA Commission/ Working Group on Toponymy, Academia Nazionale dei Lincei, Rome
Midžić F (2010) Borders and territory of Bosnia and Herzegovina from the Peace of Karlovac to the end of the First World War on old geographical maps (1699–1918). Association of Archival Workers of Una-Sana Canton
Mitin I (2018) Constructing urban cultural landscapes & living in the palimpsests: a case of Moscow City (Russia) distant residential areas. Belgeo. Revue Belge de Géographie 4:1–15. https://doi.org/10.4000/belgeo.28126
Moïsi D (2010) The geopolitics of emotion: how cultures of fear, humiliation, and hope are reshaping the world. Anchor Books, NY
Nash C (2009) Irish place names: post-colonial locations. In Berg LD, Vuolteenaho J (eds) Critical toponymies: the contested politics of place naming 2009. Ashgate, Surrey
Ndletyana M (2012) Changing place names in post-apartheid South Africa: accounting for the unevenness. Soc Dyn 38(1):87–103
NEAHF (Northeast Asian History Foundation). Dokdo in the East Sea. http://contents.nahf.or.kr/english/item/level.do?levelId=eddok_003e_0010_0030
New Zealand Geographic Board (2021) Annual report of the New Zealand Geographic Board 2020/21. https://www.linz.govt.nz/about-us/publications/new-zealand-geographic-board-maps-and-publications/annual-report-new-zealand-geographic-board
Mainnín Ó, Mícheál B (2021) Empowering multilingualism? Provisions for place names in Northern Ireland and the political and legislative context. In: Blackwood R, Dunlevy DA (eds) Multilingualism in public spaces: empowering and transforming communities. Bloomsbury, New York, pp 59–87
Ó Muraíle N (2014) Translations: the placenames work of the OS in Ireland. https://www.ria.ie/mapping-city-town-and-country-1824-ordnance-survey-ireland-0. Last Accessed 9 Nov 2021
O'Reilly G (2020) Places of memory and legacies in an age of insecurities and globalization. Springer International Publishing, Switzerland. https://www.springer.com/gp/book/9783030609818
Ormeling FJ, Kerfoot H, Zaccheddu P-G (eds) (2017) Toponymy training manual. United Nations, New York. https://unstats.un.org/unsd/ungegn/pubs/documents/Training%20Manual.pdf

Ormeling F (2017) 50 years trying for training courses in toponymy. Presentation during 11th UNCSGN/30th session UNGEGN. New York: United Nations. https://unstats.un.org/unsd/geoinfo/ungegn/docs/11th-uncsgn-docs/history_training50years.pdf
Pallasmaa J (2009) Space, place, memory and imagination. Spatial recall: memory in architecture and landscape. Routledge, New York and London, pp 16–41
Pap N, Reményi P (2020) Encounters between islam and christianity: Mohács and Kosovo Polje. In: O'Reilly Gerry (ed) Places of memory and legacies in an age of insecurities and globalization. Cham, Springer International Publishing, pp 285–305
Perko D, Ciglič R. Slovenia's regions. In: Perko D, Ciglič R, Zorn M (eds) The geography of Slovenia: small but diverse. Springer, Cham 2020, pp 227–243. https://doi.org/10.1007/978-3-030-14066-3_15
Perko D, Jordan P, Komac B (2017) Exonyms and other geographical names. Acta Geogr Slov 57(1):99–107. https://doi.org/10.3986/AGS.4891
Rose-Redwood R, Alderman D, Azaryahu M (2018) The urban landscape as political cosmos. In: Rose-Redwood R, Alderman D, Azaryahu M (eds) The political life of urban streetscapes: naming, politics and place. Routledge, London and New York, pp 1–24
Rusu MS (2021) Onoma: Journal of the International Council of Onomastic Sciences Street naming practices: a systematic review of urban toponymic scholarship. Onoma 56:269–292
Said E (2013) Orientalism. See also A. Dawson. Edward Said's imaginative geographies and the struggle for climate justice. College Liter 40(4):33–51. http://www.jstor.org/stable/24543185
Saparov A (2017) Contested spaces: the use of place-names and symbolic landscape in the politics of identity and legitimacy in Azerbaijan. Central Asian Survey 36(4):534–554
Schembri J (2019) The geographical context of the Maltese Islands. In: Landscapes and landforms of the Maltese Islands. Springer, Cham, pp 9–17
Smith B (2018) Engaging geography at every street corner: using place-names as critical heuristic in social studies. Soc Stud 109(2):112–124
Stoltman JP (2017) Geographical naming: reflections on peace, cooperation and social justice. In: The Society for East Sea (ed) Achieving peace and justice through geographical naming: proceedings of the 23rd international seminar on sea names, Seojeon Tech, Seoul
Strating R, Wallis J (2022) Maritime sovereignty and territorialisation: comparing the pacific islands and South China Sea, Marine Policy 141. https://doi.org/10.1016/j.marpol.2022.105110
Štraus I (2013) Arhitektura Jugoslavije/Architecture of Yugoslavia 1945–1990, Sarajevo Green Design Foundation, Sarajevo
Swart M (2008) Name changes as symbolic reparation after transition: the examples of Germany and South Africa. German Law J 9(2):105–121
Tent J (2015) Approaches to research in toponymy. names 63(2): 65–74. https://www.researchgate.net/publication/276158373_Approaches_to_Research_in_Toponymy
Tuan Y-F (1977, 2001) Space and place: the perspective of experience. University of Minnesota Press
UNGEGN (2002, 2007) Glossary of terms for the standardization of geographical names. NY: UN. https://unstats.un.org/unsd/ungegn/pubs/
UNGEGN (2006) Manual for the national standardization of geographical names. NY: UN. https://unstats.un.org/unsd/ungegn/pubs/
UNGEGN (UN Group of Experts on Geographical Names). https://unstats.un.org/unsd/ungegn/
United Nations Education, Scientific and Cultural Organization (UNESCO). Intangible Cultural Heritage. https://ich.unesco.org/en/
Zaccheddu P (2018) Do geographical names data play a specific role for the SDGs under a geospatial lens? UNGEGN Information Bulletin No. 54 • May 2018. SPECIAL FEATURE Geographical names supporting sustainable development
Zorn M, Mikša P (2018) The Rapallo Border between Italy and Yugoslavia after the First World War. Histoire Des Alpes 23:165–181

MIX
Papier aus verantwortungsvollen Quellen
Paper from responsible sources
FSC® C105338

If you have any concerns about our products,
you can contact us on
ProductSafety@springernature.com

In case Publisher is established outside the EU,
the EU authorized representative is:
Springer Nature Customer Service Center GmbH
Europaplatz 3, 69115 Heidelberg, Germany

Printed by Libri Plureos GmbH
in Hamburg, Germany